AA004424

MFPT 2019

Where Theory Meets Practice

Philadelphia, Pennsylvania, USA
14 - 16 May 2019

ISBN: 978-1-5108-8926-2

Printed from e-media with permission by:

Curran Associates, Inc.
57 Morehouse Lane
Red Hook, NY 12571

Some format issues inherent in the e-media version may also appear in this print version.

Copyright© (2019) by Society for Machinery Failure Prevention Technology
All rights reserved.

Printed by Curran Associates, Inc. (2019)

For permission requests, please contact Society for Machinery Failure Prevention Technology at the address below.

MFPT Society
715 Twining Road
Suite 135
Dresher PA 19025

Phone: (215) 346-2870
Fax: (215) 352-6902

https://mfpt.org

Additional copies of this publication are available from:

Curran Associates, Inc.
57 Morehouse Lane
Red Hook, NY 12571 USA
Phone: 845-758-0400
Fax: 845-758-2633
Email: curran@proceedings.com
Web: www.proceedings.com

TABLE OF CONTENTS

PRESENTATIONS

PAPERS

TUESDAY 14 May 2019

PLENARY SESSION

8:15 AM
Main Ballroom

Conference Opening Remarks & Welcome

Amy Wynn, MFPT Executive Director

Chris Nemarich (NAVSEA), MFPT Board Chairman

KEYNOTE: Reimagining The Model for Reliability and Readiness Data Analytics

Larry Burrill, President of Readiness and Sustainment Solutions Business Unit and Co-Founder McKean Defense Group

MFPT 2019 Technical Partners

Peter Gaydon, Technical Director, Hydraulic Institute

Greg Gammon, Director of Operations, Texas A&M Turbomachinery Laboratory

Bob Sayer, President, Vibration Institute

10:00 AM	**Break**			
Chairperson *Location*	**Diagnostics** L. Thomas Heiser, Ridgetop Group **Freedom I**	**Machine Learning / A.I.** Preston Johnson, Allied Reliability Group **Freedom II**	**Sensors** John Lucero, NASA **Freedom V**	**VI Track** **Freedom VI**
10:30 AM	Robot Manipulator Drive Fault Diagnostics Using Data-Driven and Analytical Modelling *Michael Lipsett; University of Alberta*	US Navy's CBM Efforts * * Featured Speaker * * *CDR Jesse Black, NAVSEA*	Universal Vibration Sensor Mount *George Zusman; Vibration Measurement Solutions, Inc.*	Vibration Institute East Coast Meeting
11:00 AM	Modern Laser Applications for Improving Machine Reliability *Ronaldo Sullivan, Hamar Laser Instruments Inc.*		Successful Vibration Monitoring of Variable Speed Equipment *Scott Dow, SPM Instrument*	
11:30 AM	Development of an Accelerated Degradation Testbed for Integrated Vibration Monitoring and Online Oil Analysis of Vehicle Differential Bearings *Student Cody Walker; University of Tennessee*	High Content, Low Bandwith, Secure Long Distance Communication of Shipboard Diagnostic Information *Eric Olson; Mechanical Solutions, Inc.*	Electrically-Improving a MEMS Sensor for Rolling Stock *James Hofmeister; Ridgetop Group, Inc.*	
12:00 PM	**Lunch & Special Speaker** **Certification in Condition Monitoring the Bindt PCN scheme** *Steve Greenfield , Eaton Aerospace, Technical Support Manager Powerplant & Rotorcraft Diagnostics*			
Chairperson *Location*	**Diagnostics** Greg Gammon, Texas A&M Turbomachinery Laboratory **Freedom I**	**Machine Learning / A.I.** Suri Ganeriwala, Spectra Quest, Inc. **Freedom II**	**Sensors** James Hofmeister; Ridgetop Group, Inc. **Freedom V**	**VI Track** **Freedom VI**
1:30 PM	Vibration Acceptance Testing of Rotodynamic Pumps - ANSI/HI 9.6.4 Standard *Peter Gaydon, Hydraulic Institute*	Jump Start Your Digital Transformation with a Failure Mode Approach to Predictive Maintenance *Preston Johnson, Allied Reliability Group*	Piezoelectric vs MEMs: Future of the Vibration Sensors *George Zusman; Vibration Measurement Solutions, Inc.*	Vibration Institute East Coast Meeting
2:00 PM	Recurrence Quantification Analysis for Identification of Cracks in Rotating Shafts *Student Foad Nazari, Villanova University*	How to Make Data Work for Your Organization *Chris MacDonald; PTC*	The Effect of Accelerometer Attachment Method on the Accuracy of Vibration Detection *William Marscher; Mechanical Solutions, Inc.*	
2:30 PM	The Diagnosis of Machinery through Use of Physics-Based Algorithms	Industrial Sensor and IIoT Data and Meta-Data Management	Low Power Wireless Triaxial Vibration Sensor Design - Prototype Review	

	William Marscher; Mechanical Solutions, Inc.	*Gopal Gopalkrishnan; OSI Soft, LLC*	*Ed Spence; The Machine Instrumentation Group*	
3:00 PM	**Break**			
Chairperson *Location*	**Diagnostics** John Tucker, Naval Research Laboratories **Freedom I**	**Machine Learning / A.I.** Chris Nemarich, NAVSEA **Freedom II**	**Exhibitor Presentations** Amy Wynn, MFPT Executive Director **Freedom V**	**VI Track** **Freedom VI**
3:30 PM	Diagnostic Applications of Motion Amplified Video *William Marscher; Mechanical Solutions, Inc.*	Driving Operational & Economic Impact with IIoT/Machine Learning Applications *AJ Alexander; ITG Technologies*	Allied Reliability Vibrant Technology, Inc. MRG Laboratories	Vibration Institute East Coast Meeting
4:00 PM	Bathtub, Failure Distribution, MTBF, MTTF, and More: They are Related *Tom Heiser; Ridgetop Group, Inc.*	Estimation of Instantaneous Angular Speed and Diagnostic of Passengerss' Coach Air-Conditioner Bearings *Dr. Renata Klein, RK Diagnostics*	Mechanical Solutions Spectra Quest, Inc.	
4:30 PM	Aerial Manipulation for Remote Robotic Machinery Diagnostics *Michael Lipsett; University of Alberta*	Maintenance Free Flow Measurements *David Weeda, Steve Nelson; Chase Defense Partners*	Check schedule outside room for updated Exhibitor presentations	
5:00 PM - 8:00PM Main Ballroom	**Exhibitor Networking Reception**			

WEDNESDAY 15 May 2019

Chairperson *Location*	**Signal Processing** Michael Lipsett; University of Alberta **Freedom I**	**Case Studies** Marc Pepi; US Army Research Laboratory **Freedom II**	**Lube Oil** Paul Howard; Paul L Howard Enterprises **Freedom V**	**VI Track** **Freedom VI**
8:30 AM	The Case of the Missing Bearing Fault Frequency *Eric Bechhoefer; GPMS, Inc.*	Selection of Thermal Isolator for a Propylene Transfer Pump- A Case Study *Mantosh Bhattacharya; Petrofac*	Introduction to Oil Debris Monitoring *Steve Greenfield; Eaton Aerospace*	Vibration Institute East Coast Meeting
9:00 AM	On Vibration Signature Analysis of Rotor/Shaft Cracks *Daming Chen; Spectra Quest, Inc.*	Full Circle Mechanical Dynamic Characterization Including Experimental Modal Analysis and Finite Element Analysis *Thomas Hazelwood; ORNL*	Oil Analysis for Condition Monitoring in an IIoT Era *Preston Johnson; Allied Reliability Group*	
9:30 AM	A Systematic Processing of a Gearbox Vibration Signal with Defective Rolling Element Bearing *Suri Ganeriwala, Spectra Quest, Inc.*	Fault Diagnosis of Excessive Pipe Vibration Due to Beating Phenomenon *Lim Meng Hee; Unversiti Teknologi Malaysia*	Case Study - Wear Debris in Industrial Gearboxes *Mark Redding; Poseidon Systems*	
10:00 AM				
Chairperson *Location*	**Bearings** John Lucero, NASA **Freedom I**	**Design for Reliability** James Hofmeister; Ridgetop Group **Freedom II**	**Lube Oil** Paul Howard; Paul L Howard Enterprises **Freedom V**	**VI Track** **Freedom VI**
10:30 AM	Revolution in Fluid Film Bearing Monitoring *Mike Ciocys, PRUFTECHNIK*	Condition Monitoring of a Cycloid Gearbox *Eric Bechhoefer; GPMS Inc.*	Advances in At-Site Detection of Debris in Lubricants *Patrick Henning, AMETEK Spectro Scientific*	

11:00 AM	An Octave Analysis Approach Based on Vibrational Data for Early Detection of Multiple Faults in Rolling Bearings *Session Cancelled*	One hour presentation: Design for Reliability and Risk Estimation and Acceptance *Dave Breslin, NAVSEA*	Reliability Centered Lubrication: Managing the Lifeblood of Machinery *Preston Johnson; Allied Reliability Group*	Vibration Institute East Coast Meeting
11:30 AM	Intelligent Fault Identification of Planet Bearings Using Discriminative Dictionary Learning Based Sparse Representation Classification Framework *Student Yun Kong; Tsingshua University*		Advances in Comprehensive Fluid Analysis *John Reintjes; KeyW Corp.*	
12:00 PM	**Lunch & Special Speaker** **Maintenance Analytics 4.0 European and Asian Achievements** *Diego Galar, Luleå University of Technology*			
Chairperson *Location*	**Systems Engineering and Prognostics** **Freedom I**	**Design for Reliability** Marc Pepi; US Army Research Laboratory **Freedom II**	**Lube Oil** Paul Howard; Paul L Howard Enterprises **Freedom V**	**VI Track** **Freedom VI**
1:30 PM	Design Example for a Robust PHM System: a Dual Set of Electromechanical Actuator Subsystems *James Hofmeister; Ridgetop Group*	Rapid Impact Testing of Any Size Structure *Mark Richardson, Vibrant Technology, Inc.*	Field Based Oil Condition Monitoring *Stuart Lunt, Parker Hannifin Manufacturing Ltd.*	Vibration Institute East Coast Meeting
2:00 PM	Evaluating the Accuracy of Prognostic Estimates: Remaining Useful Life, State of Health, Prognostic Horizon *James Hofmeister; Ridgetop Group*	Motion Amplification *Jeff Hay; RDI Technology, Inc*	Empowering the Shipboard Engineer with Timely Oil Monitoring *Thong Nguyen, US Navy Military Sea Lift Command*	
2:30 PM	How To Initiate Sensor Fusion within a Digital Environment *David Change; Dytran Instruments*	Deliberations on Foundation Design Method for Rotating Machinery *Mantosh Bhattacharya; Petrofac*	Lube Oil and Fuel Oil In-line Sensor Specification Development *Erin Babik, NAVSEA HQ*	
3:00 PM	**Break**			
Chairperson *Location*	**Signal Processing** Mark Richardson, Vibrant Technology, Inc. **Freedom I**	**Design for Reliability** Marc Pepi; US Army Research Laboratory **Freedom II**	**Lube Oil** Paul Howard; Paul L Howard Enterprises **oil Analysis for Condition Monitoring in an IIoT Era**	**VI Track** **Freedom VI**
3:30 PM	90 Minute Tutorial: Signal Processing Basics for Optimizing Benefits of The Current Vibration Analysis Programs. This tutorial will provide attendees with the fundamental principles of digital signal processing in a simplified manner without resorting to the underlying complex mathematical structure. *Suri Ganeriwala and Daming Chen; Spectra Quest, Inc.*	Devices for in Situ Tuning of Machinery Structures to Mitigate Resonance *Mantosh Bhattacharya; Petrofac*	Panel Discussion- New Technology Development Needed in Lube Oil Monitoring to Support and Enable CBM+ *Panelists TBA* *led by Paul Howard; Paul L Howard Enterprises*	Vibration Institute East Coast Meeting
4:00 PM		Defining and Determining the Appropriate Dynamic Analysis for Rotodynamic Pumps - ANSI/HI 9.6.8 Guideline *Peter Gaydon, Hydraulic Institute*		
4:30 PM		Towards Replacement of Failed Parts on the Battlefield via Casting Metals in 3D-Printed Desert Sand Molds *Marc Pepi; US Army Research Laboratory*		
8:00 - 9:00 PM	Sensors, Signal Analysis, and Diagnostics & Prognostics Focus Group Meetings Room assignments will be posted near registration			

THURSDAY 16 May 2019

Chairperson *Location*	**Bearings & Systems Engineering** Preston Johnson, Allied Reliability **Freedom I**	**Human Performance and Specialty Diagnostics** Amy Wynn, MFPT **Freedom II**	**Electrical Equipment** Jason Farmer, HII-Ingalls Shipbuilding **Freedom V**	**Freedom VI**
8:30 AM	Defect Severity Measurement in Rolling Element Bearings *Dr. Iain Epps; Mobolo Technology*	Monitoring Center of Mass Movement as an Indicator of Efficiency in Pre-Planned versus Reactive Agility Drills *Student Amanda Delaney; University of Dayton*	NSRP Electrical Technologies Panel: Overview of Electrical Safety Initiatives & Research Areas *Jason Farmer, HII-Ingalls Shipbuilding*	Focus Group Meeting Space
9:00 AM	60 Minute Tutorial: Systems Engineering *John Lucero; NASA Glenn Research Center*	Confined Space Monitoring *Zachary Kiehl, Aptima*	Fiber Optic Distributed Temperature Sensing for Real Time Monitoring of Electrical Systems Interconnections *Giovanni Tomasi; RSL Fiber Systems, LLC*	
9:30 AM		Development of an Accelerated Degradation Testbed for Integrated Vibration Monitoring and Online Oil Analysis of Vehicle Differential Bearings *Student Cody Walker; University of Tennessee*	Navy Arc Fault Detection and Prevention Systems *Brandon Whaley, NSWC Philadelphia*	
10:00 AM	**Break**			
Chairperson *Location*	**Bearing Diagnostics** Michael Lipsett; University of Alberta **Freedom I**	**Tutorial** Mantosh Bhattacharya; Petrofac **Freedom II**	**Electrical Equipment** Jason Farmer, HII-Ingalls Shipbuilding **Freedom V**	**Freedom VI**
10:30 AM	Damage Evolution in a Rolling Element Bearing: Spall Propagation *Professor Jacob Bortman*	90 Minute Tutorial: Introduction to Modal Analysis *Mark Richardson; Vibrant Technology, Inc.*	Inner and Outer Race Bearing Fault Analysis in a Permanent Magnet Motor *Suri Ganeriwala, Spectra Quest, Inc.*	Systems Engineering Focus Group Meeting
11:00 AM	Bearing Faults Detection and Identification Using Relational Data Clustering with Composite Differential Evolution Optimization *Issam Abu-Mahfouz; Penn State University Harrisburg*		Electromechanical Actuator a Case Study: Multivariate Analysis of Phase Currents to Detect Three Types of Faults *James Hofmeister; Ridgetop Group, Inc.*	Human Performance Systems Focus Group Meeting
11:30 AM	Bearing Fault Diagnosis with Improved Variational Mode Decomposition and Optimized Frequency Band Entropy *Student Dabin Jie; Chingqing University*		Interference of Variable Frequency Drive (VFD) on Rotating Machine Fault Diagnosis *Suri Ganeriwala, Spectra Quest, Inc.*	Data Management / AI Focus Group Meeting
12:00 PM	**Awards Luncheon & Conference Close** The MFPT Society recognizes an individual's contribution and excellence by awarding: *Fellows of the MFPT Society* *The Jack Frarey Memorial Award for Excellence* *The Henry and Sallie Pusey Best Paper Award* *The Hegner Gaberson Student Best Paper and Runner-Up Award*			

First Name	Last Name	Company	Title
Issam	Abu-Mahfouz	Penn State University Harrisburg	Professor of Mechanical Engineering
AJ	Alexander	ITG Technologies	Solution Architect
Garrett	Amsbaugh	MRG Labs	Research Chemist
Erin	Babik	NAVSEA HQ	
Eric	Bechhoefer	GPMS Inc	CEO/Chief Engineer
Kyle	Beckett	Military Sealift Command	
Ryan	Best	PRUFTECHNIK Inc.	Technical Sales Representative
Mantosh	Bhattacharya	PETROFAC INTL. LTD	Manager Mechanical - Rotating Equipment
Jesse	Black	USN	Deputy Marine Engineering NAVSEA 05ZB
Jacob	Bortman	BGU	Prof
Dave	Breslin	NAVSEA	
Kenneth	Burkhardt	Dupont	
Larry	Burrill	McKean Defense	President
David	Butchy	Vibration Specialty Corp	
David	Capesius	Superior Industrial Equipment	
Marty	Chamberlain	Perspecta	Senior Engineer
Dave	Change	Dytran Instruments	VP, CTO
Daming	Chen	Spectra Quest, Inc.	
Mike	Ciocys	PRUFTECHNIK	
Derek	Cruice	NSWC Philadelphia	Auxiliary System Engineering
Robert	Daniel	PROGNOST Systems, Inc.	Managing Director
Michael	Dawson	ITR LLC	
Robert	De La Garza	Allied Reliability Group	Product Manager, CM Services/SmartCBM™
Amanda	Delaney	University of Dayton	Master's Student
Xiaoxi	Ding	Chongqing University	student
Timothy	Donnelly	NSWC Philadelphia	Auxiliary System Engineering
NIRAV	DOSHI	PETROFAC INT LTD	Sr Engineer Rotating
Scott	Dow	SPM Instrument	Training Manager
Kevin	Dowdell	ABB Inc.	
Gerald	Duffy	Vibration Specialty Corp	
Iain	Epps	Mobolo Technology Ltd	Managing Director
Jason	Farmer	Huntington Ingalls	Engineer
Rob	Faverio	Bently Nevada	North Region Service Director
John	Foster	GenOn Energy	
Diane	Fricker	Naval Surface Warfare Center Philadelphia	

First Name	Last Name	Company	Title
Greg	Gammon	TEES Turbomachinery Lab	
Suri	Ganeriwala	Spectra Quest, Inc	
Piotr	Gasiorczyk	PCB/IMI	
Peter	Gaydon	Hydraulic Institute	
Michael	Golla	Structural Technologies	
Gopal	Gopalkrishnan	OSIsoft, LLC	Solutions Architect
Steve	Greenfield	Eaton Aerospace & Bindt	Chair of Bindt Technical Committee
Yuanhong	Guan	Caterpillar Inc.	Sr Engineering Specialist
Edgardo	Guevara	U.S. Coast Guard	Surface Forces Logistics Ctr
Roger	Hamby	General Electric- Bently Nevada	
Mahmoud	Hassanin	John Crane	
Jeff	Hay	RDI Technologies	CEO
Thomas	Hazelwood	ORNL	Mechanical Engineer
Tom	Heiser	Ridgetop Group Inc	President & CEO
Patrick	Henning	AMETEK Spectro Scientific, Inc.	Senior Director, R+D
Sean	Hickey	RDI Technologies	NE Regional Manager
James	Hofmeister	Ridgetop Group, Inc.	Distinguished Engineer
Paul	Howard	Paul L Howard Enterprises	President
M. David	Howard	ERBESSD Instruments	
Chao	Hu	Iowa State University	Assistant Professor
Dabin	Jie	Chongqing University	
Preston	Johnson	Allied Reliability Group	Platform Lead
Zachary	Kiehl	Aptima Inc. (Zach Kiehl)	Capability Lead & Research Engineer
John	Kingham	Baker Hughes, Bently Nevada	
Renata	Klein	R.K. Diagnostics	CEO
John	Knox	Emprise Corporation	Field Services Manager
Yun	Kong	Tsinghua University	Dr.
Meng	Li	Iowa State University	Research assistant
Roy	Li	BP	
Meng Hee	Lim	Universiti Teknologi Malaysia	Senior Lecturer
Michael	Lipsett	University of Alberta	Professor
Michael	Long	Vibration Institute	
Jacob	Loverich	KCF Technologies	CTO, VP
Hao	Lu	Iowa state university	Research Assistant
John	Lucero	NASA Glenn Research Center	AST, AERO EXP FAC & TEST TECHNIQUES

First Name	Last Name	Company	Title
Stuart	Lunt	Parker Kittiwake	
Zhusan	Luo	Praxair	
Chris	MacDonald	PTC	Head of Global Analytics CoE
Mark	Malanyn	SKF	
Justin	Malley	Flowserve	On-Site Application Engineer
Tom	Manning	Dupont	
Joe	Marino	Dytran Instruments	Eastern Regional Sales Manager
William	Marscher	Mechanical Solutions, Inc.	President
Craig	Miller	Mechanical Solutions	
Foad	Nazari	Villanova University	Post-Doctoral Research Fellow
Steve	Nelson	Emerson Automated Solutions	Senior Account Manager
Christopher	Nemarich	Naval Sea Systems Command	Engineering Manager
Thong	Nguyen	Military Sealift Command	
Ali	O Salem	Structural Technologies	
Paul	Obrien	j.m. canty inc	Engineering Manager
John	Ogren	GenOn Energy	
Eric	Olson	Mechanical Solutions, Inc.	VP
Marc	Pepi	US Army Research Laboratory	Materials Engineer
Colin	Pickett	Vibration Condition Monitoring	
Scott	Poirier	KARL STORZ, Endovision	Manager, Engineering Technologies
Bill	Pryor	PDMS	
Mark	Redding	Poseidon Systems	President
John	Reintjes	KeyW Corp	Senior Research Scientist
Mark	Richardson	Vibrant Technology, Inc.	President & CEO
Ronald	Ripley	Chevron Phillips Chemical Company	Sr. Rotating Equipment Engineer
Heather	Rowe	NAVSEA	Executive Assistant, 05ZB
Mahdi	Sadeghi	Analog Devices	Product Application Engineer
Alec	Salakovich	University of Tennessee - Nuclear Engineering Dept	Graduate Research Assistant
Bob	Sayer	Applied Structural Dynamics	
Brian	Schwarz	Vibrant Technology, Inc.	Director, Engineering Applications
Matt	Schweinberg	Emprise Corporation	Senoir Engineer
Charles	Scott	VIS Reliability Solutions	
Iram	Shaikh	Bently-Nevada	Mid-Atlantic Growth Account Manager
Peter	Sisa	Poseidon Systems	CBM Programs
Cy	Smith	PYA Analytics	Program Manager

First Name	Last Name	Company	Title
Ed	Spence	THe Machine Instrumentation Group	Managing Director
Stephen	Steen	Poseidon Systems	VP of Industrial IoT
Daniel	Stilp	Pruftechnik	
Ron	Sullivan	Hamar Laser Instruments, Inc	Brand and Distribution Manager
Giovanni	Tomasi	RSL Fiber Systems, LLC	CEO / Chief Technology Officer
John	Tucker	Naval Research Laboratory	Dr.
Guillermo	Urquizo	Sulzer Turbo Services Houston Inc.	
Jose	Vazquez	BRG Machinery Consulting	
Cody	Walker	University of Tennessee Knoxville	Graduate Student
John	Watling	Ontario Power Generation	Senior Technical Engineer
David	Weeda	Chase Defense	Government Program Manager
Brandon	Whaley	NSWC Philadelphia	
Samuel	Wick	DXP Enterprises	Director of IOT
Chris	Wilcox	m+p International Inc.	Sales Manager
Eric	Woods	Flowserve	Senior Applications Engineer
Rich	Wurzbach	MRG Laboratories	President
Jun	Yang	Spectra Quest, Inc	
George	Zusman	VMS, Inc.	President

RE-IMAGINING THE MODEL FOR RELIABILITY AND READINESS DATA ANALYTICS

Larry D. Burrill
President, McKean Technical Services
McKean Defense Group

The Challenge to do More

Spencer said he hoped to see more ***innovation*** throughout the Navy and Marine Corps – innovation in the sense of identifying everyday problems a unit faces, and solving those problems with the tools at the unit's disposal. And where additional tools are required, Spencer said he wants to know.

Innovation is achieved "not just by beating people over the head, saying, 'save money, be safe, be efficient.' It's by giving business cases and examples of what they can do, and giving them latitude," he said.

"I've got to make sure that the whole system understands ***urgency***," he added. "We have the dollars we—finally—need. We have the resources. Time is something that we can't buy, and we have to have everyone focus on that."

- Secretary of Navy Honorable Richard Spencer
USNI News

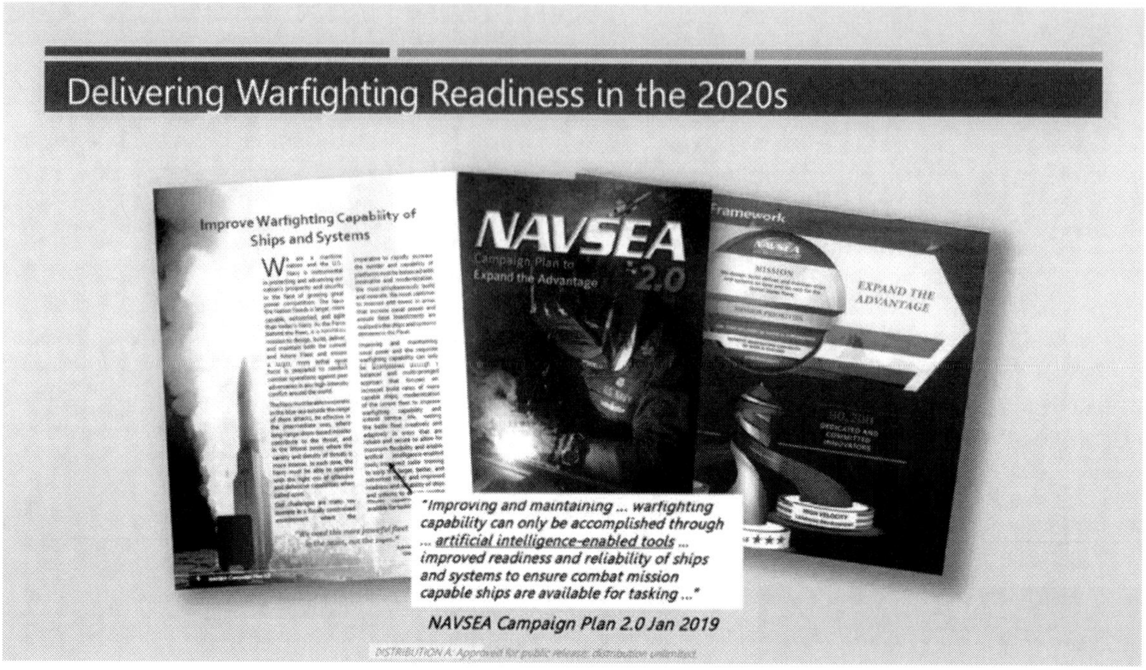
Delivering Warfighting Readiness in the 2020s
Improve Warfighting Capability of Ships and Systems
NAVSEA
Campaign Plan to Expand the Advantage
2.0
EXPAND THE ADVANTAGE
"Improving and maintaining ... warfighting capability can only be accomplished through ... artificial intelligence-enabled tools ... improved readiness and reliability of ships and systems to ensure combat mission capable ships are available for tasking ..."
NAVSEA Campaign Plan 2.0 Jan 2019
DISTRIBUTION A: Approved for public release: distribution unlimited

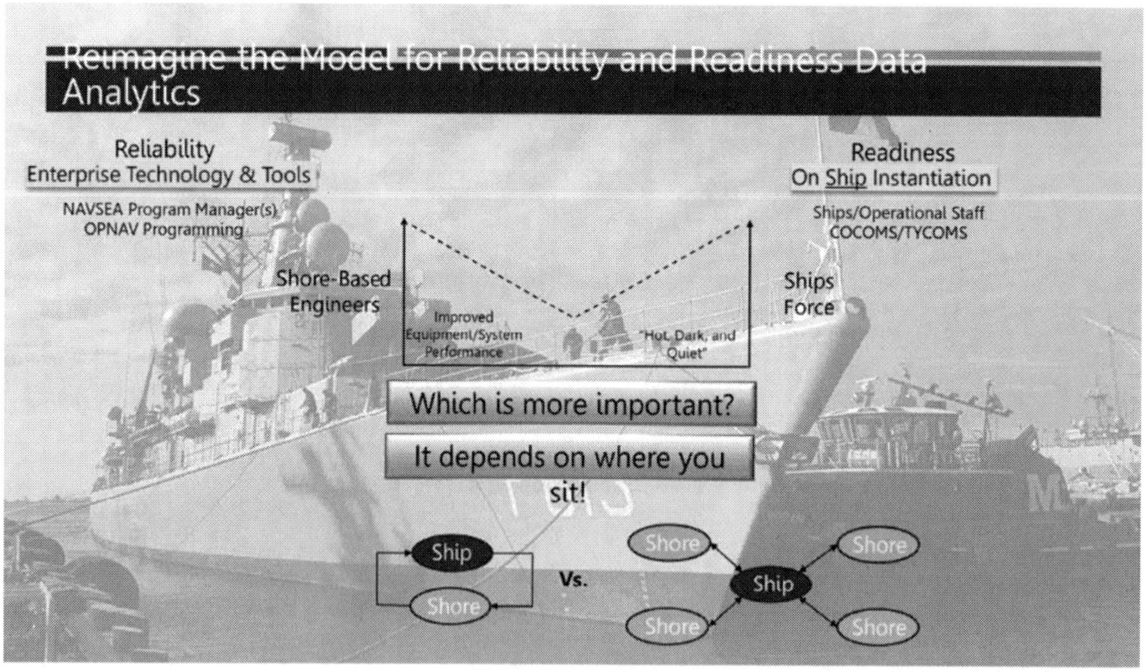
Reimagine the Model for Reliability and Readiness Data Analytics
Reliability
Enterprise Technology & Tools
NAVSEA Program Manager(s)
OPNAV Programming
Readiness
On Ship Instantiation
Ships/Operational Staff
COCOMS/TYCOMS
Shore-Based Engineers
Ships Force
Improved Equipment/System Performance
"Hot, Dark, and Quiet"
Which is more important?
It depends on where you sit!
Ship
Shore
Vs.
Shore
Shore
Ship
Shore
Shore

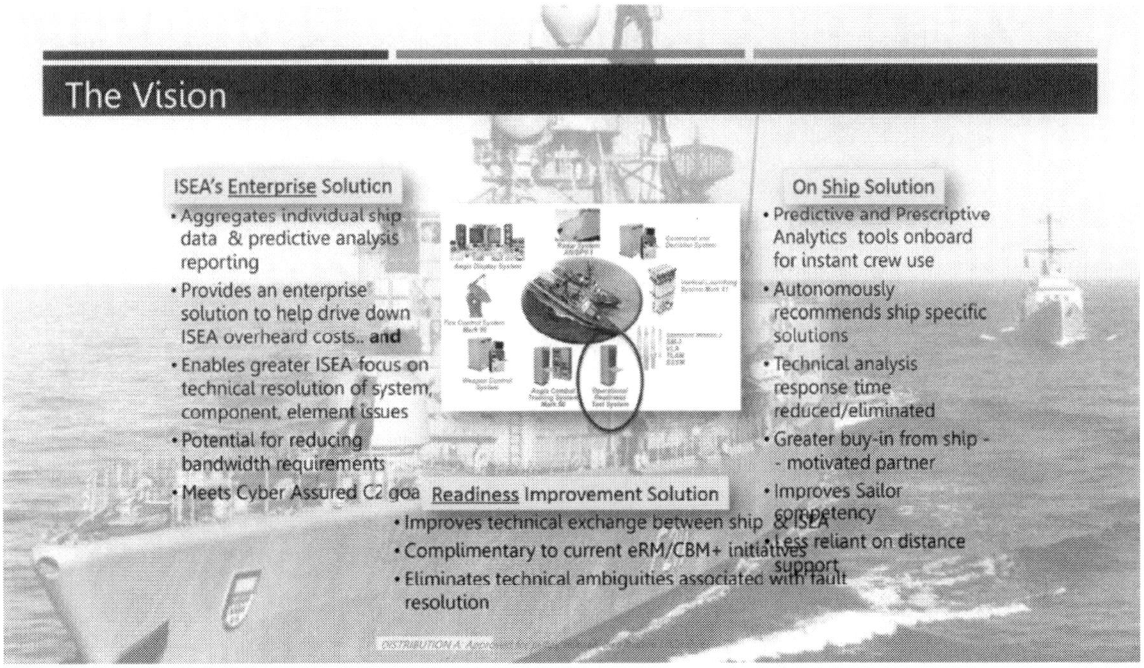
The Vision
ISEA's Enterprise Solution
• Aggregates individual ship data & predictive analysis reporting
• Provides an enterprise solution to help drive down ISEA overheard costs.. and
• Enables greater ISEA focus on technical resolution of system, component, element issues
• Potential for reducing bandwidth requirements
• Meets Cyber Assured C2 goa
On Ship Solution
• Predictive and Prescriptive Analytics tools onboard for instant crew use
• Autonomously recommends ship specific solutions
• Technical analysis response time reduced/eliminated
• Greater buy-in from ship - - motivated partner
• Improves Sailor competency
• Less reliant on distance support
Readiness Improvement Solution
• Improves technical exchange between ship & ISEA
• Complimentary to current eRM/CBM+ initiatives
• Eliminates technical ambiguities associated with fault resolution

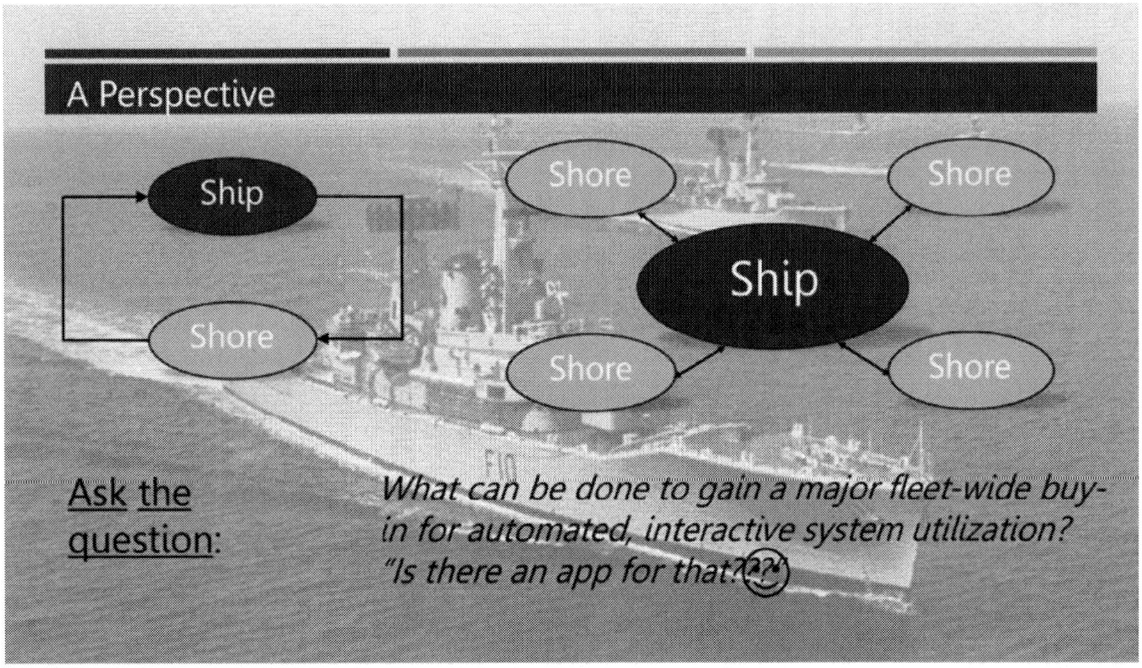
A Perspective
Ship
Shore
Shore
Shore
Ship
Shore
Shore
Ask the question:
What can be done to gain a major fleet-wide buy-in for automated, interactive system utilization?
"Is there an app for that?"

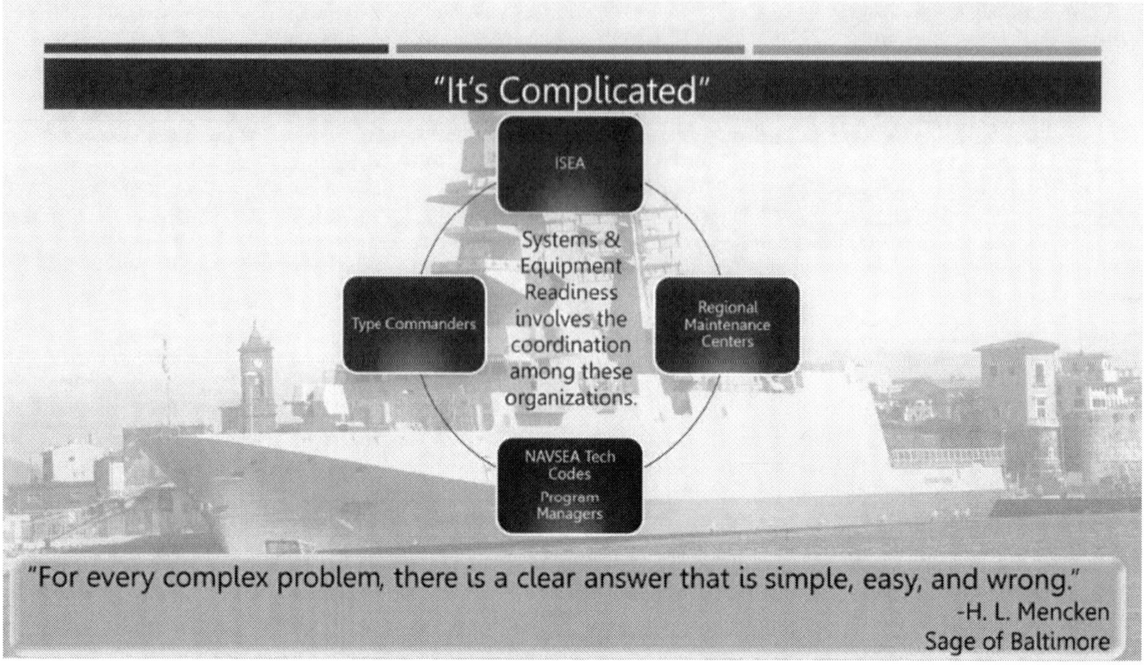
"It's Complicated"
ISEA
Type Commanders
Systems & Equipment Readiness involves the coordination among these organizations.
Regional Maintenance Centers
NAVSEA Tech Codes
Program Managers
"For every complex problem, there is a clear answer that is simple, easy, and wrong."
-H. L. Mencken
Sage of Baltimore

MFPT 2019 Where Theory Meets Practice

MAY 14-16, 2019 | PHILADELPHIA, PA

ENABLING MACHINE INTELLIGENCE

ITG

INFORMATION • AUTOMATION • INTEGRATED SOLUTIONS

- Digital Transformation Company with Deep Domain Knowledge in Automation Control & Artificial Intelligence.
- Leveraging SORBA, an AI/IIoT software Platform, for Predictive Maintenance & Process Optimization Solutions.
- Data and Time are Valuable Assets which are Underutilized and Possess Huge Economic Opportunity.
- Industrial Domain Expertise Coupled with Market Leading Technology Offers a True End-To-End IoT Solution.

SORBA

AJ Alexander | Solution Architect | 904.252.2362 | aja@itgtec.com |

MFPT 2019 Where Theory Meets Practice

MAY 14-16, 2019 | PHILADELPHIA, PA

OUR CUSTOMERS

Roche

KOCH INDUSTRIES INC

APR ENERGY. Powering your progress®

Johnson & Johnson

JEA®

OnSyte PERFORMANCE™

TISSUETECH®

OCEANEERING®

TARGET

CSX™

GERDAU

AJ Alexander | Solution Architect | 904.252.2362 | aja@itgtec.com |

sales@sorbasoft.com

MFPT 2019 Where Theory Meets Practice
MAY 14-16, 2019 | PHILADELPHIA, PA

4 KEYS FOR DIGITAL TRANSFORMATION

- ADMIT YOU HAVE A PROBLEM
- WRITE YOUR VISION ON THE WALL
- ASK FOR HELP
- WORK TOGETHER

AJ Alexander | Solution Architect | 904.252.2362 | aja@itgtec.com | sales@sorbasoft.com

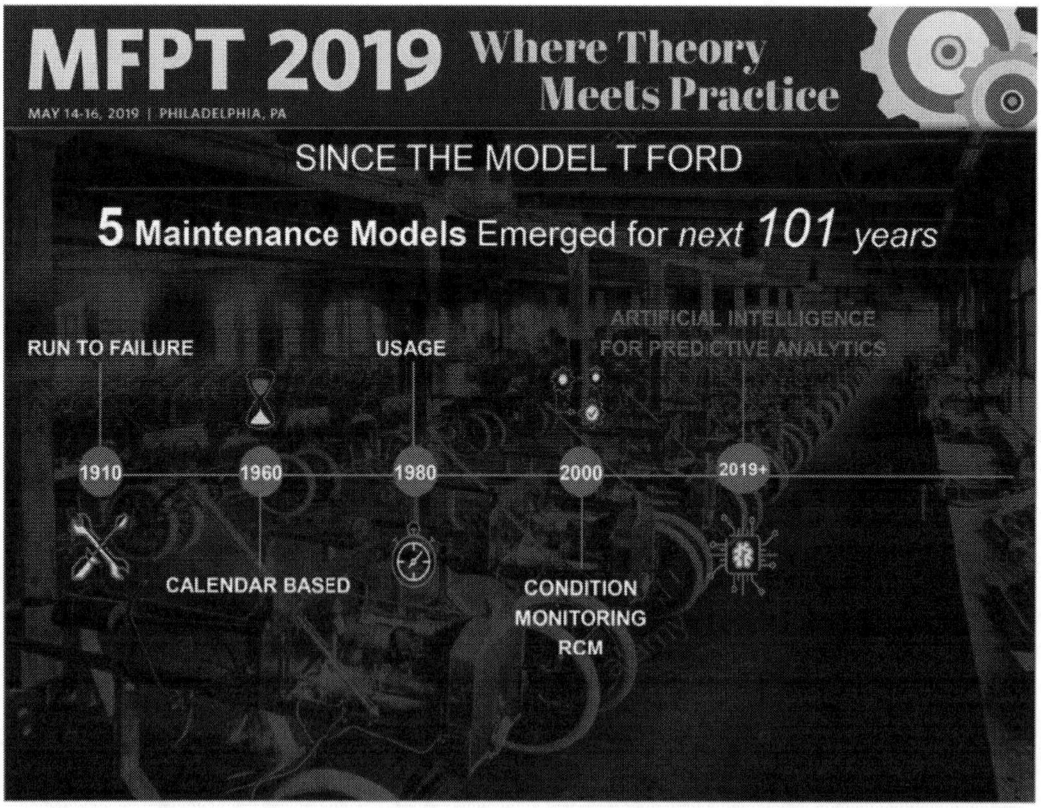

MFPT 2019 Where Theory Meets Practice
MAY 14-16, 2019 | PHILADELPHIA, PA

INEFFECTIVE MAINTENANCE/OPTIMIZATION - 100 YEARS USING SAME METHOD

"85% of all equipment failures happen on a time random basis regardless of inspection and service."

BOEING®

EMERSON™

"63% of all maintenance is unnecessary and causes more problems than it fixes."

AJ Alexander | Solution Architect | 904.252.2362 | aja@itgtec.com | sales@sorbasoft.com

MFPT 2019 Where Theory Meets Practice
MAY 14-16, 2019 | PHILADELPHIA, PA

MACHINE INTELLIGENCE TO SELF DIAGNOSE ITSELF

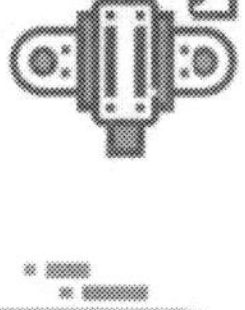

Hey Jim!
It's not the oil pressure, adjust my air value!

Hey operator!
I will have a main bearing failure in 28 days, please take me to the shop!

Operator, my quality is low! The raw materials are out of spec!

AJ Alexander | Solution Architect | 904.252.2362 | aja@itgtec.com | sales@sorbasoft.com

MFPT 2019 Where Theory Meets Practice

MAY 14-16, 2019 | PHILADELPHIA, PA

TRADITIONAL METHODS VS MACHINE LEARNING

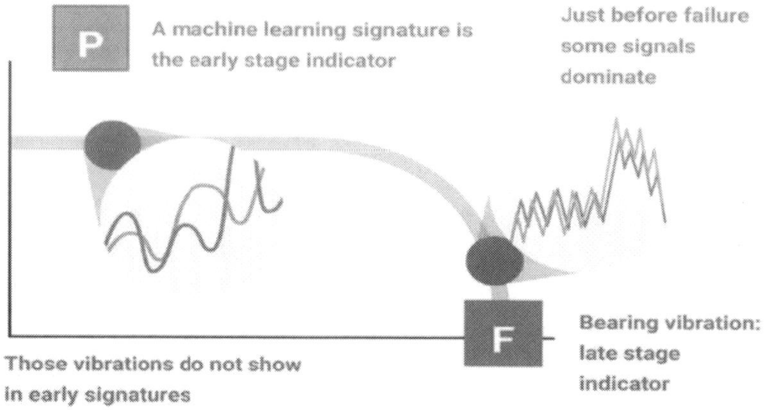

AJ Alexander | Solution Architect | 904.252.2362 | aja@itgtec.com |

sales@sorbasoft.com

MFPT 2019 Where Theory Meets Practice

MAY 14-16, 2019 | PHILADELPHIA, PA

MACHINE LEARNING IS COMPLICATED

- Businesses that want to use machine learning to solve a problem call on the cavalry.
- Simple problem requires data scientists, ML experts, and aging knowledge base.
- Process is inefficient, and takes months to get results.
- Only solves the immediate problem.
- Next problem, repeat the process all over again.

Are there AI automation tools available that enable a SME to solve these problems rapidly, confidently, and at scale???

AJ Alexander | Solution Architect | 904.252.2362 | aja@itgtec.com |

sales@sorbasoft.com

DESIGNED WITH THE END IN MIND

Benefits of increased computing power

- i.e. Time to Scan / Process 4.8 Terabytes of Data
- Traditional Relational Databases would normally take around 13 hours to process
- Versus processing in...

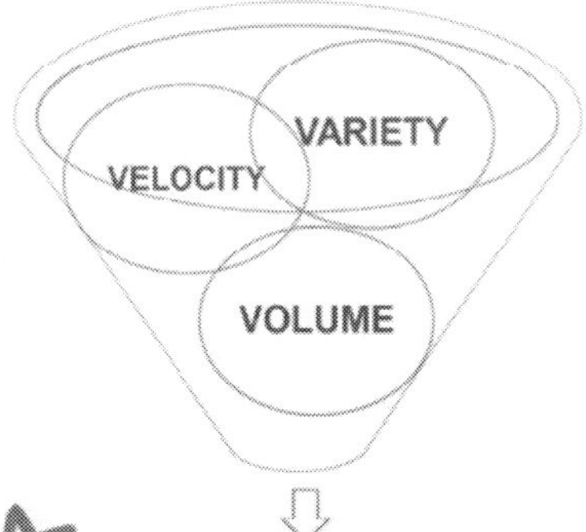

cassandra hadoop APACHE Spark

30 MINUTES

AJ Alexander | Solution Architect | 904.252.2362 | aja@itgtec.com |
sales@sorbasoft.com

MFPT 2019 Where Theory Meets Practice
MAY 14-16, 2019 | PHILADELPHIA, PA

THE AI ENTERPRISE OPERATING SYSTEM

IIoT Platform

Machine Learning For Predictive Analytics

Smart Data Collector

Centralized Management

Big Data Historian

Digital Twin For Advance Process Control

AJ Alexander | Solution Architect | 904.252.2362 | aja@itgtec.com |
sales@sorbasoft.com

IIoT PLATFORM

- **SCALABLE** and distributed Platform
- High availability no signal point of failure
- Load Balancing
- **OPEN INTEGRATION** 3rd Party Application Integration (REST API)
- **ACCESS DATA** Read/Write Real-time and Historical Data
- **MANAGE COMMUNICATIONS** Adjustable data frequency, packet size, connection thread
- **SECURE** deployment & management Cloud, On-Premise, or Hybrid

cassandra

hadoop

APACHE Spark

AJ Alexander | Solution Architect | 904.252.2362 | aja@itgtec.com |
sales@sorbasoft.com

MFPT 2019 Where Theory Meets Practice
MAY 14-16, 2019 | PHILADELPHIA, PA

SMART DATA COLLECTOR

Your IIoT Edge Gateway Interface To The Real World:

- Real-Time Data Collection
- Localized Storage
- Lightweight Historian
- Automate All ETL Data Feature Engineering
- Real-Time Analytics/ML Processing

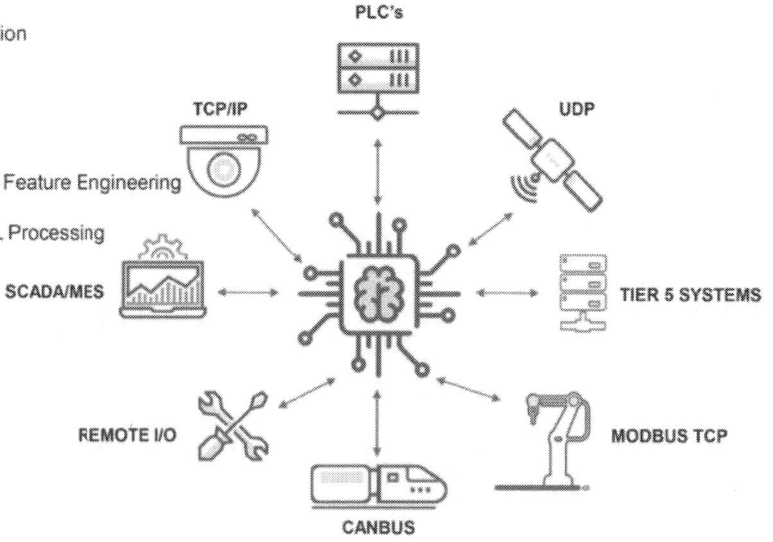

AJ Alexander | Solution Architect | 904.252.2362 | aja@itgtec.com |
sales@sorbasoft.com

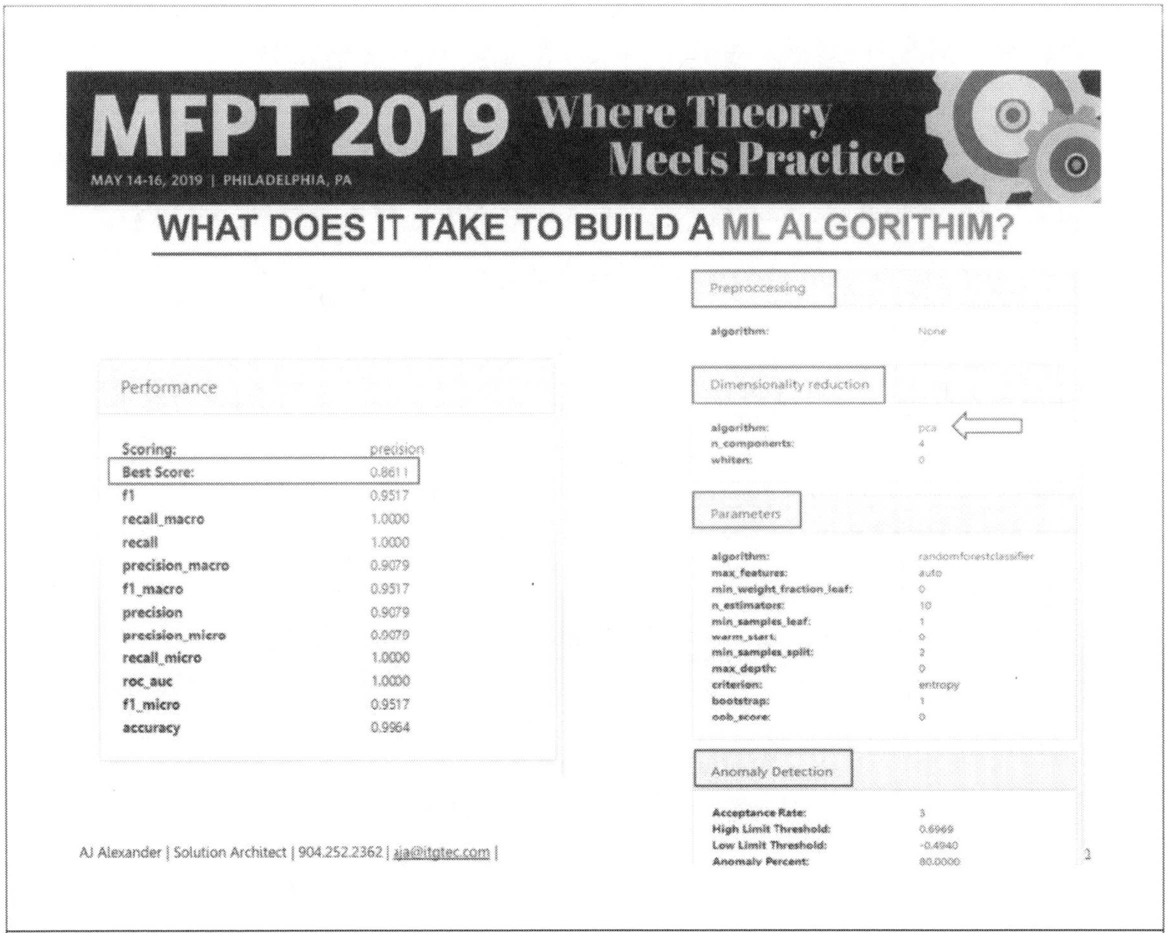
MFPT 2019
Where Theory Meets Practice
MAY 14-16, 2019 | PHILADELPHIA, PA
WHAT DOES IT TAKE TO BUILD A ML ALGORITHIM?
Performance
Scoring: precision
Best Score: 0.8611
f1 0.9517
recall_macro 1.0000
recall 1.0000
precision_macro 0.9079
f1_macro 0.9517
precision 0.9079
precision_micro 0.9079
recall_micro 1.0000
roc_auc 1.0000
f1_micro 0.9517
accuracy 0.9964
Preproccessing
algorithm: None
Dimensionality reduction
algorithm: pca
n_components: 4
whiten: 0
Parameters
algorithm: randomforestclassifier
max_features: auto
min_weight_fraction_leaf: 0
n_estimators: 10
min_samples_leaf: 1
warm_start: 0
min_samples_split: 2
max_depth: 0
criterion: entropy
bootstrap: 1
oob_score: 0
Anomaly Detection
Acceptance Rate: 3
High Limit Threshold: 0.6969
Low Limit Threshold: -0.4940
Anomaly Percent: 80.0000
AJ Alexander | Solution Architect | 904.252.2362 | aja@itgtec.com |

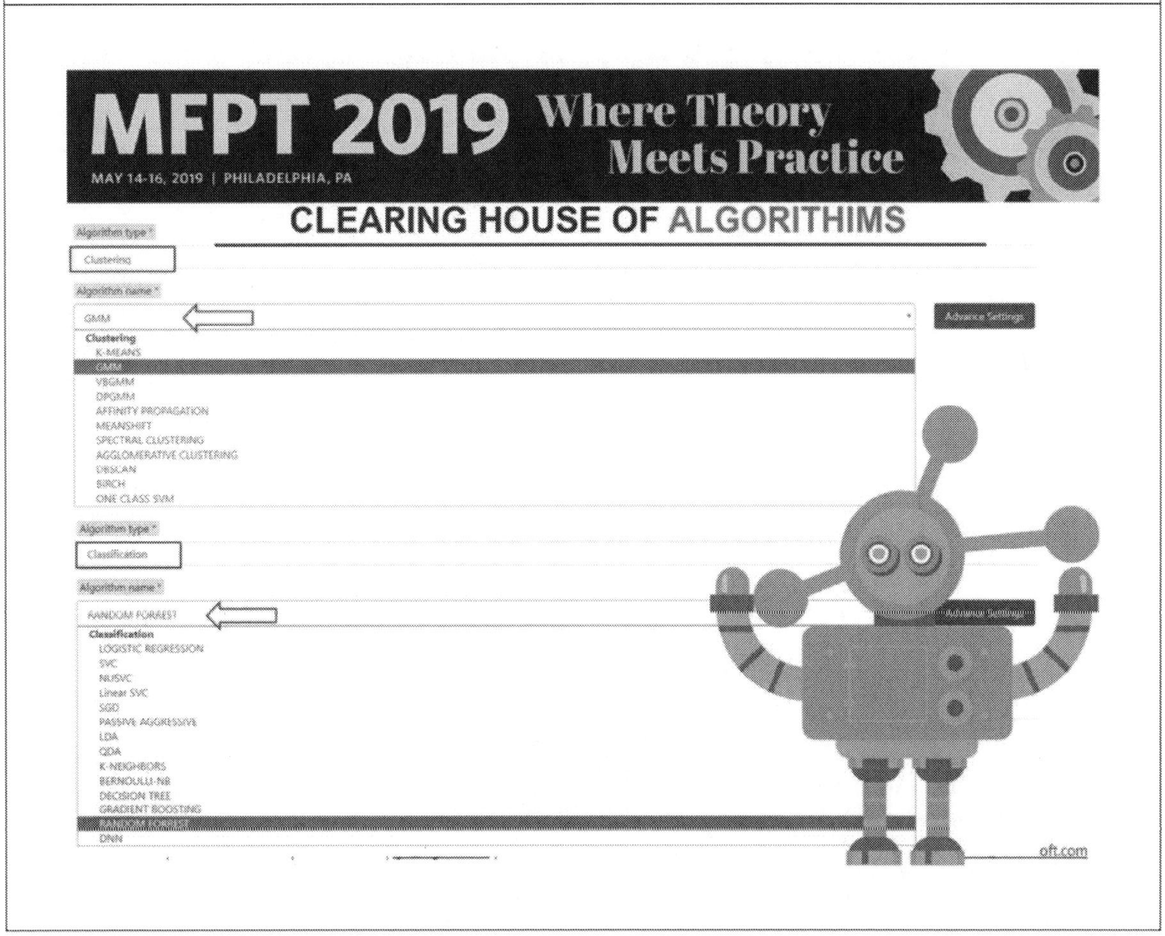
MFPT 2019
Where Theory Meets Practice
MAY 14-16, 2019 | PHILADELPHIA, PA
CLEARING HOUSE OF ALGORITHIMS
Algorithm type *
Clustering
Algorithm name *
GMM
Advance Settings
Clustering
K-MEANS
GMM
VBGMM
DPGMM
AFFINITY PROPAGATION
MEANSHIFT
SPECTRAL CLUSTERING
AGGLOMERATIVE CLUSTERING
DBSCAN
BIRCH
ONE CLASS SVM
Algorithm type *
Classification
Algorithm name *
RANDOM FORREST
Classification
LOGISTIC REGRESSION
SVC
NUSVC
Linear SVC
SGD
PASSIVE AGGRESSIVE
LDA
QDA
K-NEIGHBORS
BERNOULLI-NB
DECISION TREE
GRADIENT BOOSTING
DNN

MFPT 2019 Where Theory Meets Practice
MAY 14-16, 2019 | PHILADELPHIA, PA

AUTOMATED CLOSED LOOP ANALYTICS

- Operator Decisions are good but not optimal
- Operators base decisions on experience & training
- Look at a few variables to support decisions
- Cannot process all data

- Machine Learning Models (RNN, CNN, RL)
- Inputs are measured from the process
- Inputs are selected based on correlation analysis
- Training on historical data of known healthy states
- Retraining adjusts for dynamic changes in the process

MFPT 2019 Where Theory Meets Practice
MAY 14-16, 2019 | PHILADELPHIA, PA

REAL-TIME ADVANCE PROCESS CONTROL

PID LOOP ON STEROIDS

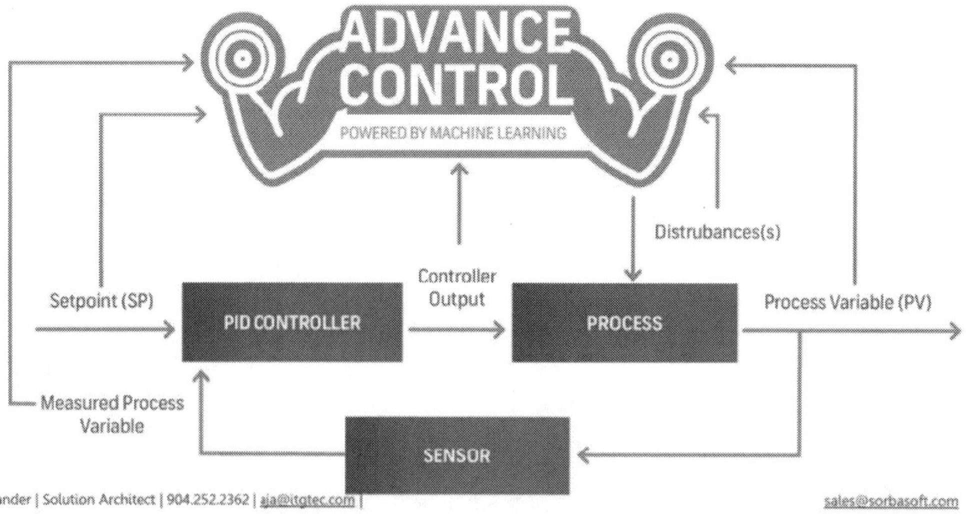

AJ Alexander | Solution Architect | 904.252.2362 | aja@itgtec.com |

sales@sorbasoft.com

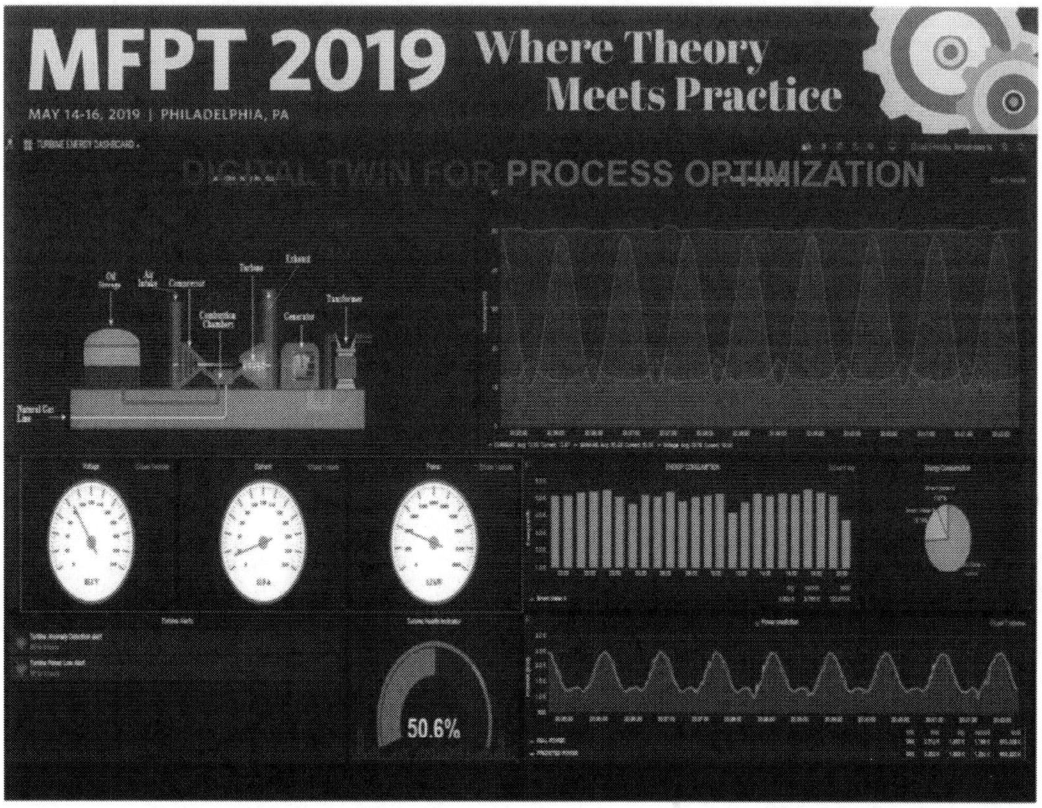

MFPT 2019 Where Theory Meets Practice

MAY 14-16, 2019 | PHILADELPHIA, PA

CENTRALIZED MANAGEMENT

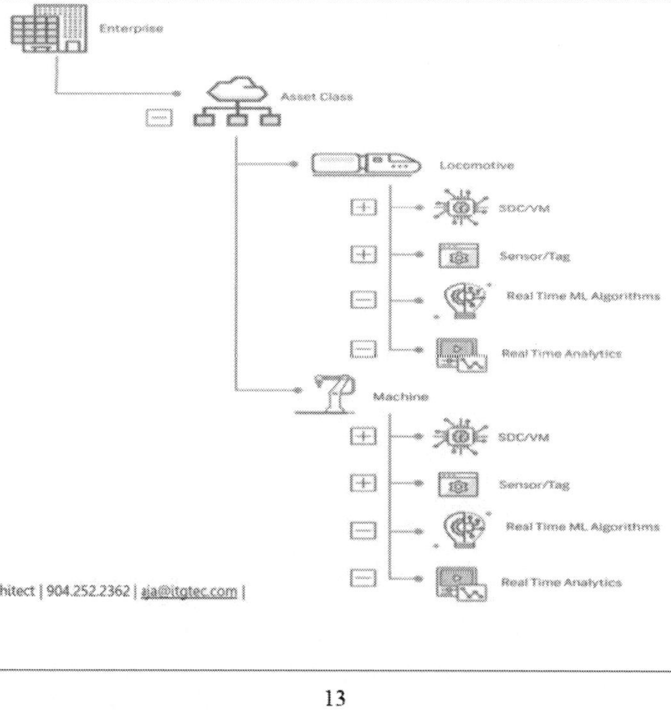

AJ Alexander | Solution Architect | 904.252.2362 | aja@itgtec.com | sales@sorbasoft.com

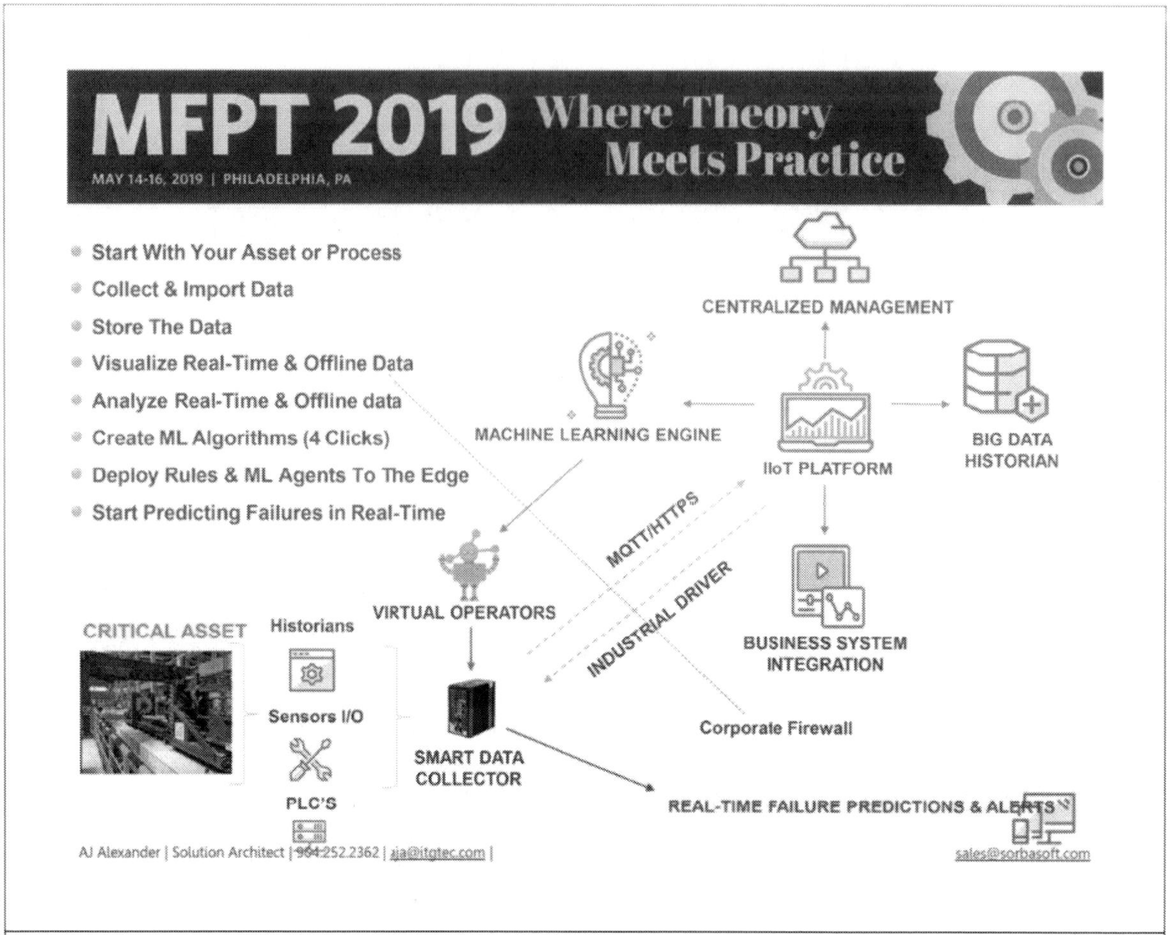
MFPT 2019 Where Theory Meets Practice
MAY 14-16, 2019 | PHILADELPHIA, PA
Start With Your Asset or Process
Collect & Import Data
Store The Data
Visualize Real-Time & Offline Data
Analyze Real-Time & Offline data
Create ML Algorithms (4 Clicks)
Deploy Rules & ML Agents To The Edge
Start Predicting Failures in Real-Time
CENTRALIZED MANAGEMENT
MACHINE LEARNING ENGINE
IIoT PLATFORM
BIG DATA HISTORIAN
MQTT/HTTPS
INDUSTRIAL DRIVER
BUSINESS SYSTEM INTEGRATION
VIRTUAL OPERATORS
CRITICAL ASSET
Historians
Sensors I/O
PLC'S
SMART DATA COLLECTOR
Corporate Firewall
REAL-TIME FAILURE PREDICTIONS & ALERTS
AJ Alexander | Solution Architect | 904.252.2362 | aja@itgtec.com |
sales@sorbasoft.com

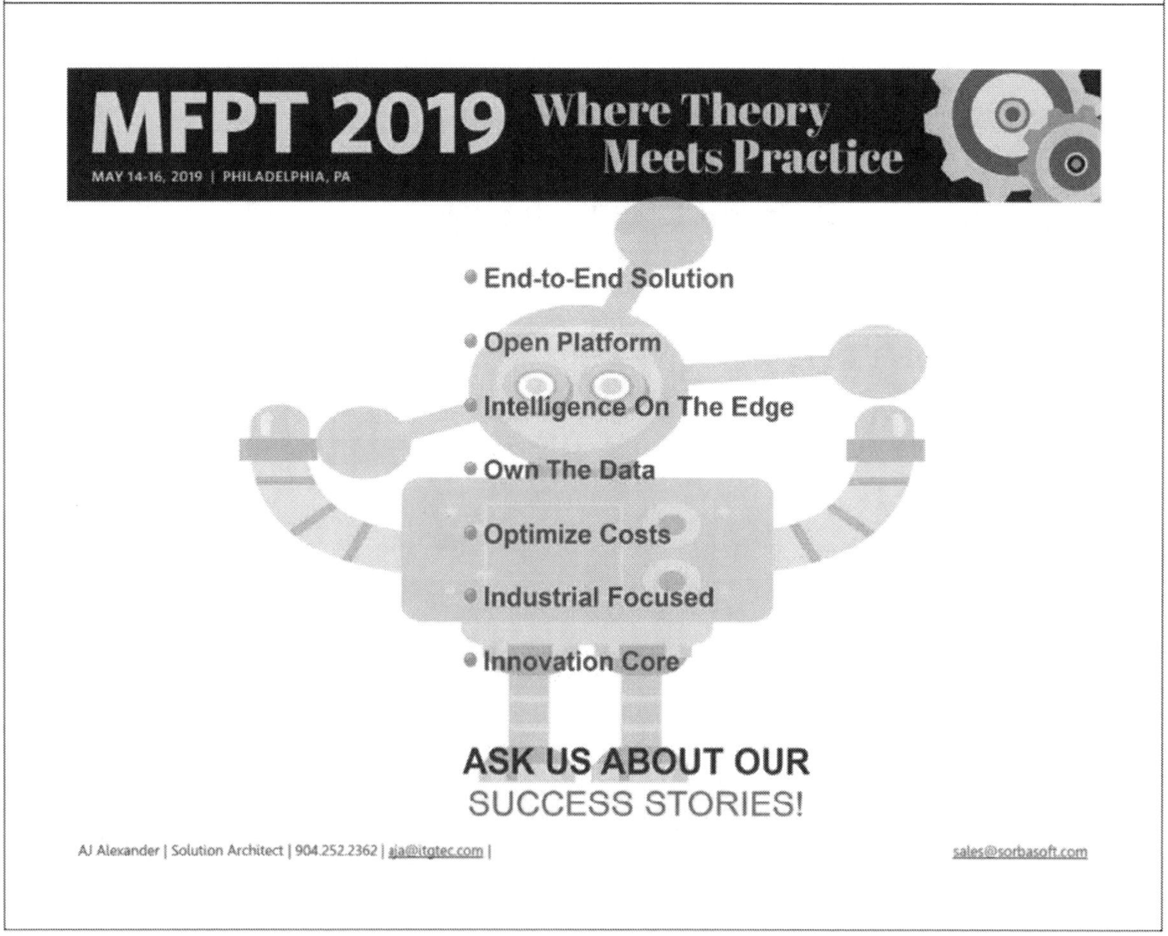
MFPT 2019 Where Theory Meets Practice
MAY 14-16, 2019 | PHILADELPHIA, PA
End-to-End Solution
Open Platform
Intelligence On The Edge
Own The Data
Optimize Costs
Industrial Focused
Innovation Core
ASK US ABOUT OUR SUCCESS STORIES!
AJ Alexander | Solution Architect | 904.252.2362 | aja@itgtec.com |
sales@sorbasoft.com

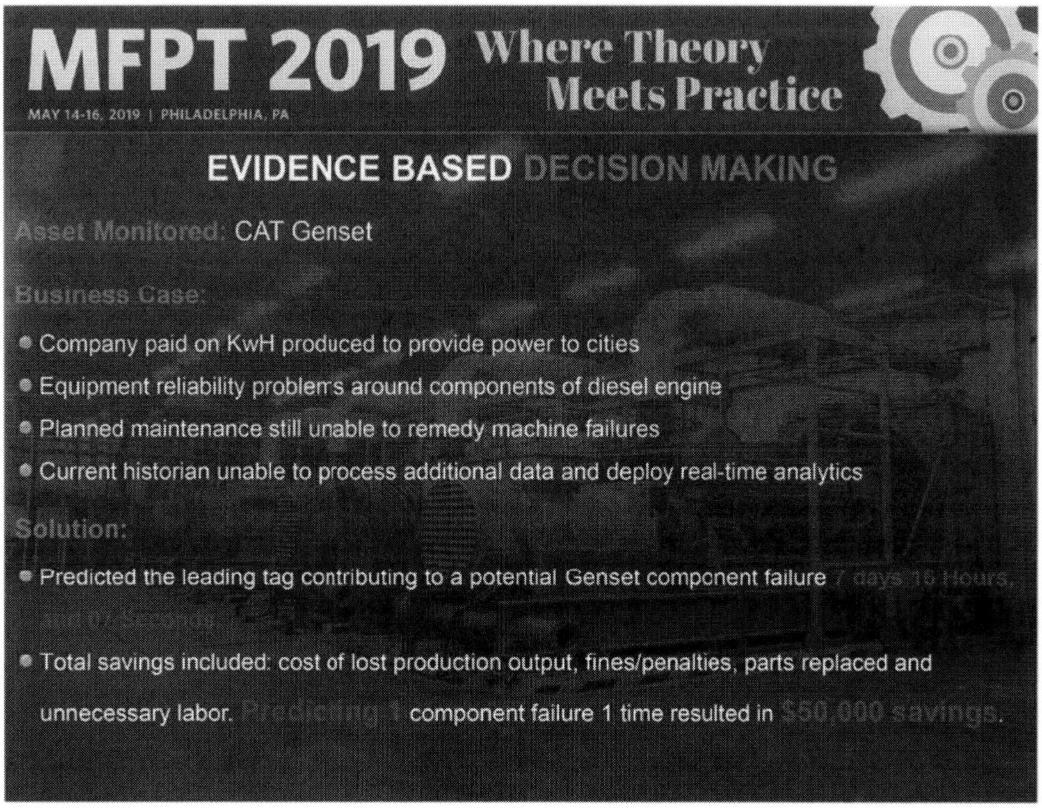
MFPT 2019
Where Theory Meets Practice
MAY 14-16, 2019 | PHILADELPHIA, PA
EVIDENCE BASED DECISION MAKING
Asset Monitored: CAT Genset
Business Case:
Company paid on KwH produced to provide power to cities
Equipment reliability problems around components of diesel engine
Planned maintenance still unable to remedy machine failures
Current historian unable to process additional data and deploy real-time analytics
Solution:
Predicted the leading tag contributing to a potential Genset component failure 7 days 16 Hours, and 07 Seconds.
Total savings included: cost of lost production output, fines/penalties, parts replaced and unnecessary labor. Predicting 1 component failure 1 time resulted in $50,000 savings.

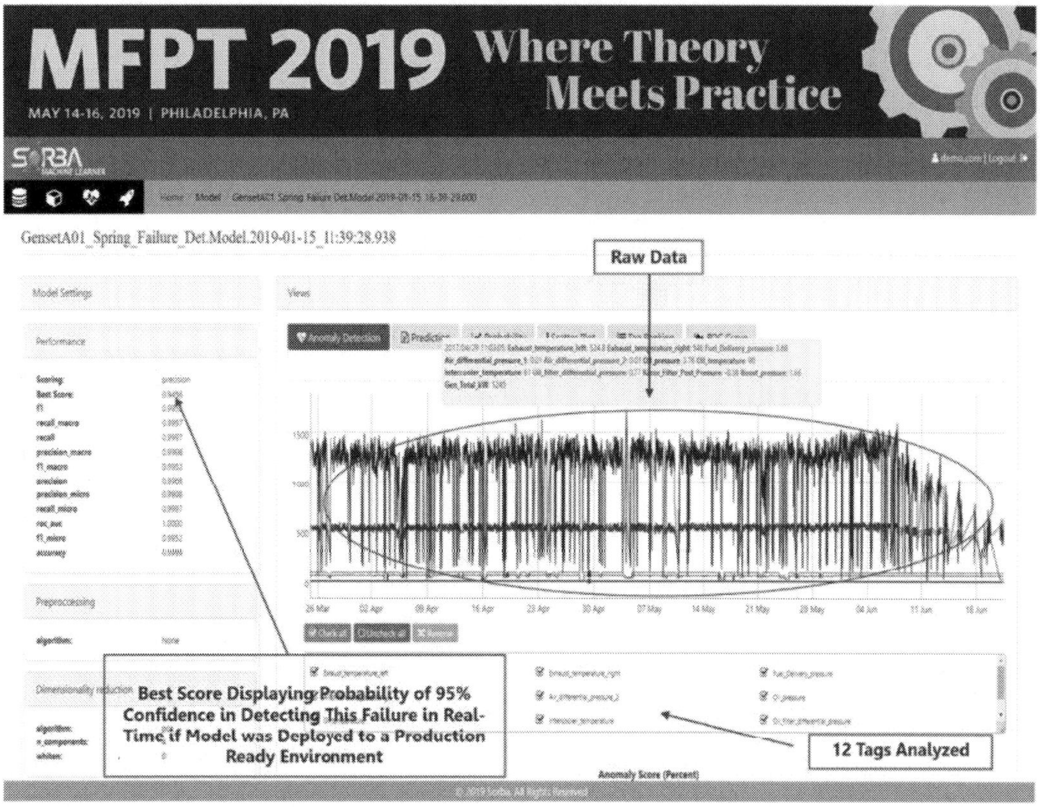
MFPT 2019
Where Theory Meets Practice
MAY 14-16, 2019 | PHILADELPHIA, PA
GensetA01_Spring_Failure_Det.Model.2019-01-15_11:39:28.938
Raw Data
Model Settings
Views
Performance
Anomaly Detection
Scoring:
Best Score:
f1
recall_macro
recall
precision_macro
f1_macro
precision
precision_micro
recall_micro
roc_auc
f1_micro
accuracy
Preprocessing
26 Mar
02 Apr
09 Apr
16 Apr
23 Apr
30 Apr
07 May
14 May
21 May
28 May
04 Jun
11 Jun
18 Jun
Best Score Displaying Probability of 95% Confidence in Detecting This Failure in Real-Time if Model was Deployed to a Production Ready Environment
12 Tags Analyzed
Anomaly Score (Percent)

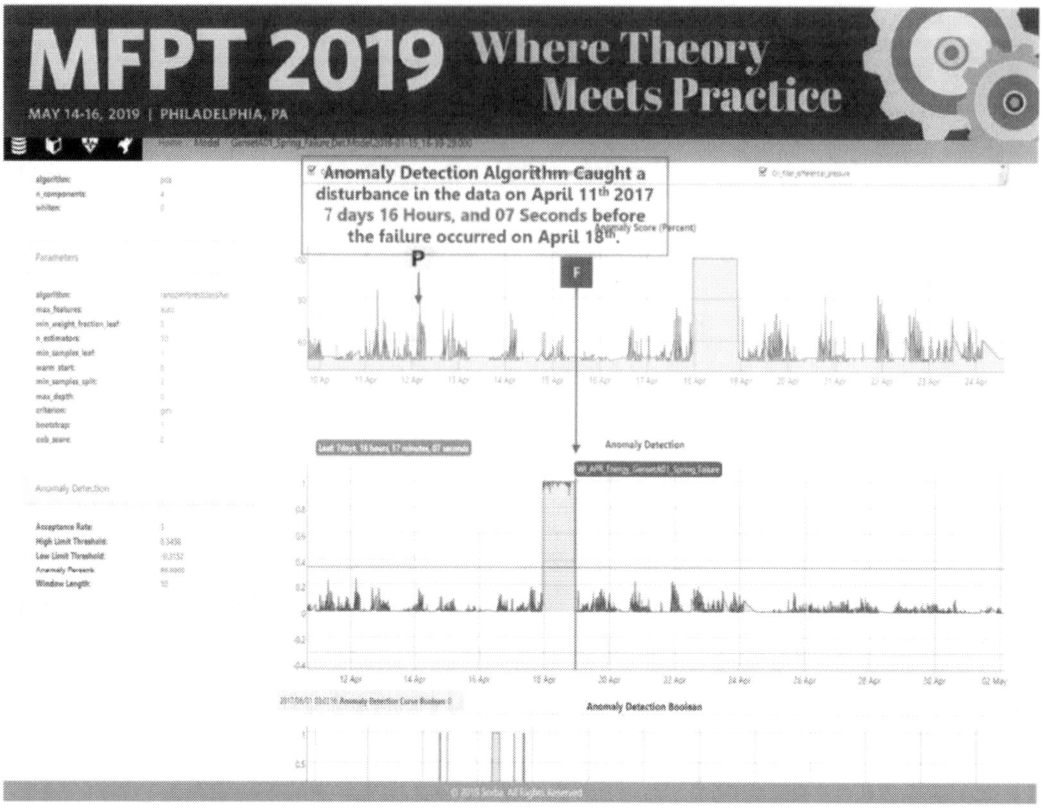
MFPT 2019
Where Theory Meets Practice
MAY 14-16, 2019 | PHILADELPHIA, PA
Anomaly Detection Algorithm Caught a disturbance in the data on April 11th 2017 7 days 16 Hours, and 07 Seconds before the failure occurred on April 18th.
P
F
Anomaly Detection
Anomaly Detection Boolean

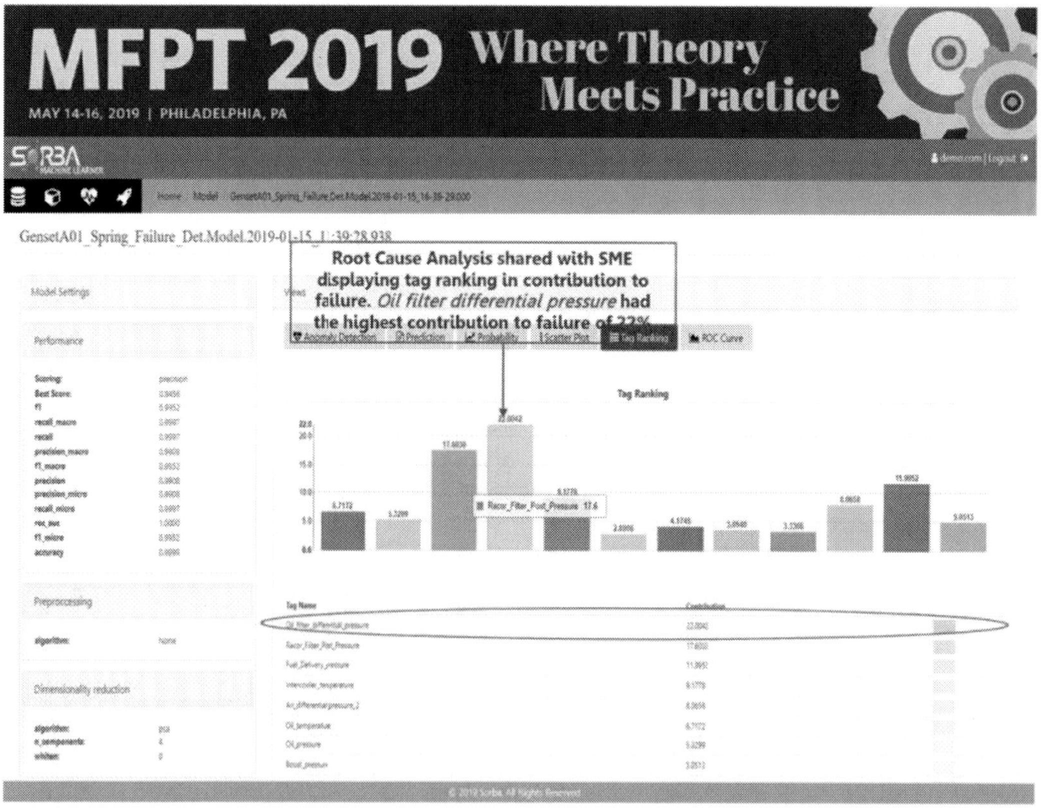
MFPT 2019
Where Theory Meets Practice
MAY 14-16, 2019 | PHILADELPHIA, PA
GensetA01_Spring_Failure_Det.Model.2019-01-15_1_:39:28.938
Root Cause Analysis shared with SME displaying tag ranking in contribution to failure. Oil filter differential pressure had the highest contribution to failure of 22%
Tag Ranking

MILITARY SEALIFT COMMAND

Empowering the Shipboard Engineer with Timely Oil Monitoring

Kyle Beckett
Senior Mechanical Engineer
Engineering Logistics and Analysis

Who is MSC?

United We Sail!

Shipboard Automated Maintenance Management

MSC's condition based monitoring program utilizes a program called SAMM (Shipboard Automated Maintenance Management). SAMM captures repair actions, machinery History, planned maintenance schedules, Test Data, and feedback to effectively monitor equipment health.

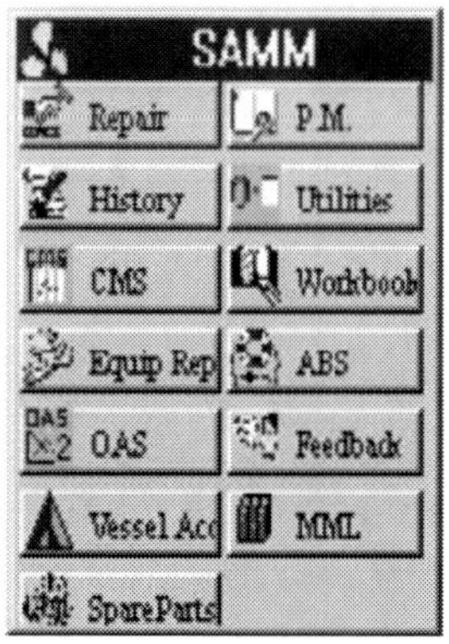

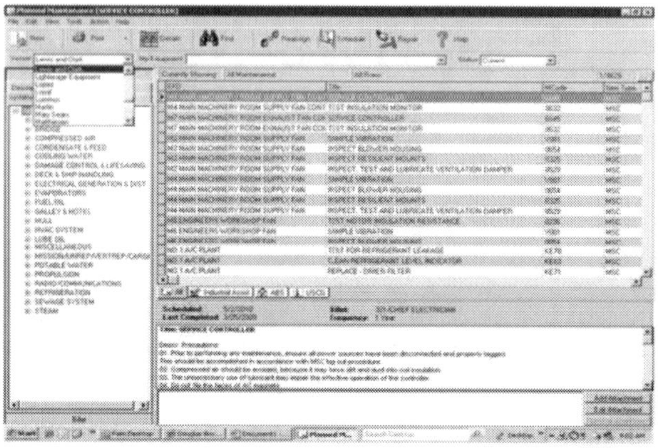

SAMM Modules

Logbook -- links directly to the Deck and Engine Logs. There is also a separate Icon on desktop for these applications.

Condition Monitoring – Repository for all Condition Monitoring test results – View Only module

Machinery History – Repository for all history records from corrective and planned maintenance actions. New records, beyond Corrective and planned maintenance completions – can be added from this module.

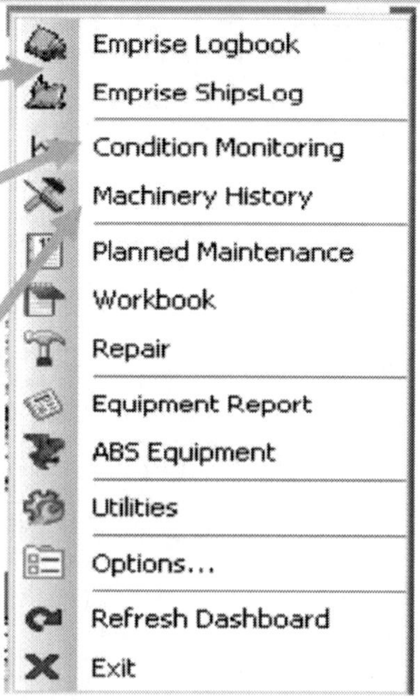

Shipboard Oil Analysis

TBN/TAN
Oxidation
Nitration
Sulfation
Antiwear additive depletion
Water
Soot
Glycol

Kinematic Viscosity @ 40C

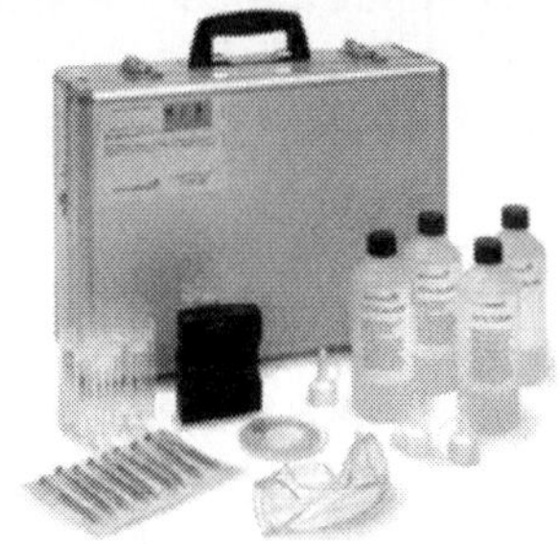

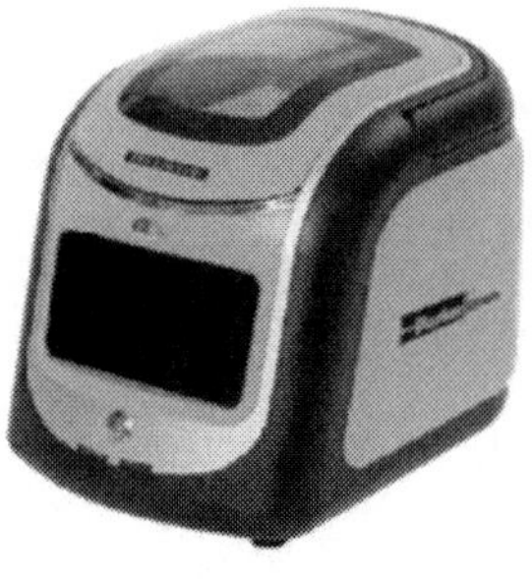

United We Sail! 5

Logbook

> Lube Oil Analysis -> Main Propulsion Engine 1

Parameter (Units)	5/8/2019 5:00:00 PM	5/7/2019 5:00:00 PM	5/6/2019 5:00:00 PM	5/5/2019 5:00:00 PM	5/4/2019 5:00:00 PM	4/29/2019 2:00:00 PM	4/22/2019 5:00:00 PM
Sulfation (Abs/mm)	18	18	18	18	18	18	17
Oil/Filter Change (Y/N)	No	No	No	No	No	No	No
Oxidation (Abs/mm)	12	12	12	12	12	12	12
Soot (%)	0.65	0.56	0.56	0.58	0.75	0.77	0.56
Total Base Number (TBN) (mgKOH/g)	14.5	14.7	14.7	14.8	13.7	13.7	14.7
Water (PPM)	15.00	15.00	15.00	0.00	0.00	0.00	113.00
Anti-Wear (Percent)	17.0	17.0	17.0	17.0	18.0	19.0	15.0
Viscosity (cSt)	144.3	142.5	142.6	143.3	145.0	142.6	144.8

Lube Oil Analysis -> Generator 1

Parameter (Units)	5/6/2019 6:00:00 PM	4/29/2019 11:30:00 AM	4/22/2019 11:30:00 AM	4/15/2019 11:30:00 PM	4/8/2019 11:30:00 PM	3/25/2019 3:30:00 PM	3/18/2019 12:00:00 PM
Soot (%)	0.01	0.00	0.04	0.00	0.00	0.00	0.01
Oil/Filter Change (Y/N)	No	No	No	No	Yes	No	No
Total Base Number (TBN) (mgKOH/g)	11.2	11.2	10.9	11.3	11.1	10.9	11.1
Water (PPM)	776.00	743.00	674.00	719.00	732.00	802.00	803.00
Viscosity (cSt)	156.0	149.0	159.0	153.0	148.0	140.0	142.0

United We Sail! 6

SAMM Modules

Planned Maintenance – Master list of all scheduled maintenance actions currently active on the vessel (s).

Workbook – Subset of Planned Maintenance – list of only scheduled maintenance actions currently due for completion. Default is current Calander month- this can be edited by Chief Engineer.

Repair – Management section for Corrective maintenance. Voyage Repairs, Ship's Force Work List, etc.

Oil Analysis Software

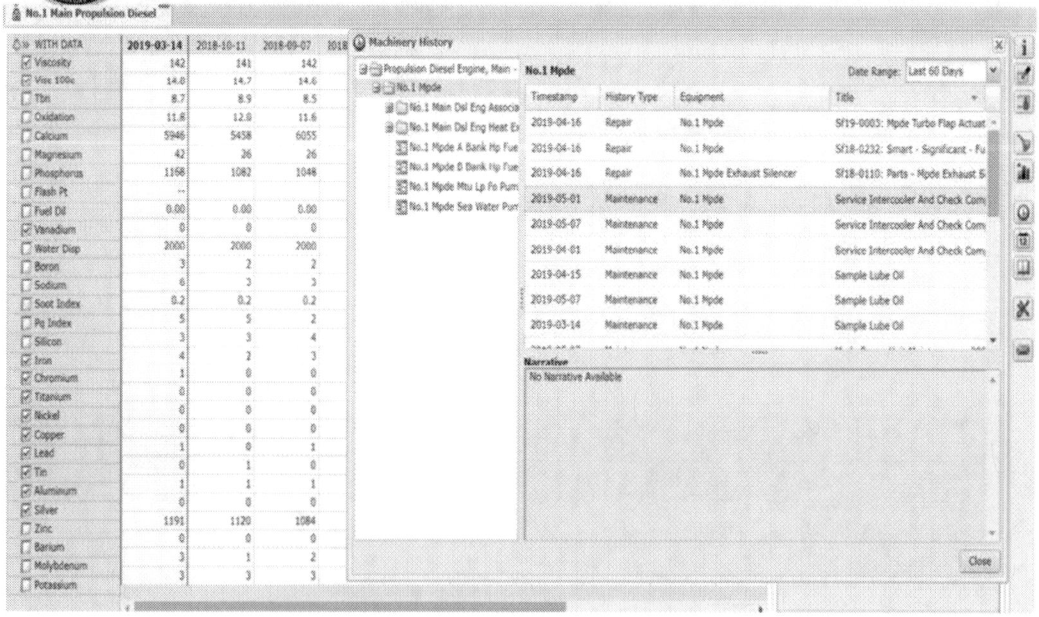

Commercial Oil Analysis

- Testing performed using industry standard methods
- Investigations performed as required
- Use of SAMM ID numbers to identify samples at lab
- Industry standard turnaround

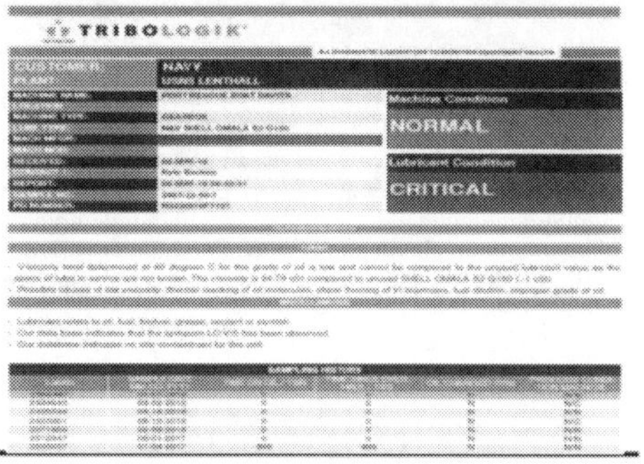

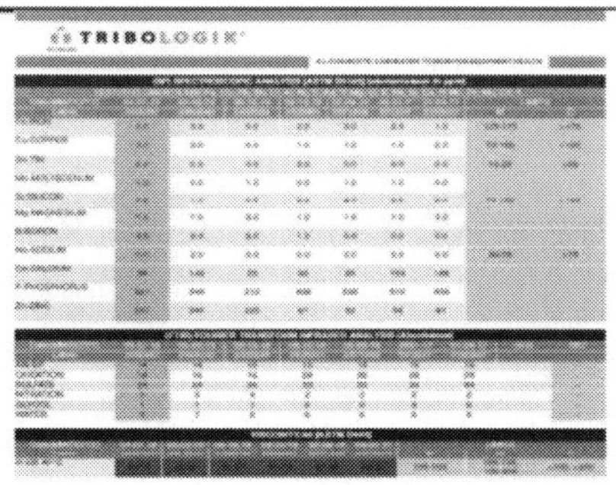

Oil Message to Ship

R 092322Z May 19
FM MSCHQ NORFO.K VA//
TO USNS BIGDA//
INFO
BT//
UNCLAS//N04700//
MSGID/GENADMIN'COMSC//
SUBJ/ USNS BIGDA LUBE OIL SAMPLE TEST RESULTS// RMKS/ 1. SAMPLES PROCESSED AT LAB 06MAY19, 07MAY19. 5 SAMPLES IDENTIFIED WITH 4 ALERTS. DATA ANALYSIS PERFORMED BY Emprise // 2. ALERTED SAMPLES:
A. EMERGENCY DIESEL ENGINE SUMP SAMPLE NUMBER: 503134019 DATE SAMPLED: 12MAR19
LUBRICANT: SHELL ROTELLA T-RIMULA SUPER 15W40
OBSERVATIONS: ALERTED FLASH PT (312.80 F), ALERTED FUEL DIL (6.14 AND INCREASING) ASSESSMENT AND RECOMMENDATIONS:
Important : Fuel Dilution (6.1%) is severely high. Decreasing Viscosity (10.8 cSt @ 100C) and Flash Point (312 F) support this high fuel reading. Check for source: injector component faults, injectors, poor timing, blow-by, heavy throttle, engine lugging, frequent short periods of operation, and excessive idling. Recommend drain the sump, replace oil filter and refill with lube chart specified lubricant.
B. NO.2 6K DIESEL FORKTRUCK TRANSMIS OIL (23-02434) SAMPLE NUMBER: 505875019 DATE SAMPLED: 16MAR19 LUBRICANT: SHELL TELLUS S2 VX 46
OBSERVATIONS: HIGH VISCOSITY (113.23 VS 46 CST AND INCREASING) ASSESSMENT AND RECOMMENDATIONS:
Desirable : Resample required. Suspect this sample was misidentified. Viscosity (113 cSt) and several other test results do not trend well with data from previous samples. This sample has properties of an engine oil. Please review your sampling procedures and resample.
3. NORMAL SAMPLES:
A. DECK AIR COMPRESSOR SAMPLE VALVE SAMPLE NUMBER: 505862019 DATE SAMPLED: 16MAR19
LUBRICANT: SHELL CORENA S4 R 68

Condition Monitoring System

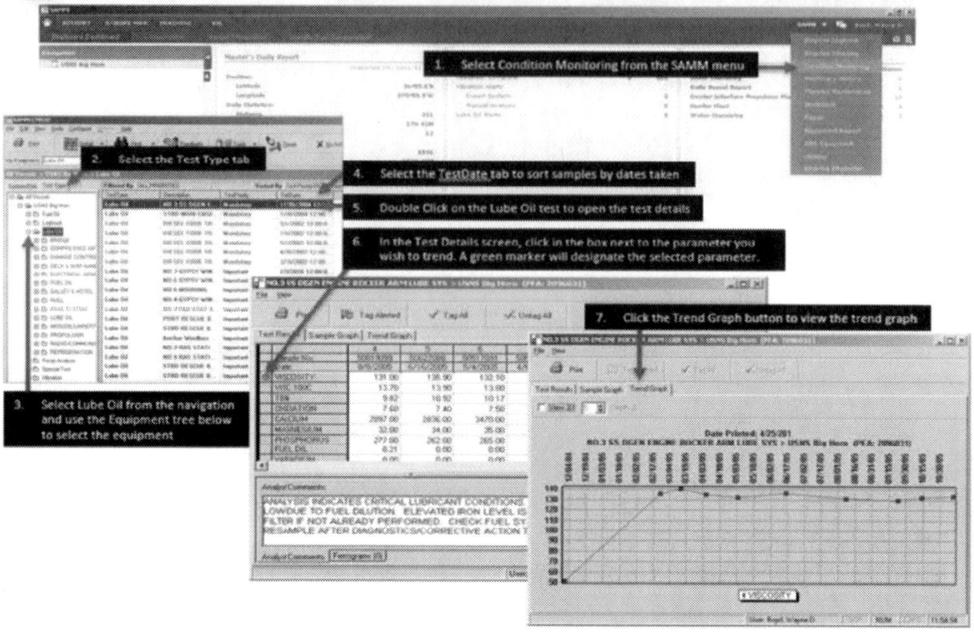

United We Sail!

Condition Based Maintenance Strategies

- **MSC Policy**
  - **Oil Analysis Policy Defined via Instructions and Standard Work Items**
  - **Use combination of MSC and OEM Approved lubricants in all equipment**
  - **Industry standards define oil selection, change out intervals, oil analysis testing.**
    - **CBM further defines any changes to periodicities or lubricants in service**
- **Centrally Managed Oil Supply Contract**
  - **Provides a single means of obtaining all required lubricants**
  - **Service focused on the needs of MSC vice product placement**
  - **Annual shipboard visits to review oil program elements**
- **Review of New Industry technologies**
  - **Determine if new technologies are align with CBM and SAMM program.**

United We Sail!

Condition Based Maintenance Strategies

- Centrally Managed Oil Analysis Contract
  - Provides MSC with ability to have routine as well as investigative analysis as required.
  - Data generated at the laboratory is submitted to SAMM daily
  - No need to rely on management of laboratory data once contract has ended
- Engineering Logistics and Analysis Contract
  - Provides MSC with: supplies needed to obtain a sample, filter systems, and analyze samples at a contract laboratory
- Civilian Mariner Engineering Training
  - Annual requirement to attend training
  - Training classes provided to Senior and Junior Engineers on the various aspects of the CBM program. Classes include:
    - Oil Analysis Training
    - Oil Supply Training
    - Bunker Fuel Training
    - Vibration Analysis
    - Chemical Analysis

United We Sail! 13

MSC Lubricants POC

- Kyle Beckett
  - Senior Mechanical Engineer
  - Lube Oil Fuel Oil SME
  - Kyle.beckett@navy.mil
  - 757-341-5535
- Christopher Lundquist
  - Lube Oil and Lube Oil Analysis ACOR
  - Christopher.lundqui1@navy.mil
  - 757-513-6457
- Michael Bigda
  - Michael.Bigda@navy.mil
  - 757-443-2239
- Thong Nguyen
  - Thong.n.Nguyen@navy.mil
  - 757-341-5536

United We Sail! 14

Ben-Gurion University of the Negev
Department of Mechanical Engineering

MFPT 2019
Where Theory Meets Practice

Damage Evolution in a Rolling Element Bearing: Spall Propagation

*Dmitri Gazizulin[1], Dr. Renata Klein[2] and Prof. **Jacob Bortman**[1]*

[1] Department of Mechanical Engineering, Ben-Gurion University of the Negev, Israel.
dmitrig@post.bgu.ac.il

[2] R.K. Diagnostics, Gilon, D. N. Misgav, Israel.
Renata.Klein@rkdiagnostics.co.il

MFPT 2019 1

Outline

- **Motivation**
- **Introduction**
- **Methodology**
- **Damage Propagation Model**
- **Experimental results**
- **Summary & Conclusions**

2

The BGU PHM Lab

Cracked shaft

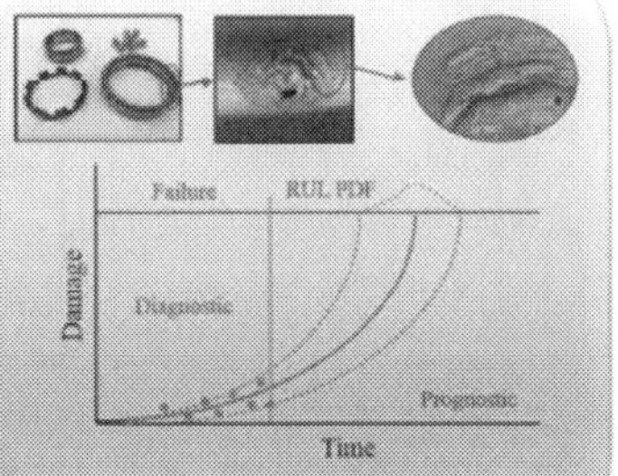

Bearing RUL

Spur gear transmission

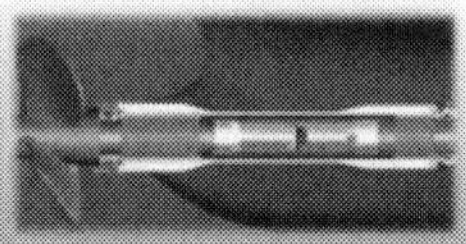

HD Bearings

Smart Bearing

3

AFRL – PHM Lab Collaboration

Advanced PHM Research for Engine Mechanical Components

AFRL
THE AIR FORCE RESEARCH LABORATORY
LEAD | DISCOVER | DEVELOP | DELIVER

- Jeremy Nickell
- Dr. Lewis Rosado

Ben-Gurion University of the Negev
Department of Mechanical Engineering

- Dmitri Gazizulin
- Ravit Ohana
- Eayl Madar
- Dr. Renata Klein
- Prof. Jacob Bortman

4

Motivation → CBM

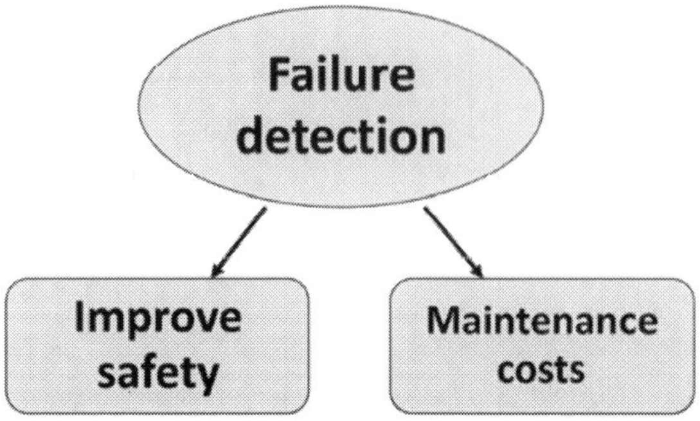

Estimation of RUL - The challenge:

- **Damage severity or size:**
  - Vibration based algorithms
  - Sensors
- **Damage growth model**
- **Future usage spectrum**

Objective

Develop a reliable **physics-based** model for **prognosis of bearing failure**

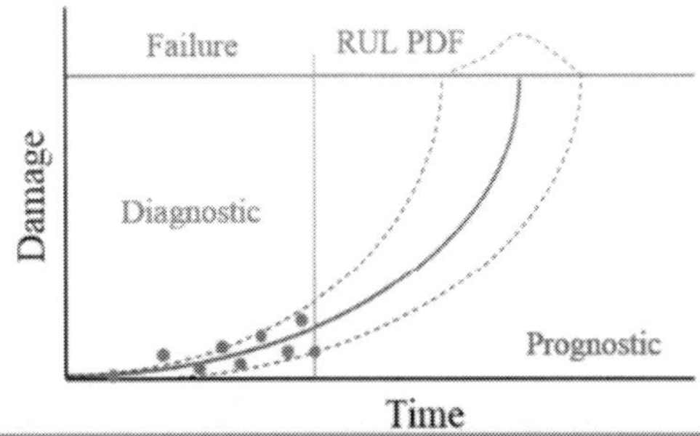

Introduction

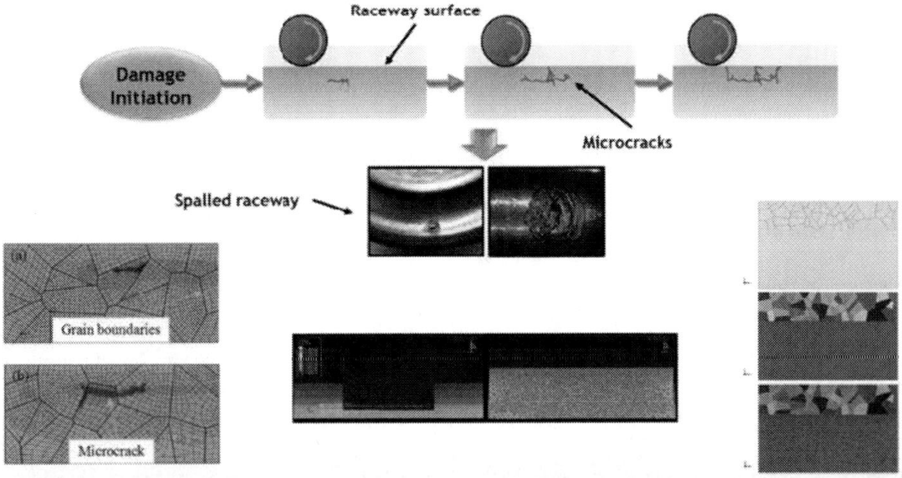

Introduction

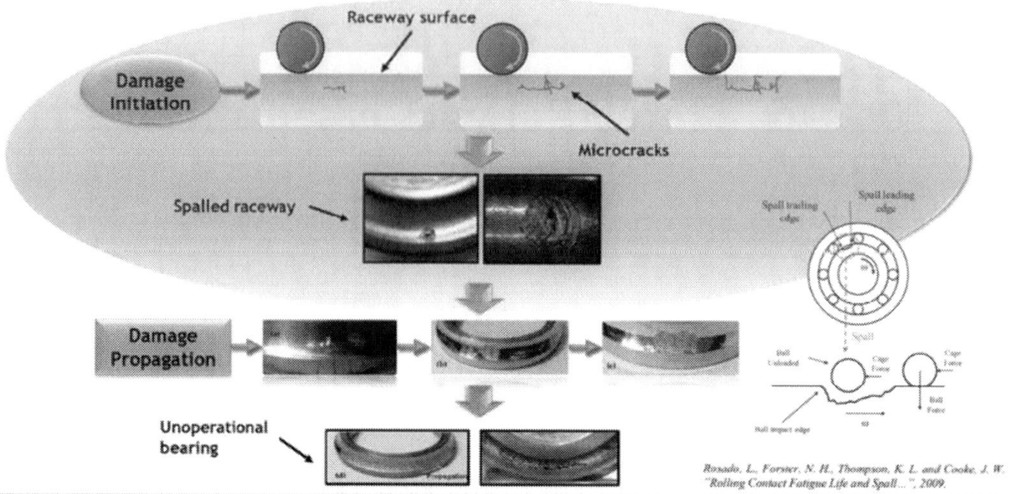

Rosado, L., Forster, N. H., Thompson, K. L. and Cooke, J. W. "Rolling Contact Fatigue Life and Spall...", 2009.

9

Damage model

- Continuum Damage Mechanics (CDM)

$$0 \le D \le D_{max}$$

$$\tilde{E} = (1 - D)E$$

$$\frac{dD}{dN} = \left[\frac{\sigma_{eff}}{\sigma_R (1 - D)} \right]^m$$

Symbol	Description
σ_{eff}	Effective stress
D	Damage variable
N	Num. of cycles
E	Modulus of elasticity
σ_R, m	Material dependent parameters

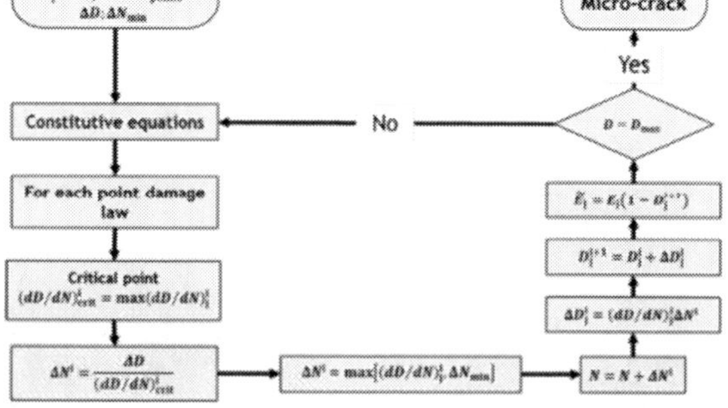

DAMAGE PROPAGATION 10

Introduction

Damage propagation models:

- Representation of the physical phenomena - **Not well based**
- Computationally **inefficient**

- *Y. Li, S. Billington, C. Zhang, T. Kurjess, S. Danyluk, S. Liang, Adaptive prognostics for rolling element bearing condition, Mechanical systems and signal processing 13 (1) (1999a) 103–113.S.*
- *Marble, B. P. Morton, Predicting the remaining life of propulsion system bearings, in: Aerospace Conference, 2006 IEEE, IEEE, 8–pp, 2006.*
- *N. K. Arakere, N. Branch, G. Levesque, V. Svendsen, N. H. Forster, Rolling Contact Fatigue Life and Spall Propagation of AISI M50, M50NiL, and AISI 52100, Part II: Stress Modeling, Tribology Transactions 53 (1) (2009) 42–51.*
- *N. Bolander, H. Qiu, N. Eklund, E. Hindle, T. Rosenfeld, Physicsbased remaining useful life prediction for aircraft engine bearing prognosis, in: Annual Conference of the Prognostics and Health Management Society, 1–10, 2009.*
- *H. Zong, H. Wang, R. Hong, Dynamic research on single-row slewing bearing with local spalling of inner ring, Advances in Mechanical Engineering 8 (2) (2016).*

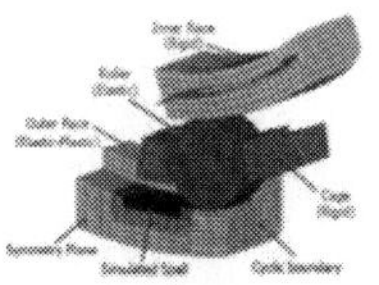

11

Endurance Test - Experimental setup

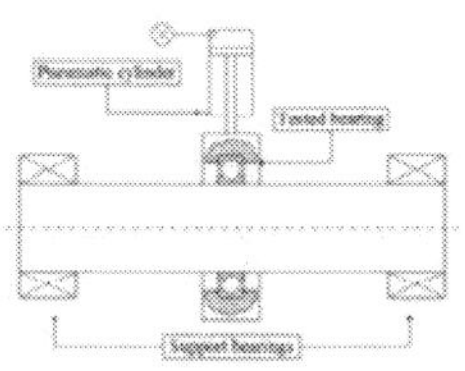

12

Spalled Bearing - Observations

- Locations of the cracks near the spall trailing edge
- Multiple ball-trailing edge impacts - material fragments detachment

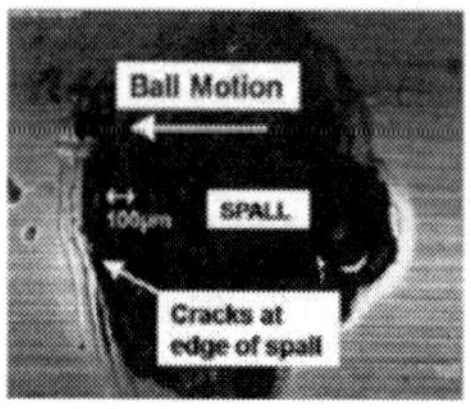

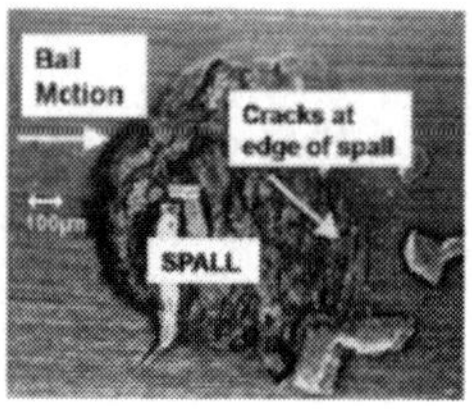

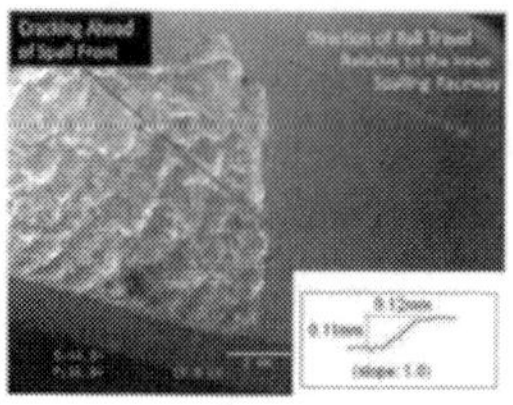

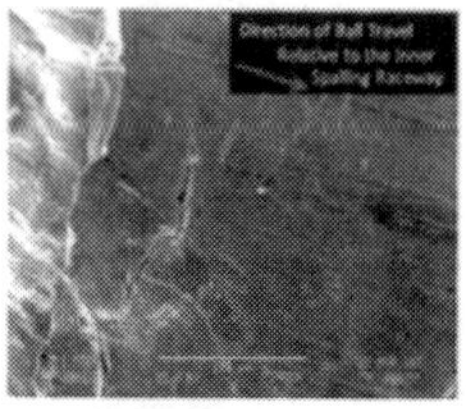

- N. A. Branch, N. K. Arakere, V. Svendsen, N. H. Forster, Stress eld evolution in a ball bearing raceway fatigue spall, Journal of ASTM international 7 (2) (2009) 118.
- M. L. French, W. M. Hannon, Angular contact ball bearing experimental spall propagation observations, Proceedings of the Institution of Mechanical Engineers, Part J: Journal of Engineering Tribology 229 (8) (2015) 902916.

Endurance Test - Damage size estimation

A section of the order spectrum at different stages of the endurance test.

Damage evolution trajectory. BPFO energy distance pattern in the envelope spectrum

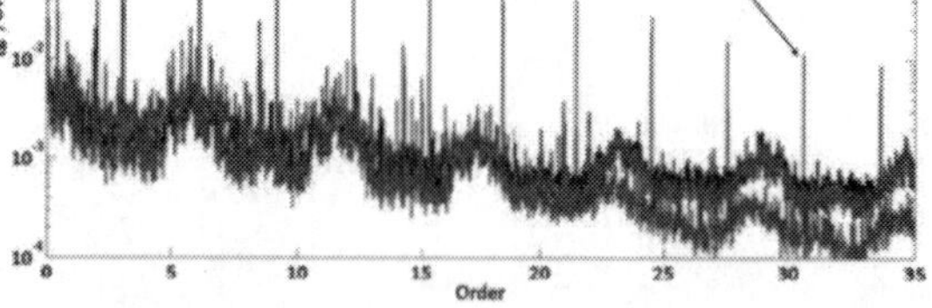

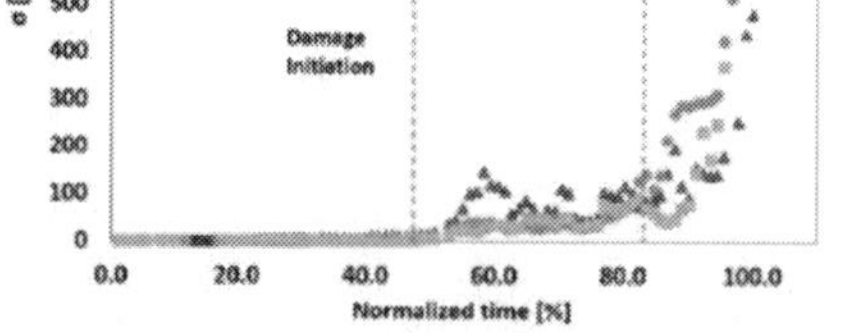

Defect size estimation - Vibrations data

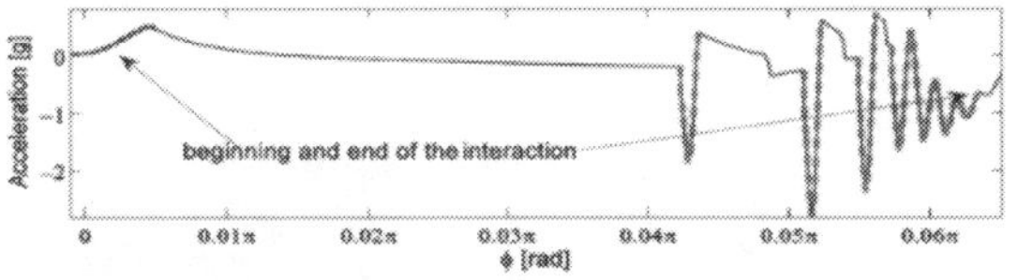

Kogan, G., J. Bortman, and R. Klein. "A new model for spall-rolling-element interaction." Nonlinear Dynamics 87.1 (2017): 219-236.

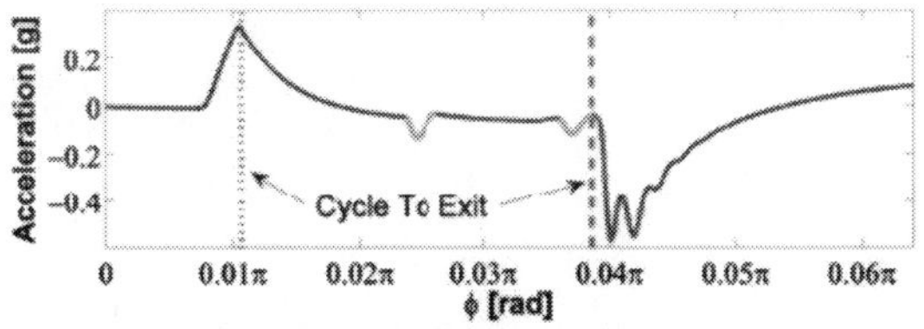

Madar, E., Kogan, G., Klein, R. & Bortman, J. (2017). An analytical model for rolling-element-spall interaction in bearing inner race. in: The First World Congress on Condition Monitoring Conference, London, England, June, 2017.

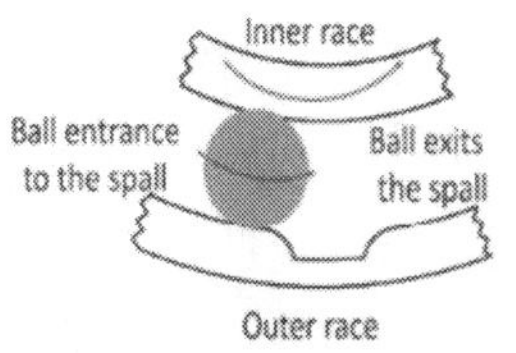

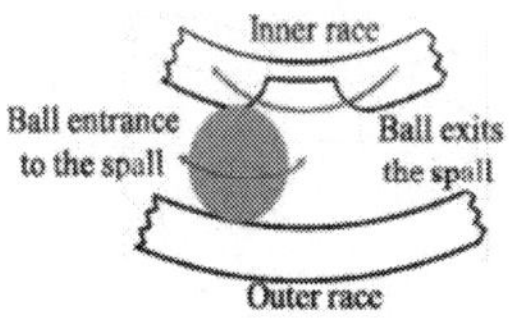

Endurance Test

- Defect size - Dynamic response

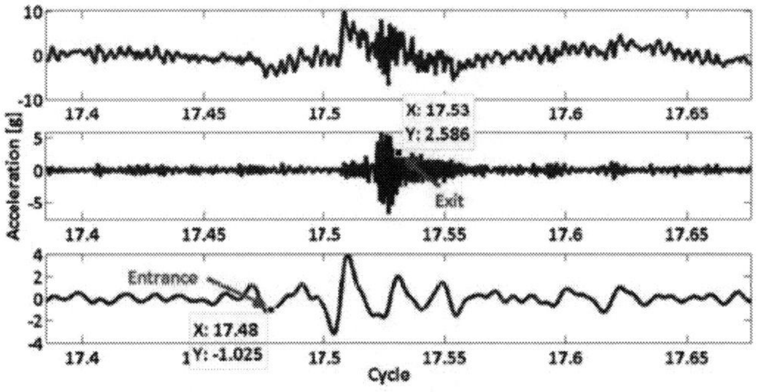

Estimated size - 3.4 *mm*

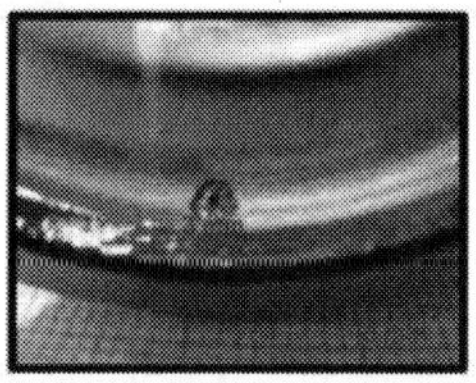

Defect size estimation - Strain data

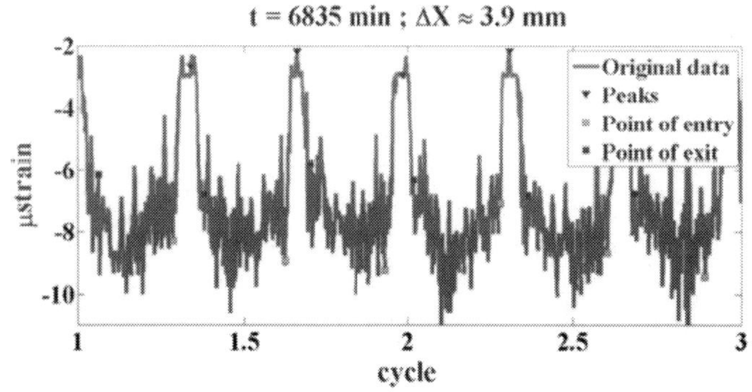

Endurance Test

- Defect size estimation - **Strain data**

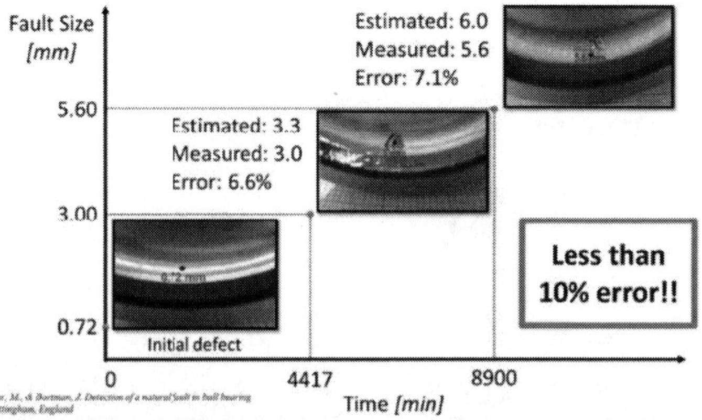

Shimoni, D., Gazizulin, D., Alian, H., Klein, R., Tur, M., & Bortman, J. Detection of a natural fault in ball bearing during endurance test by optical fiber sensors. Nottingham, England

Damage Propagation - Methodology

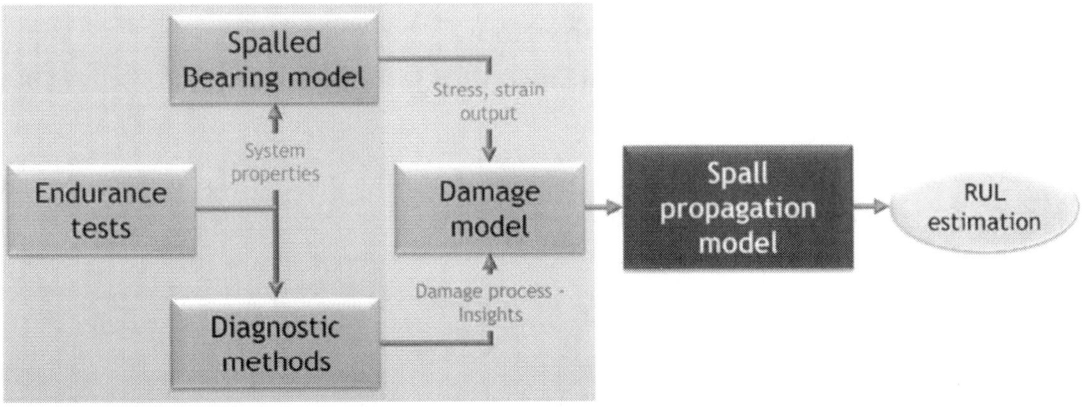

Spalled Bearing Model

- **First step towards a damage propagation model**

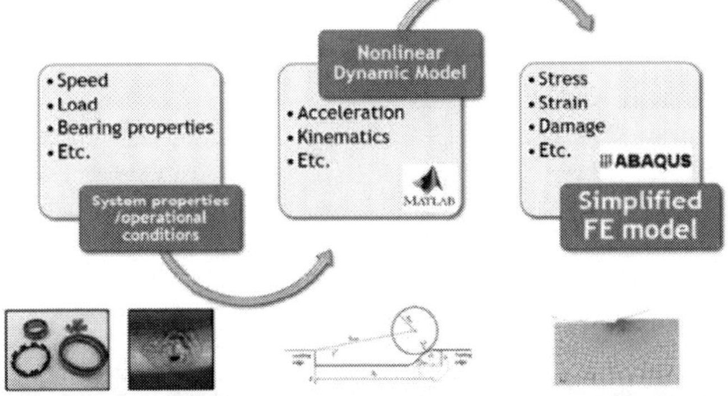

Spalled Bearing Model

Non-linear dynamic model

- Impact loaction & direction
- RE indentation into the spall edge
- Etc.

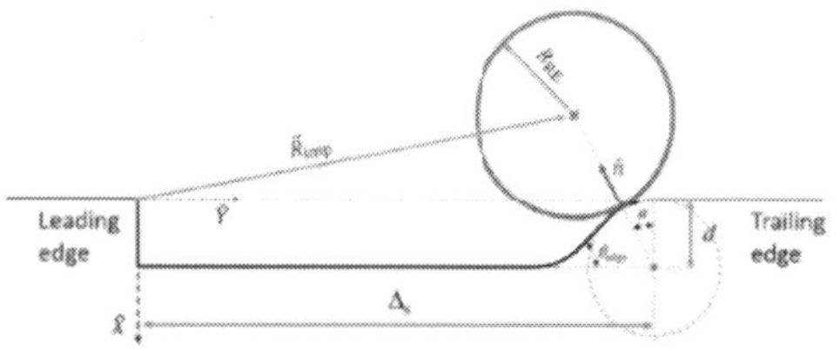

impact location

$$(x_{imp}, y_{imp}) = \left(R_{RE}\cos\alpha - h, (\mathbf{R}_{imp})_y + R_{RE}\sin\alpha\right)$$

$$h = \left|(\mathbf{R}_{imp})_x\right|$$

$$\cos\alpha = (h+d)/(R_{RE}+d)$$

impact direction

$$\hat{n}(n_1, n_2) = (\cos\alpha, \sin\alpha)$$

elastic impact model

$$F_n = k\left(R_{RE} - (n_e - n_{RE})\right)^n \left(1 + \frac{3(1-e^2)(\dot{n}_{RE} - \dot{n}_e)}{4\dot{\delta}^{(-)}}\right)$$

$$m_{RE}\ddot{n}_{RE} = -(F_n + m_{RE} g n_y)$$

$$m_e\ddot{n}_e = F_n + m_e g n_y$$

Kogan, G., Bortman, J. & Klein, R. (2016). 'A new model for spall-rolling-element interaction'. Nonlinear Dynamics pp. 1–18. doi: 10.1007/s11071-016-3037-1.

DAMAGE PROPAGATION 21

Spalled Bearing Model

FE model

- Two-dimensional, plane strain, quasi-static model
- Spall edge - elastic-plastic, anisotropic strain hardening
- RE - rigid surface
- ABAQUS/Implicit solver

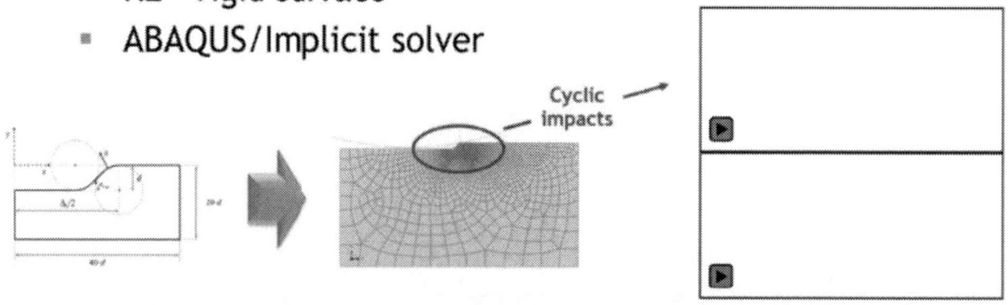

DAMAGE PROPAGATION 22

Spalled Bearing Model

Model validation

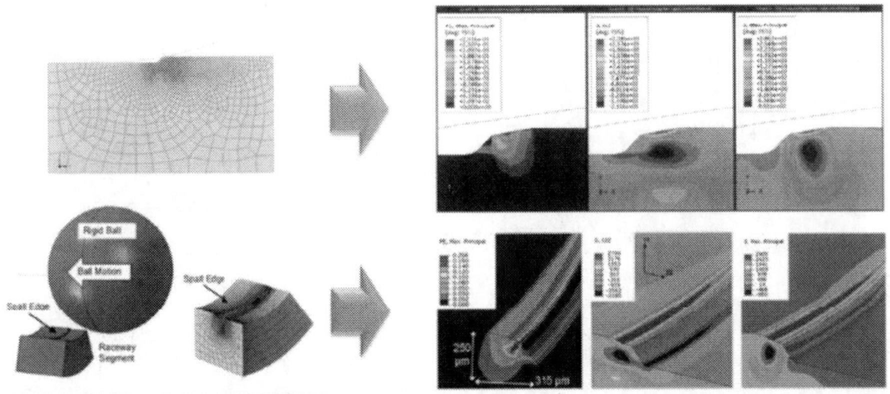

Branch et al. (2009) ,Arakere et al. (2009), ard Branch et al. (2013)

Damage model

- Continuum Damage Mechanics (CDM)

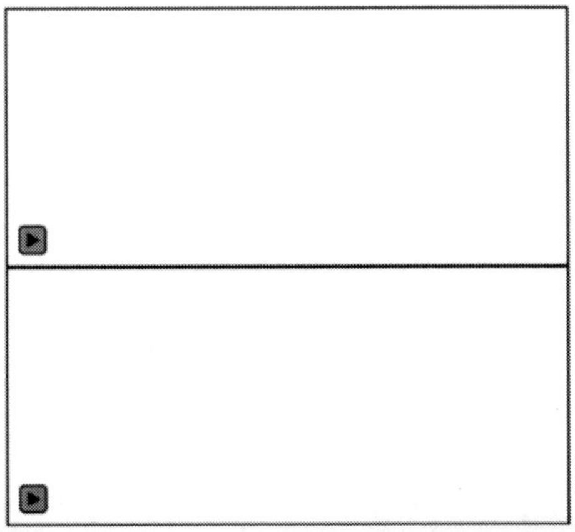

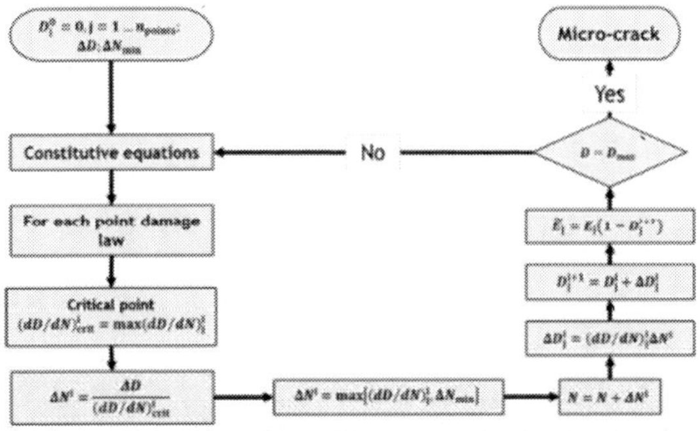

D. Gazizulin, R. Klein, and J. Bortman. Towards efficient spall generation simulation in rolling element bearing. Fatigue & Fracture of Engineering Materials & Structures, 40(9):1389–1405, 2017

Metallurgical observation

- Trailing edge

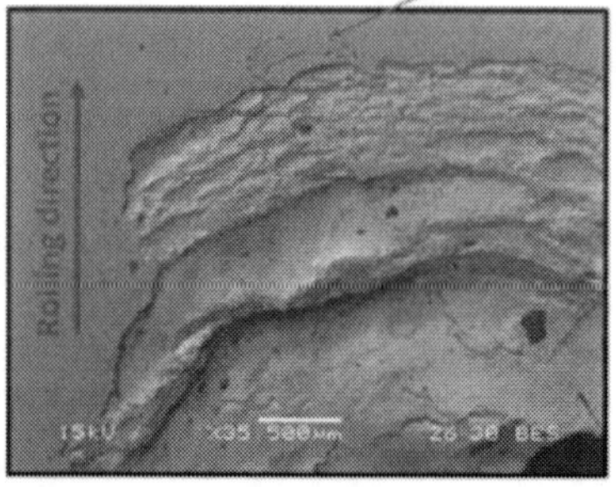

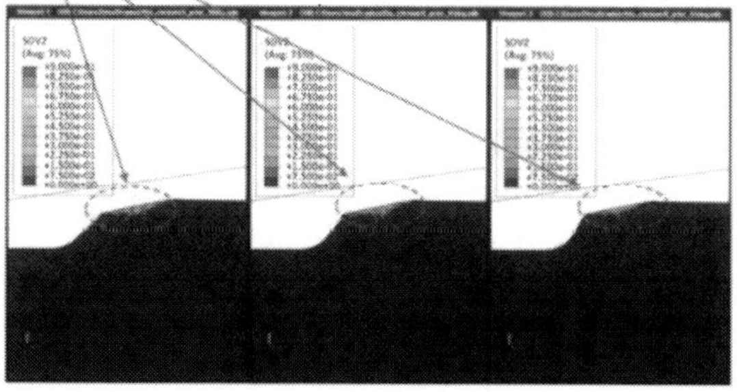

25

Spalled Bearing Model

Physical insights - Based on FE analysis, tests and literature:

- Depth
  - "Damage-free maximum shear stress" (Gazizulin et al. 2017)
- Spall front edges slopes
  - 0.8 - 2.5 (French and Hannon 2015, Branch et al. 2009)
- Crack location
  - Within the trailing edge of the spall (Literature & tests)

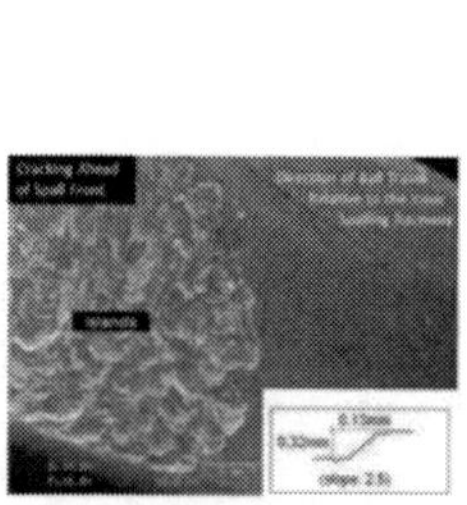

26

Summary and Conclusions

- Methodology for the REB's RUL estimation
- Development of damage severity diagnostic tools
- New damage growth modeling approach
  - Computationally efficient model
  - Potential to produce more accurate results
  - Simulations of crack evolution
- Experimental validation

27

Future work

- Continue endurance experiments
- Continue validating our model
- Develop better physical understanding of the spall propagation process
- Answer questions such as: Can we find the "knee" accurately?
- Verify the new prognostic methodology - full scale engine experiment

28

Thanks for your attention

29

Rolling Contact Fatigue Modeling

Contact Model

Microstructure Representation

Damage Model

DAMAGE INITIATION 30

RCF Modeling

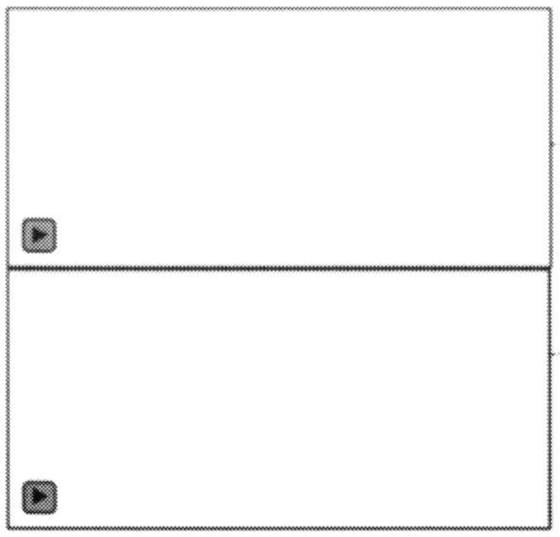

Continuum Damage Mechanics

$$0 \le D \le D_{max}$$

$$\tilde{E} = (1-D)E$$

$$\frac{dD}{dN} = \left[\frac{\Delta\tau}{\tau_R(1-D)}\right]^m$$

RCF Modeling - Results

- The general shape of the spall is in good agreement with the experimental spallation

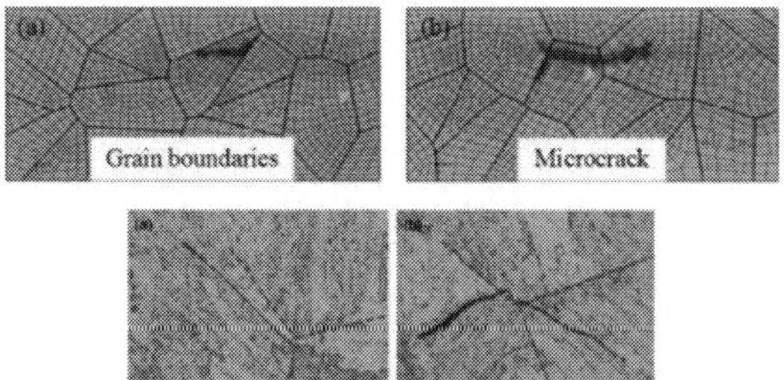

Meyer S, Brückner-Foit A, Möslang A. A stochastic simulation model for microcrack initiation in a martensitic steel. Computational Materials Science. 2003 Jan 31;26:102-10.

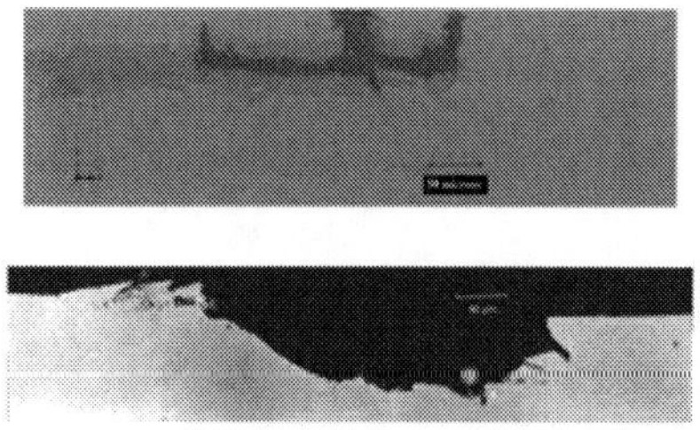

Warhadpande A, Sadeghi F, Kotzalas MN, Doll G. Effects of plasticity on subsurface initiated spalling in rolling contact fatigue. International Journal of Fatigue. 2012 Mar 31;36(1):80-95.

RCF Modeling - Results

- The simulated RCF process is dominated by the initiation stage.
- The achieved Weibull slopes are very close to the results of previously published models.

Model	Init. time Weibull slope	Total time Weibull slope
Raje et al. (2009)	-	1.61
Slack and Sadeghi (2010)	7.15	12.54
	5.11	4.08
	4.89	3.37
Jalalahmadi and Sadeghi (2005)	4.81	5.13
	4.89	6.52
Warhadpande et al. (2012)	4.81	6.58
	-	4.97
Current model	4.99	5.77

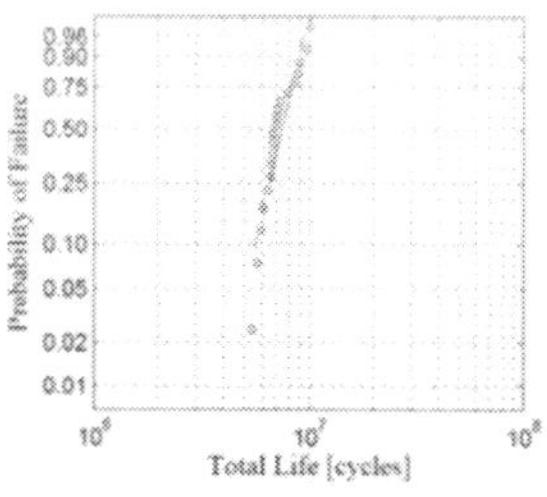

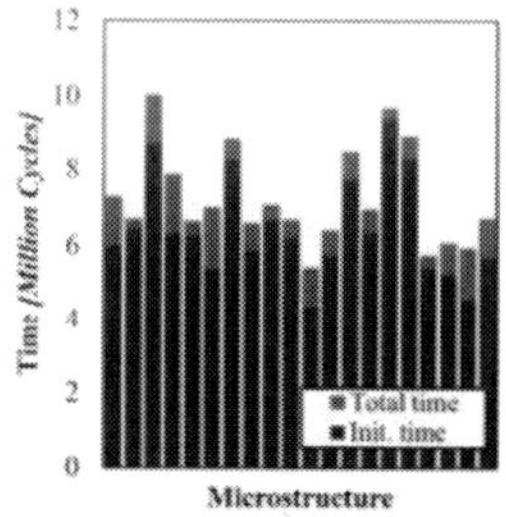

DAMAGE INITIATION 33

Introduction

Damage propagation models:

- Representation of the physical phenomena - Not well based
- Computationally inefficient

$$\frac{dD}{dt} = C_0 \left(D \right)^n$$

Y. Li, S. Billington, C. Zhang, T. Kurfess, S. Danyluk, S. Liang, "Adaptive prognostics for rolling element bearing condition", 1999.

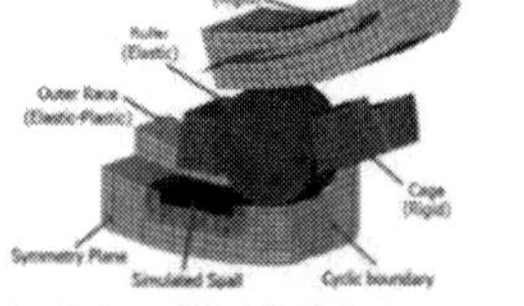

N. Bolander, H. Qiu, N. Eklund, E. Hindle, T. Rosenfeld, "Physics based ...bearing prognosis", 2009.

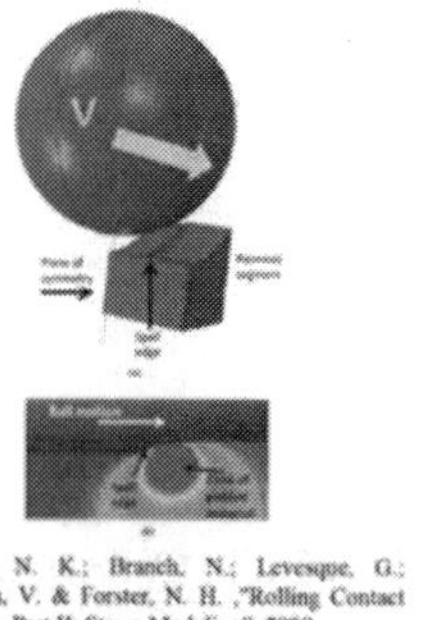

Arakere, N. K.; Branch, N.; Levesque, G.; Svendsen, V. & Forster, N. H. ,"Rolling Contact Fatigue ...Part II: Stress Modeling", 2009

34

Motivation - CBM

HUMS - Health And Usage Monitoring Systems

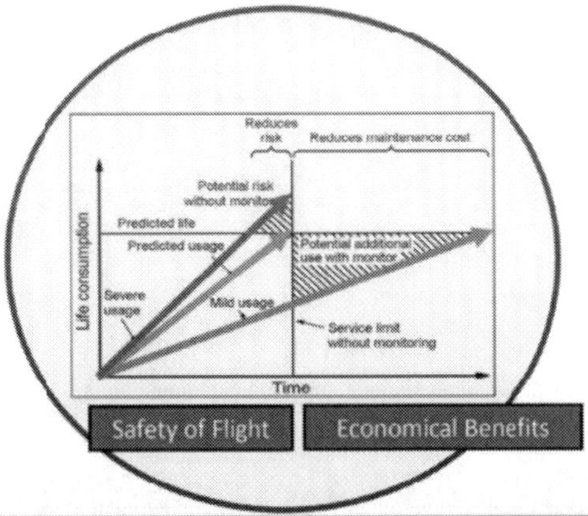

35 35

Diagnostic & Prognostic

- First defect initiation
- Diagnose current defect size
- Defect growth model - up to critical size
- Estimate RUL

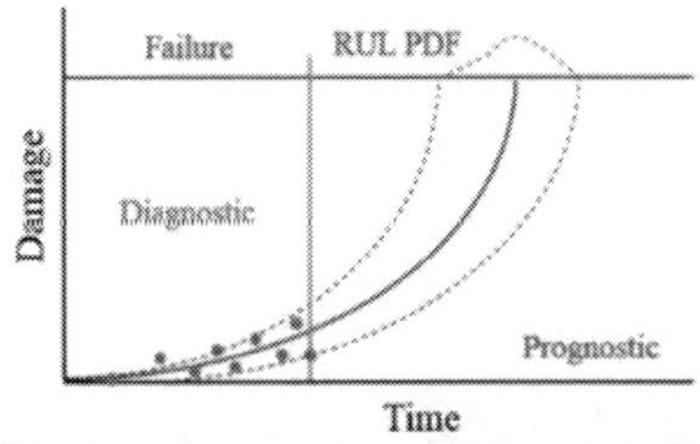

36

Diagnostic & Prognostic

- **First defect initiation**
- Diagnose current defect size
- Defect growth model - up to critical size
- Estimate RUL

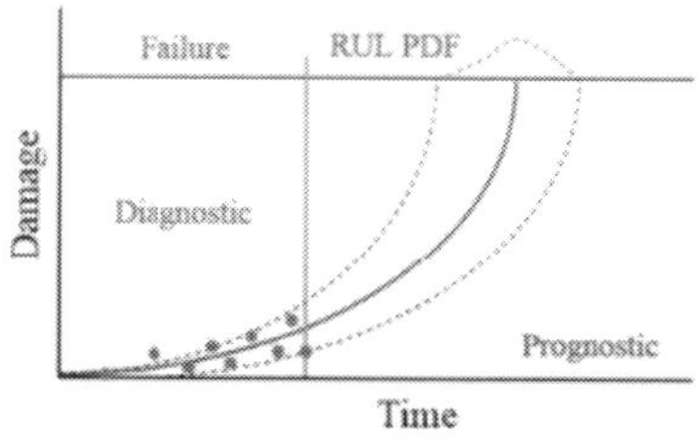

Diagnostic & Prognostic

- **First defect initiation**
- **Diagnose current defect size**
- Defect growth model - up to critical size
- Estimate RUL

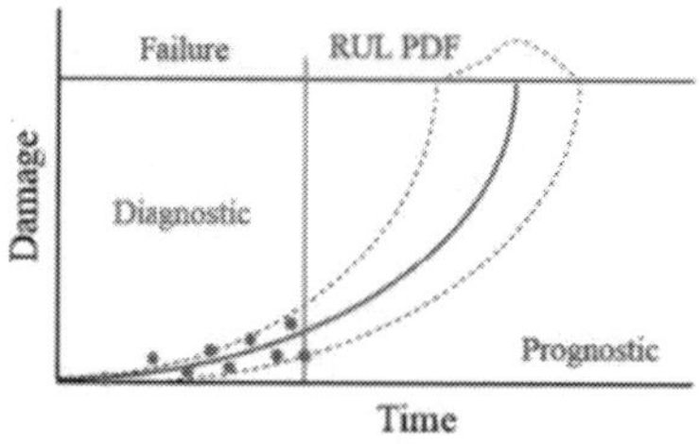

Endurance Test

- Condition Indicators

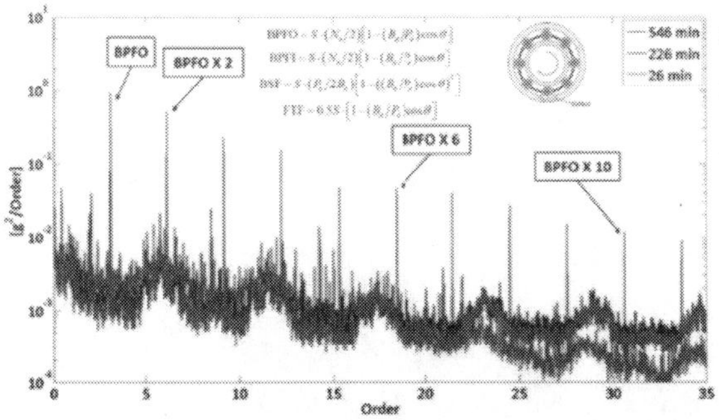

Energy of the peaks

$$E_{tot}\{f_{tone}\} = \sum_{i=1}^{I} E_i\{f_{tone}\}$$

$$E_i\{f_{tone}\} = \sum_{j=1}^{J} p_{i,j}$$

$$p_{i,j} = \sum_{x=-X}^{X} Amp(i \cdot f_{tone} + j + x \cdot \Delta f)$$

Statistical distance

$$D_{tot}[E] = \sum_{i=1}^{I} D_i[E]$$

$$D_i[E] = \frac{\sum_{j=1}^{J}(p_{i,j} - \mu_{i,j})}{\sqrt{\sum_{j=1}^{J}\sigma_{i,j}^2}}$$

$$\mu_{i,j} = \frac{1}{N}\sum_{n=1}^{N}(p_{i,j})_n$$

$$\sigma_{i,j} = \sqrt{\frac{1}{N-1}\sum_{n=1}^{N}(p_{i,j} - \mu_{i,j})_n^2}$$

39

Damage model - Simulation Results

- Simulation results - Residual stress
- Edge slope: 38°,45°, 55°
- Before crack initiation

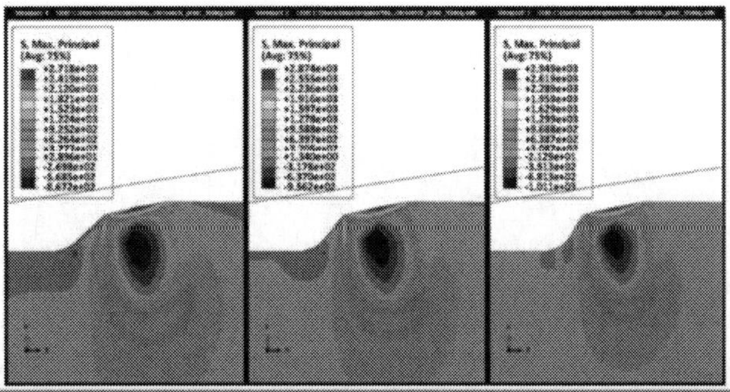

40

Damage model - Simulation Results

- Simulation results – Residual stress
- Edge slope: 38°,45°, 55°
- End of the damage simulations

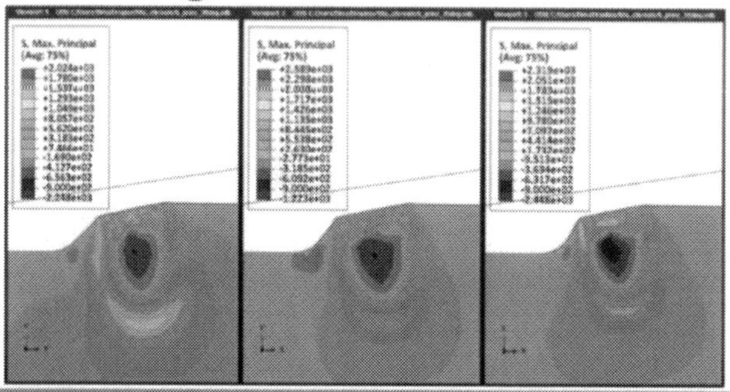

41

Damage model - Simulation Results

- Edge slope: 38°,45°, 55°
- Cracks

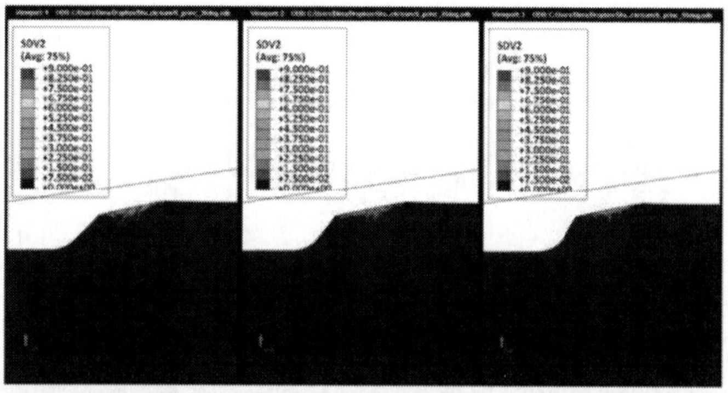

42

Metallurgical observation

- Trailing edge

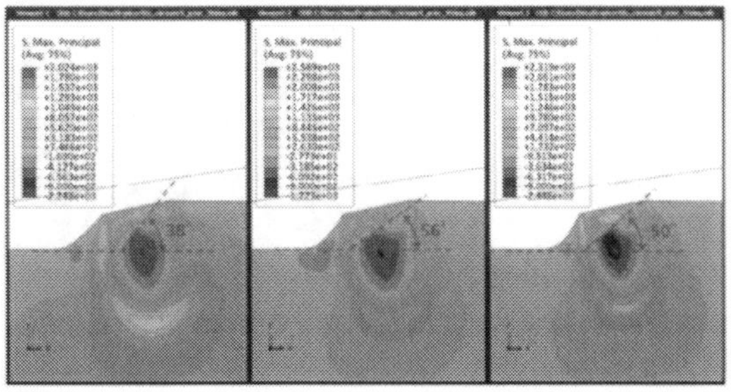

Metallurgical observation

- Trailing edge

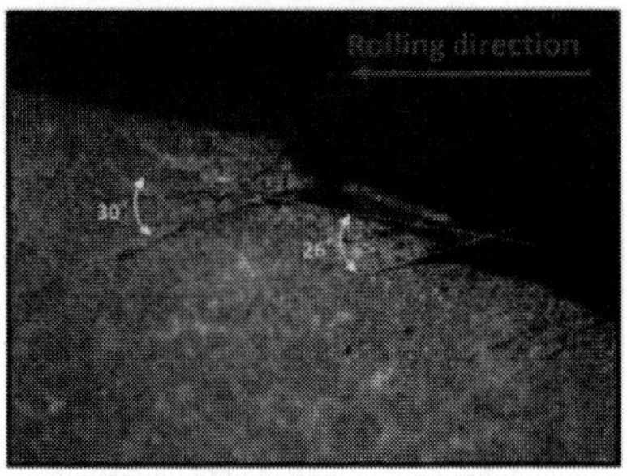

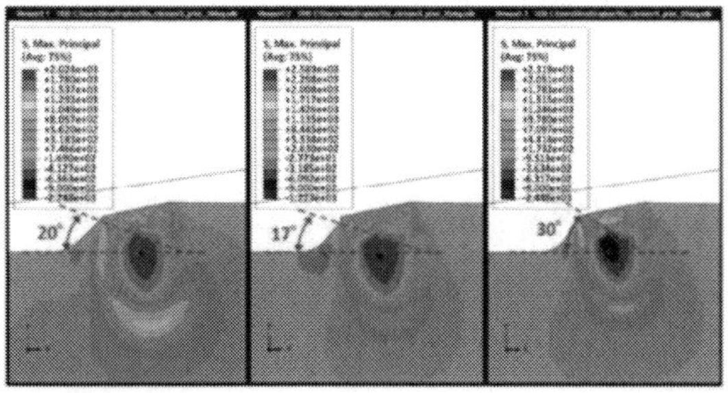

Endurance Test

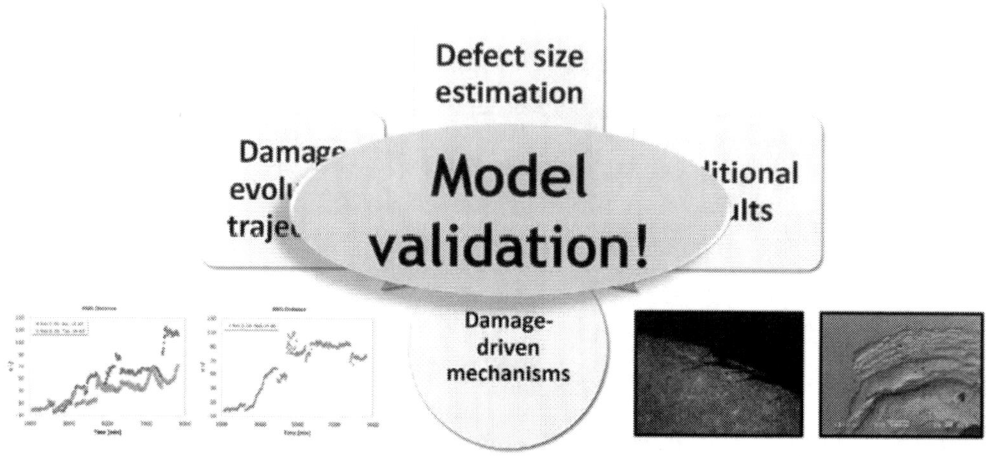

DAMAGE PROPAGATION 45

Endurance Test - Damage size estimation

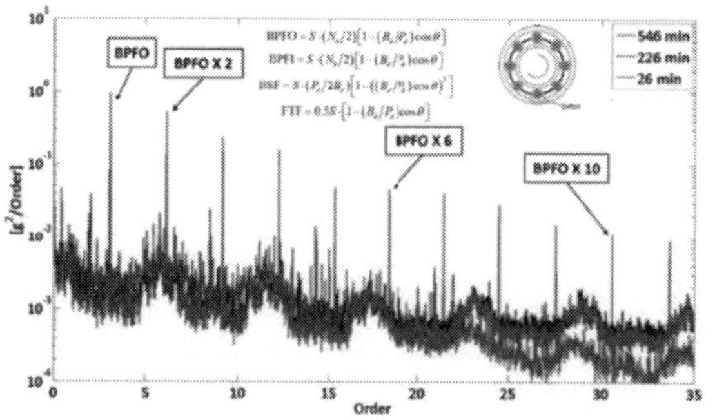

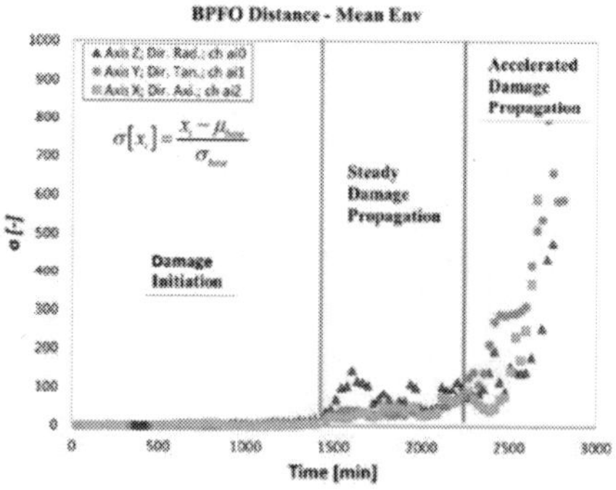

46

Spalled Bearing Model

- Simulation results – Residual stress
- Edge slope: 38°,45°, 55°

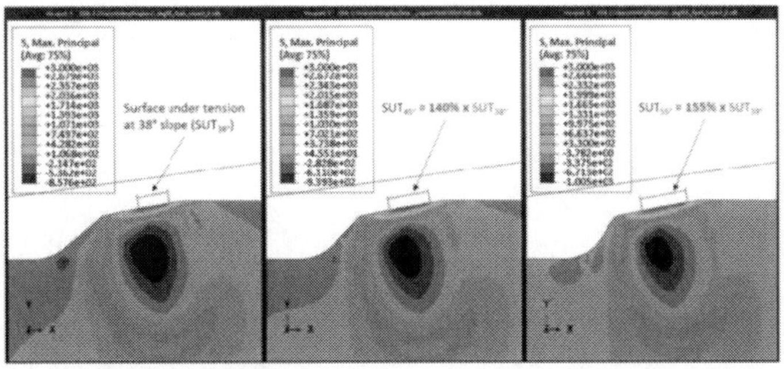

MFPT Where Theory Meets Practice

Design for Reliability…and Risk

David A. Breslin, PE
Marine Engineer, U.S. Navy
Instructor for Reliability & Risk
Adjunct Professor, Reliability Engineering & Operations Research, United States Merchant Marine Academy
Primary Contact Information :
E-Mail: David.Breslin@navy.mil
Work Telephone: 202 781 3433

Purpose

Give a high-level overview of the Design for Reliability process, as well as its relationship to risk, arguing that better products can be delivered with greater customer satisfaction and reduced cost.

Design for Reliability

A process that encompasses tools and procedures to ensure that a product meets its reliability requirements under its use environment, for the duration of its lifetime. Important parameters are reliability, R(*t*), or Availability, A.

Most often, customers use the words "reliability" and "Availability" interchangeably, and that's ok!

3

A Little Reliability Math

Many systems are designed without redundancy. In such cases, the reliability of the system, $R_{sys}(t)$, is related to the reliability of the individual components, R_i, in the following manner:

$$R_{sys}(t) = R_1 \times R_2 \times R_3 \times \ldots R_n$$

Yes, it's really that simple.

Also known as Lusser's law, named after Robert Lusser who, by the way, believed it was not possible to send humans to the Moon or Mars and return them safely to Earth.

4

Risk

The possibility of gaining or losing something of value:

Risk = Consequence x Probability

Using this form of the so-called risk equation, Risk represents the <u>*expected*</u> monetary value of some future event.

5

Risk

In this form (R = C x P), risk is an expected value, or a quantification (usually in dollars) of what the organization expects to lose.

This can help managers quantify the potential downside, which in turn helps managers decide how much money to spend on risk mitigation (redesign, redundancy, insurance, …).

6

Visual Depiction of Risk

Likelihood ↑			
High	Medium	High	Critical
Medium	Low	Medium	High
Low	Low	Low	Medium
Impact →	Low	Medium	High

Some organizations use so-called risk matrices to visualize magnitudes of risk, from low to high.

Reliability and Risk

Often enough, the risk probability is equal to the unreliability (1 – R(t)), meaning,

Risk = Consequence x (1 – R(t))

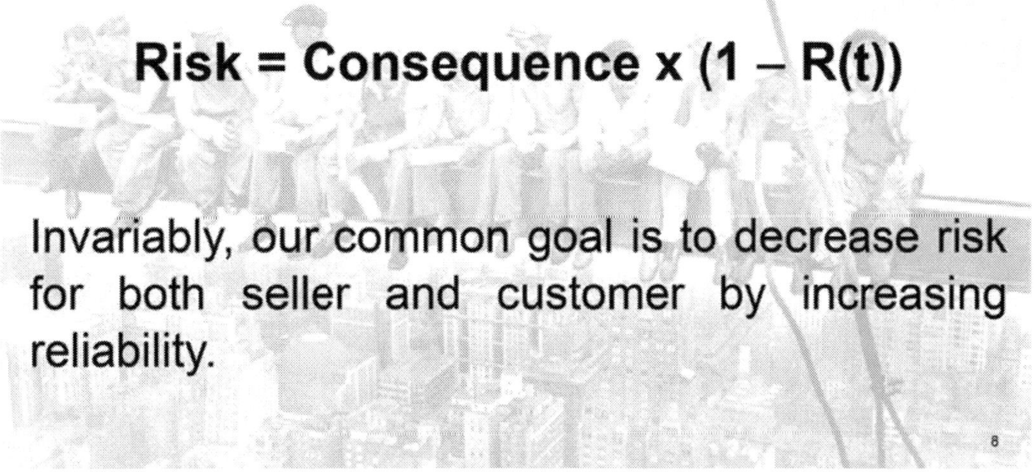

Invariably, our common goal is to decrease risk for both seller and customer by increasing reliability.

Reliability and Risk

Contrary to the popular opinion of reliability engineers, reliability is not the goal.

Reliability is a means to an end.

9

Reliability Risk

Risk to the seller:

- Liability
- Loss of market share
  - Failure to meet customer expectations
  - Failure to stay ahead of the competition

Risk to the customer:

- Liability
- Loss of revenue
  - When a system isn't working, it isn't making money
- Cost of unplanned maintenance
- Cost of premature replacement

10

The Cost Of A Bad Design

Ridiculous?
Yes, but they make a good point.

Celebrating a <u>Great</u> Design

At a total program cost of $700M, NASA's New Horizons probe was launched in 2006 and flew by Pluto in 2016. The expected mission duration was approximately 16 years, with an estimated "probability of success" of around 85%.

Questions

In 2005, what was the consequence to NASA of the event of mission failure?

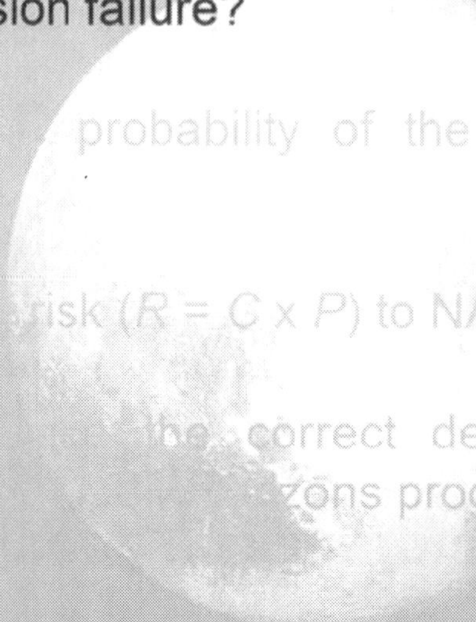

Questions

In 2005, what was the consequence to NASA of the event of mission failure?

Answer: $700M

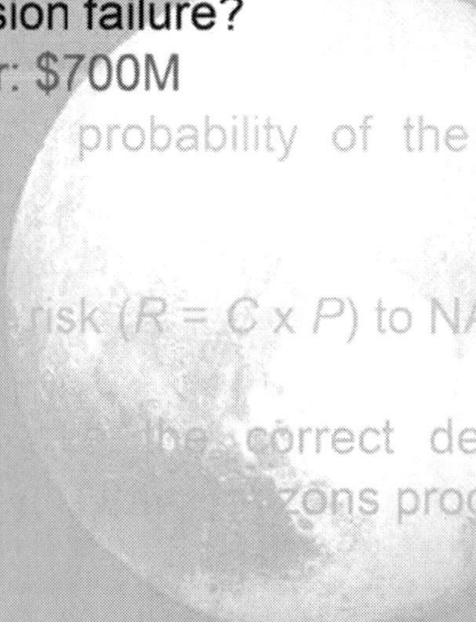

Questions

In 2005, what was the consequence to NASA of the event of mission failure?

Answer: \$700M

What was the probability of the event of mission failure?

What was the risk ($R = C \times P$) to NASA?

Did NASA make the correct decision by [illegible] forward with the New Horizons program?

15

Questions

In 2005, what was the consequence to NASA of the event of mission failure?

Answer: \$700M

What was the probability of the event of mission failure?

Answer: 15%

What was the risk ($R = C \times P$) to NASA?

Did NASA make the correct decision by [illegible] forward with the New Horizons program?

16

Questions

In 2005, what was the consequence to NASA of the event of mission failure?

Answer: $700M

What was the probability of the event of mission failure?

Answer: 15%

What was the risk (*R* = *C* x *P*) to NASA?

Did NASA make the correct decision by moving forward with the New Horizons program?

17

Questions

In 2005, what was the consequence to NASA of the event of mission failure?

Answer: $700M

What was the probability of the event of mission failure?

Answer: 15%

What was the risk (*R* = *C* x *P*) to NASA?

Answer: $105M

Did NASA make the correct decision by moving forward with the New Horizons program?

18

Questions

In 2005, what was the consequence to NASA of the event of mission failure?

Answer: $700M

What was the probability of the event of mission failure?

Answer: 15%

What was the risk ($R = C \times P$) to NASA?

Answer: $105M

Did NASA make the correct decision by moving forward with the New Horizons program?

19

Questions

In 2005, what was the consequence to NASA of the event of mission failure?

Answer: $700M

What was the probability of the event of mission failure?

Answer: 15%

What was the risk ($R = C \times P$) to NASA?

Answer: $105M

Did NASA make the correct decision by moving forward with the New Horizons program?

Answer: Absolutely!

20

Mitigation through Design

To achieve high reliability, the use of limited-life items (valves, switches, doors, memory) was kept to a minimum. When required in the design, their use was limited. For example, each "Thruster Valve" would be cycled only 200,000 times. The valve was rated for 409,000 cycles, and had only failed in prior programs after more than 500,000 cycles.

Bad Designs Are Costly

Takata Airbags, 2013
53 million cars, 11 deaths, $1B

GM Ignition Switches, 2014
30 million cars, 214 deaths, $900M

MV Sewol, 2014
304 deaths

Space Shuttle Challenger, 1986
7 lives, $2B

How Reliable?

Often, Availability is measured in "nines", as in 0.9, 0.99, 0.999, and so on.

Time, willpower, and maybe a little bit of money, are the only barriers to adding each additional "nine".

23

Quick History

In the 1950s, the ENIAC had an availability of 0.5 (it was down about half the time)

A modern data center has an availability of 0.99999, or "five nines" (5 minutes of downtime per year)

What availability will our grandchildren expect?

ENIAC: Electronic Numerical Integrator and Computer.

24

A Quick Note On Math

For the purpose of Design for Reliability as well as assessing risk, order-of-magnitude estimates are good enough…usually.

Availability: 0.9, 0.99, 0.999 …

Risk (= CxP): $10K, $100K, $1M …

Activities Flow

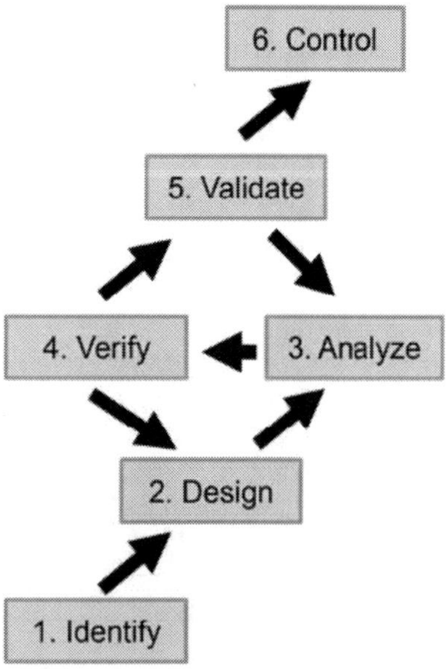

The Design for Reliability process encompasses various tools and practices and describes the order of their deployment that an organization needs to drive reliability into the products.

Depending on the industry, type of product, development cycle and other product specific factors, the Design for Reliability process can certainly vary. It can be depicted as the flow diagram to the left.

Practical Reliability Engineering, 5th Edition, by O'Connor and Kleyner.

Activities Flow

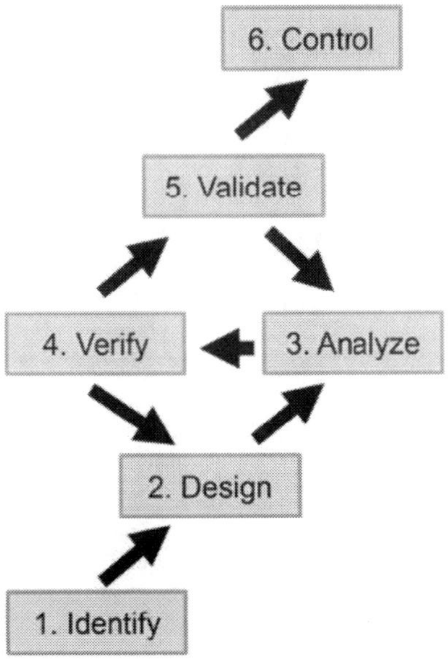

The Design for Reliability process encompasses various tools and practices and describes the order of their deployment that an organization needs to drive reliability into the products.

Depending on the industry, type of product, development cycle and other product specific factors, the Design for Reliability process can certainly vary. It can be depicted as the flow diagram to the left.

Practical Reliability Engineering, 5th Edition, by O'Connor and Kleyner.

Benefits

To the seller:

- Design cost: Less
- Development time: Shorter
- Test time: Shorter
- Manufacturing costs (parts and labor): Less
- Warranty costs: Less
- Risk: Less

To the customer:

- Satisfaction: Greater
- Value: Greater
- Risk: Less

Interesting Fact

High reliability ≠ high cost

When adjusted for inflation, the average price of new cars did not increase between 1979 and 2019. But the reliability of new cars increased **dramatically**.

In 1979, the average price of a new car was $7,000. In 2019, it was $25,000. Over the same period of time, inflation was 350%.

29

Some Tools We Use

Customer feedback
Warranty claims
Benchmarking
Quality functions
Apportionment
Accelerated life testing
Highly accelerated life testing (HALT)
Contract language
Specification language

30

Reliability Apportionment

Consider an engine. We want the engine to achieve a reliability of 0.70 over ten years of service. How do we achieve that using reliability apportionment?

Subsystems & Components

The first step is to recognize that an engine is a system comprised of subsystems and components. The reliability of an engine is dependent upon the reliabilities of those subsystems and components.

Major subsystems and components:

- Rocker lever and cam follower assembly
- Lube system
- External system
- Cylinder block
- Cylinder head
- Drive system
- Fuel system
- Cooling system
- Accessories
- Turbo system
- Miscellaneous

Engine Fault Tree

A fault tree for the major subsystems and components of the engine may look something like this. In this case, system reliability = 0.70 = $R_s = \Pi R_i$. A system reliability of 0.70 can only be achieved if each R_i has a value greater than 0.70.

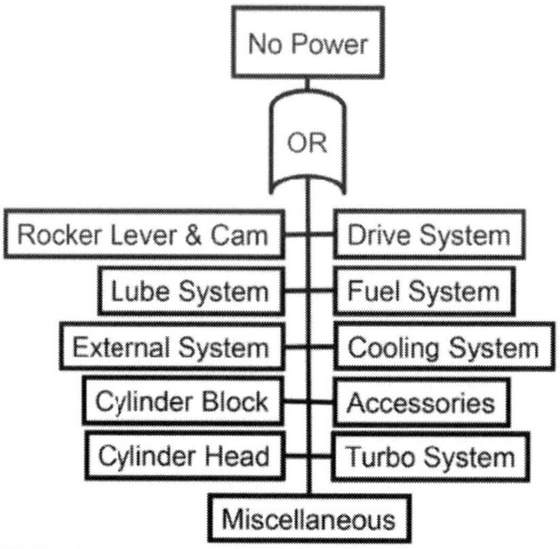

Apportionment

The reliability of each major subsystem and component is defined such that 0.70 = $R_s = \Pi R_i$. This process is repeated for each subsystem until the reliability of every component has been defined. The following figure traces reliability apportionment from the engine down to the individual components of the cam shaft. Notice the very high reliabilities of those individual components.

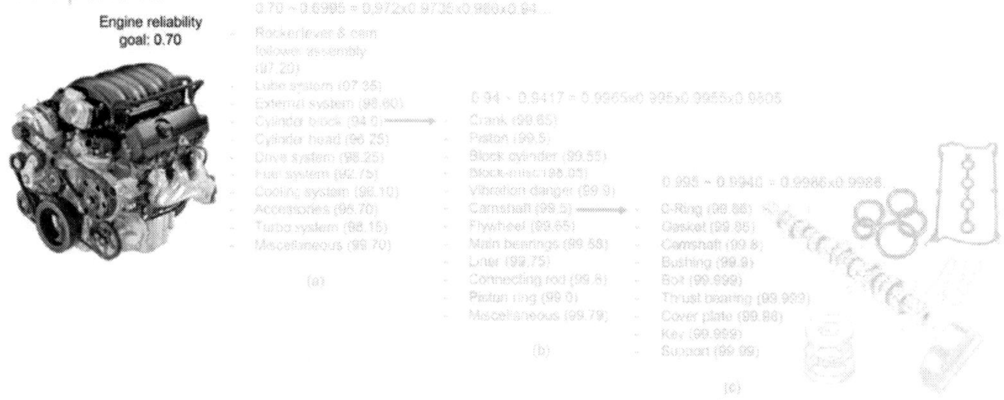

Source: Assurance Technologies by Dev G. Raheja, 1991

Apportionment

The reliability of each major subsystem and component is defined such that $0.70 = R_s = \Pi R_i$. This process is repeated for each subsystem until the reliability of every component has been defined. The following figure traces reliability apportionment from the engine down to the individual components of the cam shaft. Notice the very high reliabilities of those individual components.

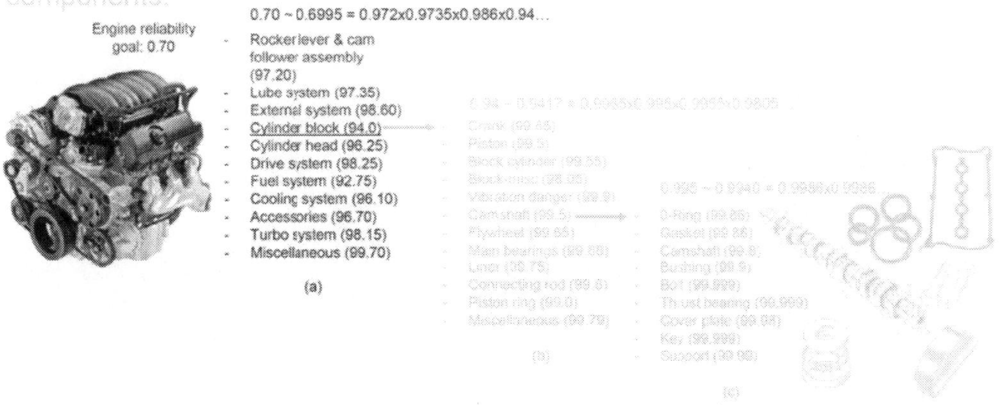

Source: Assurance Technologies by Dev G. Raheja, 1991

Apportionment

The reliability of each major subsystem and component is defined such that $0.70 = R_s = \Pi R_i$. This process is repeated for each subsystem until the reliability of every component has been defined. The following figure traces reliability apportionment from the engine down to the individual components of the cam shaft. Notice the very high reliabilities of those individual components.

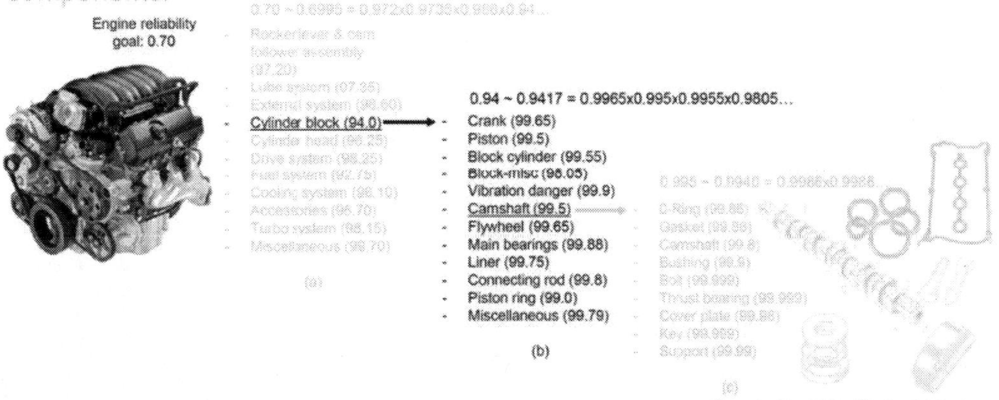

Source: Assurance Technologies by Dev G. Raheja, 1991

Apportionment

The reliability of each major subsystem and component is defined such that 0.70 = R_s = ΠR_i. This process is repeated for each subsystem until the reliability of every component has been defined. The following figure traces reliability apportionment from the engine down to the individual components of the cam shaft. Notice the very high reliabilities of those individual components.

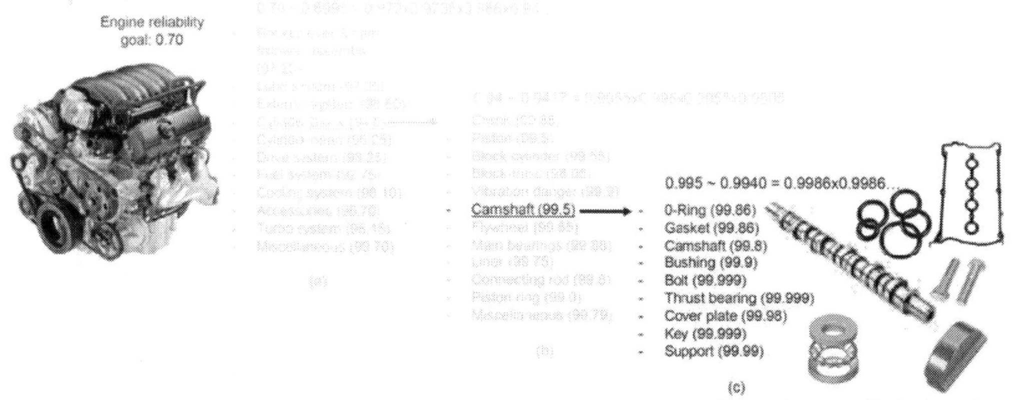

Source: Assurance Technologies by Dev G. Raheja, 1991

Design for Reliability...and Risk 37

Apportionment

The reliability of each major subsystem and component is defined such that 0.70 = R_s = ΠR_i. This process is repeated for each subsystem until the reliability of every component has been defined. The following figure traces reliability apportionment from the engine down to the individual components of the cam shaft. <u>Notice the very high reliabilities of those individual components</u>.

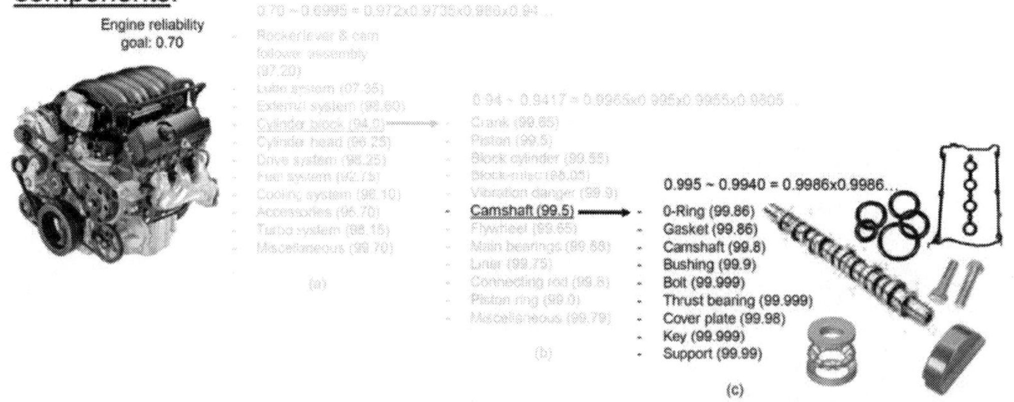

Source: Assurance Technologies by Dev G. Raheja, 1991

Design for Reliability...and Risk 38

So What?

With the reliability requirements of individual components in hand, engineers can write procurement specifications correctly, preferably in close collaboration with manufacturers and suppliers.

Note: Reliability apportionment is often an iterative process, with tradeoffs being made for weight, procurement cost, delivery times, ease of assemblage, ease of maintenance, warranty cost, and other considerations.

So What?

With the reliability requirements of individual components in hand, engineers can write procurement specifications correctly, preferably in close collaboration with manufacturers and suppliers.

Note: Reliability apportionment is often an iterative process, with tradeoffs being made for weight, procurement cost, delivery times, ease of assemblage, ease of maintenance, warranty cost, and other considerations.

Going To Mars

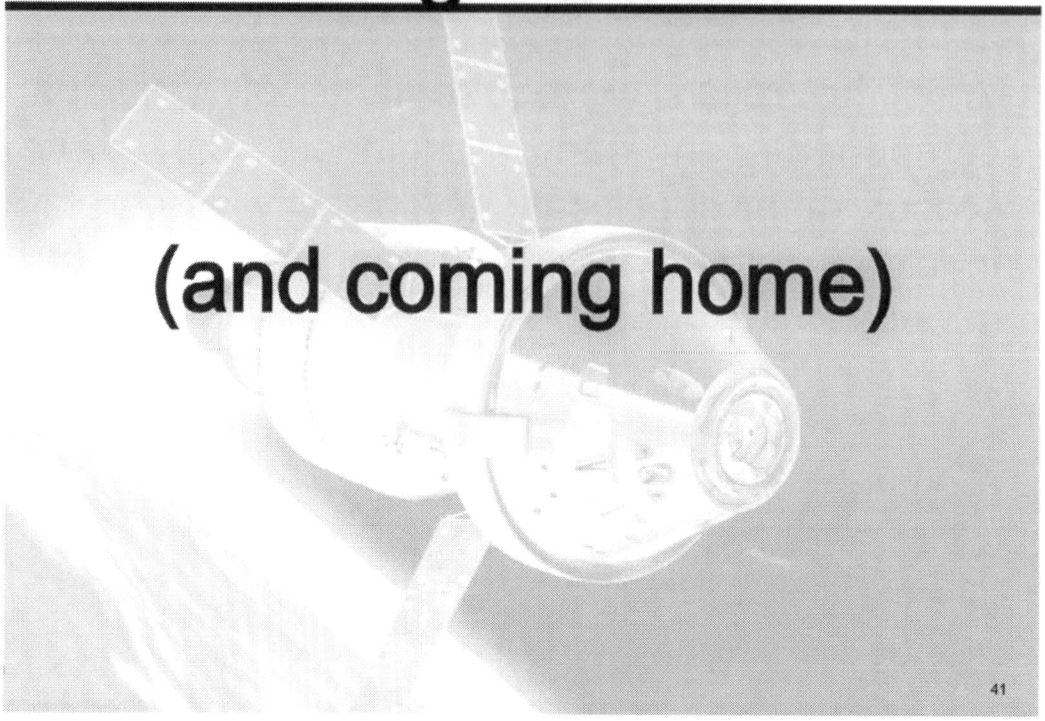

Mars

The task of delivering humans to Mars and returning them safely to Earth is not impossible…but it's really hard.

Some important points:

- Mission duration: ~1,100 days
- Maintenance depots along the way: None
- Desired probability of success: >0.999?

Source: Limitations of Reliability for Long-Endurance Spaceflight, by Owens and de Weck (MIT), 2016

42

Mean Time To Failure

Assume we want the reliability of the mission to be 0.999 (one chance in a thousand of failure). What must the MTTF of the vehicle be, assuming there will be no opportunity to conduct repairs?

$R(t) = 0.999 = e^{-\lambda t}$, where $t = 1{,}100$ days

Therefore, MTTF = $1/\lambda$ = 3,000 years (26×10^6 hrs).

Designing and building such a vehicle, however, has never been done.

43

Redundancy

The Poisson distribution is used to model rare events, and in the case of a mission to Mars can be used to make inferences about a vehicle with various levels of redundancy.

Recalling the Poisson distribution,

$$R(n, t) = 0.999 = \sum_{k=0}^{n} e^{-\lambda t}(\lambda t)^k/k!$$

where

n = the number of spares or redundant items
λ = 1/MTBF or 1/MTTF
t = 1,100 days

44

Redundancy

Redundancy, n	Required MTBF
0	3,000 years
1	68 years
2	16 years
3	7 years
4	4 years
5	3 years

Increasing levels of redundancy significantly decrease required MTBFs. However, the greater the redundancy, the greater the cost and weight. And more weight means more fuel, and even more cost.

45

What's the Solution?

Increase the MTBF of critical components; carry additional spares; have the ability to manufacture spares; or more likely do some combination of the above.

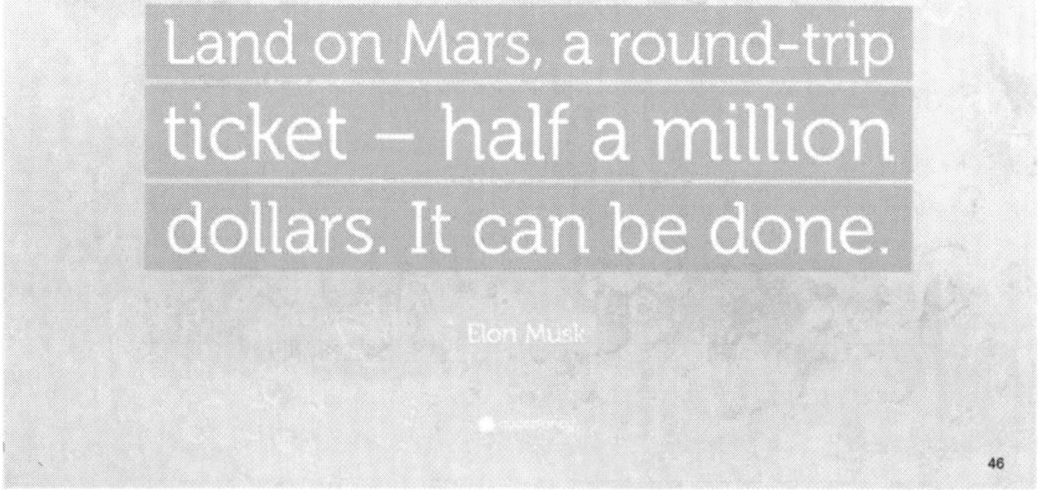

46

So What Do We Know?

The application of the principles of Design for Reliability will result in products of greater reliability and reduced cost.

Greater reliability drives down costs for both sellers and customers.

Greater reliability also reduces risk for both sellers and customers.

Design for Reliability…and Risk 47

So What Do We Know?

The application of the principles of Design for Reliability will result in products of greater reliability and reduced cost.

Greater reliability drives down costs for both sellers and customers.

Greater reliability also reduces risk for both sellers and customers.

Design for Reliability…and Risk 48

So What Do We Know?

The application of the principles of Design for Reliability will result in products of greater reliability and reduced cost.

Greater reliability drives down costs for both sellers and customers.

Greater reliability also reduces risk for both sellers and customers.

MFPT Where Theory Meets Practice

Thank You!

David A. Breslin, PE
Marine Engineer, U.S. Navy
Instructor for Reliability & Risk
Adjunct Professor, Reliability Engineering & Operations Research, United States Merchant Marine Academy
Primary Contact Information :
E-Mail: David.Breslin@navy.mil
Work Telephone: 202 781 3433

DYTRAN INSTRUMENTS, INC.

May 15 2019

How to Initiate Sensor Fusion within a Digital Environment

Dave Change
VP, CTO

Overview

01	Introduction to CAN-MD®
02	CAN-MD® Sensor Structure
03	CAN-MD® Technical Background
04	Machinery Diagnostics
05	CAN-MD® Integrations
06	Dashboard Console

ADVANCED SENSORS FOR A DYNAMIC WORLD

Introduction

CAN-MD® is a **digital sensor platform** that processes raw analog data *inside each sensor* on the machine, enabling a bussed architecture that **delivers actionable results**, not raw data!

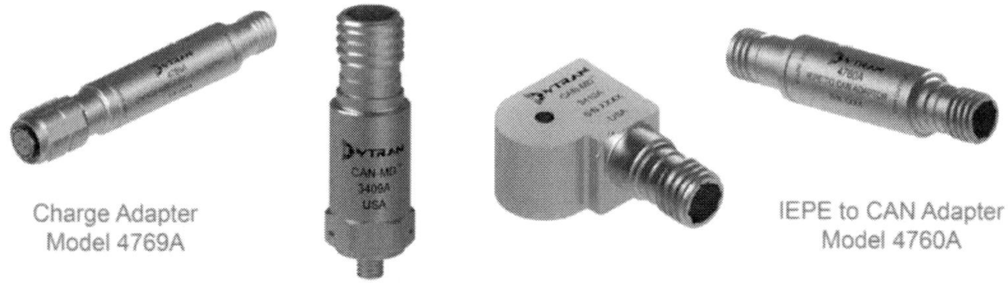

Charge Adapter
Model 4769A

Sensor
Model 3409A

Sensor
Model 3410A

IEPE to CAN Adapter
Model 4760A

ADVANCED SENSORS FOR A DYNAMIC WORLD

CAN-MD® Sensor Structure

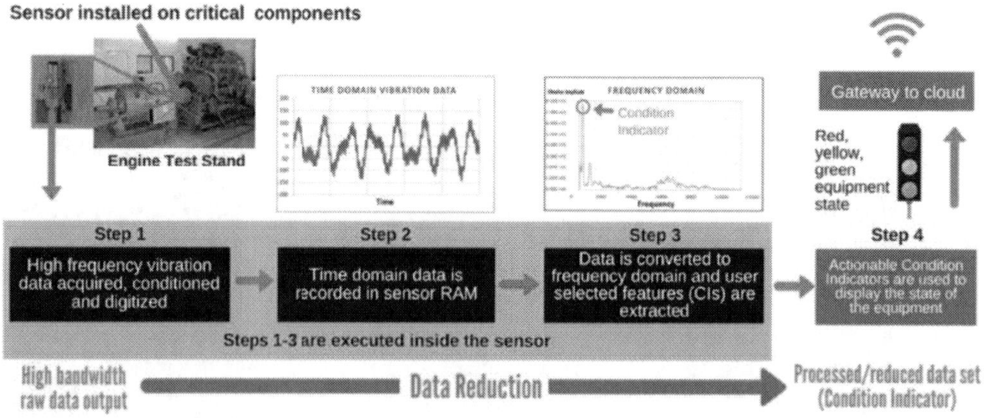

ADVANCED SENSORS FOR A DYNAMIC WORLD

CAN-MD® Data and Condition Indicators

- A Condition Indicator (CI) is the result of the processed data that is provided as an output on the data bus
- No longer is a complicated central DSP system needed to process the raw data from analog accelerometers
- In order to perform their duties, the sensors need to know some setup information.
- Each sensor can boot up and start collecting and processing CIs automatically. These CIs can later be filtered with additional information to classify "operating mode" dependent results.
- **The sensors themselves process the data!**

ADVANCED SENSORS FOR A DYNAMIC WORLD

CAN-MD® Condition Indicators

- RMS
- Skew
- Spectral Peak - Absolute
- Spectral Peak - Relative
- Spectral Energy – Relative
- Spectrum
- Time Domain
- Time Synchronous Average
- Order Domain Data Waveform
- Order Domain, Amplitude
- Order Domain, Phase

ADVANCED SENSORS FOR A DYNAMIC WORLD

CAN-MD® Functional Block Diagram

- CAN-MD® is a digital platform that allows the user to process high speed data while outputting the results using can bus, at much lower rates

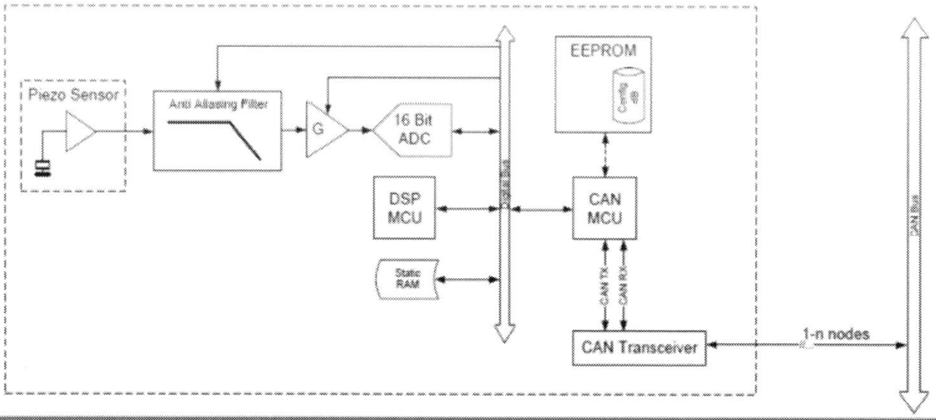

ADVANCED SENSORS FOR A DYNAMIC WORLD

Simplified Functional Diagram

Data is collected and processed within the sensor processor and output to CAN bus

or another data reduction update output, decision or alarm

ADVANCED SENSORS FOR A DYNAMIC WORLD

The Benefits of Going Digital

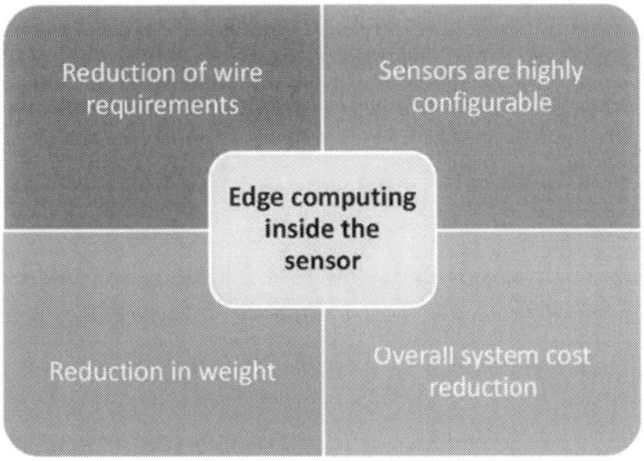

ADVANCED SENSORS FOR A DYNAMIC WORLD

Reduction of Wire Requirements

ADVANCED SENSORS FOR A DYNAMIC WORLD

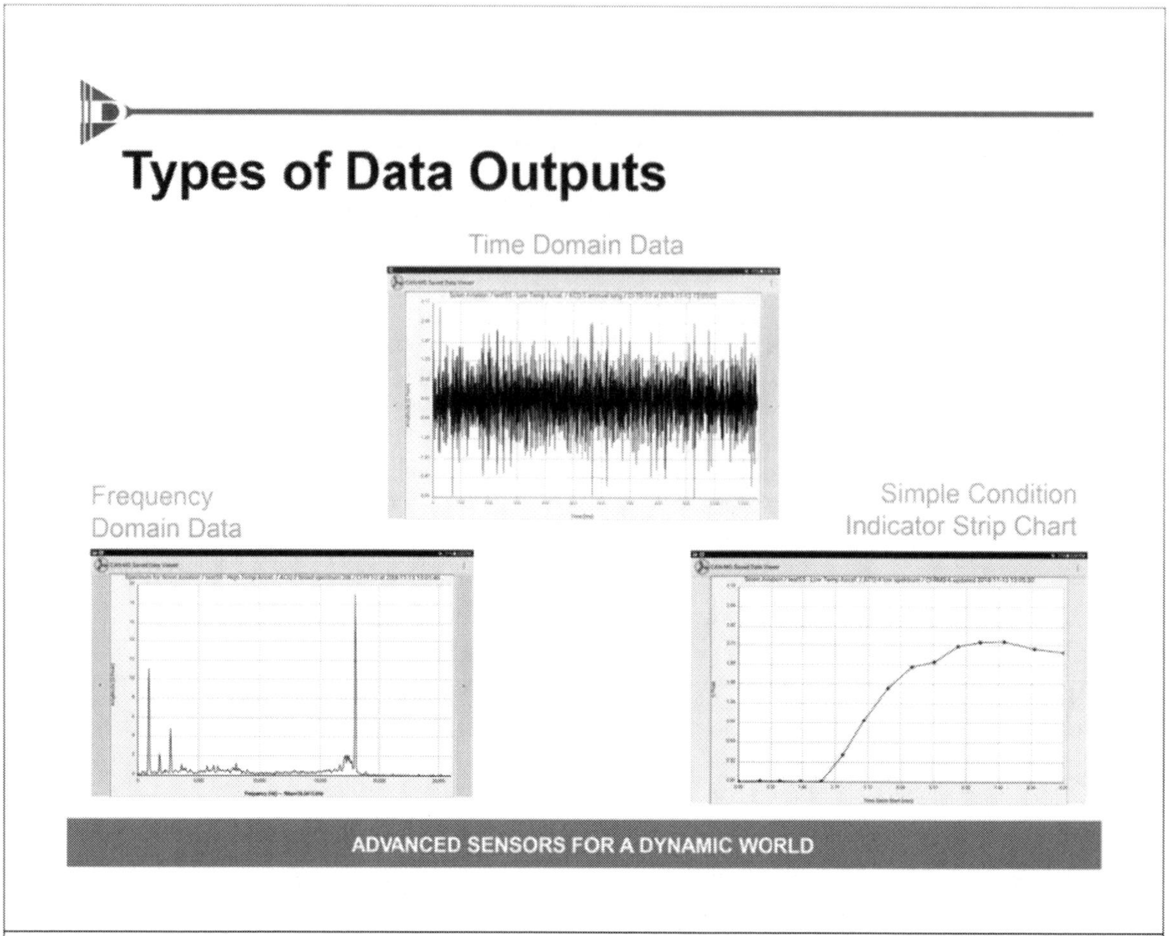
Types of Data Outputs
Time Domain Data
Frequency Domain Data
Simple Condition Indicator Strip Chart
ADVANCED SENSORS FOR A DYNAMIC WORLD

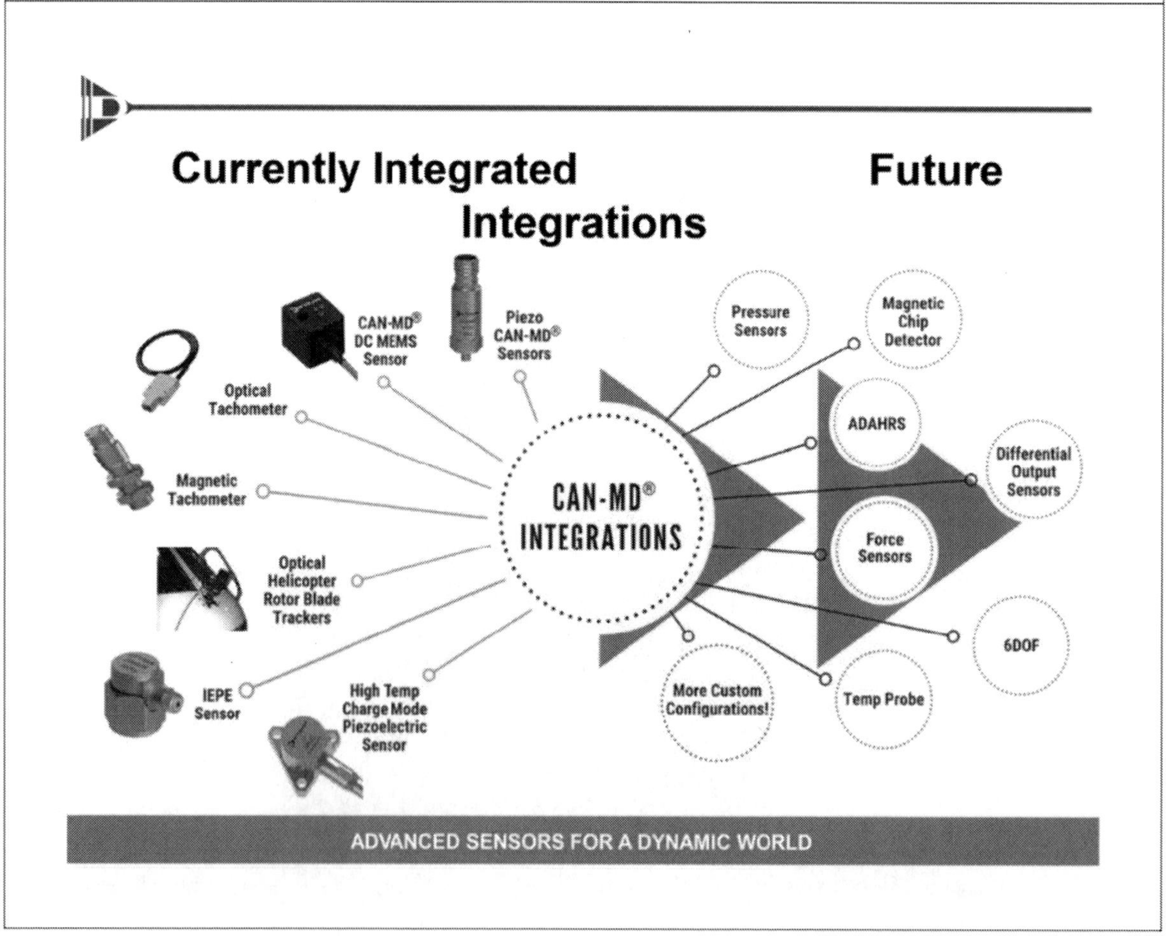
Currently Integrated
Future
Integrations
CAN-MD® DC MEMS Sensor
Piezo CAN-MD® Sensors
Optical Tachometer
Magnetic Tachometer
Optical Helicopter Rotor Blade Trackers
IEPE Sensor
High Temp Charge Mode Piezoelectric Sensor
CAN-MD® INTEGRATIONS
Pressure Sensors
Magnetic Chip Detector
ADAHRS
Differential Output Sensors
Force Sensors
6DOF
Temp Probe
More Custom Configurations!
ADVANCED SENSORS FOR A DYNAMIC WORLD

CAN-MD® Controller – New Acquisition

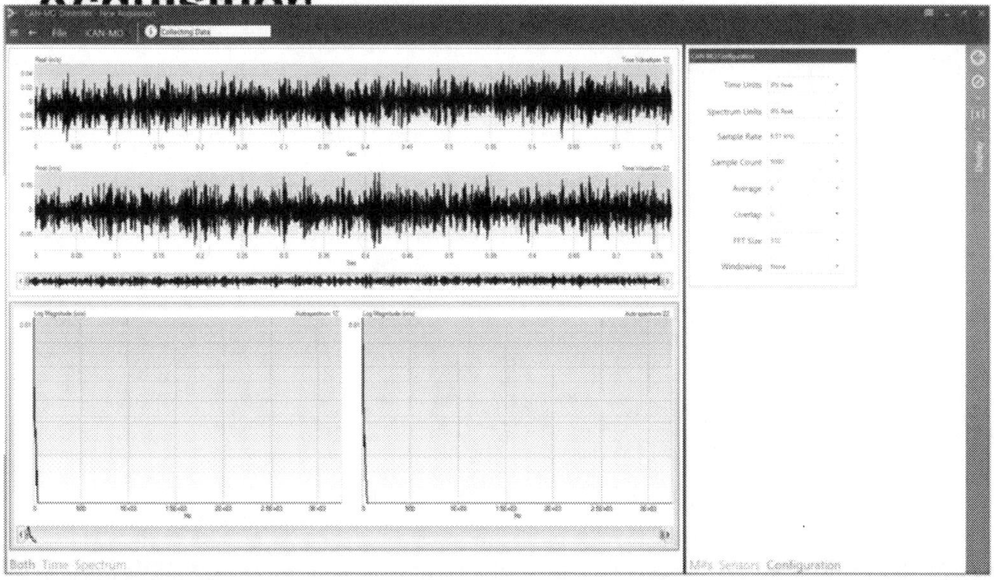

ADVANCED SENSORS FOR A DYNAMIC WORLD

CAN-MD® Machine Surveillance Console

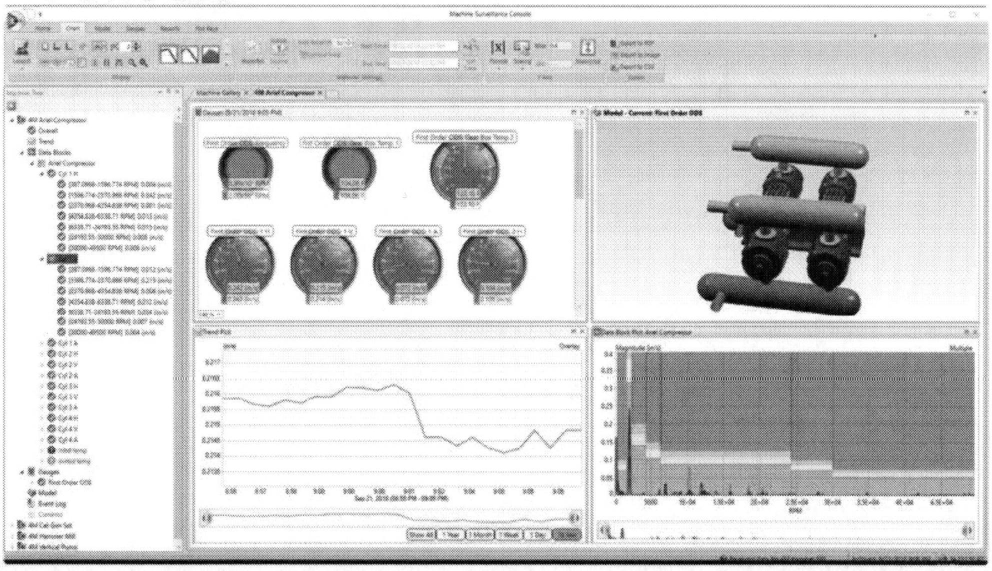

ADVANCED SENSORS FOR A DYNAMIC WORLD

CAN-MD® Machine Surveillance Console

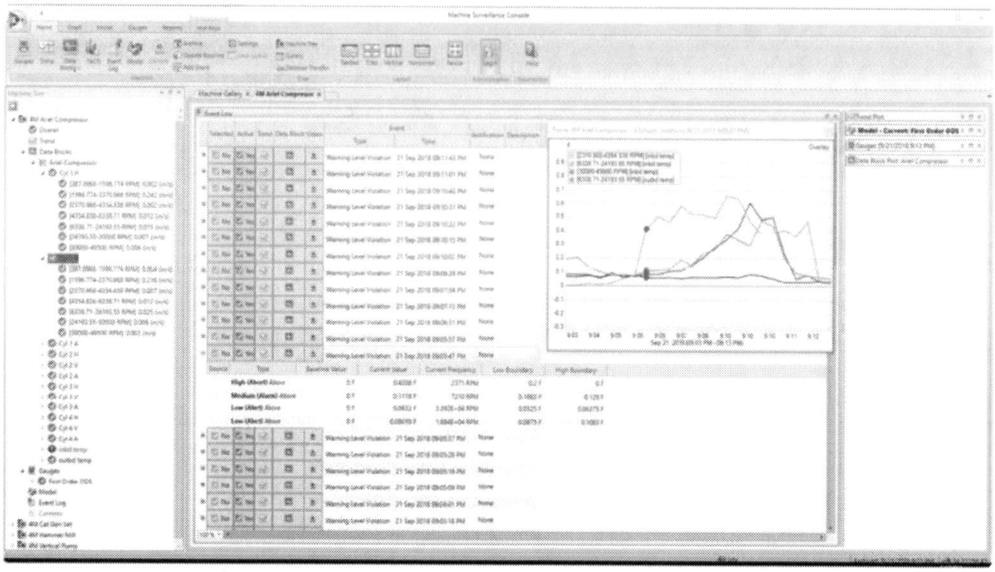

ADVANCED SENSORS FOR A DYNAMIC WORLD

Dytran Rotor Kit Model

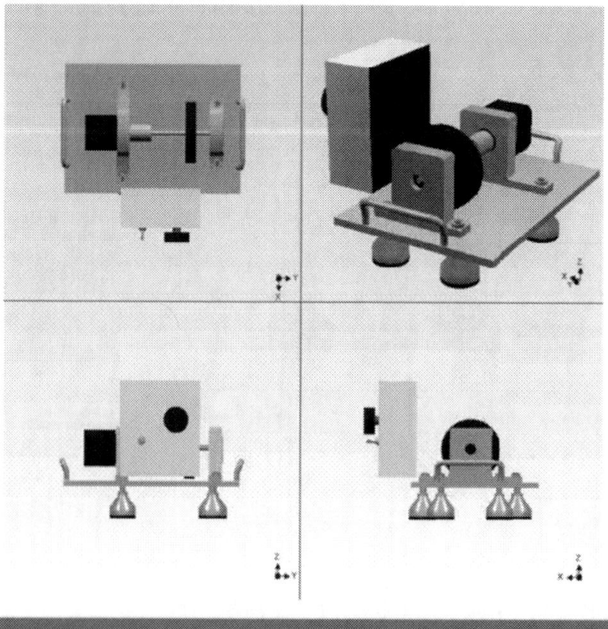

ADVANCED SENSORS FOR A DYNAMIC WORLD

Spectra Overlay with Custom Alarm Banding

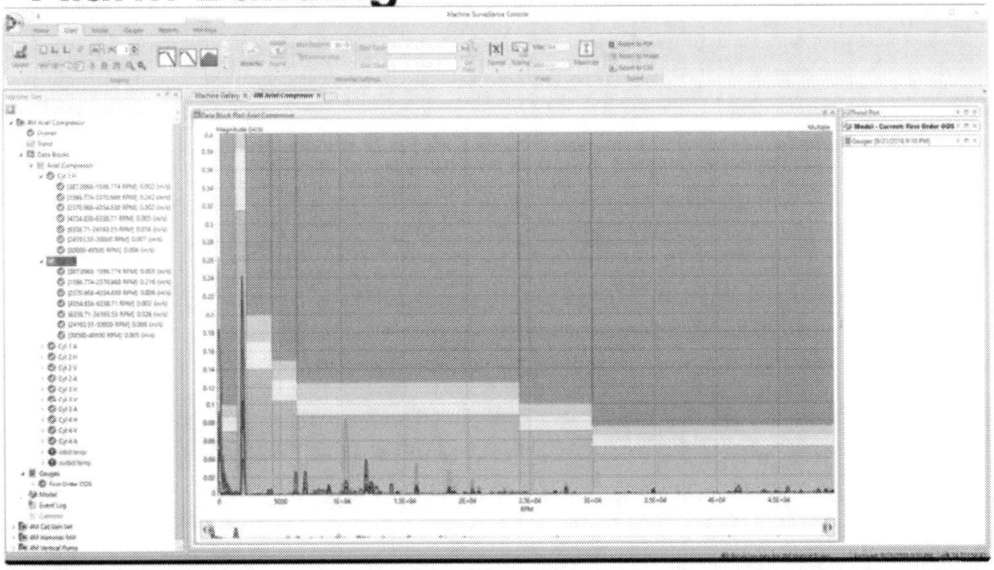

ADVANCED SENSORS FOR A DYNAMIC WORLD

Trend Plot

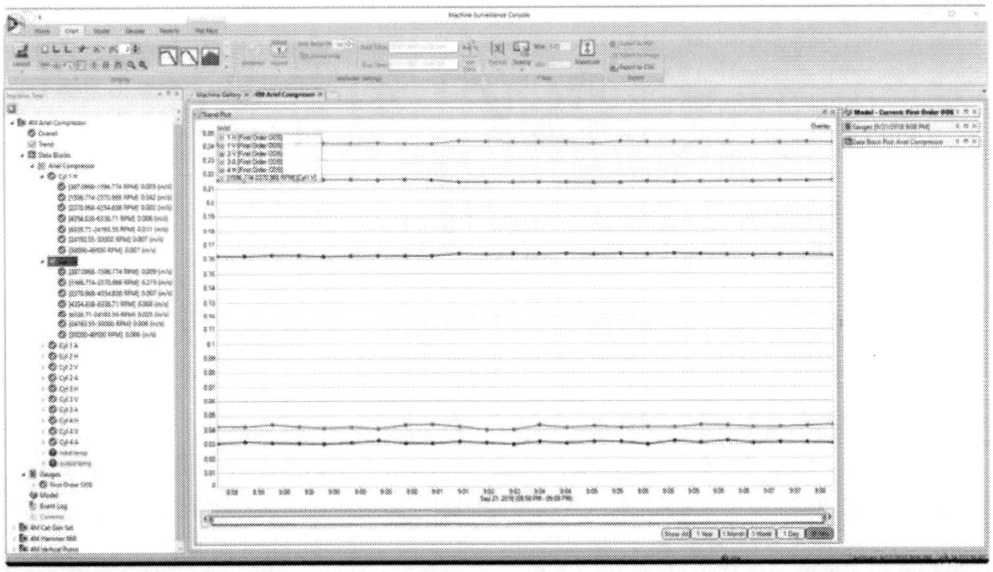

ADVANCED SENSORS FOR A DYNAMIC WORLD

Waterfall Plot

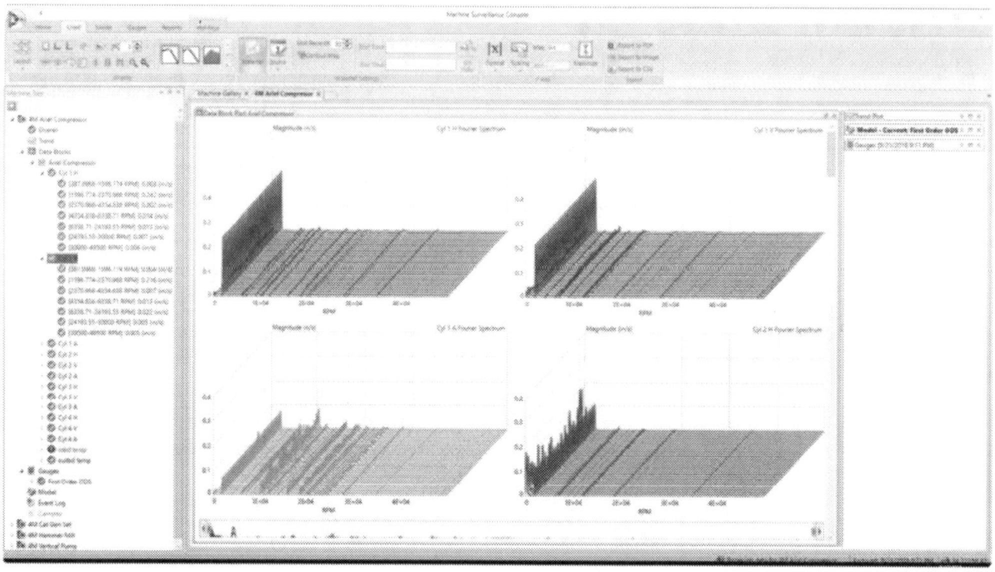

ADVANCED SENSORS FOR A DYNAMIC WORLD

CAN-MD® integration into DEWESoft CAN module and software

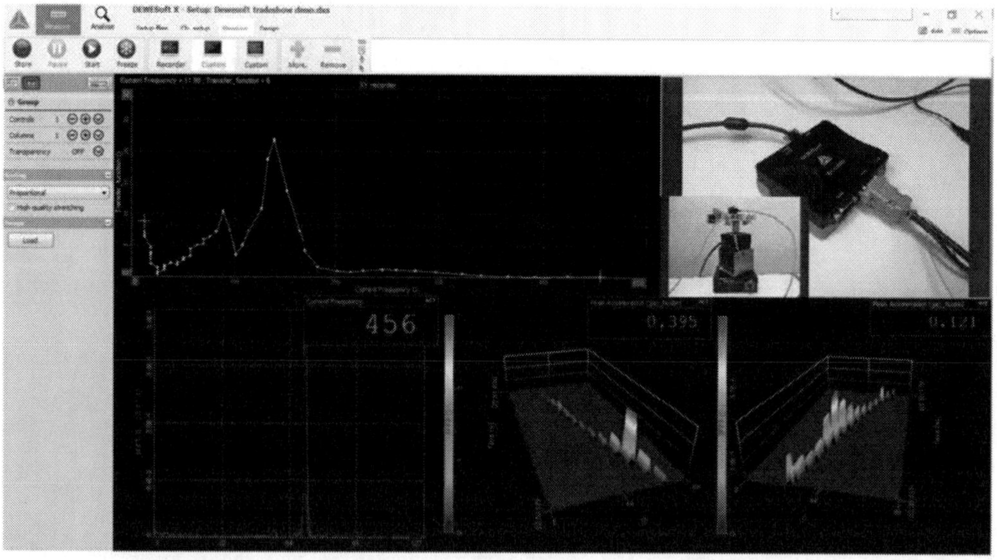

ADVANCED SENSORS FOR A DYNAMIC WORLD

Gantner Integration

- Dytran has conducted an example CAN-MD® integration utilizing the Gantner Q.station101-DT unit. The Gantner Q.station is typically found in test cells, production lines, and a variety of industrial automated environment. The architecture of the Gantner Q.series requires a Q.station controlling and additional modules for data acquisition, digital inputs/outputs, and additional communications.
- However, like many other industrial process control devices, Q.station has one independent CAN module. This makes it possible to switch the measurement chain architecture to utilize CAN-MD® technology.

ADVANCED SENSORS FOR A DYNAMIC WORLD

Gantner Integration Video

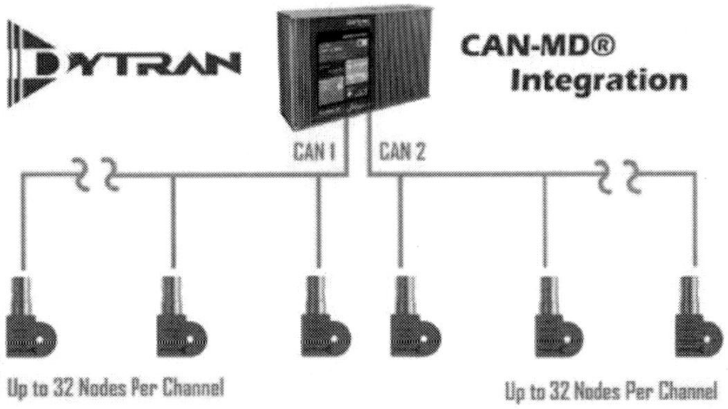

ADVANCED SENSORS FOR A DYNAMIC WORLD

Gantner Integration Video

Developer Kit Owners

GE

BOEING

ROUSH

Cummins

JOHN DEERE

AIRBUS HELICOPTERS

ADVANCED SENSORS FOR A DYNAMIC WORLD

Thank You For Attending!

Visit **Table 10** For More Information

Rolling Element Bearing Fault Analysis in a Permanent Magnet Motor

Daming Chen, Suri Ganeriwala
SpectraQuest Inc.
Richmond, Virginia, USA
suri@spectraquest.com

SpectraQuest, Inc.
8227 Hermitage Road, Richmond, VA 23228 U.S.A.
TEL +1-804-261-3300/FAX +1-804-261-3303
www.spectraquest.com

Machinery Failure Prevention Technology (MFPT) 2019
Philadelphia, May 14-16, 2019

Outline

1. Motivation
2. Methodology
3. Experiment Investigation
4. Conclusion

Motivation

- Rolling element bearing in a permanent magnet motor is most often under high load and high-speed running conditions. Hence, it is one of the most vulnerable components.
- How to enhance the bearing signal in the case of a low signal-to-noise ratio.
- How to select the optimal frequency band for envelope analysis.

Outline

1. Motivation
2. Methodology
3. Experiment Investigation
4. Conclusion

Features of Faulted Bearings

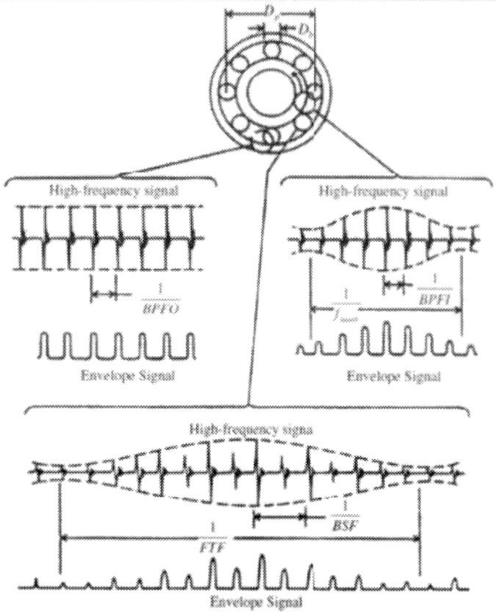

- ball pass frequency outer race (BPFO)
- ball pass frequency inner race (BPFI)
- ball spin frequency (BSF)
- fundamental train frequency (FTF)

Sawalhi, N. (2007). Diagnostics, prognostics and fault simulation for rolling element bearings. *The University of New South Wales, School of Mechanical and Manufacturing Engineering, Australia. Doctor of Philosophy thesis.*

SpectraQuest, Inc. 8227 Hermitage Rd, Richmond, VA 23228 1-804-261-3300 www.spectraquest.com 5

Minimum Entropy Deconvolution (MED) and Kurtosis

$$y(t) = x(t) * h(t) = \int_{-\infty}^{\infty} h(t-\tau)x(\tau)d\tau$$

Applying MED to find an inverse filter to remove the effect of transfer path *h(t)* and recover the original signal *x(t)*.

The coefficients of the optimal filter is obtained through formulation of an optimization problem based on kurtosis.

Endo, H., & Randall, R. B. (2007). Enhancement of autoregressive model based gear tooth fault detection technique by the use of minimum entropy deconvolution filter. *Mechanical Systems and Signal Processing, 21*(2), 906-919.

SpectraQuest, Inc. 8227 Hermitage Rd, Richmond, VA 23228 1-804-261-3300 www.spectraquest.com 6

Kurtogram

- The spectral kurtosis (SK) can indicate the nonstationary or non-Gaussian frequency components of the signal in the frequency domain.
- In the kurtogram, small values of SK can be seen at frequencies where only stationary Gaussian noise exist.
- Large values of SK can be seen at frequencies where transient occurs.
- The optimal frequency band obtained from kurtogram can be the input of envelope analysis.

Antoni, J. (2007). Fast computation of the kurtogram for the detection of transient faults. *Mechanical Systems and Signal Processing*, *21*(1), 108-124.

Envelope Analysis

$$Y(t) = y(t)\exp(-j\pi(f_l + f_u)t) \tag{1}$$

where f_l and f_u are the lower and upper limits of a selected frequency band. The second step is to lowpass filter the demodulated signal $Y(t)$ to compute the analytic signal $z(t)$. The cutoff frequency of the lowpass filter is $(f_u - f_l)/2$. The square envelope signal then can be calculated by

$$Z(t) = |z(t)|^2 \tag{2}$$

Outline

1. Motivation
2. Methodology
3. Experiment Investigation
4. Conclusion

Experimental Investigation

Drive-train diagnostics/prognostics test rig

Experimental Investigation

Rotating Speed	BPFO	BPFI	BSF	FTF
1800 RPM	91.855 Hz	148.145 Hz	60.437 Hz	11.482 Hz
3600 RPM	183.710 Hz	296.290 Hz	120.873 Hz	22.964 Hz
5000 RPM	255.153 Hz	411.514 Hz	167.880 Hz	31.894 Hz

Sampling frequency: 102,400 Hz

Outer Race Fault (BPFO)

Initial analysis (FFT)

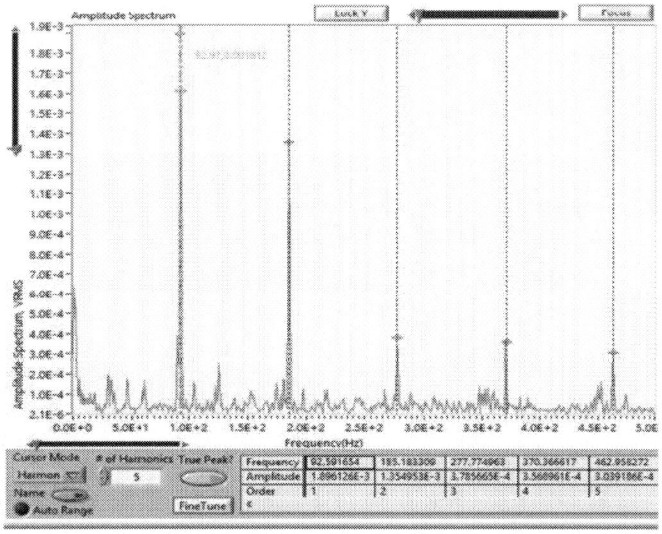

1800 RPM

Outer Race Fault (BPFO)

Initial analysis (FFT)

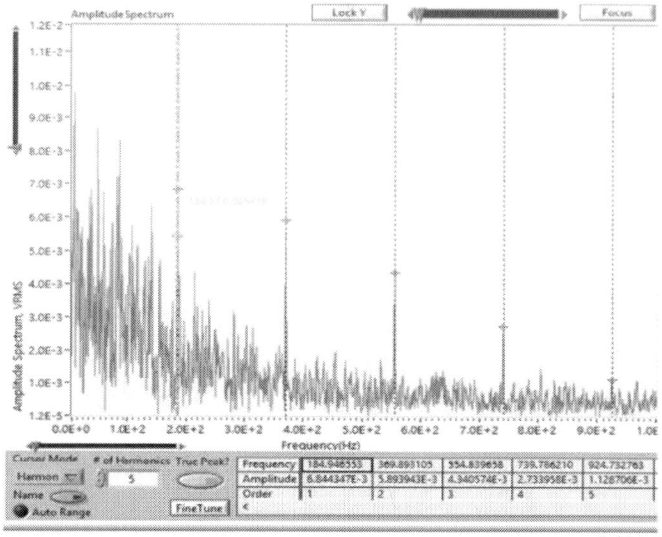

3600 RPM

Outer Race Fault (BPFO)

Initial analysis (FFT)

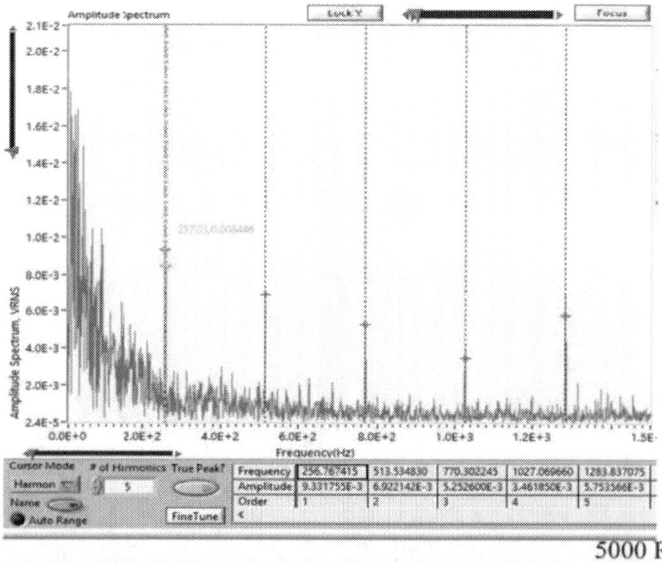

5000 RPM

Outer Race Fault (BPFO)

Raw signal at three rotating speeds | Processed signal by MED

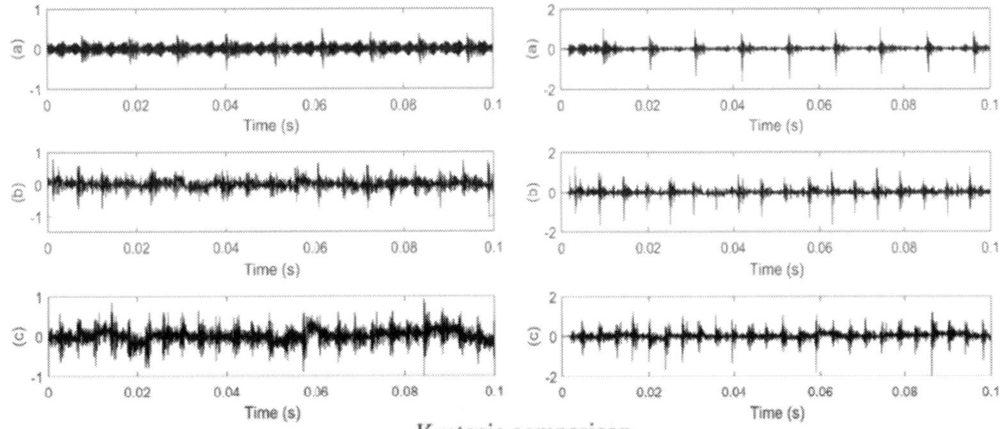

Kurtosis comparison

Rotating Speed	Before MED	After MED
1800 RPM	3.0254	42.5602
3600 RPM	3.8301	16.5063
5000 RPM	2.3856	8.7642

Kurtogram

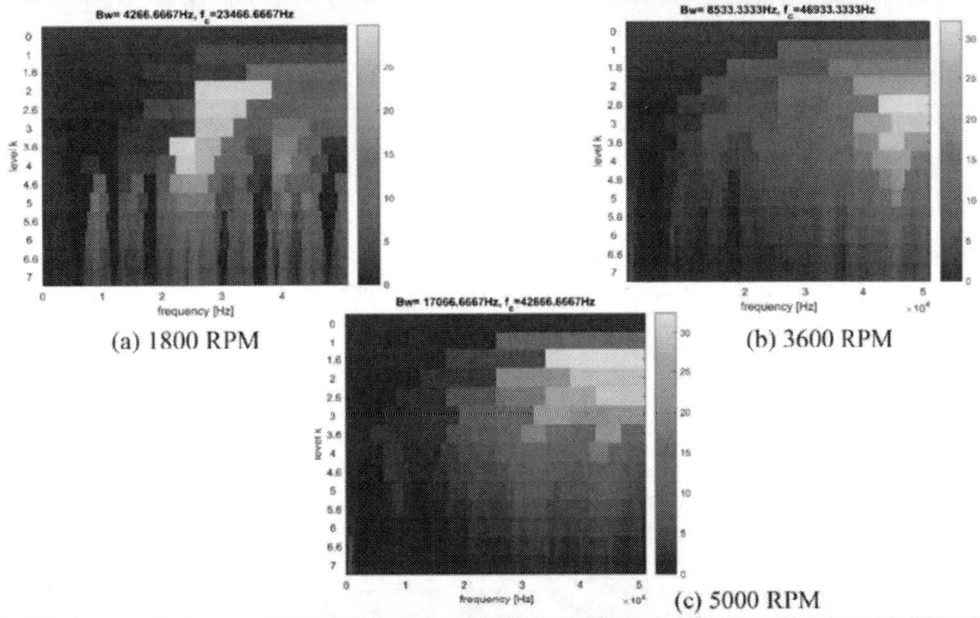

(a) 1800 RPM

(b) 3600 RPM

(c) 5000 RPM

Envelope Analysis

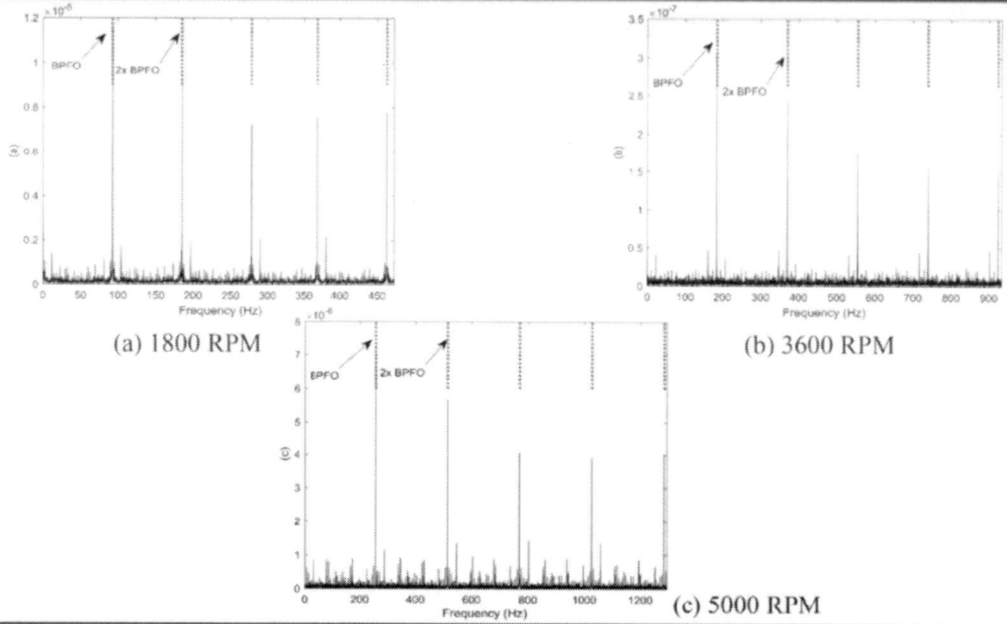

(a) 1800 RPM (b) 3600 RPM (c) 5000 RPM

Inner Race Fault (BPFI)

Initial analysis (FFT)

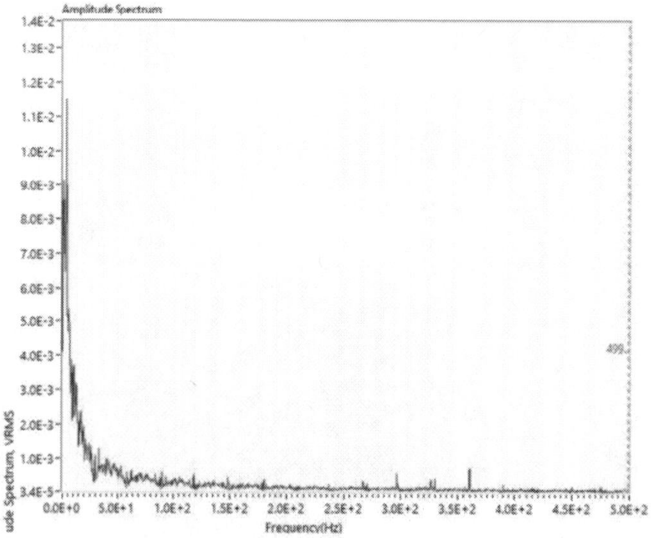

1800 RPM

Inner Race Fault (BPFI)

Initial analysis (FFT)

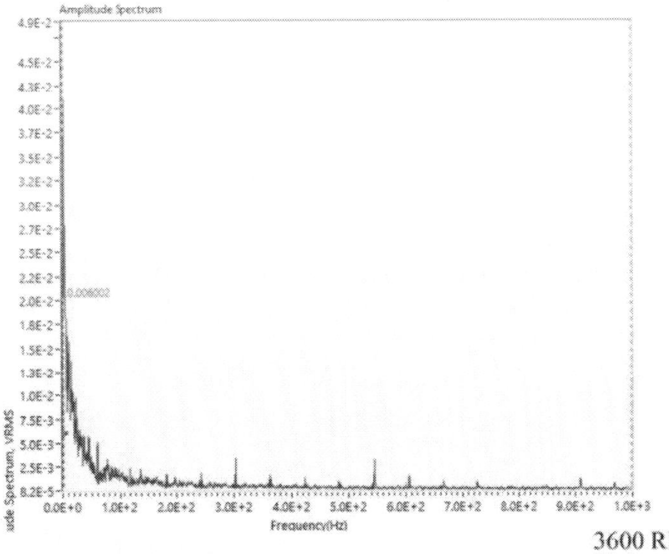

3600 RPM

Inner Race Fault (BPFI)

Initial analysis (FFT)

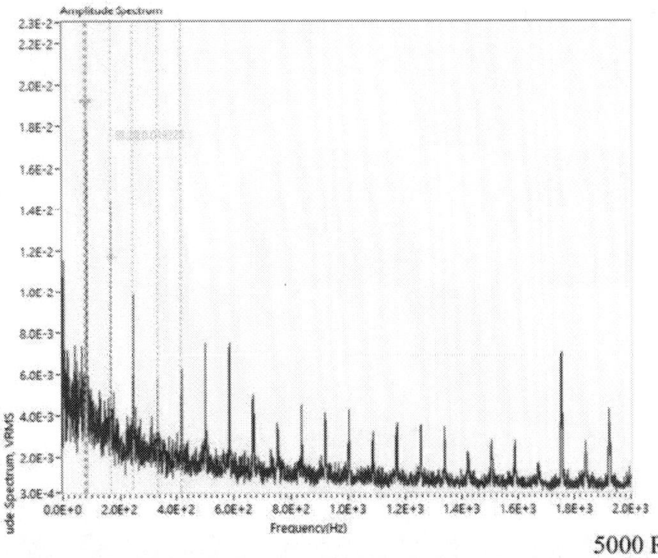

5000 RPM

Inner Race Fault (BPFI)

Raw signal at three rotating speeds | Processed signal by MED

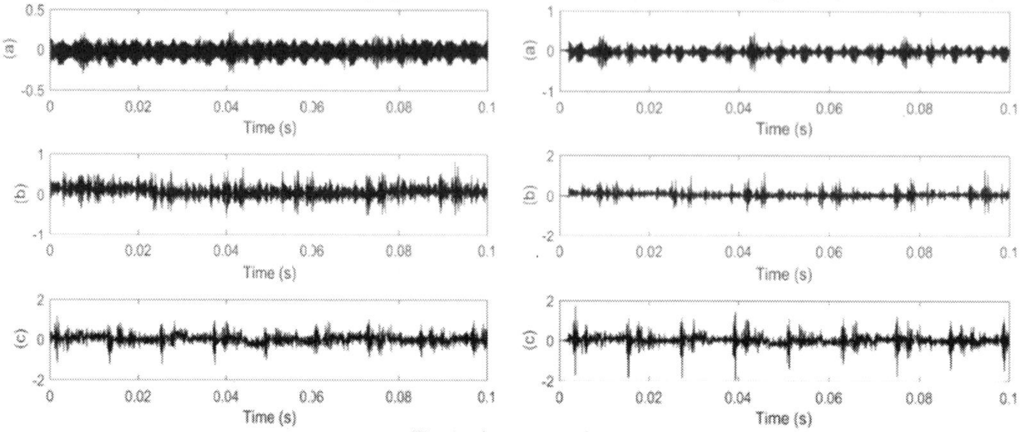

Kurtosis comparison

Rotating Speed	Before MED	After MED
1800 RPM	0.5715	4.1604
3600 RPM	1.6744	4.5812
5000 RPM	3.2212	6.8436

Kurtogram

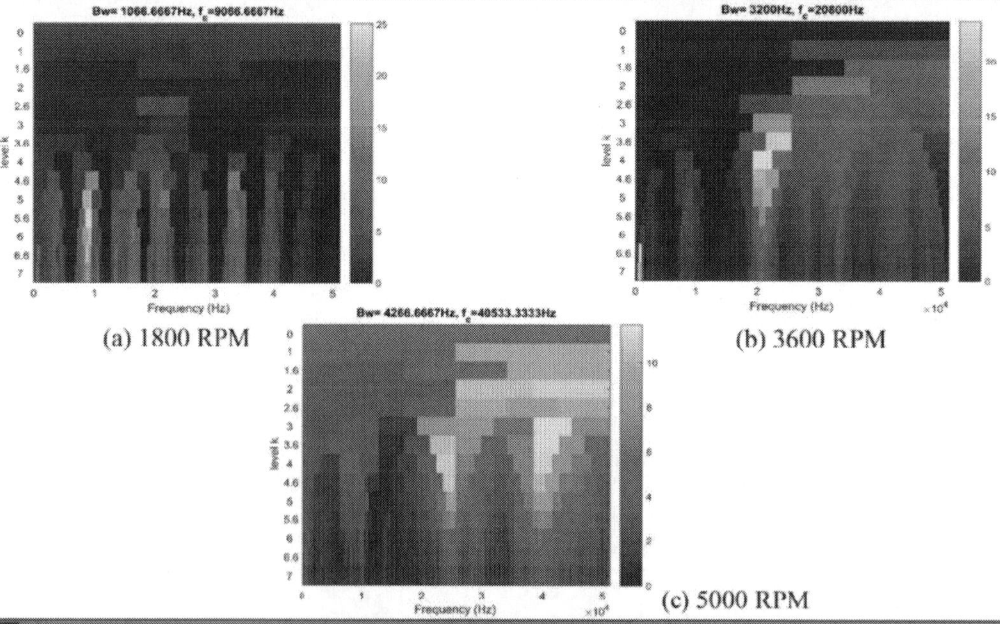

(a) 1800 RPM

(b) 3600 RPM

(c) 5000 RPM

Envelope Analysis

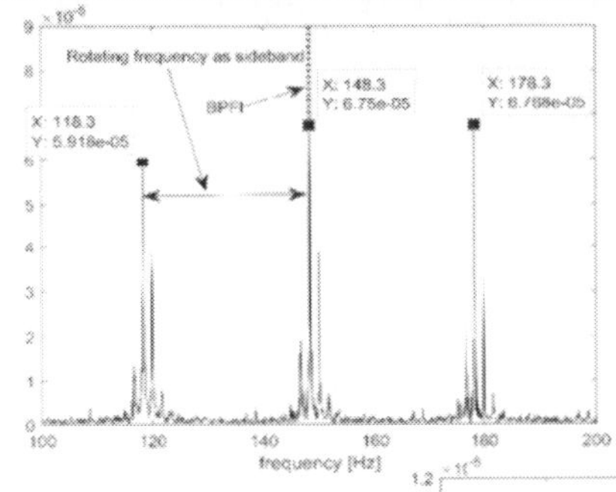

(a) 1800 RPM

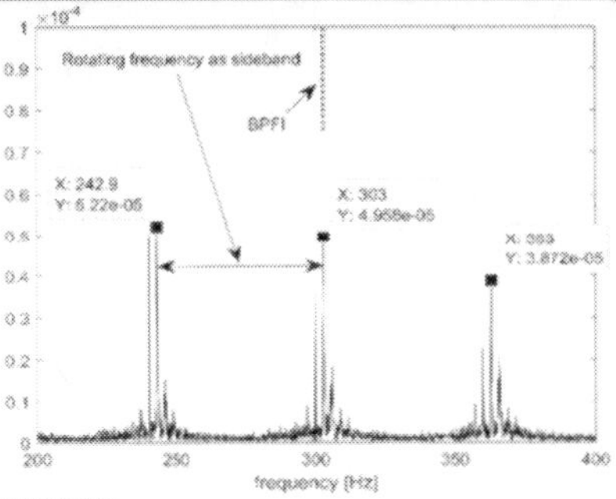

(b) 3600 RPM

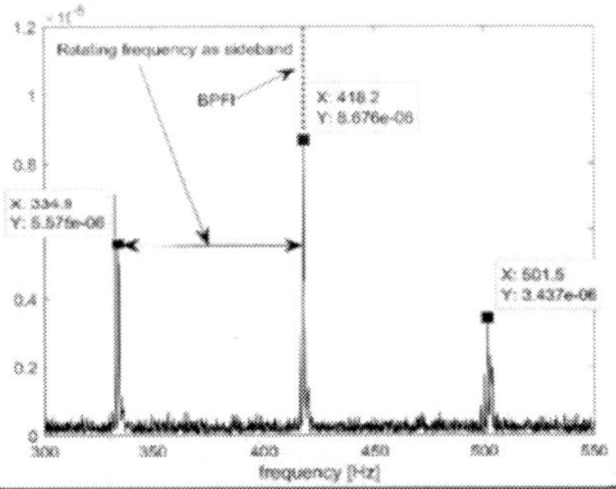

(c) 5000 RPM

Outline

1. Motivation
2. Methodology
3. Experiment Investigation
4. Conclusion

Conclusion

- By maximizing the SK of a measured signal, an inverse filter can be designed by the MED technique to reduce the effect of the transfer path on the measured signal.
- The kurtogram can help select a proper frequency band for envelope analysis.
- The combination of MED and SK shows a great potential for development of an automated diagnosis procedure to identify rolling element bearing faults.

Thank you for your attention.

MFPT 2019 Where Theory Meets Practice
MAY 14-16, 2019 | PHILADELPHIA, PA

Revolution in Fluid Film Bearing Monitoring

Insight Force Detection Sensor

Presented by Mike Ciocys

Authored by Dries Van Loon

Voyager INSTRUMENTS

PRÜFTECHNIK

MFPT 2019 Where Theory Meets Practice
MAY 14-16, 2019 | PHILADELPHIA, PA

Overview

- Intro
  - PRUFTECHNIK
- Fluid Film Bearings
  - What, Where and Why?
- Current technology for monitoring
  - Prox. Probes, Accelerometers, Temperatures, Not at all?
- FDS Sensor
  - Principle, Technology, Installation, Data Collection, Alarming, Cost, ...
- Example Data
- Conclusion

MFPT 2019 Where Theory Meets Practice
MAY 14-16, 2019 | PHILADELPHIA, PA

Intro

- Mike Ciocys
  - Started in Field Service, Application Engineer, now Technical Service & Support Manager at PRUFTECHNIK
  - Over 6 years in the organization
  - Vibration Analyst CAT II
  - Temple University graduate in Mechanical Engineering Technology
  - Specialize in online vibration projects and advanced applications

- Dries Van Loon
  - Belgium
  - 5 years in the USA
  - Electro-Mechanical Engineer
  - ISO 18436-2: Vibration Analyst CAT IV
  - 7 Years with PRUFTECHNIK (2 years Benelux, 5 years North America)
  - Application Engineer
  - Online Project Manager

MFPT 2019 Where Theory Meets Practice
MAY 14-16, 2019 | PHILADELPHIA, PA

PRÜFTECHNIK

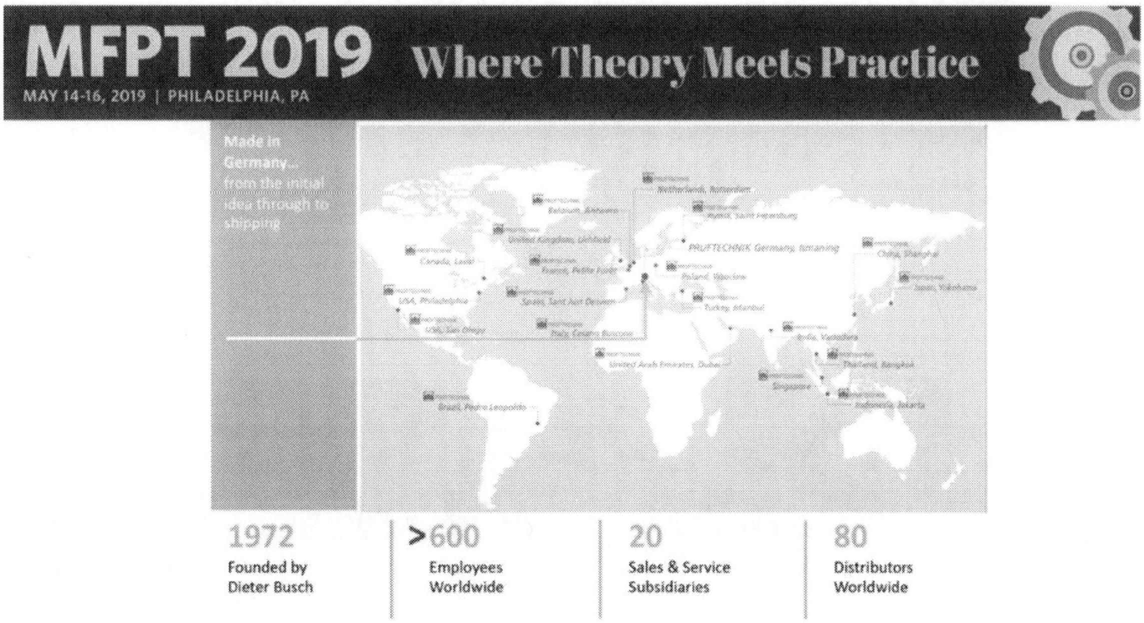

MFPT 2019
Where Theory Meets Practice
MAY 14-16, 2019 | PHILADELPHIA, PA
Made in Germany... from the initial idea through to shipping
Netherlands, Rotterdam
Canada, Laval
USA, Philadelphia
USA, San Diego
PRUFTECHNIK Germany, Ismaning
Turkey, Istanbul
United Arab Emirates, Dubai
Singapore
1972
Founded by Dieter Busch
>600
Employees Worldwide
20
Sales & Service Subsidiaries
80
Distributors Worldwide

MFPT 2019
Where Theory Meets Practice
MAY 14-16, 2019 | PHILADELPHIA, PA
Pruftechnik Montreal, QC
Pruftechnik Inc. Philadelphia, PA
Pruftechnik Inc. San Diego, CA

MFPT 2019 Where Theory Meets Practice

MAY 14-16, 2019 | PHILADELPHIA, PA

500
Patents Worldwide

250
Trademarks Worldwide

$100M
Sales 2018

MFPT 2019 Where Theory Meets Practice

MAY 14-16, 2019 | PHILADELPHIA, PA

Our Mission Complex Tasks – Easy to Use

To be a leading single source provider of premier measurement and monitoring solutions for plant maintenance and quality assurance.

MFPT 2019 Where Theory Meets Practice
MAY 14-16, 2019 | PHILADELPHIA, PA

Core Activities

Alignment Systems
Georretrical Measurements

Condition Monitoring
Vibration Analysis

Non Destructive Testing

Ultrasonic
Testing

Balancing

MFPT 2019 Where Theory Meets Practice
MAY 14-16, 2019 | PHILADELPHIA, PA

Laser Alignment Systems for Rotating Machinery

Benefits

- Longer service life
- Increased MTBF and uptime
- Lower power consumption
- Reduced temperature, vibration and mechanical repairs
- Lower operating costs
- Higher product quality

Solutions

- Shaft alignment
- Measurement of machine geometry
- Bore alignment
- Monitoring positional changes
- Roll alignment with PARALIGN
- Ultrasonic Testing

MFPT 2019 Where Theory Meets Practice
MAY 14-16, 2019 | PHILADELPHIA, PA

Condition Monitoring Vibration Analysis

Benefits

- Early detection of machine problems
- Elimination of root cause before failure occurs
- Reduced vibration and wear
- Increased machine availabilty
- Reduced operating costs
- CM feeds CbM (Condition based monitoring) and asset reliability programs

Solutions

- Portable data collectors and analyzers
- Online Condition Monitoring
- Continuous wear monitoring
- Machine protection

MFPT 2019 Where Theory Meets Practice
MAY 14-16, 2019 | PHILADELPHIA, PA

NDT - Non Destructive Testing

Benefits

- Improved product quality
- Optimized production processes
- Reduced scrap
- Improved productivity and performance

Solutions

- Eddy current testing equipment
- Flux leakage testing equipment
- Sensors and accessories
- Turnkey systems from design to commissioning

MFPT 2019 Where Theory Meets Practice
MAY 14-16, 2019 | PHILADELPHIA, PA

Worldwide Machinery Services

Laser Shaft Alignment

Turbine Alignment

Condition Monitoring

Roll Alignment

Geometric Alignment

Ultrasonic Testing

MFPT 2019 Where Theory Meets Practice
MAY 14-16, 2019 | PHILADELPHIA, PA

Fluid Film Bearings

Voyager INSTRUMENTS

PRÜFTECHNIK

MFPT 2019 Where Theory Meets Practice

MAY 14-16, 2019 | PHILADELPHIA, PA

- Plain bearings, journal bearings, or sleeve bearings
- Fundamental and critical part of rotating machinery
- **Larger**, high **speed** and high **load** equipment
  = **More Critical**
- Ideal measurement location for PdM
  - Trend Condition
  - Diagnose machine faults
  - Diffe-ent monitoring techniques than rolling element bearings

MFPT 2019 Where Theory Meets Practice

MAY 14-16, 2019 | PHILADELPHIA, PA

- Lubricant between stationary and rotating parts
- Hydrostatic
  - External pressure for lubrication
  - Potential risk of losing pressure
- Hydrodynamic
  - Rotation of shaft to pressuri-- -:l f:l--
  - No metal to metal contact
- **Theoretical Infinite Life**

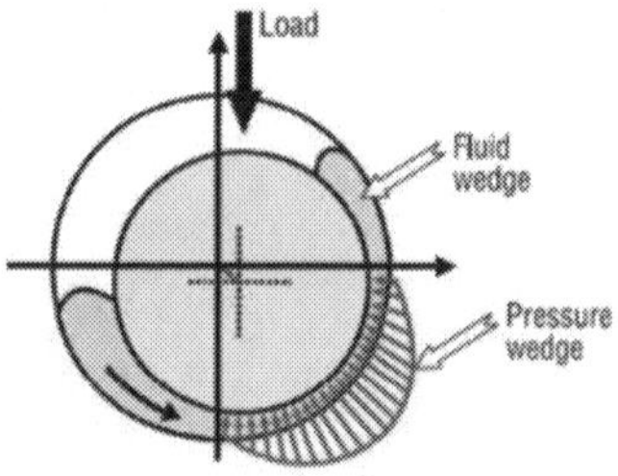

MFPT 2019 Where Theory Meets Practice
MAY 14-16, 2019 | PHILADELPHIA, PA

- Lubrication Regimes
  - Boundary lubrication
  - Mixed lubrication
  - Hydrodynamic lubrication

- Start-up and Shutdown induce most wear

- Hydrostatic lift for start-up

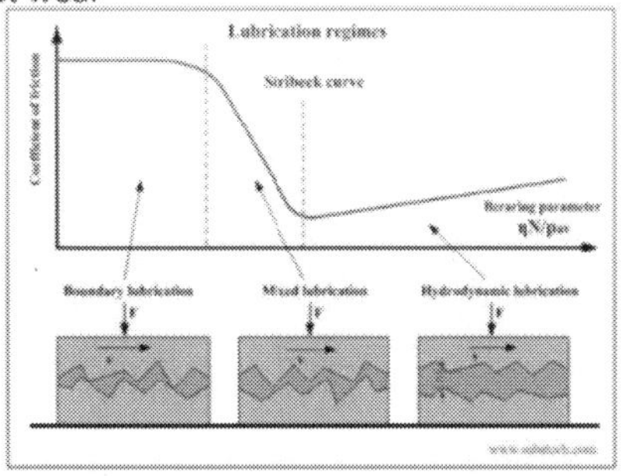

MFPT 2019 Where Theory Meets Practice
MAY 14-16, 2019 | PHILADELPHIA, PA

Monitoring Techniques

- **Proximity Probes**
  - Invented by Don Bently in the 60's
  - Eddy Current Proximitor ®
  - Relative **displacement (mils)** between bearing housing and shaft
  - Ideally X-Y (Orbit and Center Shaft Line) often only 1 probe
- Pros
  - Actual shaft position in housing (DC/gap)
  - Dynamic shaft movement (AC/orbit)
  - Easy to evaluate vs bearing clearance
  - Diagnostic possible
- Cons
  - Slow roll compensation
  - Cost (hardware >2k$)
  - Retrofit installation, drill-tap ($ >>> Hardware)

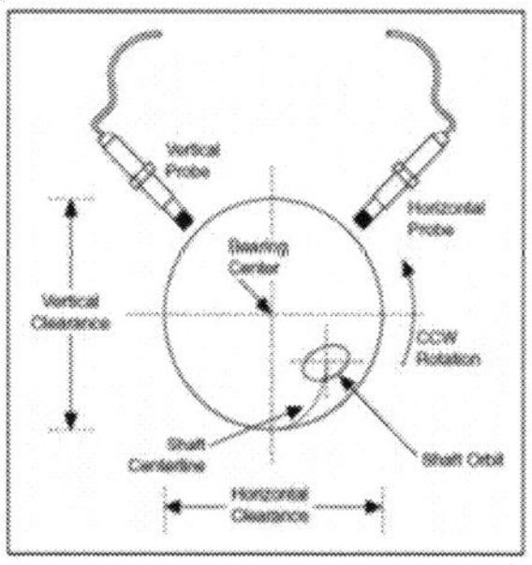

- **Temperature**
  - Bearing housing or Babbitt temperature
  - Lubricant temperature
- Pros
  - Easy to install
  - Better than no indication
- Cons
  - No diagnostic
  - Thermal Lag
  - Alarm too late (temperature after

MFPT 2019 Where Theory Meets Practice
MAY 14-16, 2019 | PHILADELPHIA, PA

- **Case Mounted Vibration**
  - Accelerometer or Velometer on casing housing
  - Casing vibration combined with relative shaft vibration (full picture)
- Pros
  - Easy to install or collect with route data coll
  - Trending and some diagnostics
  - Together with prox. probe get full picture
  - Better than nothing

- Cons
  - Transmission path not linear
  - Force from shaft to measurement locatic
  - Many harmonics hard to make diagnosti

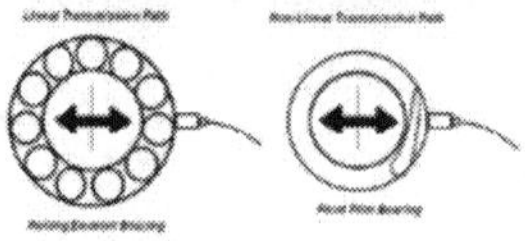

MFPT 2019 Where Theory Meets Practice
MAY 14-16, 2019 | PHILADELPHIA, PA

- **Not at all??**

- Pros
  - Cheap

- Cons
  - Very Expensive

MFPT 2019 Where Theory Meets Practice
MAY 14-16, 2019 | PHILADELPHIA, PA

- **Conclusion**
  - Proper configured X-Y proximity probes currently **best**
    - Expensive for retrofit installation (Hardware + Installation + Shutdown)
  - Other monitoring techniques not sufficient individually
  - **Need for innovation**
    - Low (retrofit) installation cost
    - Trending (protection) and diagnostic of fluid film bearings
    - New technology
  - **Insight Force Detection Sensor**

MFPT 2019 Where Theory Meets Practice
MAY 14-16, 2019 | PHILADELPHIA, PA

Insight FDS
Force Detection Sensor

Voyager INSTRUMENTS

PRÜFTECHNIK

MFPT 2019 Where Theory Meets Practice
MAY 14-16, 2019 | PHILADELPHIA, PA

- **Principle**

Use dynamic strain sensor on the housing of a fluid film bearing, installed in the load zone of the bearing (pressure wedge), to measuring indirectly the bearing forces and monitor the condition of the machine. (US Patent nr. US9841329B2)

https://patents.google.com/patent/US9841329B2/en

MFPT 2019 Where Theory Meets Practice

MAY 14-16, 2019 | PHILADELPHIA, PA

- **Kinematics**
  - Even though bearing acceleration is low, large forces are still transmitted resulting in significant strain (με)
  - Trends and alarms are direct measures of potential damage

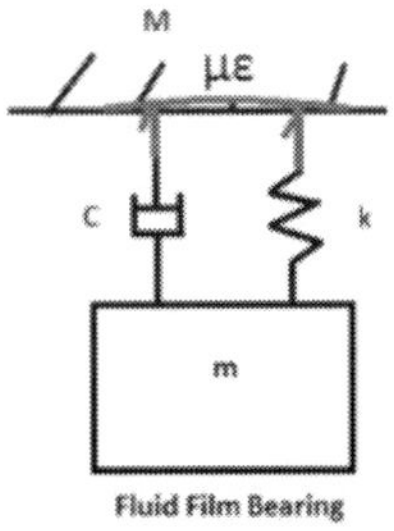

Results in microstrain (ΔL/L)

MFPT 2019 Where Theory Meets Practice

MAY 14-16, 2019 | PHILADELPHIA, PA

- **Technology**
  - Piezoelectric sensing element
  - Titanium bottom plate (engineered thickness)
  - IEPE power (same as ICP sensor)
  - Unidirectional

- **Technical Specs (PTA-FDS102-VOY)**
  - Dimensions: .425" x .6" x .180"
  - Sensitivity: 50 mV/με (± 20%)
  - Measurement Range: +-100 με
  - Frequency Range: 0.5Hz-10kHz
  - Temperature Range: -22°F to 250°F
  - Integral cable with 10-32 UNF Connector
  - Sensor and cable submersible (oil)
  - Class 1 Div. 2

MFPT 2019 Where Theory Meets Practice

MAY 14-16, 2019 | PHILADELPHIA, PA

- **Technology**
  - Piezcelectric sensing element
  - Titanium bottom plate (engineered thickness)
  - IEPE power (same as ICP sensor)
  - Unid rectional

- **Technical Specs (PTA-FDS101-VOY)**
  - Dimensions: 1.57" x .67" x .58"
  - Sensitivity: 40 mV/με (± 20%)
  - Measurement Range: +-100 με
  - Frequency Range: 0.01Hz-10kHz
  - Temperature Range: -22°F to 250°F
  - 10-32 UNF Connector

MFPT 2019 Where Theory Meets Practice

MAY 14-16, 2019 | PHILADELPHIA, PA

- **Pressure Wedge**
  - Predictable for all bearing types
  - Ideal location for FDS sensor install (BDC or 430 & 730)

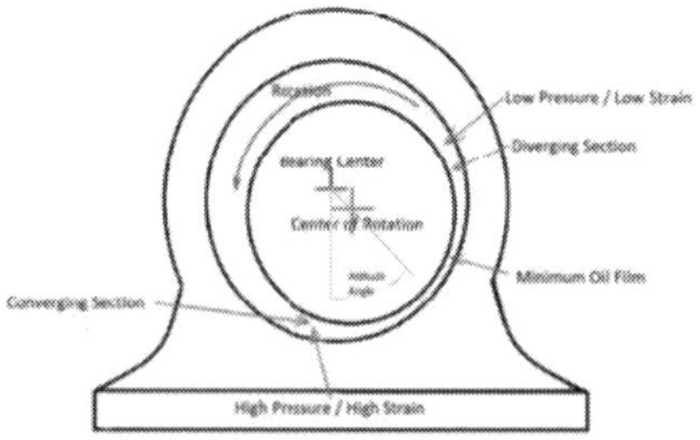

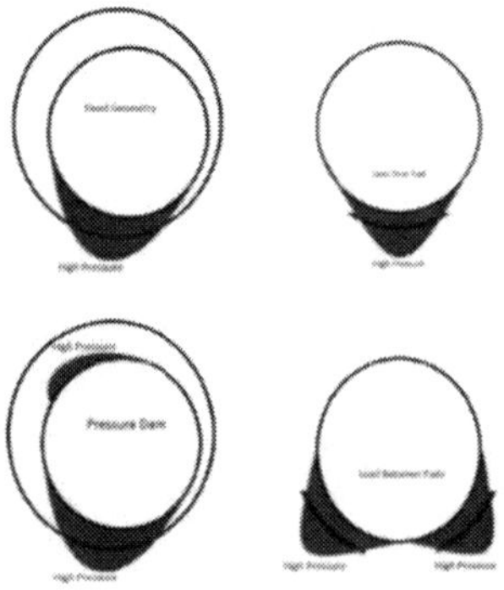

MFPT 2019 Where Theory Meets Practice

MAY 14-16, 2019 | PHILADELPHIA, PA

- **Install (<1h)**
  - Define pressure wedge location (Load?)
  - Remove loose paint, oil and solvents
  - ± Flat surface (not interested in static strain)
  - Glue sensor to bearing housing
  - Magnetic installation tool

→ ***Measure***

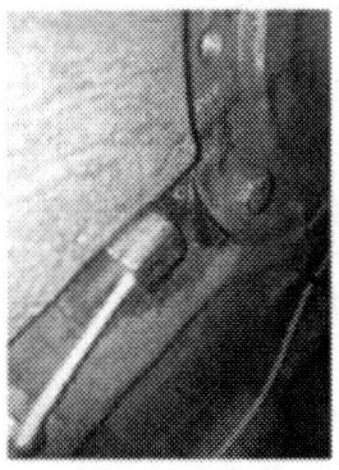

MFPT 2019 Where Theory Meets Practice

MAY 14-16, 2019 | PHILADELPHIA, PA

- **Measure (as ICP sensor)**
  - Setup sensor (50 mV/με)
  - Trend overall strain for alarming
  - Spectra and Time Waveform for detailed analysis

- **Trending in DCS**

MFPT 2019 Where Theory Meets Practice

MAY 14-16, 2019 | PHILADELPHIA, PA

- **Measure (as ICP sensor)**
  - Setup sensor (50 mV/με)
  - Trend overall strain for alarming
  - Spec:ra and Time Waveform for detailed analysis

- **Handheld data collector**

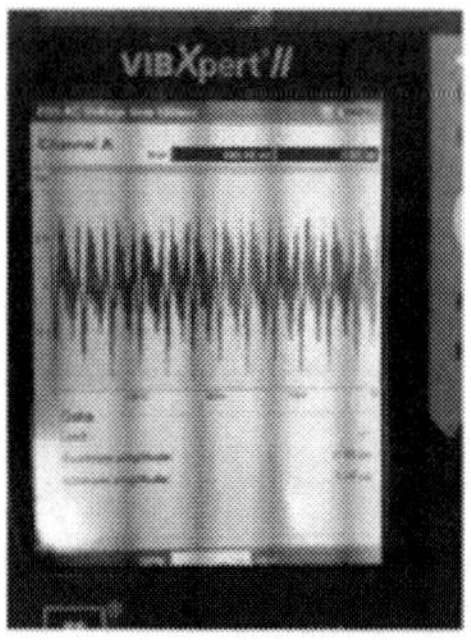

MFPT 2019 Where Theory Meets Practice

MAY 14-16, 2019 | PHILADELPHIA, PA

- **Measure (as ICP sensor)**
  - Setup sensor (50 mV/με)
  - Trend overall strain for alarming
  - Spec:ra and Time Waveform for detailed analysis

- **Continuous Monitoring**

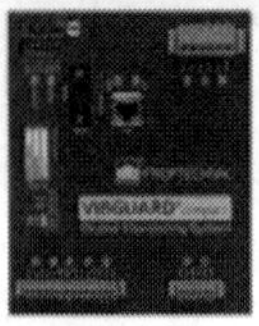

- **Alarming**
  - Evolving process (just like vibration)
  - Trending and statistical alarms (just like vibration)
  - Critical alarm level can be calculated (application specific)
    - Load and Speed
    - Babbitt material properties (elastic limits)
    - Young's Modulus
    - Alarm percentages (15-25%) of calculated max due to mounting location

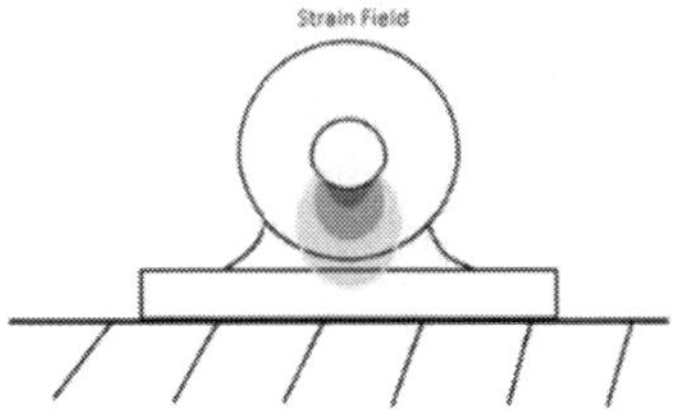

MFPT 2019 Where Theory Meets Practice
MAY 14-16, 2019 | PHILADELPHIA, PA

Example Data

Data VIBGUARD

Voyager INSTRUMENTS

PRÜFTECHNIK

MFPT 2019 Where Theory Meets Practice

MAY 14-16, 2019 | PHILADELPHIA, PA

- **Comparing data**
  - High Speed compressor
  - Prox., Accel., FDS
  - VIBGUARD portable (Cloud Connectivity)
  - OMNITREND Center Cloud Server
  - Online View (Live data and condition visualization)

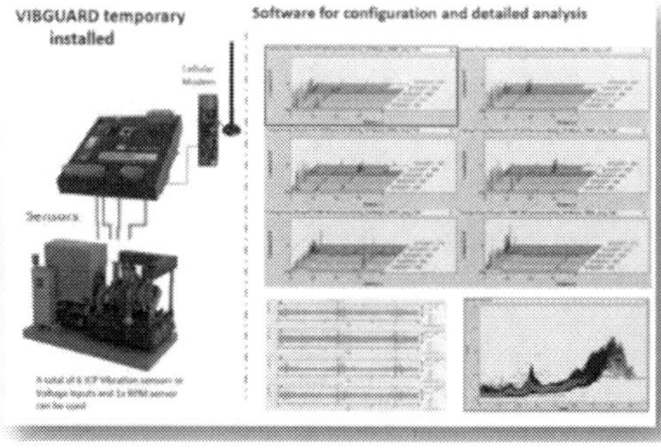

MFPT 2019 Where Theory Meets Practice

MAY 14-16, 2019 | PHILADELPHIA, PA

- Overall Pk-Pk

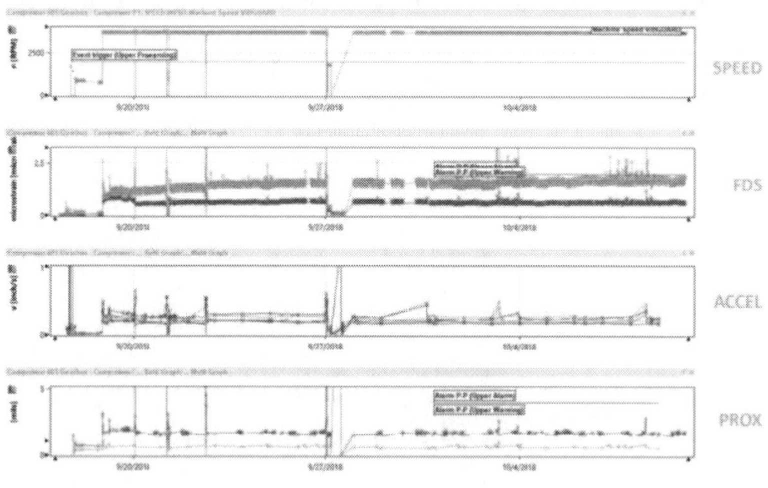

MFPT 2019 Where Theory Meets Practice
MAY 14-16, 2019 | PHILADELPHIA, PA

- Spectra

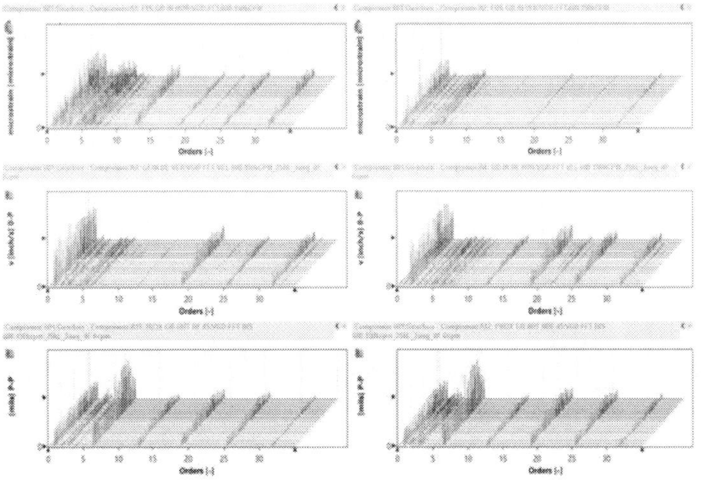

MFPT 2019 Where Theory Meets Practice
MAY 14-16, 2019 | PHILADELPHIA, PA

- Spectra - detail

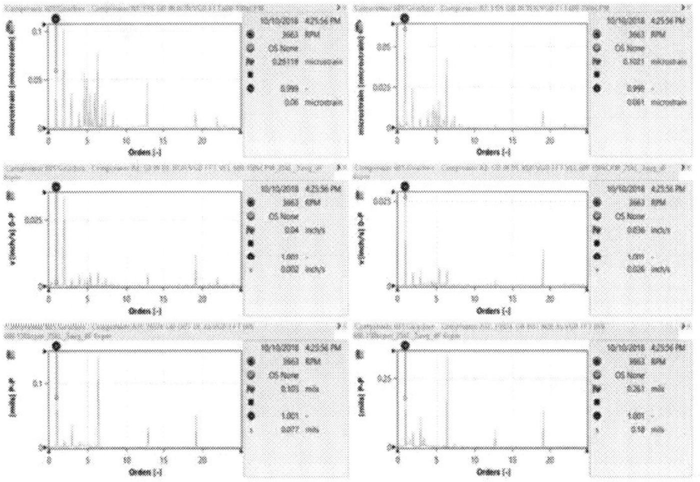

- Timewave form - detail

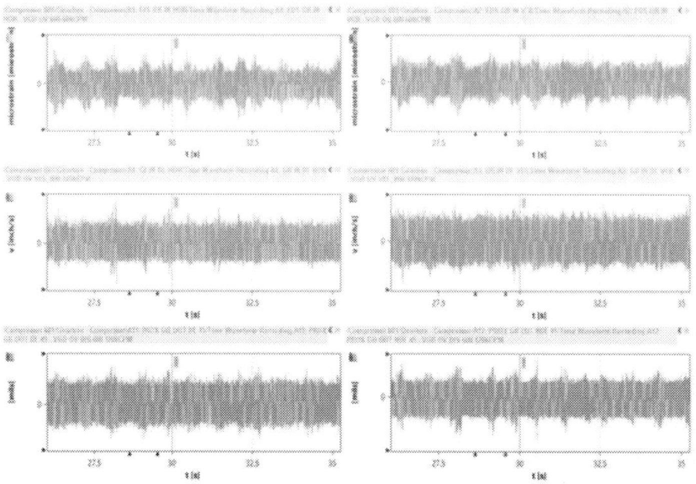

MFPT 2019 Where Theory Meets Practice
MAY 14-16, 2019 | PHILADELPHIA, PA

Lab Rotor Setup

- 100 psi bearing pressure
- 0 – 1800 RPM
- Critical Speed at 1320 RPM
- Internally designed hydrodynamic bearings
- Thick Film Lubrication at 115 RPM
- ISO VG 35 Equivalent Oil
- IB Bearing 0.00275" Clearance
- OB Bearing 0.004" Clearance
- 2" Diameter Shaft, L/D = 0.375
- Thick Babbitt 0.0625"

- Sensors:
  - Insight FDS
  - Proximity Probes
  - RTD's

MFPT 2019 Where Theory Meets Practice
MAY 14-16, 2019 | PHILADELPHIA, PA

➢ **Amplitude/pattern comparison**

- Velocity
- Dynamic Strain

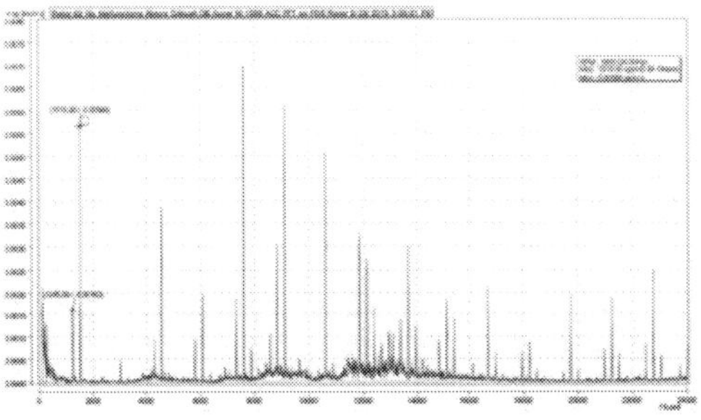

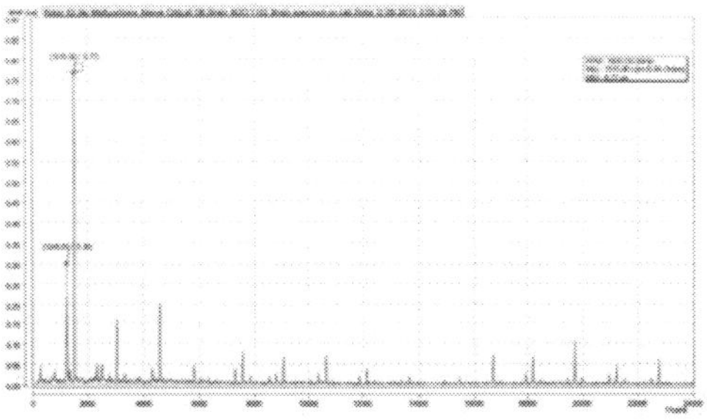

MFPT 2019 Where Theory Meets Practice
MAY 14-16, 2019 | PHILADELPHIA, PA

➢ **Amplitude/pattern comparison**

- Acceleration
- Dynamic Strain

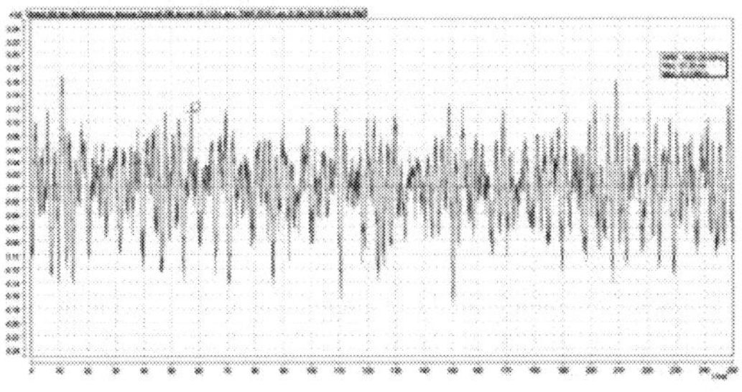

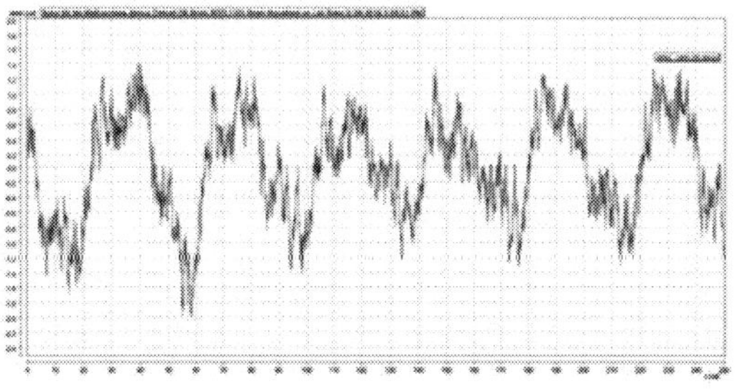

MFPT 2019 Where Theory Meets Practice
MAY 14-16, 2019 | PHILADELPHIA, PA

➢ **Amplitude/pattern comparison**

- Proximity Displacement
- Dynamic Strain

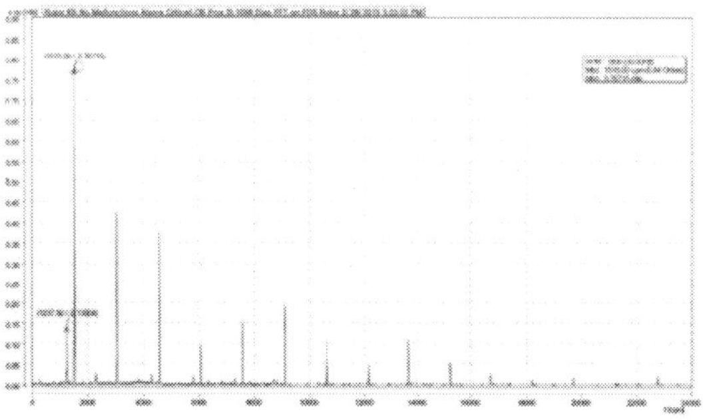

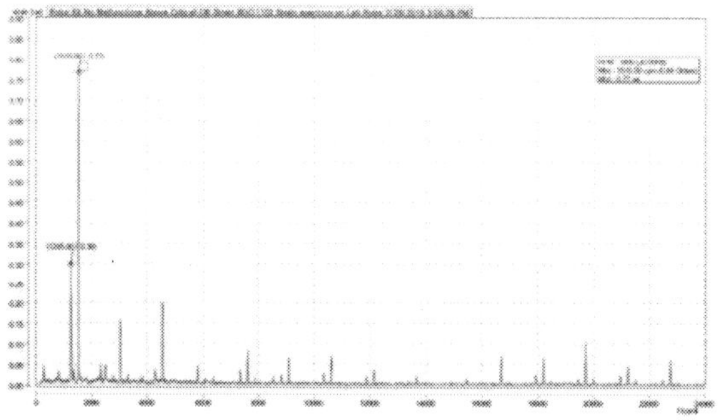

MFPT 2019 Where Theory Meets Practice
MAY 14-16, 2019 | PHILADELPHIA, PA

➢ **Amplitude/pattern comparison**

- Proximity Displacement
- Dynamic Strain

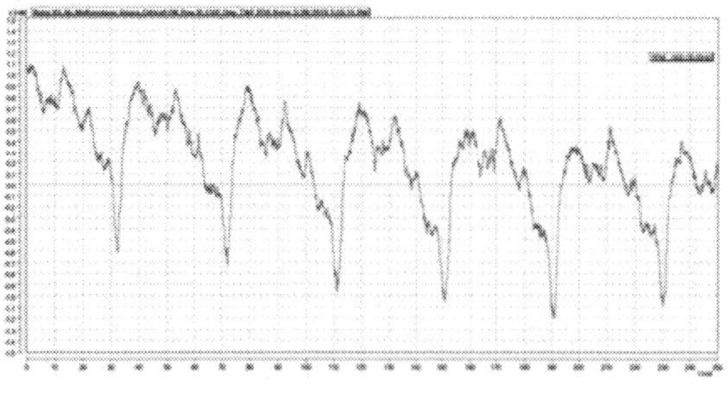

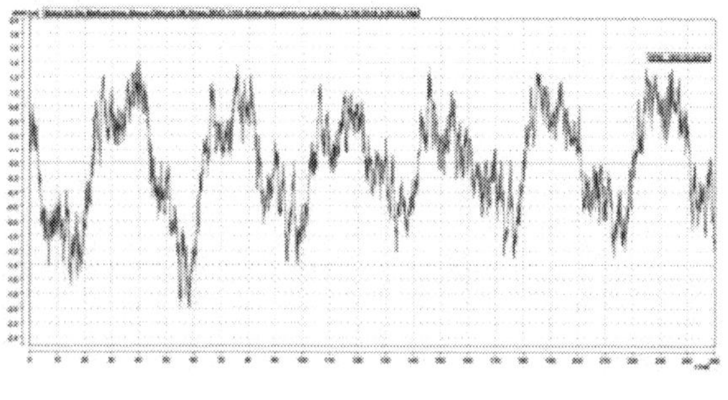

MFPT 2019 Where Theory Meets Practice

MAY 14-16, 2019 | PHILADELPHIA, PA

- **Coast Down – Bode Plot**
  - Critical Speed
  - Phase shift

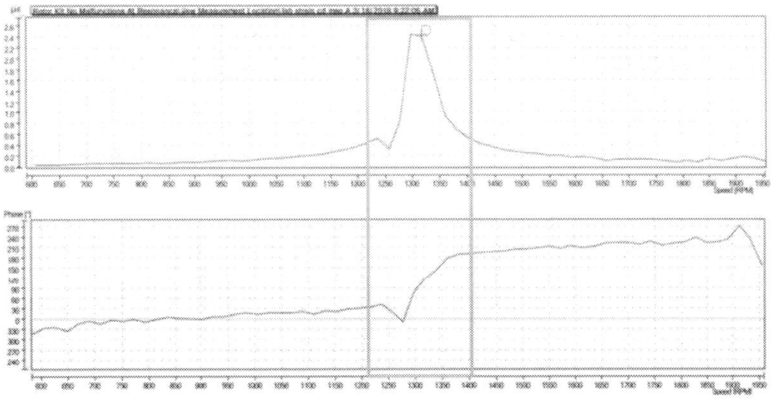

MFPT 2019 Where Theory Meets Practice

MAY 14-16, 2019 | PHILADELPHIA, PA

- **Coast Down – FFT Waterfall**
  - 1x harmonic – resonance
  - Oil Whip

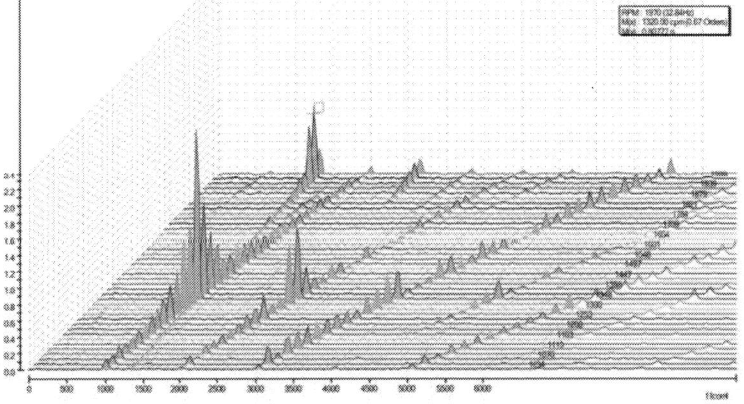

MFPT 2019 Where Theory Meets Practice

MAY 14-16, 2019 | PHILADELPHIA, PA

- **Unbalance**
  - Calculated to ISO G8.4 Unbalance
  - Applied Evenly to all Four Disks
  - Spinning at 70% of 1st Critical
    - Remains a Rigid Rotor

MFPT 2019 Where Theory Meets Practice

MAY 14-16, 2019 | PHILADELPHIA, PA

- **Unbalance**

Velocity Data – 1x = 0.0128 ips

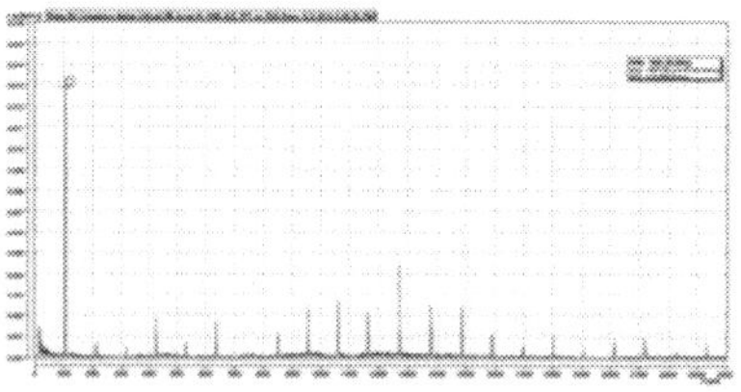

Insight FDS Data – 1x = 1.46με

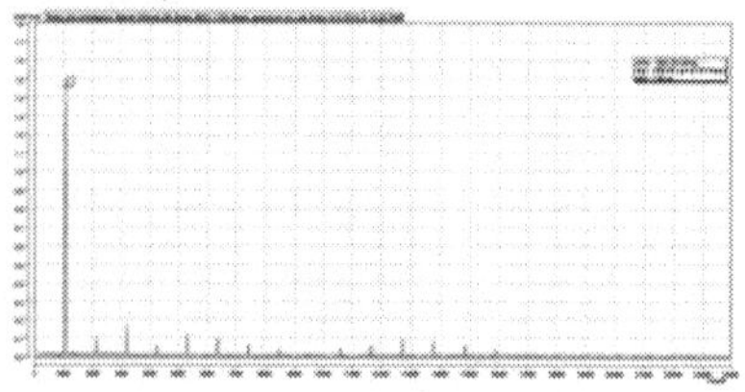

MFPT 2019 Where Theory Meets Practice
MAY 14-16, 2019 | PHILADELPHIA, PA

➢ **Unbalance**

Proximity Displacement - 1x 1.31mils

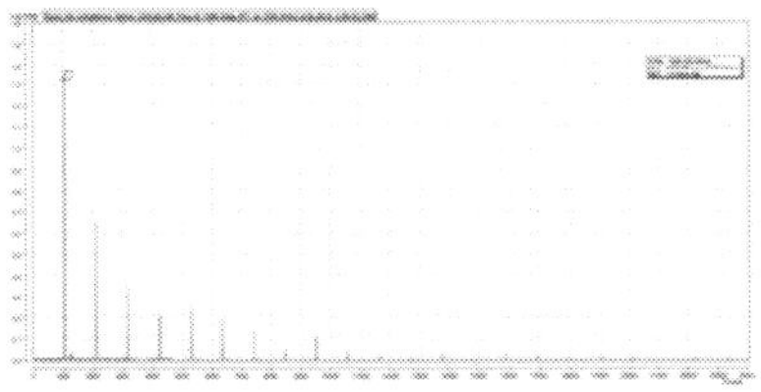

Insight FDS Data – 1x = 1.46με

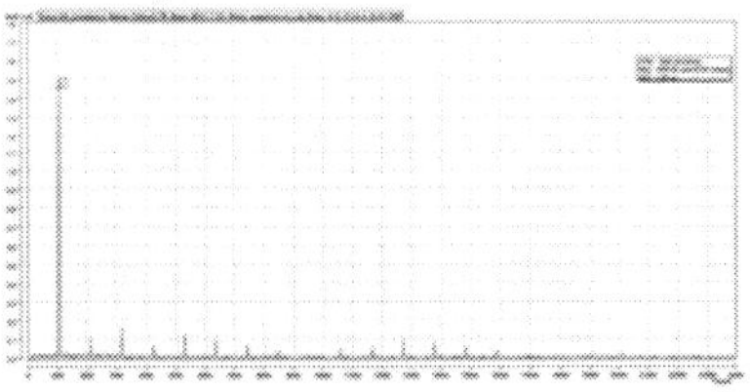

MFPT 2019 Where Theory Meets Practice
MAY 14-16, 2019 | PHILADELPHIA, PA

➢ Misalignment

- ➢ Tire Element Coupling
  - Allowable 0.0625" Misalignment
  - Set with sever Misalignment
- ➢ Used Pruftechnik Shaft Alignment Tool
- ➢ Both Proximity Probes and FDS Sensors Reflect the Malfunction Better Near the coupling
  - Similar Frequency Content

MFPT 2019 Where Theory Meets Practice
MAY 14-16, 2019 | PHILADELPHIA, PA

- **Misalignment**

Velocity

X

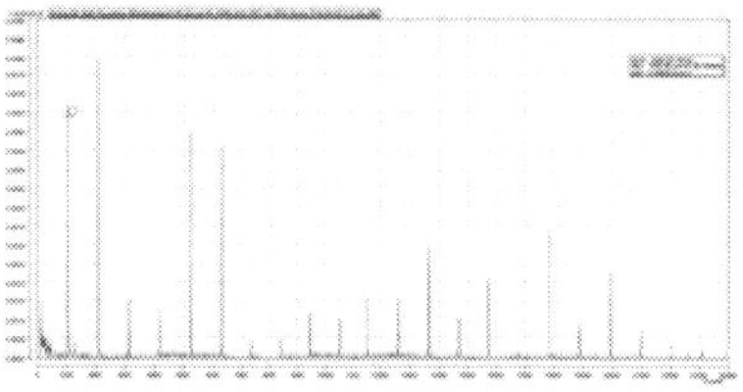

Y

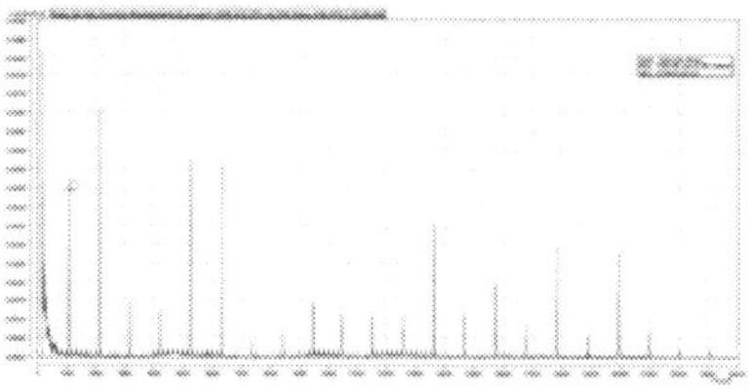

Insight FDS Data

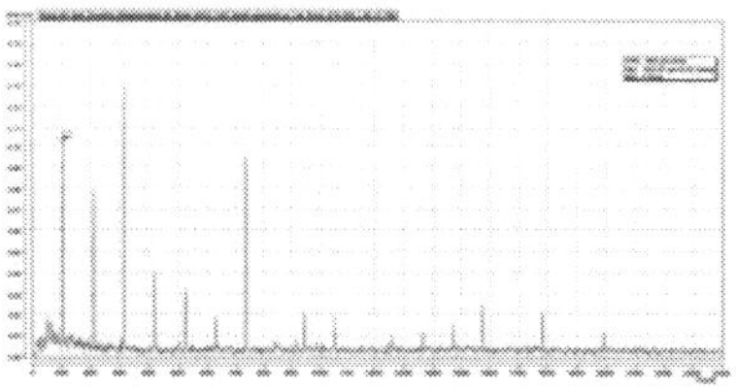

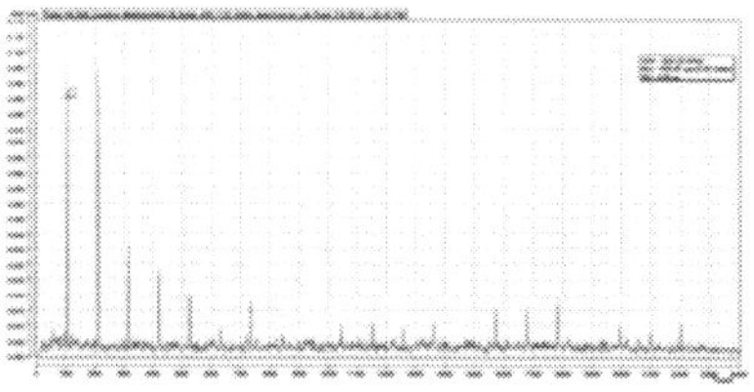

MFPT 2019 Where Theory Meets Practice
MAY 14-16, 2019 | PHILADELPHIA, PA

- **Misalignment**

Proximity

X

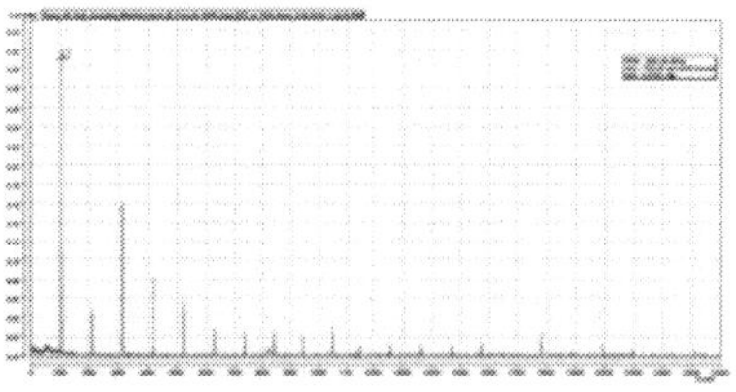

Y

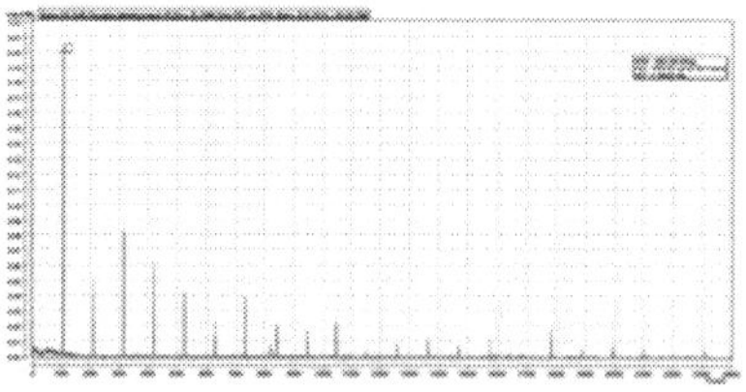

Insight FDS Data

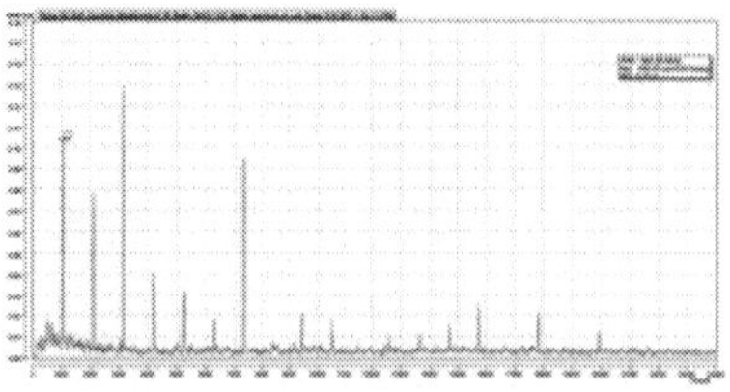

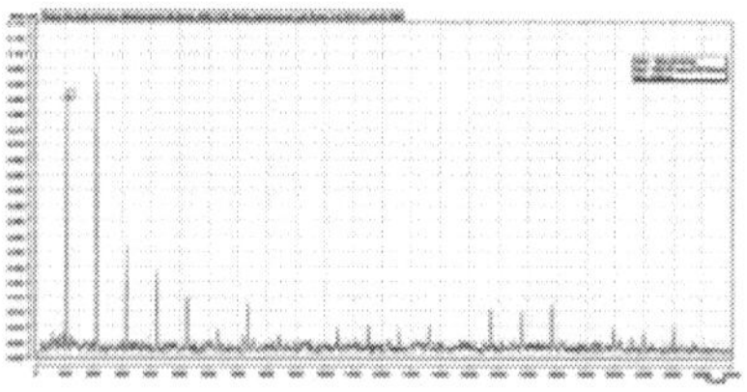

MFPT 2019 Where Theory Meets Practice
MAY 14-16, 2019 | PHILADELPHIA, PA

- **Oil Whirl**
  - Occurs in Plain Bearings at Higher Speeds and Low Eccentricity
    - Sub-synchronous
    - L/D ↑ Whirl ½ x RPM ⇒
    - Caused by Tangential Forces
  - Whirl frequency varies with eccentricity
    - Low eccentricity: whirl closer to 50%
    - High eccentricity: whirl closer to 20%
  - Can be higher than 50% with pre-swirl
  - Much lower than 50% with anti-swirl
  - Correct by increasing eccentricity

MFPT 2019 Where Theory Meets Practice
MAY 14-16, 2019 | PHILADELPHIA, PA

- **Oil Whip**
  - Occurs when shaft vibrates at a natural frequency regardless of shaft speed
    - Most commonly excited by whirl
    - Can initiate without whirl
    - Influenced by Viscosity, Temp, Load
    - 0.42 - 0.48 X RPM

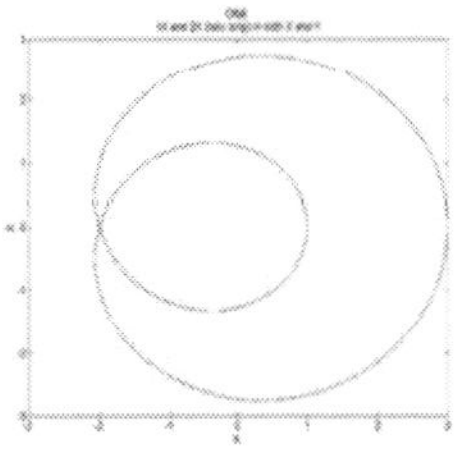

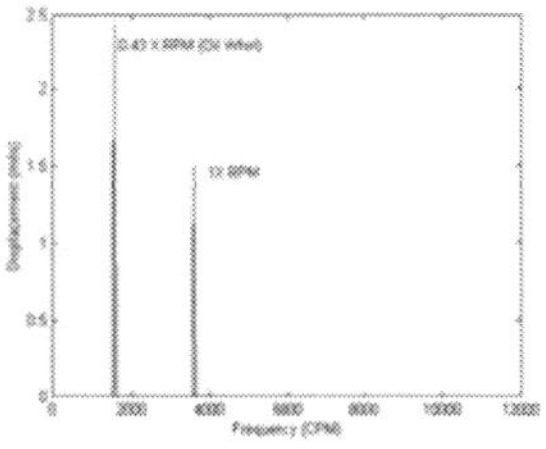

MFPT 2019 Where Theory Meets Practice
MAY 14-16, 2019 | PHILADELPHIA, PA

- **Oil Whip**
  - When Oil Whirls resonates

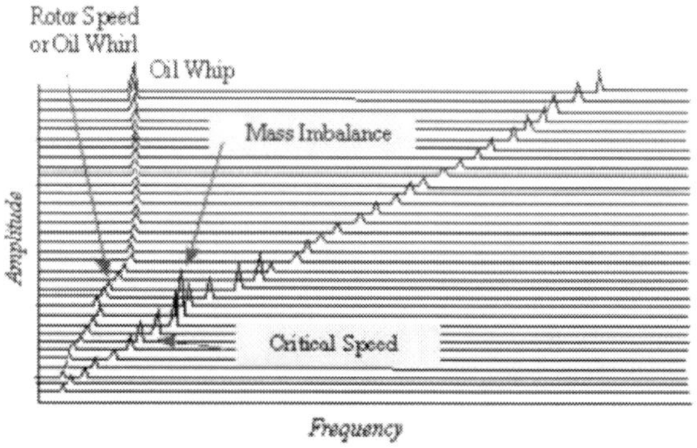

MFPT 2019 Where Theory Meets Practice
MAY 14-16, 2019 | PHILADELPHIA, PA

- **Oil Whirl / Whip**

Velocity Data

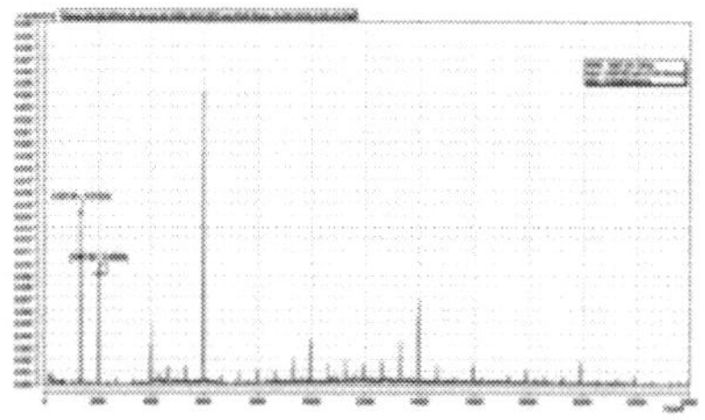

Insight FDS Data

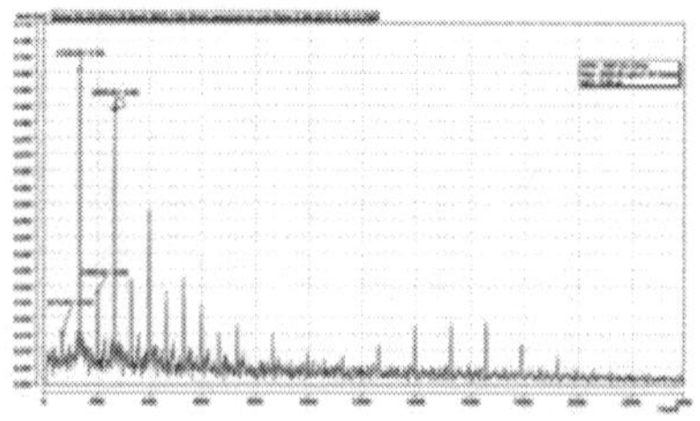

MFPT 2019 Where Theory Meets Practice

MAY 14-16, 2019 | PHILADELPHIA, PA

➢ **Oil Whirl / Whip**

Proximity Data

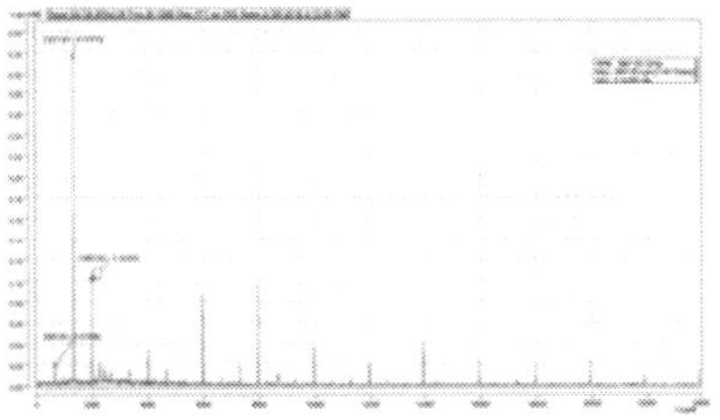

Insight FDS Data

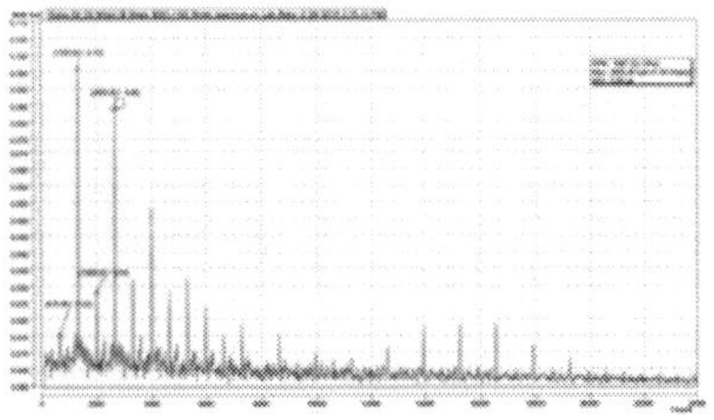

MFPT 2019 Where Theory Meets Practice

MAY 14-16, 2019 | PHILADELPHIA, PA

➢ **Loss of pressure wedge**

- ➢ Hydrodynamic Bearings Make Internal Pressure for Lift from the Babbitt
  - Function of Static Load, Dynamic Viscosity, Shaft Speed
- ➢ Lose Hydrodynamic Pressure Wedge If
  - High Static Load
    - Likely Constant
  - Low Viscosity
    - High Temperatures Most Likely Cause
  - Slow Shaft Speed
    - During Startup and Coast down

- **Loss of pressure wedge**
  - Next Slide Demonstrates Losing Pressure Wedge
    - Slowing Shaft Speed
  - Note Large Increase in Strain at Bottom Dead Center
  - Note Slight Decrease in Strain at Other Sensors
    - The Shaft is No Longer Lifted
    - Most of Strain is from Gravity Load

MFPT 2019 Where Theory Meets Practice
MAY 14-16, 2019 | PHILADELPHIA, PA

- **Loss of pressure wedge**

Before

After

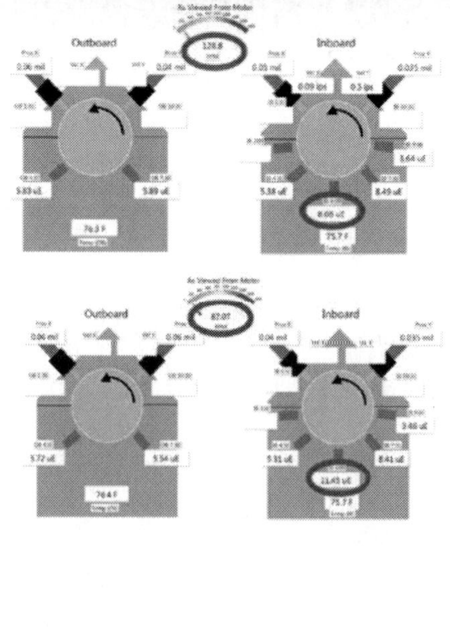

MFPT 2019 Where Theory Meets Practice
MAY 14-16, 2019 | PHILADELPHIA, PA

➢ Loss oil feed

- ➢ Hydrodynamic and hydrostatic bearings require constant oil supply to flood the bearing
- ➢ Losing oil feed can quickly lead to catastrophic failure
- ➢ Using strain gives immediate detection and shutdown options
- ➢ Note the bottom dead center strain amplitude which rises (262%) instantly with loss of feed
- ➢ Note insignificant change (10%) in proximity probe displacement

MFPT 2019 Where Theory Meets Practice
MAY 14-16, 2019 | PHILADELPHIA, PA

➢ Loss of oil feed

- Before

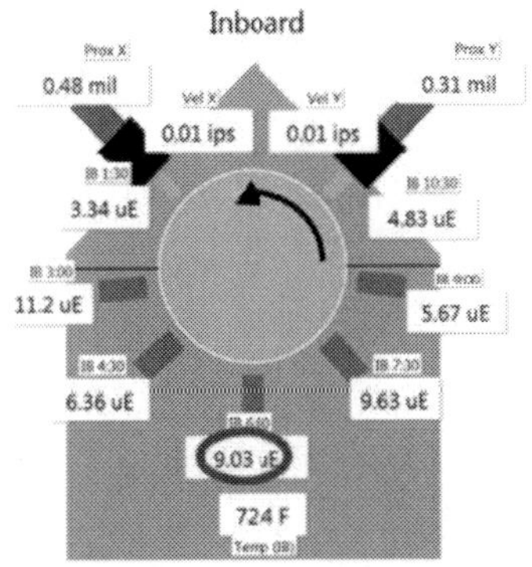

- After Cutting Feed

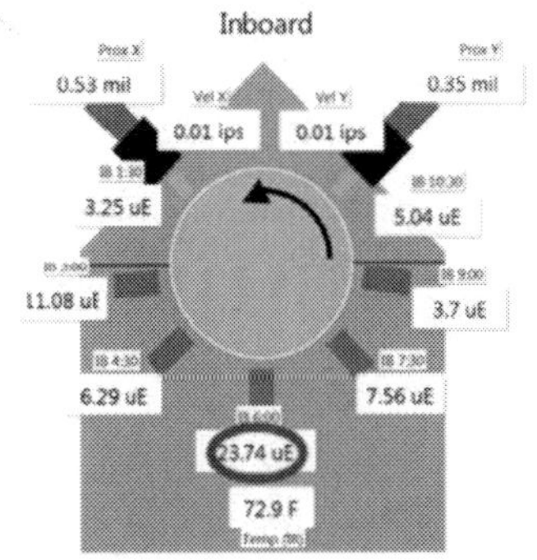

MFPT 2019 Where Theory Meets Practice
MAY 14-16, 2019 | PHILADELPHIA, PA

- **Loss of oil feed**

Velocity Data (1x from 0.003ips to 0.00472ips)

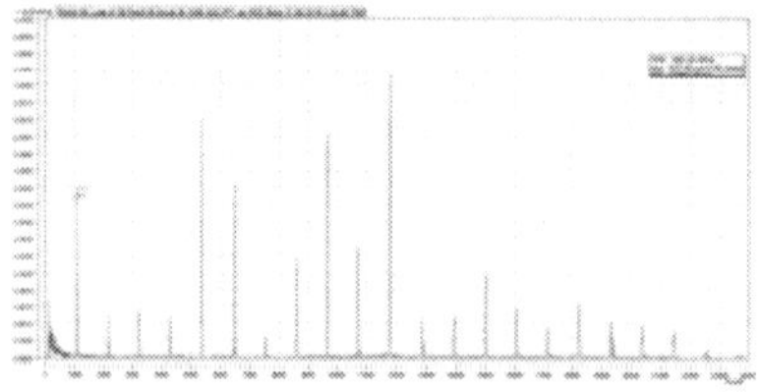

Insight FDS Data (1x from 0.32με to 1.22με)

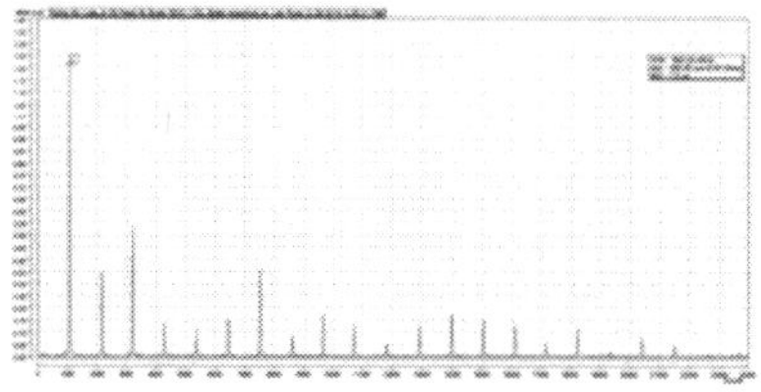

MFPT 2019 Where Theory Meets Practice
MAY 14-16, 2019 | PHILADELPHIA, PA

- **Loss of oil feed**

Proximity Data (1x from 0.19mils to 0.177mils)

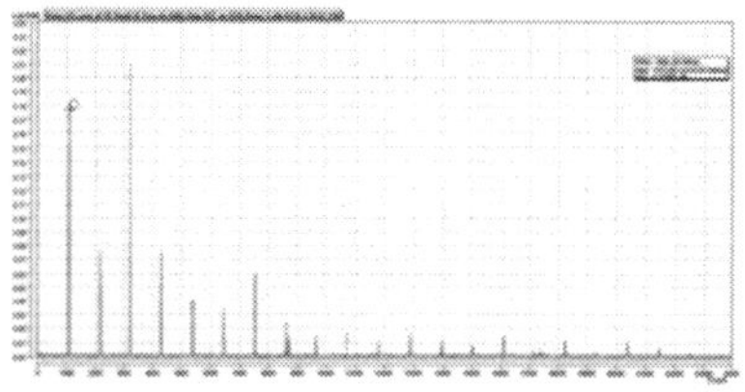

Insight FDS Data (1x from 0.32με to 1.22με)

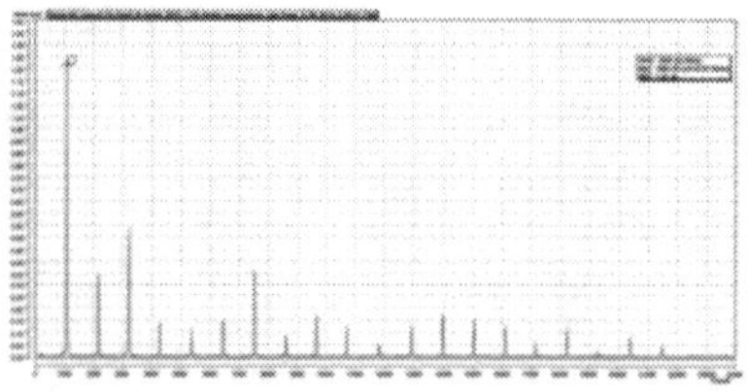

MFPT 2019 Where Theory Meets Practice

MAY 14-16, 2019 | PHILADELPHIA, PA

- **Loss of oil feed**

Acceleration Data

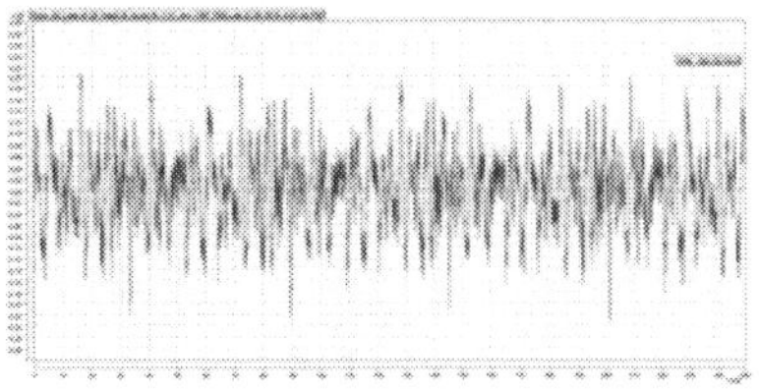

Insight FDS Data

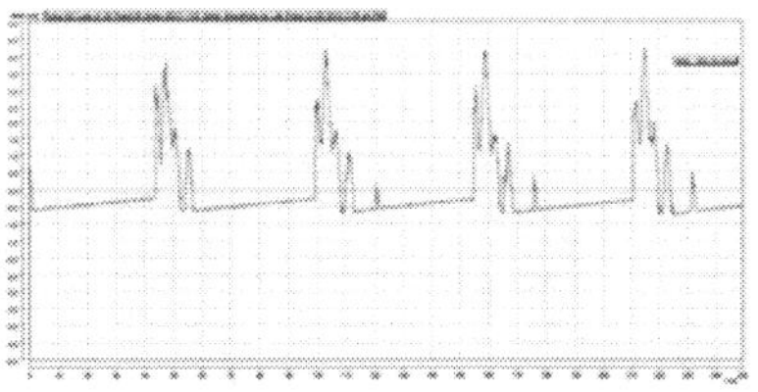

MFPT 2019 Where Theory Meets Practice

MAY 14-16, 2019 | PHILADELPHIA, PA

- **Loss of oil feed**

Proximity Data

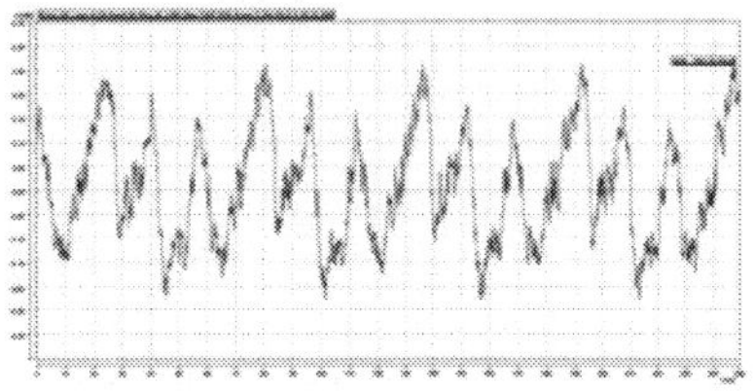

Insight FDS Data

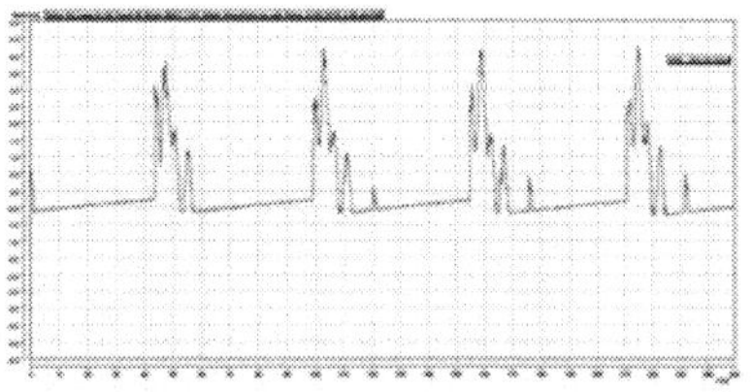

MFPT 2019 Where Theory Meets Practice
MAY 14-16, 2019 | PHILADELPHIA, PA

➢ Cocked Bearing Details

- Mounting strain sensors on both faces on the bearing can indicate a cocked bearing
  during slow roll
  - At higher speeds, many malfunctions can influence these values so only applicable at slow roll (or even turning by hand)
- Lab rotor bearing to bearing alignment checked and adjusted this way

MFPT 2019 Where Theory Meets Practice
MAY 14-16, 2019 | PHILADELPHIA, PA

➢ Cocked Bearing Details

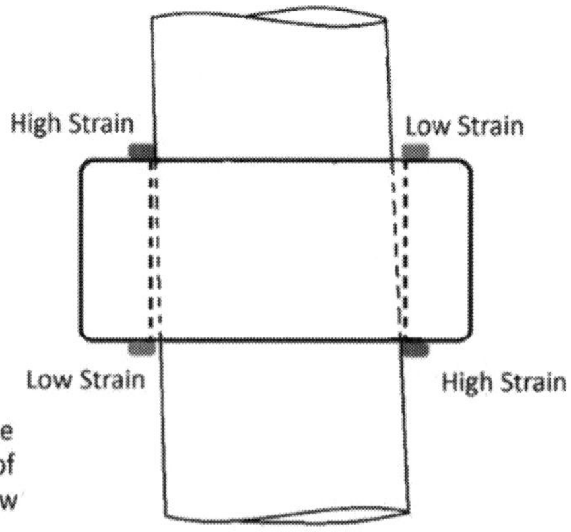

Note: Strains should be similar on both faces of the bearing during slow roll

MFPT 2019 Where Theory Meets Practice
MAY 14-16, 2019 | PHILADELPHIA, PA

➢ Rub between shaft and Babbitt

- ➢ Dwelling Near Resonance to Force a Rub
  - Mode Shape Bends the Shaft Causing the Rub

- ➢ Slide Shows Outboard Bearing (High Clearance)
  - Strain in Upper Half and Displacement in Lower Half
  - Displacement Waveform Hints at Rub
  - Strain Waveforms Clearly Present the Rub

MFPT 2019 Where Theory Meets Practice
MAY 14-16, 2019 | PHILADELPHIA, PA

➢ **Rub between shaft and Babbitt**

Velocity Data

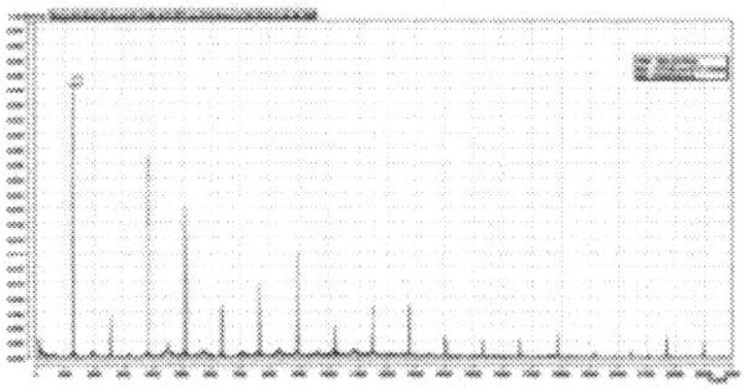

Insight FDS Data

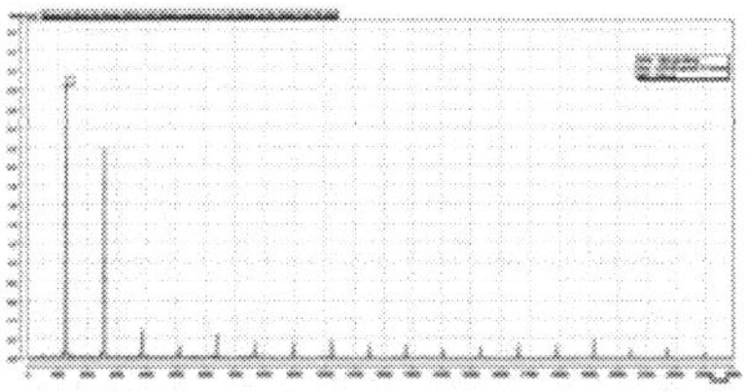

MFPT 2019 Where Theory Meets Practice

MAY 14-16, 2019 | PHILADELPHIA, PA

➢ **Rub between shaft and Babbitt**

Proximity Data

Insight FDS Data

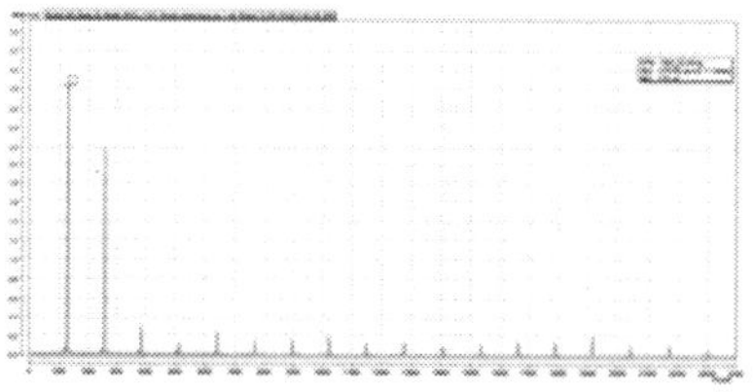

MFPT 2019 Where Theory Meets Practice

MAY 14-16, 2019 | PHILADELPHIA, PA

➢ **Rub between shaft and Babbitt**

Acceleration Data

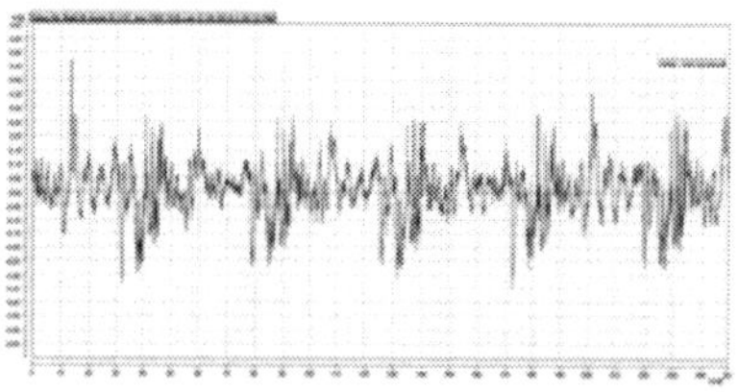

Insight FDS Data

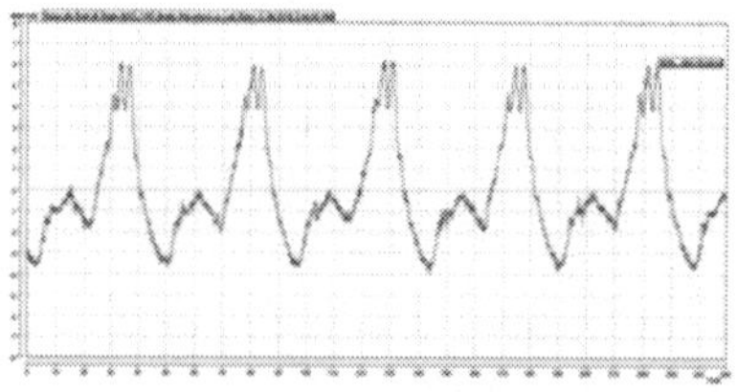

MFPT 2019 Where Theory Meets Practice

MAY 14-16, 2019 | PHILADELPHIA, PA

- **Rub between shaft and Babbitt**

Proximity Data

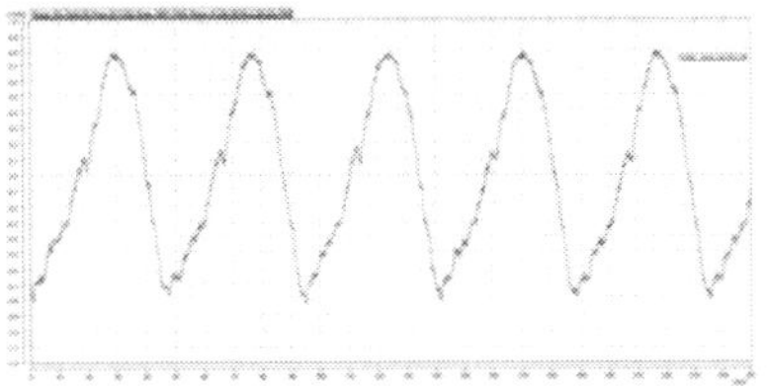

Insight FDS Data

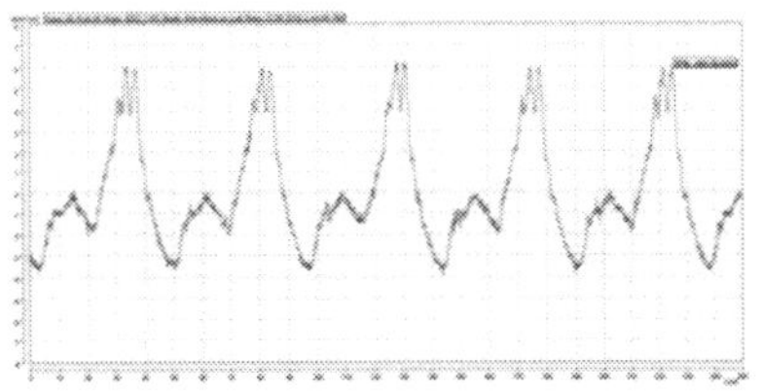

MFPT 2019 Where Theory Meets Practice

MAY 14-16, 2019 | PHILADELPHIA, PA

Field Data

Voyager INSTRUMENTS

PRÜFTECHNIK

MFPT 2019 Where Theory Meets Practice

MAY 14-16, 2019 | PHILADELPHIA, PA

- **Waukesha engine – (BP Oil)**
  - Waukesha Natural Gas Engine
  - 1st Ever Test of Dynamic Strain on Main Bearings by Pioneer Engineering
  - Used old style metal foil strain gauges for proof of concept
    - Now use IEPE powered piezoelectric strain sensors
  - No malfunctions noted
    - Examples of Clear Waveforms from Engine Main Bearings

MFPT 2019 Where Theory Meets Practice

MAY 14-16, 2019 | PHILADELPHIA, PA

- **Waukesha engine – (BP Oil)**

Installation pictures

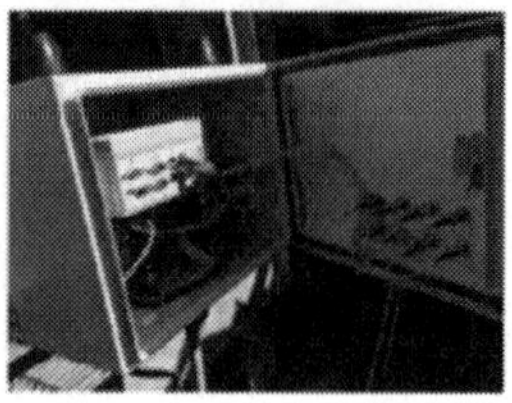

MFPT 2019 Where Theory Meets Practice

MAY 14-16, 2019 | PHILADELPHIA, PA

- Waukesha engine – (BP Oil)

Insight FDS

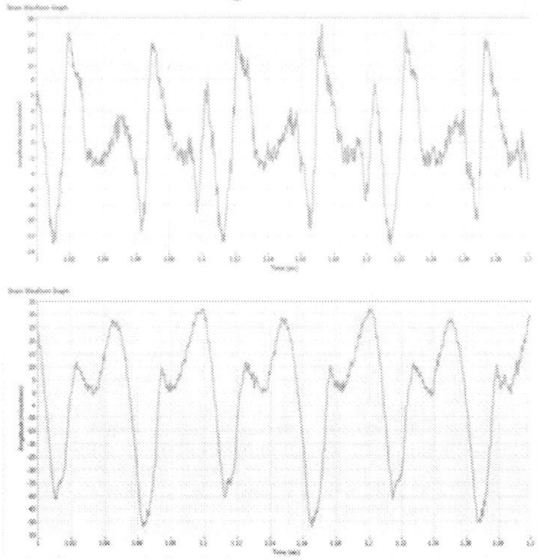

MFPT 2019 Where Theory Meets Practice

MAY 14-16, 2019 | PHILADELPHIA, PA

- Waukesha engine – (BP Oil)

Insight FDS

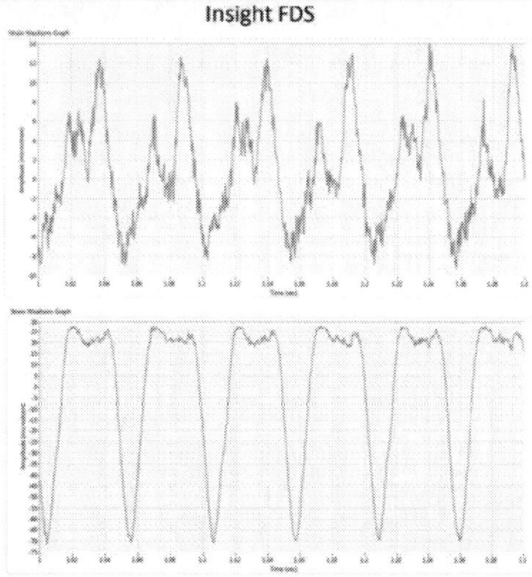

MFPT 2019 Where Theory Meets Practice

MAY 14-16, 2019 | PHILADELPHIA, PA

- ➢ **Waukesha engine – (BP Oil)**

Insight FDS

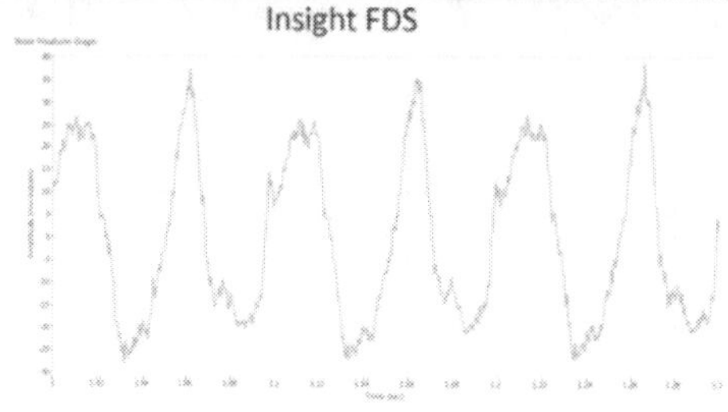

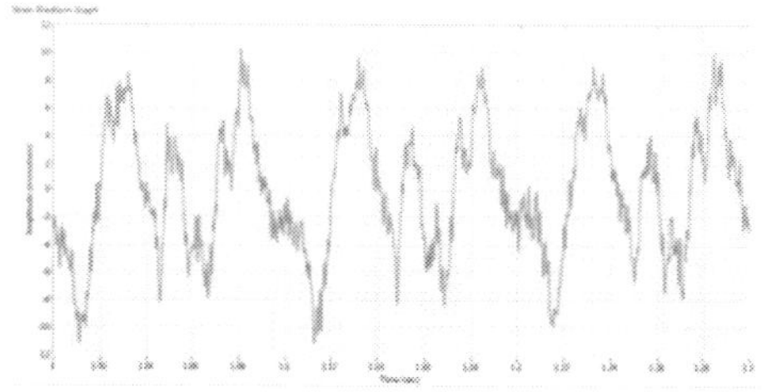

MFPT 2019 Where Theory Meets Practice

MAY 14-16, 2019 | PHILADELPHIA, PA

- ➢ **Waukesha engine – (BP Oil)**

Insight FDS

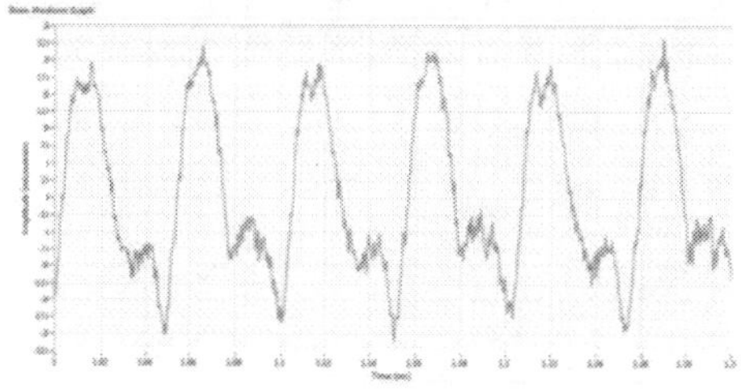

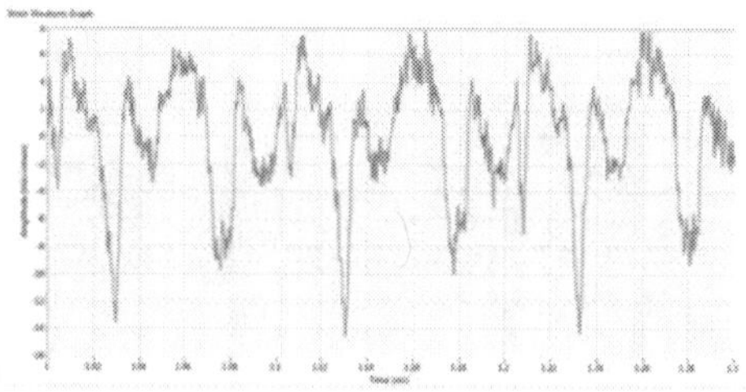

MFPT 2019 Where Theory Meets Practice

MAY 14-16, 2019 | PHILADELPHIA, PA

- **PA Fan Motor - Rotor Fault (Basin Electric)**
  - 4 pole AC induction motor (1500HP)
  - 2 Stage vane axial fan
  - Motor has a known rotor bar fault
  - Data taken on motor outboard bearing with
    Insight FDS
    Case mounted acceler
    Proximity probes
  - Prox probe data was nc
    slow roll compensated

MFPT 2019 Where Theory Meets Practice

MAY 14-16, 2019 | PHILADELPHIA, PA

- **PA Fan Motor - Rotor Fault**

Unfiltered TWF - Insight FDS

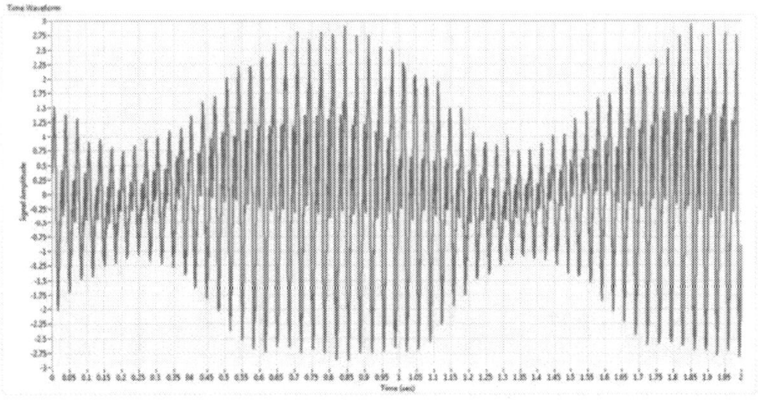

MFPT 2019 Where Theory Meets Practice
MAY 14-16, 2019 | PHILADELPHIA, PA

➢ **PA Fan Motor - Rotor Fault**

Unfiltered TWF - Accelerometer

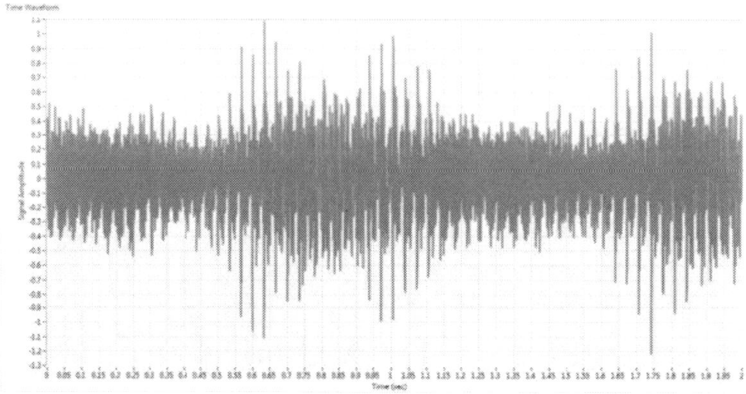

MFPT 2019 Where Theory Meets Practice
MAY 14-16, 2019 | PHILADELPHIA, PA

➢ **PA Fan Motor - Rotor Fault**

Unfiltered TWF - Prox Probe

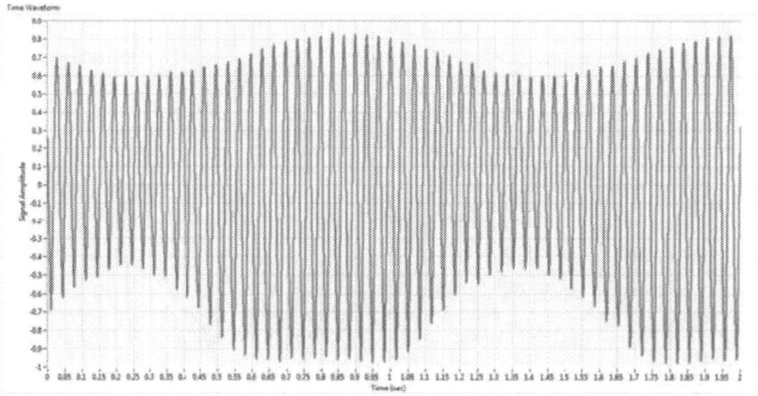

MFPT 2019 Where Theory Meets Practice

MAY 14-16, 2019 | PHILADELPHIA, PA

➢ **PA Fan Motor - Rotor Fault**

FFT - Insight FDS

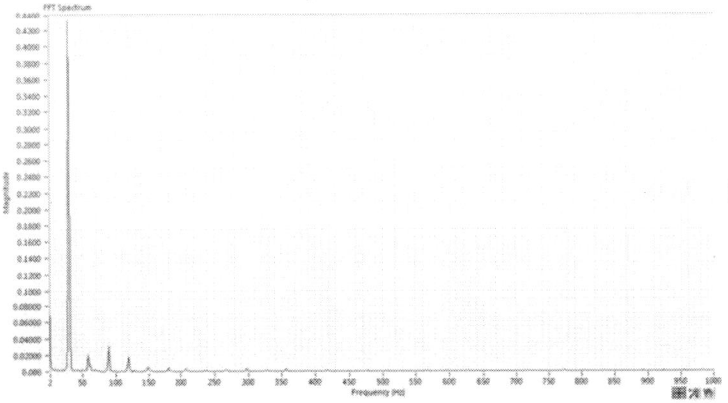

MFPT 2019 Where Theory Meets Practice

MAY 14-16, 2019 | PHILADELPHIA, PA

➢ **PA Fan Motor - Rotor Fault**

FFT - Accelerometer

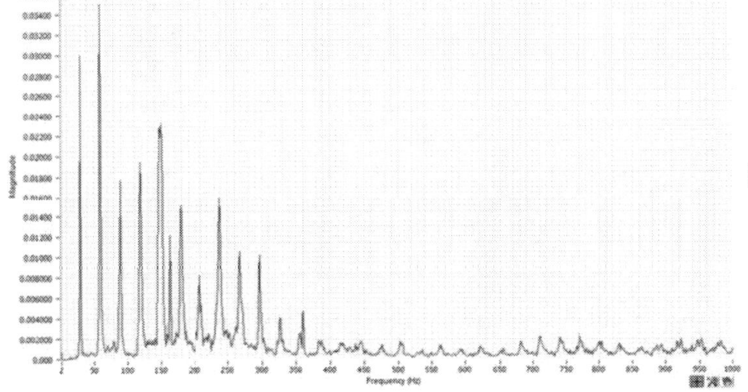

MFPT 2019 Where Theory Meets Practice

MAY 14-16, 2019 | PHILADELPHIA, PA

➢ **PA Fan Motor - Rotor Fault**

FFT - Prox Probe

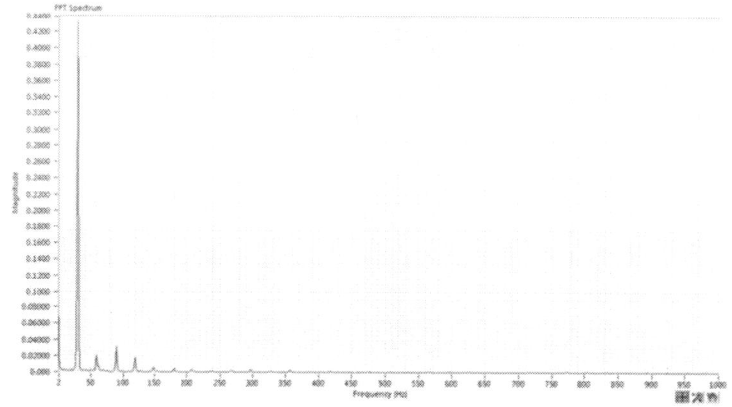

MFPT 2019 Where Theory Meets Practice

MAY 14-16, 2019 | PHILADELPHIA, PA

➢ **PA Fan Motor - Rotor Fault**

High Res. FFT - Insight FDS

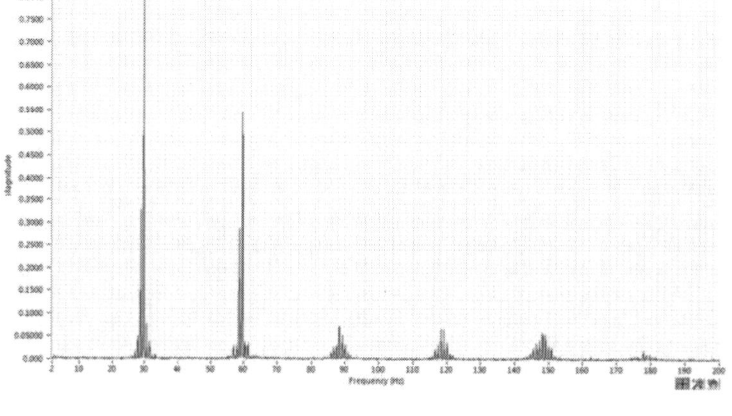

➢ **PA Fan Motor - Rotor Fault**

High Res. FFT - Accelerometer

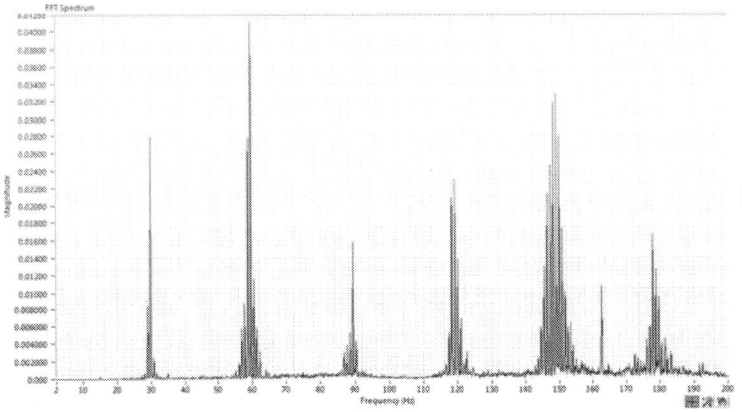

MFPT 2019 Where Theory Meets Practice
MAY 14-16, 2019 | PHILADELPHIA, PA

➢ **PA Fan Motor - Rotor Fault**

High Res. FFT - Prox Probe

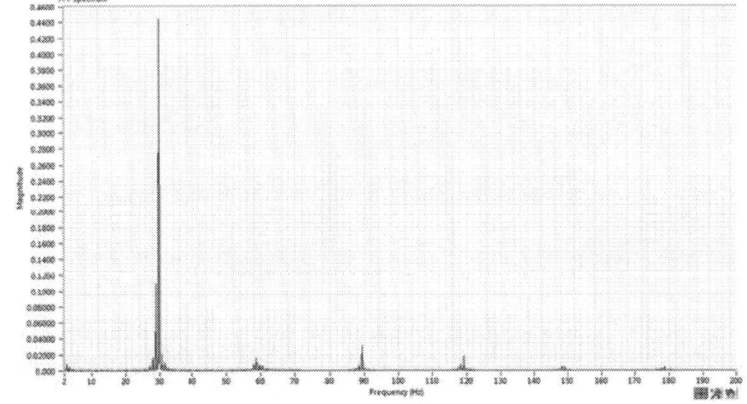

MFPT 2019 Where Theory Meets Practice

MAY 14-16, 2019 | PHILADELPHIA, PA

Conclusion

- Fluid Film Bearings on Critical Machines

- Different monitoring technologies
  - Proximity Probes – Best but very expensive
  - Temperature – To late of warning
  - Case Mounted Accelerometers – Better than nothing
  - Not at all?? – Expensive

- Insight Force Detection Sensor
  - Strain indirectly representing bearing forces
  - Easy install
  - Data quality proven

- New exciting technology available today.

MFPT 2019 Where Theory Meets Practice

MAY 14-16, 2019 | PHILADELPHIA, PA

THANK YOU

Contact: Mike.Ciocys@pruftechnik.com
Dries.VanLoon@pruftechnik.com

QUALITY MEANS TO US:

TO MAKE A DIFFERENCE EVERY DAY.

www.pruftechnik.com

Sources

- www.pruftechnik.com
- www.pioneer-engineering.com
- www.voyagerinstruments.com
- Bently Nevada
- Substech
- National Instruments

MFPT 2019 Where Theory Meets Practice

MAY 14-16, 2019 | PHILADELPHIA, PA

Defect Severity Measurement in Rolling Element Bearings

Dr Iain Epps
Managing Director, Mobolo Technology Ltd
Christchurch, New Zealand

Copyright 2019 - Mobolo Technology Ltd, New Zealand. All Rghts Reserved

mobolo

Introduction

Dr Iain Epps
Managing Director,
Mobolo Technology Ltd, Christchurch NZ

- **Mechanical Engineering at University of Canterbury NZ**
  PhD – Vibration Excited by Discrete Defects in Rolling Element Bearings (1991)
- **25 Year Career in Technology and Innovation**
  Product Development and Science – Rail Transport (UK)
  Software Development – Finance and Business Services
  Technology Start-ups – Agri-Tech and Hydro-Electricity
- **Founded Mobolo Technology in 2015**
  Research on condition monitoring of rolling element bearings
  Developing new technologies and approaches
  Focused on defect severity measurement

mobolo

Copyright 2019 - Mobolo Technology Ltd, New Zealand. All Rghts Reserved

Introduction – New Zealand (Aotearoa)

mobolo

Copyright 2019 - Mobolo Technology Ltd, New Zealand. All Rights Reserved

Overview

Defect Severity Measurement in Rolling Element Bearings

- **Severity Measurement is Important**
  Delivering on the promise of vibration monitoring
- **Severity Measurement is Difficult**
  Complexity and unknowns
- **New Approach**
  Exploring direct measurement of severity

Focusing on discrete defects (e.g. fatigue spalls) in rolling element bearing

mobolo

Copyright 2019 - Mobolo Technology Ltd, New Zealand. All Rights Reserved

Severity Measurement

Definition of Defect Severity

1) **Defect Size**
   Primarily width (dimension in direction of rolling)
   Also length/shape (or inferred by fatigue mechanics)
2) **Defect Location**
   Component – outer race, inner race or rolling element
   Position – within the load zone (if meaningful)
3) **Operating Speed and Load**
   Current operating state for the bearing

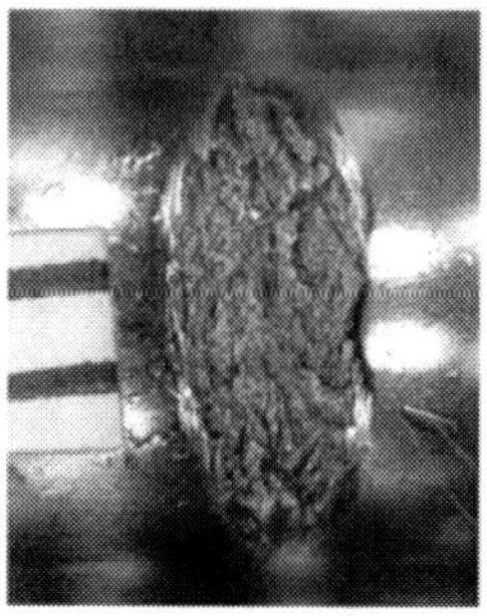

mobolo

Copyright 2019 - Mobolo Technology Ltd, New Zealand. All Rights Reserved

Severity Measurement is Important

Delivering on the Promise of Vibration Monitoring

1) **"Good News and Bad News"**
   - Detection (and diagnosis) – well understood ~100%
   - Severity (and prognosis) – much less so

2) **Business Imperatives – Economic & Environmental**
   - Maximising asset performance
   - Minimising impacts - Remanufacturing and re-use (↓ CO_2)

mobolo

Copyright 2019 - Mobolo Technology Ltd, New Zealand. All Rights Reserved

Severity Measurement is Important

Delivering on the Promise of Vibration Monitoring (cont.)

3) Beyond L_{10} and toward Prognostics

- Statistical measure... until "first sign of metal fatigue"
- Enable a better understanding of the "After Life"

$$L_{10} = \left(\frac{C}{P}\right)^p$$

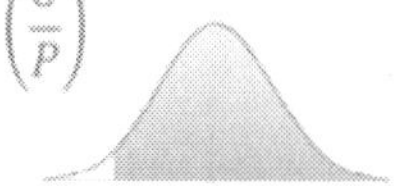

4) Feature Input for AI (rather than Feature Extraction)

- Severity as an input not an output
- Unleashing IIOT (Industry 4.0) for vibration monitoring
  - "Smarter" outputs and low bandwidth networks (LPWAN)

mobolo

Copyright 2019 - Mobolo Technology Ltd, New Zealand. All Rights Reserved

Severity Measurement is Difficult

1) Source and Response

$$x(t) \neq y(t)$$

Source $\neq$ **Response**

(what caused the vibration) (the "vibration" that is measured)

$$T\,(\,x(t)\,) = y(t)$$

Transfer Function (Source) $=$ **Response**

(what caused the response) (the "vibration" that was measured)

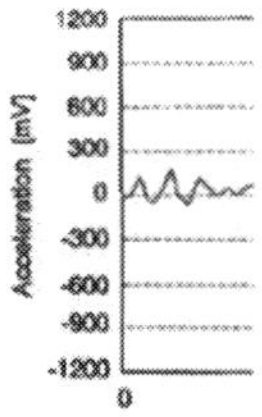

mobolo

Copyright 2019 - Mobolo Technology Ltd, New Zealand. All Rights Reserved

Severity Measurement is Difficult (cont.)

2) Response is Complex

- Interpreting the "squiggly lines"

  Maybe we don't help ourselves – more complex?

"the concept of frequency analysis…

OG Gustafsson and T Tallian (1962)
Detection of Damage in Assembled Rolling Element Bearings (ASLE)

…inappropriate"

"… an example of instrumentation technology leading rather than facilitating vibration monitoring."

SG Braun (1986)
… Applications (Academic Press)

- Few features (if any) are monotonic with severity

  Changes can be inconsistent with severity

  Historic trends may not always be reliable

"Only the size of the fault is monotonic…
…it's never going to get smaller"

…A, Chevron Corporation
MFPT 2016 Dayton OH

mobolo

Copyright 2019 - Mobolo Technology Ltd, New Zealand. All Rights Reserved

Severity Measurement is Difficult (cont.)

3) Transfer Function is Complex

- Mechanical Transmission (largely unknown)

  Structure, materials, interfaces/surfaces, sensor and mounting

  Subject to change – impacts historic trending

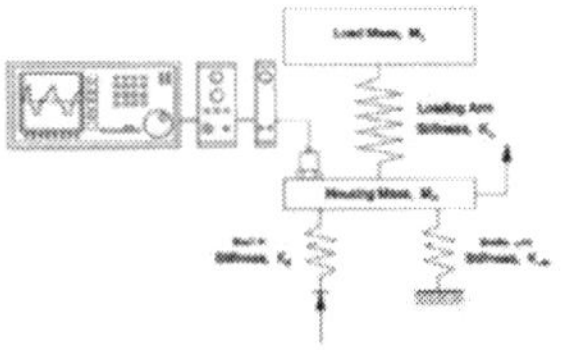

- Electrical Transmission (reasonably well known)

  Sensors, amplifiers, filters, interference, anti-aliasing, digitisation

  Can be hidden – preparatory processes

mobolo

Copyright 2019 - Mobolo Technology Ltd, New Zealand. All Rights Reserved

Severity Measurement is Difficult (cont.)

4) Source is simple?

- Universal assumption (circa 1988)
  "the source of vibration is a repetitive simple impulse"

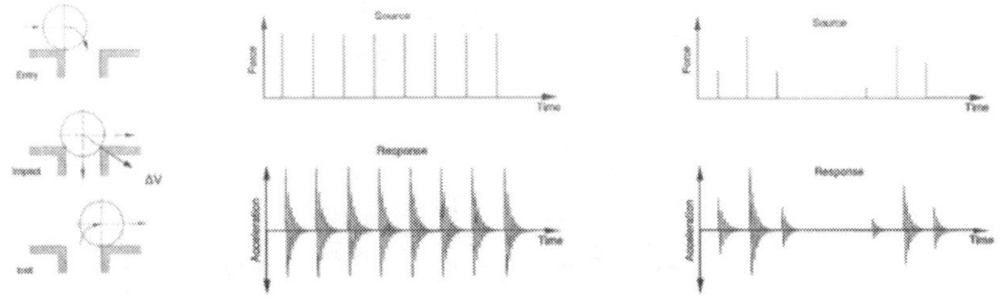

- Assumption tested by Epps PhD (1991)

Copyright 2019 - Mobolo Technology Ltd, New Zealand, All Rights Reserved

Severity Measurement is Difficult (cont.)

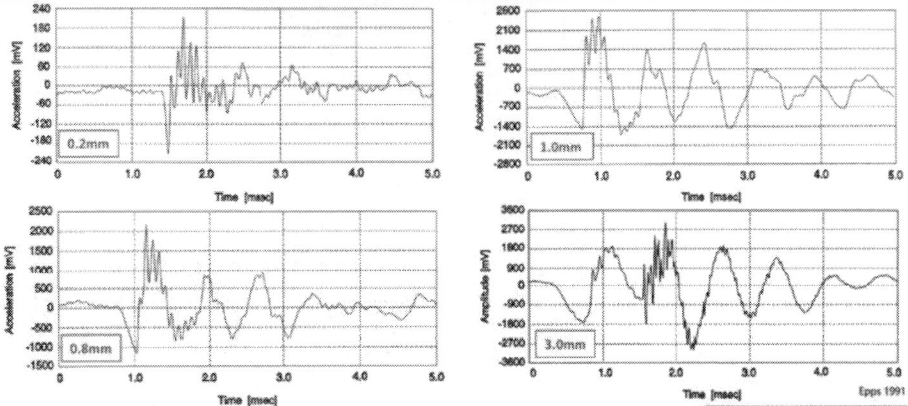

Epps 1991

Copyright 2019 - Mobolo Technology Ltd, New Zealand, All Rights Reserved

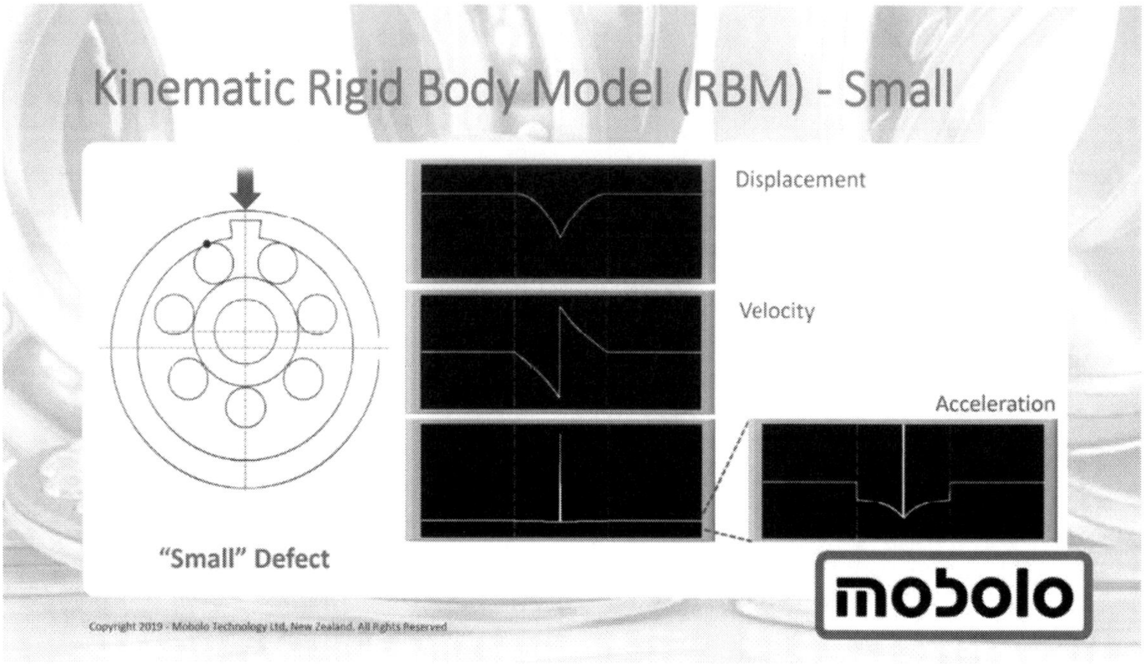

Kinematic Rigid Body Model (RBM) - Small
Displacement
Velocity
Acceleration
"Small" Defect
mobolo
Copyright 2019 - Mobolo Technology Ltd, New Zealand. All Rights Reserved

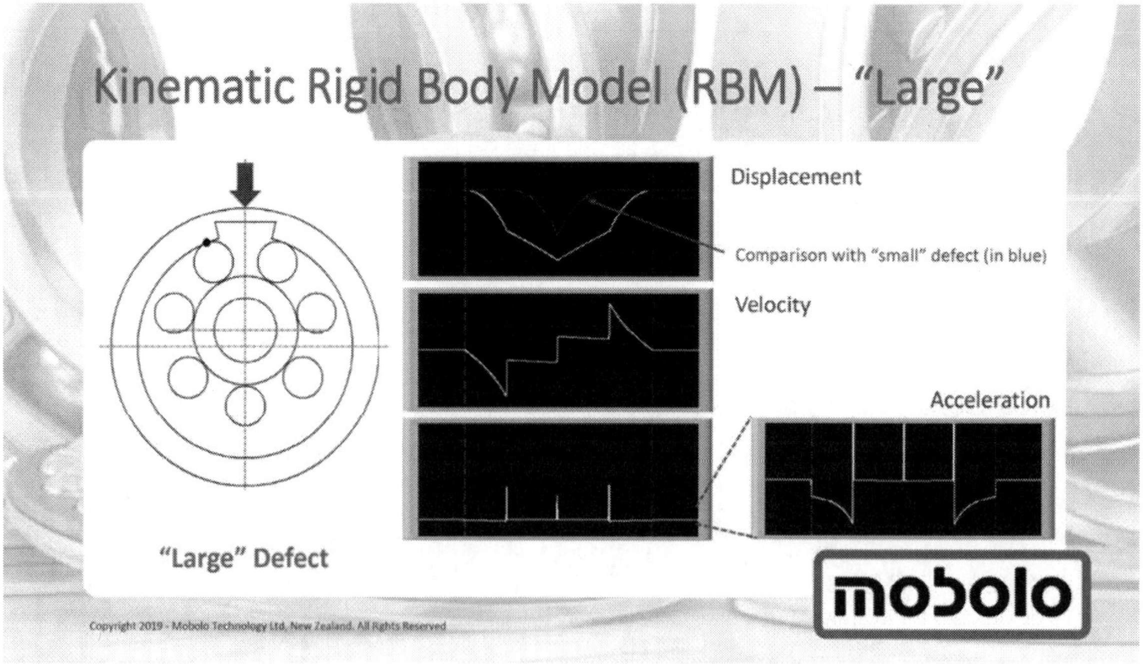

Kinematic Rigid Body Model (RBM) – "Large"
Displacement
Comparison with "small" defect (in blue)
Velocity
Acceleration
"Large" Defect
mobolo
Copyright 2019 - Mobolo Technology Ltd, New Zealand. All Rights Reserved

Severity Measurement is Difficult (cont.)

4) Source is ~~Simple~~ Complex

- **Source is just not a simple impulse**
- **There are multiple events per impact**
  - Minimum of two (can be many more)
- **The timing changes between events**
  - With both speed and defect width

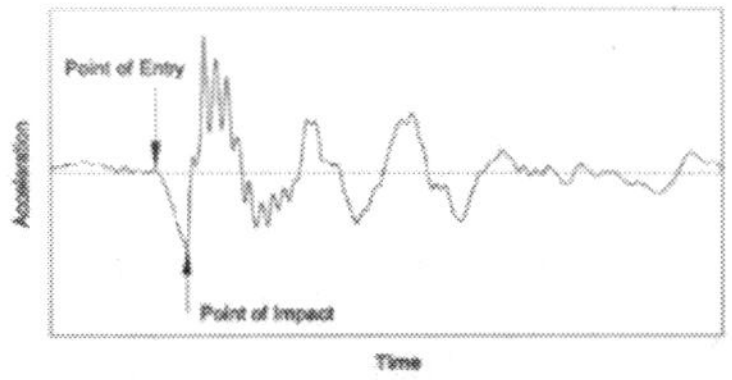

mobolo

Copyright 2019 - Mobolo Technology Ltd, New Zealand. All Rights Reserved

Exploring a New Approach

Although more complex, the new model for the Source provides a method for direct measurement of defect size

Step 1. Find Point of Entry

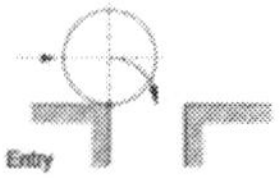

Step 2. Find Point of Impact

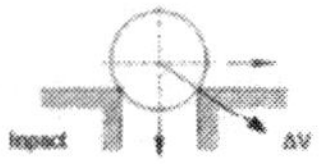

Step 3. Difference α width
Time to Impact (TTI)

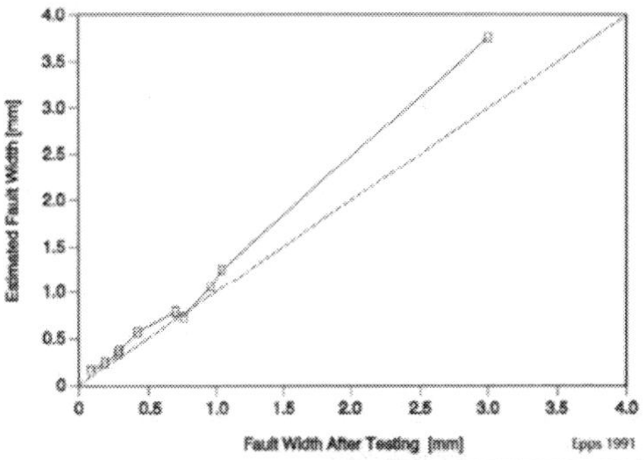

mobolo

Copyright 2019 - Mobolo Technology Ltd, New Zealand. All Rights Reserved

Exploring a New Approach (cont.)

Practical application – two key challenges:

1) **Identifying impulsive events related to bearing defects**
   Many sources of noise and other impulsive events (e.g. gear mesh and looseness)

2) **Identifying the Point of Entry**
   "Low energy step-change" in acceleration response

MFPT 2019 Where Theory Meets Practice

mobolo

Copyright 2019 - Mobolo Technology Ltd, New Zealand. All Rights Reserved

Exploring a New Approach (cont.)

Synchronous Averaging – powerful technique

Enhance event features (coherent) and reduces noise (non-coherent)
Need to synchronise with each ball pass

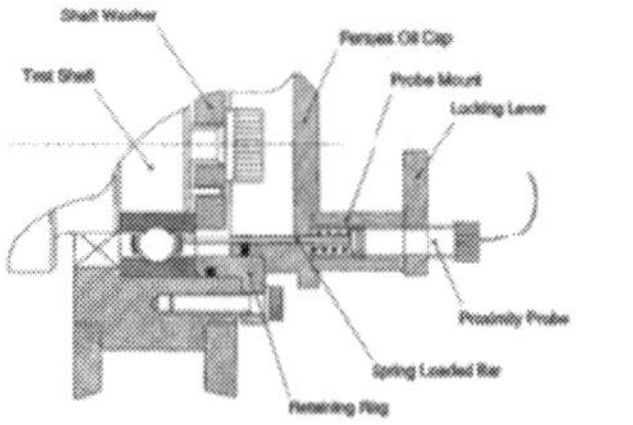

Cage Sensor (Eddy Current Proximity Probe)

Ball Pass Sensor (Laser)

mobolo

Copyright 2019 - Mobolo Technology Ltd, New Zealand. All Rights Reserved

PhD Research – Problem with Data

Initial experimental data had underlying "noise"

Consistent and regular sine wave

- Frequency = outer race ball pass (f_o or BPFO)
- Amplitude α applied load
- Visible whether defect present or not

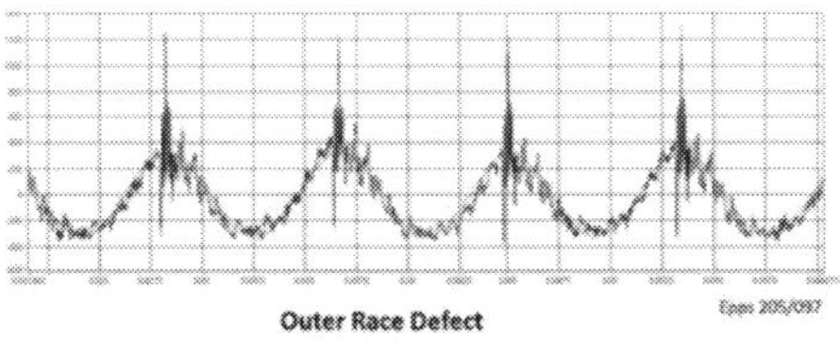

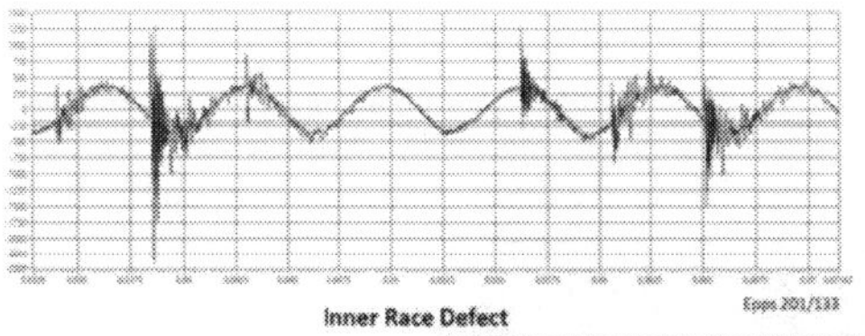

mobolo

Copyright 2019 - Mobolo Technology Ltd, New Zealand. All Rights Reserved

PhD Research – Cause

Underlying "noise" was caused by accelerometer base strain

- Occurs in simple compression type piezoelectric accelerometers (~1960's)

Caused by bending of piezoelectric element

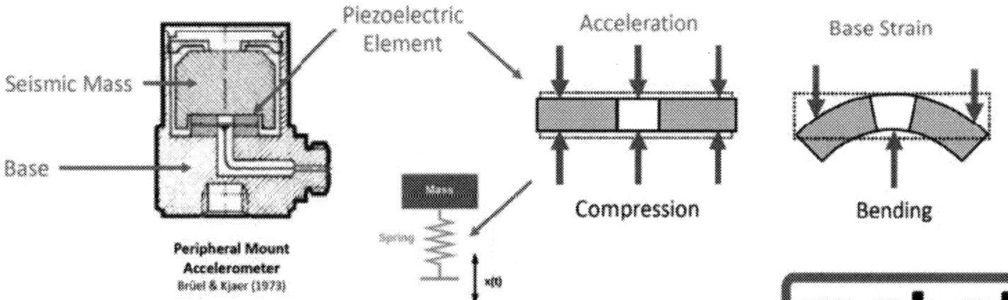

mobolo

Copyright 2019 - Mobolo Technology Ltd, New Zealand. All Rights Reserved

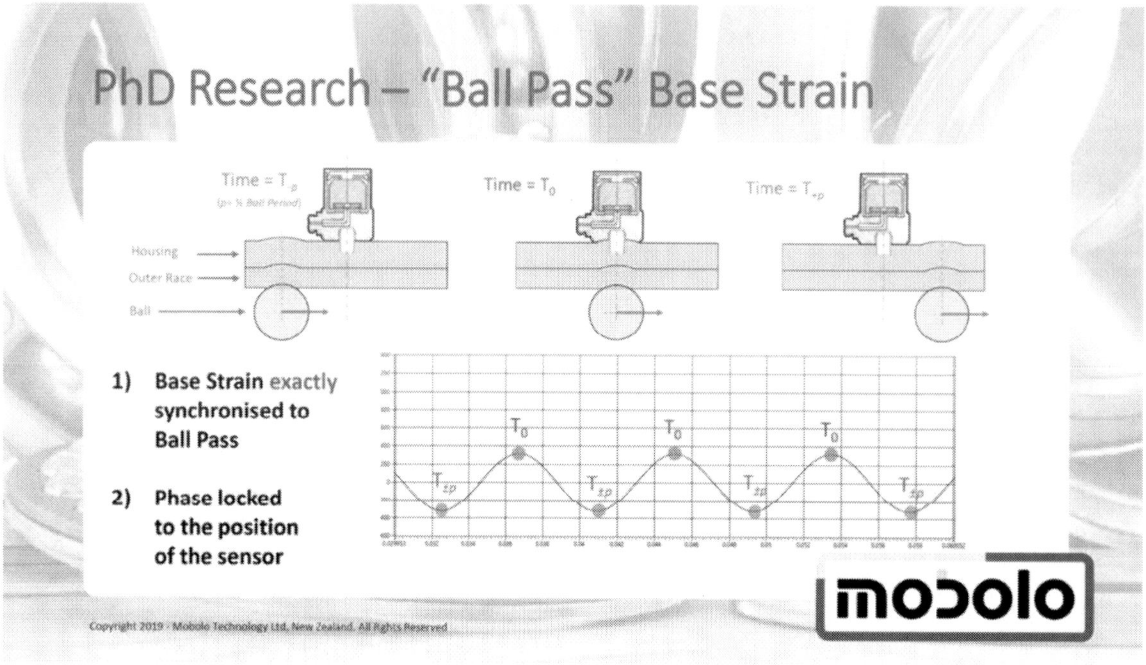
PhD Research – "Ball Pass" Base Strain
Time = T_0
Housing
Outer Race
Ball
1) Base Strain exactly synchronised to Ball Pass
2) Phase locked to the position of the sensor
mobolo
Copyright 2019 - Mobolo Technology Ltd, New Zealand. All Rights Reserved

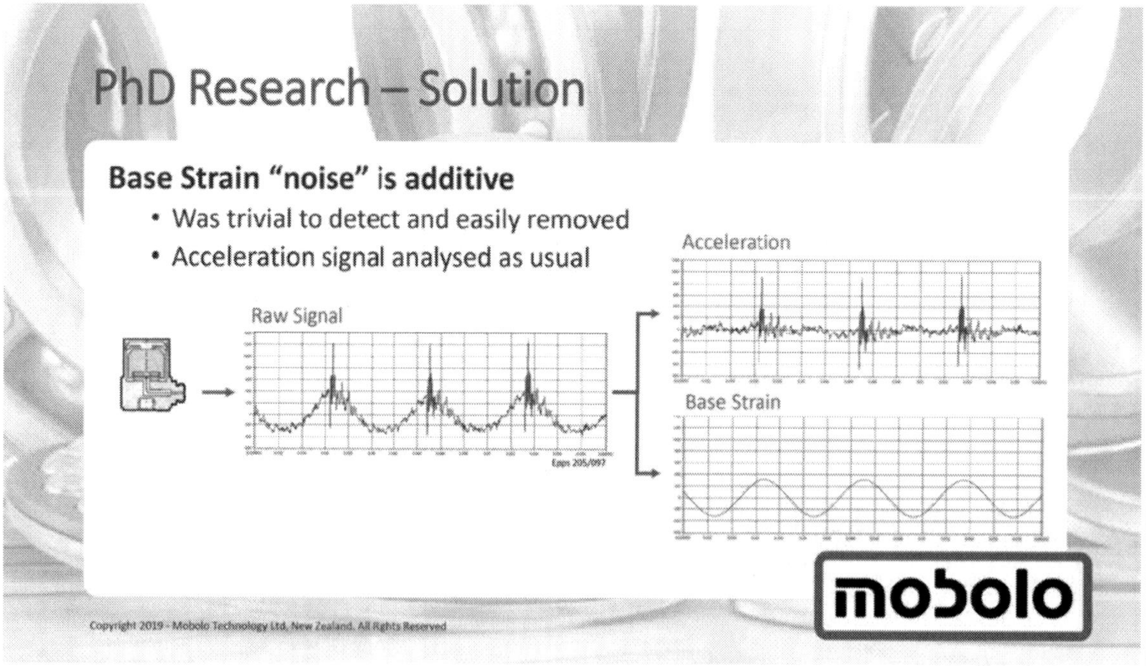
PhD Research – Solution
Base Strain "noise" is additive
• Was trivial to detect and easily removed
• Acceleration signal analysed as usual
Raw Signal
Acceleration
Base Strain
mobolo
Copyright 2019 - Mobolo Technology Ltd, New Zealand. All Rights Reserved

"Discovery consists of seeing what everyone else has seen, and thinking what no one else has thought"

Albert Szent-Gyorgyi
Nobel Prize for Medicine 1937

mobolo

Copyright 2019 - Mobolo Technology Ltd, New Zealand. All Rights Reserved

Exploring a New Approach (cont.)

Base strain is widely regarded as "noise" and "undesirable"

- Modern sensors (accelerometers) are designed to eliminate it

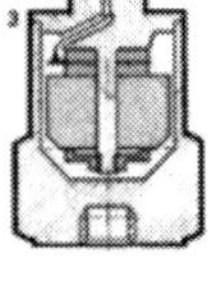

Inverted Centre Mount Accelerometer
Brüel & Gjaer (1973)

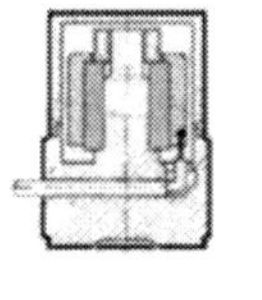

Shear Type Accelerometer
Brüel & Kjaer (:973)

- Alternatively... sensors can be designed to enhance it (and have)

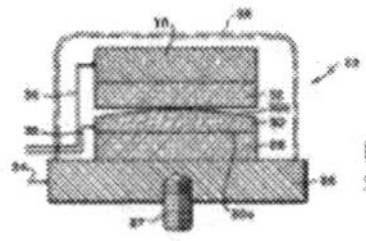

Dynamic Curvature Sensor
Sheridax et al (1968)

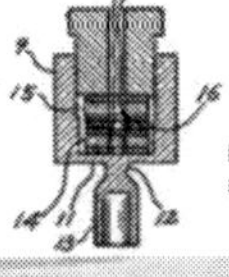

Impulse Sensor
Lew (1989)

mobolo

Copyright 2019 - Mobolo Technology Ltd, New Zealand. All Rights Reserved

New Approach - Operating Principle

Ball Pass Strain (BPS) Enhanced Acceleration

1) New sensor designed to accentuate base strain from ball pass
2) Outputs both acceleration and ball pass strain (BPS) signals
3) Ball pass strain then used to enhance acceleration

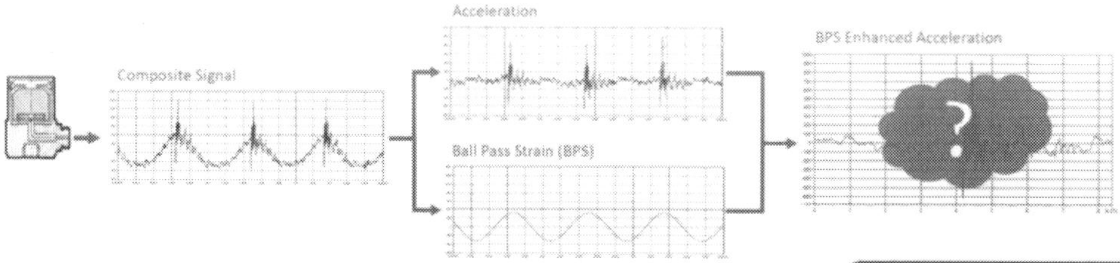

mobolo

Copyright 2019 - Mobolo Technology Ltd, New Zealand. All Rights Reserved

#1 – Synchronous Averaging

Acceleration can be segmented and averaged at ball pass

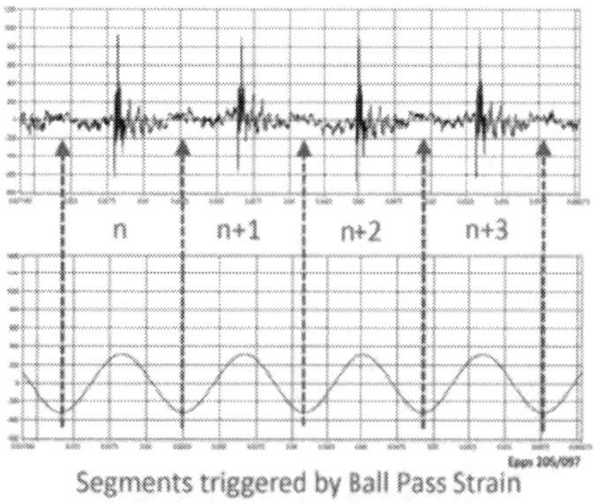

Segments triggered by Ball Pass Strain

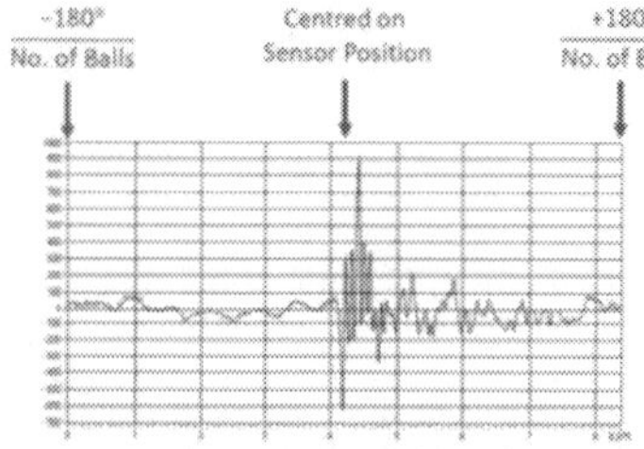

Segments Combined
(e.g. Averaged)

mobolo

Copyright 2019 - Mobolo Technology Ltd, New Zealand. All Rights Reserved

#1 – Synchronous Averaging (cont.)

Benefits

1) **Accentuates events related to bearing defects**
   - Noise reduction including other impulsive events (non-coherent)

2) **Identification of defective component**
   - Outer race directly by ball pass (BPFO) – inner race (and others) via scaling

3) **Accentuates the Point of Entry**
   - Low energy event, otherwise difficult to detect (c.f. impact)

Copyright 2019 - Mobolo Technology Ltd, New Zealand. All Rights Reserved

#1 – Synchronous Averaging (cont.)

Benefits (cont.)

4) **Measurement unaffected by changes in speed**
   - Variable and continuously variable speeds (e.g. road and rail)
   - Random changes (e.g. slippage)

5) **Single "self-synchronising" sensor**
   - No separate encoder and/or trigger source required

Copyright 2019 - Mobolo Technology Ltd, New Zealand. All Rights Reserved

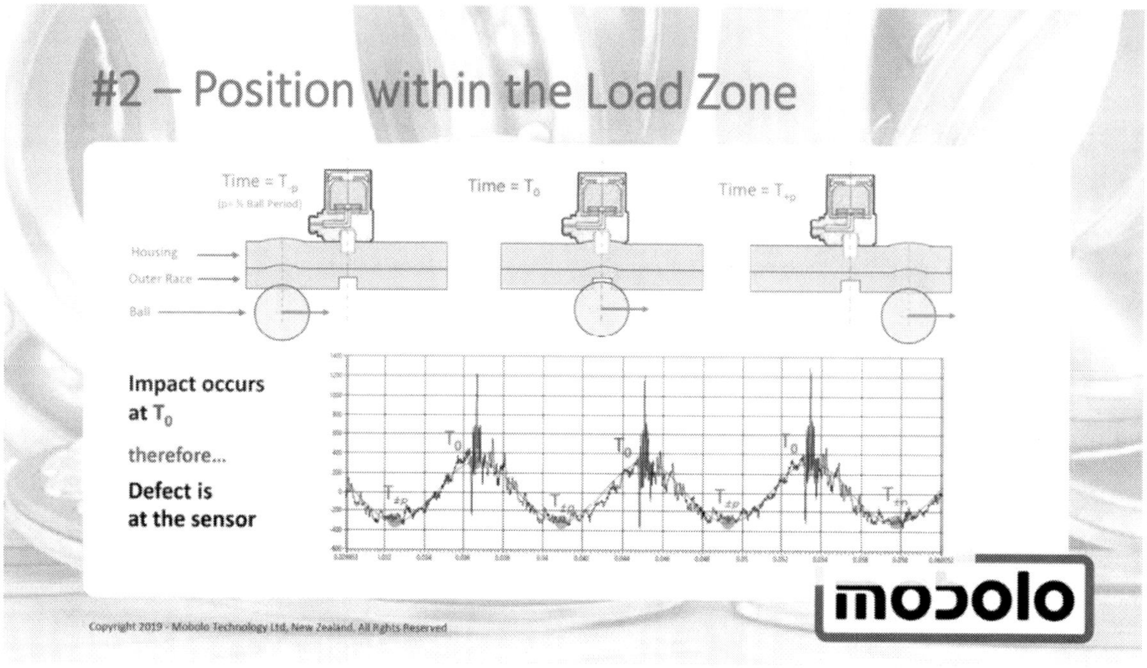
#2 – Position within the Load Zone
Time = T_{-p}
Time = T_0
Time = T_{+p}
Housing
Outer Race
Ball
Impact occurs at T_0
therefore...
Defect is at the sensor
T_0
T_0
T_0
Copyright 2019 - Mobolo Technology Ltd, New Zealand. All Rights Reserved
mobolo

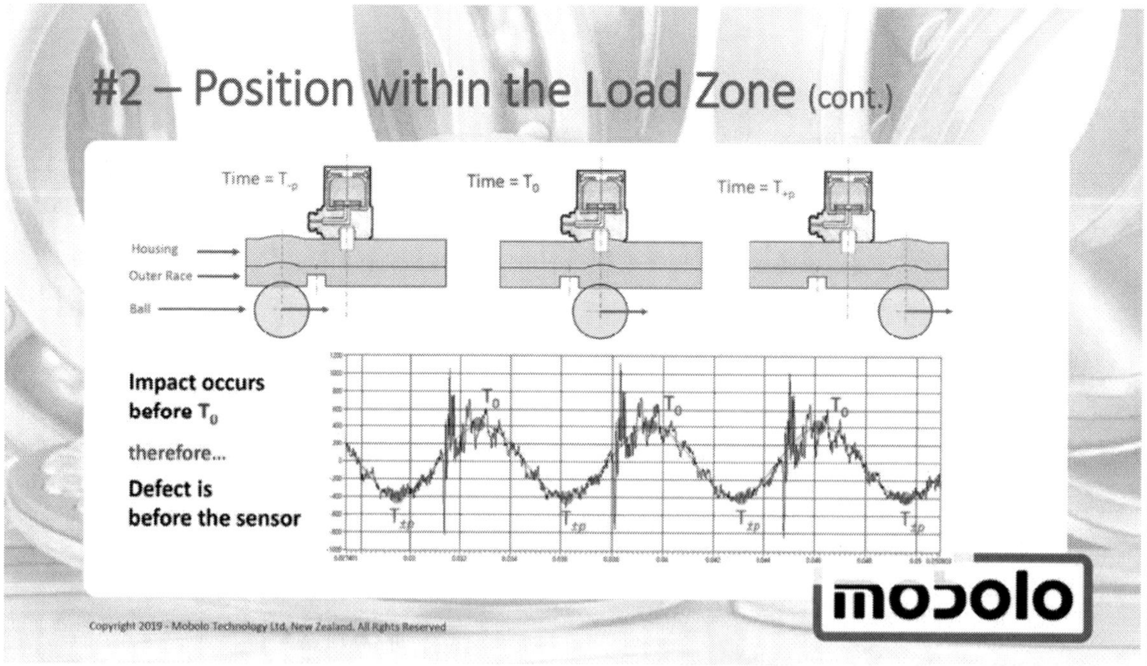
#2 – Position within the Load Zone (cont.)
Time = T_{-p}
Time = T_0
Time = T_{+p}
Housing
Outer Race
Ball
Impact occurs before T_0
therefore...
Defect is before the sensor
T_0
T_0
T_0
Copyright 2019 - Mobolo Technology Ltd, New Zealand. All Rights Reserved
mobolo

#2 – Position within the Load Zone (cont.)

Angular position given by phase relative to the sensor

Position within the load zone → estimate load distribution

Defect positioned 10° **before** the sensor

Defect positioned 10° **after** the sensor

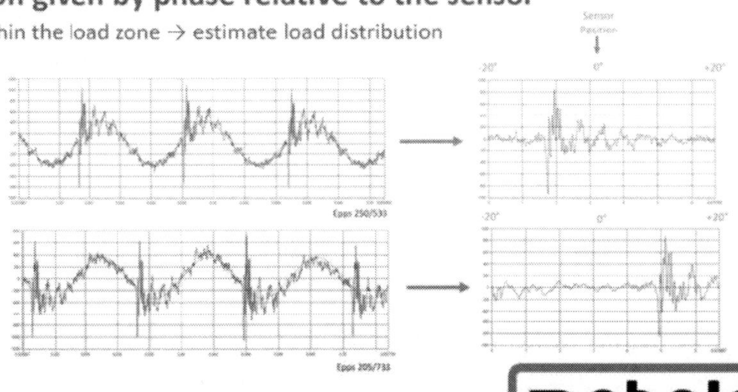

mobolo

Copyright 2019 - Mobolo Technology Ltd, New Zealand. All Rights Reserved

#3 – Operating Speed and Load

Operating state detected by the same sensor

- Frequency of Ball Pass Strain → Speed
  Shaft speed from bearing geometry (assumed zero slippage)
- Amplitude of Ball Pass Strain → Load
  Calibrated by correlation or computation (FEA)

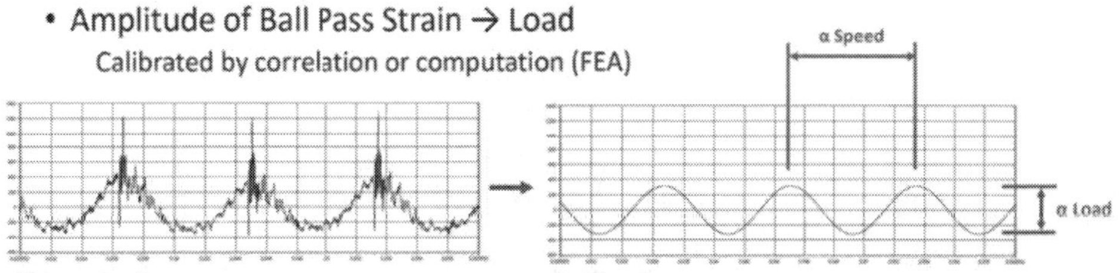

Composite Signal

Ball Pass Strain

mobolo

Copyright 2019 - Mobolo Technology Ltd, New Zealand. All Rights Reserved

New Approach - Summary

1. **Severity Measurement** ✓
   - #1 - Synchronous Averaging
   - #2 - Position within the Load Zone
   - #3 - Operation Speed and Load

   → • Defect Size • Defect Location • Operating Speed and Load

2. **Direct Measurement of Severity** ✓
   - Single reading – no trending or data storage required

3. **Measurement is Independent of Speed** ✓
   - Including continuously variable speed bearings (e.g. road & rail)

from a
Single Sensor

using a
Simple Process

Copyright 2019 - Mobolo Technology Ltd, New Zealand. All Rights Reserved

New Approach – Prototype IOT Edge Device

Simple process means a simple device (aka low cost)

- Uses well known technologies
- Processor could be within the sensor

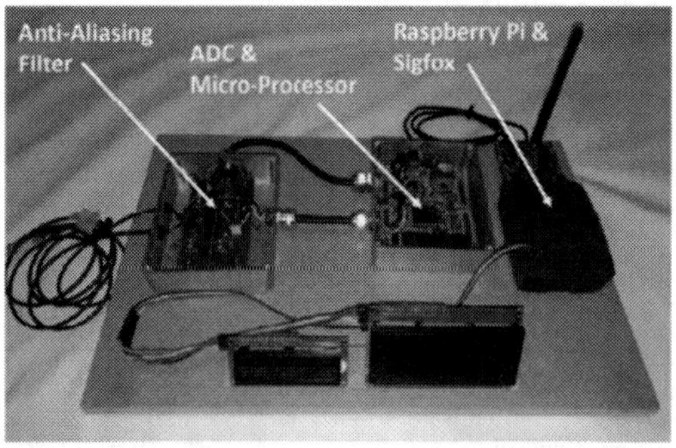

IOT Edge Device
- 2x Channel Proof of Concept
- Single Micro-Processor
- Raspberry Pi for Comms / Supervision
- Sigfox LPWAN

Copyright 2019 - Mobolo Technology Ltd, New Zealand. All Rights Reserved

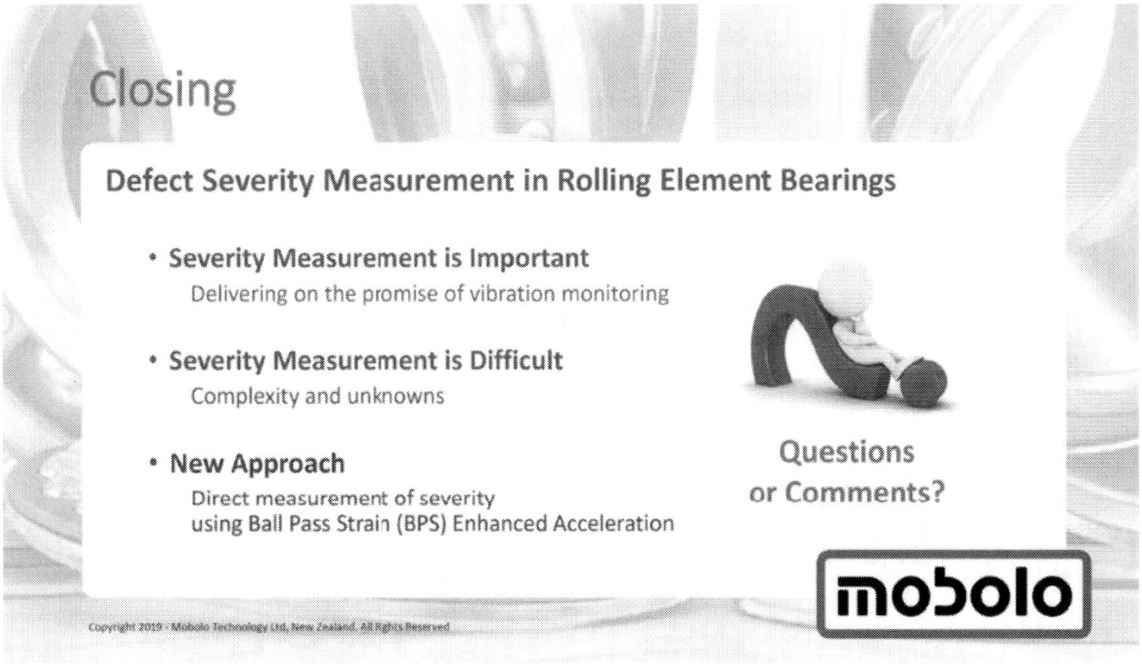
Closing
Defect Severity Measurement in Rolling Element Bearings
• Severity Measurement is Important
Delivering on the promise of vibration monitoring
• Severity Measurement is Difficult
Complexity and unknowns
• New Approach
Direct measurement of severity
using Ball Pass Strain (BPS) Enhanced Acceleration
Questions
or Comments?
mobolo
Copyright 2019 - Mobolo Technology Ltd, New Zealand. All Rights Reserved

MFPT 2019
Where Theory Meets Practice
MAY 14-16, 2019 | PHILADELPHIA, PA
For further information
Original Research:
PhD Thesis: "An Investigation into the Vibrations Excited by Discrete Faults in Rolling Element Bearings". Iain K. Epps, University of Canterbury NZ, June 1991.
Current Research:
Mobolo Technology Ltd, Christchurch NZ, August 2015 – May 2019
iain.epps@mobolo.co.nz • +64 (27) 544 6070 • www.mobolo.co.nz
mobolo
Copyright 2019 - Mobolo Technology Ltd, New Zealand. All Rights Reserved

Interference of Variable Frequency Drive (VFD) on Rotating Machinery Diagnostics

Suri Ganeriwala, Ph.D.
&
Jun Yang

SpectraQuest Inc.
Richmond, Virginia, USA
suri@spectraquest.com

SpectraQuest, Inc.
8227 Hermitage Road, Richmond, VA 23228 U.S.A.
TEL +1-804-261-3300/FAX +1-804-261-3303
www.spectraquest.com

Machinery Failure Prevention Technology (MFPT) 2019
Philadelphia, May 14-16, 2019

MFPT2019

This presentation will discuss considerations that must be taken when analyzing data collected from a machine controlled by a Variable Frequency Drive (VFD) and using it as a diagnostics tool

1. Basics of VFD operation

2. Experimental setup

3. Discussion of results

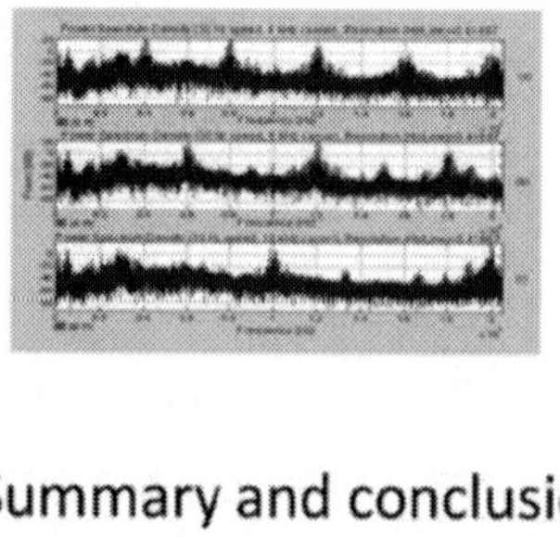

4. Summary and conclusions

SpectraQuest, Inc.

MFPT 2019

Induction motor rotational speed and torque can be controlled using a VFD

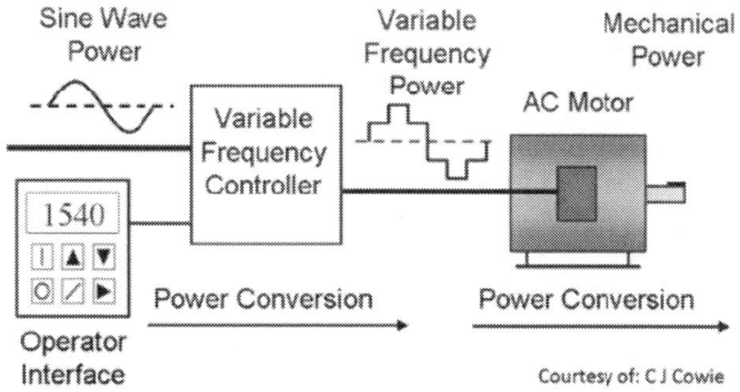

- Input: grid power with constant frequency (60Hz)
- Output: variable frequency power
- Frequency and voltage magnitude controls motor rotating speed and torque

SpectraQuest, Inc. MFPT 2019

A VFD controls the rotational speed and torque of an induction motor

$$n_{sync}(rpm) = \frac{120 * f}{p}$$

$$n_{actual}(rpm) = \left(1 + n_{sync}\right) * s$$

f=frequency
s=motor slip

n_{actual} = *actual motor speed*
n_{sync} = *synchronous speed*

- Input: grid power with constant frequency (60Hz)
- Output: variable frequency power
- Frequency and voltage magnitude controls motor rotating speed and torque

SpectraQuest, Inc. MFPT 2019

A VFD is comprised of three main systems

1. Rectifier: converts AC voltage (60 Hz) to DC
2. DC-link: smooth and flatten the rectified voltage to a constant DC
3. Inverter: switches generate voltage at the desired frequency

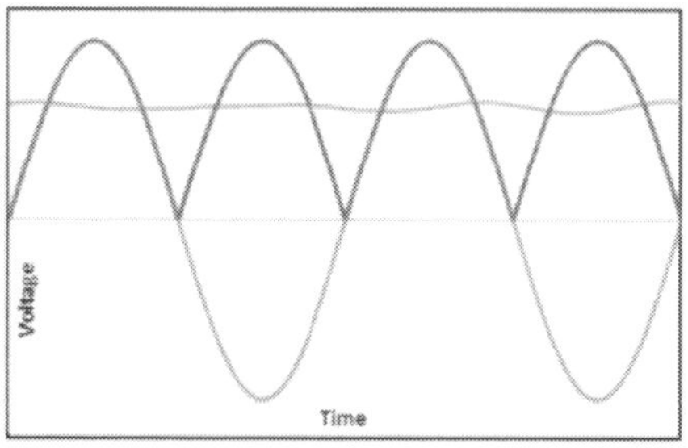

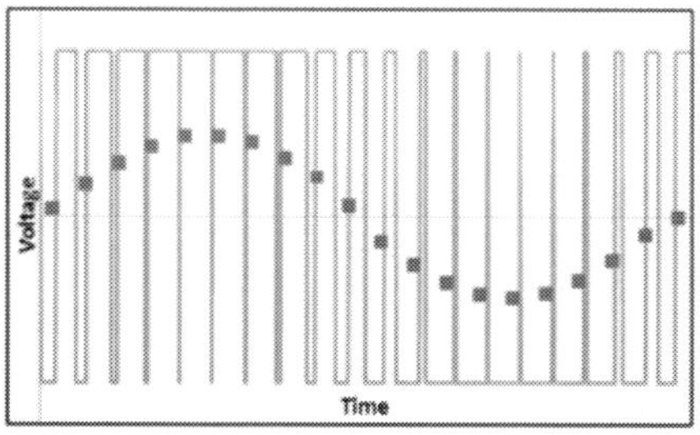

Pulse Width Modulation (PWM) is the most common approach used to generate a pseudo-line frequency

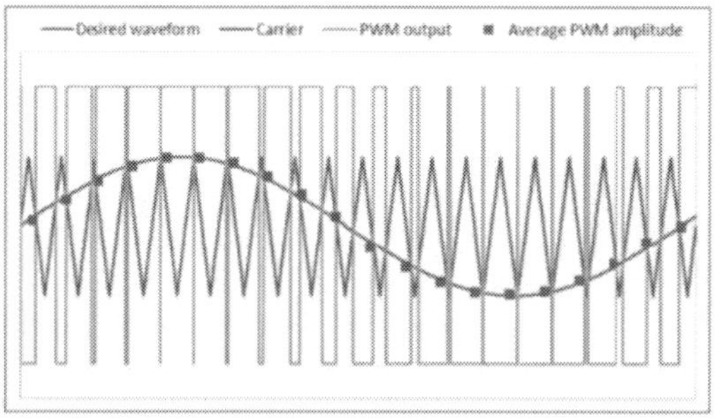

- Superimpose triangle (or sawtooth) wave on a sine wave at the desired output frequency
- Comparator varies the duty cycle of switches (typically IGBT)
- Results in a pseudo line frequency that is output to the motor
- "Carrier frequency" is that of the triangle wave

A Machine Fault Simulator (MFS) was used to generate all data

- Data collection and analysis: VibraQuest Pro, MATLAB, National Instruments PCI-4472
- Motors:1 HP Marathon
- VFD: Lenze AC Tech SMVector
- Accels: PCB IMI 608 A11
- Disabled reciprocating mechanism
- Disabled gearbox
- Two inertial discs for loading bearings

SpectraQuest, Inc. MFPT 2019

Drivetrain Diagnostics Simulator

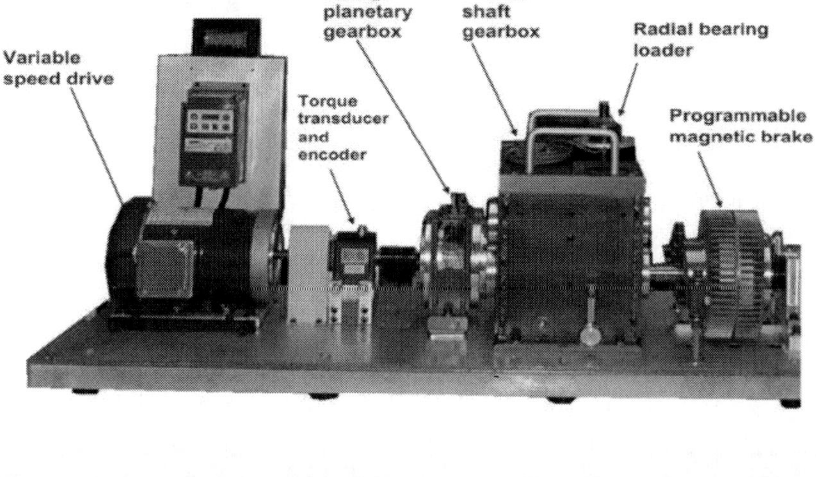

SpectraQuest, Inc. MFPT 2019

Variables used during different experiments were as follows:

- VFD output frequency: 15-60 Hz
- Carrier frequencies: 6, 8, 10 kHz
- Types of motor (2 and 4 poles)
- Loaded/unloaded
- Coupled/uncoupled shafts

SpectraQuest, Inc. MFPT 2019

Table 1 Experiment variables

VFD frequency (Pseudo line)	2-Pole motor (speed Hz)	4-pole motor (Hz)
15	14.97	7.49
30	29.89	14.89
60	57.88	29.4

SpectraQuest, Inc. MFPT 2019

The carrier frequency and its harmonics are distinguished by families of peaks centered at those frequencies

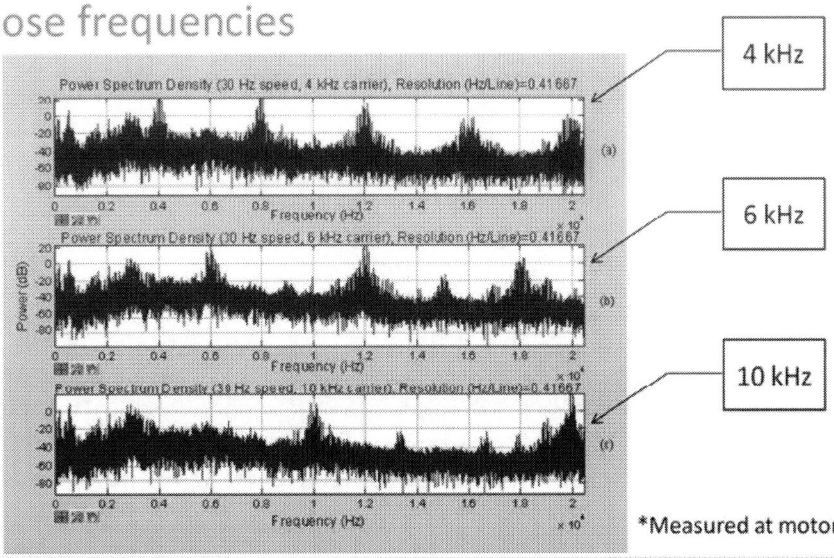

*Measured at motor

SpectraQuest, Inc.

MFPT 2019

Carrier frequency does not appear to have any influence on the lower frequencies

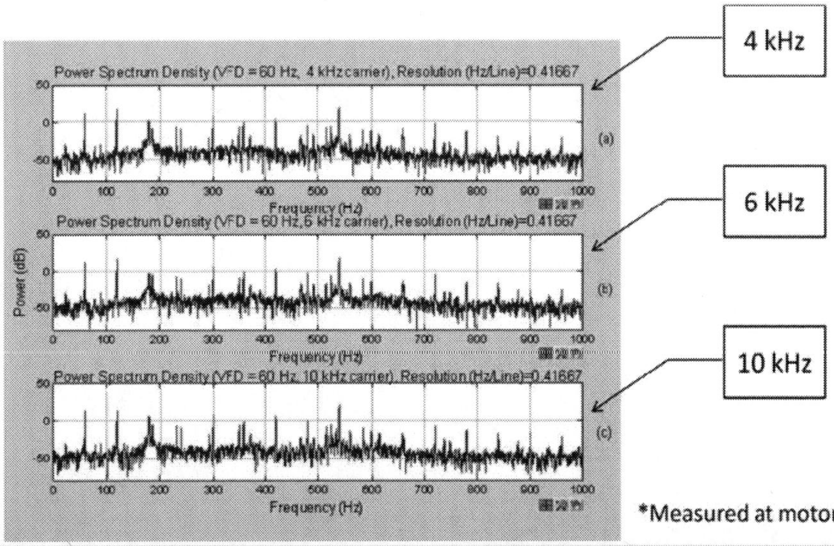

*Measured at motor

SpectraQuest, Inc.

MFPT 2019

The time domain accelerometer signals on 4-pole (top) and 2-pole (bottom) motors have different characteristics

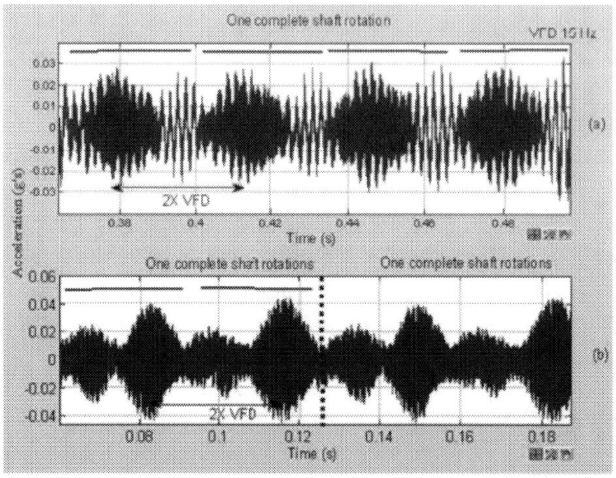

- Noticeable lower freq. component in 4-pole motor (480 Hz)
- High frequencies are present in both cases due to carrier frequency

*Measured at motor

SpectraQuest, Inc.

MFPT 2019

The sidebands about the carrier are spaced at both the pseudo line frequency as well as shaft rotation speed

Using a 4 pole motor allows the pseudo line frequency to be easily distinguished from shaft speed as compared to a 2 pole motor

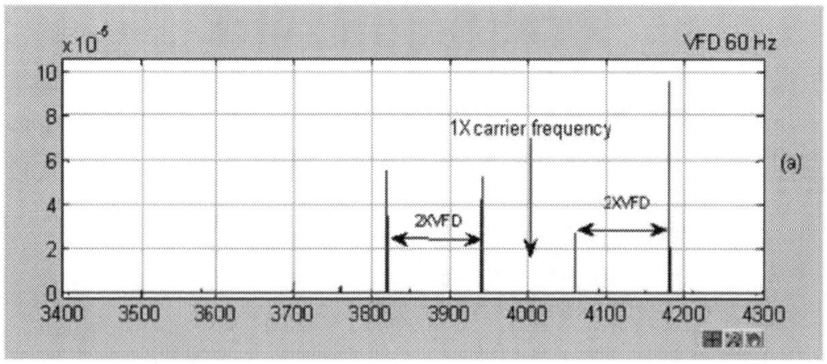

- Pseudo line frequency: 60 Hz
- Shaft speed: 30 Hz

*Measured at motor

SpectraQuest, Inc.

MFPT 2019

The sidebands about the carrier are spaced at both the pseudo line frequency as well as shaft rotation speed

Using a 4 pole motor allows the pseudo line frequency to be easily distinguished from shaft speed as compared to a 2 pole motor

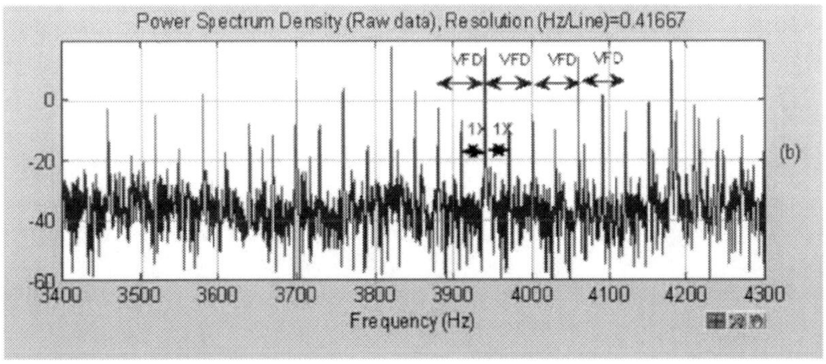

- Pseudo line frequency: 60 Hz
- Shaft speed: 30 Hz

*Measured at motor

MFPT 2019

The sidebands about the carrier are spaced at both the pseudo line frequency as well as shaft rotation speed

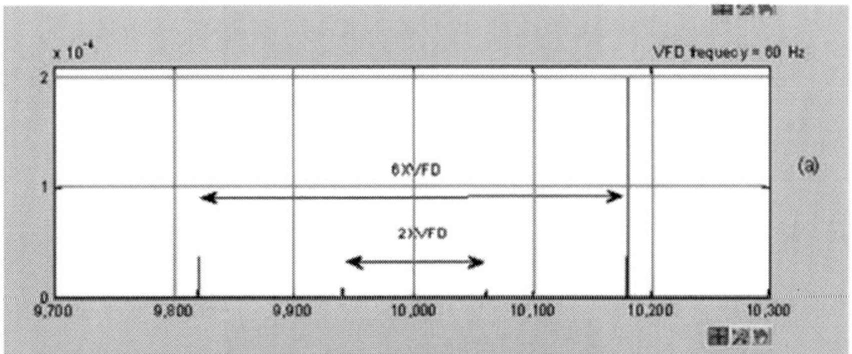

- Pseudo line frequency: 60 Hz
- Shaft speed: 30 Hz

*Measured at motor

MFPT 2019

The sidebands about the carrier are spaced at both the pseudo line frequency as well as shaft rotation speed

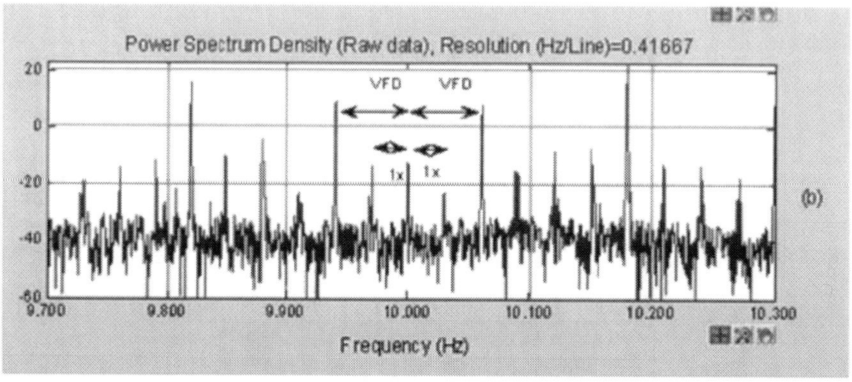

- Pseudo line frequency: 60 Hz
- Shaft speed: 30 Hz

*Measured at motor

Broken Bar :Current Signature around Carrier

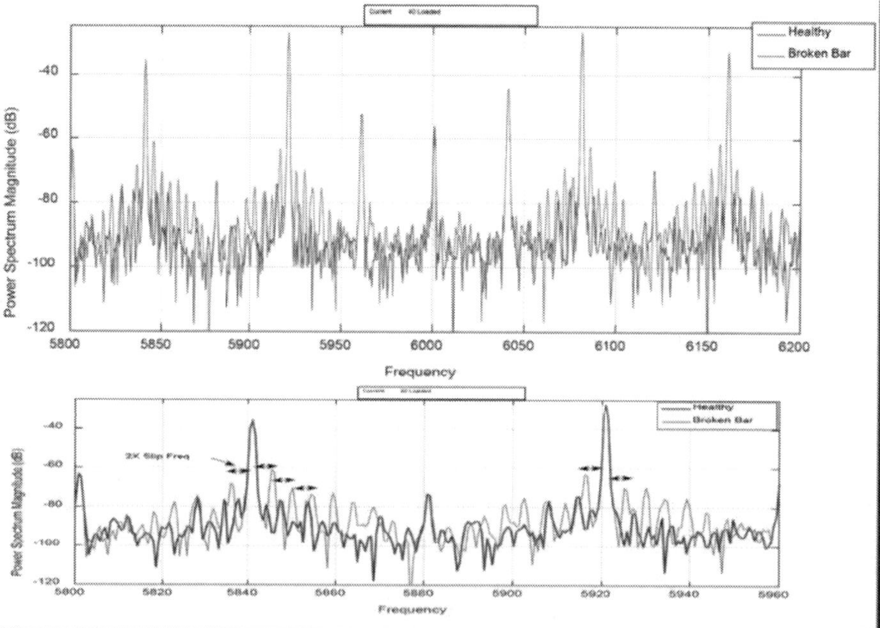

Vibration: Broken Bar 2
Diagnosis Information around the VFD Carrier Frequency

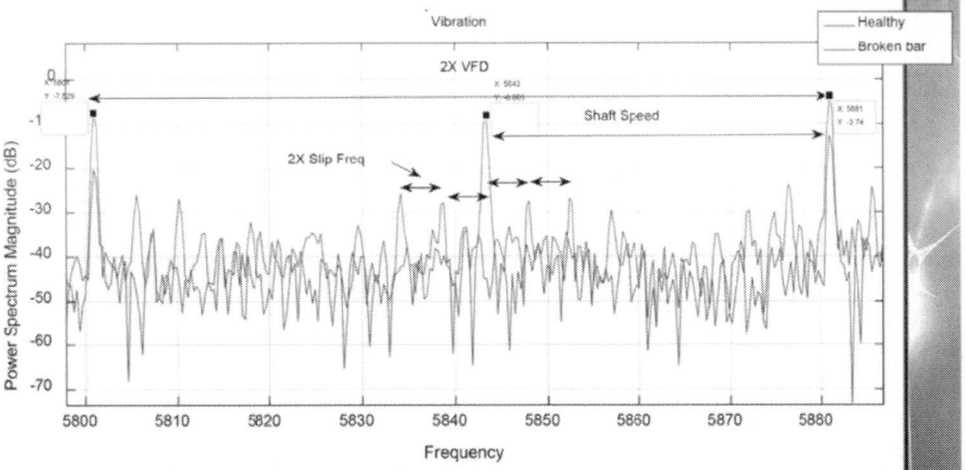

SpectraQuest, Inc.

MFPT 2019

Broken Bar :Current Signature around Carrier

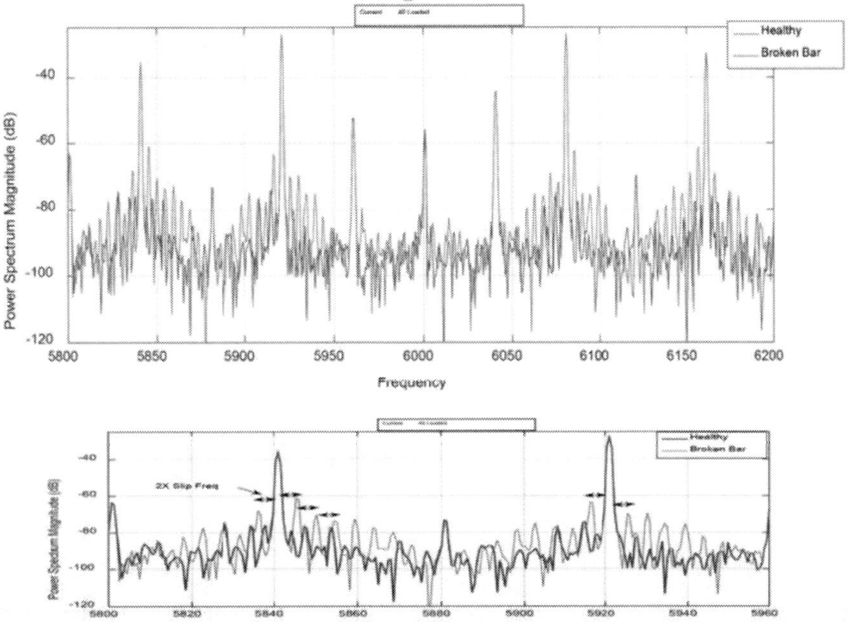

SpectraQuest, Inc.

MFPT 2019

Sidebands Around the VFD Carrier Frequencies

Data show modulation by the intermediate shaft speed (9.577 Hz) is identified

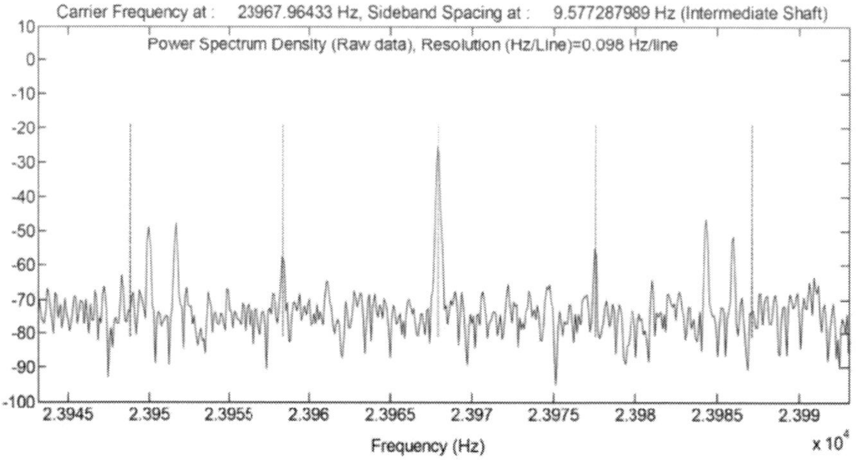

SpectraQuest, Inc. MFPT 2019

Sidebands Around the VFD Carrier Frequencies

Data show modulation by the Input shaft speed (high speed)
Sideband cursors showing the speed of input shaft (16.32 Hz)

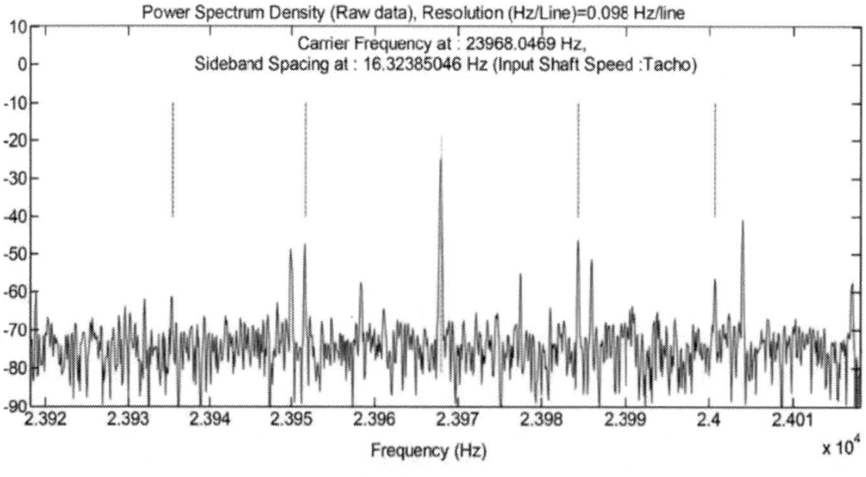

SpectraQuest, Inc. MFPT 2019

VFD Frequency and Slippage Calculation

- Note VFD (18 Hz) and 2XVFD sidebands
- Slippage % can be calculated ((17.975-16.323)/17.975)*100%=9.19 %

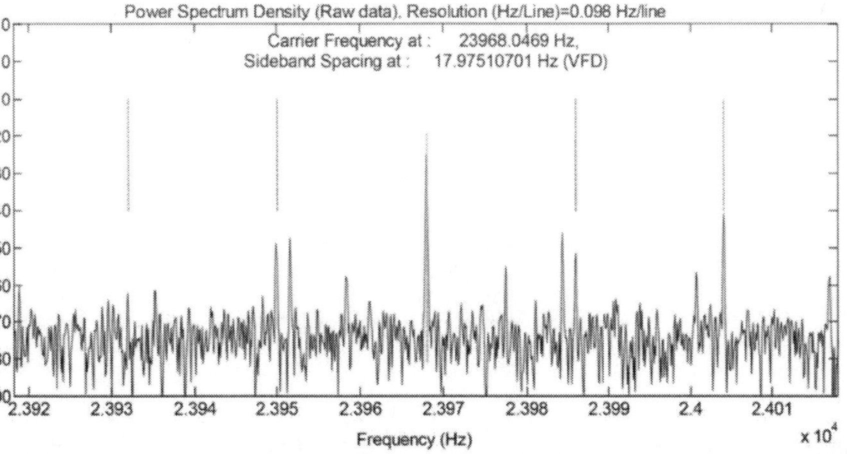

SpectraQuest, Inc. MFPT 2019

Conclusions

- Three types of Motor faults (Mechanical) were examined in this work (Broken bar, bearing fault and unbalance).
- Diagnosis using vibration data and Motor current signature analysis were attempted.
- Vibration data was found to have the features of the faults and in particular around the VFD carrier frequency.
- Current power spectra were very useful only in the case of broken bar and showed little change for bearing and unbalance.
- A proposed algorithm to separate electrical and mechanical signatures was proposed and the squared envelope spectrum of the residual whitened signal was analyzed.
- The envelope spectra were found to be different in each case with clear diagnosis in the case of broken bar
- More work and refinement are still in need to utilize the approach and make it beneficial for fault diagnosis

SpectraQuest, Inc. MFPT 2019

Questions?

SpectraQuest, Inc.

SpectraQuest, Inc. MFPT 2019

Limit Field Vibration with Dynamic Analysis

ANSI/HI 9.6.8 Rotodynamic Pumps:
Dynamics of Pumping Machinery

Peter Gaydon – Technical Director, Hydraulic Institute

HYDRAULIC INSTITUTE

Rotodynamic Pumps: ANSI/HI 9.8 Guideline for the Dynamics of Pumping Machinery

New American National Standard in 2014

- Initial use period and gaining market awareness for specification of dynamic analysis.
- Basis of today's discussion on the mitigation of vibration by the use of appropriate dynamic analysis

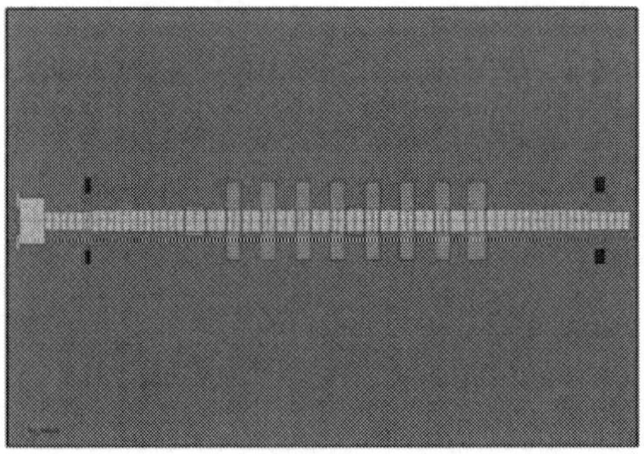

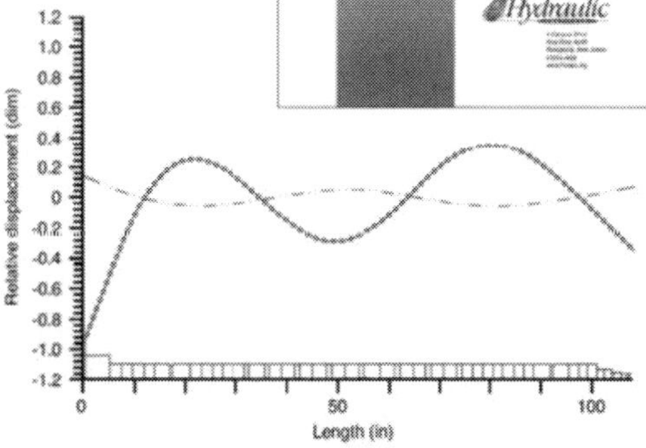

Introductory Terminology to be discussed

- Natural frequency
- Excitation Frequency (forcing frequency)
- Resonance
- Amplification Factor
- Damping
- Separation Margin
- Campbell Diagram

- Natural Frequency - frequency at which a rotor and its support system can theoretically execute free vibration indefinitely without the need for any external energy input, after being excited.
- Critical Speed- shaft rotational speed at which the rotor/bearing/support system is in a state of resonance with a natural frequency with an amplification factor greater than 3.33

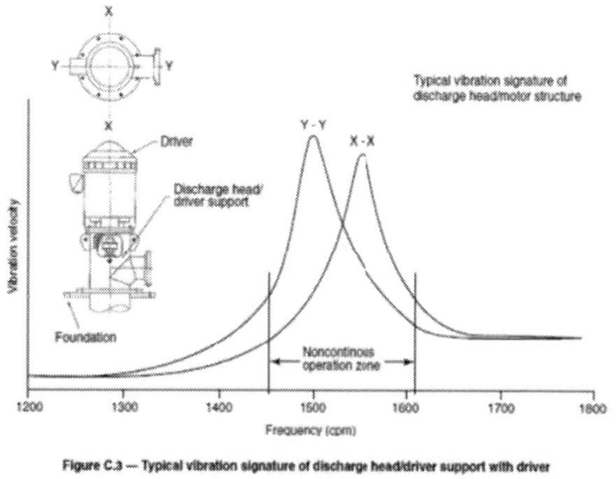

Figure C.3 — Typical vibration signature of discharge head/driver support with driver

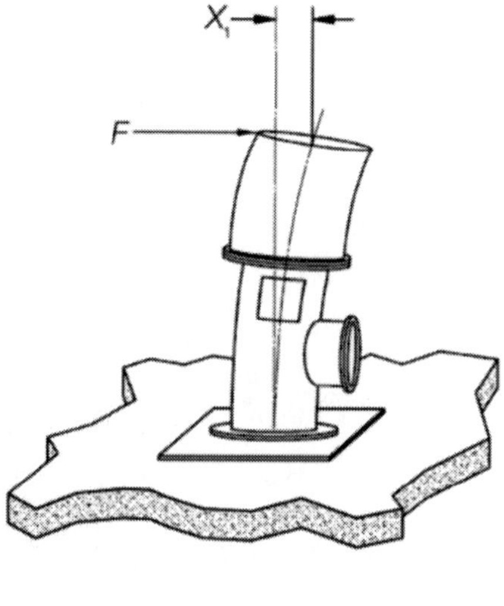

Excitation Frequency (Forcing Frequency)

- Forces sourced in the pump system that occur at a frequency (i.e. imbalance 1XRPM)

Imbalance

Vane Pass

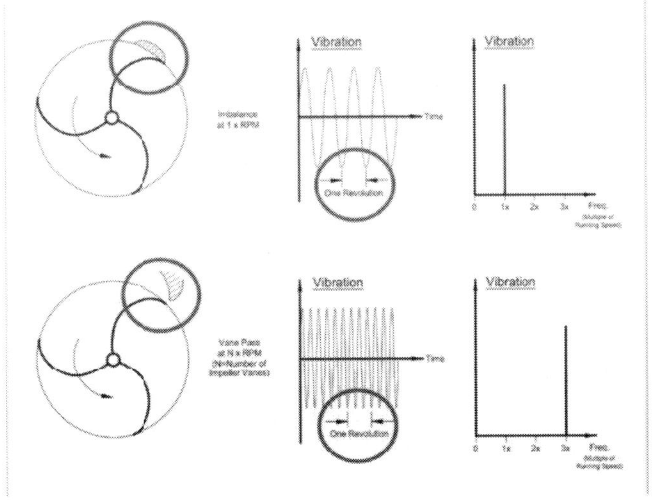

HYDRAULIC INSTITUTE

Amplification Factor

- Ratio of the response level to a given excitation force applied at the resonant frequency, to the response that would be obtained if the excitation force was applied statically.

Damping

- Any effect that tends to reduce the amplitude of oscillations in an oscillatory system, typically through either viscous or Coulomb frictional energy dissipation.

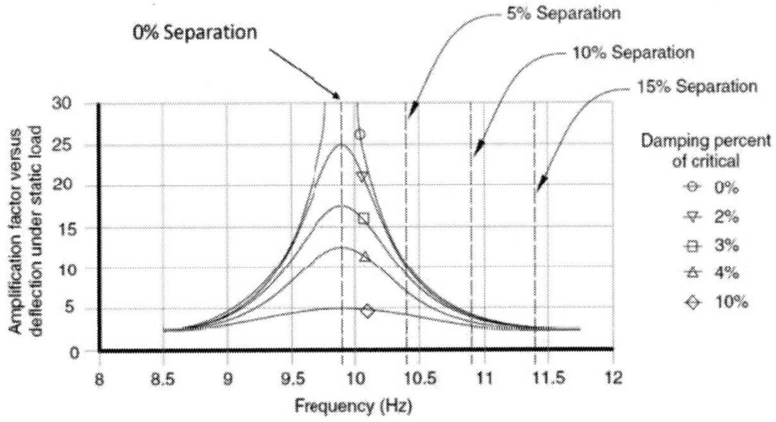

Figure E.1 — Amplification chart

HYDRAULIC INSTITUTE

Separation Margin

- Margin, defined as a percentage of the operating speed, between a critical speed that is outside of the operating speed range and the operating speed or other specified excitation frequency closest to it.

Campbell Diagram

- A plot of natural frequencies and excitation frequencies as a function of running speed. The intersections of the excitation and natural frequency lines provide approximate locations of resonance.

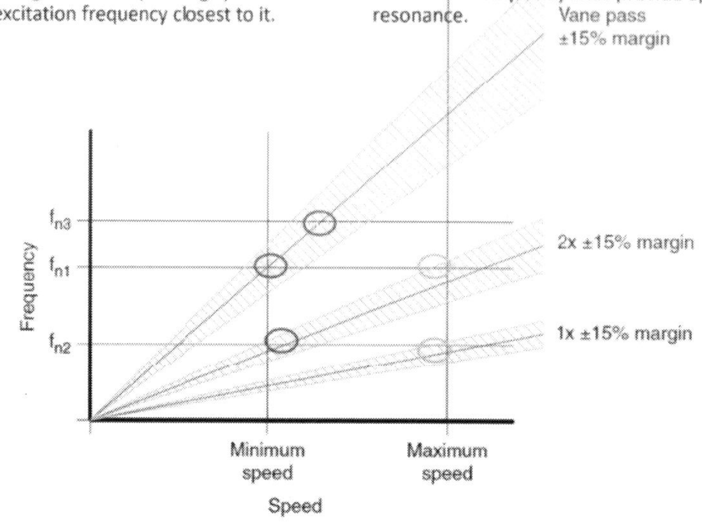

HYDRAULIC INSTITUTE

Purpose of Dynamic Analysis & Guideline

- To define dynamic analysis levels, and evaluate pumping machinery construction attributes and relevant site characteristics to determine the effects of dynamic performance on equipment life
- Why: Mitigate risk of resonance prior to installation or remediate resonance after installation so installed vibration is per ANSI/HI 9.6.4 Allowable
- Guideline Provides:
  - types of analyses,
  - level of analyses,
  - required inputs for analyses, and
  - sample specifications so the "non-expert" can determine and specify the level of analysis and separation margin appropriate for the application

HYDRAULIC INSTITUTE

What is Dynamic Analysis?

Dynamic analysis is the:

- evaluation of forces
- their frequency input to the pumping system compared the pump system's natural frequencies and
- the consequence of their interaction.

Guideline defines dynamic analyses into three categories, and each category includes three levels of complexity resulting in successively less uncertainty

1. Lateral rotordynamic analysis
   - Level I, II, & III
2. Torsional rotordynamic analysis
   - Level I, II, & III
3. Lateral structural dynamic analysis
   - Level I, II, & III

HYDRAULIC INSTITUTE

Analysis Types Defined (Simple)

There are three main types of dynamic analyses for pumps and pump trains

- *Lateral Vibration- vibration of a rotor in any plane that is perpendicular to the axis of rotation*
- *Torsional Vibration- vibration of a rotor in a manner that twists about the axis of rotation*

1. Rotor Lateral

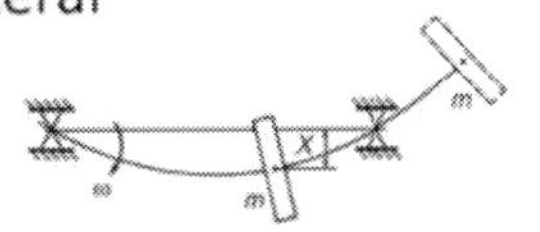

$$N_c = \sqrt{\frac{k}{m}} \text{ rad per unit time} = \left(\frac{60}{2\pi}\right)\sqrt{\left(\frac{g}{d}\right)},$$

2. Rotor Torsional

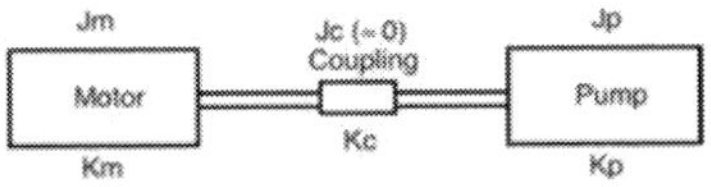

$$\frac{1}{Ke} = \frac{1}{Km} + \frac{1}{Kc} + \frac{1}{Kp}$$

$$Nc_t = \left[\frac{60}{2\pi}\right]\left[\frac{Ke(Jm + Jp)g}{Jm\,Jp}\right]^{0.5}$$

HYDRAULIC INSTITUTE

Analysis Types Defined (Simple)

3. Lateral Structural

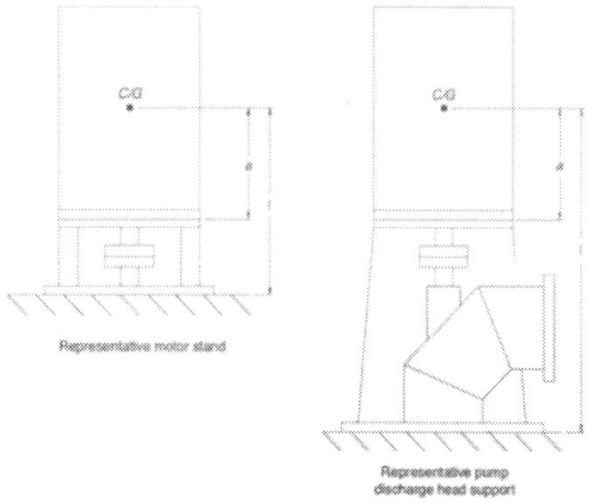

Figure C.1 — Vertical motor structures

$$RCF = \left(\frac{60}{2\pi}\right)\sqrt{\frac{g}{d}}$$

$$RCF = \left(\frac{60}{2\pi}\right)\sqrt{\frac{g}{\left[y_{static} + \frac{W(l^3 - a^3)}{3EI}\right]}}$$

RCF = the reed critical frequency of the structure (cpm)

g = acceleration of gravity = 9.81 m/s^2 (386 in/s^2)

y_{static} = the static deflection of the motor CG, in m (in.), supplied by the motor vendor

a = the motor CG height from the bottom of the motor, in m (in.), supplied by the motor vendor

l = the motor CG height from the foundation support, in m (in.)

W = force due to motor weight, in newtons (N) (pounds [lb]), supplied by the motor vendor

E = the modulus of elasticity of the material of the motor stand or pump discharge head, in N/m^2 (pound per square inch [psi])

I = the effective moment of inertia of the motor stand or pump discharge head, in m^4 (in.^{4})

HYDRAULIC INSTITUTE

Analysis Types Defined (Intermediate / Complex)

Advanced Finite Element Analysis (FEA) methods can be used to calculate the frequencies and modes with more certainty

- Rotor lateral

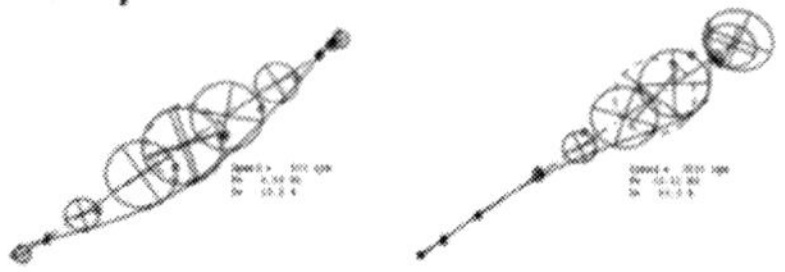

- Rotor torsional

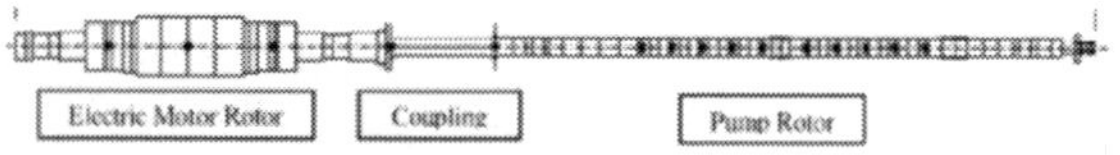

- Structural analysis

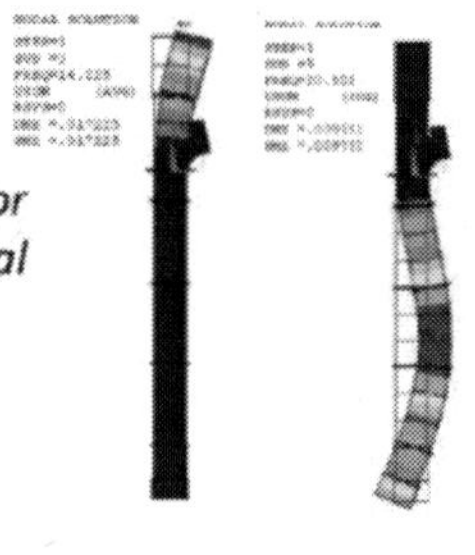

<u>Finite Element Analysis (FEA)</u>- a mathematical technique for analyzing stress or deflection, which breaks down a physical structure into substructures called finite elements

HYDRAULIC INSTITUTE

Introduction to Analysis Levels

- Level 1 - Simple calculations performed using a hand calculator and standard equations with significart simplifying assumptions.
- Level 2 - Intermediate methods involving basic mass elastic modeling using commercially available software tools such as finite element analysis.
- Level 3 – Advanced methods involving multiple specialty programs and complex methodologies, typically undertaken by specialists experienced in resolving the attendant complexities.

Expected Inaccuracies			
Analysis Level	1 - Simple Calculations	2 - Intermediate Methods	3 - Advanced Methods

Typical Level 1 and Level 2 Analyses

- **Typical Level 1 Analyses**
  - Rotor lateral frequency, rotor torsional frequency, or vertical structure reed frequency using simple equations
- **Typical Level 2 Analyses – Using general FEA or specific Rotordynamic code**
  - **Lateral Rotordynamic Analyses**
    - Undamped Natural Frequency Analysis
    - Damped Natural Frequency Analysis
    - First Dry Critical Speed of Vertically Suspended Pumps
  - **Torsional Rotordynamic Analyses**
    - Undamped Torsional Natural Frequency Analysis
  - **Structural Analyses**
    - Horizontal Pump Shaft-end Deflection Analysis (Static Structural FEA)
    - Horizontal Pump & Baseplate Natural Frequency Analysis (Modal FEA)
    - Vertically Suspended Pump Structural Natural Frequency Analysis (Modal FEA)

HYDRAULIC INSTITUTE

Typical Level 3 Analyses

Typical Level 3 Analyses **(Using general FEA or specific Rotordynamic code)**

- Lateral Rotordynamic Analysis
  - Forced response analysis of horizontal pump
  - Vertical pump dynamics addressed as a combined rotordynamic and structural analysis
- Torsional Rotordynamic Analyses
  - Forced response analysis of Driver-Coupling-Pump shaft-train
  - Transient forced response analysis for Synchronous motor – coupling – pump shaft trains
- Structural Analyses
  - Forced Response analysis of horizontal pump, driver and baseplate
  - Forced response analysis of vertical pump including flexible elements such as piping supports or other flexible mounting elements

- Recommendations for procedures for Level 2 and Level 3 analyses are provided in Appendix H

HYDRAULIC INSTITUTE

When and What to Analyze – Tools & Complexity

- The appropriate type of dynamic analysis **(lateral, torsional, structural)** depends on the pump type configuration, and orientation

- Besides what type of analysis to perform, some big questions are:
  - **When** should one do an analysis?
  - **Why** should a given analysis tool or tools be used?
  - **What** Level of analysis should be conducted and what should be considered in an analysis?

- To help answer these questions the guideline provides a
  - **Risk ("R")** and **Uncertainty ("U")** assessment,
  - for determining the **RUN value**.

Step 4 - Using the calculated "RUN" value from step 3 for each analysis type (lateral, torsional, or structural), determine the suggested level of analysis for each type of analysis from the guidelines below.	
RUN value from step 3	Suggested level of analysis
≤ 15	None Required
> 15, ≤ 20	Level 1
> 20, ≤ 50	Level 2
> 50, ≤ 160	Level 3
> 160	Level 3 +Validation*

Step 1 – Quantify Uncertainty

Uncertainty Values Table – Rigid/Non-rigid rotors & VS pumps

Pump Type	Lateral Rotordynamic Analysis	Torsional Rotordynamic Analysis	Structural Dynamic Analysis
Types OH & BB Pumps with Rigid Rotor Designs	Maximum speed > 3800 rpm, U = 2 Fly wheel driven, U = 2 Drive shaft driven, U = 2 Variable speed driven, U = 2 Power > 30 kW (40 bhp) and < 375 kW (500 bhp), U = 1 Power > 375 kW (500 bhp) and < 750 kW (1000 bhp), U = 2 Power > 750 kW (1000 bhp), U = 3 No. of vanes = 3 or fewer, U = 3	Trains with three or more elements, U = 1 Synchronous motor driven, U = 2 Fly wheel driven, U = 2 Drive shaft driven, U = 2 Internal combustion engine driven, U = 2 Variable speed driven, U = 3 Power > 30 kW (40 bhp) and < 375 kW (500 bhp), U = 1 Power > 375 kW (500 bhp) and < 750 kW (1000 bhp), U = 2 Power > 1000 bhp, U = 3	Flexible foundations, U = 1 Variable speed driven, U = 3 Power > 30 kW (40 bhp) and < 375 kW (500 bhp), U = 1 Power > 375 kW (500 bhp) and < 750 kW (1000 bhp), U = 2 Power > 750 kW (1000 bhp), U = 3 No. of vanes = 3 or fewer, U = 3 NOTE: For vertically oriented OH & BB pump types, use type VS pump values.
	System Configuration Total U _______ (Sum)	System Configuration Total U _______ (Sum)	System Configuration Total U _______ (Sum)
Types OH & BB Pumps with Flexible Rotor Designs (See Appendix A for definition of flexible rotor.)	Maximum speed > 3800 rpm, U = 2 Specific gravity < 0.7, U = 2 Fly wheel driven, U = 2 Drive shaft driven, U = 2 Variable speed driven, U = 3 Power > 30 kW (40 bhp) and < 375 kW (500 bhp), U = 1 Power > 375 kW (500 bhp) and < 750 kW (1000 bhp), U = 2 Power > 750 kW (1000 bhp), U = 3 No. of vanes = 3 or fewer, U = 3	Trains with three or more elements, U = 1 Synchronous motor driven, U = 2 Fly wheel driven, U = 2 Drive shaft driven, U = 2 Internal combustion engine driven, U = 2 Variable speed driven, U = 3 Power > 30 kW (40 bhp) and < 375 kW (500 bhp), U = 1 Power > 375 kW (500 bho) and < 750 kW (1000 bhp), U = 2 Power > 750 kW (1000 bnp), U = 3	Flexible foundations, U = 1 Variable speed driven, U = 3 Power > 30 kW (40 bhp) and < 375 kW (500 bhp), U = 1 Power > 375 kW (500 bhp) and < 750 kW (1000 bhp), U = 2 Power > 750 kW (1000 bhp), U = 3 No. of vanes = 3 or fewer, U = 3 NOTE: For vertical structures for OH & BB pump types, use type VS pump values.
	System Configuration Total U _______ (Sum)	System Configuration Total U _______ (Sum)	System Configuration Total U _______ (Sum)
Type VS Pumps	Maximum speed > 3800 rpm, U = 2 Specific gravity < 0.7, U = 2 Fly wheel driven, U = 2 Drive shaft driven, U = 2 Variable speed driven, U = 2 Power > 30 kW (40 bhp) and < 375 kW (500 bhp), U = 1 Power > 375 kW (500 bhp) and < 750 kW (1000 bhp), U = 2 Power > 750 kW (1000 bhp), U = 3 No. of vanes = 3 or fewer, U = 3	Trains with three or more elements, U = 1 Synchronous motor driven, U = 2 Fly wheel driven, U = 2 Drive shaft driven, U = 2 Internal combustion engine driven, U = 2 Variable speed driven, U = 3 Power > 30 kW (40 bhp) and < 375 kW (500 bhp), U = 1 Power > 375 kW (500 bho) and < 750 kW (1000 bhp), U = 2 Power > 750 kW (1000 bnp), U = 3	Drivers supported separately, U = 1 Drivers supported by pumps, U = 2 Flexible foundations, U = 2 Variable speed driven, U = 3 Power > 30 kW (40 bhp) and < 375 kW (500 bhp), U = 1 Power > 375 kW (500 bhp) and < 750 kW (1000 bhp), U = 2 Power > 750 kW (1000 bhp), U = 3 No. of vanes = 3 or fewer, U = 3
	System Configuration Total U _______ (Sum)	System Configuration Total U _______ (Sum)	System Configuration Total U _______ (Sum)

Typical Uncertainty Values - Zoomed

Pump Type	Lateral Rotordynamic Analysis
Types OH & BB Pumps with Rigid Rotor Designs	Maximum speed > 3800 rpm, U = 2 Fly wheel driven, U = 2 Drive shaft driven, U = 2 Variable speed driven, U = 2 Power > 30 kW (40 bhp) and < 375 kW (500 bhp), U = 1 Power > 375 kW (500 bhp) and < 750 kW (1000 bhp), U = 2 Power > 750 kW (1000 bhp), U = 3 No. of vanes = 3 or fewer, U = 3
	System Configuration Total U _______ (Sum)

- The table includes "U" values for various system features.
- Add up the "U" value scores to yield the total "U" value for each of the three analysis types
- The uncertainty number value "U" increases with increased uncertainty as related to the various application factors

HYDRAULIC INSTITUTE

Decision Matrix - Quantification of Risk

Step 2 – Quantify risk using the values provided

Step 2 - Determine and enter risk value "R" from suggested values below.			Enter selected R value
RISK NUMBER, R	Unknown, new design with no field experience.	20	
	Significant modifications to standard product or similar design - no experience in field.	10	
	Minor modifications to standard product or similar design proven in field.	4	
	Identical or standard product, proven field history.	2	

Note - "R" indicated in step 2 increases with increased risk

HYDRAULIC INSTITUTE

Decision Matrix – RUN Number

Decision Matrix Continued

- Step 3 – Determine the product of "R" and "U" ("RUN" number)
- The "RUN" number is an **INDICATOR** of the combined effects of risk and uncertainty
- Use the value to determine the appropriate analysis level that should be considered

Step 3 - Multiply the "R" values from step 2 times the risk value "U" selected in step 1 for each type of analysis. These are the "RUN" values.		
Lateral	Torsional	Structural
Products of R x U, or RUN numbers		

Step 4 - Using the calculated "RUN" value from step 3 for each analysis type (lateral, torsional, or structural), determine the suggested level of analysis for each type of analysis from the guidelines below.	
RUN value from step 3	Suggested level of analysis
≤ 15	None Required
> 15, ≤ 20	Level 1
> 20, ≤ 50	Level 2
> 50, ≤ 160	Level 3
> 160	Level 3 +Validation*

*Validation testing by the manufacturer is recommended prior to shipment.

Work Examples - Existing Pump Upgrade

- Vertically Suspended
- Pump Type VS3
- Cooling Water Circulation Pump
  - Existing pump has lower reliability
  - Operating back on curve
  - Requires two pump operation to meet demand
- Upgrade each pump to meet demand
  - New Impeller/bell
  - Increased head and capacity
  - Increased power requires a new larger 2000 hp synchronous motor

Table 9.6.8.4 — Analysis complexity levels

Analysis Complexity	Includes	Expected Inaccuracies
Level 1 Simple Calculations	1. Simple calculations using standard equations	Highest
Level 2 Intermediate Methods	1. Lateral rotordynamic analyses 1.1 Undamped natural frequency analysis - OH or BB pumps with one or two stages 1.2 First dry critical speed calculation - VS pumps 1.3 Damped natural frequency analysis - OH or BB pumps with more than two stages 2. Torsional rotordynamic analyses 2.1 Undamped torsional natural frequency analysis 3. Structural analyses 3.1 Shaft deflection analysis (static structural FEA) - OH and BB pumps 3.2 Pump baseplate or support frame, natural frequency analysis (modal FEA) - OH and BB pumps 3.3 Pump structural natural frequency analysis (modal FEA) VS (and vertically mounted OH and BB pump types)	
Level 3 Advanced Methods	1. Lateral rotordynamic analysis 1.1 Forced response analysis OH and BB pumps 1.2 Vertically suspended pump rotordynamics addressed as a combined rotor and structural analysis 2. Torsional rotordynamic analyses 2.1 Transient forced response analysis of driver-coupling-pump shaft-trains (driver: electrical motors, typically induction [= asynchronous] motors). Also trains operated through gears and fluid-couplings. 3. Structural analyses 3.1 Forced response analysis of driver and baseplate or support frame - OH and BB pumps 3.2 Forced response analysis of VS pump including flexible elements 3.3 Forced response analysis of pump, driver, and support structure – OH, BB, and VS pumps	Lowest

Work Examples - Determining Uncertainty

Pump Type	Lateral Rotordynamic Analysis	Torsional Rotordynamic Analysis	Structural Dynamic Analysis
VS Pumps	Maximum speed > 3800 rpm, U = 2	Trains with three or more elements, U = 1	Drivers supported separately, U = 1
	Specific gravity < 0.7, U = 2	Synchronous motor driven, U = 2	Drivers supported by pump, U = 2
	Fly wheel driven, U = 2	Fly wheel driven, U = 2	Flexible foundations, U = 2
	Drive shaft driven, U = 2	Drive shaft driven, U = 2	Variable speed driven, U = 3
	Variable speed driven, U = 2	Internal combustion engine driven, U = 2	Power > 30 kW (40 bhp) and < 375 kW (500 bhp), U = 1
	Power > 30 kW (40 bhp) and < 375 kW (500 bhp) U=1	Variable speed driven, U = 3	Power > 375 kW (500 bhp) and < 750 kW (1000 bhp), U = 2
	Power > 375 kW (500 bhp) and < 750 kW (1000 bhp), U = 2	Power > 30 kW (40 bhp) and < 375 kW (500 bhp), U = 1	Power > 750 kW (1000 bhp), U = 3
	Power > 750 kW (1000 bhp), U = 3	Power > 375 kW (500 bhp) and < 750 kW (1000 bhp), U = 2	No. of vanes = 3 or fewer, U = 3
	Number of Vanes ≤ 3, U = 3	Power > 1000 bhp, U = 3	
	System Configuration Total U = 3 (Sum)	System Configuration Total U = 5 (Sum)	System Configuration Total U = 7 (Sum)

HYDRAULIC INSTITUTE

Work Examples - Determining Risk & RUN value

Step 1 - Determine and enter Uncertainty Values "U"			
Analysis	Lateral rotordynamic analysis	Torsional rotordynamic analysis	Structural dynamic analysis
Uncertainty (U)	3	5	7

Step 2 - Determine and enter risk value "R" from suggested values below			Enter selected "R" Value
RISK NUMBER "R"	Unknown, new design with no field experience	20	10*
	Significant modification to standard product or similar design - no field experience	10	
	Minor modification to standard product or similar design with proven field experience	4	
	Identical or standard product, with proven field history	2	

Note: After review with manufacturer it was determined that 2000 hp synchronous motor had never been proven on this discharge head and the pump impeller & bell were a new design; however they have been proven on similar designs

Step 3 - Multiply the "R" value in step 2 by the "U" value in step 1		
Lateral	Torsional	Structural
3 X 10 = 30	5 X 10 = 50	7 X 10 = 70
Products of R X U, or RUN numbers		

Step 4 - Determine Level of analysis based on the calculated RUN value	
RUN value	Recommended Analysis Level
≤ 15	None
> 15 but ≤ 20	Level 1
> 20 but ≤ 50	Level 2
> 50 but ≤ 160	Level 3
> 160	Level 3 + Validation

HYDRAULIC INSTITUTE

Sample Specification for Recommended Analysis

Level 3, structural analysis, VS pumps, driver baseplate structural system

To determine the potential for a critical structural natural frequency occurring within the normal operating speed range of the pump, a level 3 structural dynamic analysis shall be performed in accordance with ANSI/HI 9.6.8 *Rotodynamic Pumps Guideline for Dynamics of Pumping Machinery,* Table 9.6.8.4.

The structural analysis shall include foundation mass and stiffness within a radial distance (measured from the center of the pump base) at least equal to the height of the top of the motor relative to the level of attachment of the baseplate to the floor and piping details important to modal mass and stiffness (including enclosed fluid) within a spherical zone of radius (relative to the centerline) equal to twice the height of the top of the motor relative to the level of attachment of the baseplate to the floor.

The pump structure shall be subject to a natural frequency analysis (modal FEA) in accordance with Section 9.6.8.6.2.3.3. The pump should be considered complete with driver, baseplate, and structural system. The minimum frequency separation margin obtained by analysis shall be ± _____ (to be completed by specifier). Pumps shall also be subject to a forced response analysis in accordance with Section 9.6.8.7.2.3.2. The forced response analysis of the structure shall demonstrate compliance with the vibration acceptance criteria as per ANSI/HI 9.6.4 *Rotodynamic Pumps for Vibration Measurements and Allowable Values.*

HYDRAULIC INSTITUTE

Input Data Requirements for VS Level 3

Structural Analysis – VS pumps

1. Motor drawing/motor representation (weight, cg, reed frequency, feet location)
2. Discharge head drawings/solids model
3. Motor stand (if applicable)
4. Suction bell, diffuser casing, column, mounting plate, sole plate drawing/solids models
5. Foundation information and/or support pillars (typical of vertically mounted sewage pumps)
6. Discharge piping info to first constraint or expansion joint

1. Required separation margins and/or specifications to be met
2. Level 3 only – Agreement on loads (imbalance or other) to be input for the forced response analysis

HYDRAULIC INSTITUTE

Vertical Motor Reed Critical Freq. Considerations

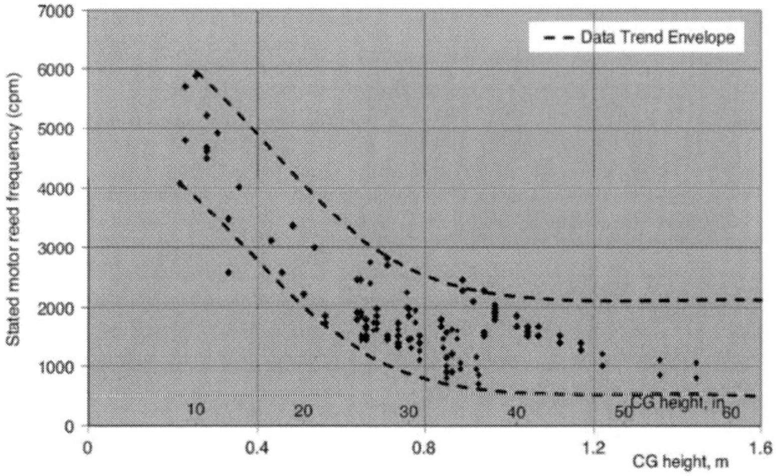

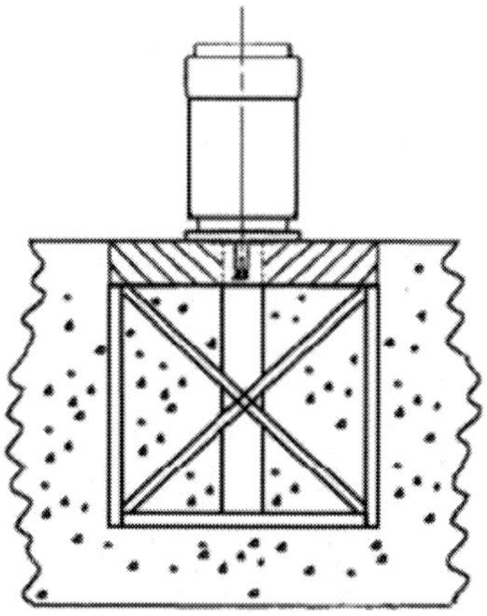

Figure D.5 — Typical range of stated motor reed frequency versus CG height (data trend envelope indicated)

HYDRAULIC INSTITUTE

Sample Specification – Vertical Motors

F.4 Instructions to motor purchasers

When specifications such as presented herein apply, they are best provided to the motor vendor in the quotation phase of the project so as to allow for the assessment of any commercial considerations, such as potential impact on pricing and delivery, to be included in the motor vendor's quotation and proposal.

F.5 Sample specification

1) In the proposal phase, the expected reed critical frequency (RCF), mass, and location of the center of gravity of the motor shall be determined by the vendor and provided to the purchaser as part of the tendering documentation.

2) The accuracy of th
obtained by verifica
dicular planes when
massive as the mot
as applicable.

3) After manufacture of the motor and before shipment, the motor RCF, mass, and center of gravity properties shall be verified at the motor manufacturer's facility. In the case of RCF, this involves an impact test in two perpendicular planes, to be denoted relative to the conduit box location.

4) A report of the impact test results (as-built RCF) in both directions (to be denoted relative to the conduit box location), the as-built mass of the motor, and the as-built center of gravity location shall be provided to the purchaser of the motor, before shipment of the motor.

5) If the as-built values of reed critical frequency, mass, and center of gravity location are outside of the allowed tolerances, then the vendor shall inform the purchaser of the motor and ensure that the matter is resolved with the purchaser, before shipment of the motor.

Overall Summary

- Methodology applied to improve the reliability of pumping systems through specification or the proper analysis and level of analysis
  - Applied to new systems
  - Applied to existing systems
- Defined Analysis types
  - Lateral Rotor
  - Torsional Rotor
  - Lateral Structure
- Defined Analysis levels
  - Level 1 (Simple methods) for initial sizing or refinement of a design
  - Level 2 (Intermediate methods) for higher level accuracy need to prove a case of moderate complexity or risk
  - Level 3 (Advanced methods) for highest level of accuracy and greatest degree of mitigation of risk and uncertainty for highly complex systems.
- Sample specification(s) improve communication of the needs of the analysis
- Field testing supports model calibration

HYDRAULIC INSTITUTE

Questions?

Contact

Peter Gaydon
Technical Director, Hydraulic Institute
862-242-5679
pgaydon@pumps.org
www.pumps.org

Special Acknowledgements

ANSI/HI 9.6.8 Standard Committee

ANSI/HI 9.6.8 Education Committee

- **Jack Claxton,** Patterson Pump Company
- **Mick Cropper,** Sulzer Pumps
- **Bill Marscher, Paul Boyadjis, Eric Olson,** Mechanical Solutions Inc.

PCN CERTIFICATION OF CONDITION MONITORING

Working Group Secretary & Certification Office Manager – Jennifer Cook

CM Administrator – Matthew Calver

CM Administrator – Leanne Hemsley

©The British Institute of Non-Destructive Testing

BINDT are based in Northampton England, which is roughly 70 miles away from central London. The institute moved in to Midsummer House in 2016 and have since grown in to the extra space due to continued growth of the institute.

We look forward to the continued growth of the institute in the future.

©The British Institute of Non-Destructive Testing

What is PCN independent (third party) certification?

Independent central (or third party) certification means that personnel are required to pass examinations which are devised and set by professional examiners authorised by a Certifying Body which has overall control over the certification process which it operates.

The main advantage of such a system is that independently awarded verifiable qualifications carry wider recognition and acceptance.

The British Institute of Non-Destructive Testing (BINDT) has developed the internationally recognised PCN scheme.

©The British Institute of Non-Destructive Testing

....Continued

BINDT is a certification body whose PCN scheme is accredited by the United Kingdom Accreditation Service (UKAS), to provide Condition Monitoring certification of personnel, the resulting certification will comply with the requirements of international standard ISO 18436.

The purpose of a PCN certificate is to provide assurances of independence, impartiality and conformity to international standards of proficiency in CM. The proficiency element is indicative of competence, though it is the employer's responsibility to authorise their employees to undertake given tasks.

©The British Institute of Non-Destructive Testing

An Industry Led Initiative

The management and administration for the PCN Scheme is provided by BINDT from its Northampton headquarters.

The CM Technical Committee (CMTC) delegates PCN scheme technical functions to sub-committees called 'Working Groups', also representative of industry, each with a published terms of reference.

The role of these Working Groups is to provide support/guidance with any technical matters and industry demand relating to the certification of personnel in the CM methods.

The Terms of Reference and Method of Working for CM Working Groups are approved by the CMTC, whose terms of reference is approved by the Council of the BINDT.

©The British Institute of Non-Destructive Testing

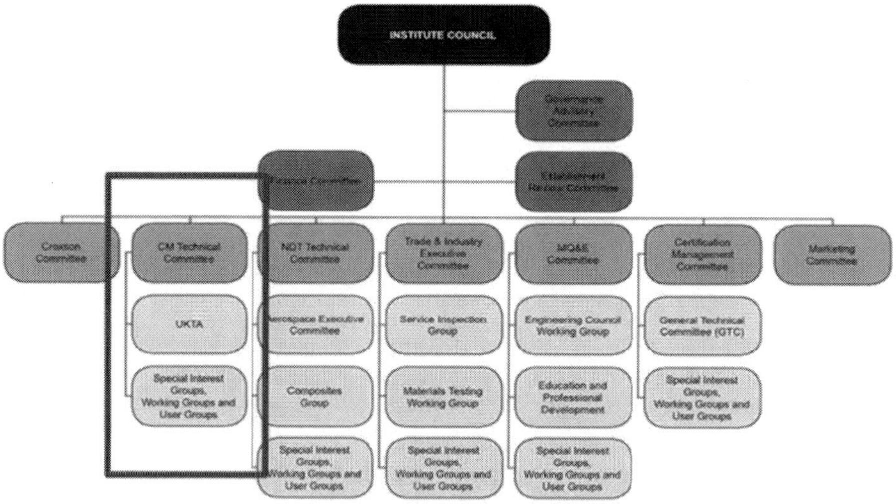

Figure 1. Committee structure

©The British Institute of Non-Destructive Testing

....Continued

The Working Groups are constituted by nominated representatives (volunteers) from organisations and industrial companies with specific interests and expertise in the CM method.

This is an excellent opportunity to receive feedback from industry in order to influence everything from certification examinations to international standards.

Current active PCN scheme working groups:

- Vibration Analysis (VA)
- Infrared Thermography (IRT)
- Acoustic Emissions (AE)
- Lubrication Analysis (LA)
- Ultrasonic Testing (UT) Under development.

©The British Institute of Non-Destructive Testing

Approved Training Organisations

We have a worldwide network of Approved Training Organisations (ATO's) that offer training in Vibration Analysis, Lubrication Analysis, Infrared Thermography and Acoustic Emission.

These ATO's are approved by BINDT to offer training that complies with the required standards and are audited annually to ensure compliance with these requirements is maintained.

We currently have 26 ATO's around the world offering CM training, a list of which can be found on the BINDT website.
http://www.bindt.org/downloads/CM-MARINE-APPROVED-TRAINING-ORGANISATIONS.pdf

©The British Institute of Non-Destructive Testing

Examinations

BINDT examinations are offered through our exam centres and can be taken worldwide. These exams are created, maintained and reviewed with the help of the Working Groups.

New questions are created/reviewed by the working groups to ensure they meet BINDT's requirements and are relevant to the technology/level the question is intended for.

Once reviewed/approved these questions are compiled in to exam papers with the relevant examiners help and are then offered to candidates.

©The British Institute of Non-Destructive Testing

....Continued

PCN offers CM certification for the following CM methods:

- Vibration Analysis (Category 1,2,3 and 4)
- Lubrication Analysis (Category 1,2 and 3)
- Infrared Thermography (Category 1,2 and 3)
- Acoustic Emission (Category 1, 2 and 3)
- UT under development

We also have some availability for taking PCN exams in other languages to help provide greater availability for non English speaking candidates.

©The British Institute of Non-Destructive Testing

PCN Statistics

To date there are 2,062 certificate holders holding 2,438 valid CM certificates. We are seeing a steady increase year on year.

This chart shows the increase of certificates issued per year since 2009

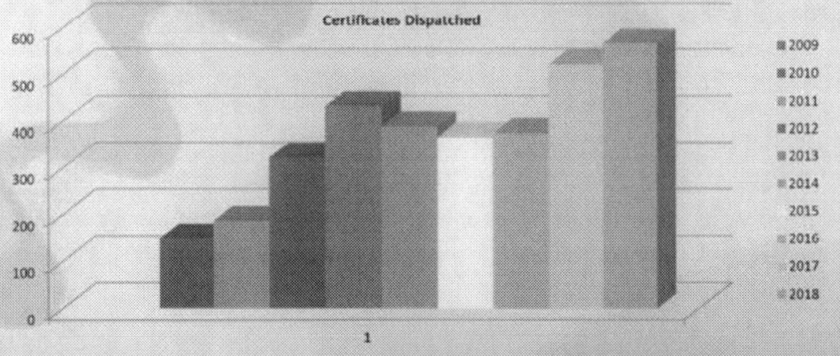

©The British Institute of Non-Destructive Testing

PCN Condition Monitoring documents

By visiting the BINDT website you can find and download all the relevant PCN CM documents covering requirements and eligibility for certification, syllabus requirements, application forms and more:
http://www.bindt.org/Certification/pcn-exam-requirements-and-document-download/Condition-Monitoring-Documents/

You can also contact PCN directly at the following email address:
cm.admin@bindt.org

©The British Institute of Non-Destructive Testing

PCN Certification Renewal

Each certificate issued by BINDT is valid for a period of 5 years and can be renewed for a further 5 year period each time.

After the first 5 year period the certificate can be revalidated without examination for a further 5 years. To show this the individual completes the CP-16 CM form to show that they have continued their professional development.

After the second 5 year period the certificate will be revalidated through re-examination. Once recertified through re-examination the certificate will be valid for a further 5 years and can be renewed again by completing CP-16 CM.

©The British Institute of Non-Destructive Testing

Membership with the Institute

Membership of the British Institute of Non-Destructive Testing is a recognition of an individual or organisation's professional status within this vital sector of the engineering industry and is recognised internationally.

Membership of the British Institute of Non-Destructive Testing is open to anyone or any company working in or having an interest in NDT, condition monitoring, diagnostic engineering, or materials and quality testing in general.

Individual members range from students embarking on engineering, physics, metallurgy or a wide range of other courses, practitioners in the hands-on world, managers and academics, to research scientists.

Company members come from an equally diverse range of industries – aerospace, power generation, oil and gas, petrochemical, universities to name but a few – as well as the equipment manufacturers and providers of specialist services, and consultancies. Company members are automatically members of the Institute's Industry Group.

To find out more please visit the BINDT website: http://www.bindt.org/membership/

©The British Institute of Non-Destructive Testing

Membership Benefits

Some of the many benefits of becoming a member with the institute are:

- Status – to demonstrate your standing in the engineering profession;
- Education – to further your knowledge and enhance your career prospects;
- Information – industry and employment news;
- Free copy of Insight, the Insitute's monthly Journal;
- *NDT News* monthly – Branch bulletin and certification news;
- Free *BINDT Yearbook*;
- Free Branch membership;
- Personal legal helpline;
- Discounts on books, membership items and other goods;
- Discounts for attendance at Institute conferences and seminars;
- Engineers' registration opportunities – CEng, IEng and EngTech;
- Opportunity to participate in industry committees.

©The British Institute of Non-Destructive Testing

Apprenticeships

The NDT apprenticeships have been developed by an employer group, led by Rolls-Royce plc and supported by the British Institute of Non-Destructive Testing (BINDT). The apprenticeships are considered to be a major benefit to employers of NDT personnel and provide a significant career opportunity to aspiring NDT apprentices, whether they come from the existing NDT community, a related engineering community or straight from school.

Three separate NDT apprenticeships have been developed:

- NDT Operator Apprenticeship
- NDT Engineering Technician Apprenticeship
- NDT Engineer Apprenticeship

Work has now also commenced on developing a similar scheme for Condition Monitoring for which government funding has already been secured.

©The British Institute of Non-Destructive Testing

Events

A comprehensive series of events covering a wide range of subjects of interest to members of the NDT and condition monitoring community is arranged each year.

BINDT's own events include:

- NDT Conference
- CM Conference
- Materials Testing exhibition
- Aerospace
- Structural Health Monitoring

The Institute also co-sponsor's events organised by other societies and from time to time arrange/manage conferences, seminars, exhibitions on behalf of other organisations.

©The British Institute of Non-Destructive Testing

Events

The sixteenth International Conference on Condition Monitoring and Asset Management

CM2019 in co operation with MFPT

Tuesday 25 to Thursday 27 June

The Principal Grand Central Hotel

Glasgow

Scotland

WWW. cm-mfpt.org

Email cm_mfpt@bindt.org

©The British Institute of Non-Destructive Testing

Publications

The British Institute of Non-Destructive Testing produces a wide range of publications covering all aspects of non-destructive testing, condition monitoring, diagnostic engineering and all other materials and quality testing disciplines.

These publications include:

- Insight – The institutes journal and covers NDT, Condition Monitoring and other associated sectors.
- ICNDT Journal – The official newsletter of the International Committee for Non-Destructive Testing.
- IJCM – The international journal of condition monitoring.
- NDT News – The newspapers for NDT, Inspection, Condition Monitoring and quality practitioners.
- Condition Monitor – Monthly publication of CM related news
- The BINDT Yearbook – Regarded as the industry "bible" and is extensively used by those working in NDT and CM.

©The British Institute of Non-Destructive Testing

Questions

? ? ?

©The British Institute of Non-Destructive Testing

MFPT Bindt 2019 Philadelphia
Introduction to On-Line oil debris monitoring Technology

Steve Greenfield
Technical Support Manager
Engine & Rotorcraft Heath Monitoring
(Eaton's Aerospace Business)

Presentation schedule

1. How do EFD-(Early Failure Detection) Systems for lubrication systems work?
2. Requirements to an "ideal" EFD-System
3. Progression of Failure / Efficiency
4. Particle Capture
5. Particle Probes
6. Principlal circuitry of a modern EFD-system example shown a Jet Engine.
7. Outlook
8. Questions and Answers

Principle of Operation

EFD-system, based on the inspection of ferromagnetic particles

1. Metallic (ferromagnetic) particles floating in the lubricant circulation are used as an indicator for system condition.

2. These particles (debris) are collected by means of a magnetic probe.

3. Debris will be inspected right away or later in a laboratory.

4. One must decide between "normal wear" (even in high quantity) and failure-relevant particles.

5. Evidence about actual condition of drive train can be obtained from number, shape and size of particles.

An "ideal" Early Failure Detection (EFD) – system

1. Operates in real time
2. Does not create any false indications
3. Contains integrated self-test feature
4. Detects a beginning failure in time, before secondary damage or catastrophic failure of drive train will occur.
5. Will evaluate all captured particles on magnetic probe
6. Recognizes the "Trend" of an upcoming failure.
   This is an important aspect of failure detection.
   The capture of just single particles is not necessarily indicating the condition of the drive system.
7. Should be able to operate "on-condition" and avoid unnecessary replacement of healthy components (cost!)

An "ideal" Early Failure Detection (EFD) – system

1. Is compatible to on board monitoring systems (HUMS)
2. Isolates and indicates the location of failure.
3. Monitors all relevant components and covers all possible failure modes.
4. Takes care that all failure relevant particles will be captured and indicated by EFD-sensor.

In addition: Ultra fine filtration (3µ absolute) can enhance the function of the EFD-system. By ultra fine filtration basic contamination of the lubricant will be reduced significantly. Thus the quality of failure prediction can be improved.

Failure-relevant Particles, Definition

- Particle size in normal operation up to 100 µm
- Failure relevant particles bigger than 100 µm
- Definition particle size:
  Multiangular chips, longest dimension corresponds to indication of given particle size. Thickness usually 1/10 to 1/5 of diameter.

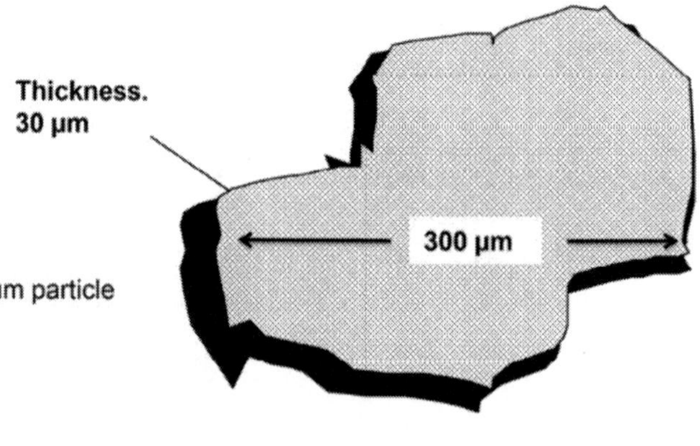

A typical 300µm particle

Typical Progression of a Failure (Bath tub curve)

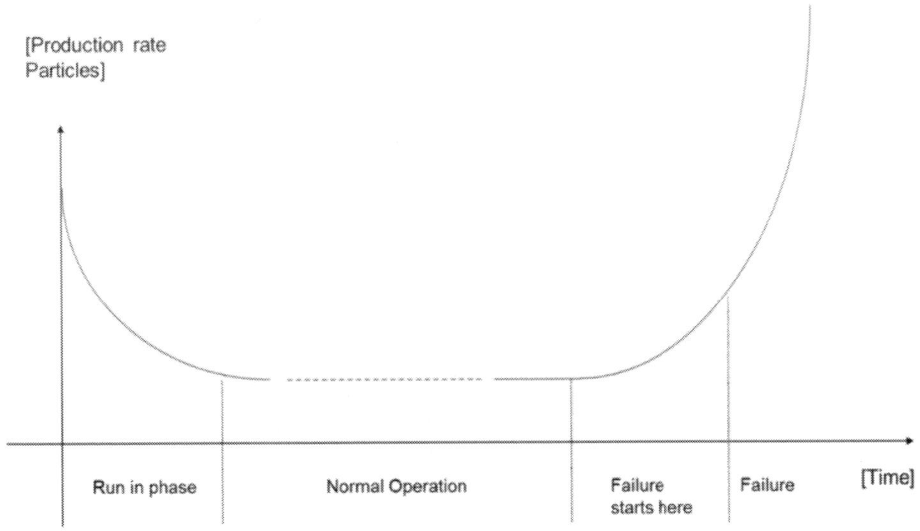

Typical Progression to a Failure

- During run in of the drive train the particle production rate will decrease continuously. It will stabilize at a certain level.
- That's where normal operation begins, characterized by constant particle production. Particle size is well below a critical threshold.
- At the begin of a failure particle production rate will increase continuously, sometimes even dramatically. Not only production rate increases, but also size of generated particles.
- At this point a reliable EFD system must indicate and report the problem.
- Only by timely indication of a developing failure, secondary damage and catastrophic failure can be avoided

Capture Efficiency of an EFD-System

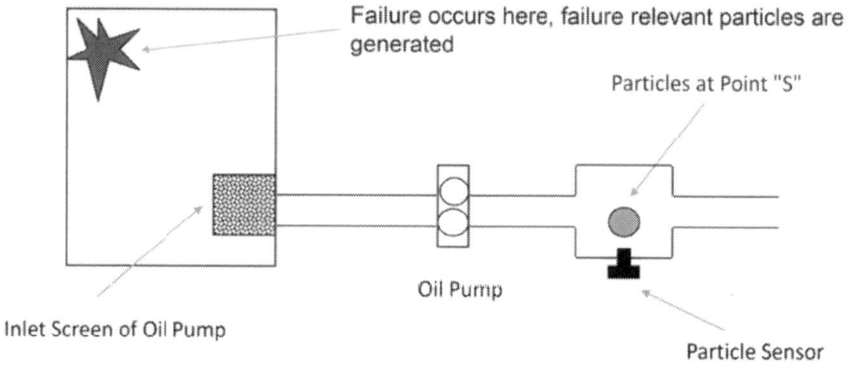

Efficiency [Transport] ηt = Particles at Point "S" / Number of generated Particles

Efficiency [Captured] ηc = Particles on Sensor / Particles at Point"S"

Efficiency [Indicated] ηi = Particles indicated / Particles on Sensor

Efficiency [System] $\eta s = \eta t * \eta c * \eta i$ = Particles indicated / Number of generated Particles

Capture Efficiency of an EFD-System - Concept

- Best Efficiency [Transport] ηt
  = Particles at Point "S" / Number of generated Particles

- Optimisation:
- When designing a lubrication system care must be taken that all failure relevant particles can reach the location of the EFD- probe.
- Scavenge Oil Flow must be possible without any obstacles or traps for particles.
- Avoid mesh or filters upstream from particle sensor.

- Note: A good Efficiency [Transport] ηt is challenge and task for the designer of the lubrication system.

Capture Efficiency of an EFD-System - Concept

Best Efficiency [Captured] ηc
= Particles on Sensor / Particles at Point "S"

Optimisation:

- As many particles as possible that reach the location of the particle sensor shall be captured by this probe.
- This can be achieved by design measures:
  * Reduce speed of oil flow
  * Introduce curves or "knees" to oil pipe
  * Use Strainers or metal mesh
  * Integrate Cyclonic particle separator (Lubriclone)

- A good Efficiency [Captured] ηc is obtained by design of oil flow and use of appropriate hardware.

Improved Capture Efficiency of an EFD-System - Methods

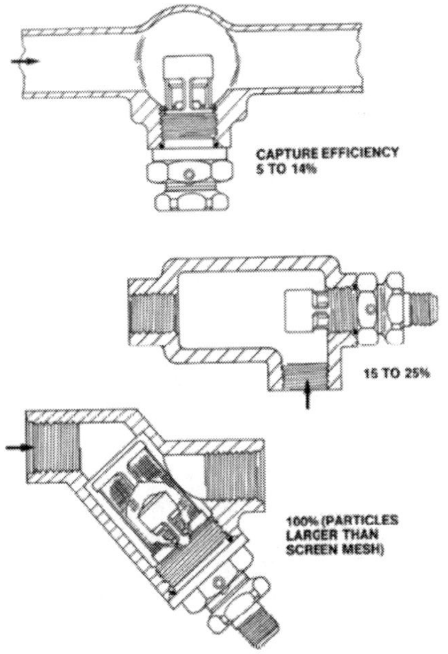

- Slow down oil flow by increasing diameter of oil line.
- Sharp bend in direction of flow results in vortex around sensor.
- Metal Grid around sensor (Strainer) captures all particles larger than mesh size.

Cyclonic Particle Separator Lubriclone™, Principle

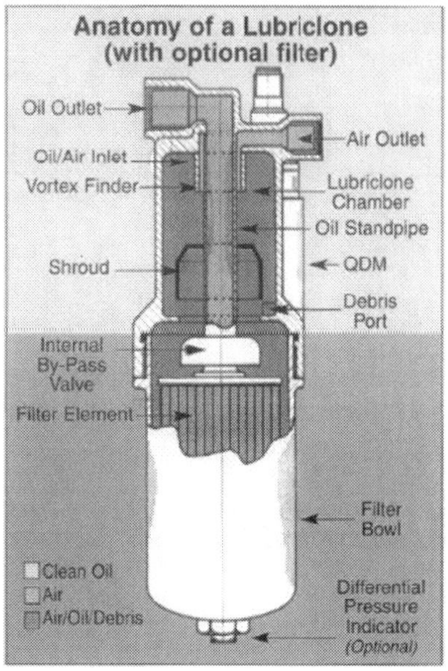

- Oil enters tangentially into Lubriclone and starts rotating.
- All particles with specific weight higher than the one of lubricant are forced to the outside of vortex by centrifugal force.

  They will finally end up on the surface of particle sensor.
- Lightweight air will concentrate in the center of vortex and can leave Lubriclone vertically. Also a good method to de-aerate the lubricant.

Cyclonic Particle Separator Lubriclone™, Data

Oil-air mixture entering the unit
will be separated by centrifugal forces.

All particles larger than target size, say 500µm, will be driven out tangentially.
also 70% of all particles larger than 100µm.

99% of the air mixed with lubricant can be vented to the outside.
By de-aerating the volume of the lubricating oil will be reduced, resulting in a drop of system "back pressure".

Capture Efficiency of an EFD-System - Concept

Best Efficiency [Indicated] ηi
= Particles indicated / Particles on Sensor
Optimisation:

- As many particles as possible, captured by the magnetic probe, shall be recognised and indicated. This is the base for predicting an upcoming failure.
- Captured particles can either be collected for evaluation in a laboratory (off line) or analyzed by the probe and / or connected electronics (on line).
- A good Efficiency [Indicated] ηi can be obtained by the optimum magnetic probe configuration.

Particle Sensors: Magnetic Plugs - Function

- Magnetic Plug (Chip Collector)

- A magnetic Probe is located in lubrication circuit or in oil sump.
- Ferromagnetic particles will be attracted by the probe's magnet and captured on it's surface.
- For easy access and optimum performance the probe is located inside a body with self closing valve.
  (Avoid oil leakage when inspecting probe)
- The Chip Collector will be removed frequently and inspected for debris. This can be visual inspection or analysis in a laboratory.
- Result is the basis for further operation of the machine.

Particle Sensors: Magnetic Plugs

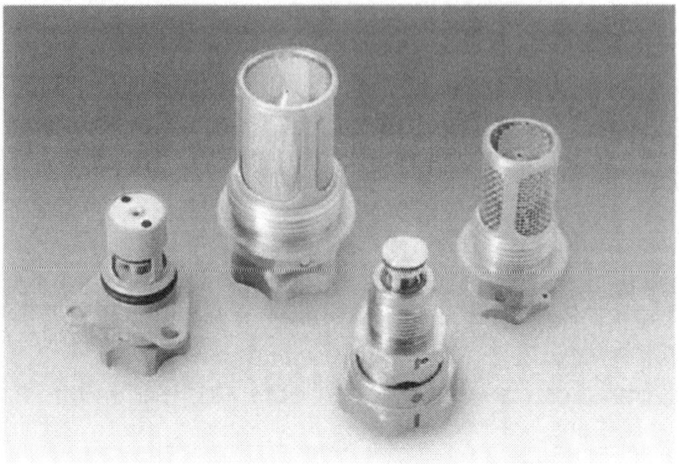

Magnetic Plugs with Self Closing Valve

Particle Sensors – Magnetic Plugs, Properties

- PROs:
  - Proven technology *
  - Cost-effective

* Result of a 5 years study initiated by Rolls Royce.

Percentage of Failures detected by the following Technologies:

Magnetic Plug	27 %
High Oil Consumption	14 %
Vibration	12 %
Air Temperature at Power Stage	10 %
Exhaust Temperature	10 %
Optical Inspection	8 %
Thermography	3 %

CONs:

- *No remote indication*
- *Inspection Intervals depending on Experience. No real time Results*
- *Improper installation can cause problems, i.e. false or no indication at all.*
- *Interpretation of debris is very subjective.*
- *Requires regular maintenance*
- *No link to on board electronics possible.*

Particle Sensors : Electric Chipdetector, Function

- <u>Chip Collector with Remote Indication</u>

- A magnetic probe with electrically insulated magnet, is located in the lubrication circulation or in the oil sump. The magnet is electrically wired via an indicator light (called chip light) to a power supply. The housing is grounded.

- Particles or metal chips captured by probe line up along the magnetic field lines. Thus they can bridge the oil/air gap between magnet and body and close the electric circuit. As a result the chip light comes on.

Particle Sensors : Electric Chipdetector, Details

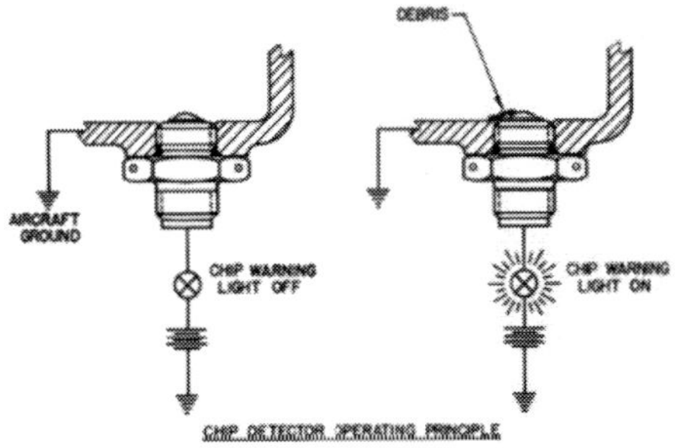

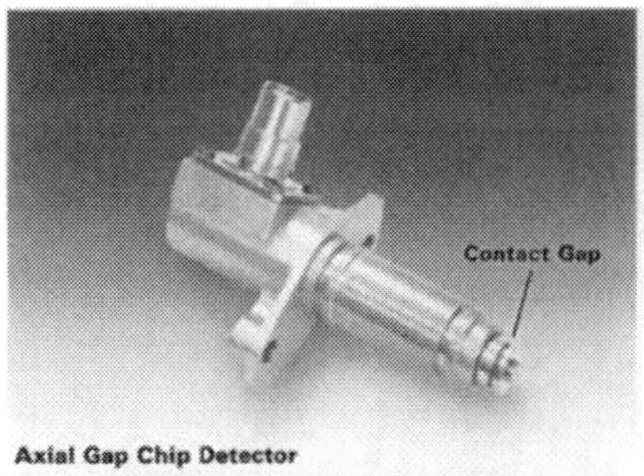

Axial Gap Chip Detector

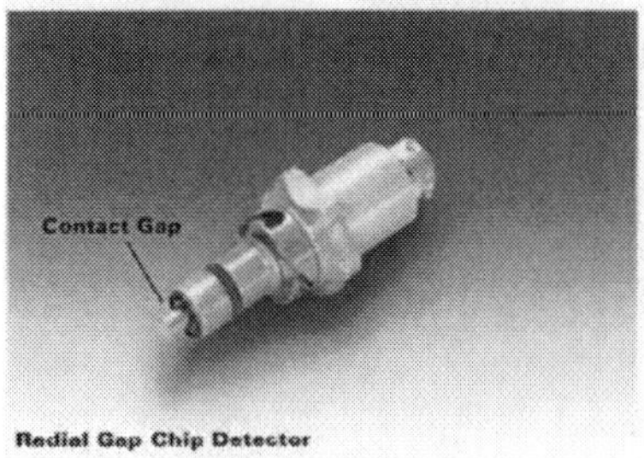

Radial Gap Chip Detector

Particle Sensors: Electric Chipdetector, Properties

- PROs:

- Proven technology, working in many helicopters, gearboxes and engines.
- Very cost-effective.
- No regular inspection required.

CONs:

- *Nuisance indications possible.*
  Cannot distinguish normal debris from failure relevant chips.
- *Not possible to monitor the trend of a failure.*

Particle Sensors: Zapper™ Pulsed-Chipdetector, Function

- <u>Electric Chip detector with Pulse-Network</u>

- Similar to Electric Chip Detector, but with built-in capacitor network.
- When air gap is bridged by metal particles, capacitor will discharge and current flows over the conductive bridge. Discharge can be triggered manually or automatically.
- Depending on current, small particles will be melted, bigger chips will resist and remain on the sensor.
- Discharge current is proportional to the capacitance of the capacitor network.
  In combination with air gap dimension, one can easily define a threshold for critical particle size. All particles bigger than critical size will not melt due to discharge.

Particle Sensors: Zapper™ Pulsed-Chipdetector, Details

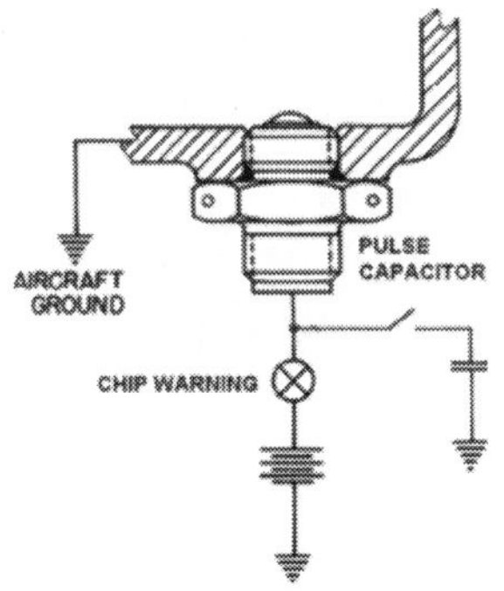

Zapper™ circuit

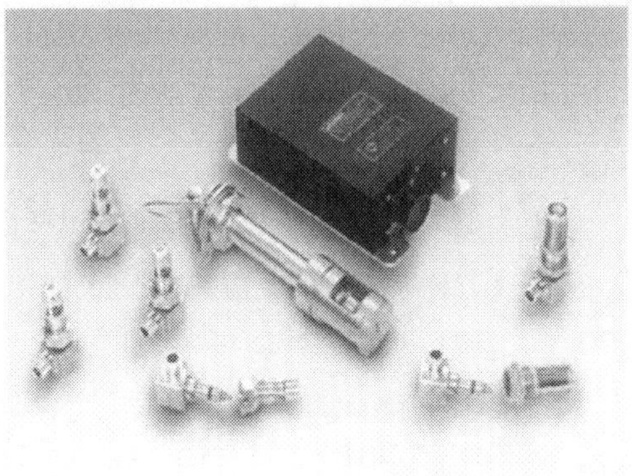

Zapper™ with control electronics

Particle Sensors: Zapper™ Pulsed-Chipdetector, Properties

- PROs:

- Well proven Helicopter technology
- A current pulse will eliminate normal wear, while large particles remain unchanged
- Practically zero nuisance indications
- Automatic or manual mode.

CONs:

- *Electronic interface required*
- *Limited capability to monitor the trend of a failure – by counting successful pulse events.*
- *Large quantities of normal debris stay on the surface and can contaminate the sensor.*

Particle Sensors: "Smart Zapper" Pulsed-Chipdetector
(Function)

- <u>Improved Enhancement of Zapper</u>

- Universal data bus interface.
- Strength and shape of pulse can be configured in field.
- Free programmable threshold for critical particle production.
- Built-in-test feature.

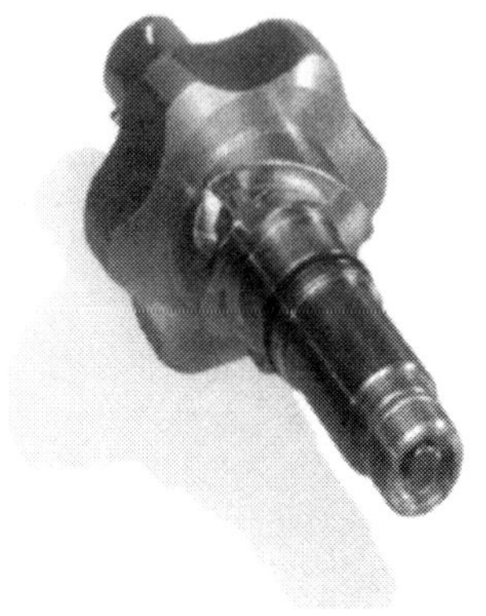

Smart Zapper

Particle Sensors: "Smart Zapper" Pulsed-Chipdetector
(Function)

PROs:

- All advantages of Zapper-technology
- Built-in self test (BIT)
- Micro processor technology
- Interface to data bus and HUMS possible
- Only marginal extra cost compared to conventional Zapper
- Trend of failure can be monitored

CONs:

- *Large quantities of debris are not removed automatically*

Particle Sensors: "QDM" Quantitative Debris Monitor
(Function)

Magnetic Plug with integrated Induction Coil

- Particles hitting the surface cause a current pulse in the electric coil. This pulse is proportional to the size or mass of the particle
- The current pulse is processed in the signal interface electronics and converted into a digital signal.
- All particles will be registered and counted.
  They can be classified by size and mass.
  Example:
  12,5 µg relates to a 250 µm-particle,
  100 µg relates to a 500 µm-particle.
- Depending on size and quantity of particles, a two stage warning signal can be displayed:
  Service warning or
  Early failure warning

Particle Sensors: QDM™ Quantitative Debris Monitor

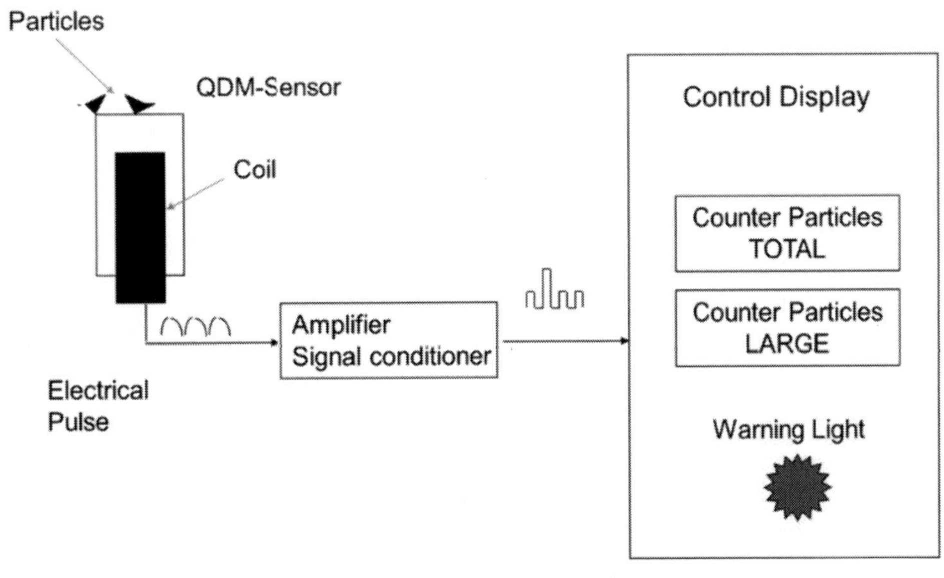

Particle Sensors: QDM™ Quantitative Debris Monitor

QDM-Sensor

QDM-Sensor and Lubriclone GE90

Particle Sensors: QDM™ Quantitative Debris Monitor

PROs:

- Collects and keeps debris for later analysis
- Can count single particles and categorize by size
- Can easily monitor the trend of a proceeding failure

CONs:

- *Large amounts of debris on probe can reduce sensitivity of sensor However quantity required is several times greater than severe failure levels*

Particle Sensors: QDM™ Quantitative Debris Monitor

- QDM Selected for large fan aero engines on all the latest aircraft, 777, A380, Dreamliner, A350
- Whatever technology is selected the debris is always easily available for human and spectrographic analysis

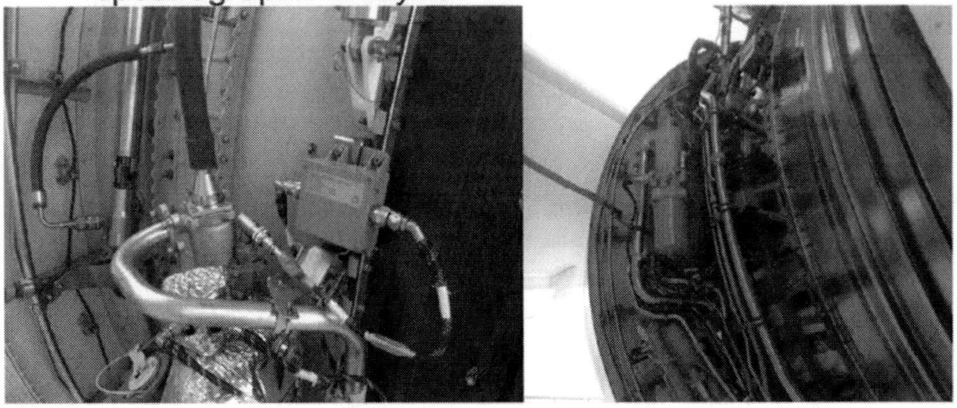

Optimal EFD System for a complex machine

(Gas Turbine)

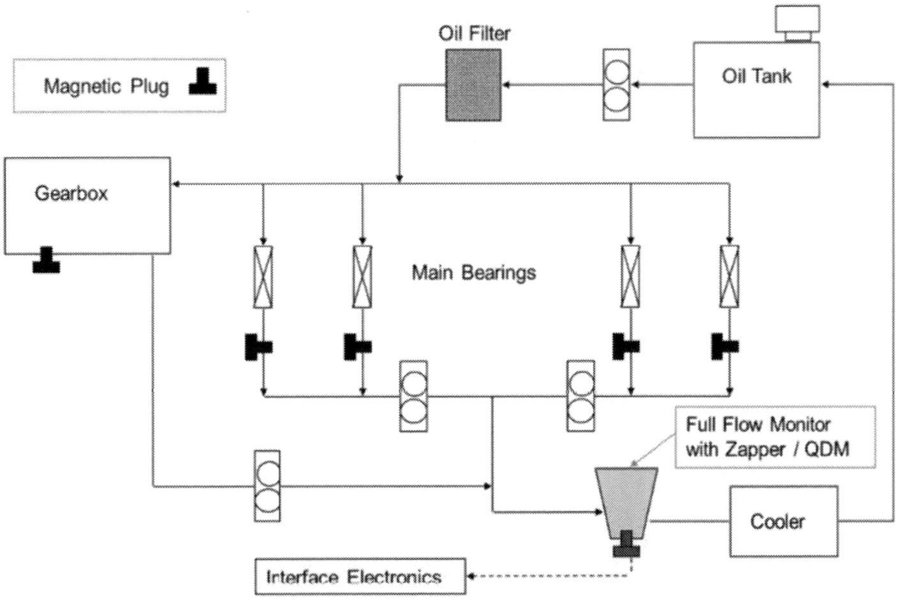

Outlook – Future Systems

- Future generations of gas turbines will feature ceramic bearings.
- New materials require new technologies for early Failure Detection.
- EATON are developing three new technologies, for particle analysis.

  * Optical
  * Acoustic
  * Engine main shaft seal wear sensors

Questions?

EATON
Powering Business Worldwide

MFPT 2019 Where Theory Meets Practice

MAY 14-16, 2019 | PHILADELPHIA, PA

Industrial sensor & IIoT data and meta-data management

Machinery Failure Prevention Technologies
MFPT 2019
https://mfpt.org/
Machine Learning/AI track, May 2019

Gopal GopalKrishnan, P.E.
Solutions Architect
gopal@osisoft.com
OSIsoft, LLC.

OSIsoft. ©2019 OSIsoft, LLC

Industrial sensor & IIoT data and meta-data management

Summary
Industrial sensor data is Big Data and largely of time-series nature for asset intensive industries such as oil & gas, chemicals, paper & pulp, pharma, mining etc., including utility companies in electricity, gas and water production/generation and transmission/distribution. Attend this session as we review the challenges in sensor and IIoT (Industrial Internet of Things) data and meta-data management. This involves working with large volumes of high velocity data from automation and control systems with hundreds of thousands to millions of sensors; each sensor collecting measurements every minute, some every second or even few microseconds. The data may also be from CMS (condition monitoring systems) and several other disparate line-of-business data sources such as planning, work management, quality, weather, web pages etc. IIoT and edge devices now bring in even more data with their own new challenges during data collection, and the need to merge this data with legacy sensor data sets.

This presentation will focus on the lessons learnt during our 35+ years of working with data and information management in manufacturing operations (https://en.wikipedia.org/wiki/Manufacturing_operations_management). And, we will cover how data analytics (with machine learning when appropriate) and visualization are used in customer use cases in energy reduction, predictive maintenance and reliability, process yield improvement, product quality, and others.

A data infrastructure approach with a faceted and application-adapted digital model allows flexible and extensible methods for data collection, analysis, and visualization that are fit-for-purpose and finally, actionable. Also see OSIsoft.

OSIsoft. ©2019 OSIsoft, LLC

Takeaways

Failure Prevention & Performance Monitoring

Industrial sensor & IIoT data and meta-data management

A single data/meta-data store
- SCADA/process data
- Machine condition data
- Collect, Contextualize, Analyze, Visualize, Share, Act

IIoT = Industrial Internet of Things

OSIsoft. ©2019 OSIsoft, LLC

IIoT enabled Operations

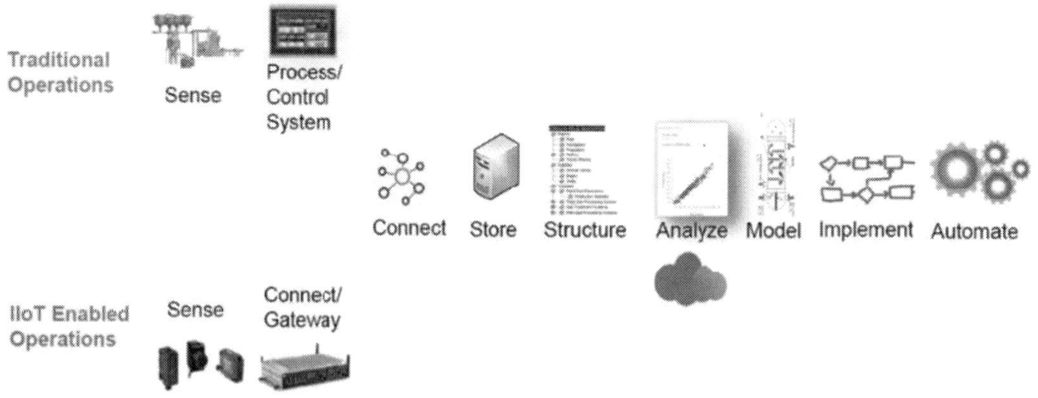

OSIsoft. ©2019 OSIsoft, LLC

OSIsoft

OSIsoft. PI System

1980
Founded

127+	19,000+	1.5B+	65%
countries installed in	sites deployed	licensed sensor-based data streams	of the Industrial Fortune 500 use us

PI System data infrastructure is widely deployed

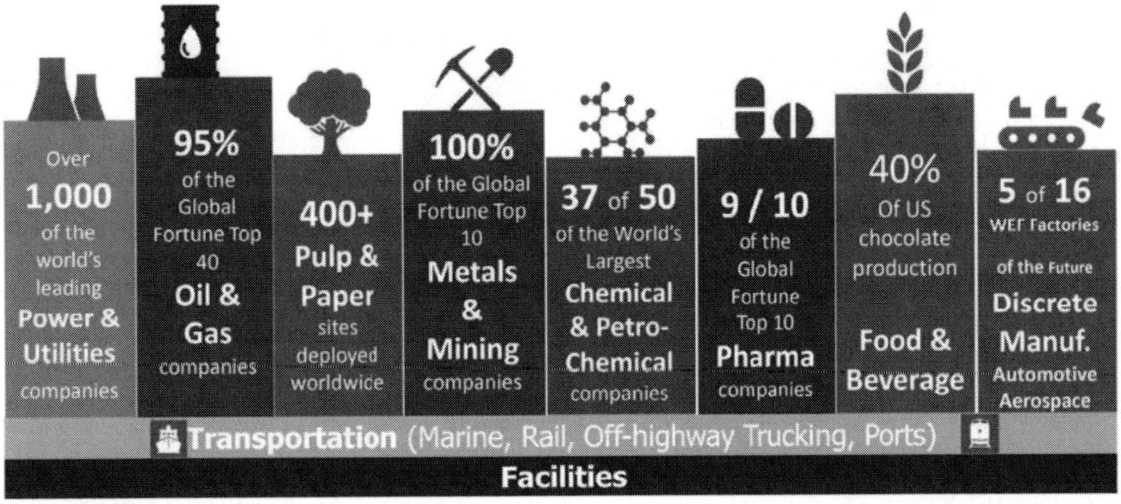

Takeaways

Failure Prevention & Performance Monitoring

Industrial sensor & IIoT data and meta-data management

A single data/meta-data store

- SCADA/process data
- Machine condition data
- Collect, Contextualize, Analyze, Visualize, Share, Act

IIoT = Industrial Internet of Things

MFPT 2019 Philadelphia, PA May 2019

OSIsoft. ©2019 OSIsoft, LLC

Process data vis-à-vis Machine Condition data

- Process data - continuous online measurements
  - Data source - Control, Safety, Environmental, Process monitoring, Trips/Interlocks
  - SCADA, PLCs, other instrumentation, IIoT ...
  - Process measurements - Pressure, temperature, flowrate, level, amperage, voltage
  - Equipment states - Open/Close, Running/Stopped,...
  - Operator rounds
  - ....
- CM data - may not be on-line and may not be continuous
  - Vibration
  - Infrared (Thermography)
  - Acoustic (Ultrasound)
  - Oil sampling - motor oil, transformer oil...
  - Motor current analysis
  - Strain-gauge (coke drums in a refinery)
  - Machinery inspection, surveillance, operator rounds, ...
  - ...

MFPT 2019 Philadelphia, PA May 2019

OSIsoft. ©2019 OSIsoft, LLC

AMEREN Process data + Vibration data

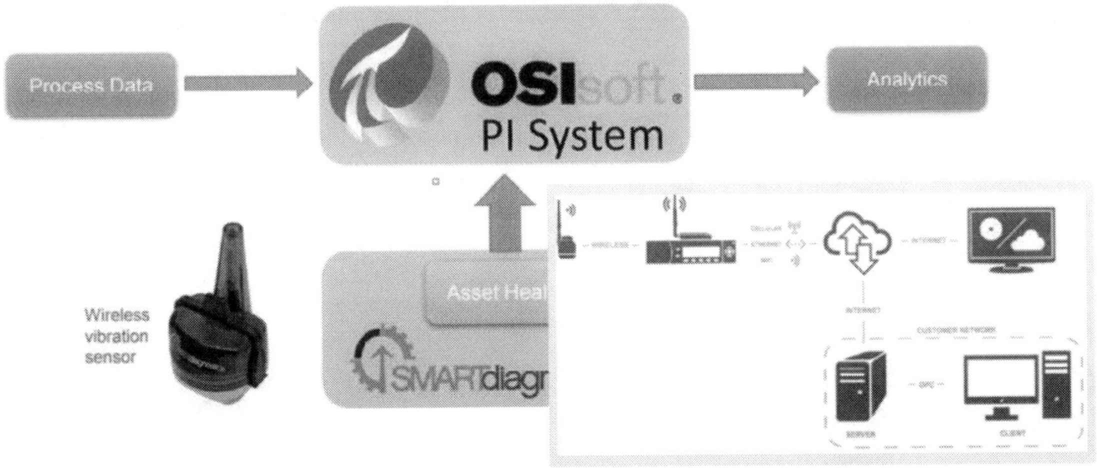

https://www.osisoft.com/Presentations/Journey-to-a-new-paradigm-in-utility-condition-based-maintenance--Amerenx/

Vibration - accelerometers - continuous monitoring

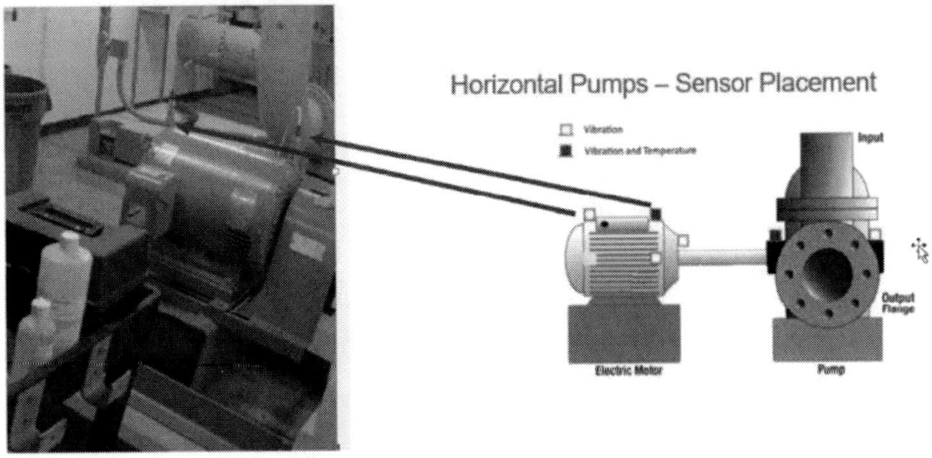

IoT can create Data Silos...

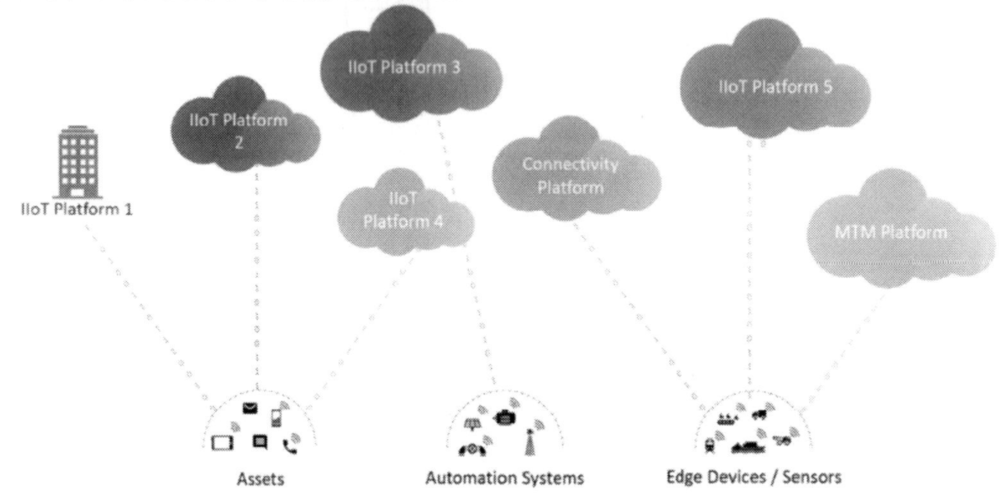

OSIsoft. ©2019 OSIsoft, LLC

IoT with Operations data infrastructure

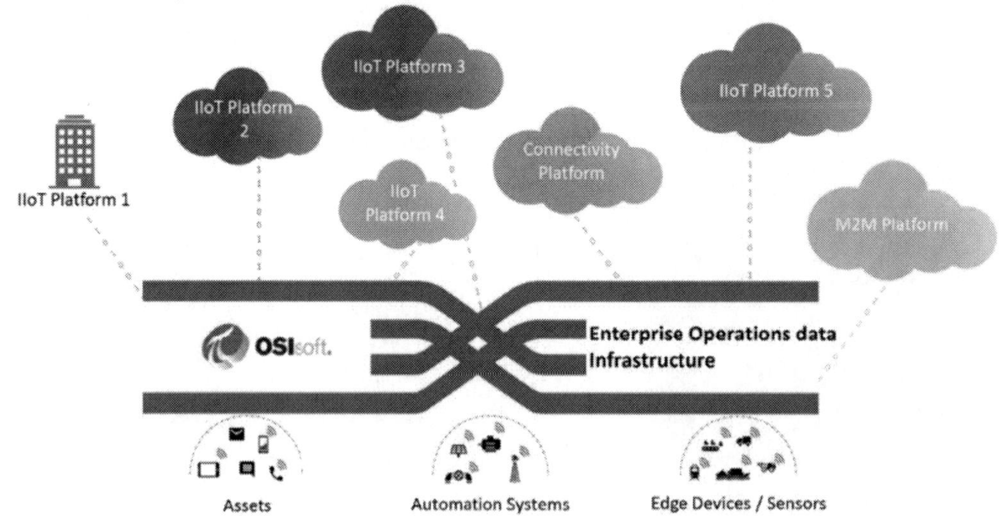

OSIsoft. ©2019 OSIsoft, LLC

PI System's role in Failure Prevention

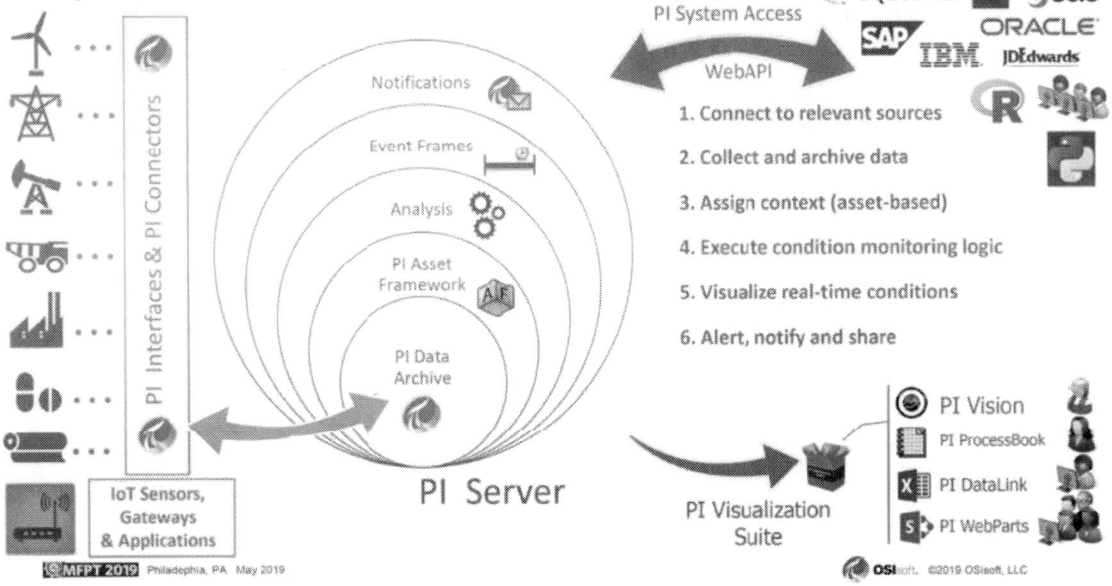

MFPT 2019 Philadelphia, PA May 2019 OSIsoft. ©2019 OSIsoft, LLC

Singe datastore for Process data and Machine condition data

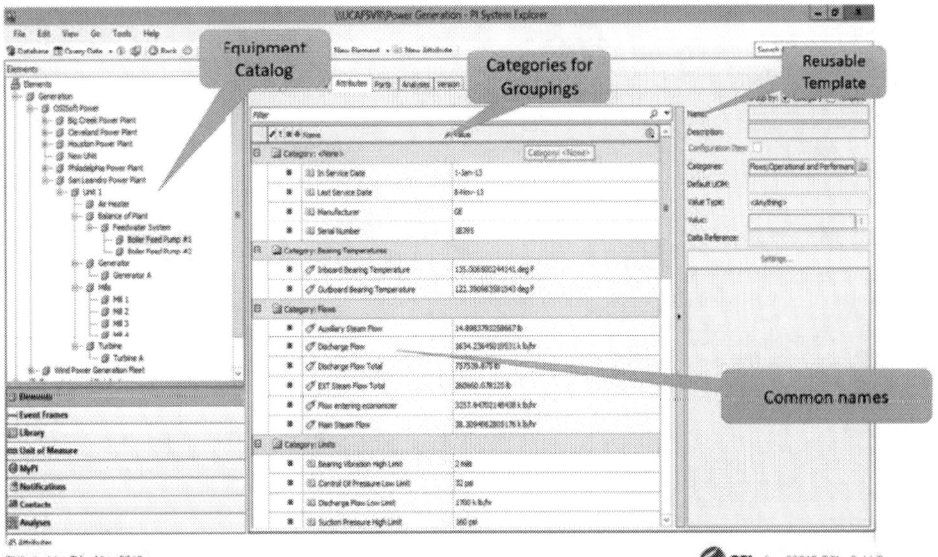

MFPT 2019 Philadelphia, PA May 2019 OSIsoft. ©2019 OSIsoft, LLC

Event Detection – process events and machine events

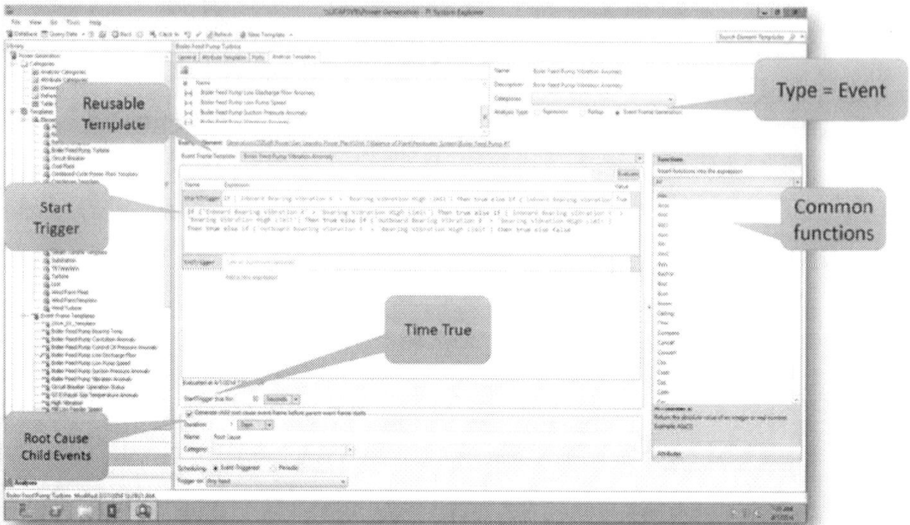

OSIsoft. ©2019 OSIsoft, LLC

Monitoring Asset Conditions

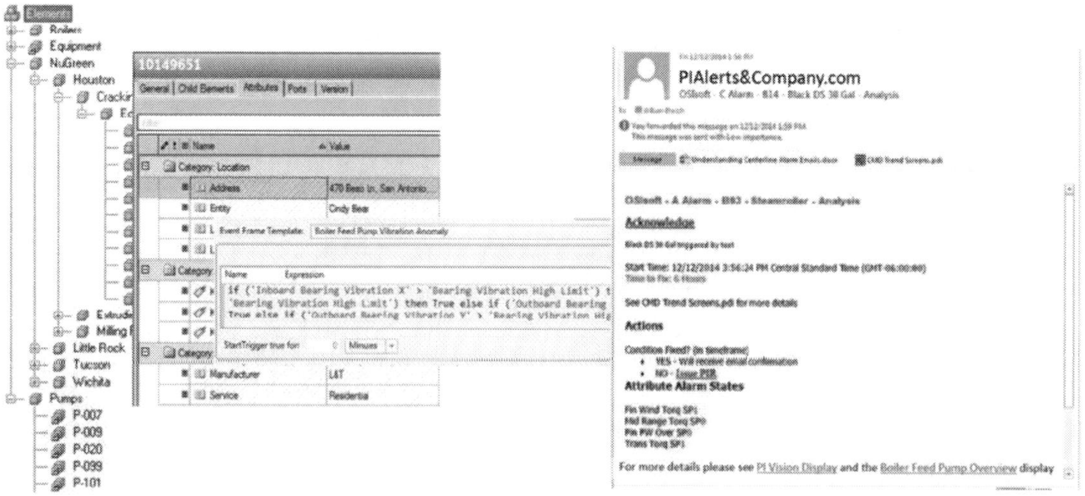

OSIsoft. ©2019 OSIsoft, LLC

Rotating Equipment Analytics

System Solution Architecture

PETRONAS

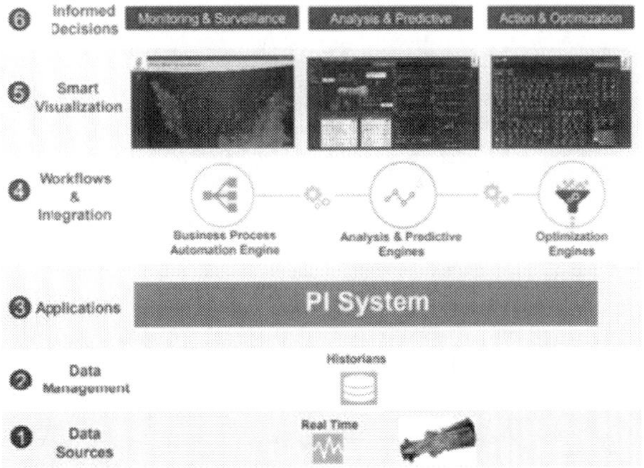

© 2017 Petroliam Nasional Berhad (PETRONAS)

https://www.osisoft.com/Presentations/PROTEAN---Rotating-Equipment-Analytics-in-PI-AF-at-Petronas/

OSIsoft. ©2019 OSIsoft, LLC

Rotating Equipment Analytics

Initial Problem Statement

PETRONAS

Over 130 pieces of gas turbine driven equipment

Numerous manufacturers

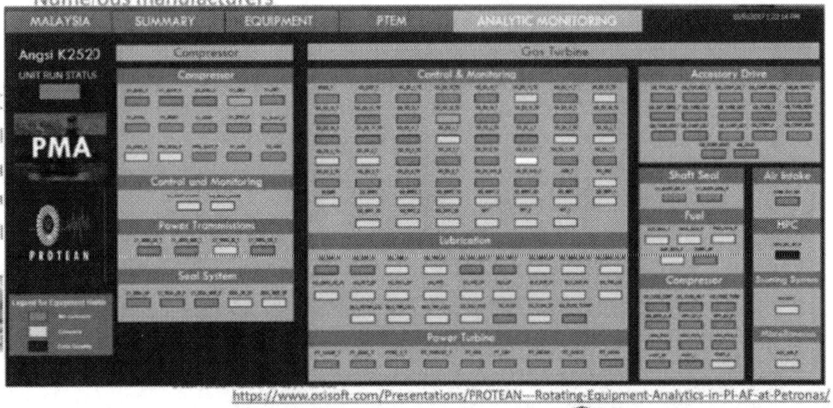

https://www.osisoft.com/Presentations/PROTEAN---Rotating-Equipment-Analytics-in-PI-AF-at-Petronas/

OSIsoft. ©2019 OSIsoft, LLC

Why should I care

- Industrial sensor data, ~ 40 years, 2 billion+ sensors, 20,000+ installs
- IIoT data, ~ 5 years
- Data and meta-data
  - Single datastore for process data & machine condition data
  - Collect, Contextualize, Analyze, Visualize, Share, Act - stack is extensible & API enabled
- Scalable - single equipment, several equipment (in a plant), several plants (organization), community, …
- Scalable - few sensors to millions, 100s of thousands events/sec
- Hybrid - on-prem & cloud
- …
- …

MFPT 2019 Philadelphia, PA May 2019

OSIsoft. ©2019 OSIsoft, LLC

Vibration (extracted features) data

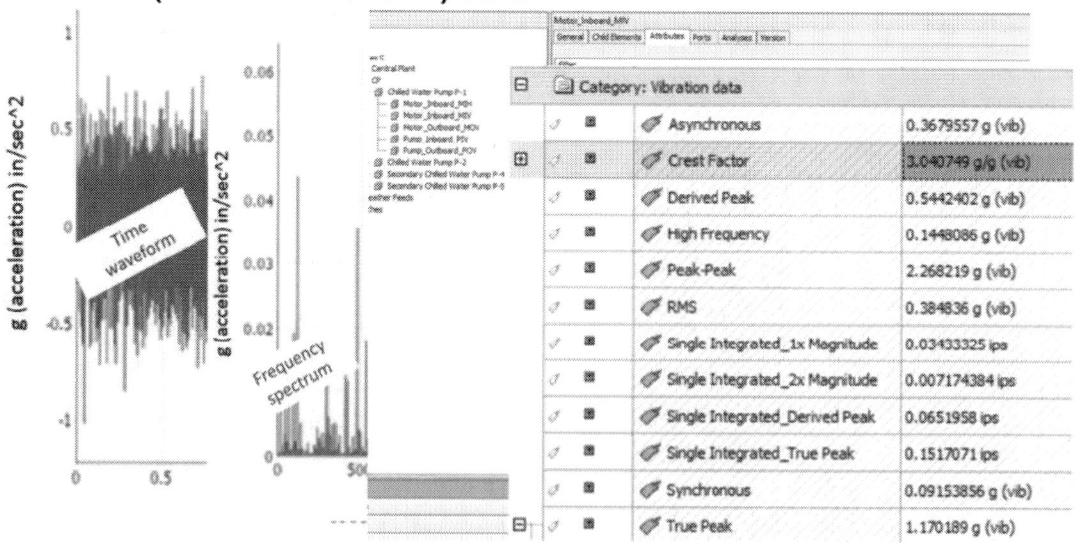

MFPT 2019 Philadelphia, PA May 2019

OSIsoft. ©2019 OSIsoft, LLC

Vibration (features) data in PI

Fluke 805 FC – Vibration Meter

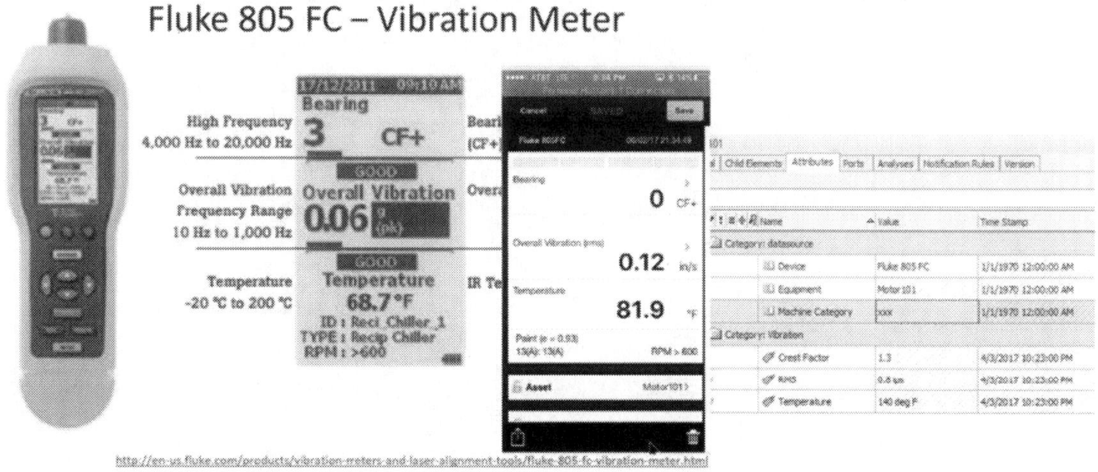

http://en-us.fluke.com/products/vibration-meters-and-laser-alignment-tools/fluke-805-fc-vibration-meter.html

OSIsoft. ©2019 OSIsoft, LLC

Fluke Connect Modules:

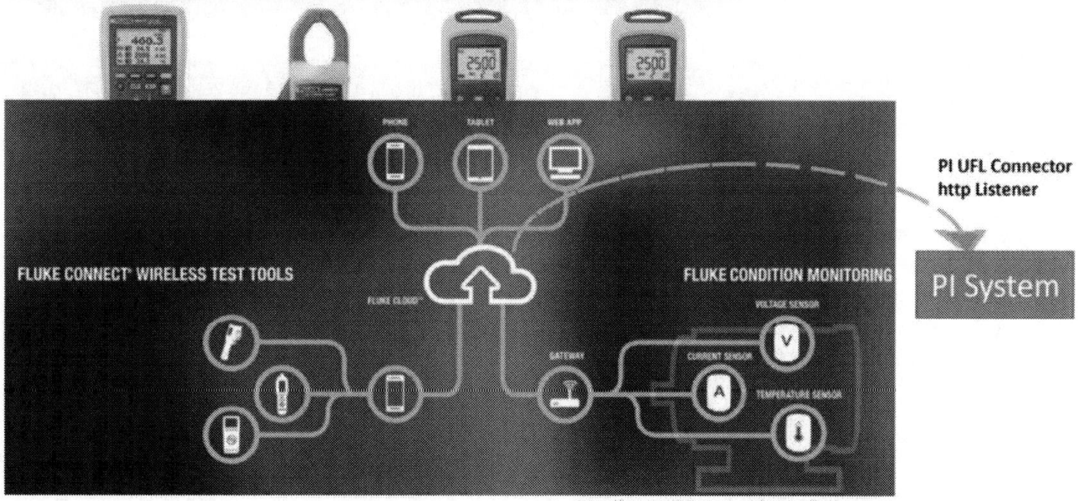

https://www.myflukestore.com/content/fluke_connect

OSIsoft. ©2019 OSIsoft, LLC

Portland Gas & Electric - Turbine startup - trend overlays

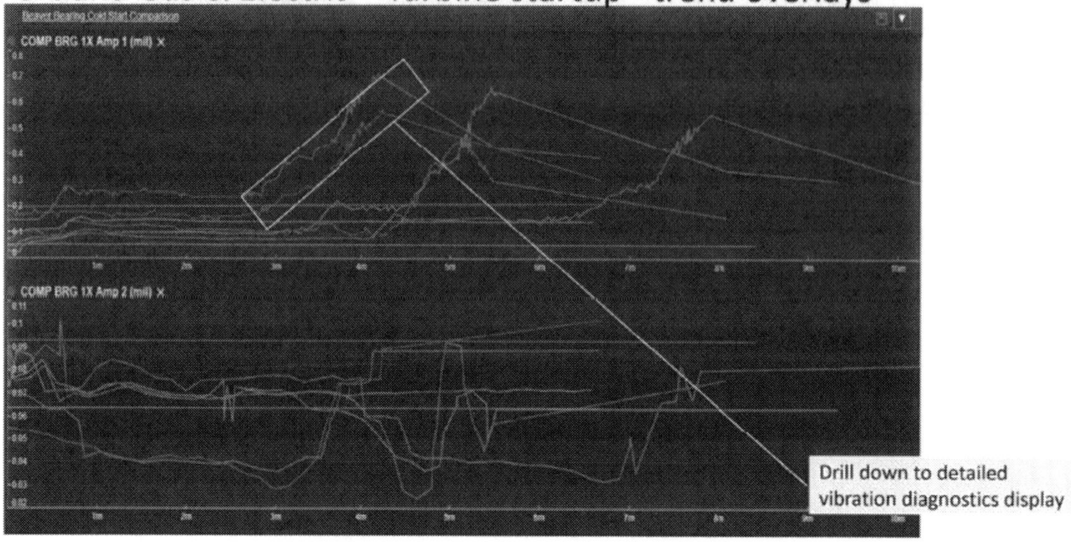

https://www.osisoft.com/Presentations/Experience-using-the-PI-System-for-Full-Featured-Vibration-Monitoring-at-Portland-General-Electric/

MFPT 2019 Philadephia, PA May 2019

OSIsoft. ©2019 OSIsoft, LLC

Vibration waveform and spectrum displays

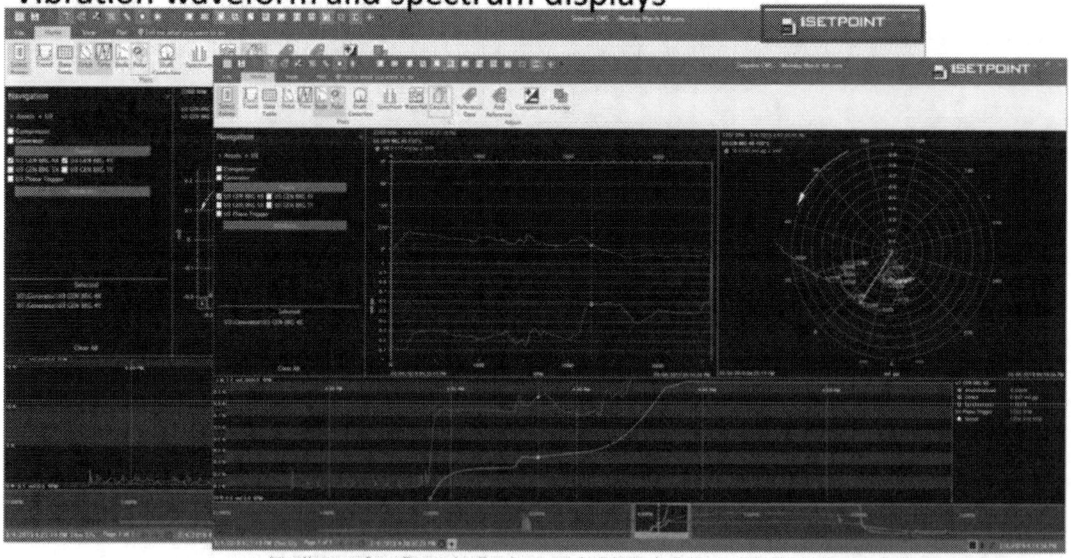

https://www.osisoft.com/Presentations/Experience-using-the-PI-System-for-Full-Featured-Vibration-Monitoring-at-Portland-General-Electric/

MFPT 2019 Philadephia, PA May 2019

OSIsoft. ©2019 OSIsoft, LLC

Thermography - Infrared

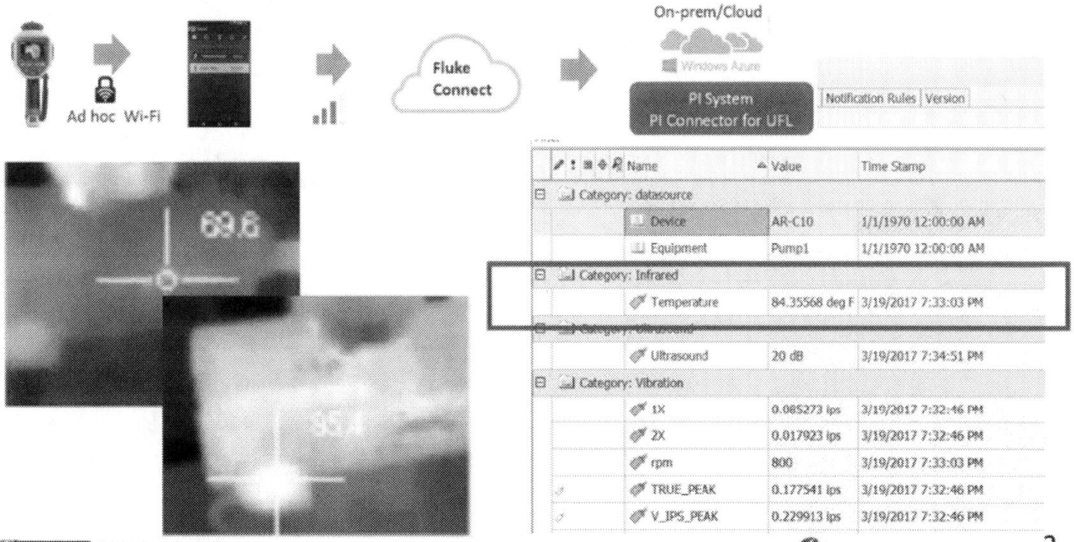

OSIsoft. ©2019 OSIsoft, LLC

Ultrasound (UE devices)

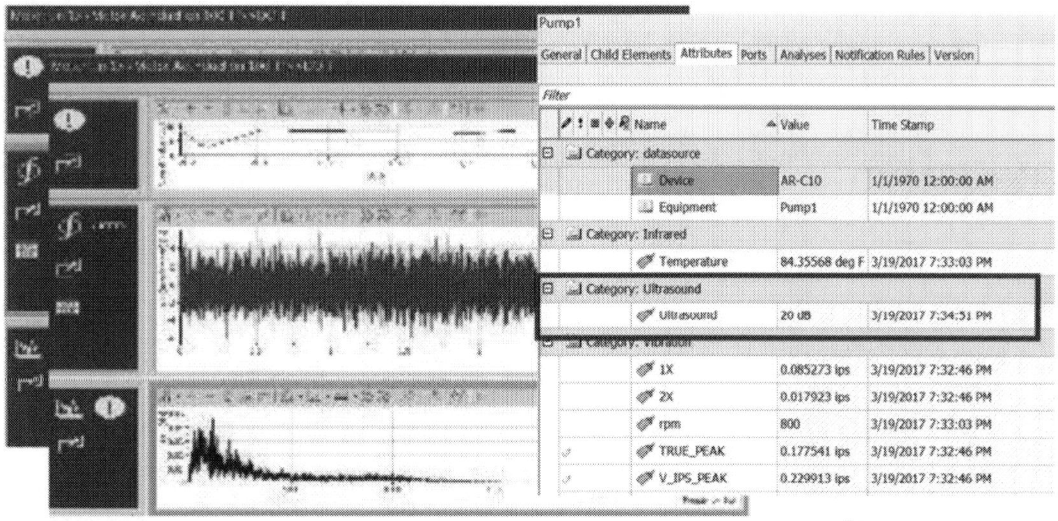

OSIsoft. ©2019 OSIsoft, LLC

Oil analysis results

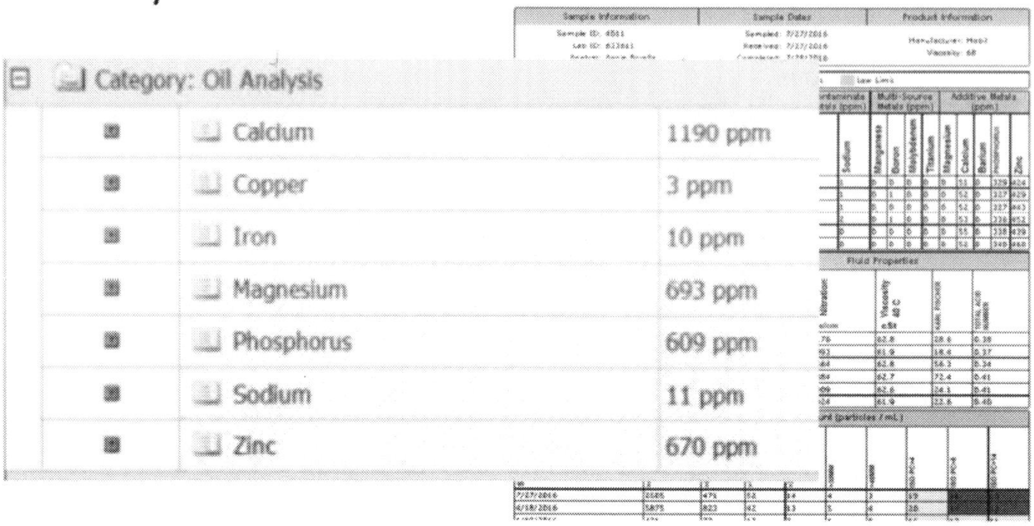

OSIsoft. ©2019 OSIsoft, LLC

Electric Motor Current Analysis

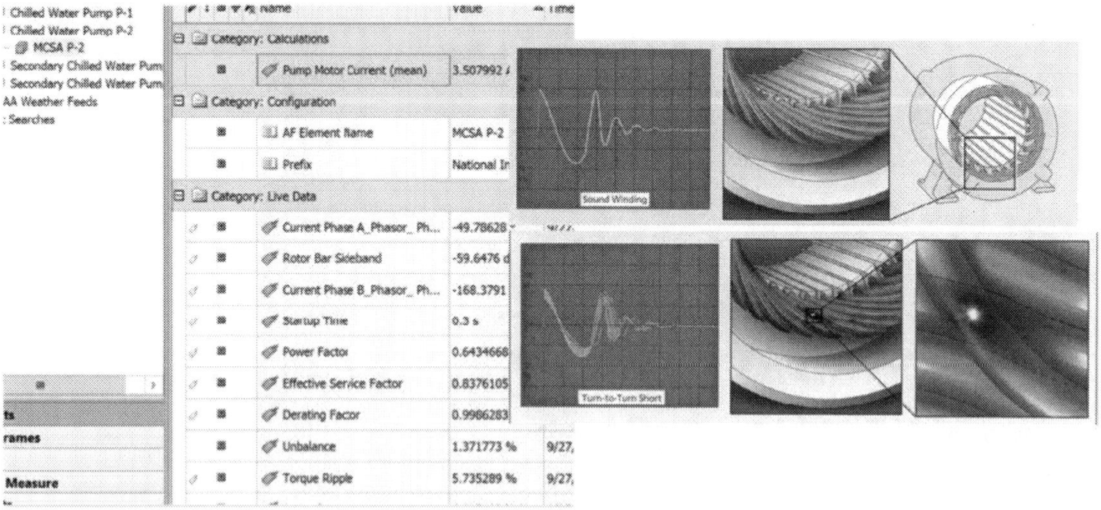

OSIsoft. ©2019 OSIsoft, LLC

Process data and Machine data working together

- Combined process and MCI data in a single data model for analytics and display
- Condition assessments can use both process and MCI data
- MCI alerts use dynamic thresholds (instead of static thresholds) based on process conditions
- Confirm MCI analysis with process data e.g.
  - ➢ Non-sync high frequency peaks indicating pump cavitation confirmed by low suction pressure
  - ➢ Increase in overall vibration on a wind turbine indicating increase in load confirmed by pitch angle
  - ➢ Increase in high frequency vibration indicating lubrication issue on a dryer can confirmed by high lubricant temperature (lowered viscosity)
  - ➢ Increase in barring frequency indicating combination of moisture, basis weight, and machine speed is outside of optimal
  - ➢ Increased overall vibration on a fan indicating increased load confirmed by the damper position
  - ➢ ...
  - ➢ ...

MCI = Machine Condition Indicator

OSIsoft. ©2019 OSIsoft, LLC

Pump Real-time Condition Monitoring

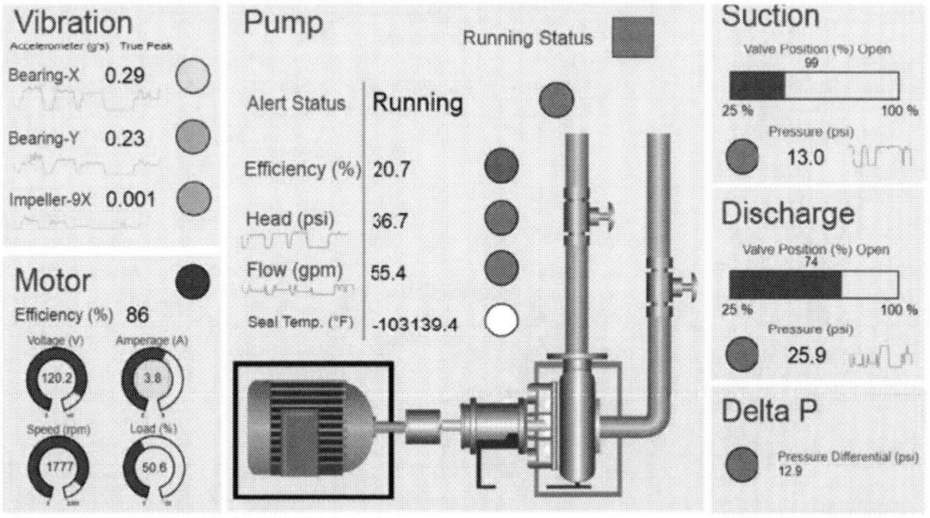

MFPT 2019 Philadelphia, PA May 2019

OSIsoft. ©2019 OSIsoft, LLC

PSE&G – Consolidate & Correlate Data

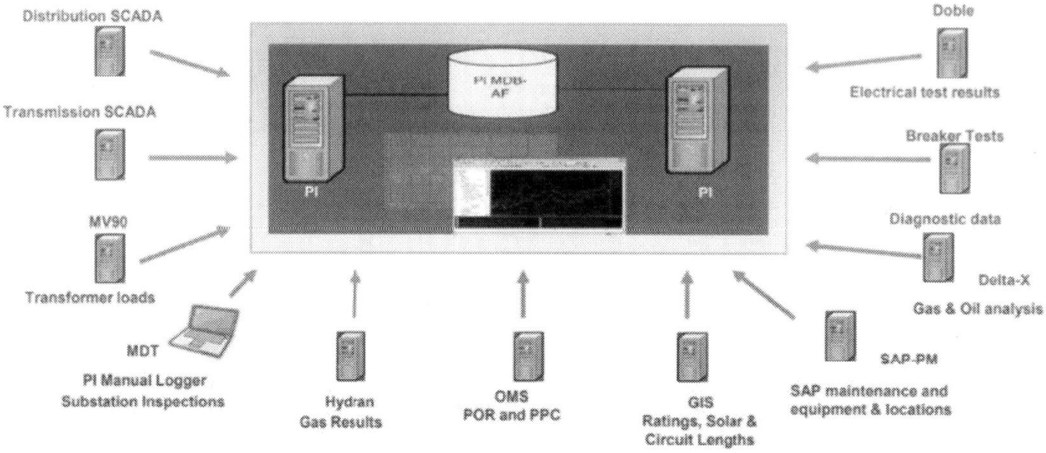

https://www.osisoft.com/Presentations/Condition-Based-Maintenance/

Asset Watchlist with health score

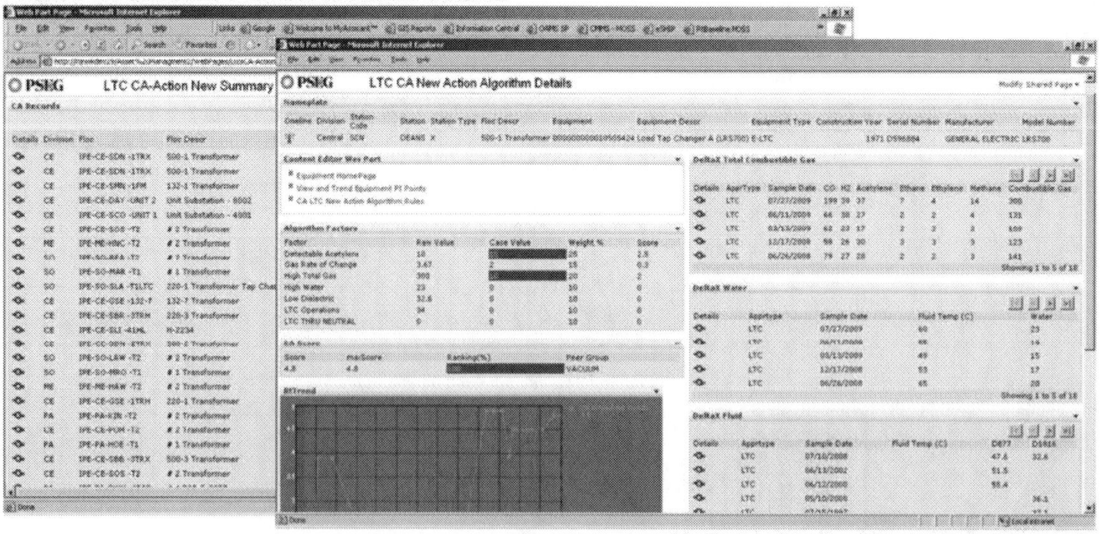

https://www.osisoft.com/Presentations/Condition-Based-Maintenance/

Analytics - simple predictive - air heater fouling

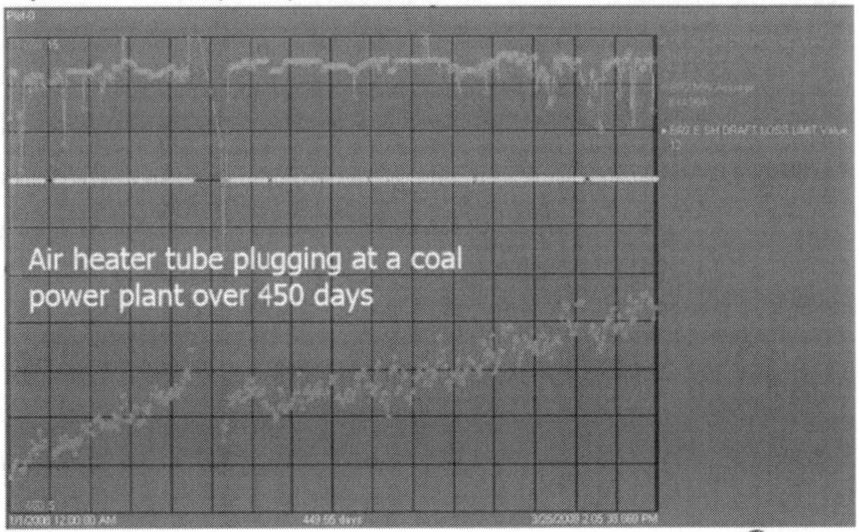

OSIsoft. ©2019 OSIsoft, LLC

Analytics - advanced predictive - fault detection

VISTRA ENERGY

ECG Predict-IT APR - Advanced Pattern Recognition

Early identification of high generator winding temperature

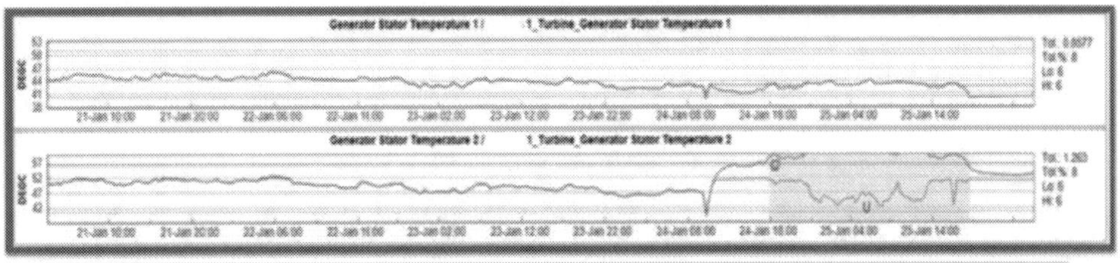

Action Taken	**Plant notification and subsequent investigation**
Corrective Action	**Hydrogen Cooler leak repaired**
Consequence Avoided	**Internal inspection and repair**

https://www.osisoft.com/Presentations/Vistra-Energy--Cloud-Transformation-in-a-Nationwide-Power-Producer/

OSIsoft. ©2019 OSIsoft, LLC

Takeaways and Q&A

Failure Prevention & Performance Monitoring

Industrial sensor & IIoT data and meta-data management

A single data/meta-data store

- SCADA/process data
- Machine condition data
- Collect, Contextualize, Analyze, Visualize, Share, Act

IIoT = Industrial Internet of Things

OSIsoft. ©2019 OSIsoft, LLC

Resources

Portland General Electric - power generation Experience-using-the-PI-System-for-Full-Featured-Vibration-Monitoring
Ameren - nuclear power generation A-new-paradigm-in-utility-condition-based-maintenance
Vistra Energy (Luminant) - power Cloud-Transformation-in-a-Nationwide-Power-Producer
Barrick Gold - metals & mining Mine-Haul-Truck-Health-Monitoring-System
EQT - natural gas pipeline Reliability-Centered-Maintenance-Program
King County - waste water treatment PI-as-a-Maintenance-Tool
PSE&G - electricity distribution - transformers Condition-Based-Maintenance-Case-Study
PSE&G - electricity distribution - circuits Extending-Condition-Based-Maintenance-Program-to-Underground-Networks/
Petronas - oil & gas PROTEAN---Rotating-Equipment-Analytics-in-PI-AF
Silicon Valley Power - power generation Optimizing-Predictive-Maintenance-by-Integrating-Vibration-Data-and-Process-Data
SNCF - railways Using-Operational-Data-for-the-Future-Maintenance-of-the-French-Rail-Infrastructure
Alyeska Pipeline - oil & gas Achieving-Reliability-Centered-Maintenance-and-Diagnostics-with-the-PI-System
ADNOC - oil & gas Improving-Equipment-Reliability-and-Availability-through-Real-time-Data
Las Vegas Valley Water District - water treatment Evaluating-Efficiency-and-Predicting-Maintenance-with-Existing-Data-Using-PI-AF-and-Asset-Analytics
Chevron – oil exploration Condition-Based-Maintenance-and-Smart-Monitoring-in-Frade-FPSO-with-the-PI-System
Vattenfall – hydro power Deploying-a-Condition-Based-Maintenance-Strategy-in-the-Hydro-Power-Business
PowerStream - electricity distribution Equipment-Failure-is-Not-an-Option
GenOn Energy - power Proactive-Maintenance-Data-Gateway-Built-on-the-OSIsoft-PI-System-Infrastructure
Uniper - power How-OSIsoft-PI-supports-Unipers-Maintenance-Strategy-Planning
ESB - power Using-the-PI-System-in-Power-Generation-Operation-and-Maintenance-to-Optimise-Commercial---Environment
MOL – oil refinery Supporting-Operational-Availability-by-Connecting-the-PI-System-and-SAP-PM
OSIsoft Introduction-to-Asset-Monitoring-and-Condition-based-Maintenance-with-the-PI-System
OSIsoft Getting-started-with-iiot-sensors (may need to register – free)

OSIsoft. ©2019 OSIsoft, LLC

AMETEK® | Spectro Scientific
Confidence in knowing

Advances in At-Site Detection of Debris in Lubricants

Patrick Henning, Thomas Barraclough, Robert Yurko, Eric Olson, Peter Loges, Matthew Fratkin, Randi Price, Daniel Walsh, and Yuegang Zhao

05/15/19

MFPT 2019

AMETEK® | Spectro Scientific
Confidence in knowing

Machinery Health ->What Debris Should We Measure and How?

- Large (>20 microns) and small particles
- Steel alloys – ferrous is just about everywhere – both oxidized and shiny
- Dirt/Sand/Fibers – silicon, etc.
- Non-ferrous metals indicative of failure (eg. pump housing)

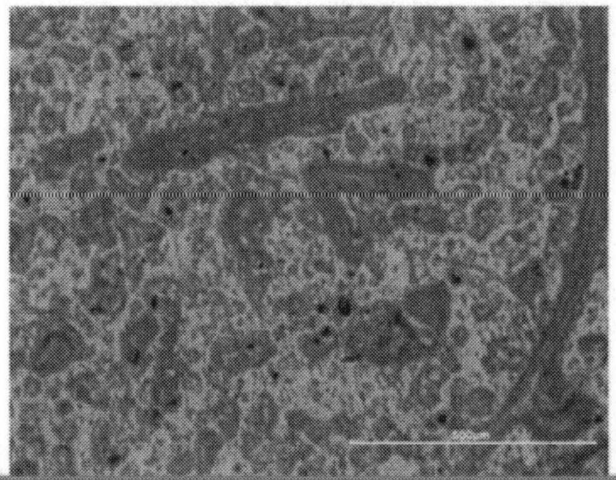

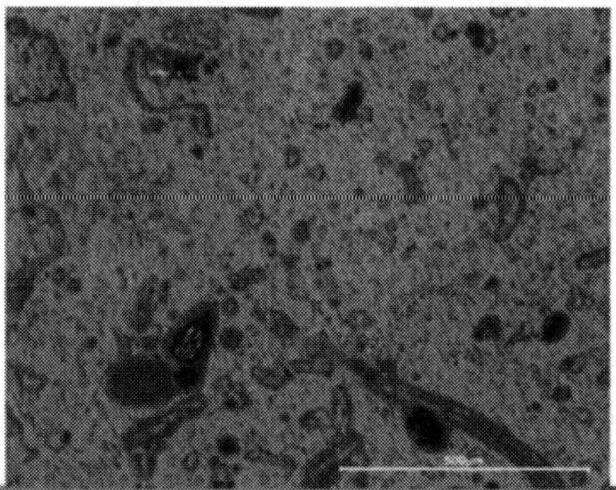

Machinery Health ->What Debris Should We Measure and How?

- For every chunk of debris seen, I want to answer those questions
- The oil may be sooty, dark or very clean
- It may be grease, or have widely varying viscosity
- No sample preparation for all oils, fast answers (minutes)

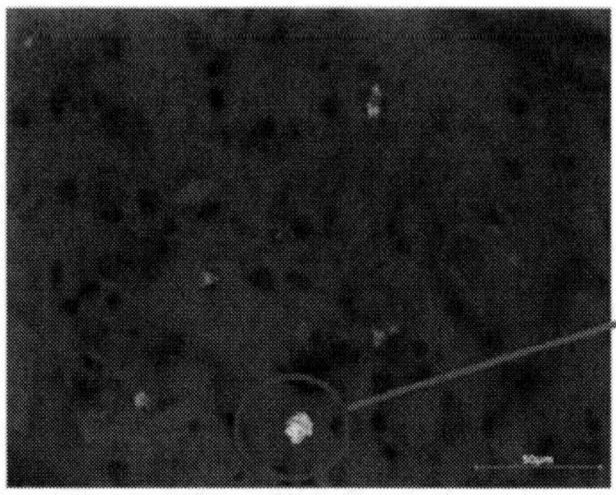

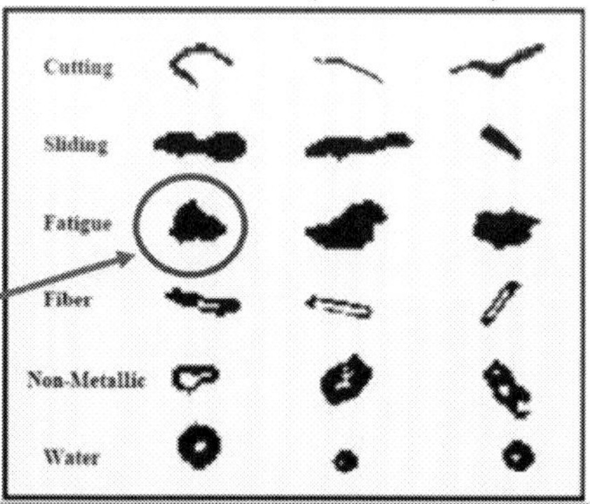

Traditional Techniques To Do These Tasks

- Ferrography
- Elemental Analysis
- Acid Digestion+Elemental Analysis
- Particle Counters
- Ferrous meters
- Microscopy

- All can only do a few things

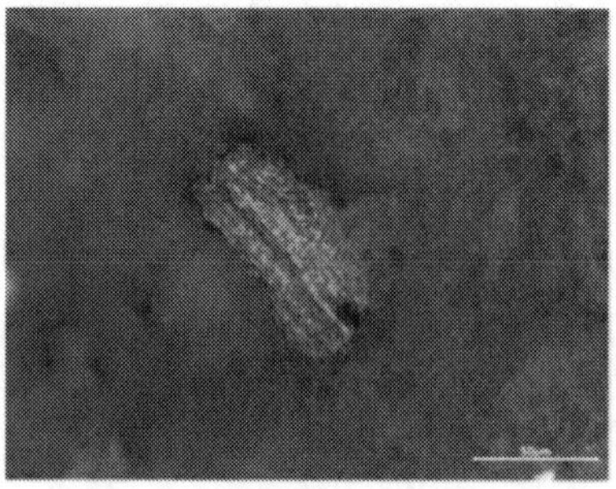

AMETEK® | Spectro Scientific
Confidence in knowing

What Debris Should We Measure and How?

- LaserNet Fines Technology – Particle Analysis Using Imaging
  - Particle Count Statistics (> 4 microns)
  - Image Map and Individual Particle Classification (20 microns and above)

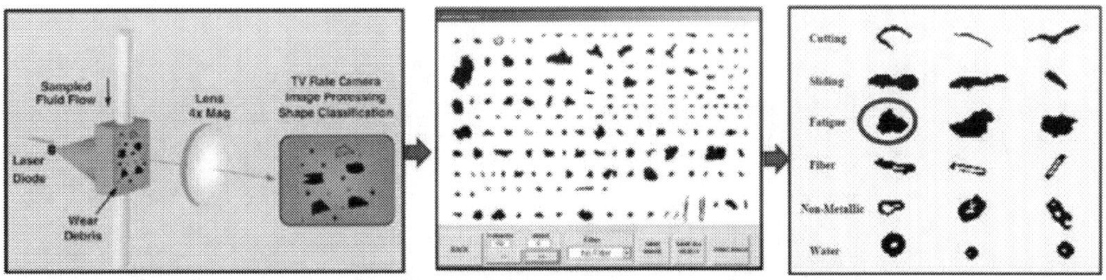

AMETEK® | Spectro Scientific
Confidence in knowing

Imaging and Sensing Particulate – LaserNet 230

- Small Coils: A – single magnetic particles > 25 microns
- Large Coils: B – total ferrous
- Flow Cell: C - imaging

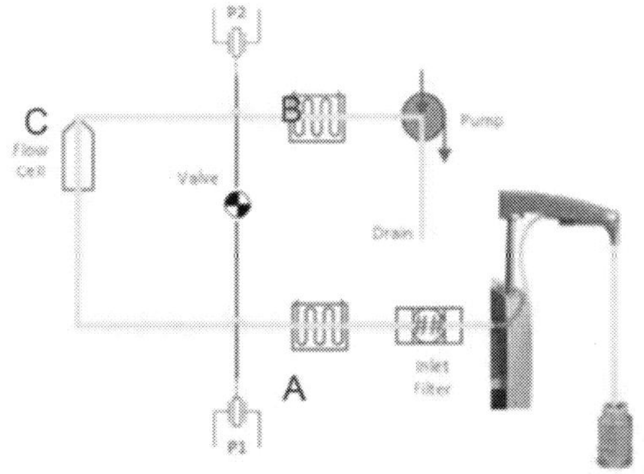

AMETEK® | Spectro Scientific
Confidence in knowing

What Debris Should We Measure and How?

- Large (>20 microns) and small particles ☑
- Steel alloys – ferrous is just about everywhere – both oxidized and shiny ☑
- Dirt/Sand/Fibers – silicon, etc. ☑
- Non-ferrous metals indicating failure, eg. pump housing ☑

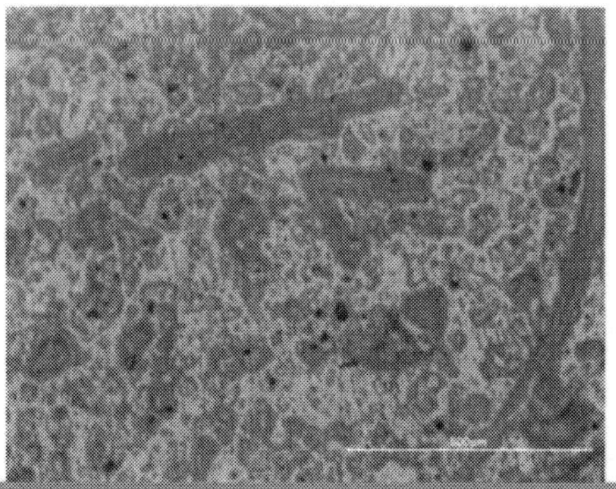

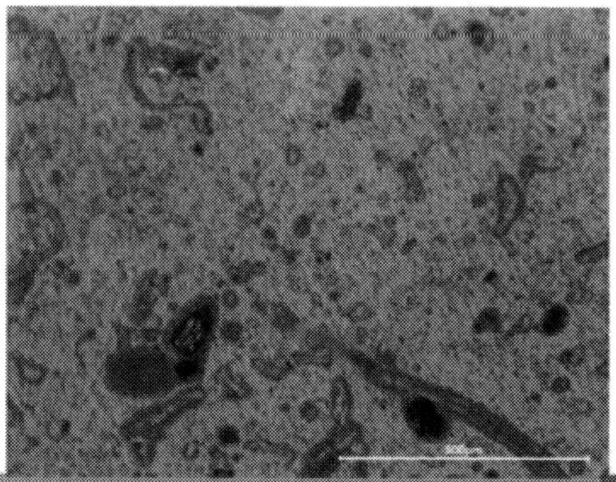

MFPT 2019 7

AMETEK® | Spectro Scientific
Confidence in knowing

LaserNet 230 Provides Quantitative Ferrographic Outputs

$$\%\text{large particles} = \left(\frac{\text{Concentration (ppm)}}{\text{Total Ferrous Concentration (ppm)}} \right) * 100$$

$$\text{Wear Severity Index} = \text{Concentration (ppm)} * \text{Total Ferrous Concentration (ppm)}$$

MFPT 2019 8

AMETEK® | Spectro Scientific
Confidence in knowing

Simple process-answers the critical questions of oil analysis

Spectroil Elemental Analyzer
- 24 Elements in PPM
- 31 with extended metals option

LaserNet 230 particle counter, ferrous monitor and wear classifier
- ISO and other Cleanliness Codes
- Automatic wear particle shape classification and count

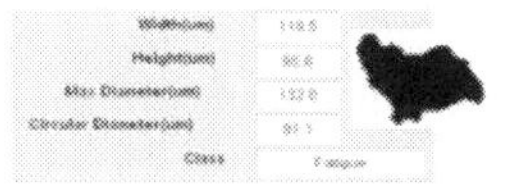

- Total Ferrous Particle Content (PPM)
- Large Ferrous Particle Count and size distribution

MiniVisc Kinematic Viscometer
- Viscosity @ 40C
- Visc 100 calculation from VI

Fluidscan® Infrared (IR) Oil Analyzer
- Oxidation
- Total Acid Number
- Water
- Additional parameters for engine oils

- **Sample preparation-test-clean in < 15 minutes**
- **ASTM methods and NIST traceable standards for all instrumentation**

AMETEK® | Spectro Scientific
Confidence in knowing

Why Onsite Testing?

- Simple, standard Process improves speed and quality of data
- Fewer people involved simplifies process
- Faster information improves the workflow
- Single Information management system eliminates human errors
- Freshly collected sample is most representative of machinery
- Immediate retest to confirm results
- Alarms, diagnostics, recommended actions- Intelligence individualized to your assets
- Better Economics – branching out is easy (incoming inspection,…)
- Crosscheck (with other CBM technologies)

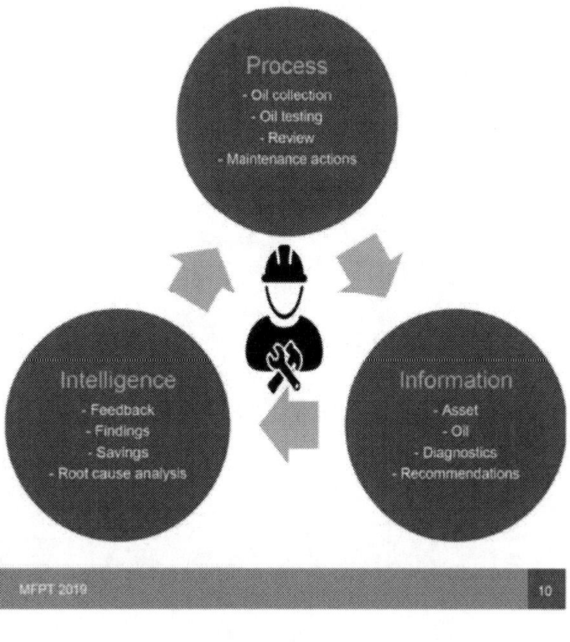

Putting it together…a simple and holistic approach to Oil Analysis – TruVu 360

AMETEK® | Spectro Scientific — Confidence in knowing

Sample & test → Review / edit software generated diagnostic and recommended actions → Release & distribute report →

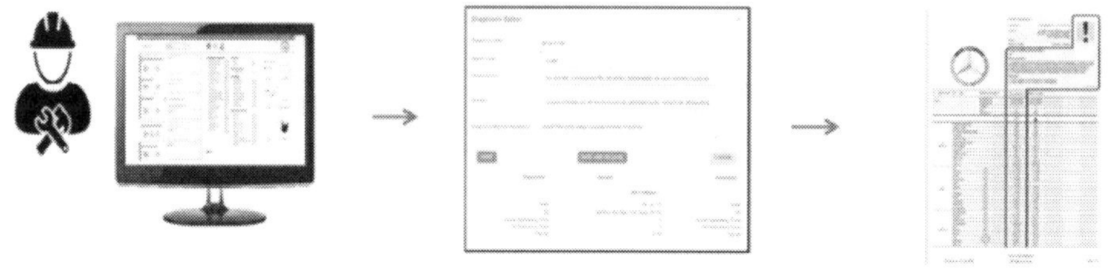

Work Order request and execution in company CMMS/EAM → After work order executed, log maintenance findings and cost savings

Adaptive Rules Engine to update asset alarms, diagnostics and maintenance actions

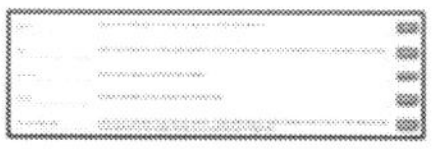

Track asset, plant and corporate metrics for oil program

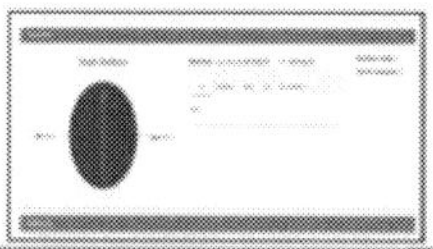

AMETEK® | Spectro Scientific — Confidence in knowing

Case Study - Benefit of On-Site Fluid Analysis

- Reduce fluid cost
- Increase machine uptime and availability
- Extend component lifetime
- Reduce cost of failure and repair

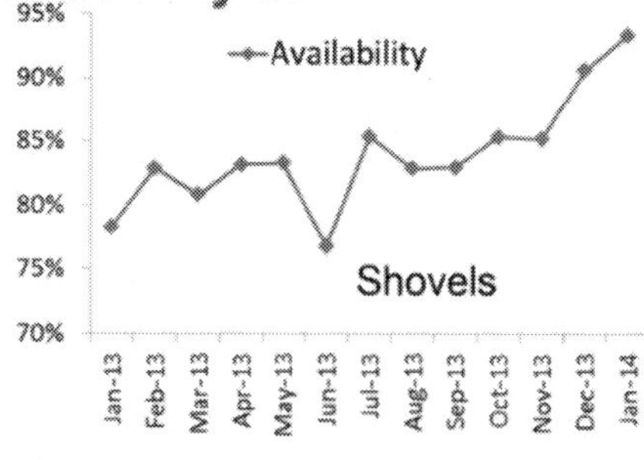

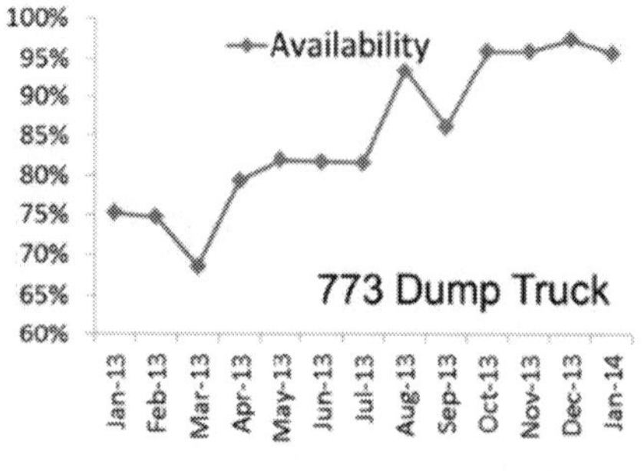

AMETEK | Spectro Scientific
Confidence in knowing

Case Studies of Onsite Oil Analysis

USCG
https://www.spectrosci.com/knowledge-center/gallery/customer-success-stories/case-studies/lubes-n-greases-magazine-cutters-cut-costs-with-oil-analysis/

Simmons Feed
https://www.machinerylubrication.com/Read/31304/simmons-lubrication-program

NASA /Jacobs
https://www.machinerylubrication.com/Read/31385/jacobs-oil-analysis

Nissan North America
https://www.machinerylubrication.com/Read/30871/lubrication-excellence-journey
https://www.machinerylubrication.com/Read/29457/nissan-lubrication-program

Eli Lilly
http://www.efficientplantmag.com/2017/04/reliability-changes-lives/.

MFPT 2019 13

AMETEK | Spectro Scientific
Confidence in knowing

Importance of Large Particles...PPM and Count

Wear Particle Weight

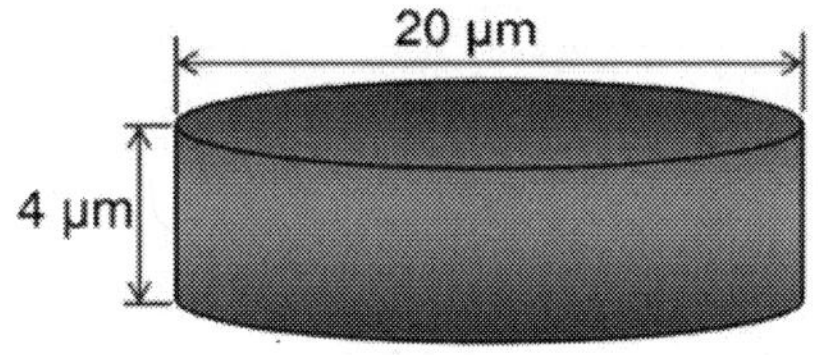

$V = \pi d^2 h/4$

$V = \pi(20 \times 10^{-4})^2 (4 \times 10^{-4})/4$

$V = 1.26 \times 10^{-9}$ cm^3

Weight = Density x Volume

Weight = 8 g/cm^3 x 1.26×10^{-9} cm^3

Weight = 10×10^{-9} g = 10 ng

PPM (mass) vs. Particle concentration (quantity) on the FPQ filter

Just 30 of these could indicate bearing failure!

Based on weight, 1 ml of oil would need 100 particles similar to the platelet above to increase the concentration by 1 ppm

100 particles as above = 100 x 10 ng
= 1,000 ng
= 1 µg

MFPT 2019 14

AMETEK® | Spectro Scientific Confidence in knowing

Importance of Large Particles...PPM of what?

High hardness wear particles induce further wear

-The more you can distinguish between harmful and benign particles, the better

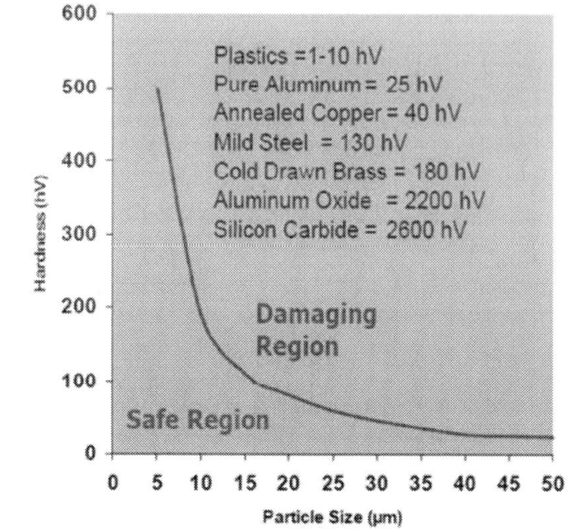

AMETEK® | Spectro Scientific Confidence in knowing

Back to the Debris...Grease...Extreme Oils...FPQ/XRF

- SEM/EDX has been known for a long time
- Pore-Blockage particle counting is also well-established
- Idea: Combine aspects of these into a battery-operated, portable system

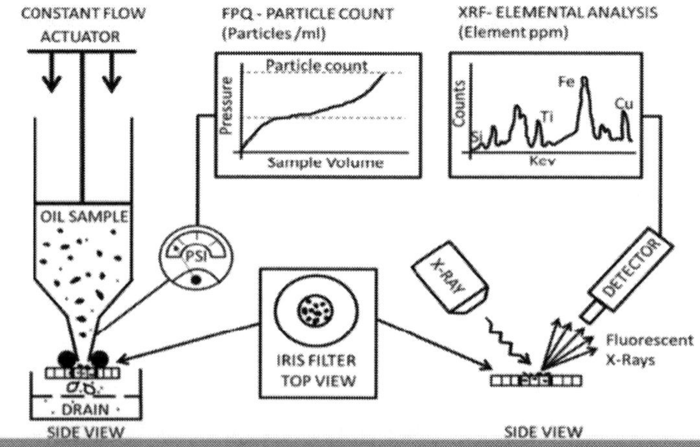

AMETEK® | Spectro Scientific
Confidence in knowing

Back to the Debris...Grease...Extreme Oils...FPQ/XRF

- FPQ/XRF is utilized in the FieldLab 58M
- Battery-operated
- No liquids or chemicals
- No sample prep
- Greases, sooty oil, dark oil
- Answers anywhere, immediate

AMETEK® | Spectro Scientific
Confidence in knowing

Back to the Debris...Grease...Extreme Oils...FPQ/XRF

- Sees the large particles that typical elemental analysis can't see
- Demonstrated elemental correlation via acid digestion

	Before Acid Digestion -ICP (ppm)						After Acid Digestion - ICP (ppm)					
Sample	A	B	C	D	E	F	A	B	C	D	E	F
Ag	0	0	0	0	0	0	0	0	0	0	0	0
Al	0	0	0	10	10	21	0	0	0	0	13	28
Cr	6	0	0	0	6	7	6	0	0	0	6	8
Cu	0	0	0	0	11	11	0	0	0	0	10	10
Fe	10	7	0	0	33	67	11	10	0	0	35	86
Mo	0	0	0	0	0	0	0	0	0	0	0	0
Ni	0	0	0	0	0	0	0	0	0	0	0	0
Pb	0	0	0	0	0	0	0	0	0	0	0	0
Sn	0	0	0	0	0	0	0	0	0	0	0	0
Ti	0	0	0	0	0	0	0	0	0	0	0	0
V	0	0	0	0	0	0	0	0	0	0	0	0
Total ppm	16	7	0	10	60	106	17	10	0	0	64	132

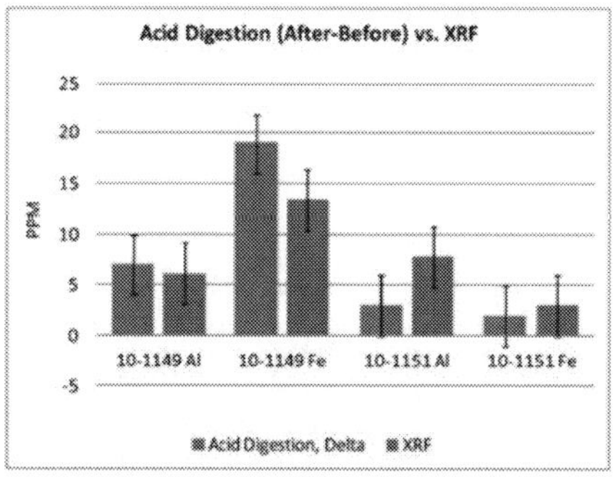

AMETEK® | Spectro Scientific Confidence in knowing

Advances in At-Site Detection of Debris in Lubricants - Conclusions

- At-Site, immediate detection and quantification of debris via, morphology, magnetism, and elemental content has many documented benefits

- New technologies exist which enable this for the first time

- Thank you! (MFPT, and everyone)

- Questions?

MFPT 2019 19

MFPT 2019 Where Theory Meets Practice
MAY 14-16, 2019 | PHILADELPHIA, PA

Bearing Fault Diagnosis with Improved Variational Mode Decomposition and Ensemble Kurtosis

Dabin Jie, Yimin Shao*, Liming Wang and Xiaoxi Ding
State Key Laboratory of Mechanical Transmission, Chongqing University

Reporter: Dabin Jie
MAY 16, 2019

重庆大学
CHONGQING UNIVERSITY

MFPT 2019 Where Theory Meets Practice
MAY 14-16, 2019 | PHILADELPHIA, PA

OUTLINE

1 Introduction

2 Theory Analysis

3 Experimental Descriptions

4 Result Analysis

5 Conclusion

2

MFPT 2019 Where Theory Meets Practice
MAY 14-16, 2019 | PHILADELPHIA, PA

1 Introduction

Bearings are important parts in mechanical equipment. There are many important applications of them in wind turbines, railway equipment, machine tools, engineering machinery and other fields.

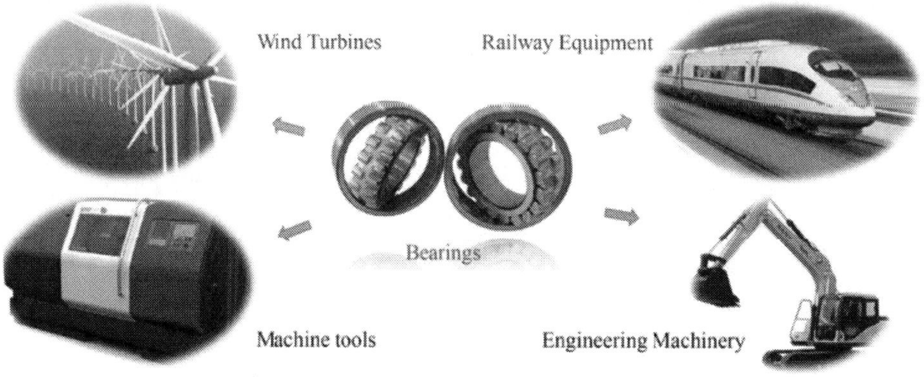

3

MFPT 2019 Where Theory Meets Practice
MAY 14-16, 2019 | PHILADELPHIA, PA

1 Introduction

However, bearing failure can cause significant economic losses. Even many people lost their precious lives.

Catastrophe I

Bearing fault

Catastrophe II

Bearing fault diagnosis is critical for an safety and reliable operating process in the application of rotating machinery.

4

1 Introduction——Challenges / Research gaps

- ✓ Vibration signal analysis has been proved to be a key technology for fault diagnosis. Many signal processing methods have been applied to the field of bearing fault diagnosis.
- ✓ Two crucial issues:
  - ◆ Heavy background noise
  - ◆ Complex Modulation effect

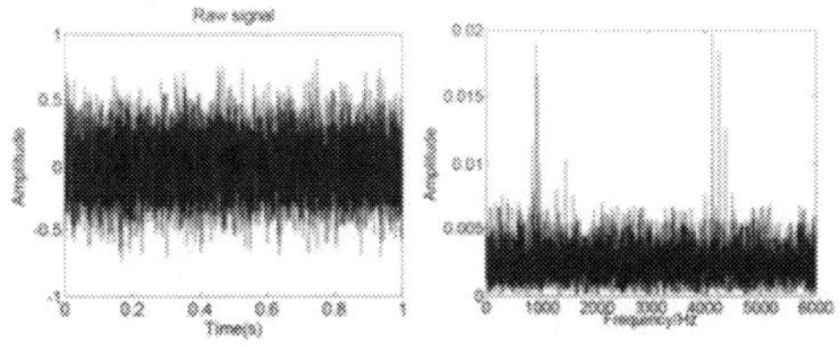

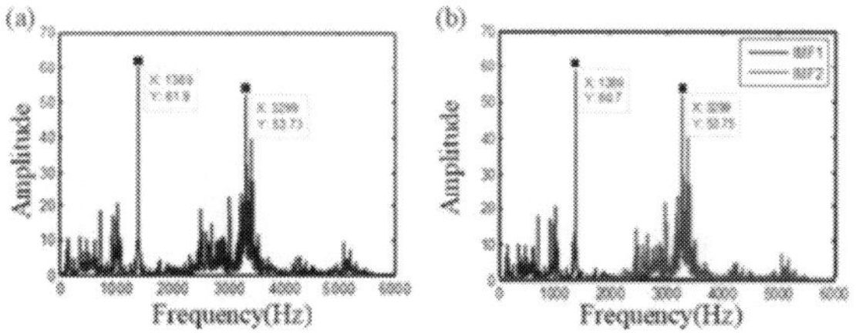

5

MFPT 2019 Where Theory Meets Practice
MAY 14-16, 2019 | PHILADELPHIA, PA

2 Theory analysis

2.1 Denoising based on a weighted EMD method

Firstly, the signal is decomposed into N different intrinsic functions (*IMFs*),

$$sig_{matrix} = f_{emd}(x) = \left[IMF_1, IMF_2, \ldots, \ IMF_N\right]^T$$

Then, Pearson correlation coefficient method is used to identify noise-free and noise-dominant IMFs,

$$\rho(x, c_n) = \frac{\sum_{k=1}^{K}[x(k) - \bar{x}]\,[IMF_N(k) - \overline{IMF_N}]}{\sqrt{\sum_{k=1}^{K}[x(k) - \bar{x}]^2 \sum_{k=1}^{K}[IMF_N(k) - \overline{IMF_N}]^2}}$$

Therefore, the correlation coefficient vector between all intrinsic functions and the raw signals can be obtained,

$$\rho_{matrix} = [\rho_1, \rho_2, \ldots \rho_N]^T$$

6

2 Theory analysis

2.1 Denoising based on a weighted EMD method

In order to suppress noise, intrinsic mode functions are divided into noise-dominant and noise-free IMFs based on threshold. Then, weighted signal reconstruction is performed to suppress noise, as follows,

$$x_{rec} = \sum_{i=1}^{M} w_i \times IMF_i$$

The expression of *wi* is as follows,

$$w_i = \begin{cases} \rho_i / max(\rho_{matrix}), & \rho_i \geq \rho_m \\ 0, & \rho_i < \rho_m \end{cases}$$

threshold

7

2 Theory analysis

2.1 Denoising based on a weighted EMD method

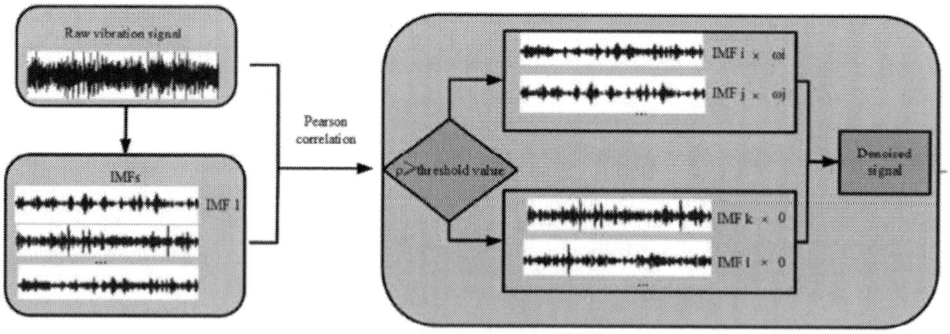

The flow diagram of denoising process

8

2 Theory analysis

2.2 IVMD based on SDA

Recently, a new self-adaptively method, VMD, is proposed. The goal of VMD is to decompose a real valued input signal into a discrete number of sub-signals (modes). That have specific sparsity properties while reproducing the input. Here, the sparsity prior of each mode is chosen to be its bandwidth in spectral domain. In order to assess the bandwidth of a mode, the resulting constrained variational problem is the following,

$$\min_{\{u_k\},\{w_k\}}\left\{\sum_k \left\|\partial_t\left[\left(\delta(t)+\frac{j}{\pi t}\right)*u_k(t)\right]e^{-jw_kt}\right\|_2^2\right\}\quad s.t.\quad \sum_k u_k = f$$

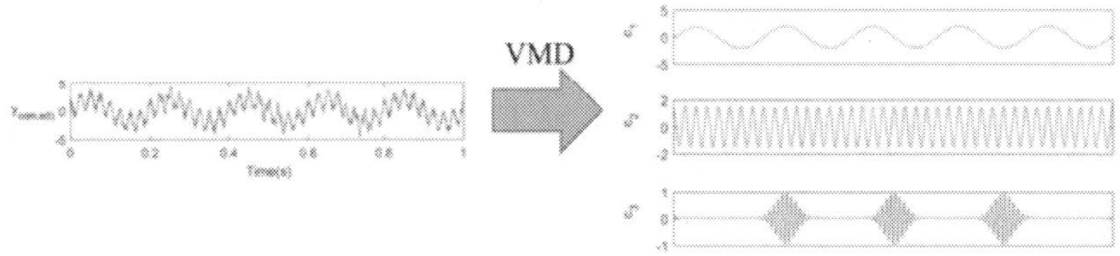

9

2 Theory analysis

2.2 Adaptive VMD based on SDA for an optimal level *K*

Before decomposing the signal with VMD method, we must choose the appropriate mode number. Because the choice of K has an important influence on the results of VMD. Usually, we choose the mode number empirically. So a mode number determination method is proposed by using signal difference average (SDA) for the VMD method. SDA can reflect the similarity between IMF and original signal. And the smaller the SDA, the higher the similarity.

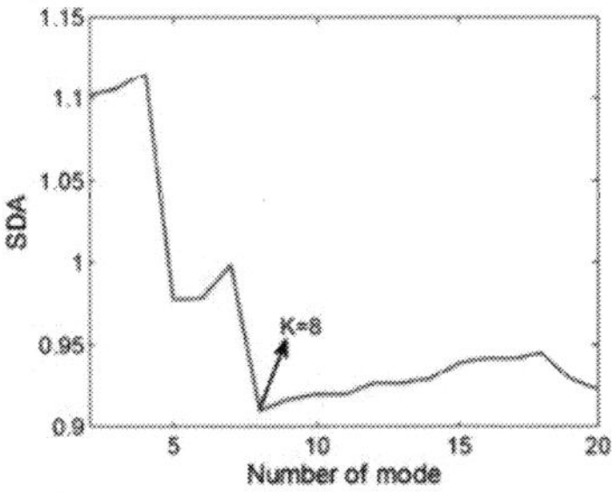

$$\text{SDA} = \frac{1}{N}\sum_{i=1}^{n}\left(y_{IMF_s} - y_s\right)$$

10

2 Theory analysis

2.2 Adaptive VMD based on SDA for an optimal level *K*

Before decomposing the original signal with VMD method, we must choose the appropriate mode number. Because the choice of K has an important influence on the results of VMD. Usually, we choose the mode number empirically. So a mode number determination method is proposed by using signal difference average (SDA) for the VMD method SDA can reflect the similarity between IMF and original signal. And the smaller the SDA, the higher the similarity.

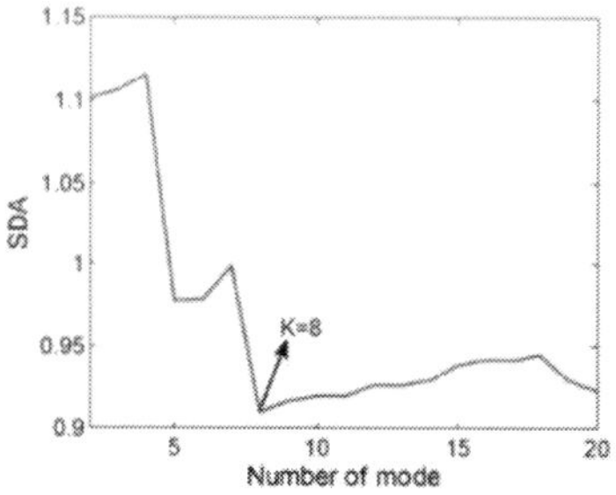

IVMD

Optimal level $K = 8$

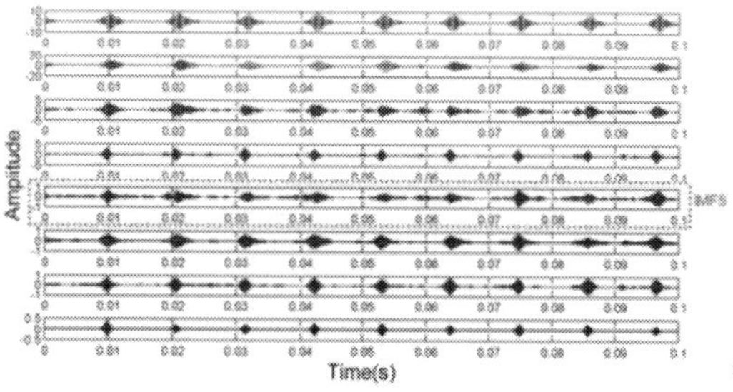

11

MFPT 2019 Where Theory Meets Practice
MAY 14-16, 2019 | PHILADELPHIA, PA

2 Theory analysis

2.3 Optimal component output based on EK

The appropriate evaluation criteria of the fault feature are rather significant for the machinery fault diagnosis. So a index, named ensemble kurtosis (EK), is proposed, which is constructed by the virtue of kurtosis and envelope spectrum kurtosis. It is defined as follows,

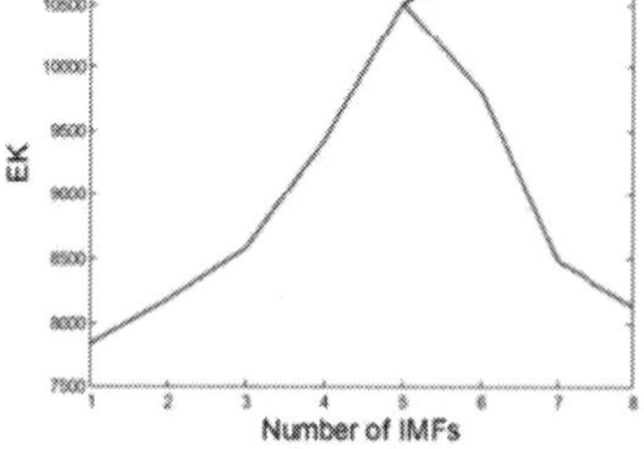

$$EK = ESK \cdot K$$

$$ESK = \frac{\sum_{p=1}^{P} |\overline{SE(p)}|^4}{\left(\sum_{p=1}^{P} |\overline{SE(p)}|^2\right)^2}$$

12

2 Theory analysis

2.3 Optimal component output based on EK

The appropriate evaluation criteria of the fault feature are rather significant for the machinery fault diagnosis. So a index, named ensemble kurtosis (EK), is proposed, which is constructed by the virtue of kurtosis and envelope spectrum kurtosis. It is defined as follows,

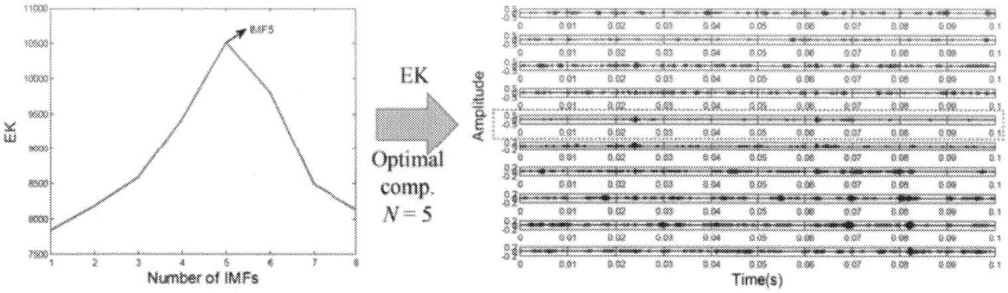

13

MFPT 2019 Where Theory Meets Practice
MAY 14-16, 2019 | PHILADELPHIA, PA

2 Theory analysis

2.3 Optimal component output for frequency diagnosis

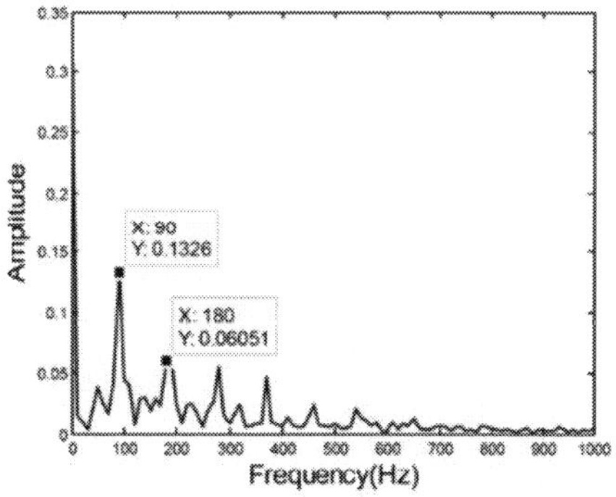

Envelop spectrum of selected component

14

MFPT 2019 Where Theory Meets Practice
MAY 14-16, 2019 | PHILADELPHIA, PA

2 Theory analysis

2.4 Procedure of the proposed IVMD

The proposed method can be divided into three parts, Part 1 is the procedure of denoising based on weighted EMD. Part 2 is the principle of improved VMD. Part 3 is envelop analysis of the selected component from IVMD. And the detailed steps of the proposed method are shown in right figure.

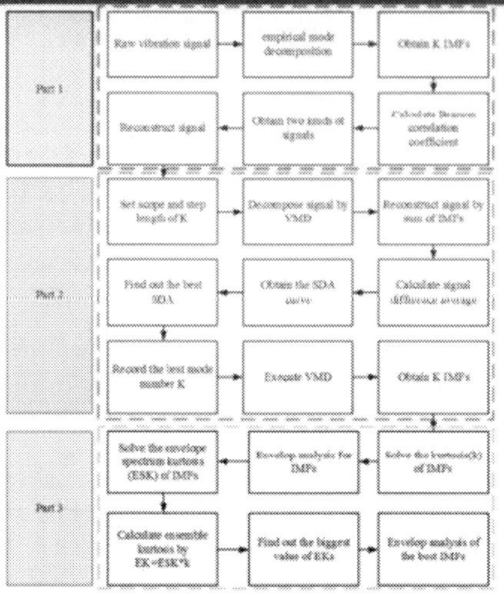

The flow diagram of proposed method

15

MFPT 2019 Where Theory Meets Practice
MAY 14-16, 2019 | PHILADELPHIA, PA

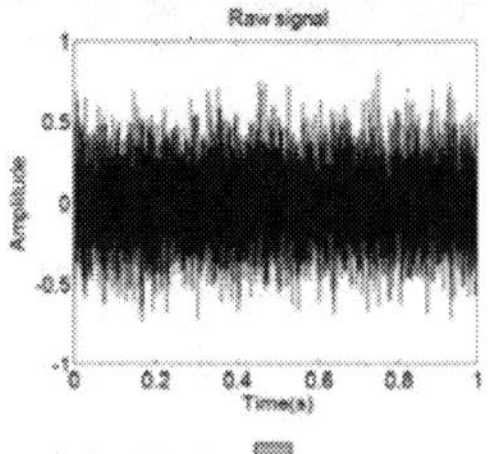

Weight EMD Denoising

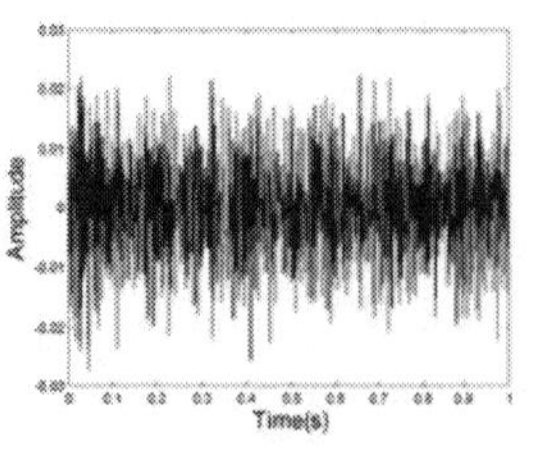

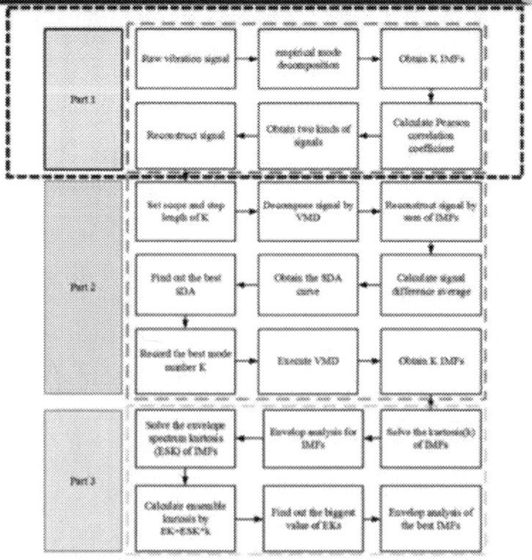

The flow diagram of proposed method

16

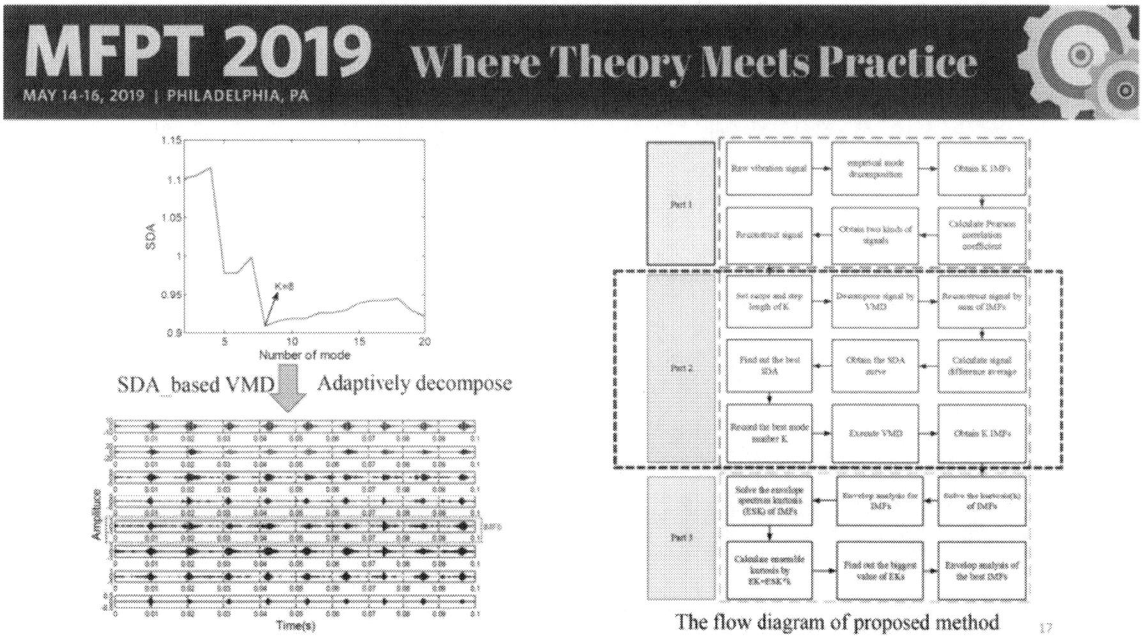
MFPT 2019
Where Theory Meets Practice
MAY 14-16, 2019 | PHILADELPHIA, PA
SDA
Number of mode
K=8
SDA_based VMD
Adaptively decompose
Amplitude
Time(s)
The flow diagram of proposed method
17

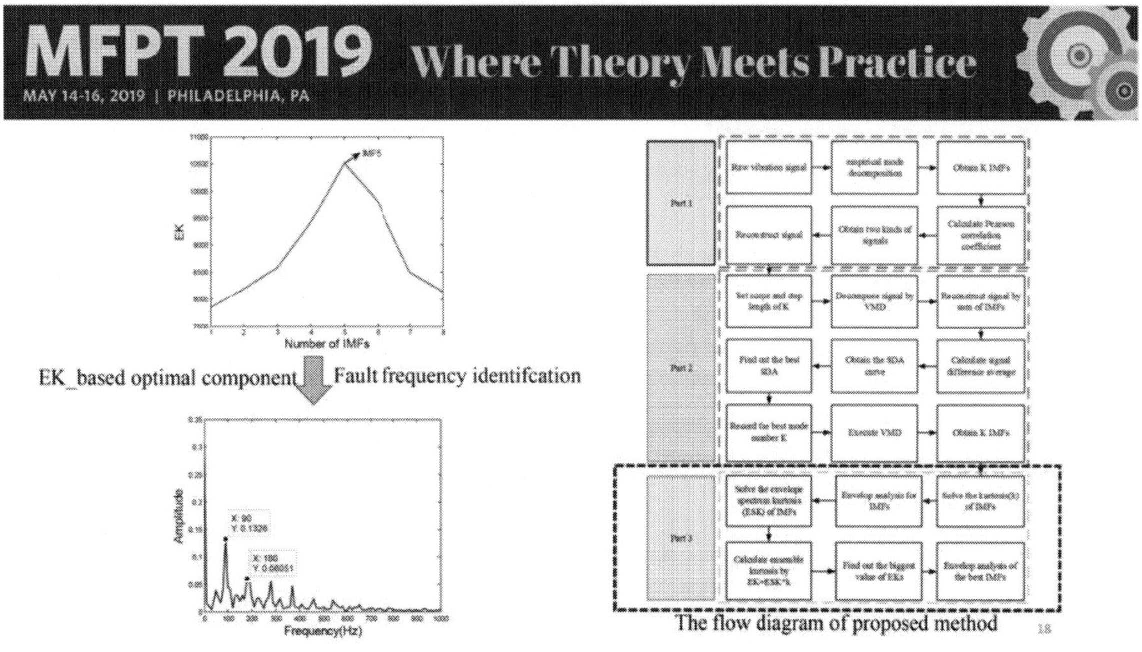
MFPT 2019
Where Theory Meets Practice
MAY 14-16, 2019 | PHILADELPHIA, PA
EK
Number of IMFs
IMF5
EK_based optimal component
Fault frequency identifcation
Amplitude
Frequency(Hz)
X: 90
Y: 0.1326
X: 180
Y: 0.08051
The flow diagram of proposed method
18

3 Experimental descriptions

- ✓ The faulty signal provided from our experimental setup is very clear, and the fault feature can be well defined by all the conventional methods.
- ✓ However, as mentioned in the two issues, once the operating conditions were complex, the signal will be submerged by heavy noise and complex modulated inferences.
- ✓ So we analyzed another data from the Case Western Reserve University Bearing Data Center

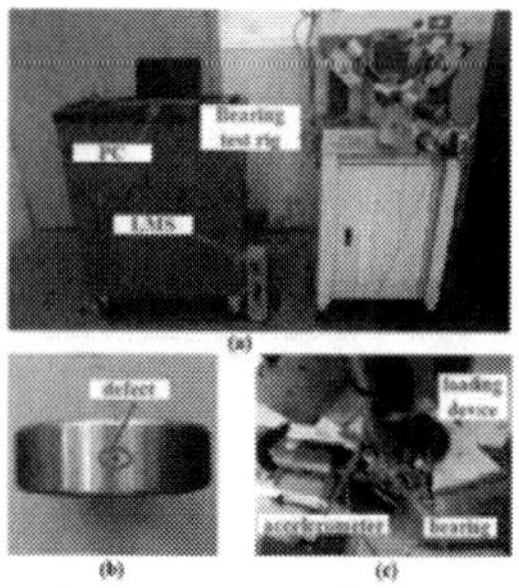

Our experimental setup

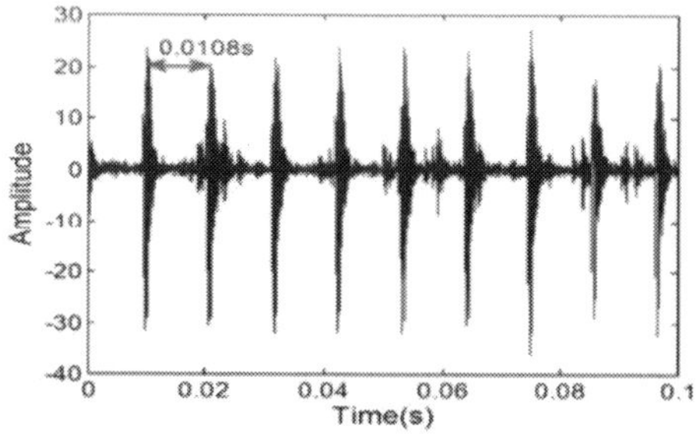

Obvious !

19

MFPT 2019 Where Theory Meets Practice
MAY 14-16, 2019 | PHILADELPHIA, PA

3 Experimental descriptions

- ✓ The faulty signal provided from our experimental setup is very clear, and the fault feature can be well defined by all the conventional methods.
- ✓ However, as mentioned in the two issues, once the operating conditions was complex, the signal will be submerged by heavy noise and complex modulated inferences.
- ✓ So we analyzed another data from the Case Western Reserve University Bearing Data Center Website

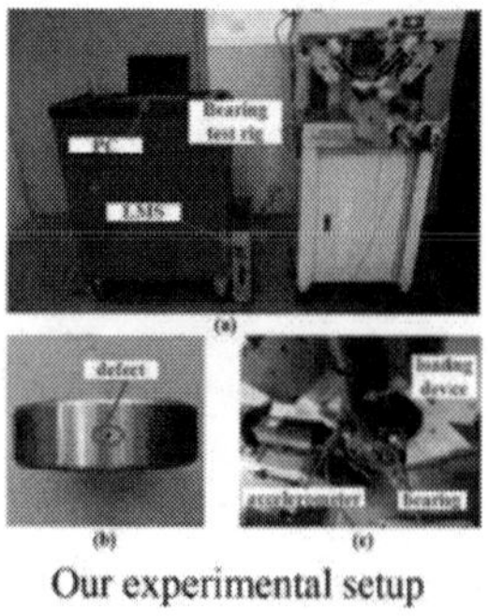

Our experimental setup

Setup from the CWRU

20

MFPT 2019 Where Theory Meets Practice
MAY 14-16, 2019 | PHILADELPHIA, PA

3 Experimental descriptions

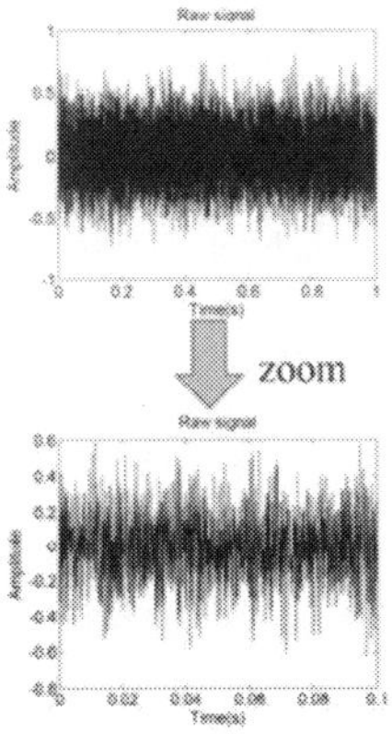

Time domain signal and its zoomed signal

Signals with inner ring faults are analyzed. The fault characteristic frequency of inner ring is 162 Hz. The other parameters of bearings are shown in the table below.

Parameters of the experiment	
Bearing type	6205-2RS JEM SKF, deep groove ball bearing
Length of signal (s)	1
FCF (Hz) of inner ring	162
Motor speed (rpm)	1796
Sampling frequency (Hz)	12K

21

MFPT 2019 Where Theory Meets Practice
MAY 14-16, 2019 | PHILADELPHIA, PA

4 Results

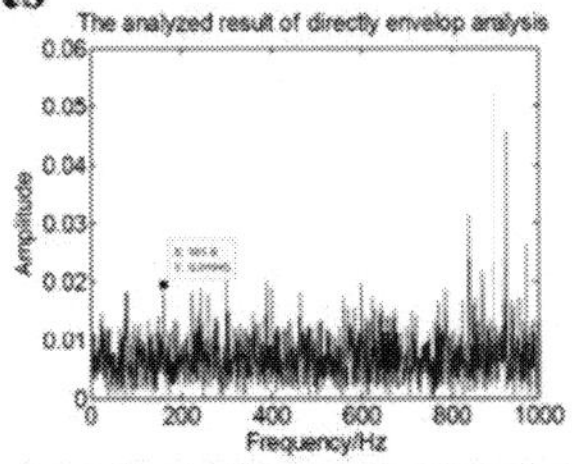

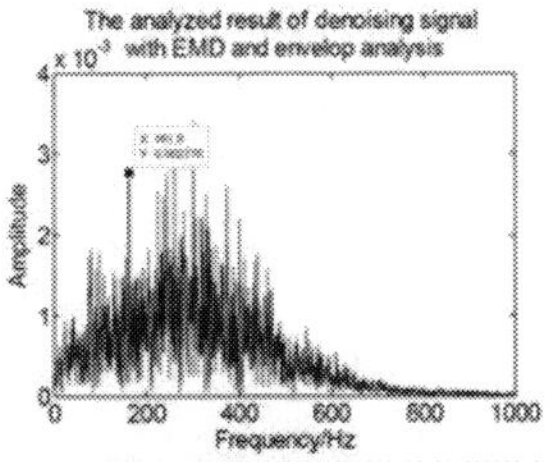

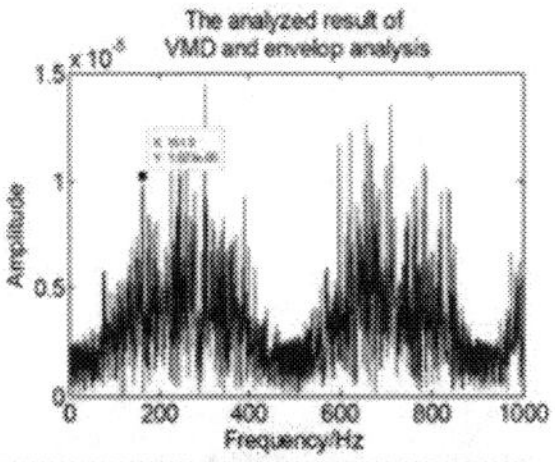

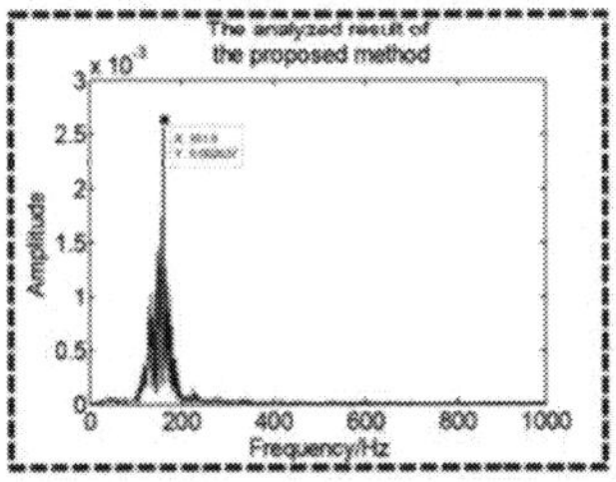

Methods	Denoising effect	Fault frequency identification
Directly envelop	×	×
EMD	√	×
VMD	√	×
Proposed method	√	√

MFPT 2019 Where Theory Meets Practice
MAY 14-16, 2019 | PHILADELPHIA, PA

5 Conclusion

- ✓ Variational mode decomposition (VMD) is an effective method to detect bearing fault if the signal is clear and the reasonable mode number *K* can be effectively pre-set.
- ✓ Weighted EMD can effectively remove the noise with the desired feature remained.
- ✓ With rule of SDA and EK selection, IVMD achieves a sound optimal component output for bearing fault feature in an adaptive way.
- ✓ **Comparisons show that the proposed IVMD method can identify the bearing fault frequency successfully.**

23

MFPT 2019 Where Theory Meets Practice
MAY 14-16, 2019 | PHILADELPHIA, PA

Thank you!

24

MFPT 2019 Where Theory Meets Practice
MAY 14-16, 2019 | PHILADELPHIA, PA

Reliability Centered Lubrication:

Managing the Lifeblood of Machinery

Preston Johnson, Allied Reliability

MFPT 2019 Where Theory Meets Practice
MAY 14-16, 2019 | PHILADELPHIA, PA

Why Reliability Centered Lubrication?

MFPT 2019 Where Theory Meets Practice
MAY 14-16, 2019 | PHILADELPHIA, PA

Field Lubricant Analysis – ISO 18436-4 Category I

1. Maintenance Strategies
2. Lubrication Fundamentals
3. Lubricant Selection
4. Lubricant Application
5. Lubricant Management
6. Contamination Control
7. Lubricant Sampling
8. Lubricant Health Monitoring
9. Wear Debris Analysis

Serving the Lubrication Community
ICML
International Council for Machinery Lubrication

MFPT 2019 Where Theory Meets Practice
MAY 14-16, 2019 | PHILADELPHIA, PA

Why Machines Fail

Dr. Ernest Rabinowicz of MIT found that 70% of the loss of usefulness of material objects is due to surface deterioration...

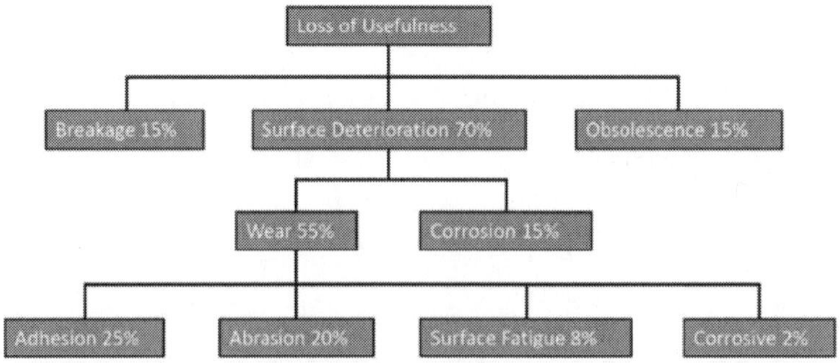

...most of which can be caused by lubrication failure!

MFPT 2019 Where Theory Meets Practice
MAY 14-16, 2019 | PHILADELPHIA, PA

Why Machines Fail

- In engines, approximately 70% of in-service failures are due to contamination, of which 50% are due to wear-related problems.
- Over 50% of rolling element bearings fail due to improper lubrication, while 10% fail at normal fatigue life limit.
- Approximately 75% of gear failures are due to in-service problems. Of these, half are due to wear and the rest to overload.
- Over 75% of hydraulic systems fail due to contamination.

Source: Pall

MFPT 2019 Where Theory Meets Practice
MAY 14-16, 2019 | PHILADELPHIA, PA

Why Machines Fail

A 1997 study by BHP Research found that 63% of gearbox repairs and failures at Pt. Kembla Steelworks were lubrication-related.

Source: BHPB

MFPT 2019 Where Theory Meets Practice
MAY 14-16, 2019 | PHILADELPHIA, PA

Costs of Inadequate Lubrication

Mechanism	Losses ($B US)
Wear	170
Friction	17
Lubrication	3.4
Electrical Contacts	3.5
Adhesion	0.05
Frictional Electricity	0.05
Total	**194**

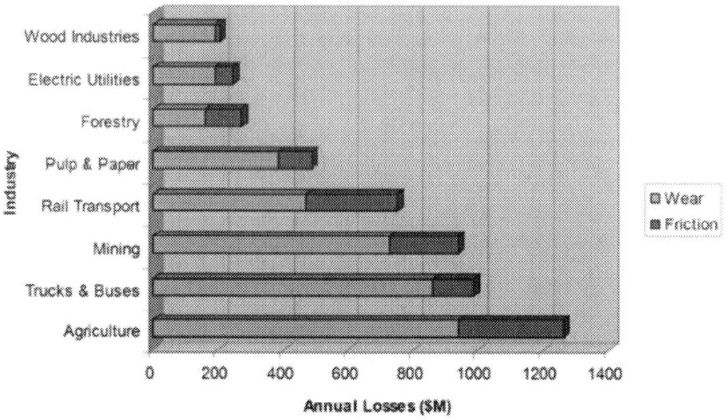

Source: Leugner

MFPT 2019 Where Theory Meets Practice
MAY 14-16, 2019 | PHILADELPHIA, PA

Benefits of Improved Lubrication

- Through improved lubrication, in 5 years Kawasaki Steel achieved:
  - 96% reduction in hydraulic component failures
  - 80% reduction in oil consumption

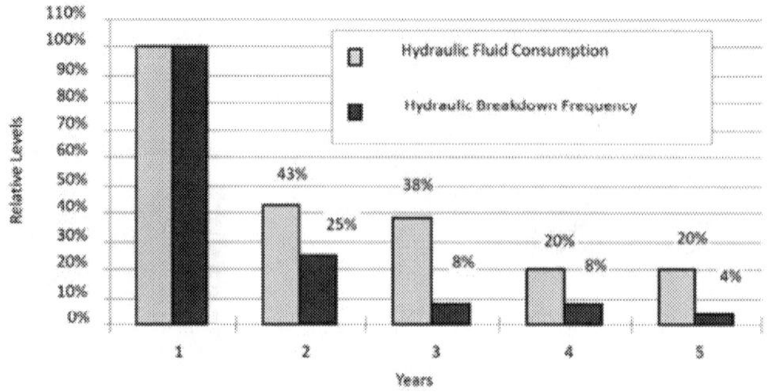

Source: Kawasaki

MFPT 2019 Where Theory Meets Practice

MAY 14-16, 2019 | PHILADELPHIA, PA

Benefits of Improved Lubrication, cont.

- Through improved contamination control, in 6-7 years, NSC achieved:
  - 50% reduction in bearing purchases plant-wide
  - 80% reduction in hydraulic pump replacement
  - 83% reduction in oil consumption
  - 90% reduction in pump overhauls
  - 90% reduced frequency of lubrication failures

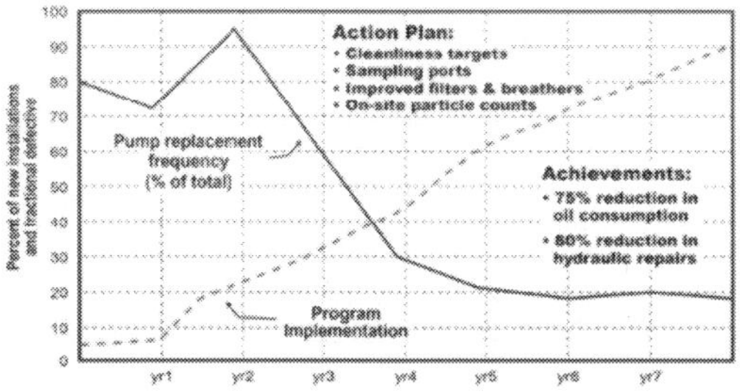

Source: NSC

MFPT 2019 Where Theory Meets Practice

MAY 14-16, 2019 | PHILADELPHIA, PA

Benefits of Improved Lubrication, cont.

- Through filtration system improvements, in 3 years, the BHP Newcastle Rod Mill achieved:
  - 92% sustained reduction in bearing failure rates
  - 140% mill speed increase

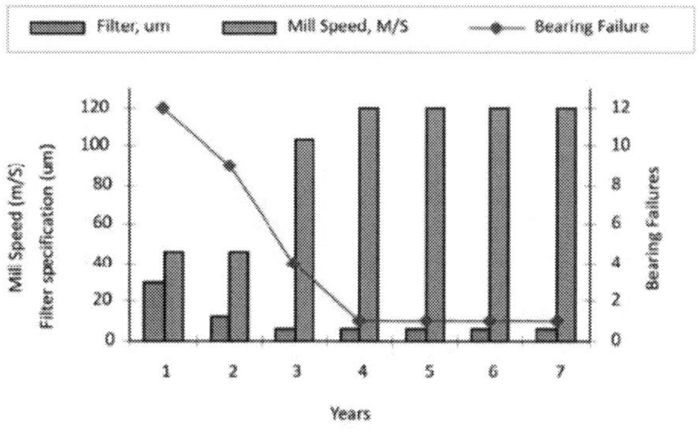

Filters: 30 → 12 → 6 → 3 μm
Bearing failures: 12 → 9 → 4 → 1 /yr
Mill speed: 44 → 105 → 117 m/s

Source: BHPB

So Why Should You Lubricate Equipment Properly?

- Maximize component and lubricant life
- Reduce energy consumption
- Reduce maintenance and operating costs
- Improve plant reliability and availability

"If you won't pay for quality lubrication, you'll pay for poor lubrication!"

"There is nothing that has more impact on equipment reliability than lubrication."

"There is nothing more dangerous to equipment reliability than an untrained lube tech or a poorly designed lube program."

MFPT 2019 Where Theory Meets Practice
MAY 14-16, 2019 | PHILADELPHIA, PA

Lubrication Strategies

MFPT 2019 Where Theory Meets Practice
MAY 14-16, 2019 | PHILADELPHIA, PA

Lubrication Maintenance Strategies

- A process used to determine the lubrication and contamination control requirements of any physical asset in its operating context, allowing for means of condition monitoring, thus having the ability to impact the overall asset health of the lubricated component.

MFPT 2019 Where Theory Meets Practice
MAY 14-16, 2019 | PHILADELPHIA, PA

Introduction: Lubricant Functions

Lubricate	prevent or minimize wear by separating metal-to-metal contacts
Protect	surfaces against corrosive substances
Cool	absorb and transfer excess heat to a cooler region
Clean	transports contaminants to filters/separators
Seal	e.g. at piston rings to prevent blow-by
Insulate	in equipment such as transformers and switch gear
Transmit Power	in hydraulic systems, etc.

MFPT 2019 Where Theory Meets Practice
MAY 14-16, 2019 | PHILADELPHIA, PA

Determining Bearing Re-grease Volumes, cont.

- If OEM data is not available, use the SKF formula:

Grease quantity (g) = 0.005 x Brg OD (mm) x Brg width (mm)

Grease quantity (oz) = 0.114 x Brg OD (in) x Brg width (in)

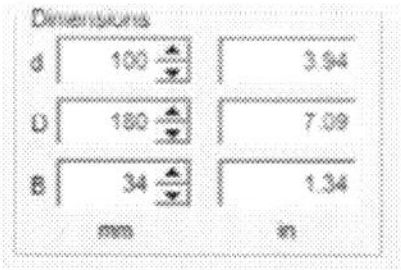

MFPT 2019 Where Theory Meets Practice
MAY 14-16, 2019 | PHILADELPHIA, PA

Determining Oil Feed Rates

LUBRICATING CALCULATIONS

If the amount of lubrication needed is not specified by the equipment manufacturer, this table can be used as a guide.

Illustration	Application	Oil Required in cc/hour
	Ball Bearings	.10 x bearing diameter (in.) x number of rows or bearings
		.04 x bearing diameter (cm.) x number of rows or bearings
	Plain Bearings	.15 x bearing shaft diameter (in.) x bearing length (in.)
		.023 x bearing shaft diameter (cm.) x bearing length (cm.)
	Flat Slides	.04 x [slide length (in.) + travel (in.)] x slide width (in.)
		.006 x [slide length (cm.) + travel (cm.)] x slide width (cm.)
	Cylindrical Slides	.15 x [slide length (in.) + travel (in.)] x slide diameter (in.)
		.023 x [slide length (cm.) + travel (cm.)] x slide diameter (cm.)
	Ball Bearing Slides	.03 x slide length (in.) x number of slides or rows
		.012 x slide length (cm.) x number of slides or rows
	Cams	.08 x surface area (sq. in.)
		.013 x surface area (sq. cm.)
	Gears	.30 x gear pitch diameter (in.) x face width (in.)
		.046 x gear pitch diameter (cm.) x face width (cm.)
	Chains	.05 x length of chain (in.) x width (in.)
		.008 x length of chain (cm.) x width (cm.)

MFPT 2019 Where Theory Meets Practice
MAY 14-16, 2019 | PHILADELPHIA, PA

Lubricant Storage and Management

Section 4

MFPT 2019 Where Theory Meets Practice
MAY 14-16, 2019 | PHILADELPHIA, PA

Causes of Lubricant Contamination

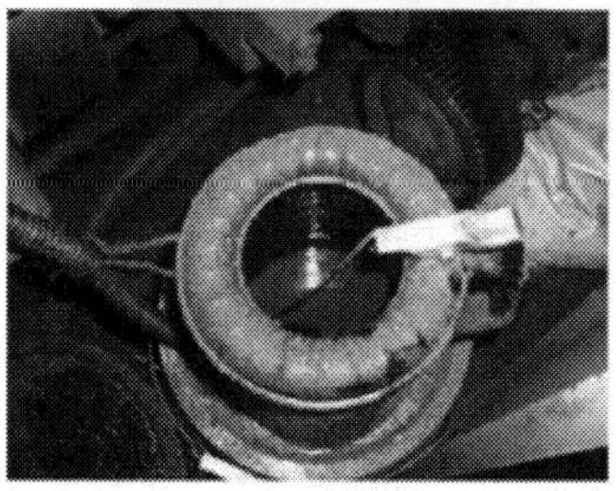

MFPT 2019 Where Theory Meets Practice
MAY 14-16, 2019 | PHILADELPHIA, PA

Lube Storage

- Cool, clean, and dry indoor area
- Storage temperatures should remain moderate at all times
  - Extreme hot or cold can cause chemical degradation
- Containers should be rotated and used on a first in, first out basis

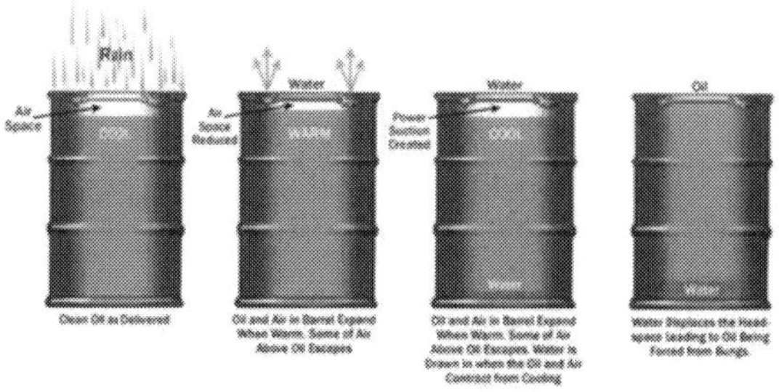

MFPT 2019 Where Theory Meets Practice
MAY 14-16, 2019 | PHILADELPHIA, PA

Storage

- Grease

- Store in clean, cool, dry indoor area
- Utilize first in, first out practice
- Keep containers closed
- Wipe off lid prior to opening
- Acclimatize grease prior to use
- Store grease tubes and loaded grease guns vertically for less bleed

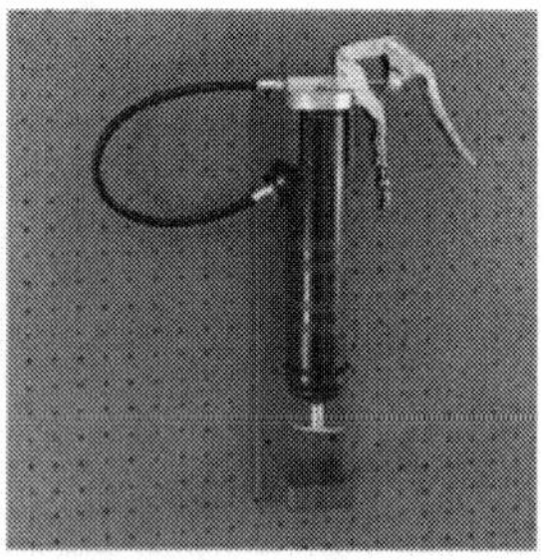

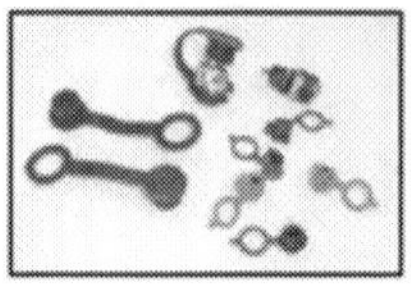

MFPT 2019 Where Theory Meets Practice
MAY 14-16, 2019 | PHILADELPHIA, PA

Lubricant Labeling

- Use combination of color coding and lube specification
- Reduces probability of misapplication
- Allows after hours correct application
- Assists in training new technicians/maintenance

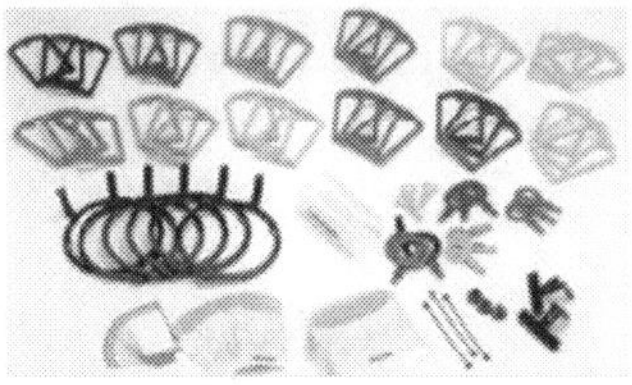

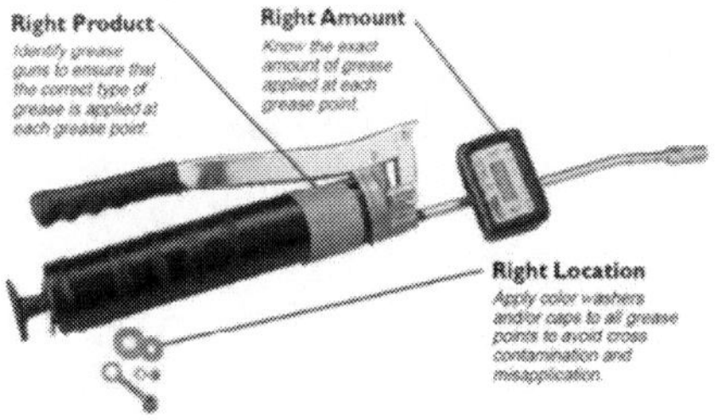

MFPT 2019 Where Theory Meets Practice
MAY 14-16, 2019 | PHILADELPHIA, PA

Contamination

- How Important Is It?

The majority of lubrication problems are caused by:

Solid Particulate. Moisture. Air

"Reliability and maintainability are a function of contamination control and contamination control leads to long life" - General Electric

"Contamination is the greatest single cause of lubricant malfunction and subsequent excessive wear of parts" - Mobil

"Bearings can have an infinite life when particles larger than the lubricant film are removed" - SKF

Particle Contamination

- Particle contamination can:
  - Degrade the wetted surfaces of the system
    - Three-body abrasion
    - Particle impingement erosion
  - Interfere with the movement of surfaces or fluid
    - Silt Lock or Dynamic Jam
  - Serve as catalyst in breaking down or reformulating the fluid and additives

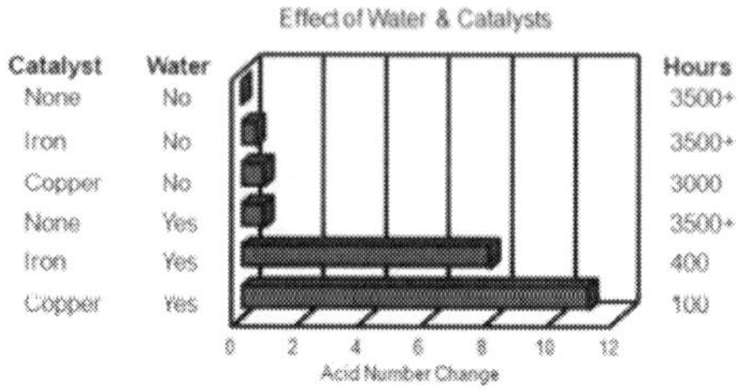

Source: E Fitch

MFPT 2019 Where Theory Meets Practice
MAY 14-16, 2019 | PHILADELPHIA, PA

Particle Contamination

- Why It Matters

How much of a difference will cleaning your oil from ISO 23/21/17 to ISO 18/16/13 really make?

Amount of "Dirt" in Oil	Amount of "Dirt" Flowing Through Pump in 1 Year
ISO 18/16/13	9 Kg/yr
ISO 21/18/15	36 Kg/yr
ISO 23/21/18	288 Kg/yr

Source: Lubrosoft

MFPT 2019 Where Theory Meets Practice
MAY 14-16, 2019 | PHILADELPHIA, PA

Water Contamination

- States of Existence
  - Dissolved
    - Water is dispersed on a molecular level throughout the oil
    - Water is held below the saturation point
  - Free
    - Water that is fully separated from the oil
  - Emulsified
    - Components suspended as tiny globules within one another

Typical oil saturation levels:
Hydraulic - 200-400 ppm (0.02-0.04%)
Lubrication - 200-750 ppm
Transformer - 30-50 ppm

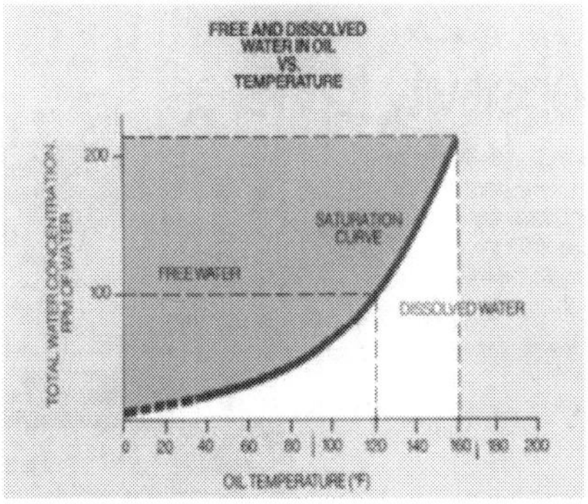

Source: Lubrosoft

MFPT 2019 Where Theory Meets Practice
MAY 14-16, 2019 | PHILADELPHIA, PA

Air Contamination

- Forms of air in hydraulic and lubrication fluids:
  - Free or emulsified air
  - Dissolved air
- Sources of air contamination:
  - Ingression through reservoir and system vents
  - Change in fluid temperature/pressure in the system
  - Leakage through seals
  - Fluid top-off

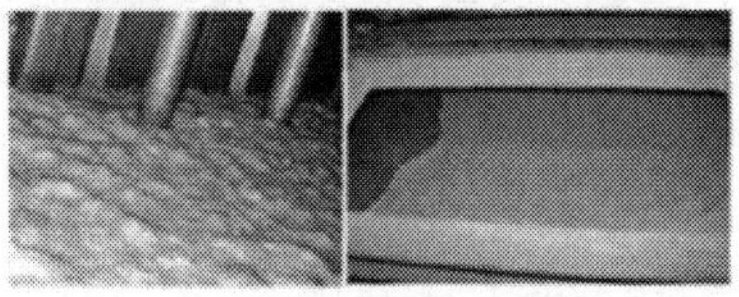

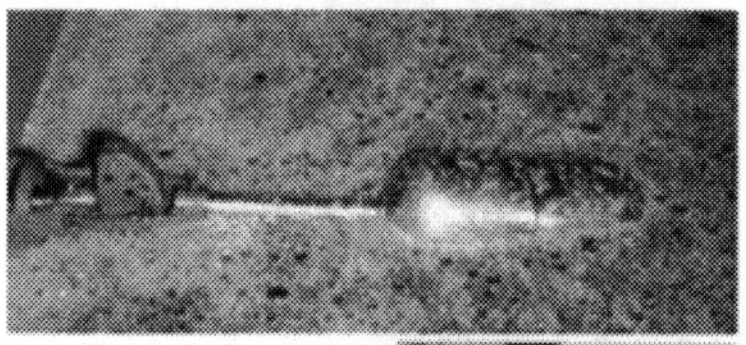

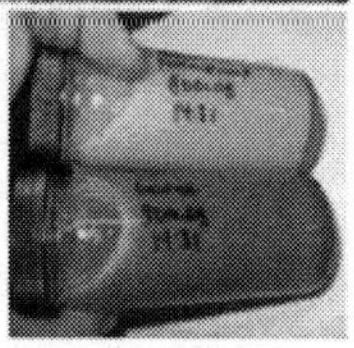

Fluid samples taken upstream and downstream of a purifier

Lubricant Sampling

MFPT 2019 Where Theory Meets Practice
MAY 14-16, 2019 | PHILADELPHIA, PA

Failure Modes Detectable with Oil Analysis

Often referred to as the life blood of equipment, Lubrication and Oil Analysis can provide a great deal of information on the condition of equipment, as well as serve as a leading indicator of the future condition of the equipment. Allied's approach to ensuring that all failure modes are covered through appropriate technologies allows for a well-documented approach to not only Oil Analysis, but to overall Reliability Centered Lubrication (RCL).

When used in conjunction with other CBM technologies, Lubrication Analysis can help enable the Proactive Workflow Model by identifying defects in advance. This allows corrective work to be planned and scheduled generating higher reliability gains.

- Incorrect Lubricant Use/Selection
- Over Extended Oil Drains
- Component Fatigue
- Sliding Wear
- Abrasive Wear
- External Contamination
- Poor Handling/Sampling Practices
- Inadequate Contamination Control Measures
- Wear Particles from bearing and gear wear
- Additive content loss
- Viscosity
- Humidity and Temperature

	Contaminates		Fluid Properties			
Sample Date	Water % Vol	[illegible]	Oxidation abs/cm	Nitration abs/cm	Viscosity 40 C cSt	[illegible]
[illegible]	Nil	[illegible]	[illegible]	[illegible]	[illegible]	[illegible]
[illegible]	Nil	[illegible]	[illegible]	[illegible]	[illegible]	[illegible]
[illegible]	Nil	[illegible]	[illegible]	[illegible]	[illegible]	[illegible]
[illegible]	Nil	[illegible]	[illegible]	[illegible]	[illegible]	[illegible]
[illegible]	Nil	[illegible]	[illegible]	[illegible]	[illegible]	[illegible]
[illegible]	Nil	[illegible]	[illegible]	[illegible]	[illegible]	[illegible]

MFPT 2019 Where Theory Meets Practice
MAY 14-16, 2019 | PHILADELPHIA, PA

What Is Oil Analysis?

We take oil samples…

"To obtain a clear indication of the operating state of the equipment"

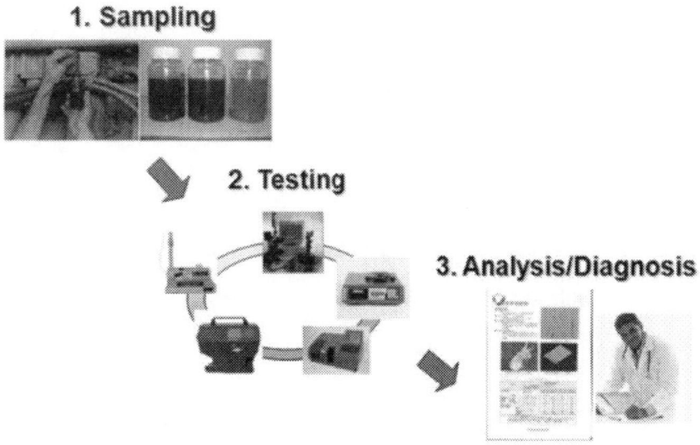

Source: Lubrosoft

MFPT 2019 Where Theory Meets Practice
MAY 14-16, 2019 | PHILADELPHIA, PA

Lubricant Sampling

- In order to provide value as a PdM/PM tool, Oil Sampling must be:
  - Repeatable
    - Same characteristics regardless of who collects the sample
  - Accurate
    - Must reflect current lubricant and machine condition
  - Properly Focused
    - What are we looking for?
      - Oil chemistry?
      - Wear debris?
      - Contaminants?

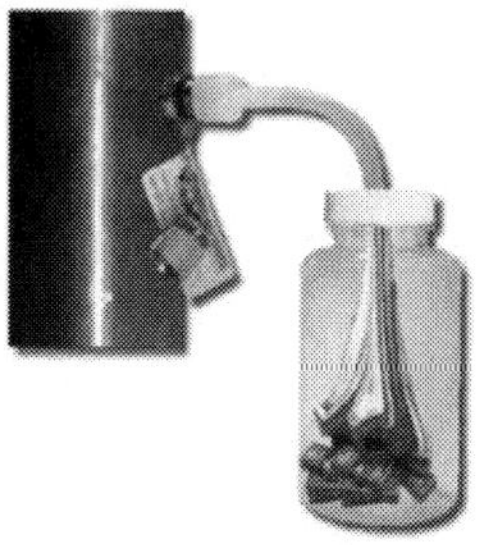

Source: AMRRI

Sampling Golden Rules

- To achieve those objectives, the sampling process design should address:

Three Key Factors

1. Location: Where
2. Consistency: Who and Procedure
3. Correct hardware/fittings

Seven Points

4. Time: when
5. Frequency: how often
6. Clean tools and sampling ports
7. Drain dead oil out: 5-10X pathway volume
8. Bottle Sample volume: 60-75% full
9. Labelling: hours/make up...
10. Flushing sampling bottle, cap, tubes, etc.

Oil sampled must be as close as possible to *critical load zone* to be most representative of what the component is actually exposed to.

Source: TRICO/Lubrosoft

Default Sampling Frequency

- Consider 5 frequency adjustment factors
  1. Economic Penalty of Failure: downtime costs, repair costs, and business interruption penalty
  2. Fluid Environment Severity Pressures, load, temperature, speed, contaminants in oil, and duty cycle
  3. Machine Age:
     - Infant machines: going through break-in
     - Old age machines: showing symptoms of distress
  4. Oil Age:
     - Infant oils: just being changed or <10% into expected life
     - Old age oils: showing symptoms of degradation
  5. Target Tightness:
     - Tight: oil properties extremely close to targets and limits
     - Loose: trend well within targets and limits

Source: Lubrosoft

Bearings	500 hrs
Chillers	500 hrs
Compressors	500 hrs
Differentials	300 hrs
Engine, Aviation Recip	50 hrs
Engine, Diesel	150 hrs
Final Drives	300 hrs
Gearing, Aviation	150 hrs
Gearing, High Speed Industrial	300 hrs
Gearing, Low Speed	1000 hrs
Hydraulics, Aviation	150 hrs
Hydraulics, Industrial	700 hrs
Hydraulics, Mobile	250 hrs
Transmissions	300 hrs
Turbines, Aviation	100 hrs
Turbines, Gas	500 hrs
Turbine, Steam	500 hrs

MFPT 2019 Where Theory Meets Practice
MAY 14-16, 2019 | PHILADELPHIA, PA

Reliability Centered Lubrication

Summary

MFPT 2019 Where Theory Meets Practice
MAY 14-16, 2019 | PHILADELPHIA, PA

Top Considerations

- Lubrication practices have a significant impact on machinery and machinery component life.
- Providing the right lubrication in the right amounts at the right time is critical for proper lubrication
- Proper lubrication storage and management prevents contamination of lubrication, and insures the right lubrication is applied
- Sampling of lubrication helps identify machine part wear, and lubrication breakdown

MFPT 2019 Where Theory Meets Practice
MAY 14-16, 2019 | PHILADELPHIA, PA

Oil Analysis for Condition Monitoring

In an IIoT ERA

Preston Johnson, Allied Reliability

MFPT 2019 Where Theory Meets Practice
MAY 14-16, 2019 | PHILADELPHIA, PA

Failure Prevention is Multifaceted

PEOPLE

Work Execution Management
- Planning & Scheduling
- Job Plan Improvement
- Develop Work Procedures
- Spare Parts
- BOMs

Routine Equipment Care

Recruiting
- Talent Sourcing
- Talent Screening
- Direct/Contract Hire

Workforce Development
- Training
- Coaching & Mentoring

PROCESS

Asset Health Assurance
- Asset Catalog
- Criticality
- Doc & BOMs
- FMEA (AHA + RCM)
- PM Writing
- EMP Design/Implementation

Reliability Engineering
- PM Evaluation/ PM Optimization
- Root Cause Analysis
- Reliability Centered Maintenance
- Failure Reporting and Corrective Action System
- Spare Parts

TECHNOLOGY

Full Suite of CBM Applications /Technologies
- Vibration
- Infrared
- Ultrasound
- Oil Analysis
- Motor Circuit Analysis
- Advanced Diagnostics
- Sensory Inspection
- Inline Monitoring

Engineered Products & Services
- Lubrication Systems
- Emissions
- Controls
- Oil and Fluid Filtration

Reliability Management System (iReliability™)

Change Management & Leadership Alignment

MFPT 2019 Where Theory Meets Practice

MAY 14-16, 2019 | PHILADELPHIA, PA

Oil Analysis and Machinery Failure Modes

MFPT 2019 Where Theory Meets Practice

MAY 14-16, 2019 | PHILADELPHIA, PA

Failure Modes Detectable with Oil Analysis

Often referred to as the life blood of equipment, Lubrication and Oil Analysis can provide a great deal of information on the condition of equipment, as well as serve as a leading indicator of the future condition of the equipment. Allied's approach to ensuring that all failure modes are covered through appropriate technologies allows for a well-documented approach to not only Oil Analysis, but to overall Reliability Centered Lubrication (RCL).

When used in conjunction with other CBM technologies, Lubrication Analysis can help enable the Proactive Workflow Model by identifying defects in advance. This allows corrective work to be planned and scheduled generating higher reliability gains.

- Incorrect Lubricant Use/Selection
- Over Extended Oil Drains
- Component Fatigue
- Sliding Wear
- Abrasive Wear
- External Contamination
- Poor Handling/Sampling Practices
- Inadequate Contamination Control Measures
- Wear Particles from bearing and gear wear
- Additive content loss
- Viscosity
- Humidity and Temperature

	[illegible]		Fluid Properties			
Sample Date	Water % Vol	[illegible]	Oxidation abs/cm	Nitration abs/cm	Viscosity 40 C cSt	[illegible]
[illegible]	[illegible]	[illegible]	[illegible]	[illegible]	[illegible]	[illegible]
[illegible]	[illegible]	[illegible]	[illegible]	[illegible]	[illegible]	[illegible]
[illegible]	[illegible]	[illegible]	[illegible]	[illegible]	[illegible]	[illegible]
[illegible]	[illegible]	[illegible]	[illegible]	[illegible]	[illegible]	[illegible]
[illegible]	[illegible]	[illegible]	[illegible]	[illegible]	[illegible]	[illegible]
[illegible]	[illegible]	[illegible]	[illegible]	[illegible]	[illegible]	[illegible]

MFPT 2019 Where Theory Meets Practice
MAY 14-16, 2019 | PHILADELPHIA, PA

Oil Analysis Failure Modes and Technologies

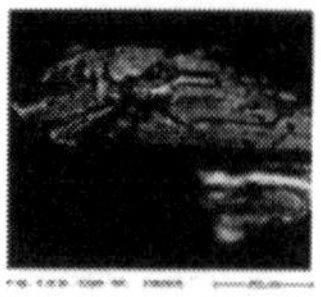

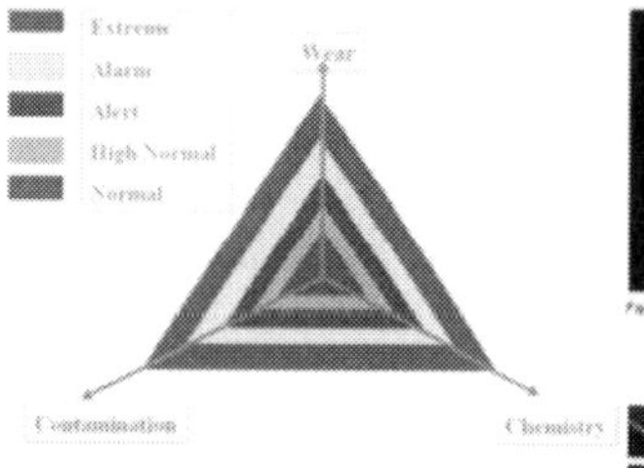

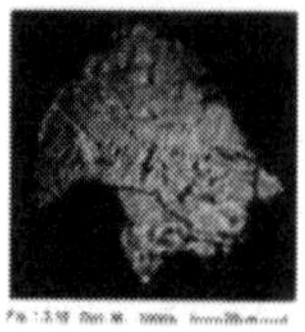

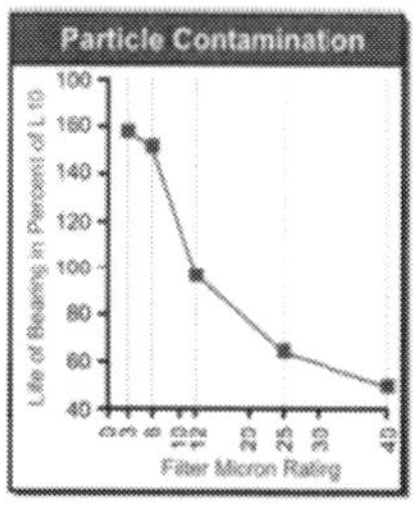

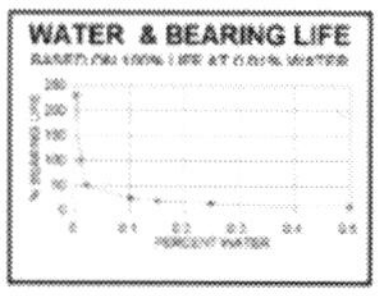

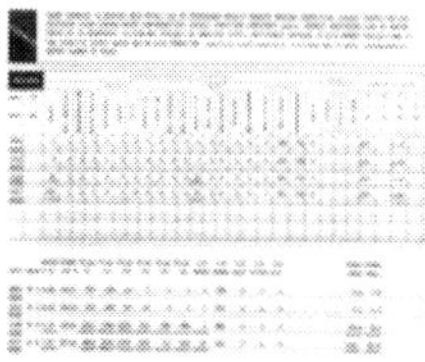

MFPT 2019 Where Theory Meets Practice
MAY 14-16, 2019 | PHILADELPHIA, PA

Oil Analysis
Failure Modes Detected

- Component Failure Modes
  - Compromised Seal
  - Bearing Fault
  - Gear Problem
  - Seal Damage
  - Clutch Damage
- Lubricant Failure Modes
  - Oxidation
  - Nitration
  - Additive Depletion
  - Soot Loading
  - Viscosity Change
  - Poor Handling Practices
  - Fuel Dilution
  - Water

MFPT 2019 Where Theory Meets Practice
MAY 14-16, 2019 | PHILADELPHIA, PA

Value of Oil Analysis

- Must be used to initiate corrective actions
  - Filtering
  - Oil changes
  - Equipment repairs and improvements
- Extends oil life
  - Reduce or eliminate time based oil changes

MFPT 2019 Where Theory Meets Practice
MAY 14-16, 2019 | PHILADELPHIA, PA

Chemistry Tests

- Test the physical properties of the lubricant to determine if the proper lubricant is being used and the levels of lubricant degradation
  - Viscosity
  - Dielectric Strength
  - Total Acid Number (TAN)
  - Total Base Number (TBN)
  - Spectroscopy (metals < 10 microns)
  - Fourier Transform Infrared Spectroscopy (FTIR) Spectroscopy

M & E page - 8

MFPT 2019 Where Theory Meets Practice
MAY 14-16, 2019 | PHILADELPHIA, PA

Contamination Tests

- Test the lubricant for contaminants that can lead to lubricant degradation or particles that indicate wear
  - Water Content
  - Direct Ferrography or Ferrous Density
  - Particle Count
  - Analytical Ferrography

MFPT 2019 Where Theory Meets Practice
MAY 14-16, 2019 | PHILADELPHIA, PA

Water Contamination

Water	Contaminants
Dissolved	**Appears invisible** **Similar to humidity**
Emulsified	**Appears cloudy** **Similar to fog**
Free	**Separates to bottom** **Water layer visible** **Similar to rain**

* Emulsified and Free water cause most problems

MFPT 2019 Where Theory Meets Practice
MAY 14-16, 2019 | PHILADELPHIA, PA

Oil Analysis Common Traps

- Belief that "free" oil analysis is the answer to lubrication analysis program
- Lack of understanding of lubricant testing profiles
- No understanding of the limitations of each testing technique
- Only sampling on an "as needed" basis
- Extended time interval between sampling and analysis
- Improper sampling techniques

MFPT 2019 Where Theory Meets Practice
MAY 14-16, 2019 | PHILADELPHIA, PA

Oil Sampling and Analysis in Practice

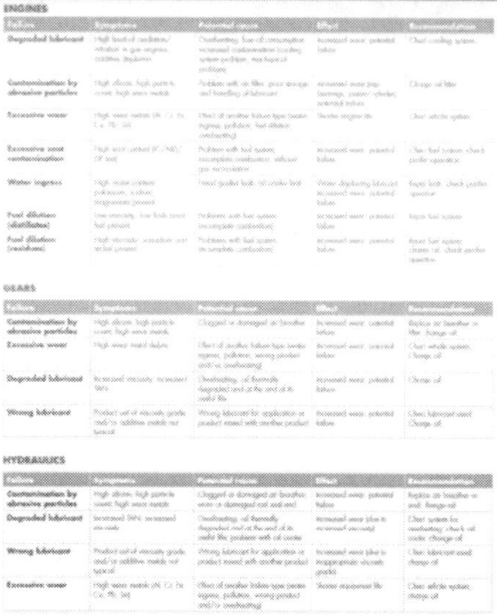

Typical Failures Based on Component Type

Abrasive Particles are common across all component types and contribute to the Wear Chain as well as lubricant degradation

MFPT 2019 Where Theory Meets Practice
MAY 14-16, 2019 | PHILADELPHIA, PA

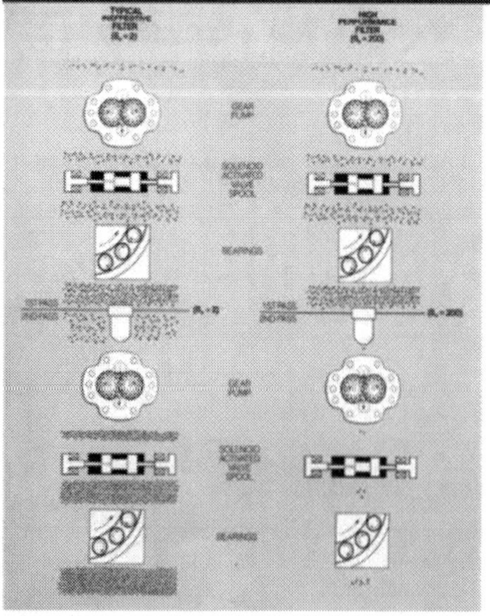

The Presence of Particles Creates More Particles

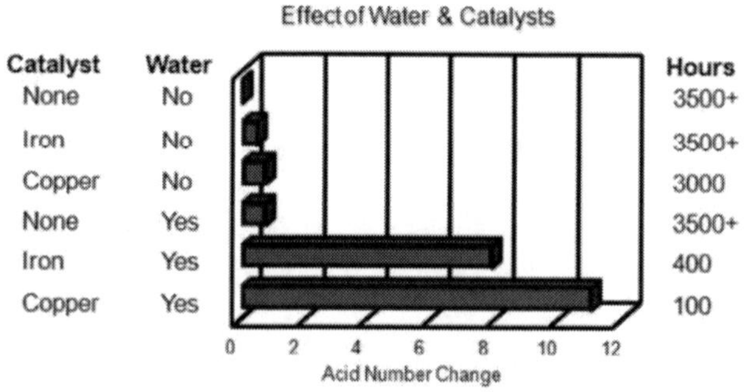

MFPT 2019 Where Theory Meets Practice
MAY 14-16, 2019 | PHILADELPHIA, PA

Oil Analysis as a Missed Opportunity

- Most common frequency for sampling industrial oils: Quarterly
- Most common frequency for sampling engine oils: Drain Schedule

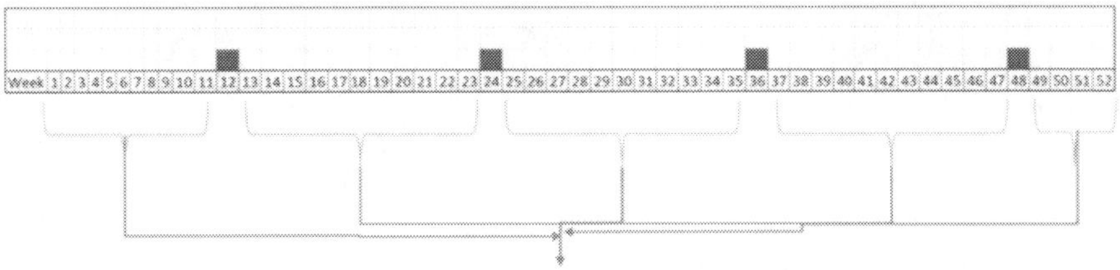

Failure Mode Coverage: 0

Multi-Failure Mode Coverage Occurrence: 4

Cost of Coverage: $100-$200 annual
Per Occurrence: $25-$50

MFPT 2019 Where Theory Meets Practice
MAY 14-16, 2019 | PHILADELPHIA, PA

Monthly Sampling

- Full failure mode coverage requires a comprehensive test slate for each sample tested

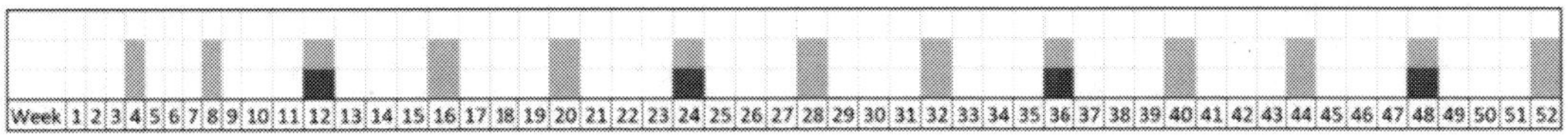

Multi-Failure Mode Annual Coverage Occurrence: 12

Cost of Coverage: $300-$540 annual
Per Occurrence: $25-$45

Comprehensive Testing Options

On-site Testing Option

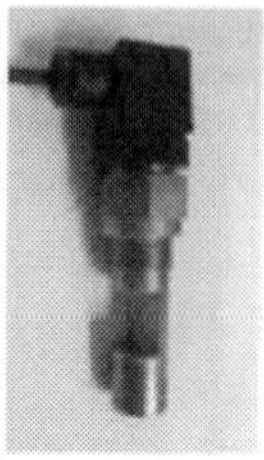

Inline Testing Option

- Using on-site and/or inline drives
  - CONDITION BASED LAB ANALYSIS
- Primary focus lies in critical systems (top 2-5% of criticality)

MFPT 2019 Where Theory Meets Practice
MAY 14-16, 2019 | PHILADELPHIA, PA

Example Oil Online Monitoring

- Poseidon Systems: http://www.poseidonsys.com/products/
- Macom: http://www.macom.co.uk/index6a46.html?option=com_content&task=view&id=79&Itemid=111
- Hy-Daq: http://www.hydac.com.au/Condition-Monitoring.aspx
- Tan Delta Systems: http://www.tandeltasystems.com/products/

MFPT 2019 Where Theory Meets Practice
MAY 14-16, 2019 | PHILADELPHIA, PA

Understanding Your Data

Limited View

Columns of data:

Operations
Machine Condition

Full View

MFPT 2019 Where Theory Meets Practice
MAY 14-16, 2019 | PHILADELPHIA, PA

Oil Analysis Features into Common AI Rules Engine

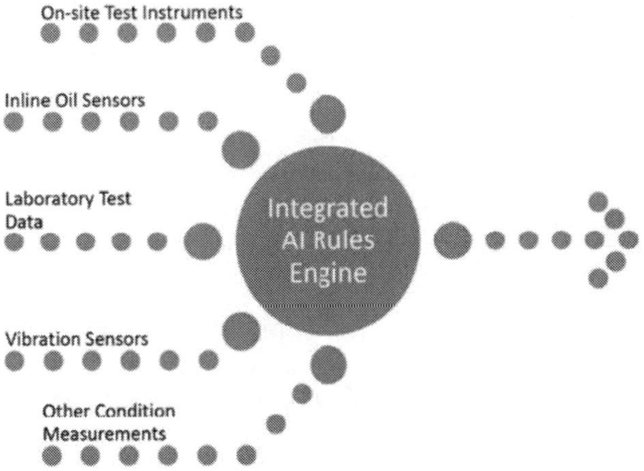

Specific Problem identification with integrated confidence factor

MFPT 2019 Where Theory Meets Practice
MAY 14-16, 2019 | PHILADELPHIA, PA

Failure Mode Mapping to Oil Analysis Testing

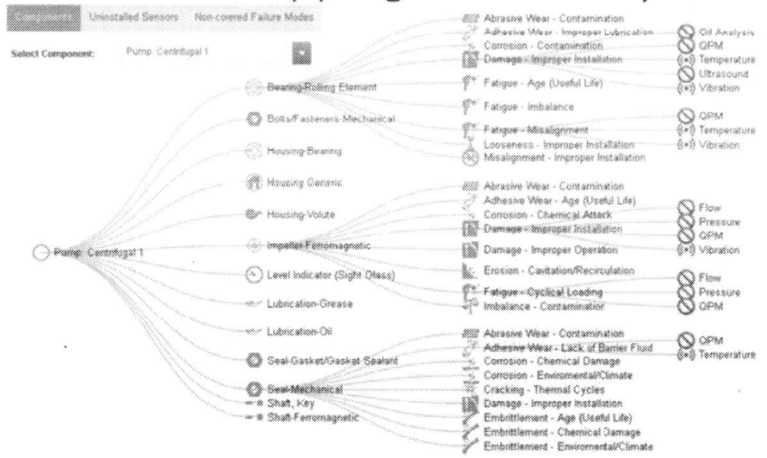

MFPT 2019 Where Theory Meets Practice
MAY 14-16, 2019 | PHILADELPHIA, PA

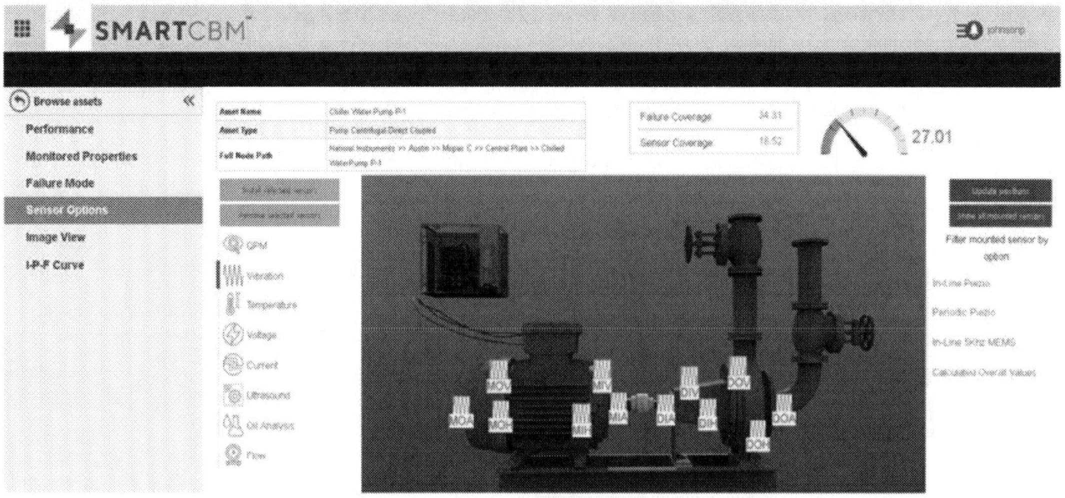

MFPT 2019 Where Theory Meets Practice

MAY 14-16, 2019 | PHILADELPHIA, PA

Data Integration to AI and Machine Learning

MFPT 2019 Where Theory Meets Practice

MAY 14-16, 2019 | PHILADELPHIA, PA

Inline Monitoring with AI Integration

- Inline monitoring covers primary causes of failure:
  - Particle Contamination
  - Moisture Contamination
  - Overheating Conditions

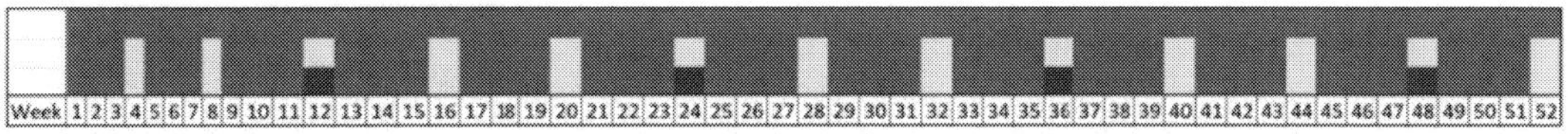

Multi-Failure Mode Coverage Occurrence: 2,190

Cost of Coverage, Year 1: $5540; Year 2: $540

Per Occurrence, Year 1: $2.53; Year 2: $0.25

Oil Analysis and the P-F Curve

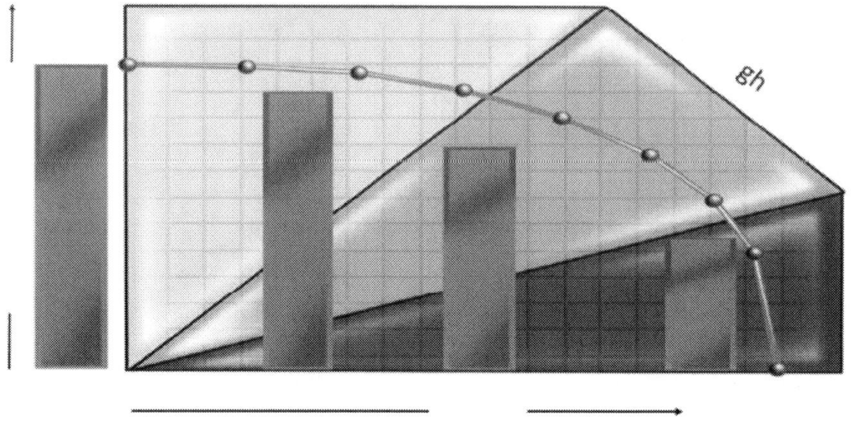

MFPT 2019 Where Theory Meets Practice
MAY 14-16, 2019 | PHILADELPHIA, PA

Future of Machine Learning

- Machine and failure mode specific action recommendations
- Early adoption will result in sensor price reduction
- Continued use of commercial lab testing

MFPT 2019 Where Theory Meets Practice
MAY 14-16, 2019 | PHILADELPHIA, PA

Jump Start Your Digital Transformation with a Failure Mode Approach to Predictive Maintenance

Preston Johnson
Platform Lead: Intelligent Monitoring and IIoT
May 9, 2019

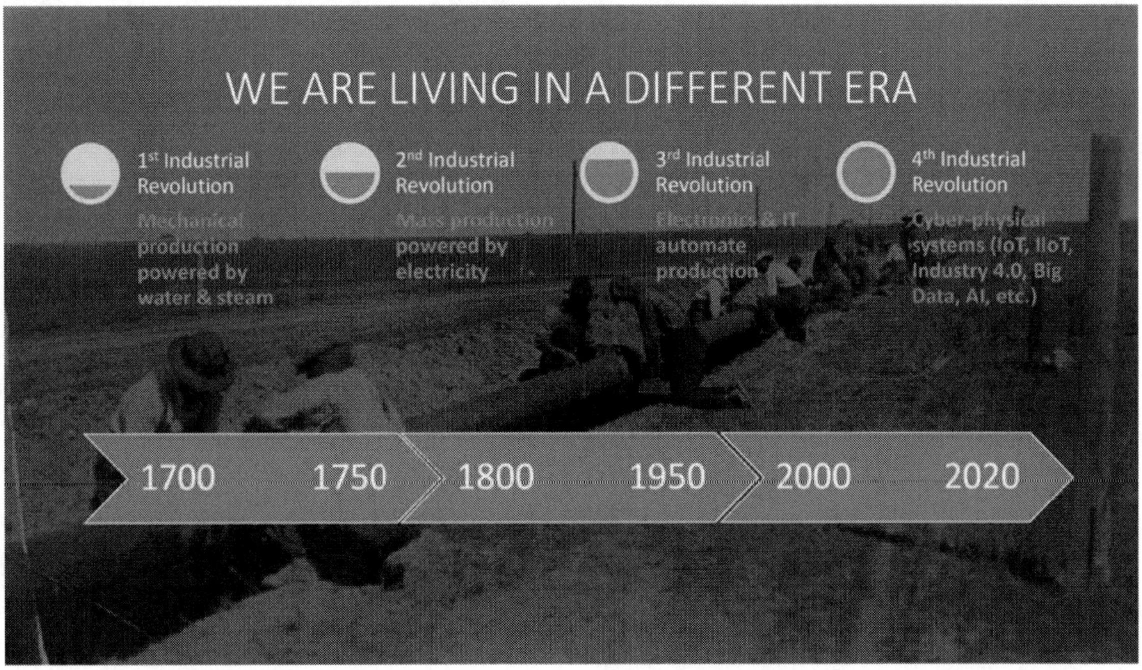

Common Themes

MFPT 2019 Where Theory Meets Practice
MAY 14-16, 2019 | PHILADELPHIA, PA

LNS IIoT and Big Data Analytics

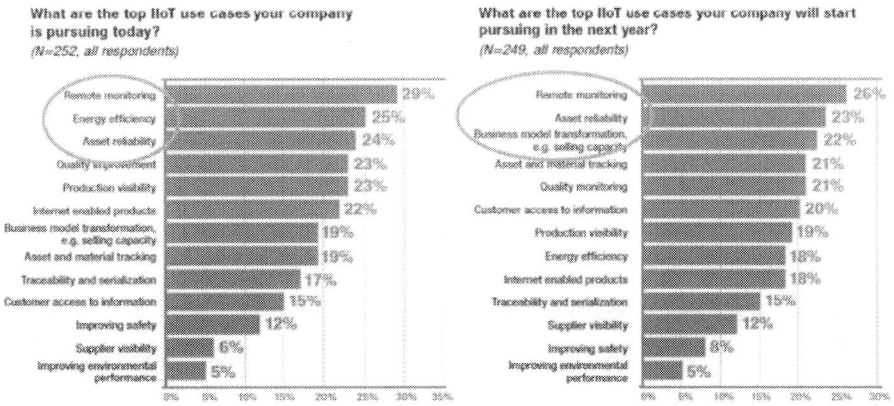

Credit: ***lnsresearch.com***, **"IIoT AND BIG DATA ANALYTICS: How Manufacturing System Architecture Is Being Transformed "**

MFPT 2019 Where Theory Meets Practice
MAY 14-16, 2019 | PHILADELPHIA, PA

Online Monitoring Solution, Power Generation

10,000 30,000 2,300 60 1

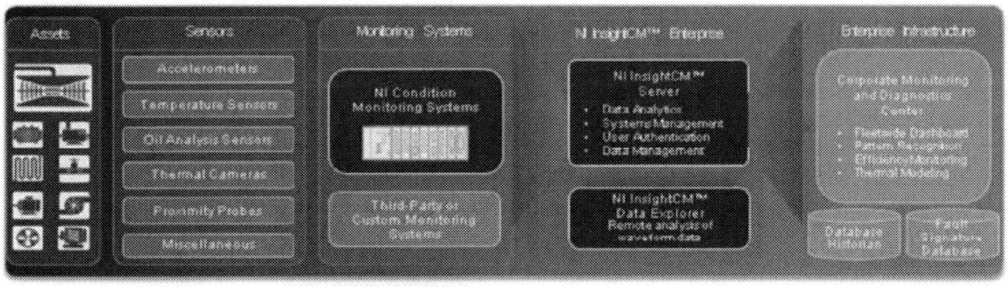

Subject Matter Expertise

Data Acquisition and Analysis

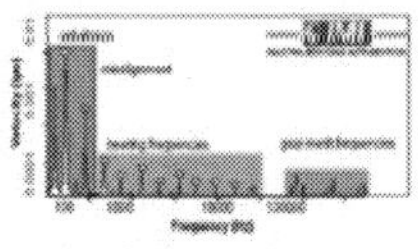

Condition Indicators Alarming

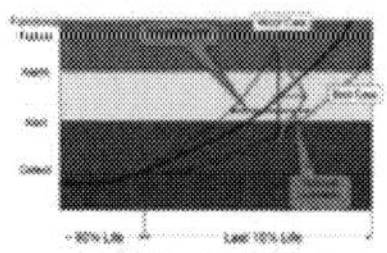

Predictive Analytics

Images: www.ni.com, NIWeek 2014

https://reliabilityweb.com/assets/uploads/documents/61431174_duke_energy_case_study.pdf

Best Practices Predictive Maintenance

A properly designed and implemented condition monitoring program delivers productivity:

↓ 28% reduction in annual maintenance spend

↑ 20% increase in OEE in 3 years

↓ 30% reduction in parts inventory

↑ 100% increase in production over 3 years

Performance data represens a multitude of plants, across different industries and scopes of work.

PdM Assessments + Action = ↑ OEE
(Major Tire Manufacturer)

Targeting Failure Modes + Action = ↓ Mtnc Costs
(Major Surface Mining Plant)

MFPT 2019 Where Theory Meets Practice
MAY 14-16, 2019 | PHILADELPHIA, PA

What might a Digital Transformation Look Like?

Also known as Manufacturing 4.0, Industry 4.0, Smart manufacturing, and China 2025

MFPT 2019 Where Theory Meets Practice

MAY 14-16, 2019 | PHILADELPHIA, PA

Industrial Innovation Platform

3) Augmented Reality makes human labor 30-50% more efficient

4) Combines data collection, analytics, visualization, and integration with the enterprise

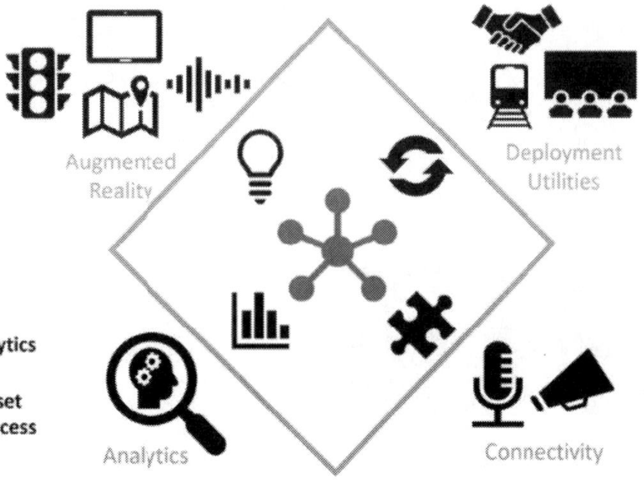

2) Predictive analytics drives 10-30% improvement in asset availability and process optimization

1) Industrial IoT drives IT/OT convergence across the value chain and increases automation and productivity up to 25%

MFPT 2019 Where Theory Meets Practice

MAY 14-16, 2019 | PHILADELPHIA, PA

The Point of Detection to failure curve (P-F curve)

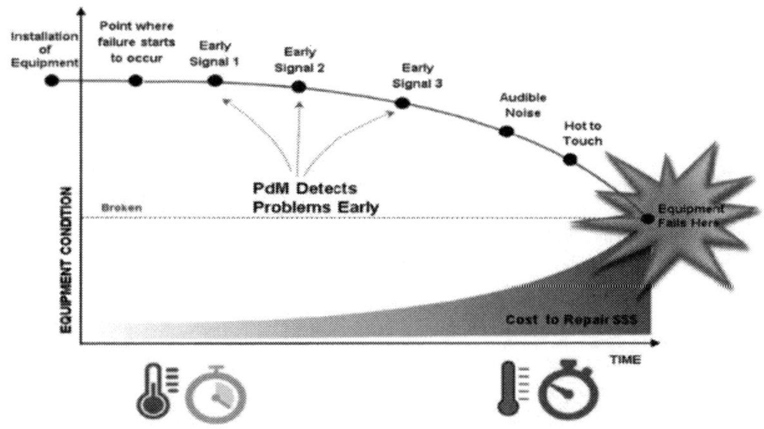

Real-Time IIoT

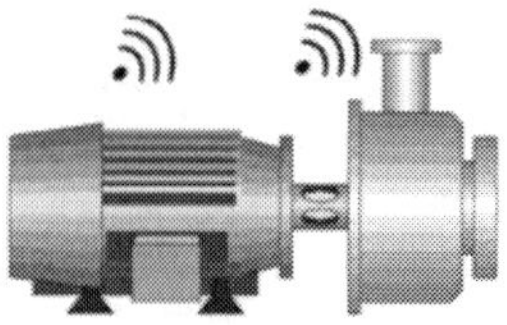

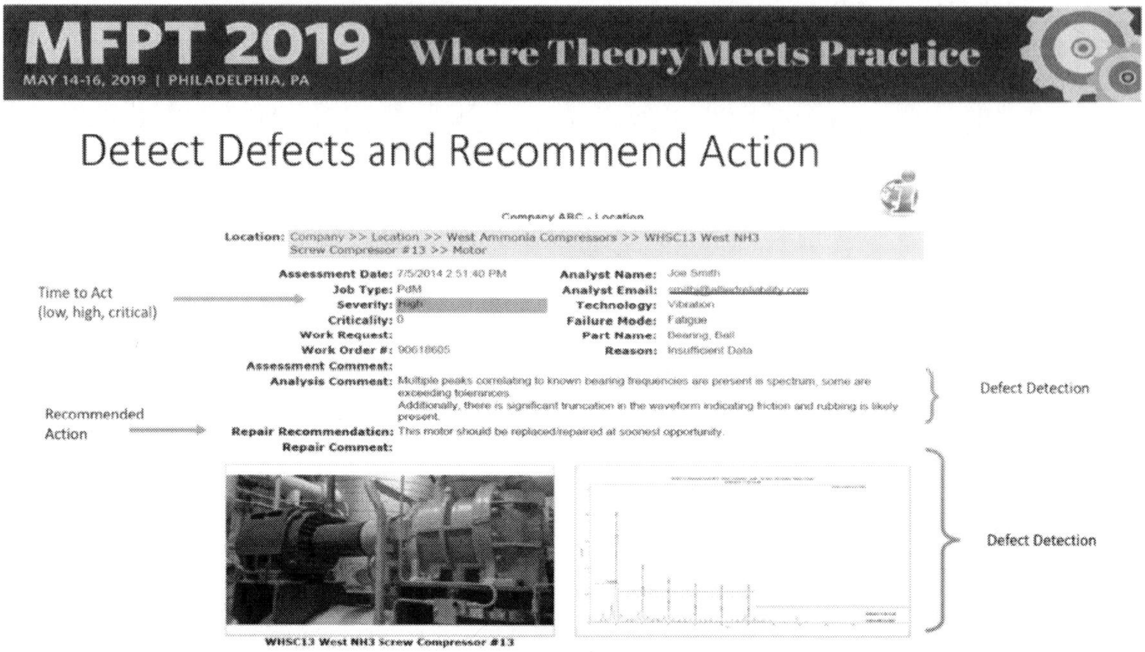
MFPT 2019
MAY 14-16, 2019 | PHILADELPHIA, PA
Where Theory Meets Practice
Detect Defects and Recommend Action
Company ABC - Location
Location: Company >> Location >> West Ammonia Compressors >> WHSC13 West NH3 Screw Compressor #13 >> Motor
Assessment Date: 7/5/2014 2:51:40 PM
Job Type: PdM
Severity: High
Criticality: 0
Work Request:
Work Order #: 90618605
Analyst Name: Joe Smith
Technology: Vibration
Failure Mode: Fatigue
Part Name: Bearing, Ball
Reason: Insufficient Data
Assessment Comment:
Analysis Comment: Multiple peaks correlating to known bearing frequencies are present is spectrum, some are exceeding tolerances.
Additionally, there is significant truncation in the waveform indicating friction and rubbing is likely present.
Repair Recommendation: This motor should be replaced/repaired at soonest opportunity.
Repair Comment:
Time to Act
(low, high, critical)
Recommended
Action
Defect Detection
Defect Detection
WHSC13 West NH3 Screw Compressor #13

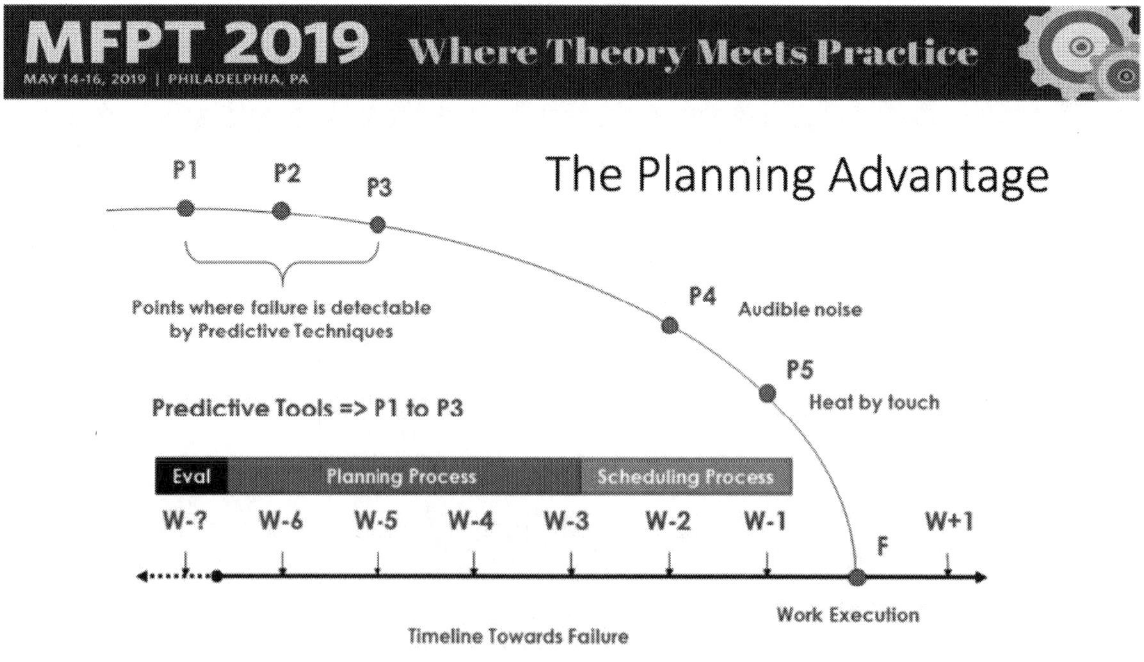
MFPT 2019
MAY 14-16, 2019 | PHILADELPHIA, PA
Where Theory Meets Practice
The Planning Advantage
P1
P2
P3
Points where failure is detectable
by Predictive Techniques
P4
Audible noise
P5
Heat by touch
Predictive Tools => P1 to P3
Eval
Planning Process
Scheduling Process
W-?
W-6
W-5
W-4
W-3
W-2
W-1
F
W+1
Work Execution
Timeline Towards Failure

MFPT 2019 Where Theory Meets Practice
MAY 14-16, 2019 | PHILADELPHIA, PA

Path to Augmented Work Instructions and Aides

MFPT 2019 Where Theory Meets Practice
MAY 14-16, 2019 | PHILADELPHIA, PA

Understanding Your Data

Columns of data:

Operations
Machine Condition

MFPT 2019 Where Theory Meets Practice
MAY 14-16, 2019 | PHILADELPHIA, PA

Dashboards of Success: Specific KPIs

- Production Count (leading)
- Production Rate (leading)
- Task / Cycle Time (leading)
- Asset Health % (leading)
- Overall Equipment Effectiveness (OEE) (lagging)

- Reject Ratio (lagging)
- Downtime (lagging)
- Total Cost Per Unit (lagging)
- Production Losses (lagging)

MFPT 2019 Where Theory Meets Practice
MAY 14-16, 2019 | PHILADELPHIA, PA

Failure Modes Approach

Preventing Failures is the best reliability and maintenance strategy

An SME's Approach to "Digital Transformation"

Building on decades of Domain Expertise:

Human "Supervised" Reliability Centered Machine Learning

Machine Components at Risk

Time to Failure

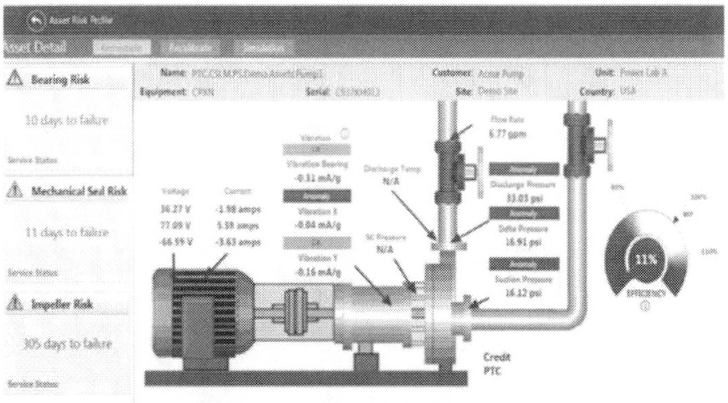

SME's Advise on Sensor Placement / Inclusion Based on Component Failure Modes

SME Commissions System

MFPT 2019 Where Theory Meets Practice
MAY 14-16, 2019 | PHILADELPHIA, PA

Centrifugal Pump and Failure Modes

- **Hydraulic Failure Modes**
  - **Cavitation.**
  - **Pressure Pulsation.**
  - **Pump Recirculation.**
  - **Radial and Axial Thrust.**
- **Mechanical Failure Modes**
  - **Shaft Seizure or Break**
  - **Bearing Failure**
  - **Seal Failure**
  - **Vibrations**
  - **Fatigue**
  - **Misalignment**

- **Lubrication**
  - **Viscosity**
  - **Water**
  - **Wear particles**
- **Motor**
  - **Rotor bar damage**
  - **Eccentricity**
  - **Mechanical looseness**

MFPT 2019 Where Theory Meets Practice

MAY 14-16, 2019 | PHILADELPHIA, PA

Failure Mode Analysis

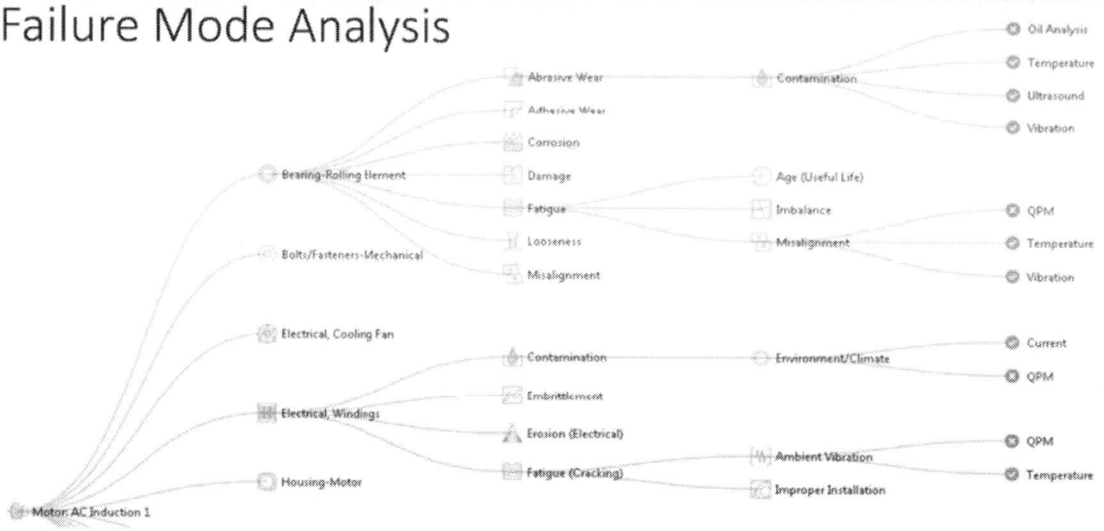

MFPT 2019 Where Theory Meets Practice

MAY 14-16, 2019 | PHILADELPHIA, PA

Organized for Reliability

What happens if 2 shafts are misaligned?

- Increased Power Requirements
- Damaged Bearings
- Damaged Coupling
- Broken Shaft
- Broken Machine Base

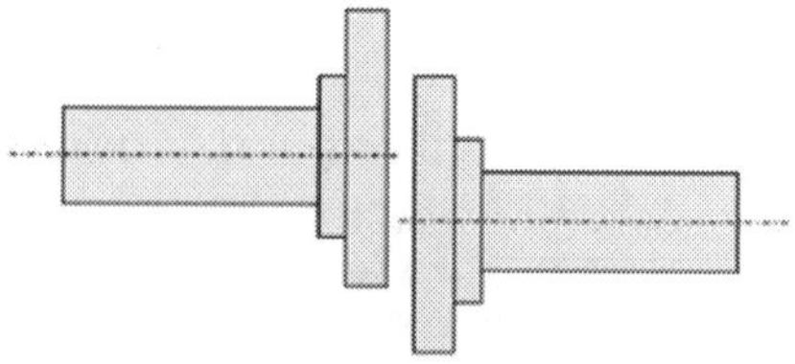

MFPT 2019 Where Theory Meets Practice
MAY 14-16, 2019 | PHILADELPHIA, PA

Failure Modes Detectable with Vibration

Imbalance	Cracked or broken gear teeth
Misalignment	Chain looseness
Bent shaft	Rolling element bearing defects
Bearing issues	Resonance
Structural looseness	Sheave run-out (Belts, cables)
Gear wear	Fluid and airflow problems
Cavitation	

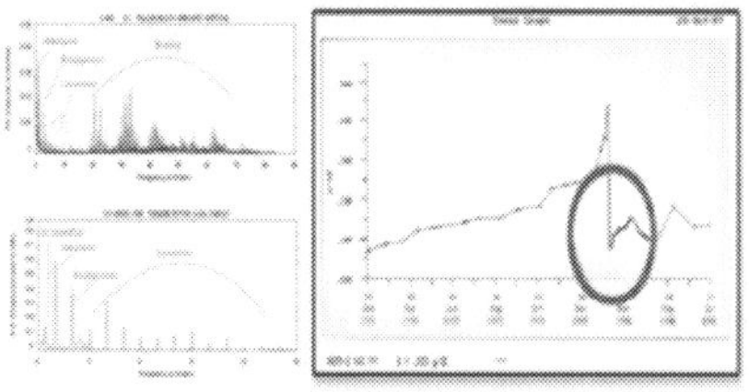

Vibration Analysis

Vibration Analysis is the most frequently deployed CBM technology on the market today. Why? As Art Crawford, the foremost expert on Vibration once stated, ***"There is no single measurement available that contains more information about the mechanical condition of a machine than the vibration signature."***
Simply put, Vibration Analysis can detect a large number of failure modes across a multitude of equipment types.

All rotating equipment vibrates to some degree. As components and parts begin to fail, they begin to vibrate more dramatically and in distinct ways.

MFPT 2019 Where Theory Meets Practice
MAY 14-16, 2019 | PHILADELPHIA, PA

Overall value sensors have a role, yet...

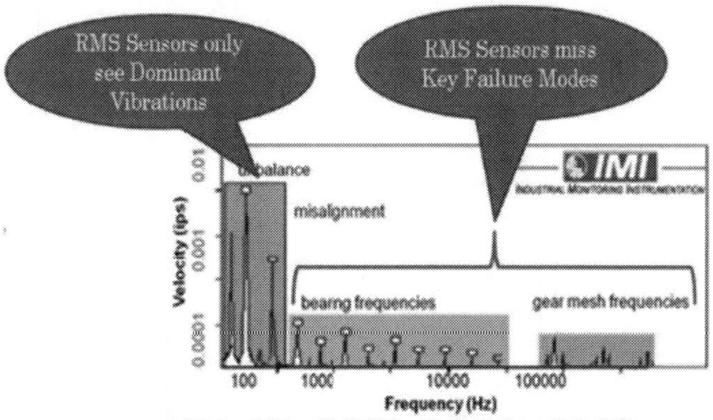

1X, 2X, BPFI, BPFO, GMF

- Include SME in technologies
- Does the sensor provide diagnostic information?
- Does the sensor have the frequency range?
- Is an "engine light" warning appropriate?

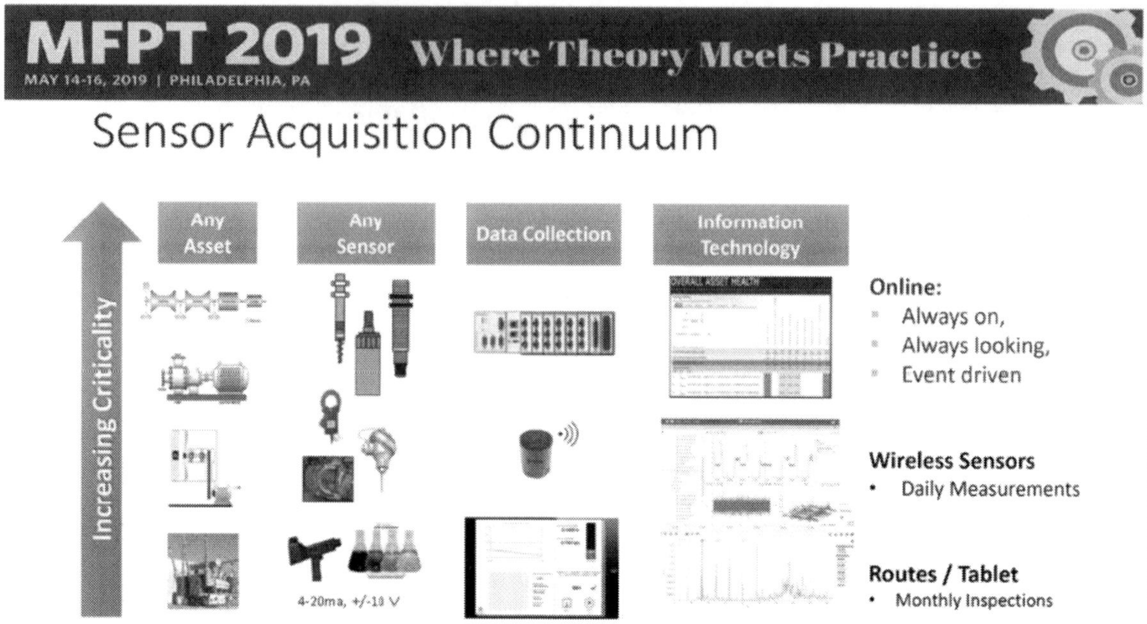
MFPT 2019 Where Theory Meets Practice
MAY 14-16, 2019 | PHILADELPHIA, PA
Sensor Acquisition Continuum
Increasing Criticality
Any Asset
Any Sensor
Data Collection
Information Technology
4-20ma, +/-10 V
Online:
Always on,
Always looking,
Event driven
Wireless Sensors
Daily Measurements
Routes / Tablet
Monthly Inspections

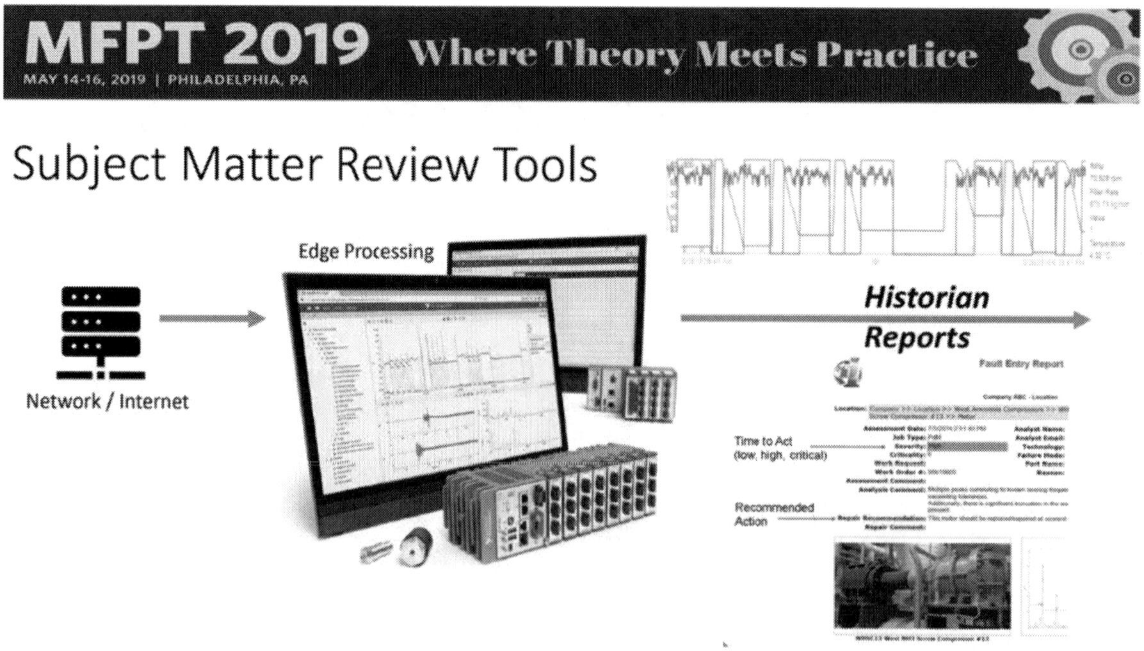
MFPT 2019 Where Theory Meets Practice
MAY 14-16, 2019 | PHILADELPHIA, PA
Subject Matter Review Tools
Edge Processing
Network / Internet
Historian
Reports
Time to Act (low, high, critical)
Recommended Action

MFPT 2019 Where Theory Meets Practice
MAY 14-16, 2019 | PHILADELPHIA, PA

Getting to Digital

First and second steps in implementing your digital transformation for predictive maintenance

MFPT 2019 Where Theory Meets Practice
MAY 14-16, 2019 | PHILADELPHIA, PA

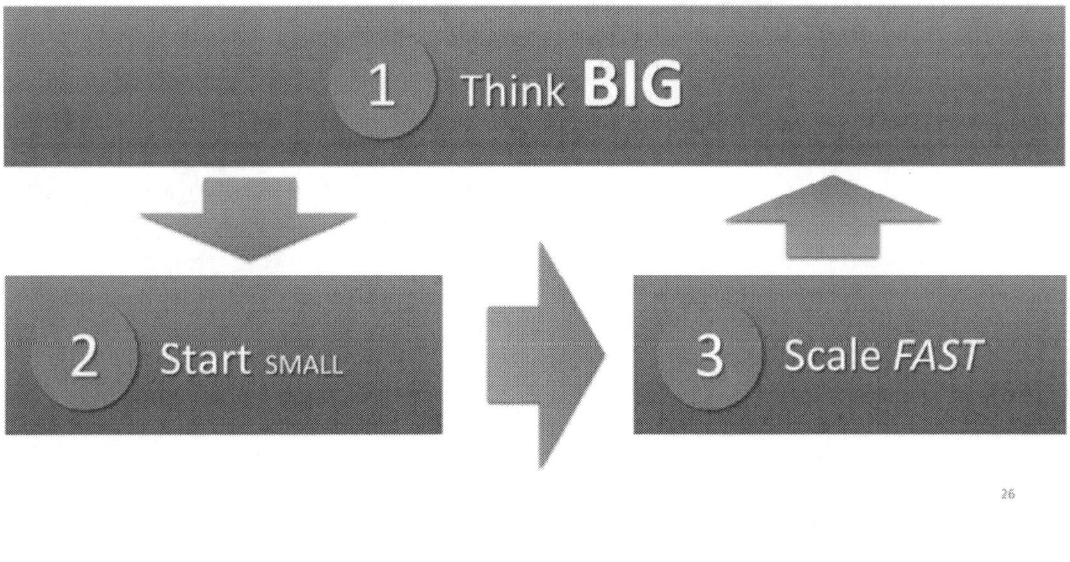

26

Digital Transformation Iterative Process

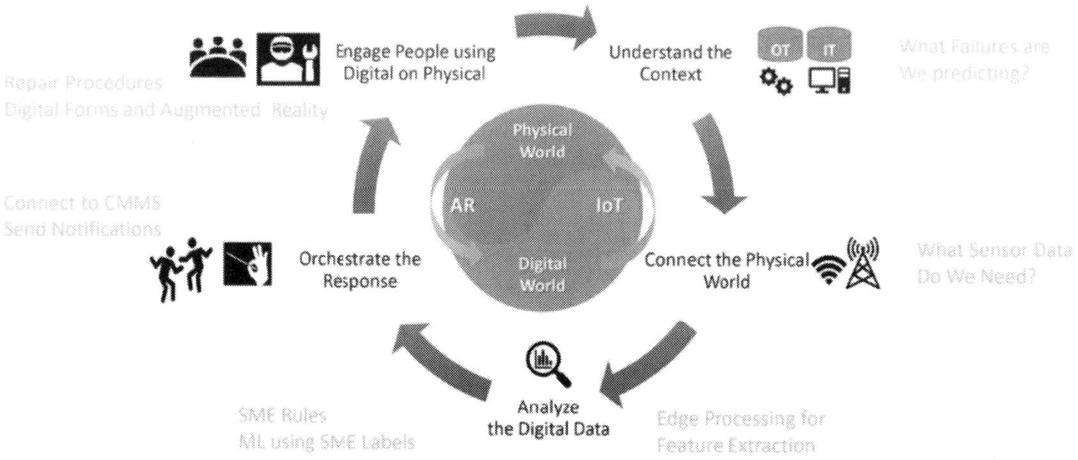

Reliability IIoT Is a Team Sport

- Reliability Experts
  - Equipment maintenance and productivity experts
- Information Technology
  - Computers, networking, historian, CMMS
- Integration Partners
  - Connecting client enterprise applications
  - Machine learning
- CBM Experts
  - Diagnostics, repair recommendation
- Leadership
  - Alignment and KPIs

Must partner with multiple companies to get all needed items.

Business Impact Dashboards

- Determine KPIs
- Organize by user roles

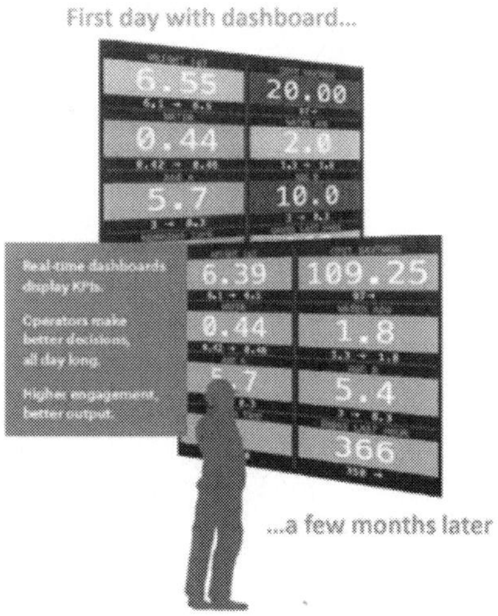

MFPT 2019 Where Theory Meets Practice
MAY 14-16, 2019 | PHILADELPHIA, PA

Deployment Architecture

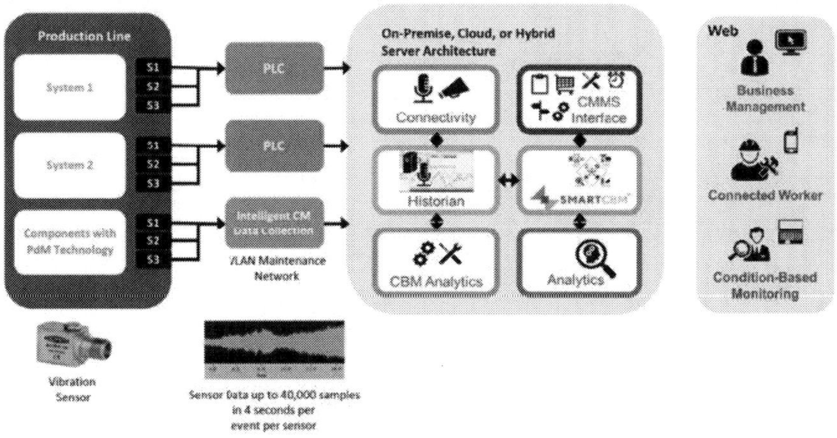

Case Studies

MFPT 2019 Where Theory Meets Practice
MAY 14-16, 2019 | PHILADELPHIA, PA

Snack Foods, Fryer

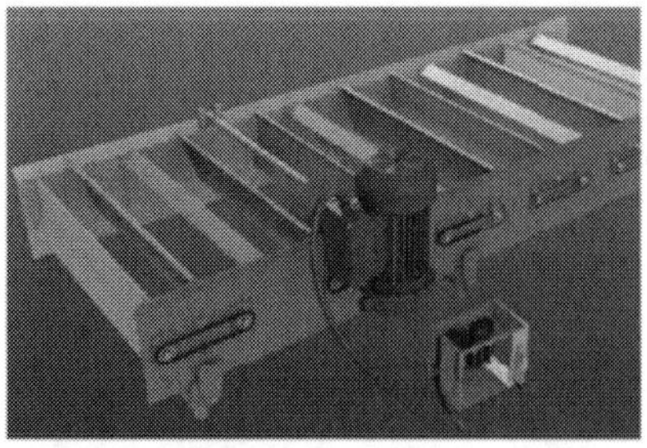

Improve Equipment Productivity
Optimizing Maintenance
Implement Digital Transformation

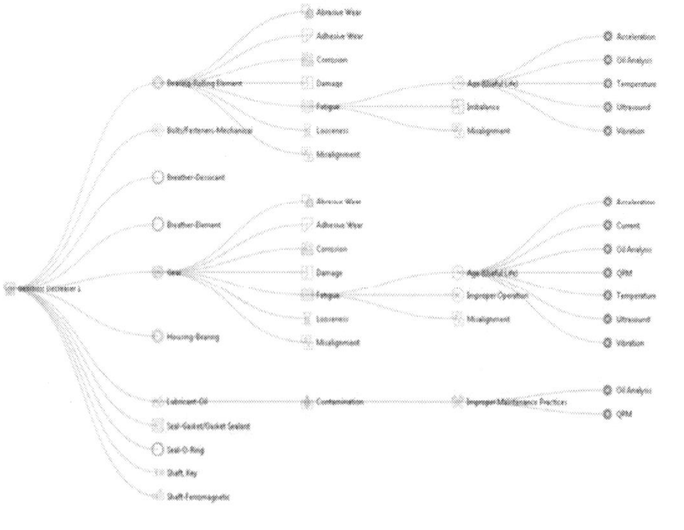

MFPT 2019 Where Theory Meets Practice
MAY 14-16, 2019 | PHILADELPHIA, PA

Steel Plant, Hydraulic Pumps

Improve Equipment Productivity
Lower Maintenance Costs (in 6 months, 3 early IDs)
Implement Digital Transformation

MFPT 2019 Where Theory Meets Practice
MAY 14-16, 2019 | PHILADELPHIA, PA

Pharmaceutical, Air Handling Unit

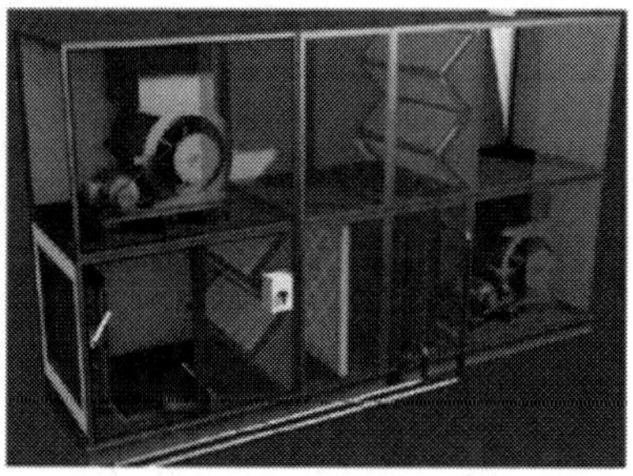

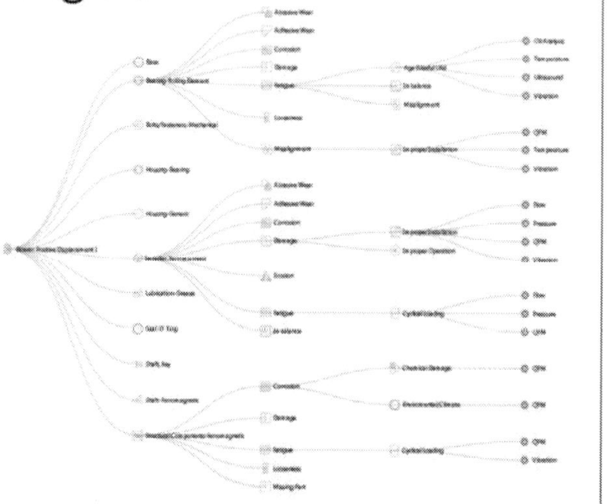

Prove Technology for Digital Transformation
Early identification of defects

Personal Care: AC Servo Gear Motor

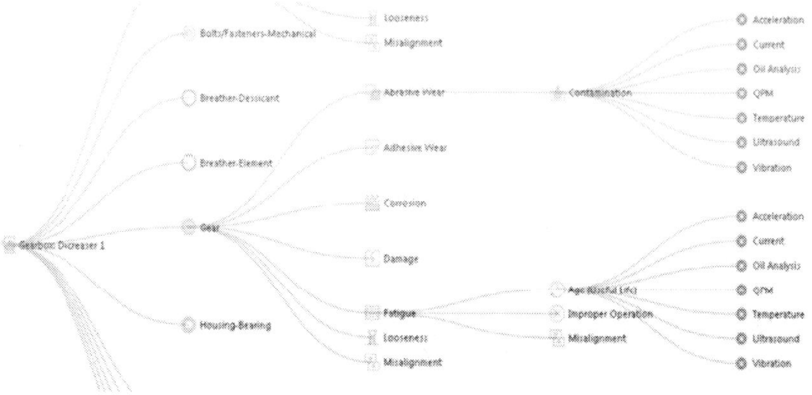

Prove Technology
Early identification of defects
Multi-site roll out

MFPT 2019 Where Theory Meets Practice
MAY 14-16, 2019 | PHILADELPHIA, PA

Refinery of the Future

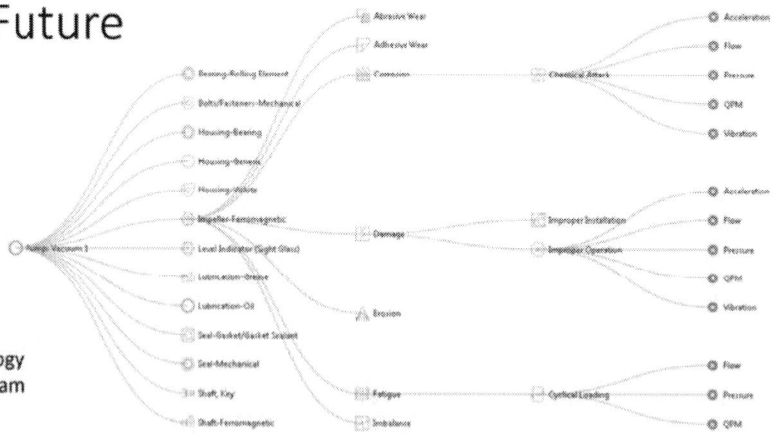

Prove Digital Transformation technology
Reference sight for transformation team
Improve plant capacity

MFPT 2019 Where Theory Meets Practice
MAY 14-16, 2019 | PHILADELPHIA, PA

Midstream: Pipeline Pumping

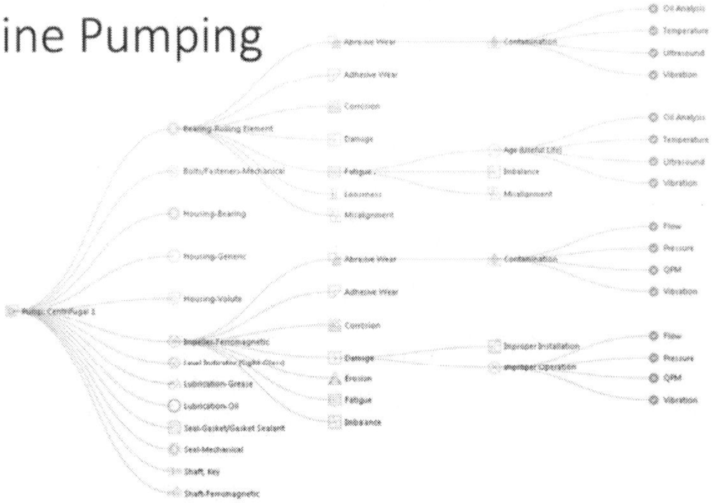

Visibility to remote assets
Reduce manpower to monitor
Lower Maintenance costs
Prove technology

MFPT 2019 Where Theory Meets Practice
MAY 14-16, 2019 | PHILADELPHIA, PA

Summary

MFPT 2019 Where Theory Meets Practice
MAY 14-16, 2019 | PHILADELPHIA, PA

How Do We Get Started?

How do we get there?

- Start with small set of assets
- Understand the failure modes
- Map Sensing technology
- Online, real-time reporting
- Think **BIG**, start small, scale *fast*

Define your purpose

- Lower maintenance costs
- Increase uptime
- Improve visibility
- Prove the technology
- ???

Estimation of Instantaneous Angular Speed and Diagnostics of Train Coach A/C Fan Bearings

Gabriel Davidyan, Renata Klein and Jacob Bortman

PHM Lab, BGU
R.K. Diagnostics

MFPT 2019 Where Theory Meets Practice
MAY 14-16, 2019 | PHILADELFHIA, PA

Ref: 201909-0351

Objectives

- Algorithm for diagnostics of bearings in the A/C of train cars
- Estimation of instantaneous angular speed based on the vibration signal

Outline

- Motivation
- Instantaneous Angular Speed (IAS) estimation
- IAS accuracy evaluation
- Experimental analysis results
- Summary

Ref: 201909-0351 May 2019 Diagnostics of REB & IAS estimation 3

Introduction

- Some train cars suffer from damaged bearings in the A/C.
  - Long maintenance time and increased maintenance costs.
- The A/C is located on the cars' roof
- Not feasible to install a rotating speed sensor

Ref: 201909-0351 May 2019 Diagnostics of REB & IAS estimation 4

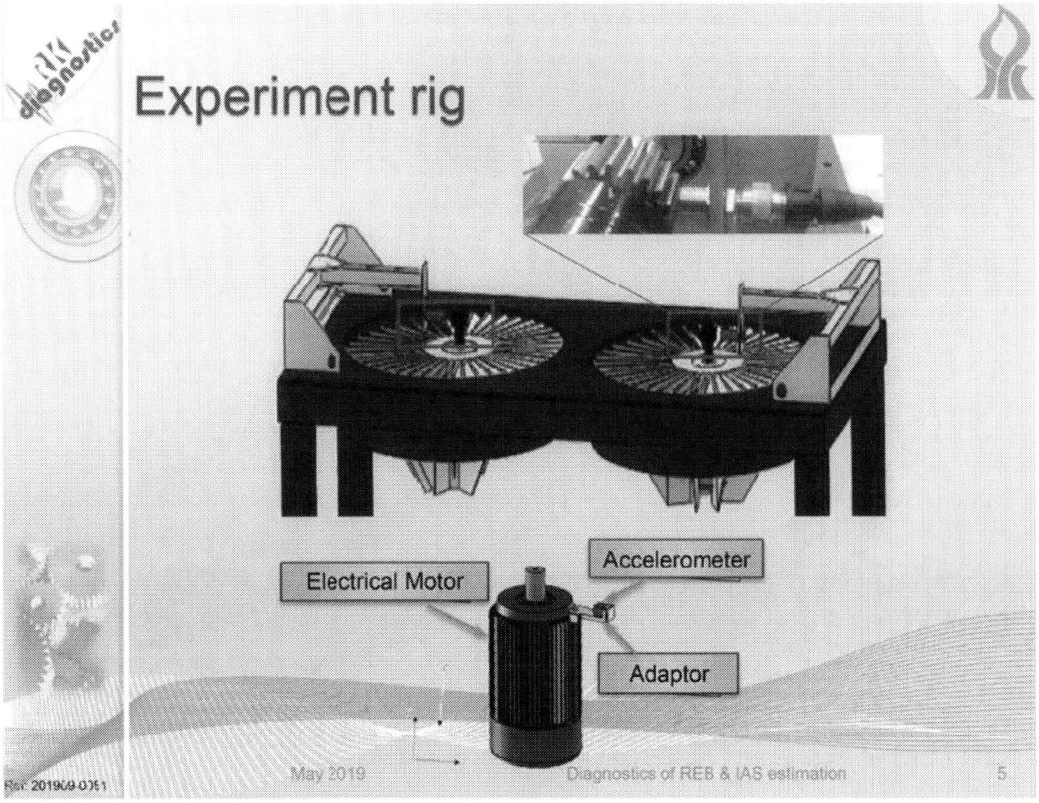
Experiment rig
Accelerometer
Electrical Motor
Adaptor
May 2019
Diagnostics of REB & IAS estimation
5

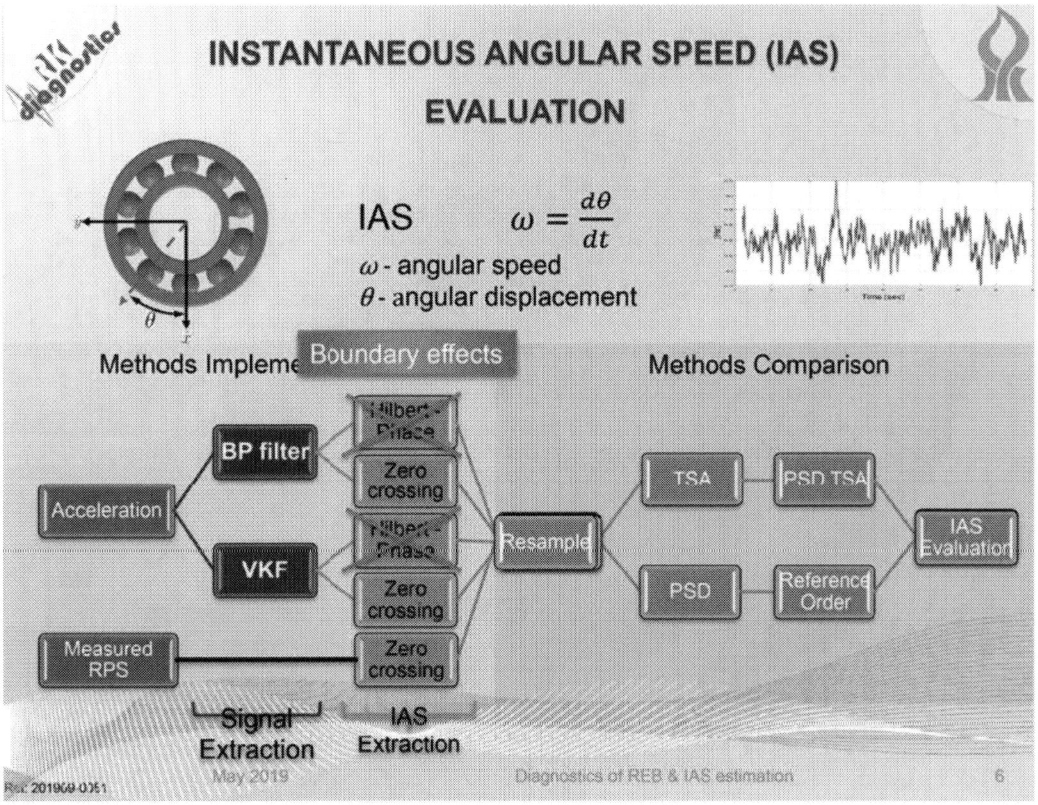
INSTANTANEOUS ANGULAR SPEED (IAS)
EVALUATION
IAS $\omega = \frac{d\theta}{dt}$
ω - angular speed
θ - angular displacement
Methods Impleme
Boundary effects
Methods Comparison
BP filter
Acceleration
VKF
Measured RPS
Zero crossing
Zero crossing
Zero crossing
Resample
TSA
PSD TSA
PSD
Reference Order
IAS Evaluation
Signal Extraction
IAS Extraction
May 2019
Diagnostics of REB & IAS estimation
6

ANGULAR SPEED EVALUATION

$$EL_{SA} = 10 \sum_i \log_{10} \frac{S_{O,i}}{S_{SA,i}}$$

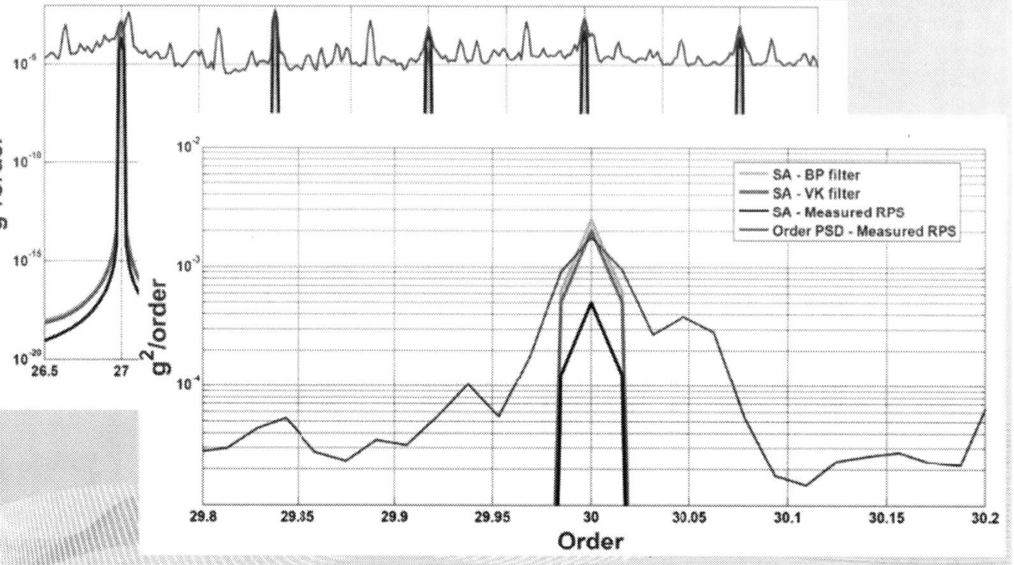

Band Pass Filter

- Butterworth filter

Attenuation (dB)	Ripple (dB)	Filter Order	Filter Cascades	Bandpass (Hz)	Harmonic	Nominal RPS (Hz)
10	3	Auto	20	[20 30]	1	24.6
				[70 80]	3	
				[290 300]	12	

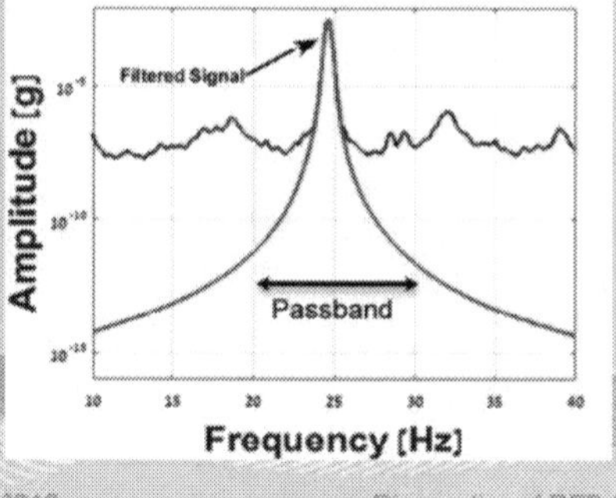

VKF - Vold Kalman Filter

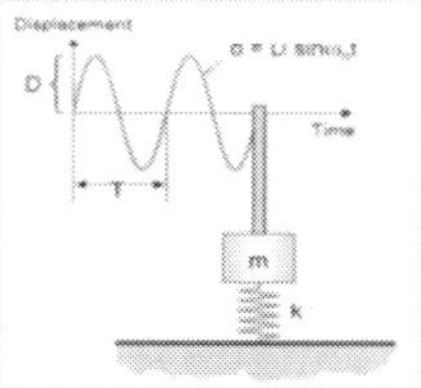

- Data equation:
  $$x(n\Delta t) - 2\cos(\omega\Delta t)x\big((n-1)\Delta t\big) + x\big((n-2)\Delta t\big) = \varepsilon(n)$$
- Structural equation: $y(n) = x(n) + \varepsilon(n)$
  - *x(n)* - filter output, harmonic signal of a specified frequency
  - *ω* - angular frequency
  - *y(n)* - measured signal.
  - *ε(n)* - error term

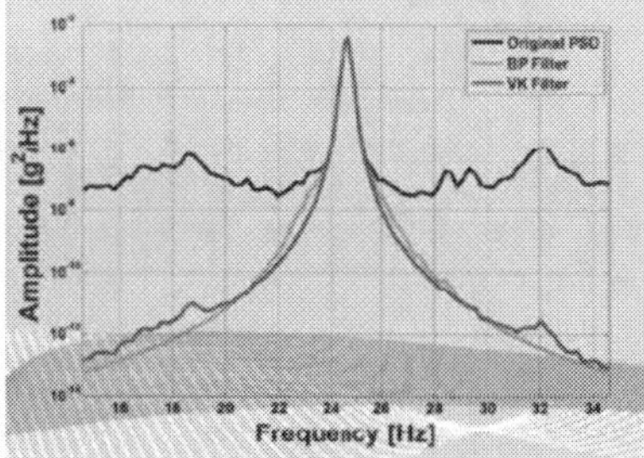

Bandpass (Hz)	Harmonic	Nominal Frequency (Hz)
2.46	1	24.6
	3	
	12	

METHODS COMPARISON

$$EL_{SA} = 10\sum_i \left(log10\frac{S_{O,i}}{S_{SA,i}}\right)$$

- 1st harmonic provided best results at high orders
- VKF and BPF have very similar results
- ✓ **VKF** was selected because of its efficiency

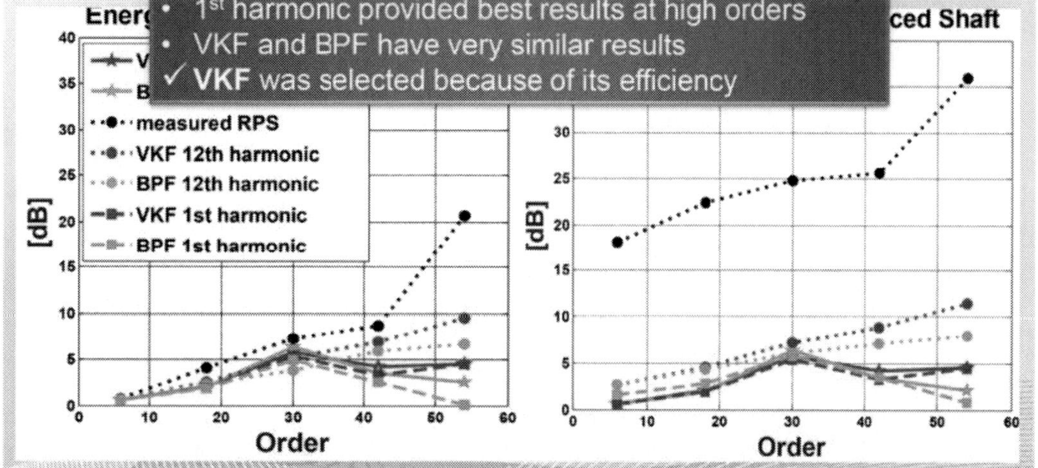

Experiment

- 5 bearings 6205 TVH with spalls of width:
  - 0 – healthy
  - 1.5, 2, 2.5, 3.5 mm
- RPS 24.6 Hz
- 25 repetitions

Bearing Tone	Order
BPFI	5.418
BPFO	3.578
FTF	0.4
BSF	4.707

Data Analysis

Vibrations
- IAS – 1st order VKF

→

Angular Resampling
- PSD
  - Order

↓

Envelope Analysis
- Envelope – Hilbert transform
  - PSD
    - Order

←

Feature Extraction
- Bearing CIs:
  - Energies and statistical distances @ BPFO & SB

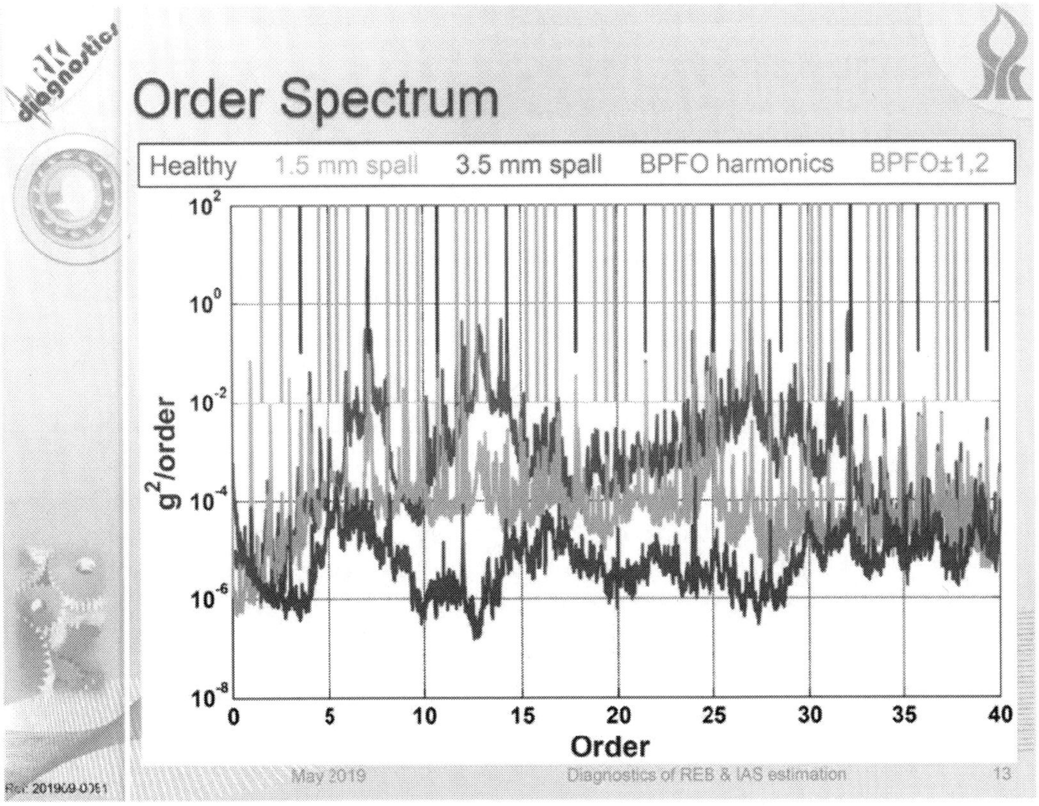
Order Spectrum
Healthy
1.5 mm spall
3.5 mm spall
BPFO harmonics
BPFO±1,2
g²/order
10²
10⁰
10⁻²
10⁻⁴
10⁻⁶
10⁻⁸
0
5
10
15
20
25
30
35
40
Order
Ref: 201909-0351
May 2019
Diagnostics of REB & IAS estimation
13

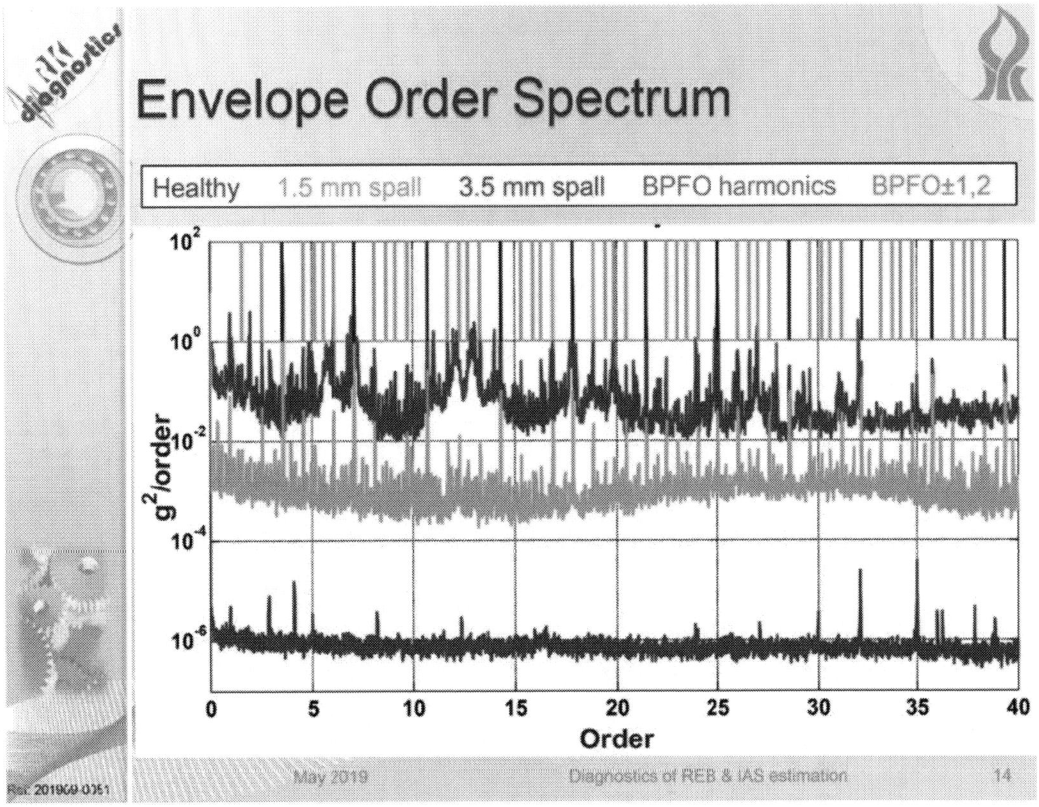
Envelope Order Spectrum
Healthy
1.5 mm spall
3.5 mm spall
BPFO harmonics
BPFO±1,2
g²/order
10²
10⁰
10⁻²
10⁻⁴
10⁻⁶
0
5
10
15
20
25
30
35
40
Order
Ref: 201909-0351
May 2019
Diagnostics of REB & IAS estimation
14

Summary

- Both methods for rotating speed signal extraction from the vibrations, i.e. cascade of BP filters and Vold Kalman filter provided good results.
- The angular speed estimation based on zero crossings was found superior to phase estimation via Hilbert transform.
- A new method for RPS accuracy evaluation was proposed and implemented.
- Analysis of experiments with seeded faults validated the concept of extracting RPS from the vibration signal, and demonstrated the abilities to detect bearing faults and to rank their severity.
- Based on the above findings, an algorithm for diagnostics of a bearing without a rotating speed sensor was developed for the case of A/C in train cars.

Questions?

1ST VS 12TH HARMONIC

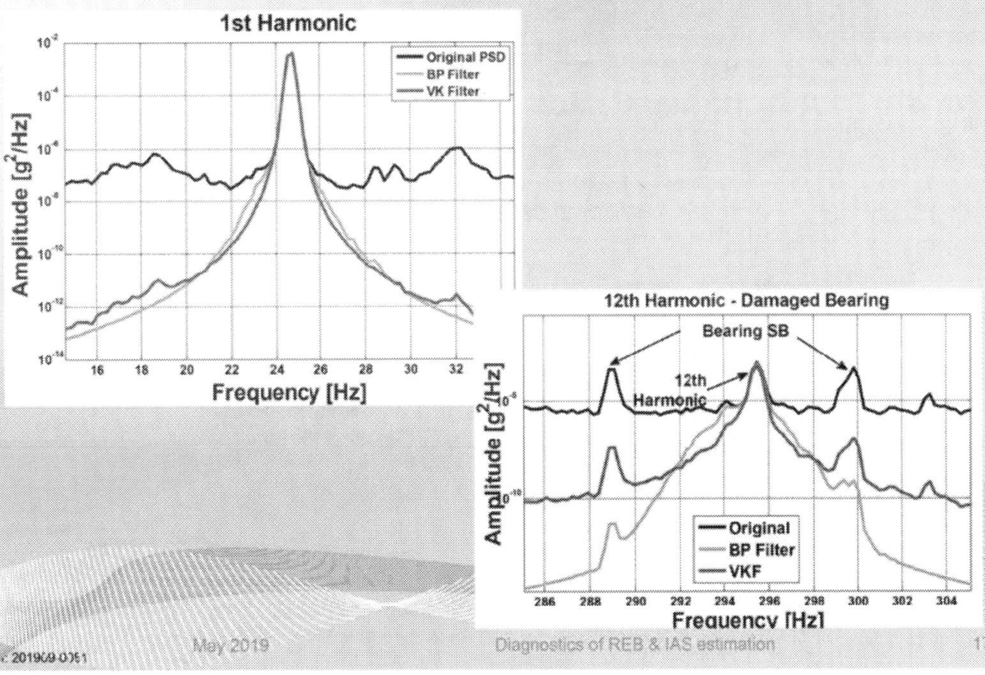

Reference Order Selection

- Energy Leakage $EL = \frac{\sum_i S_{O,i}}{\sum_j S_{O,j}}, \forall i \in \mathbb{Z}_O, j \neq i$
  - Large EL represents the order that has the least energy leakage, i.e. the order with the narrowest and highest peaks.
- Peak Energy Concentration $PEC = \frac{\sum_i S_{O,i}}{RMS}$
  - PEC measures the energy percentage that is concentrated in the peaks

	EL	PEC	EL	PEC	EL	PEC
Harmonic	1st		3rd		12th	
VK Filter	0.562	1.862	0.56	1.87	0.542	1.908
BP Filter	0.56	1.645	0.558	1.75	0.521	1.535
Measured	0.564	2.135	0.564	2.135	0.564	2.135

Transforming Work Through Human Sensing: A Confined Space Monitoring Application

Machinery Failure Prevention Technology 2019
Human Performance and Specialty Diagnostics

May 13th, 2019

Presenter: Zach Kiehl

Co-authors: Kevin Durkee, Kent Halverson, Ph.D., James Christensen, Ph.D., & George Hellstern

© 2019 Aptima, Inc.

Acknowledgments and Disclaimers

This material is based upon work supported by the U.S. Air Force Sustainment Center under Contract No. FA8501-16-C-0020. The following opinions, findings, recommendations, and conclusions expressed in this material are those of the authors and do not necessarily reflect the views of the U.S. Air Force.

A special thank you to the Air Force Research Laboratory for their help and support.

© 2019 Aptima, Inc.

APTIMA®
Human-Centered Engineering®

Confined space work operations are (1) prevalent across many domains and (2) relatively dangerous

About 2.1 million workers enter permit-required confined spaces annually

In 2017 and 2016, there were 166 and 144 workplace deaths involving confined spaces

Approximately 60 percent of confined-space fatalities are rescuers

© 2019 Aptima, Inc.

3

APTIMA®
Human-Centered Engineering®

Rigid safety practices have been implemented due to the precariousness nature of confined space entry

Confined space monitoring is a practice mandated by OSHA to protect worker health & safety

Current monitoring practices requires manual oversight (e.g., 1:1 ratio) and frequent verbal communication

Current monitoring practices are arguably somewhat ineffective and unequivocally costly

© 2019 Aptima, Inc.

4

APTIMA
Human-Centered Engineering®

Common confined space entry protocol

© 2019 Aptima, Inc.

5

APTIMA
Human-Centered Engineering®

Can we improve the safety and efficiency of current confined space operations for aircraft maintenance?

© 2019 Aptima, Inc.

6

Confined Space Monitoring System (CSMS): A cloud- & IoT-based architecture for remote entrant monitoring

- **Unobtrusive sensor suite** to remotely monitor worker health signals, atmospheric hazards, & location in confined spaces

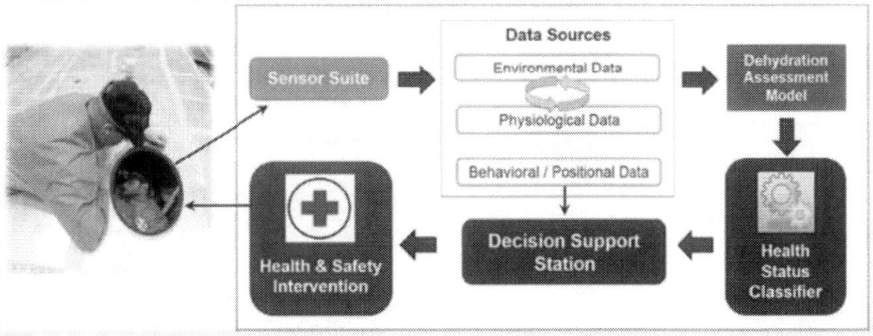

- **Integrated decision support** for alerting & intervention
  - Single person can monitor many confined spaces concurrently
  - Continuously assess health status and safety risks
  - Early warning alerts & accelerated emergency response

© 2019 Aptima, Inc.

APTIMA®
Human-Centered Engineering®

Operational view of the CSMS

© 2019 Aptima, Inc.

Technical components of the CSMS

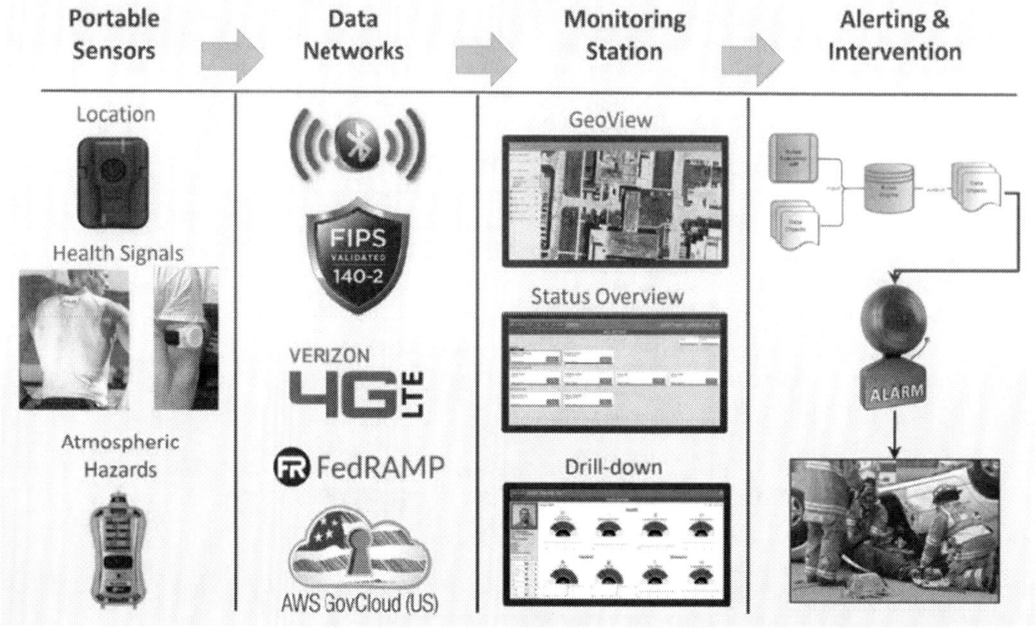

© 2019 Aptima, Inc.

System architecture of the CSMS

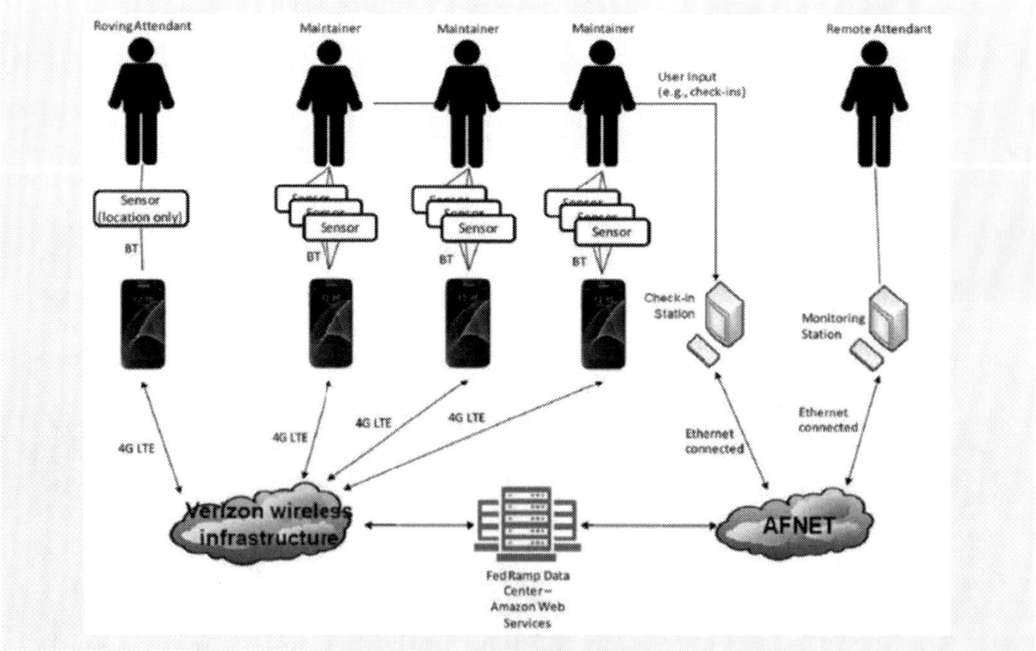

© 2019 Aptima, Inc.

APTIMA®
Human-Centered Engineering®

Example physiological data streams

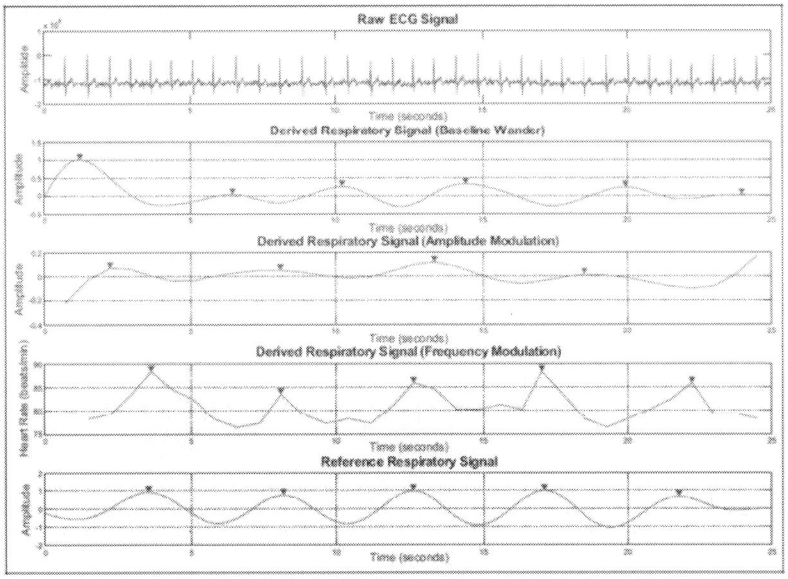

© 2019 Aptima, Inc.

Potential adverse health states of confined space entrants

State or condition	Signal or metric	Presenting symptom	Factors used to distinguish between dangerous and safe conditions
1. Tachycardia	ECG or HR in BPM	Elevated resting HR	Activity levels via accelerometry; high levels of motion can increase HR
2. Bradycardia	ECG or HR in BPM	Slow resting HR	Activity levels via accelerometry; low levels of motion can decrease HR
3. Hyperventilation	Respiration waveform or RR in breaths per minute	Elevated respiration rate	Activity levels via accelerometry; high levels of motion can increase RR
4. Hypoventilation	Respiration waveform or RR in breaths per minute	Slow respiration rate	Activity levels via accelerometry; low levels of motion can decrease RR
5. Hyperthermia	Skin or core temperature	Elevated body temperature	Ambient temperature; extreme temperatures can increase or reduce core body temperature
6. Hypothermia	Skin or core temperature	Low body temperature	Ambient temperature; extreme temperatures can increase or reduce core body temperature
7. Acute injury	ECG or HR, RR, accelerometer signals	Elevated HR, elevated RR, sudden motion or posture change	User input from technician via smartwatch
8. Panic attack (e.g. due to claustrophobia or agoraphobia)	ECG or HR, RR, accelerometer signals	Elevated HR, elevated RR, increased motor activity	User input from technician via smartwatch
9. Hypoxia	SpO2, RR, Environmental oxygen	Reduced SpO2, elevated RR, decrease atmospheric oxygen saturation	Ambient environmental oxygen saturation
10. Hemorrhage	ECG, RR, PPG	Elevated HR, reduced RR, reduced amplitude of PPG waveform	User input from technician via smartwatch
11. Apnea (cessation of breathing)	RR and SpO2	Reduced RR and SpO2	Flat spirogram, atmospheric composition, and activity via accelerometry
12. Cardiac arrest	ECG or HR, PPG, accelerometer signals	Reduced HR, flat PPG signal, reduced movement	User input from technician via smartwatch

ECG: electrocardiography; HR: heart rate; BPM: beats per minute; PPG: photoplethysmogram; SpO2: blood oxygen saturation.

© 2019 Aptima, Inc.

Remote attendants provide enhanced safety and labor efficiency

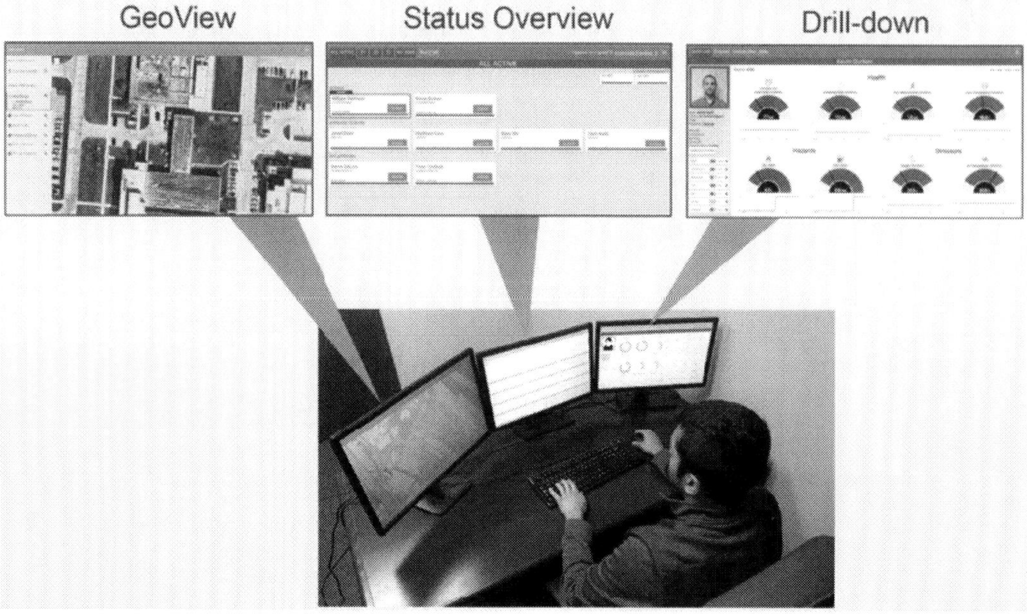

© 2019 Aptima, Inc.

APTIMA®
Human-Centered Engineering®

Risks, mitigations, & concerns of the CSMS

User acceptance	Data dropouts & alarm sensitivity	Return on Investment
• Met with labor union & OSHA representatives • Allowed user to try on sensor suites • Met with working groups	• Conducted ecologically valid dropout testing • Conducted pilot testing at maintenance depots • Decommissioned wing test	• Estimated payback period of 3 years • Estimated annual savings of >$1 million per year • Scalability

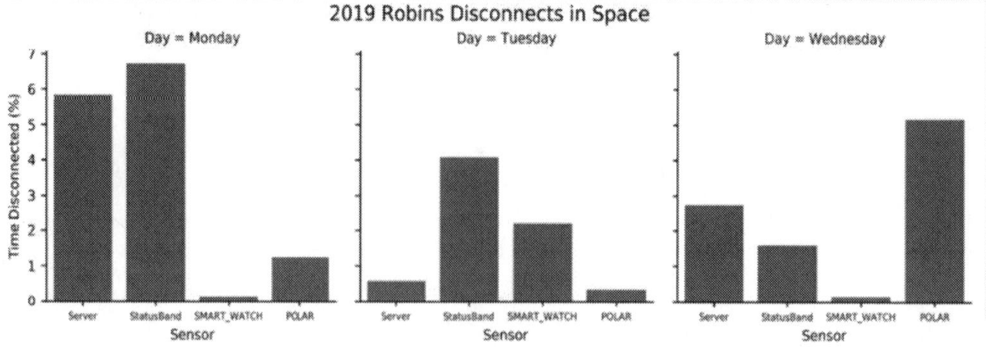

APTIMA
Human-Centered Engineering®

Future research plans and applications

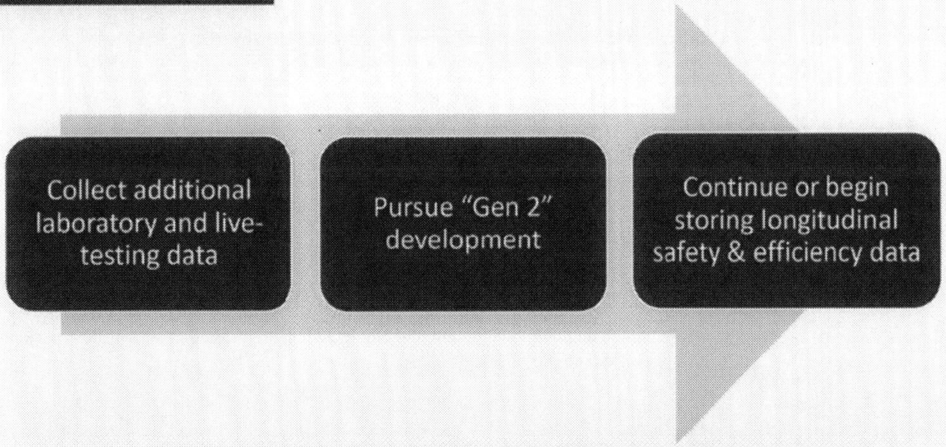

Kiehl, Z. A., Durkee, K. T., Halverson, K. C., Christensen, J. C., & Hellstern, G. F. (2019). Transforming work through human sensing: a confined space monitoring application. *Structural Health Monitoring*, 1475921719840994.

© 2019 Aptima, Inc.

15

APTIMA
Human-Centered Engineering®

Zach Kiehl, Capability Lead for Human-System Performance

2555 University Blvd., Suite 300, Fairborn, OH 45324
zkiehl@aptima.com
Direct: (937) 490-8028
Mobile: (937) 606-4354

www.aptima.com

© 2019 Aptima, Inc.

Systems Engineering Awareness

MFPT 2019

John Lucero
NASA Glenn Research Center
King of Prussia, PA
May 16, 2019

Presenter

John Lucero

(32 years exp.)
Facilities Testing Division

NASA Glenn Research Center
Brook Park, Ohio 44135

Office 216 433 2684
Cell 440 915 9277
John.M.Lucero@nasa.gov

Question to audience

- Why are we sitting here?

Possible Answer

- Seeking greater success in project endeavors.

Proposed solution

- Performing activities defined by Systems Engineering methods will greatly increase project success.

Introduction

History– Wikipedia

1. Systems Engineering can be traced back to Bell Telephone Laboratories in the 1940s
2. The development and identification of new methods and modeling techniques
3. International Council on Systems Engineering (INCOSE) 1995
4. The conception, design, development, production and operation of physical systems

Definitions and Terms

System:

- NASA: A set of interrelated components which interact with one another in an organized fashion toward a common purpose.

- DOD: An integrated composite of people, products, and processes that provide a capability to satisfy a stated need or objective.

- INCOSE: A construct or collection of different elements that together produce results not obtainable by the elements alone.

MFPT 2019 – Systems Engineering Awareness

Definitions and Terms

Systems Engineering:

- MIL-STD-499B: An interdisciplinary approach encompassing the entire technical development effort to evolve and verify an integrated and life-cycle balanced set of system, people, product and process solutions that satisfy customer needs. Systems engineering encompasses:
  - Technical effort (develop, manufacture, verify, operate, etc)
  - Define and manage the system configuration
  - Translate the system to a work breakdown structure
  - Develop information for decision making.

Definitions and Terms

Systems Engineering:

- INCOSE: An interdisciplinary approach and means to enable the realization of successful systems, focusing on customer needs early in the design cycle, document requirements and proceed to design synthesis and system validation. Consider the complete problem: operations; performance; test; mfging; $ and time; training/support; disposal.

Definitions and Terms

Systems Engineering:

- NASA: Requires the application of a systematic, disciplined engineering approach that is quantifiable, recursive, iterative, and repeatable for the development, operation, maintenance and disposal of the systems integrated into a whole throughout the life cycle of a project or program. The emphasis is on achieving stakeholder functional, physical, and operational performance requirements in the intended use environments over the system's planned life within cost and schedule.

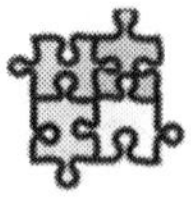

Definitions and Terms

Systems Engineering is:

- Multidisciplinary
- Systematic
- Processes
- Development/operation
- Complex product
- Limited resources
- Constraints

Process:

- Complex work activity
- Transforms inputs into outputs
- Consumes resources

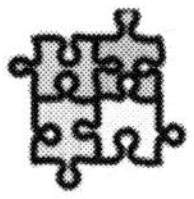

Definitions and Terms

Three Types of Systems Engineering:

1. Product Systems Engineering (PSE) is the focus on the design of physical systems consisting of hardware and software.

2. Enterprise Systems Engineering (ESE) pertains to the view of enterprises, that is, organizations or combinations of organizations, as systems.

3. Service Systems Engineering (SSE) has to do with the engineering of service systems.

Benefit of SE

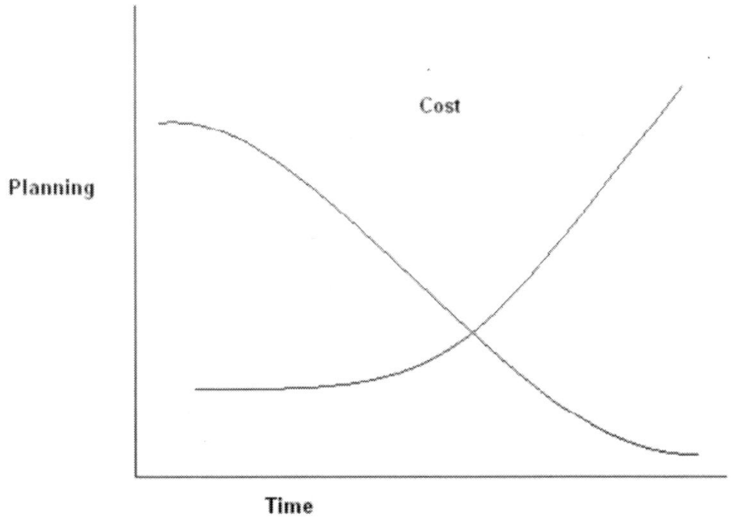

Definitions and Terms

Systems Engineering Leadership:

- Knowledge base
  - What do you know about the subject?
- Engineering judgment
  - Experience, successes
- Effective Communication
  - Written, verbal, diplomacy
- Systematic Processes
  - Company policy
  - Examples: CMMI, ISO, NPR 7123.1A

Definitions and Terms

Interaction of PM, SE, and TL = WBS creation:

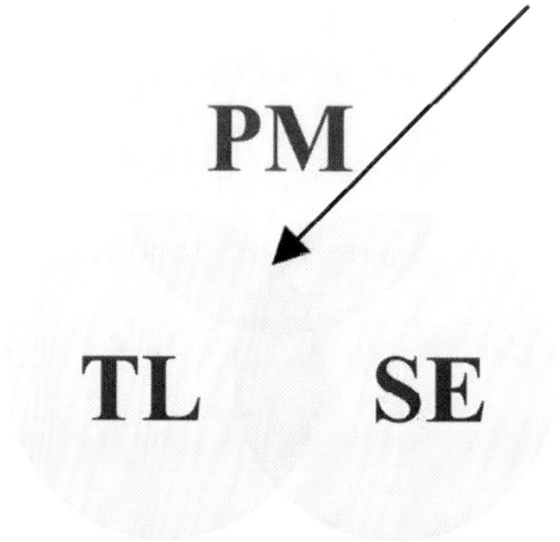

SE Dilemma

MUST BALANCE

Risk (R)! Performance (P)! Cost (C)!

C*R = P

1. Reduce C at constant R = P drops
2. Reduce R with constant C = P drops
3. Reduce P at constant C = Risk drops
4. Reduce P at constant R = Cost drops
5. Reduce C at constant P = Risk increase
6. Reduce R at constant P = Cost increase
7. Etc.

Design Engineering Dilemma

<u>Engineers want a solution NOW!</u>

- General statements lead to detailed design on first step.
- Difficult to step back and look at BIG picture
- Up front requirements definition and systems engineering planning are PARAMOUNT before designs start getting built.
- Late design changes cost $$$$$
- Communication is KEY.

NASA Procedural Reqmt. 7123.1A

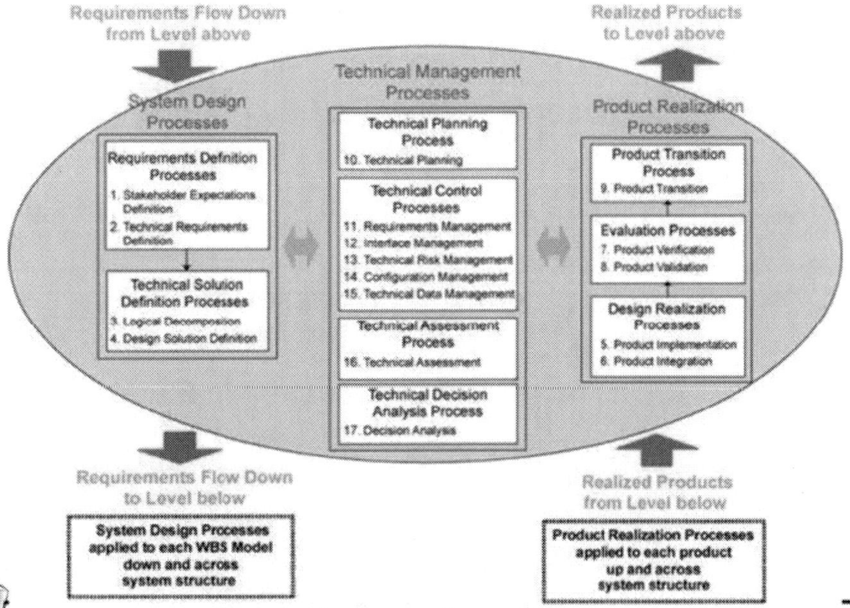

This handbook consists of six core chapters:

(1) Systems engineering fundamentals discussion
(2) the NASA program/project life cycles
(3) systems engineering processes to get from a concept to a design
(4) systems engineering processes to get from a design to a final product
(5) crosscutting management processes in systems engineering
(6) special topics relative to systems engineering

CDR	Critical Design Review	PLAR	Post-Launch Assessment Review
CERR	Critical Events Readiness Review	PRR	Production Readiness Review
DR	Decommissioning Review	P/SDR	Program/System Definition Review
FRR	Flight Readiness Review	P/SRR	Program/System Requirements Review
KDP	Key Decision Point	PSR	Program Status Review
MCR	Mission Concept Review	SAR	System Acceptance Review
MDR	Mission Definition Review	SDR	System Definition Review
ORR	Operational Readiness Review	SIR	System Integration Review
PDR	Preliminary Design Review	SRR	System Requirements Review
PFAR	Post-Flight Assessment Review	TRR	Test Readiness Review
PIR	Program Implementation Review		

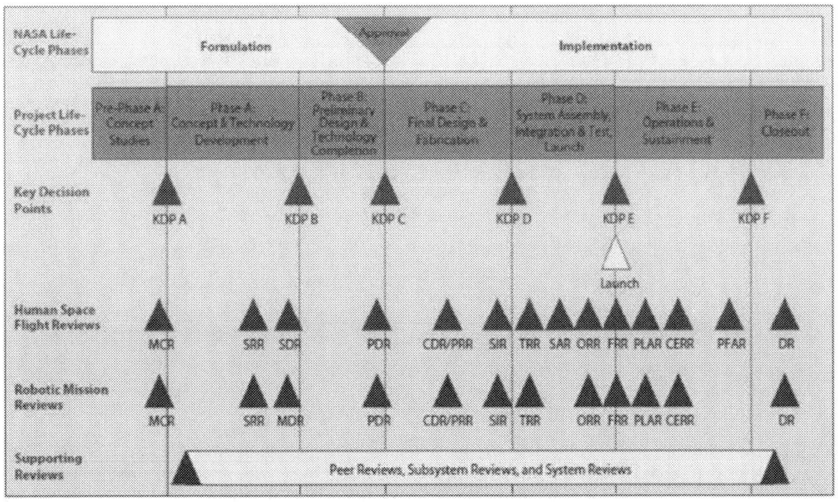

Purpose of PDR

- To determine the feasibility and desirability of a suggested new major system and establish an initial baseline compatible with strategic plans.
- Develop final mission concept, system-level requirements, and needed system structure technology developments.
- Mature requirements for all products in the developing product tree, develop ConOps, preliminary designs, and perform feasibility analysis of the verification and validation concepts to ensure the designs will likely be able to meet their requirements.

MFPT 2019 – Systems Engineering Awareness

Output of PDR

End products in the form of mockups, trade study results, specification and interface documents, and Prototypes.

Systems Engineering Focus Group

21

MFPT 2019 – Systems Engineering Awareness

How to get to PDR

- Concept of Operations
- Develop High Level Requirements
- Identify Key Driving Requirements
- Define Verification Methods
- Identify Design Solutions
- Perform Trade Studies
- Develop a Work Break Down Structure (Product Based)
- Cost and Schedule
- Risk Management and Mitigation
- Technical Performance Measures
- Configuration Management

Systems Engineering Focus Group

22

Most “key” valuable lessons

- Techniques on calling, holding, and archiving meetings/action items
- Human interface, stakeholder education for synthesis of requirements document (design by requirements is bad)
- Functional Analysis (gives insight, interfaces, WBS, PBS based on product architecture)
- Clarity (no vague requirements wording)
- Plan to iterate
- Diplomacy
- One shall per requirement
- Validate – use it
- Verify – feasible and affordable

Most “key” valuable lessons

- Golden Rule of Requirements – who is going to pay?
- Integration and test – plan what to test, test it, plan early (test what you fly, fly what you test)
- Trade Study Methodology– always iterate
- WBS – product oriented, based on architecture
- PLAN EARLY !!! System Engineering Management Plan (SEMP)

MFPT 2019 – Systems Engineering Awareness

Keys to Success

- Strong SE effort early in project
- Plan and Baseline at each step
- Iterate often
- Strong Risk Management Plan
- Keep an eye on Technical Performance Measures

Systems Engineering Focus Group

MFPT 2019 – Systems Engineering Awareness

Conclusions

- Proper application of SE methods upfront in a project life cycle will greatly increase success.
- Paying close attention to the Stakeholder Analysis will greatly increase project success.
- Performing activities defined by SE methods will increase chances of obtaining funding to complete your project.
- Verification and Validation is key at every level in the project.

Systems Engineering Focus Group 26

FIELD BASED OIL CONDITION MONITORING

Avoiding Premature Failure

Stuart Lunt
Research Manager, Parker Hannifin, UK.
Presented at MFPT 2019, Philadelphia, USA.

ENGINEERING **YOUR** SUCCESS.

Overview

1. Oil Analysis Programs
2. Lubricating Oil Condition Parameters
3. Chemical Test Kits
4. Portable Offline instruments
5. Online Sensors
6. Case Study - Marine Diesel Engines
7. Concluding Remarks

2

Oil Analysis Programs

Laboratory oil analysis programs for condition based maintenance are well established.

- Samples pulled at defined intervals.
- Shipped to laboratory for analysis.
- Results (and sometimes advice!) by return.

OK for stationary and/or accessible equipment.

What about remote or no-access machinery? e.g.

- Marine vessels.
- Wind turbines.
- Nuclear installations.

Solution: Field testing or online sensing.

3

Parker

Oil Condition Parameters

- **Physical properties**
  - ***Viscosity****
  - ***Density****
  - ***Temperature****
- **Contamination**
  - ***Water****
  - ***Coolant****
  - ***Fuel Dilution****
  - ***Wear Debris****
  - ***Insolubles/Soot/Dirt****
- **Condition/Quality**
  - AN
  - ***BN****
  - Additive depletion
  - ***Oxidation****
  - Sulphonation
  - Nitration

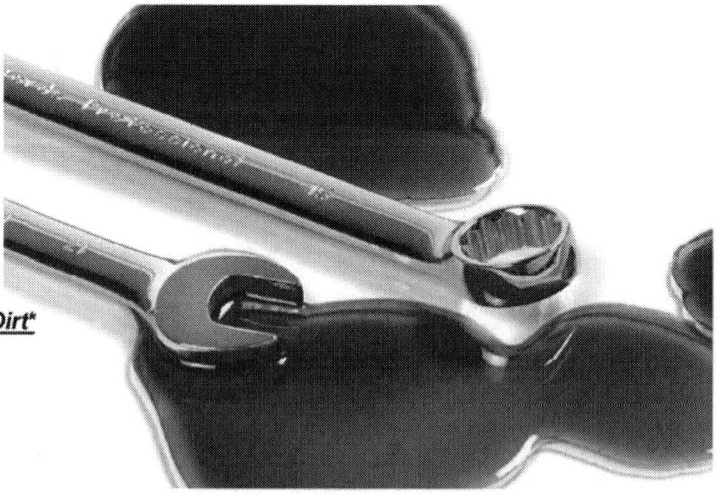

4 ***Sensor available****

Parker

Chemical Test Kits

Pros

- Simple, easy to use.
- Inexpensive.
- Portable

Cons

- Reagents:
    - Consumable items.
    - Environmental issues.
- Time consuming
- Precision sometimes limited.

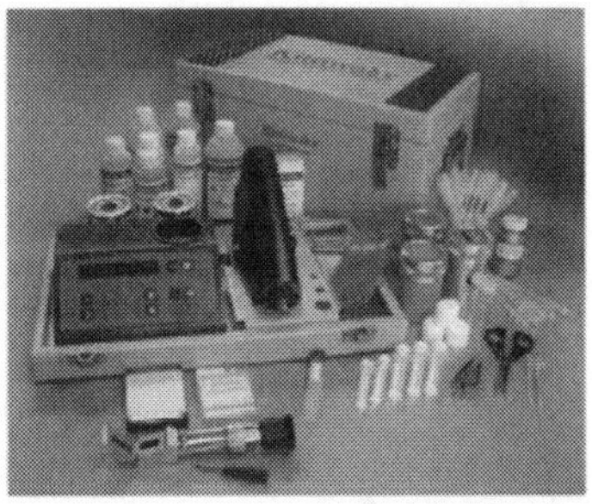

5

Chemical Test Kit Examples - I

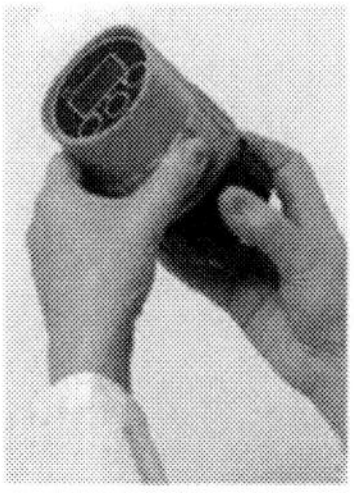

DIGICell

- Gas tight reaction cell
- Defined oil sample volume + reagent
- Free water content determination:
    - Calcium Hydride reaction – H_2 generated.
- Base Number (BN) determination:
    - Acid reaction with carbonate alkaline additive – CO_2 generated.

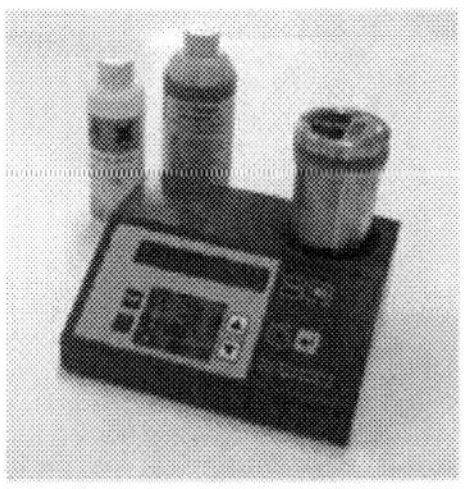

TAN - Titration by dropper bottle

- Acid Number (AN) by simplified IP139 test

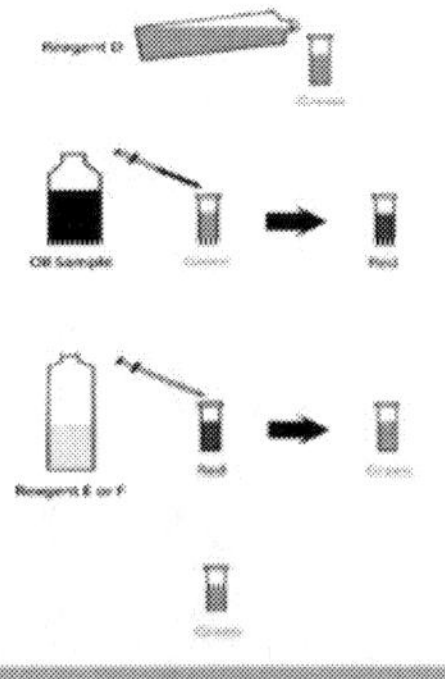

6

Chemical Test Kit Examples - II

Cold Corrosion Test Kit
(CCTK) for cylinder drain oil analysis.

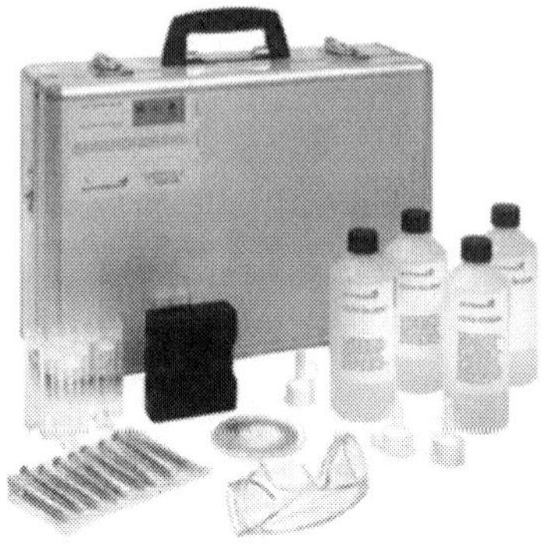

- Determines iron in corrosion by-products.
  - Iron sulphates and oxides.
- Fe^{2+} and Fe^{3+} but not Fe.
  - Fe is from normal wear.

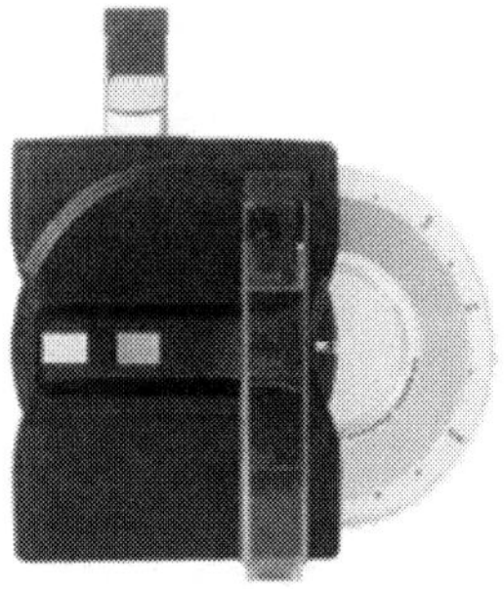

Compounded iron content determined by a chemical reagent which forms a coloured complex only with the oxidized state.
Two ranges: 0 - 400 ppm
And 0 – 800 ppm
Test time is < 5 mins

7

Portable Offline Instruments

Single parameter
- Wear debris – magnetometry for iron/steel
- Viscosity – falling or rolling ball

Multi parameter
- Spectroscopy based
  - IR-ATR
  - XRF

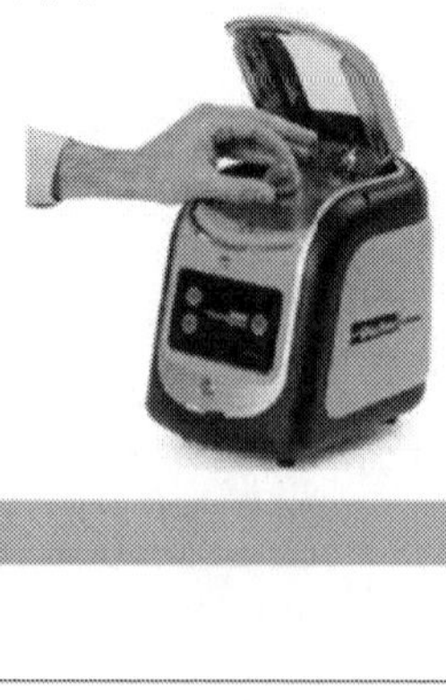

8

Online Sensors

Fluid Condition
- AC Conductivity – aging, fuel dilution, contamination.
- Relative Permittivity – soot contamination, oxidation, incorrect top up.
- Relative Humidity – water ingress.
- Temperature – measurement compensation.
- Pressure – filter plugging, leakages etc.

Wear Debris – Particle Counting
- Inductive sensing
- Ferrous debris – iron/steel
- Non-ferrous debris – white metal, bronze cages etc.

Iron Content
- Ferrous (metallic) iron only

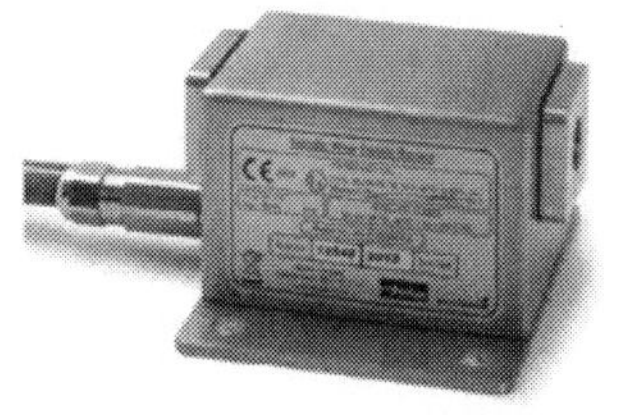

9

Online Sensor Packages – One Example

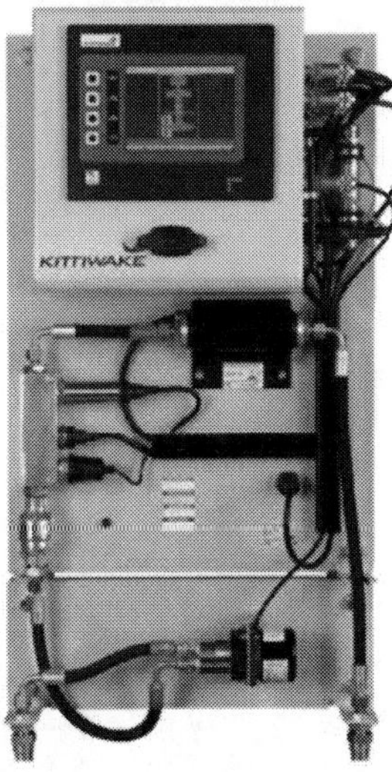

Azimuthal Thruster Monitoring Package:
1. Gearbox Oil &
2. Drive Oil
   - Debris
   - Oil condition – dielectric measurements
   - Moisture content
3. Other Possible Inputs
   - Vibration
   - Temperature
4. Remote master HMI
5. Multiple Slaves
   - Daisy Chained

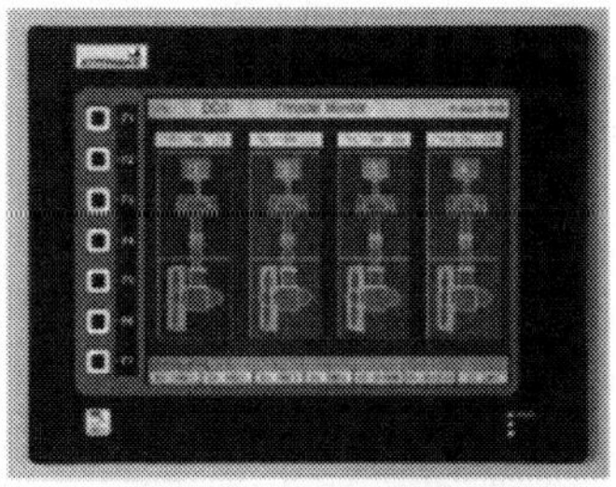

10

Case Study - Marine Diesel Engines

Slow Speed Two Stroke Cylinder Lubrication

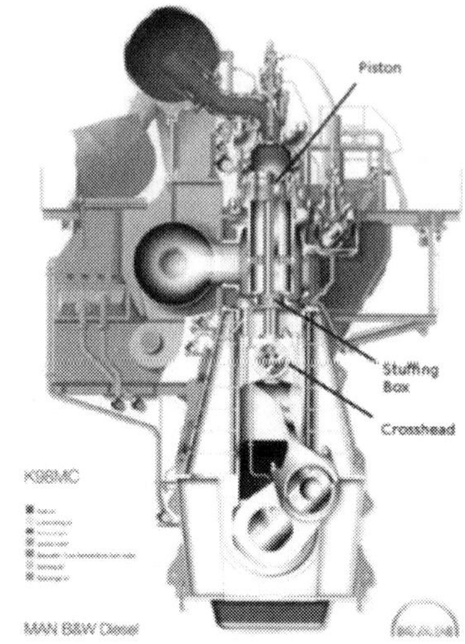

Piston Ring – Cylinder Liner Interface in Two-Stroke Marine Diesel Engines is lubricated on a one-time basis with a specially formulated Cylinder Lube Oil.

Two functions:

1. Lubrication
2. Neutralisation of acidic sulphur combustion by-products

Residual used oil is “scraped down” into a drain line for collection and subsequent disposal.

11

Two Stroke Cylinder Lubrication

1. Lubrication
   - Formulation based on usual considerations!
2. Acid Neutralisation
   - Alkaline rich additive package.
   - Overbased metallo-organic carbonates are the most common additive.
   - Base Number requirements from 25 to >100 depending on Fuel Sulphur content.
   - IMO 2020 sulphur cap (0.5%) will impact fuel quality.
   - Fuel switching - non ECA to ECA - will require Cylinder Oil switching.

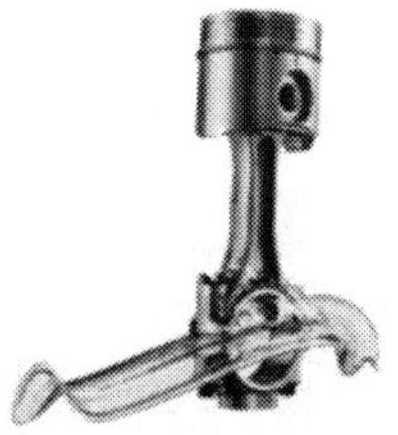

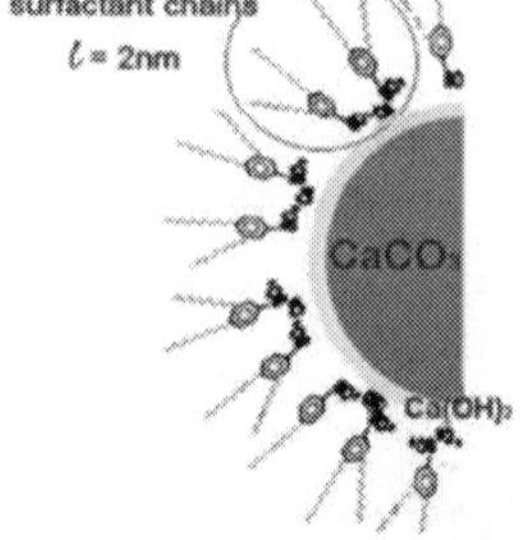

12

Two Stroke Cylinder Lubrication

Monitoring of used scrapedown oil gives information on cylinder liner health:

	Condition Parameter
Adequate lubrication - "normal" abrasive wear. Under lubrication - increased abrasive wear	Ferrous iron content
Sufficient neutralisation of acidic combustion by-products – alkaline additive is depleted but not exhausted. Insufficient neutralisation – corrosive wear.	Base Number

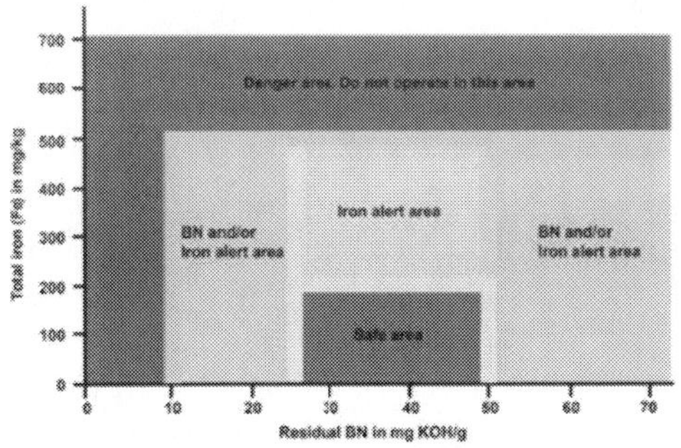

13

Two Stroke Cylinder Lubrication Monitoring with Test Kits

1. Base Number Monitoring
   - Pressure cell measures CO_2 generation from reaction of carbonate component with acidic reagent.
   - Calibrate cell with a sample of unused oil of known BN.
   - Reduction in pressure from this value for a used sample = reduction in BN as carbonate is depleted.

Pros
- Simple, reliable test
- Quick – approximately 3 minutes per test

Cons
- Manual test with sampling required
- Resource intensive for multiple cylinders and frequent testing regimes

14

Two Stroke Cylinder Lubrication Monitoring with Test Kits

2. Iron Content (non ferrous)
    - Newly developed reagent (patented)
    - Extracts Iron(II) and Iron(III) from oil sample into an aqueous phase
    - Converts any Iron(III) to Iron(II)
    - Forms coloured complexes with all of the now Iron (II)
    - Colour intensity related to original Iron(II) + Iron(III) content

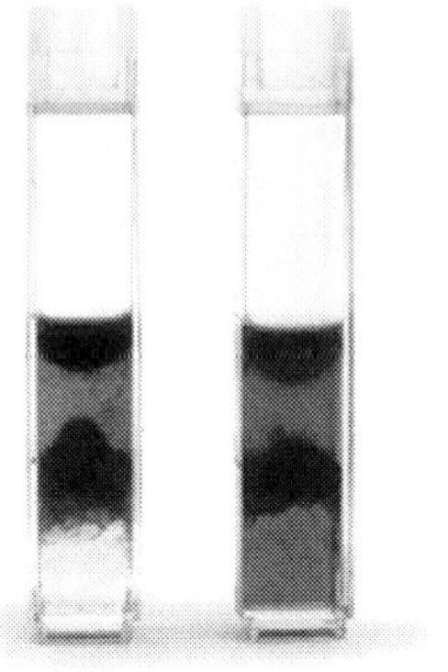

Pros
- Simple, quantitative test
- Quick – approximately 5 minutes per test
- Determines Iron compounds not metallic Iron

Cons
- Manual test with sampling required
- Resource intensive for multiple cylinders and frequent testing regimes

NB: No metallic Iron – requires an instrument

15

Two Stroke Cylinder Lubrication Monitoring with Test Kits & Instruments

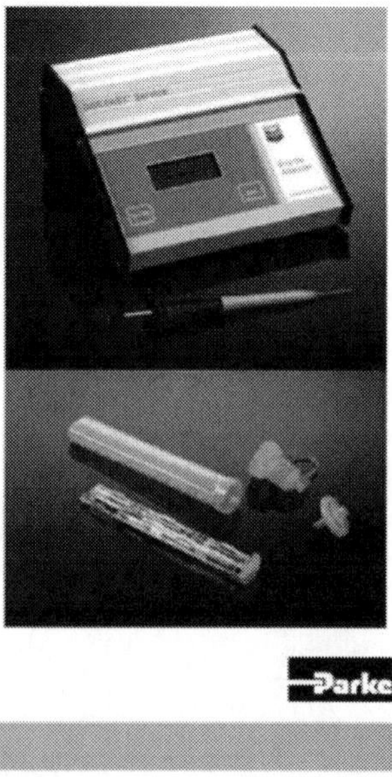

3. Total Iron Content
    - Acid digestion followed by colorimetry
    - E.g. Chevron DotFAST
    - Determines Total Iron content – metallic and compounded.

Pros
- Quantitative test

Cons
- No differentiation between wear and corrosion.
- Time consuming – 4 + hours
- Hazardous reagents
- Manual test with sampling required
- Resource intensive for multiple cylinders and frequent testing regimes

16

Two Stroke Cylinder Lubrication Monitoring by Offline Instruments

1. Metallic Iron by Magnetometry
   ppm or PQ[1] Index

   Pros
   - Accurate, reliable test
   - Quick – approximately 2 minutes per test

   Cons
   - Sampling required
   - Resource intensive for multiple cylinders and frequent testing regimes

Note: Laboratory iron measurements are total iron content.

Metallic iron + Corroded iron = Total iron

FWM (Metallic iron)   CCTK (Corroded iron)

17

[1] PQ is a registered trademark of Parker Hannifin

Two Stroke Cylinder Lubrication Monitoring by Offline Instruments

2. BN by IR absorption – ATR sampling interface. Calibrate against known samples.

   Pros
   - Accurate, reliable test
   - Quick –2 minutes per test

   Cons
   - Sampling required
   - Resource intensive for multiple cylinders and frequent testing regimes
   - Oil specific

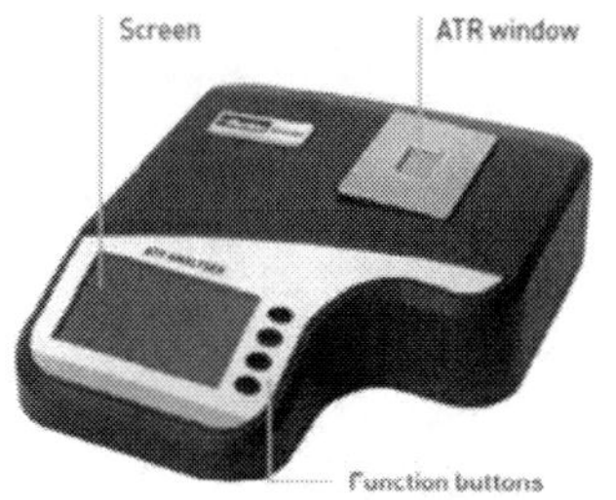

18

Two Stroke Cylinder Lubrication Monitoring by Offline Instruments

2. BN by IR absorption

- Monitor alkaline additive depletion
- Monitor sulphate uptake
- BN by chemometric modelling

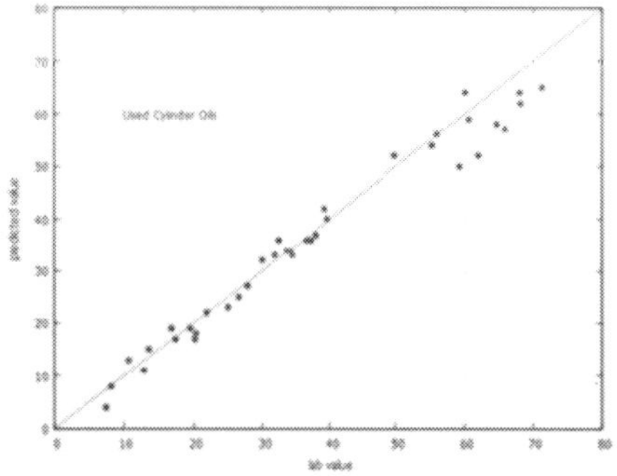

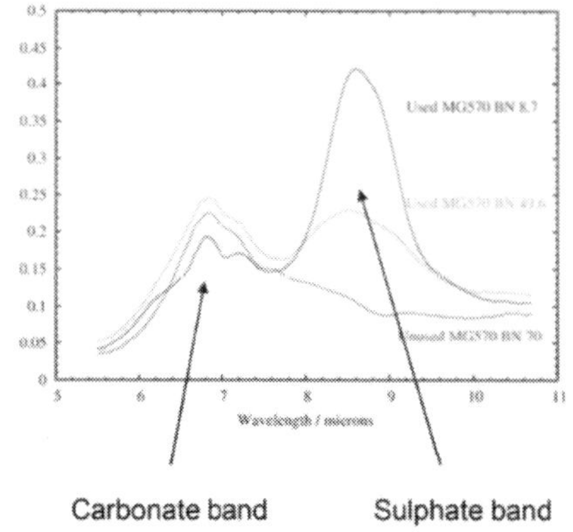

19

Two Stroke Cylinder Lubrication Monitoring by Online Sensors

1. Metallic Iron
   Ferrous iron content by magnetometry
   Periodic measurement cycle with selectable interval period
   Iron content determined as ppm

Pros
   Continuous monitoring
   No sampling required
   Direct reporting to control room
   Feedback to oil supply system
Cons
   Capital cost

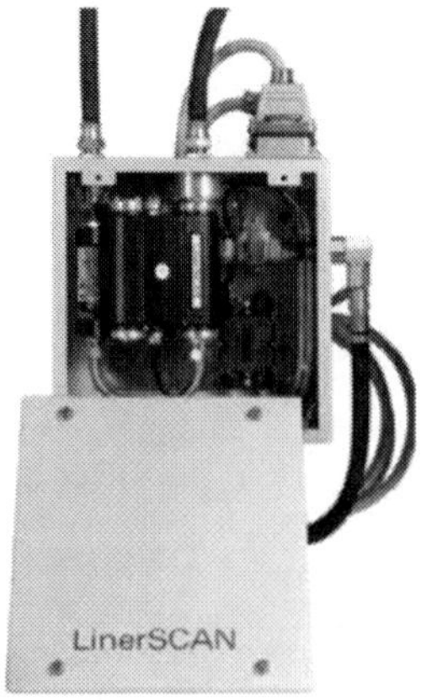

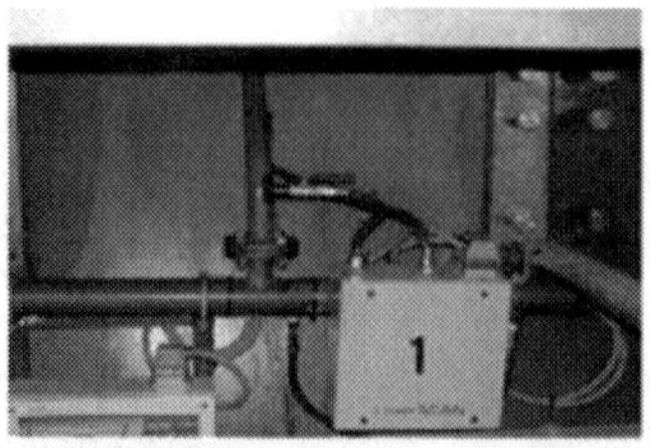

20

Two Stroke Cylinder Lubrication Monitoring by Online Sensors

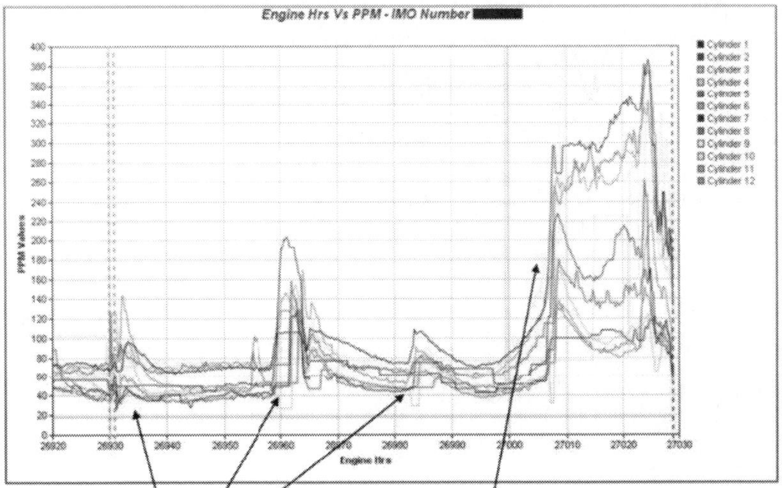

24 hr sweep tests

Catfine ingress on fuel switchover

21

Parker

Concluding Remarks

- Field based conditioning monitoring provides:
  - Rapid screening at point of use
  - Quick feedback on oil quality
  - Early warning of potential issues
  - Real time optimisation data
- Field based condition monitoring does not:
  - Replace in-depth laboratory testing

22

Parker

LEARNING HOW TO MAKE DATA WORK FOR YOUR ORGANIZATION

Chris MacDonald
Head of Global Analytics CoE

May 14 2019

ptc

THE TIME TO TRANSFORM IS NOW

ptc

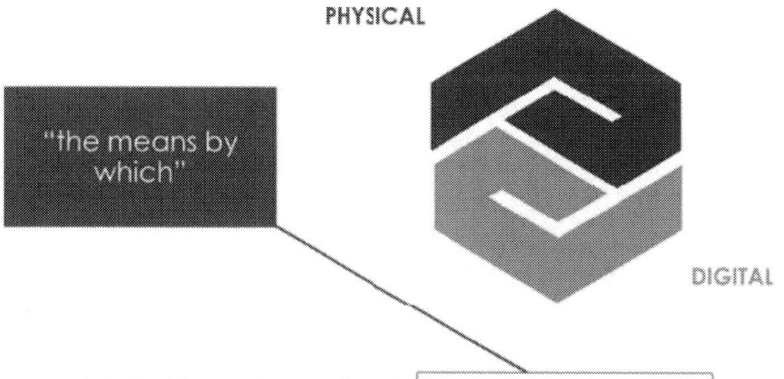

Digital Transformation is the means by which industrial companies capitalize on the **PHYSICAL DIGITAL** convergence that is at the core of **INDUSTRIE 4.0**

2

ptc

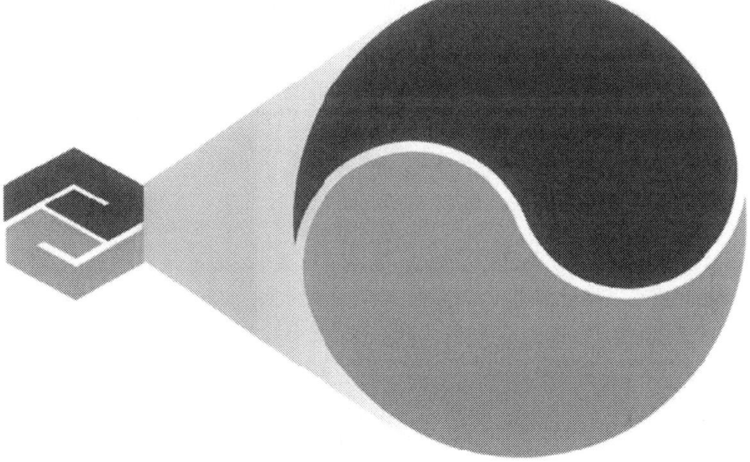

3

ptc

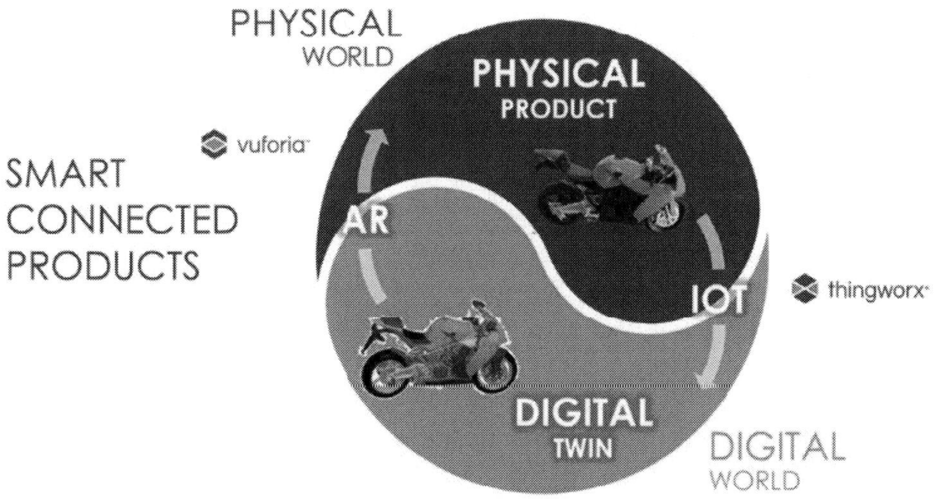
PHYSICAL
WORLD
PHYSICAL
PRODUCT
vuforia
SMART
CONNECTED
PRODUCTS
AR
IOT
thingworx
DIGITAL
TWIN
DIGITAL
WORLD

4

ptc

SMART
CONNECTED
OPERATIONS

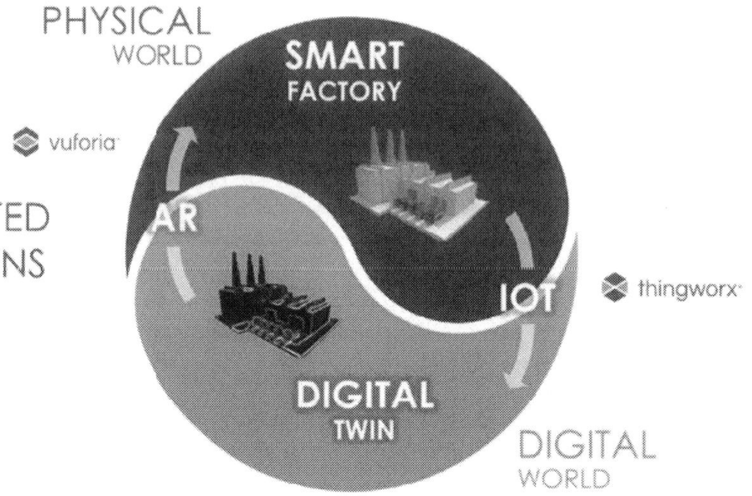

5

TRANSFORM ASSETS, PROCESSES, AND PEOPLE

ptc

6

DRIVE BUSINESS OUTCOMES

ptc

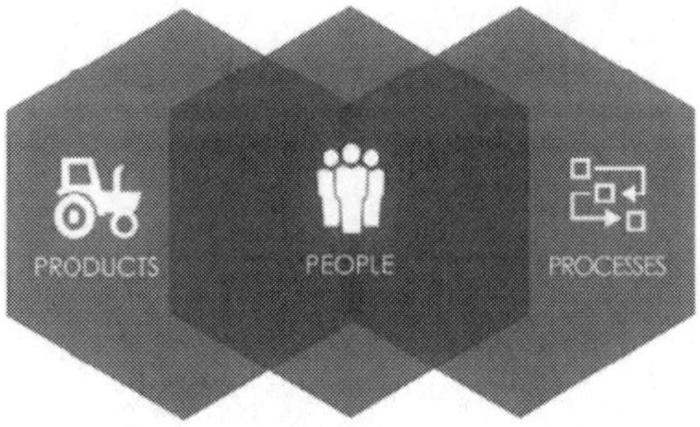

Smart Connected Products Drive Critical Product and Service Differentiation

Smart Connected Processes Enable Breakthroughs in Operational Effectiveness

Smart Connected People Makes Workforce Productivity and Quality Soar

7

DIGITAL TRANSFORMATION SOLUTION SUITE

ptc

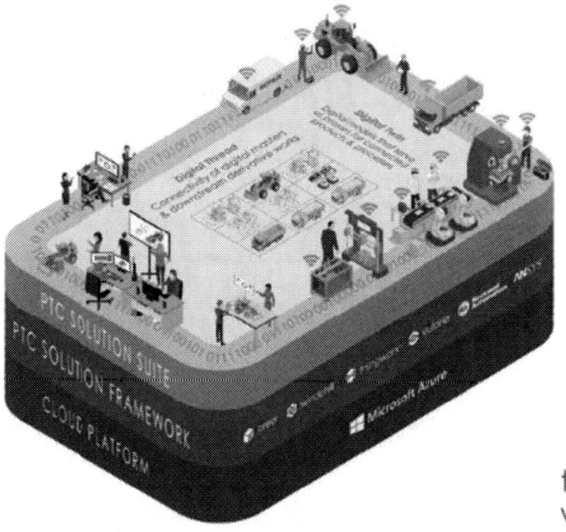

the means by which...

8

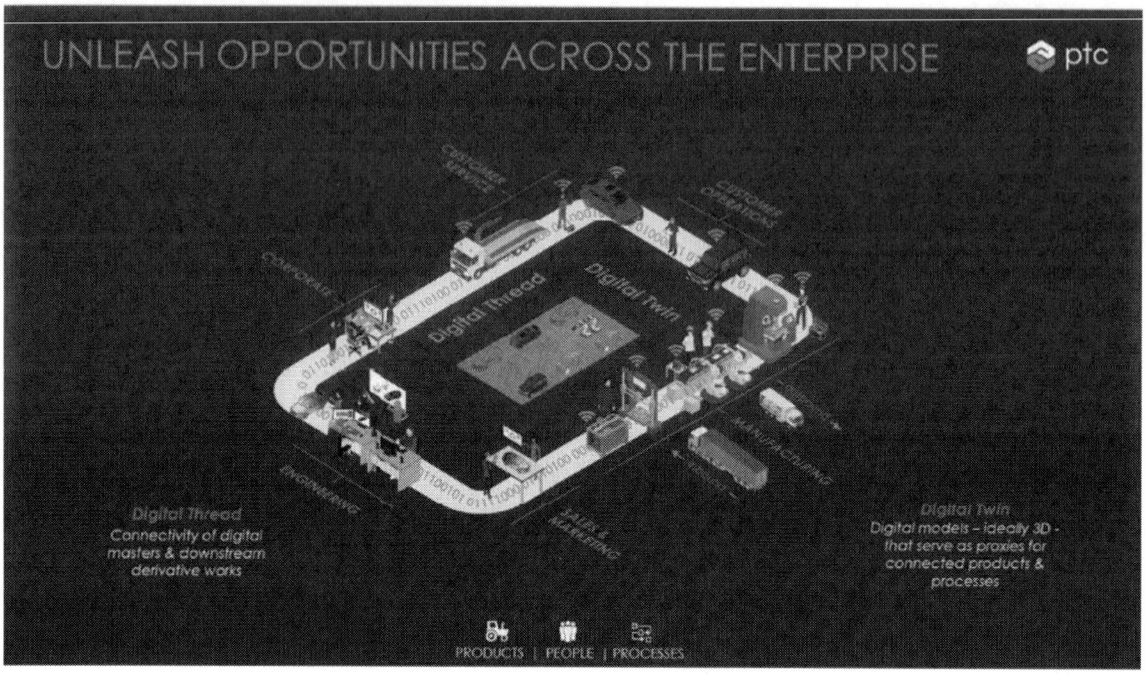
UNLEASH OPPORTUNITIES ACROSS THE ENTERPRISE
ptc
CUSTOMER SERVICE
CUSTOMER OPERATIONS
CORPORATE
Digital Thread
Digital Twin
MANUFACTURING
ENGINEERING
SALES & MARKETING
Digital Thread
Connectivity of digital masters & downstream derivative works
Digital Twin
Digital models – ideally 3D - that serve as proxies for connected products & processes
PRODUCTS | PEOPLE | PROCESSES

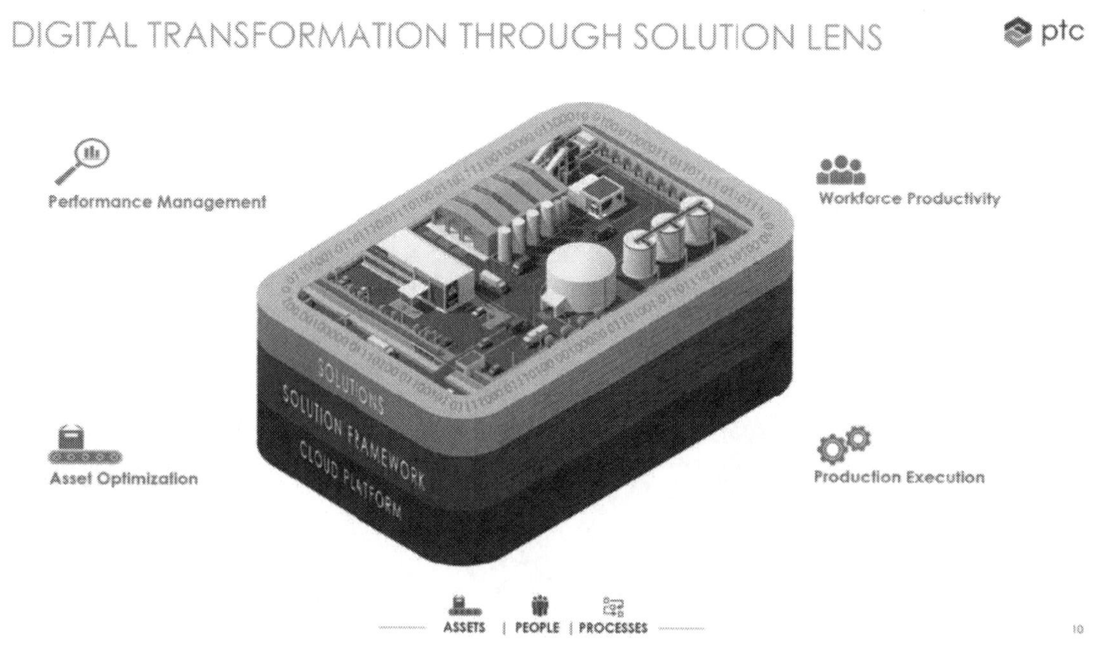
DIGITAL TRANSFORMATION THROUGH SOLUTION LENS
ptc
Performance Management
Workforce Productivity
SOLUTIONS
SOLUTION FRAMEWORK
CLOUD PLATFORM
Asset Optimization
Production Execution
ASSETS | PEOPLE | PROCESSES
10

ANALYTICS ISN'T "EASY"

ptc

- Lets accept this upfront
- Demystify buzzwords
- It takes more than technology
- Requires organizational change
  - People
  - Process
  - Culture
- Why do organizations continue to mature their analytics capacities knowing it's complex?

BECAUSE DATA DRIVEN INSIGHTS ARE TRANSFORMING ORGANIZATIONS

PRODUCTS | PEOPLE | PROCESSES

12

ASKING THE RIGHT DATA ANALYSIS QUESTIONS

- ✓ What exactly do you want to find out?
- ✓ What standard KPIs will you use that can help?
- ✓ Where will your data come from?
  - Which scales apply to your different data sets?
- ✓ How can you ensure data quality?
- ✓ Which statistical analysis techniques do you want to apply?
- ✓ Who are the final users of your analysis results?
- ✓ What data visualizations should you choose?
- ✓ How can you create a data-driven culture?

COMPREHENSIVE APPROACH TO ANALYTICS

ptc

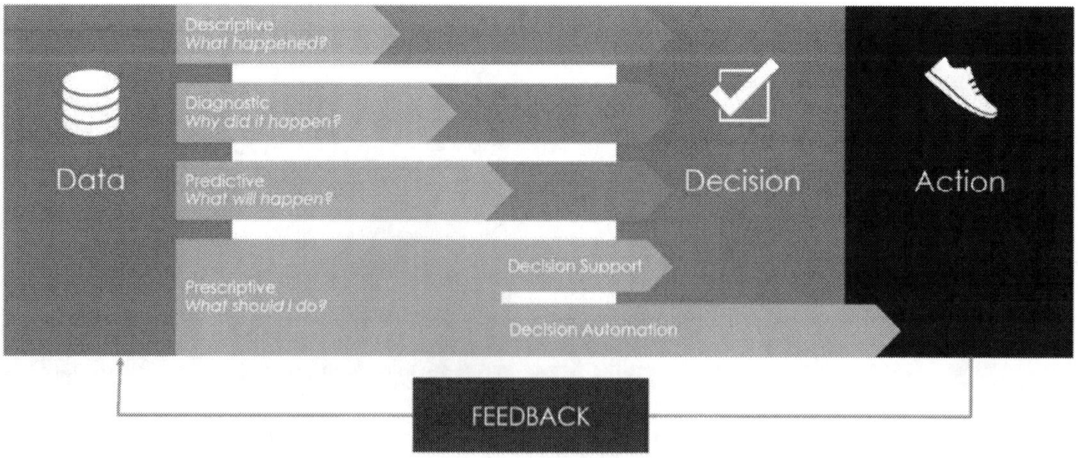

14

TECHNOLOGY ENABLED EVOLUTION OF MAINTENANCE

ptc

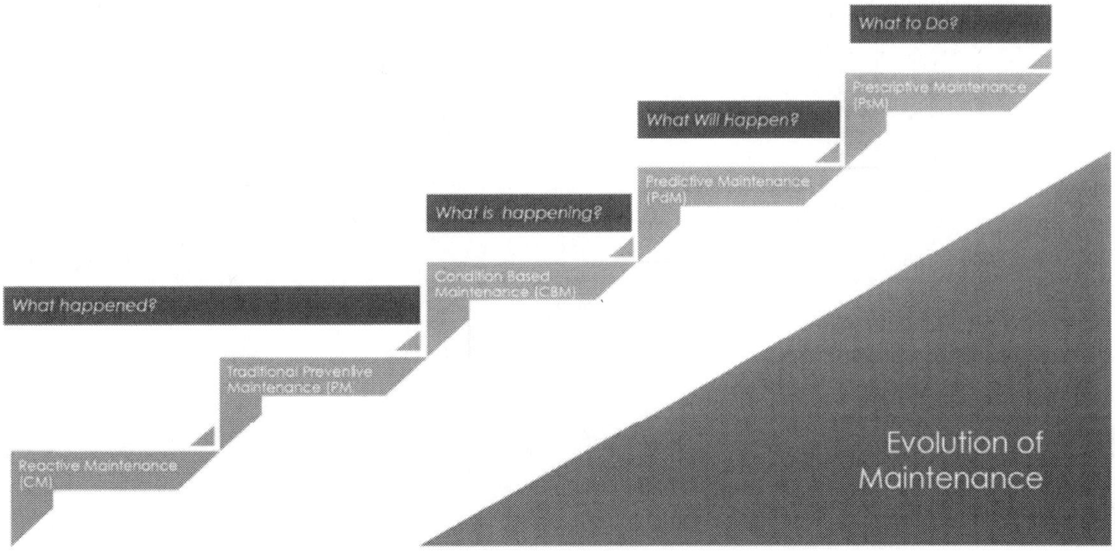

15

DESCRIPTIVE AND DIAGNOSTIC ANALYTICS

IMMEDIATE VALUE AND A FOUNDATION FOR AI/ML

ptc

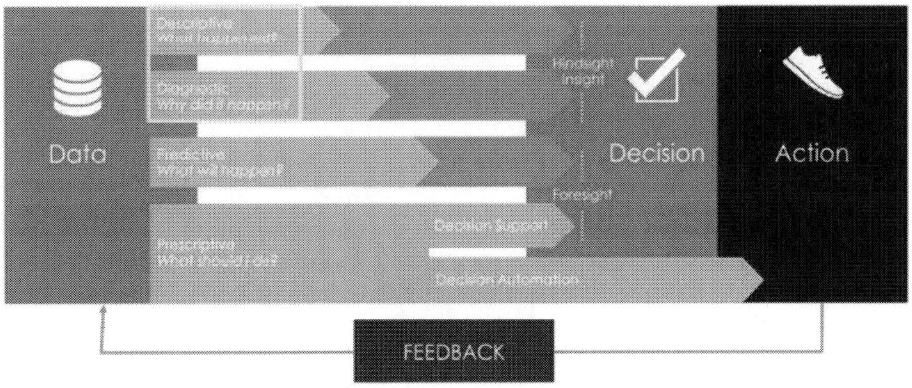

- **Introductory data analysis** and SPC capabilities.
- Rules defined by domain experts applied to individual variables to **alert when variables fall out of tolerance.**
- More advanced analytics applied **to identify root causes and patterns** that correlate to important events like Asset Downtime, Process Performance Reductions or Quality Issues.
- Take KPIs and Patterns from descriptive and diagnostic analytics to **create enriched, historical asset and process data schema.**

16

MACHINE LEARNING AND DECISION OPTIMIZATION

ENABLE SMARTER CONNECTED OPERATIONS

ptc

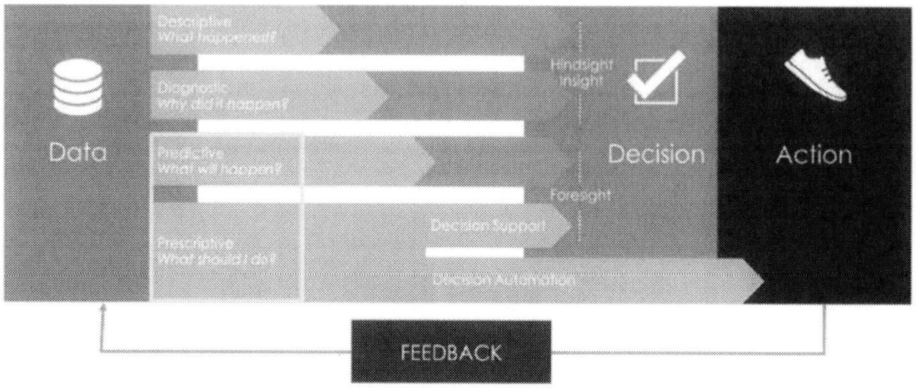

- **Apply predictive models trained with machine learning** algorithms and historical data into ThingWorx Applications.
- Proactively identify important (**such as process inefficiency**) issues with enough time horizon to schedule a preventative action.
- **Prescribe preventative measures** on production equipment or processes to improve performance or avoid downtime.

17

---

ptc

> I would not give a fig for the simplicity this side of complexity, but I would give my life for the **simplicity on the other side of complexity**.
>
> \- Oliver Wendell Holmes Jr.

18

WHY DATA SCIENCE FEELS INACCESSIBLE...

ptc

Interconnection among the components of Data Science

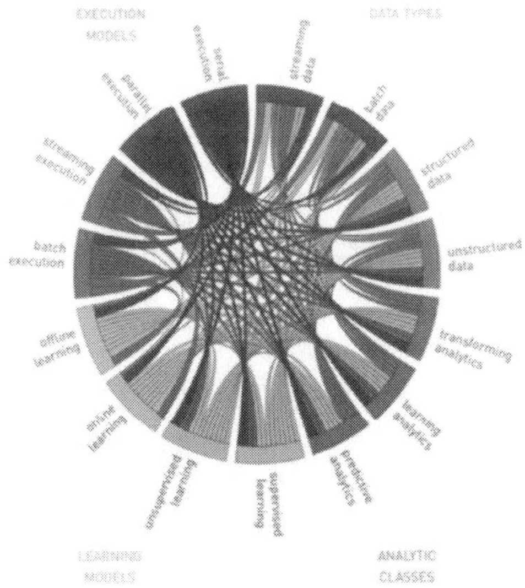

Source: Booz Allen Hamilton

DATA SCIENCE, SIMPLIFIED

ptc

- Data Science is the art of turning data into actions.
- This is accomplished through the creation of data products, which provide actionable information without exposing decision makers to the underlying data or analytics.
- Examples: buy/sell strategies for financial instruments, a set of actions to improve product yield, or steps to improve product marketing.

PREDICTIVE ANALYTICS PROCESS

ptc

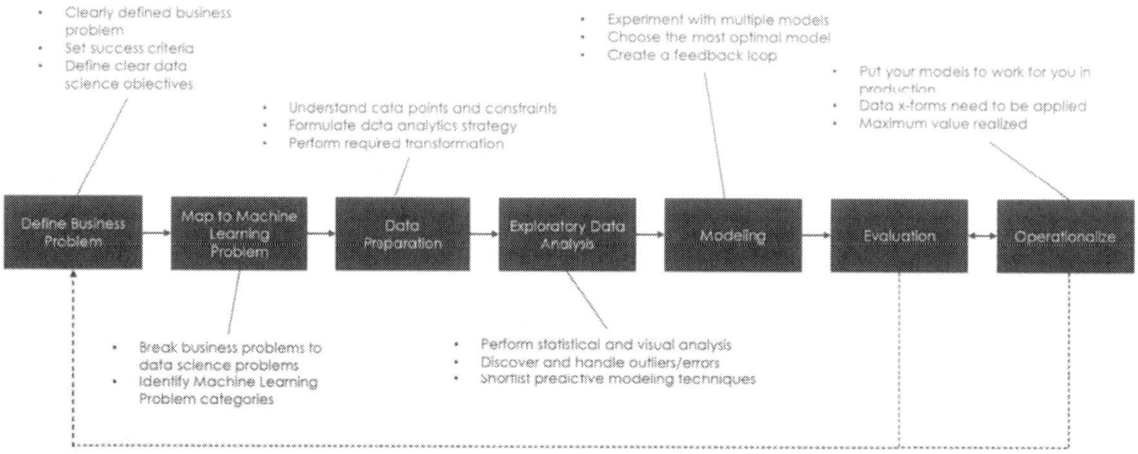

21

CLEARLY DEFINE THE BUSINESS PROBLEM

- Critical to **Framing the Analytics Question**
- Remember: be careful *how* you ask
- You may get exactly what you wish for
- "A million bucks?"

Define Business Problem

FIVE QUESTIONS

ptc

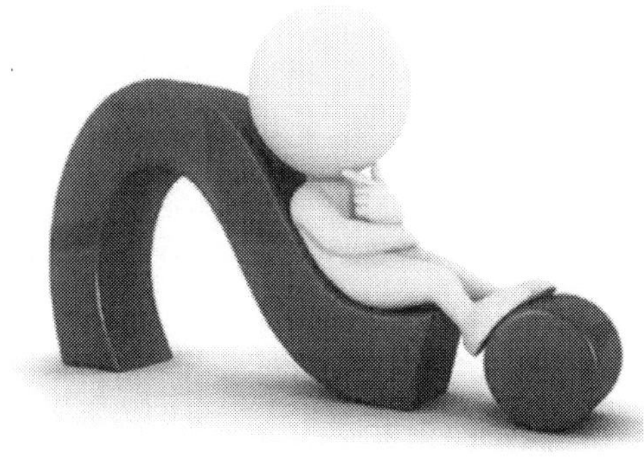

Map to Machine Learning Problem

- Is it this or that?
  - Classification algorithms
  - Supervised
- How many?
  - Regression algorithms
  - Supervised
  - Numerical predictions
- Is this (in any way) organized?
  - Clustering algorithms
  - Unsupervised
  - No one right answer
- Is it different?
  - Not normal behavior
  - The class of problems known as anomaly detection
- What should I do next?
  - Reinforcement learning algorithms
  - Learn from trial and error

23

USING THE RIGHT DATA

Data Preparation

- Understand your data and how it relates to the problem you are trying to solve
- Transformation (feature engineering) will almost always be needed
- That transformation must be replicable in operation

24

ptc

IS YOUR DATA READY?

Better Data = Better Results

- Relevant
  - Is it [potentially] relevant to the outcome (or internally consistent)
- Connected
  - No gaps in the data
  - The more missing data, the harder it is to provide insights
- Accurate
  - Tight Groupings or distributions of data, but precise
  - Imprecise can be accurate if imprecision is representative of the problem
- Enough
  - More data potentially means more detailed/precise decisions

Exploratory Data Analysis

25

MACHINE LEARNING APPLIED

ptc

	Process	Assets	Quality	Energy
Cloud and Data Center	Determine when where to run	Reduce machine failures	"Golden Batch"	Optimize product profit
	Manage setup time	Improve performance	Improve yield	Reduce energy cost
Edge	Optimize output to specification	Increase uptime	Predict product grade	Determine when or where to run
	Dynamically adjust set points	Extend tool life	Reduce scrap and rework	Reduce energy use

REQUIRED CAPABILITIES FOR MACHINE LEARNING SUCCESS

ptc

Prepare	Explore	Model	Operationalize
Data ingestion and normalization	Statistical analysis	Train predictive model	Codify data transformations
Create training data set	Feature discovery	Feature engineering	Tune model based on new data
	Identify & manage outliers		Directed action in application
SME defined business problem	Visual data analysis	Vetting model by SME	Scoring streaming analytics at the edge

ADVANCED ANALYTICS IS A HUMAN EXERCISE

ptc

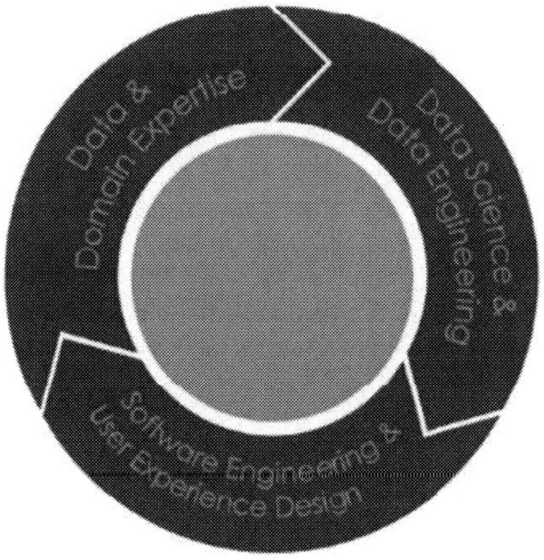

A human is always accountable for the insight and the call to action

- ✓ Understand the accountability
- ✓ Spend time framing the question
- ✓ Limit "black boxing" solutions
- ✓ Look at business success, not simply the data science

28

ptc

MFPT 2019 Where Theory Meets Practice
MAY 14-16, 2019 | PHILADELPHIA, PA

Physics-Based Machinery Diagnosis

Bill Marscher

Mechanical Solutions, Inc.

MFPT Annual Meeting

Philadelphia PA

May 14-16, 2019

Mechanical Solutions, Inc.
Engineering Analysis, Test & Technology

11 Apollo Drive, Whippany, New Jersey 07981, USA
E-Mail: wdm@MechSol.com Website: www.MechSol.com

MFPT 2019 Where Theory Meets Practice
MAY 14-16, 2019 | PHILADELPHIA, PA

Experience of Author's Company with Pump Diagnostics

Foundation: Decades of troubleshooting, design, and analysis experience to understand pump physics and system interactions.

Diagnostics and Prognostics System Development:

- 2006 - High Speed Continuous Blade Health Monitoring for Turbine Engines (NAVAIR)
- 2005-8 - Active Vibration and Physics Based Health Management (USAF AEDC)
- 2008 - Blade Vibration Measurement System (NASA)
- 2010 - Valve Health Monitoring System (USAF AEDC)
- 2010 - Blade Vibration Measurement System for Turbine Engines (USAF WPAFB)
- 2010 - Blade Vibration Measurement System for Unducted Fan Engines and Rotorcraft Blades (NASA Glenn)
- 2012-4 - High Temperature Blade Health Measurement System for Adaptive Engines (USAF AEDC)
- 2013-5 – Pump & Motor Health Monitoring (USAF AEDC)

2

What are the goals of machinery automated diagnostic systems?

- Protect capital assets
  - Catch problems early
    - Avoid collateral damage, through actionable information
    - Minimize warranty expenses
    - Alert connected OEM to spare parts opportunities
  - Avoid "misses" leading to catastrophic failure events
- Maintain end-user production, improve ROI, enhance safety
  - Avoid unexpected downtime
  - Avoid human injuries or environmental leaks
  - Avoid "false positives" encouraging unnecessary shut-down
- "Internet of Things" collaboration
  - Facilitates system optimization
  - Facilitates reliable health assessment and/ or troubleshooting

MFPT 2019 Where Theory Meets Practice
MAY 14-16, 2019 | PHILADELPHIA, PA

What are the system sensors and input information required?

Typical:

- Accelerometer(s)
- Inlet Pressure
- Discharge Pressure
- Flow Rate
- Thermocouples or RTD's
- RMS Voltage
- RMS Electric Current

Useful if Available:

- Triax Accelerometers
- Proximity Probes
- Tachometers
- Microphones
- Dynamic Pressure Transducers
- Seal RTD's & Flow Rate
- Voltage FFT
- Electric Current FFT

MFPT 2019 Where Theory Meets Practice
MAY 14-16, 2019 | PHILADELPHIA, PA

Typical vibration measurement hardware

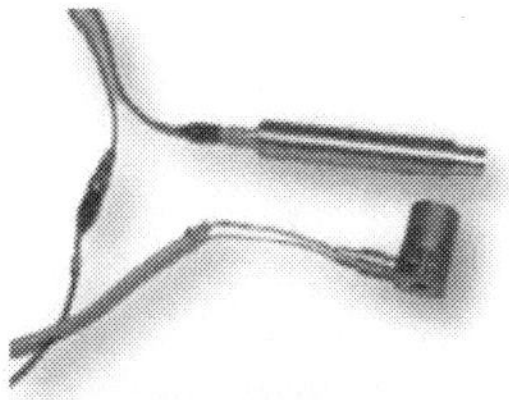

- **Accelerometers**
- Can vary in construction depending on the application. For high temperature applications there are accelerometers with the signal conditioning electronics built in a separate device from the actual sensor, minimizing its exposure to the temperature.
- There are special accelerometers for submersible applications with an integrated sealed cable connection.

MFPT 2019 Where Theory Meets Practice
MAY 14-16, 2019 | PHILADELPHIA, PA

Instrumentation options wish list

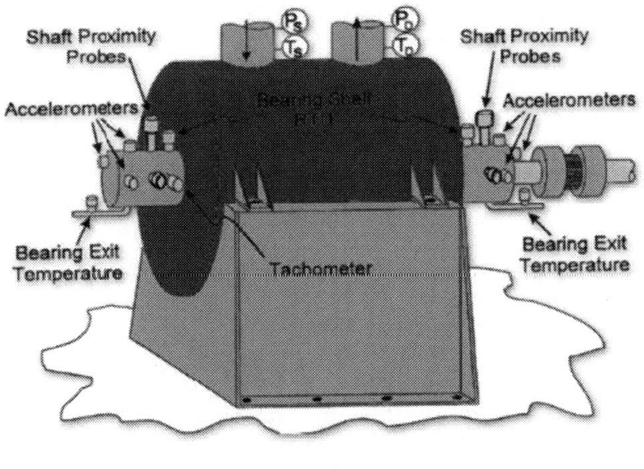

There are two fundamental approaches to interpret the signals these sensors provide:

- Statistics or "Data-Driven" methods
  - Take as much data as possible
  - Take data over an extended time period
  - Learn to see patterns
  - From this, determine anomalies and problems, estimate time-to-failure
- "Physics-Based" or "Physiology-of-Pumps" methods
  - Put data in perspective of speed, load, pump type
  - Assess where the pump is on its H-Q curve
  - Use pre-knowledge of the pump design to assess if anomalies are problems
- Both methods can/ should use ANSI/HI, API, and ISO specs as a back-stop

MFPT 2019 Where Theory Meets Practice
MAY 14-16, 2019 | PHILADELPHIA, PA

Example fault "condition indicator" (CI) trending/ diagnosis/ prognosis

Faults Addressed:

- Imbalance/Misaligrment
- Running Off BEP
- Hydraulic Issues: Recirculation/Stall
- Pump Cavitation
- Resonant Conditiors
- Seal Damage
- Roller Bearing Deterioration
- Journal Bearing Rub
- Oil Whip/Whirl
- Vane Pass Issues: Obstructions, Damaged Volute
- Soft Foot
- Looseness
- Casing and Foundation Structural Problems
- Motor Rotor Bar Cracked
- Motor Static/Dynamic Air Gap Eccentricity
- Gearbox Deterioration
- Belt Drive Faulty
- Acoustic Resonance

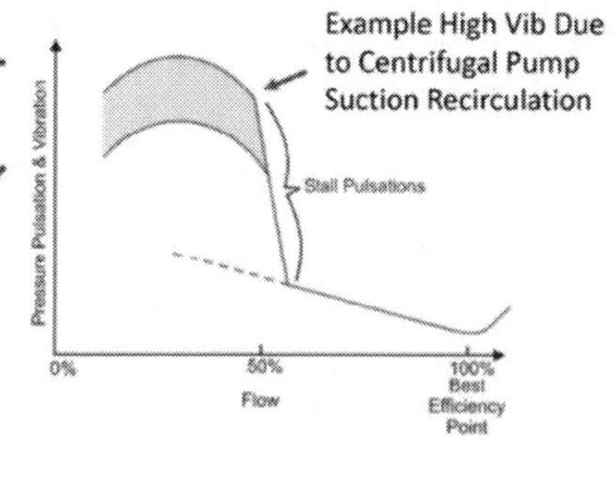

MFPT 2019 Where Theory Meets Practice

MAY 14-16, 2019 | PHILADELPHIA, PA

Real-life pumps are complex!

Example: Screw Pump Assembly

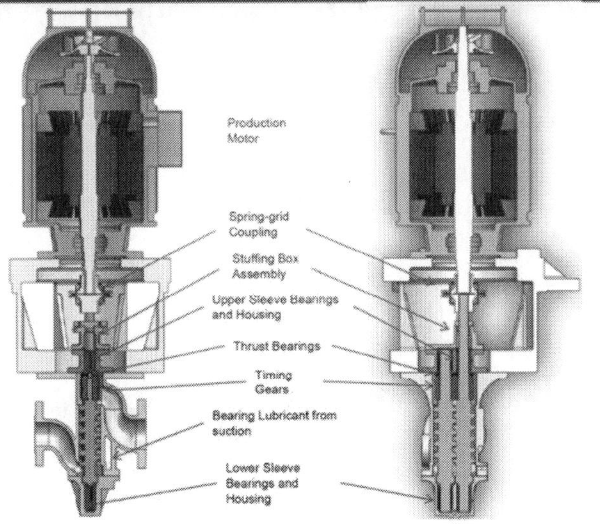

9

MFPT 2019 Where Theory Meets Practice

MAY 14-16, 2019 | PHILADELPHIA, PA

What user interface options work best?

- Example: Marine fuel oil screw pump:

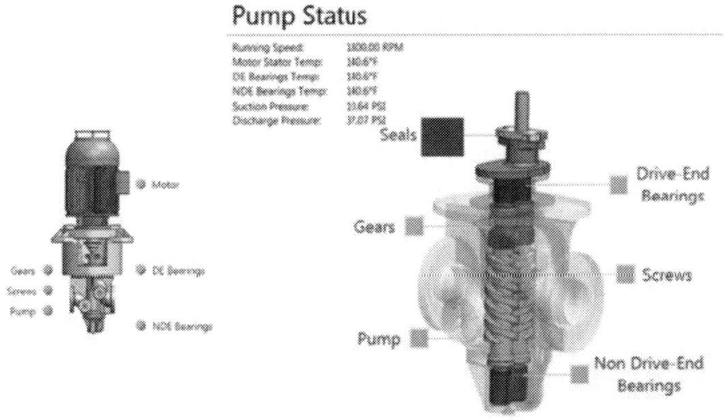

First & Foremost:

Red/ Yellow/ Green

Is there a problem? If no, just get a Green light.

If Red or Yellow: Why? Which machine/ component?

If Red: Critical to operation? How long can it run?

MFPT 2019 Where Theory Meets Practice
MAY 14-16, 2019 | PHILADELPHIA, PA

Examples of "drill down" information:

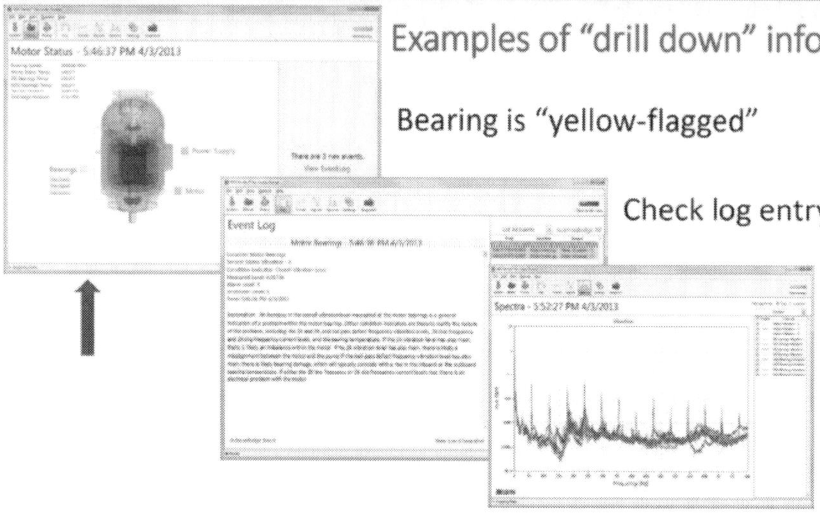

Bearing is "yellow-flagged"

Check log entry

Check (or not) data like bearing housing FFT Spectrum.

Send to expert?

MFPT 2019 Where Theory Meets Practice
MAY 14-16, 2019 | PHILADELPHIA, PA

Ideas for a user-friendly software environment

- A directory "tree" to list machinery and instrumentation
- A main display area to provide easy access to signals, spectra, live condition indicator values, historical condition indicator trends, and more
- A timeline to show historical operating status, recordings, user notes, and maintenance events
- Playback controls to allow for easy recall of historical data and return to live data

MFPT 2019 Where Theory Meets Practice
MAY 14-16, 2019 | PHILADELPHIA, PA

Some ideal qualities for a diagnostic approach

1. Gives Customers actionable info, not data overload
2. Gives Customers a birds-eye view of the "whole system"
3. When problems are diagnosed, predicts root causes, MTTF
4. Employs a combination of physics and data statistics
5. Determines & accounts for operating point (e.g. low flow off-BEP)
6. If data is available, tracks efficiency in pursuit of optimization
7. Dovetails into existing plant systems (e.g. Pi, Bently System One)
8. Less expensive if developed with "other people's money", e.g. Gov

MFPT 2019 Where Theory Meets Practice
MAY 14-16, 2019 | PHILADELPHIA, PA

Typical failures in centrifugal pumps

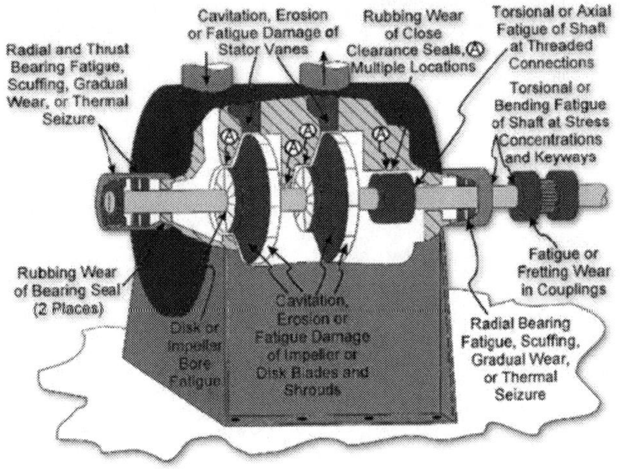

MFPT 2019 Where Theory Meets Practice
MAY 14-16, 2019 | PHILADELPHIA, PA

Damaging forces in centrifugal pumps

Both Data and Physics-based methods look for tell-tales of these

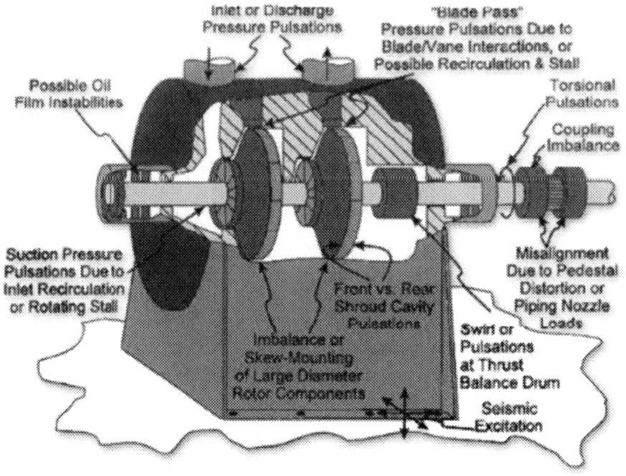

MFPT 2019 Where Theory Meets Practice
MAY 14-16, 2019 | PHILADELPHIA, PA

Various rigs for "seeded fault" testing

20HP Multistage Compressor

150 HP Screw Pump

Tabletop Gearbox Rig

20 HP Centrifugal Pump

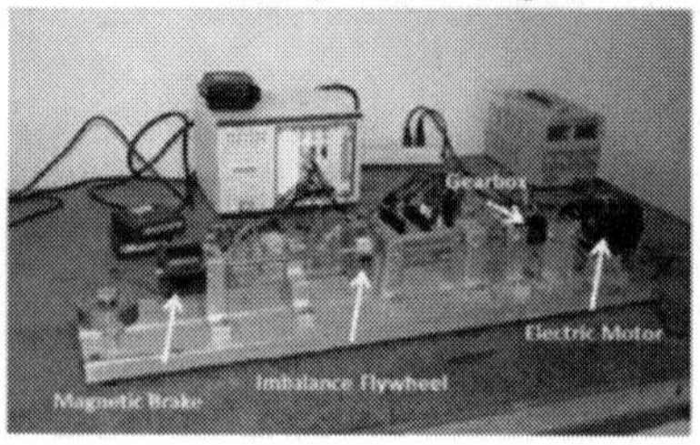

16

MFPT 2019 Where Theory Meets Practice
MAY 14-16, 2019 | PHILADELPHIA, PA

Example: Cavitation in an axially split case double suction pump

These pumps are used when suction pressure is very low.

Therefore, suction recirculation & cavitation have been issues

Other Possible Issues:

- Double suction impeller typically equalizes thrust side-to-side, but at off-design is prone to axial shuttling
- 180 degree "drip pocket" allows bearing housing and stuffing box to flex downward due to pressure pulsations, e.g. at 1x or vane pass

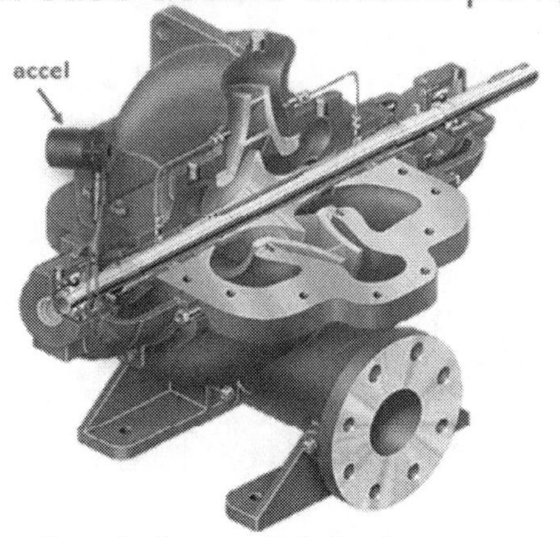

Example of a properly-designed pump

MFPT 2019 Where Theory Meets Practice
MAY 14-16, 2019 | PHILADELPHIA, PA

Cavitation issue in pumps: pressure depression at inlet

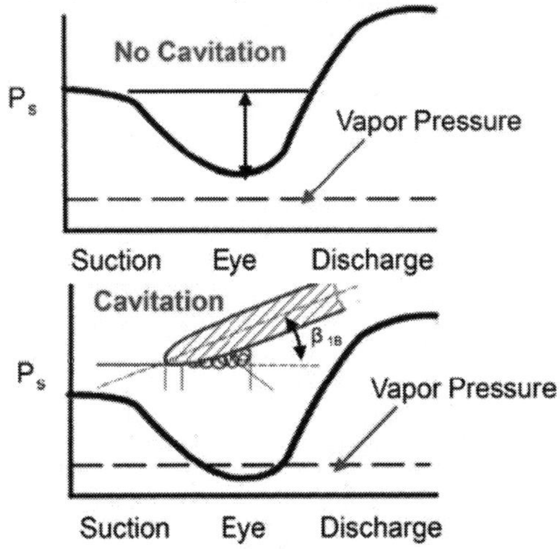

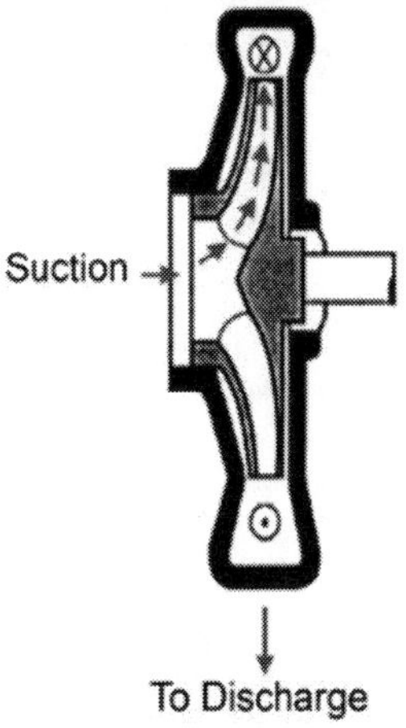

MFPT 2019 Where Theory Meets Practice
MAY 14-16, 2019 | PHILADELPHIA, PA

Centrifugal pump cavitation and damage

MFPT 2019 Where Theory Meets Practice
MAY 14-16, 2019 | PHILADELPHIA, PA

Scope trace of cavitation event in double suction pump

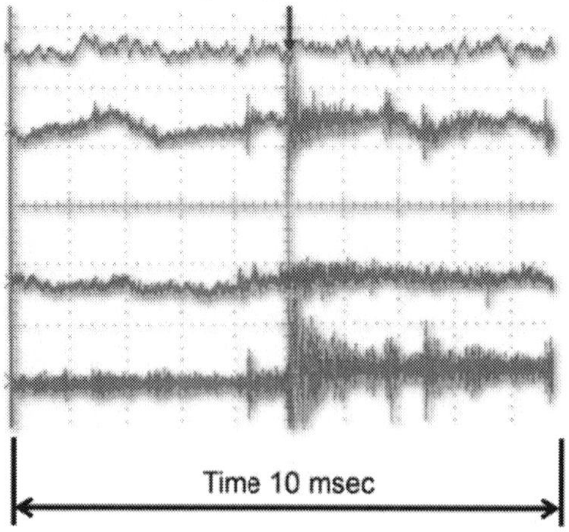

Physics-basis example: Misalignment (per ODS)

- An ODS is created by collecting vibration data and using specialized processing. It shows exaggerated motion (but to scale) at specific frequencies.
- Left - animated Operating Deflection Shape (ODS) of typical angular misalignment.

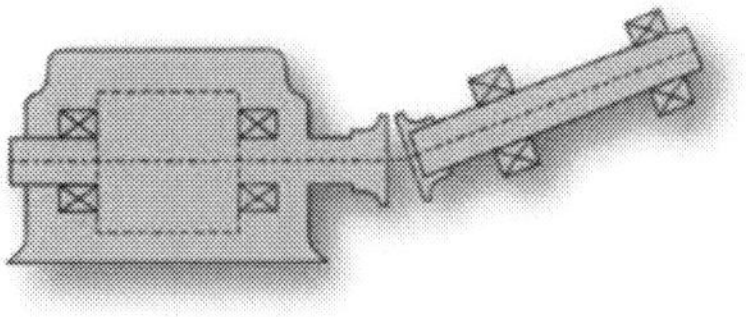

MFPT 2019 Where Theory Meets Practice
MAY 14-16, 2019 | PHILADELPHIA, PA

Physics-basis example: Misalignment (per Motion-Magnified Video System)

Physics-basis example: Bearing Problems

$FTF = \frac{N}{2}(1-\gamma)$ Fundamental Train Frequency

$BSF = \frac{D_p}{D_b} * \frac{N}{2}(1-\gamma^2)$ Ball Spin Frequency

$BPFO = N_b * \frac{N}{2}(1 - \gamma)$ Ball Pass Frequency Outer Race

$BPFI = N_b * \frac{N}{2}(1 + \gamma)$ Ball Pass Frequency Inner Race

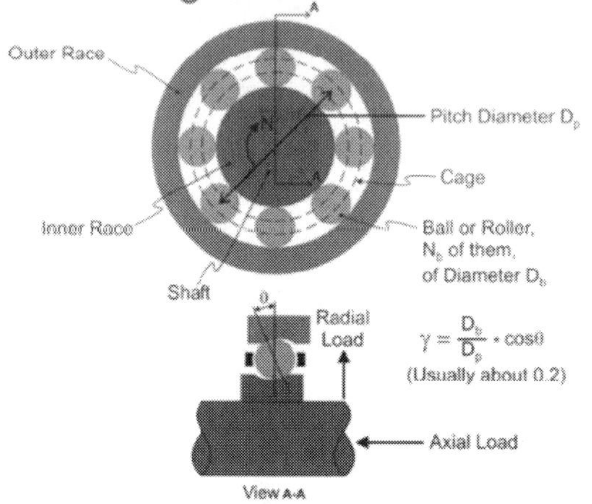

MFPT 2019 Where Theory Meets Practice
MAY 14-16, 2019 | PHILADELPHIA, PA

So ... What should we do?

- If you have data, try to make sense of it.
- However ... if you get a serious disease, will you go to a physician or a statistician to get your diagnosis?
- But ... how about a top physician that also considers human disease statistics? Good idea!

MFPT 2019 Where Theory Meets Practice
MAY 14-16, 2019 | PHILADELPHIA, PA

Evaluation of Alternative Accel Attachments

Bill Marscher

Mechanical Solutions, Inc.

MFPT Annual Meeting

Philadelphia PA

May 14-16, 2019

Mechanical Solutions, Inc.
Engineering Analysis, Test & Technology
11 Apollo Drive, Whippany, New Jersey 07981, USA
E-Mail: wdm@MechSol.com Website: www.MechSol.com

MFPT 2019 Where Theory Meets Practice
MAY 14-16, 2019 | PHILADELPHIA, PA

Vibration Measurement Using Accelerometers

- Vibration measurement accuracy and repeatability depends upon how well the base of the accelerometer is able to follow the true movement of the surface being measured.

- This requires a stiff attachment, one that does not allow the accelerometer surface to part from the measured surface.

- Also an attachment should not allow the natural frequency of the mounted accelerometer to drift down into the intended measurement frequency range.

MFPT 2019 Where Theory Meets Practice
MAY 14-16, 2019 | PHILADELPHIA, PA

Typical Vibration Measurement Hardware

Types of accelerometers

- Single-axis, double-axis, and tri-axis accelerometers.
- As small as 0.2 inches long (5.1 mm) and weighing 0.007 oz (0.2 grams). The typical accel used for troubleshooting purposes is about 1 inch (25 mm) long and 0.75 inches (19 mm) Hex and weighs about 1 oz (30 grams) without the magnetic mount.

MFPT 2019 Where Theory Meets Practice
MAY 14-16, 2019 | PHILADELPHIA, PA

- ➢ The ideal attachment is metal-to-metal contact. This may be obtained by a screw or bolt that enforces significant compression on the interface.
- ➢ Depending upon the mass of the accelerometer and the stiffness of the mounting surface, the bolted joint method can result in accurate and repeatable measurements to order of 50 kHz.
- ➢ Finite element analysis provides the insight that, for flat-surface-on-flat-surface, a bolt may be torqued enough to tightly compresses the two surfaces together sufficiently so that the separation force from acceleration F=ma, remains below the spring-like compression force across the surface, F=kx
- ➢ The author's testing used the screw attachment method as the standard against which all other methods were compared.

4

MFPT 2019 Where Theory Meets Practice

MAY 14-16, 2019 | PHILADELPHIA, PA

- Screw-on attachment of an accelerometer requires tapping of the measured surface, which may not be permitted, or may not be practical (e.g. thin surface walls).
- In other instances, screw connections may be excessively time-consuming and/or inconvenient, such as a typical Operating Deflection Shape (ODS) test, when hundreds of measurements on many types of surfaces are required.
- Screw connections are also the case for walk-around routes in large plants, where technicians must measure key locations on a large number of machines in a brief period of time, as they follow their routes.
- Therefore, other methods of attachment are typically used. One of the simplest and oldest is the use of hand-held pressure, by holding the accelerometer sensor base (sometimes through a needle point "stinger") firmly against the measurement location.

5

MFPT 2019 Where Theory Meets Practice

MAY 14-16, 2019 | PHILADELPHIA, PA

- In the authors' experience, as demonstrated by the laboratory testing reported in this presentation, hand-held measurements are typically more accurate than the other long-time method of a magnetic base firmly screwed onto the accelerometer base.
- Good accuracy and repeatability to 5 kHz is possible with good technique.
- A problem with the magnetic base is that at moderate frequency the magnetic surface can temporarily lose contact with the magnetic target surface, especially at higher acceleration G levels, and particularly at higher frequencies.

6

- Other attachment methods include adhesively "sticking" the accelerometer base on the target surface. These include:
- A thin layer of special sticky wax (only works for surfaces with temperatures to about 100 F, or 38 C)
- A thin layer of glue/ cement (typically cyano-acrylic "superglue")
- Double-sided sticky tape on the order of 7 mils (0.175 mm) thick.
- Studies have been performed by various organizations, and opinion-based curves of adequate ranges for each technique are published by such standards organizations as ISO (ISO 5348), Hydraulic Institute (ANSI/HI 9.6.4), and ANSI (ANSI S2.61).

7

Accelerometer Attachment Experiments

- A 12 inch long, 2 inch wide, 2 inch thick cantilever beam was attached to a stiff support over a 4 inch length at one end, firmly bolted to a massive slotted machine foundation plate.
- The beam retained 8 inches of cantilevered free length, as shown in the figure on the next slide.
- The modal characteristics of the final assembly were determined by impact testing and an Operating Deflection Shape test, assisted by 3-D finite element analysis.
- Many natural frequency modes were detected in the range of 100 Hz to 10 kHz (unavoidable in a real-life laboratory set-up), . The test results were evaluated in the context of these modes rather than assuming infinite stiffness of the attachment of the beam "fixed end" to ground, or the base of the exciter to ground.

8

MFPT 2019 Where Theory Meets Practice
MAY 14-16, 2019 | PHILADELPHIA, PA

- A screwed-on reference accelerometer was attached to the bottom of the free end of the beam, 2.5 inch from the beam end.
- The tested accelerometer/ attachment method in each test was attached to the beam, on the top side of the beam, precisely on the opposite side of the reference accelerometer.

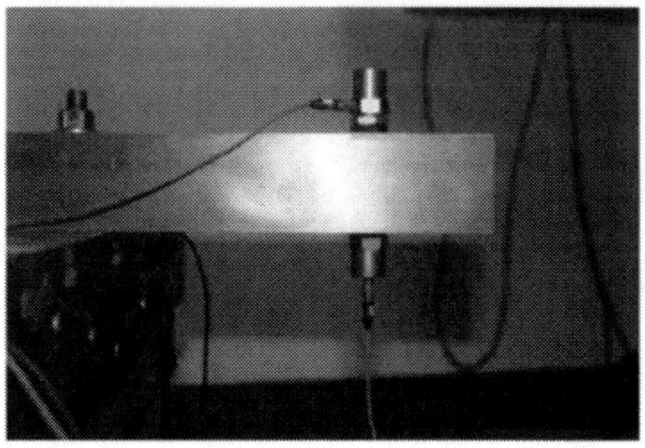

9

MFPT 2019 Where Theory Meets Practice
MAY 14-16, 2019 | PHILADELPHIA, PA

- The tests consisted of exciting the end of the beam with an electrodynamic shaker, with an accelerometer on its head to detect the spectrum input to the base of a stinger.
- For all acceleration measurements in the experimental set-up, the authors used off-the-shelf accelerometers from a major supplier.
- Oscillating force was provided by a Labworks ET-132-2 shaker, driven by a power amplifier PA-138, with LabView based software control custom-programed by the authors' group.
- Data acquisition equipment consisted of a National Instruments (NI) data acquisition system chassis model number NI PXIe-1073 with two cards of model NI PXIe-4497.

10

MFPT 2019 Where Theory Meets Practice
MAY 14-16, 2019 | PHILADELPHIA, PA

- The FFT system was set up to scan and read a frequency span of 100 Hz to over 10 kHz (after anti-aliasing), a resolution of ¼ Hz, an overlap of about 90%, and an update time of 0.375 sec.
- The analyzer was set to Peak Averaging mode throughout a frequency sweep at a constant rate of typically 20 Hz/s.
- Excitation was performed at a nominally constant amplitude of 0.1 G at the base of the shaker stinger, where the stinger attached to the shaker head.
- The shaker possessed system resonances at about 4800 Hz and 8500 Hz, which resulted in increased acceleration input to the stinger of order a 1 G when the shaker was operating within several percent of those frequencies.
- The presence of the resonant frequencies was not observed to affect the conclusions.

11

MFPT 2019 Where Theory Meets Practice
MAY 14-16, 2019 | PHILADELPHIA, PA

- A 100 mV/g reference accelerometer was stud mounted to the underside of the beam.
- For each attachment-tested accelerometer, after acquiring data at 0.1 G excitation across the frequency range, the excitation amplitude was increased by increasing the current fed to the power amplifier. Three amplitudes were tested (0.1 G peak, 0.3 G peak, and 1.0 G peak) for each tested attachment and each tested accelerometer type.
- Specifically, the procedures were repeated with different accelerometers to ensure repeatability of the various attachment methods, or to investigate the effect of accelerometer mass on the result. Two different serial number (but same model) 100mV/g accelerometers were tested, and were then replaced with a roughly 1/10th mass 10 mV/g accelerometer.
- Each specific test was repeated four times to ensure consistency. The results were found to be quite consistent, and within the range of frequency and amplitude investigated, the accelerometer size had no measurable effect.

12

MFPT 2019 Where Theory Meets Practice

MAY 14-16, 2019 | PHILADELPHIA, PA

- The following figures show only representative results from the large number of tests.
- The initial four plots show the raw frequency spectra, plotting the tested accelerometer and attachment as a blue line, and the reference accelerometer as the red line.
- These plots are log amplitude plots, which have the advantage of showing low amplitude and high amplitude responses on an equal basis, but can give the false impression of better percentage agreement than is actually present.
- The results are re-plotted in the remaining figures based on a statistical average of the repeatability testing
- The repeatability results are provided in terms of the ratio of the tested accelerometer reading to that of the reference accelerometer, along with any phase difference between the two, over the tested range of frequency. Beware of a false impression of gross error, because of "divide by zero" resolution at low amplitude anti-resonance "valleys".

13

MFPT 2019 Where Theory Meets Practice

MAY 14-16, 2019 | PHILADELPHIA, PA

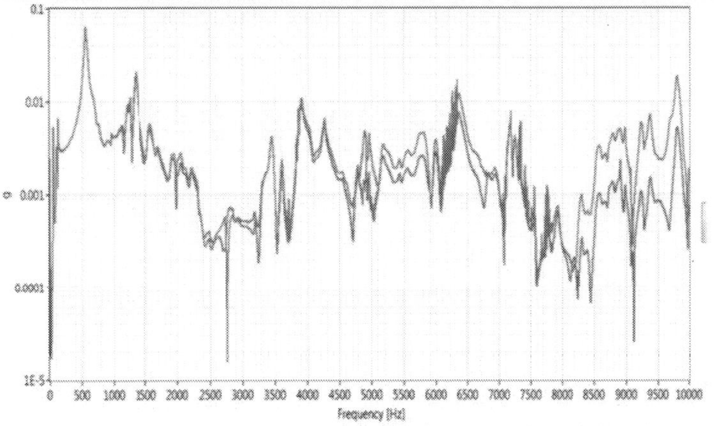

Acceleration attachment test raw spectra, handheld-only attachment; blue is the signal of the tested accelerometer & attachment, red is the reference accelerometer signal

14

MFPT 2019 Where Theory Meets Practice
MAY 14-16, 2019 | PHILADELPHIA, PA

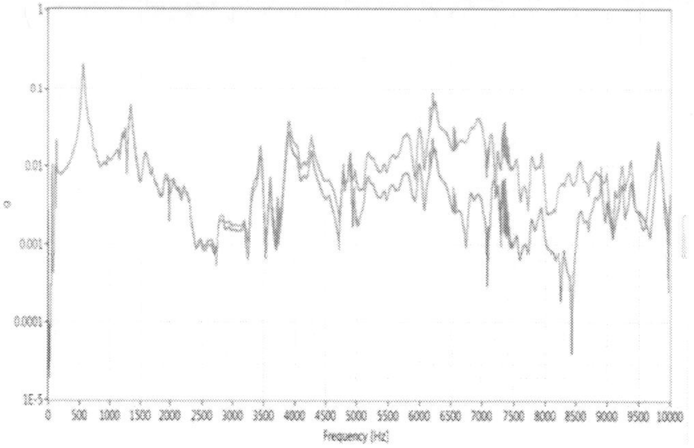

Acceleration attachment test raw spectra, magnet-only attachment; blue is the signal of the tested accelerometer & attachment, red is the reference accelerometer signal

15

MFPT 2019 Where Theory Meets Practice
MAY 14-16, 2019 | PHILADELPHIA, PA

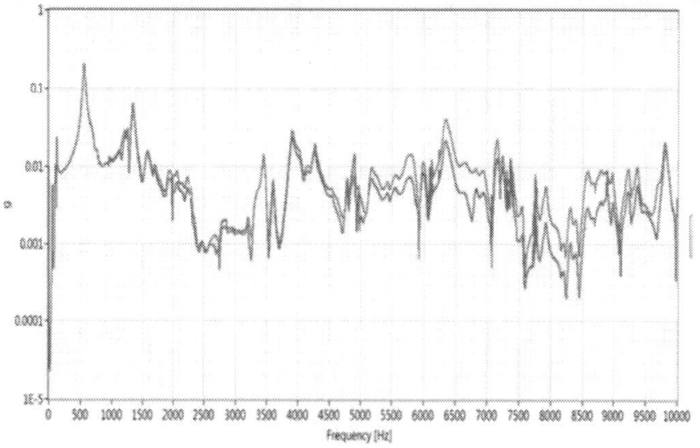

Acceleration attachment test raw spectra, wax-only attachment; blue is the signal of the tested accelerometer & attachment, red is the reference accelerometer signal

16

MFPT 2019 Where Theory Meets Practice

MAY 14-16, 2019 | PHILADELPHIA, PA

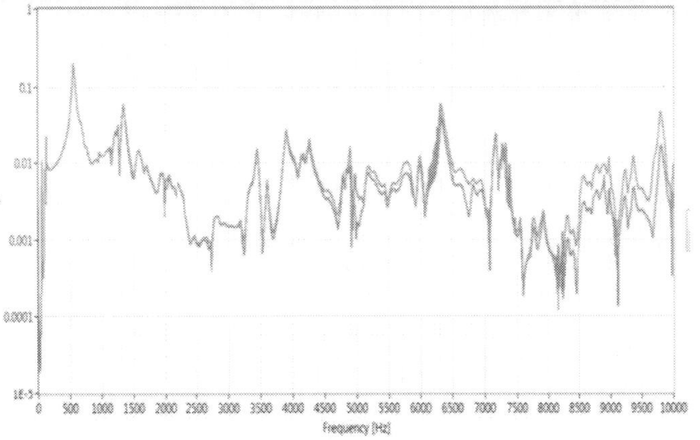

Acceleration attachment test raw spectra, magnet-plus-wax attachment; blue is the signal of the tested accelerometer & attachment, red is the ref. accelerometer signal

17

MFPT 2019 Where Theory Meets Practice

MAY 14-16, 2019 | PHILADELPHIA, PA

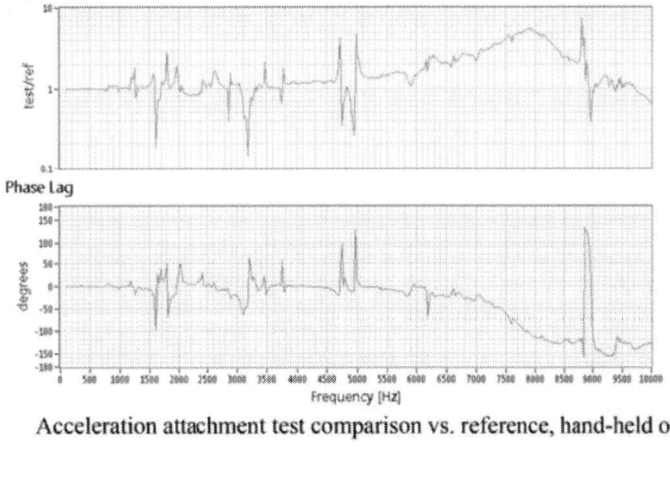

Acceleration attachment test comparison vs. reference, hand-held only

18

MFPT 2019 Where Theory Meets Practice

MAY 14-16, 2019 | PHILADELPHIA, PA

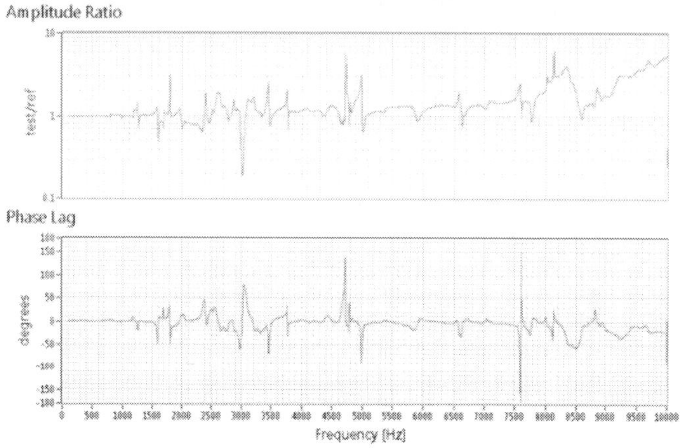

Acceleration attachment test comparison vs. reference, magnetic mount only

19

MFPT 2019 Where Theory Meets Practice

MAY 14-16, 2019 | PHILADELPHIA, PA

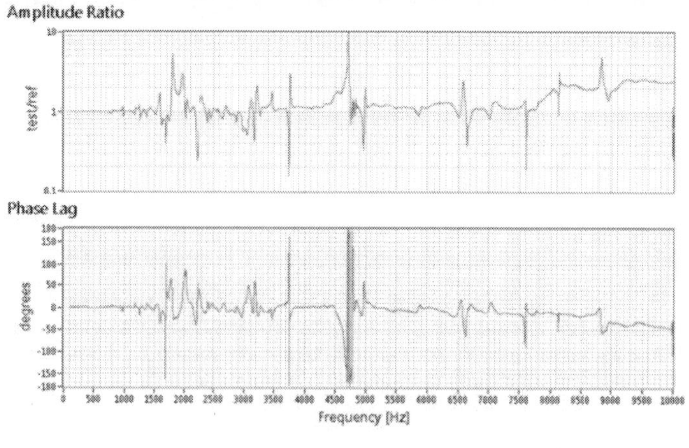

Acceleration attachment test comparison vs. reference, wax only

MFPT 2019 Where Theory Meets Practice
MAY 14-16, 2019 | PHILADELPHIA, PA

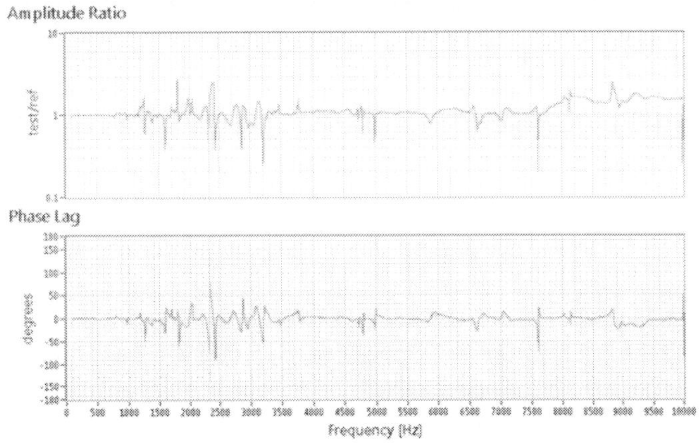

Acceleration attachment test comparison vs. reference, wax plus magnet mount

MFPT 2019 Where Theory Meets Practice
MAY 14-16, 2019 | PHILADELPHIA, PA

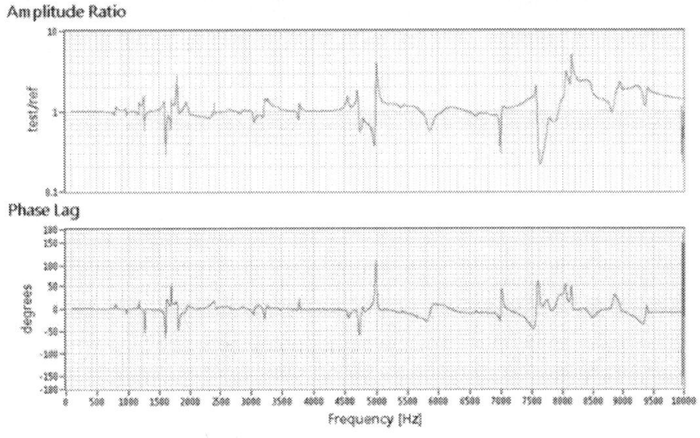

Acceleration attachment test comparison vs. reference, 3 mil thick tape

MFPT 2019 Where Theory Meets Practice

MAY 14-16, 2019 | PHILADELPHIA, PA

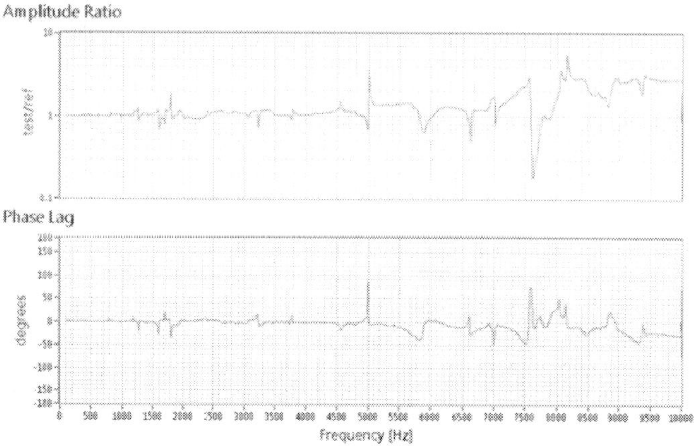

Acceleration attachment test comparison vs. reference, 7 mil thick tape

MFPT 2019 Where Theory Meets Practice

MAY 14-16, 2019 | PHILADELPHIA, PA

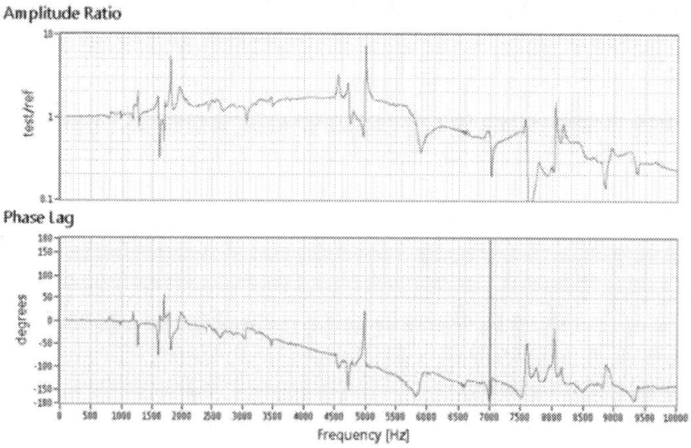

Acceleration attachment test comparison vs. reference, 15 mil thick tape

MFPT 2019 Where Theory Meets Practice
MAY 14-16, 2019 | PHILADELPHIA, PA

- In the opinion of the authors, a screw compression attachment of flat-surface-against-flat-surface is the "gold standard" of accelerometer attachment methods. This is consistent with the recommendations of international standards such as ISO and ANSI/HI.

- All of the tested attachment techniques were found to provide accurate and repeatable results for frequencies from 1 Hz to 1.5 kHz, indistinguishable on a practical basis.

MFPT 2019 Where Theory Meets Practice
MAY 14-16, 2019 | PHILADELPHIA, PA

- The best non-damaging technique of accelerometer attachment was found to be moderate thickness (7 mil, or 0.175 mm) double-sided fiber-reinforced paper tape.

- 50% thinner tape did not produce as high fidelity results, nor did 2x thicker tape. The 7 mil thick tape attachment method provided reasonable fidelity from 1 Hz to 7 kHz, apparently being thick enough to better fill in surface undulations than the 3 mil thick tape, but not be so thick as to lose significant stiffness though the thickness of the tape as in the case of the 15 mil thick tape.

- Some fidelity was lost at specific natural frequency peaks and anti-resonance "zeros", such that the reference accelerometer might report higher or lower vibration amplitude within +/-1% of such zones.

- The use of wax by itself provided results that were inferior to wax plus a magnet.

- The "lowest fidelity method" label was shared by magnet-alone, hand-held alone, and wax-alone.

- The magnet-alone was dependent on the flatness of the attachment surface.

- Handheld-alone was generally superior, but dependent on the firmness and steadiness of the hand holding the accelerometer. Generally, any unsteadiness in the hand holding affected frequencies at 30 Hz and below.

- The wax attachment was dependent to some extent on the thickness of the wax coating between the accelerometer base and the target surface. Less than 3 mils thickness is recommended, and can be achieved by a twist during the attachment process.

NAVSEA
NAVAL SEA SYSTEMS COMMAND

NAVSEA CBM+

May 2019

File Name: SNA 2019 CBM+ Rev2
Distribution Statement A: Approved for public release; Distribution unlimited

NAVSEA
NAVAL SEA SYSTEMS COMMAND

NAVSEA
NAVAL SEA SYSTEMS COMMAND

Senior Leadership Direction

ADM Richardson has made the Digital transformation a foundation in his Design 2.0

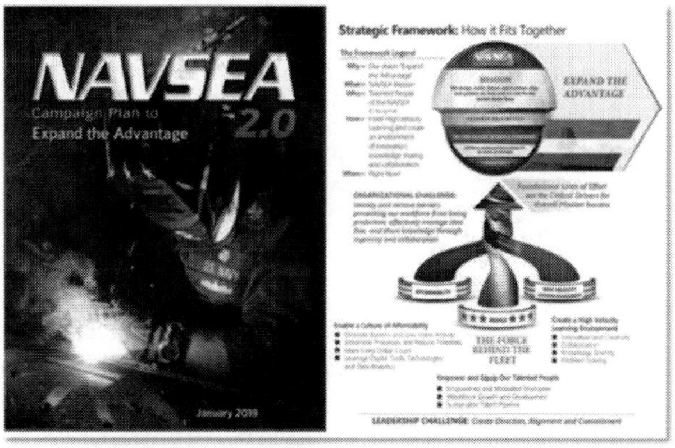

ADM Grady at AFCEA-USNI West 2019 Conference Feb 2019:

"With Predictive analytics in play, we will find ourselves in a place where we are making data-informed decisions, maximizing the return on investment for taxpayers"

File Name: SNA 2019 CBM+ Rev2
Distribution Statement A: Approved for public release; Distribution unlimited

2

NAVSEA

CBM

DoD Instruction 4151.22 of 16 October 2012

Notes: AIT=Automatic Identification Technology, SIM=Serialized Item Management, IETMs=Interactive Electronic Tech Manuals, IUID = Item Unique Identification, AISs=Automated Information Systems, ERPs=Enterprise Resource Planning, PMAs=Portable Maintenance Aids

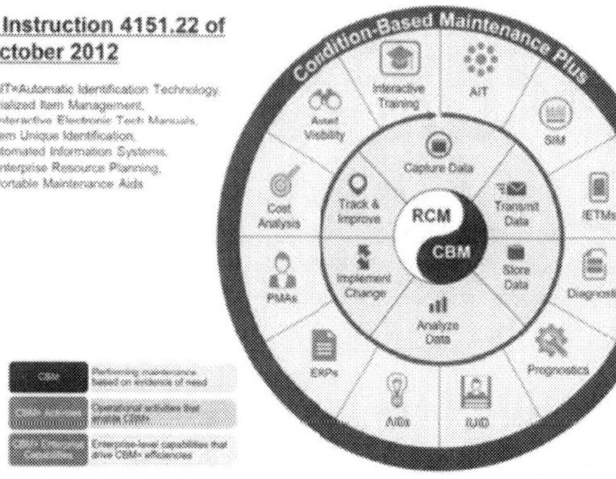

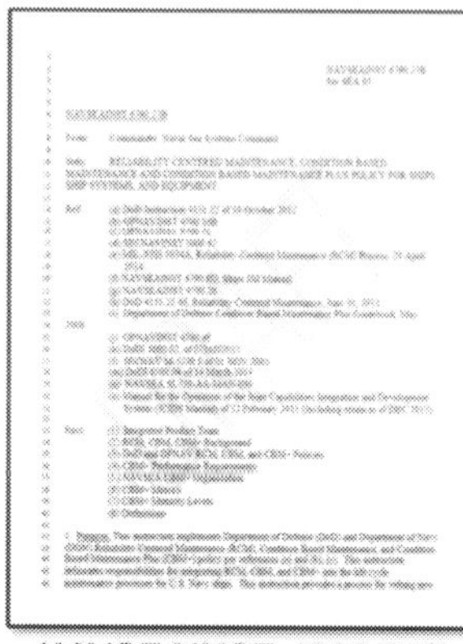

Reliability Centered Maintenance (RCM)
- Since 2001 over 10K trained and certified
- FY19 Plan has 66 RCM Courses and 6 Fleet Maintenance Effectiveness Reviews which will review ~9000 maintenance Procedures
- RCM is the feedback mechanism with the Fleet operators and maintainers

Condition Based Maintenance (CBM)
- Over ten years of investment in CBM
- Recent improvements to CBM+ enablers are going to make vast improvements in our Operational Availability, reduce system down time and also save money
- This new information will be fed back into the RCM process to drive further improvements

NAVSEAINST 4790.27B
- Final draft in review with NAVSEA 04 and NAVSEA 05
- Release Date in May 2019

eRM and CMAS

Integrated Condition Assessment System (ICAS)
- Has been NAVSEA's common tool to support Condition Based Maintenance (CBM) for the past 20 + years

Enterprise Remote Monitoring (eRM)
- Increased speed, scalability, cyber posture
- Able to support rapid development for
  - New algorithms
  - Digital twin
  - Linkage to existing systems and sensors

Consolidated Machinery Assessment System ashore (CMAS)
- Diagnostics, parametric trend analysis and providing condition alerts

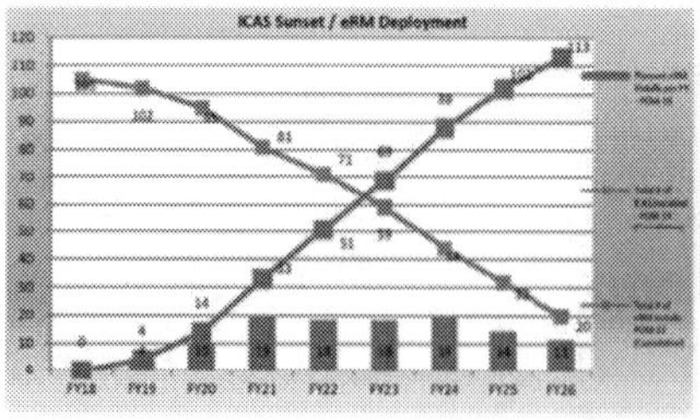

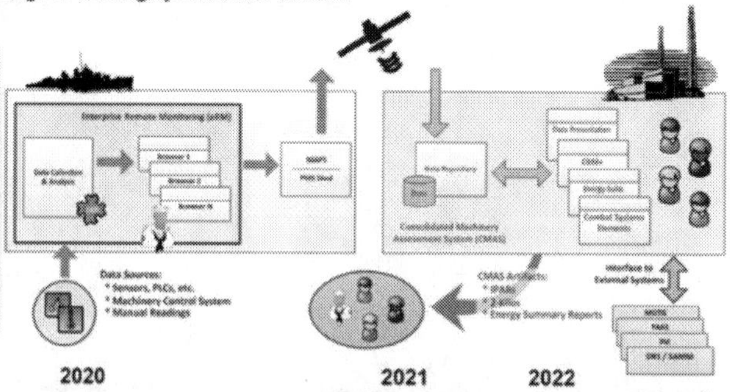

2018 2019 2020 2021 2022

eRM First in class: CG, LCS, DDG, LSD, LPD, LHD

Advanced Analytics: Gas Turbines, Drive Trains, Diesels / Air Cond & Reefers / Air Compressors, Others →

Software: eRM v3.1 to v3.3; CMAS v1.2 to v1.3

CMAS is in The Early Stages of Realizing Digital Twin Capability for CBM+

NAVSEA
NAVAL SEA SYSTEMS COMMAND

How CBM+ fits in the BIG picture

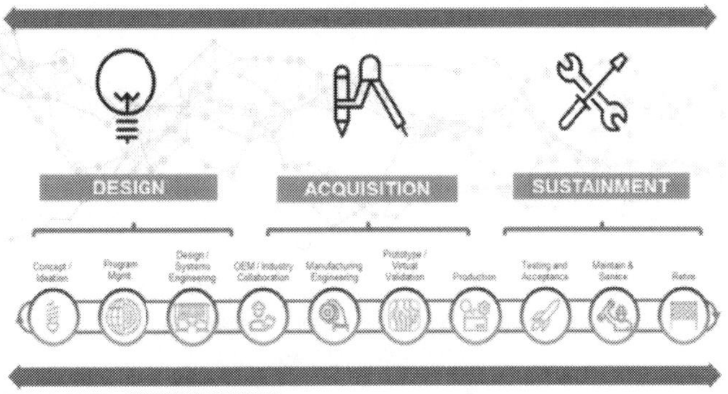

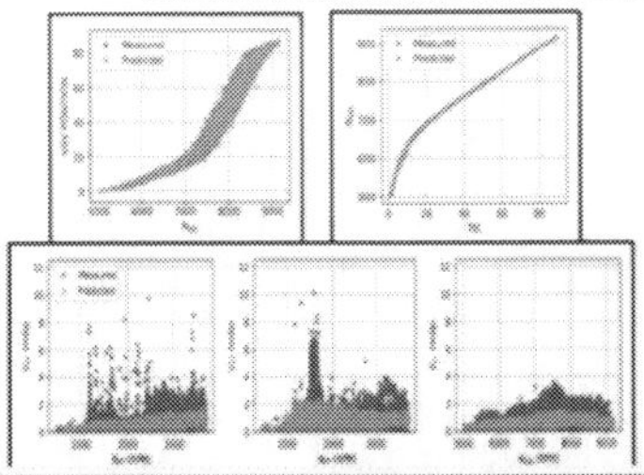

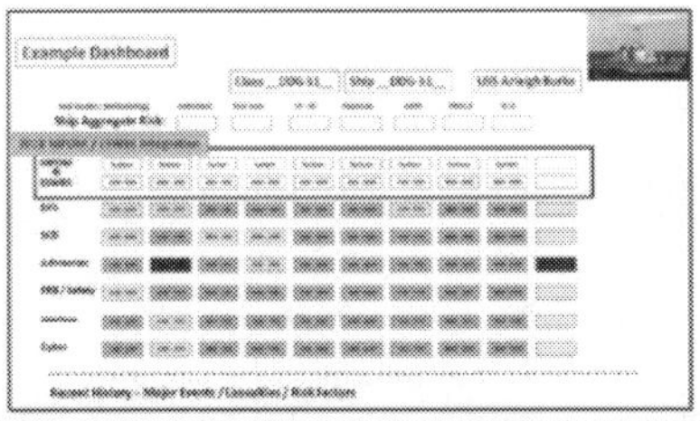

CBM+ is a Critical Enabler to Close the Digital Thread

File Name: SMA 2019 CBM+ Rev2
Distribution Statement A. Approved for public release: Distribution unlimited
5
NAVSEA

High Content, Low Bandwidth Secure Long Distance Communication of Shipboard Diagnostics Information

Distance Communication Maintenance System (DCoMS)

Presented by: Eric Olson
Vice President, Mechanical Solutions, Inc.
973-326-9920 ext 125
Eric.olson@mechsol.com

May 14, 2019

NAVSEA RIF: RIF-2017-000112
Contract: N00024-18-C-4003

Mechanical Solutions, Inc.
Engineering Analysis, Test & Technology

Distribution Statement A. Approved for Public Release

"Over the Horizon" Communication Telemedicine

Practice:

- Telemedicine is the use of telecommunication and information technology to provide health care over great distances.
- It has been used to overcome distance barriers and to improve access to medical services that would not often be available

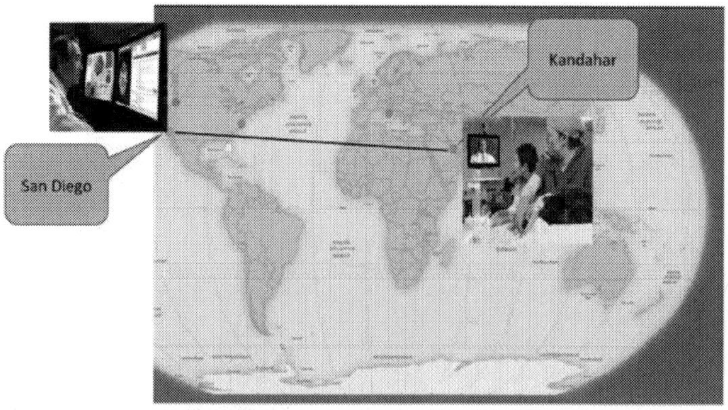

Distribution Statement A. Approved for Public Release

"Over the Horizon" Communication Telemaintenance

Concept:

The Distance Communication Maintenance System (DCoMS) enables shore-site Subject Matter Experts (SMEs) to remotely support more timely shipboard maintenance and troubleshooting activities without days of separation by travel.

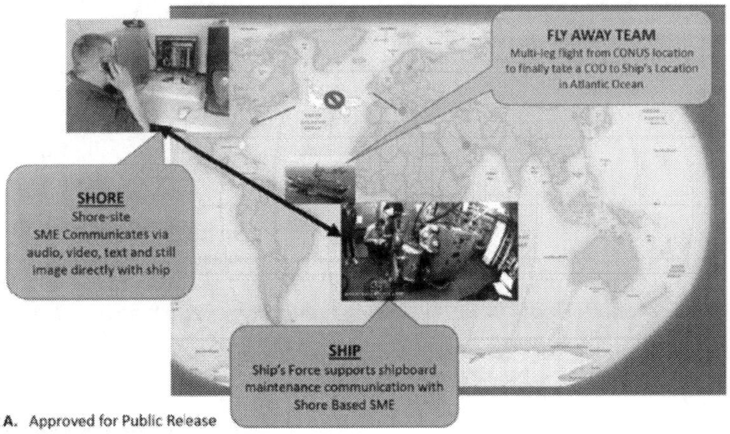

Distribution Statement A. Approved for Public Release

DCoMS - Issue/Benefit/Objectives

- Connects shore-site Subject Matter Expert (SME) to shipboard maintenance personnel.
  - **Eliminates unnecessary travel**
  - **Multiplies the effectiveness of the on call SME force.**
  - **Increases overall fleet readiness**
- Uses Consolidated Afloat Networks and Enterprise Services (CANES) or Integrated Shipboard Network System (ISNS) for secure ship-to-shore data transmittal.
  - **DOES NOT interface with any other shipboard system. CANES or ISNS for data transport only.**
- System integrates a Historical Archive Library (HAL), which is a lessons learned database.
  - **Shared resource available to all DCoMS users**

- **Rapid Innovation Fund (RIF) Efforts**
  a. Finalize development of DCoMS v1.0 (completed)
  b. Interim Authority to Test (IATT) (ECD June 2019)
  c. System Integration Test (SIT) of v1.0 – July to August 2019 time frame (preparations in process)
  d. FLEX process for DCoMS demonstration test aboard a carrier in September or October 2019 (completed)
  e. NPC for three kit trial period (completed)
  f. RIF Goal: initial deployment on PROD Network after trial period (on schedule with seminal transition event in March 2020)

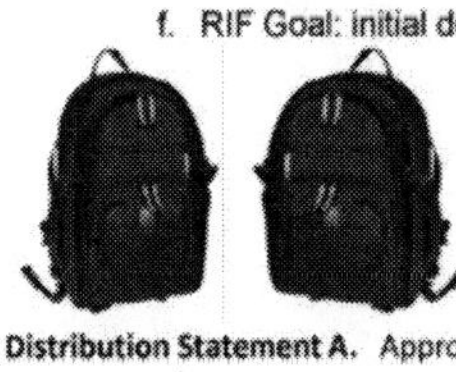

DCoMS Kit is:
- 2 backpacks
- Weigh less than 25 lb. ea.
  - **Worn equipment weighs only 2.3 lb.**
  - Laptop weighs 4 lb.

Distribution Statement A. Approved for Public Release

Testing Onboard the Self Defense Test Ship - 1
November 2015 Test

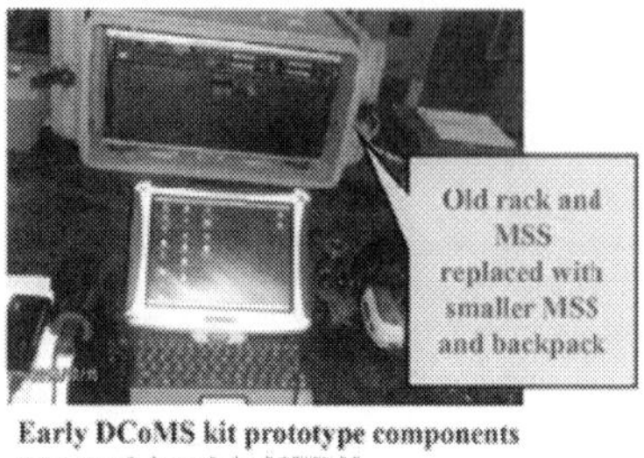

Early DCoMS kit prototype components on a work bench in MER II

Above is the still frame of the video at the start of a walk-around inspection recorded for testing HAL. Bulkhead mounted camera view. The shore-site SME remotely controls the camera pan and tilt.

Successfully demonstrated remote troubleshooting technology in a shipboard environment!

Other early testing demonstrated effective audio, video, and text communications at bandwidths as low as <u>128Kbps</u>

Distribution Statement A. Approved for Public Release

5

Testing Onboard the Self Defense Test Ship - 2
November 2015 Test

Example of what the shore-site SME or ISEA sees on their NMCI computer screen

Bulkhead camera view

Headset camera view

Inspection camera view

Distribution Statement A. Approved for Public Release

6

DCoMS Shipboard Kit Hardware

Device #	Device
1	Machinery Space Server (MSS)
2	Machinery Space Laptop (MSL)
3	Noise Cancelling Headset
4	Cameras (3 per kit) - one with ambient noise microphone
5	Inspection camera lense
6	Bags (x2 per kit)
7	Pan/Tilt Clamp
8	Pan/Tilt Assembly & Controller
9	Cables, batteries, Wireless Antennas, Battery Packs

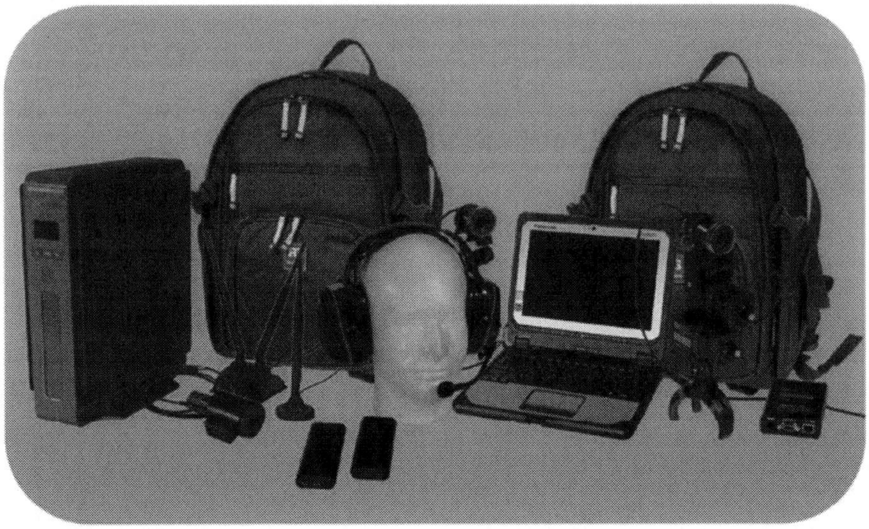

Distribution Statement A. Approved for Public Release

RIF Project Schedule to Include 3rd Year Transition

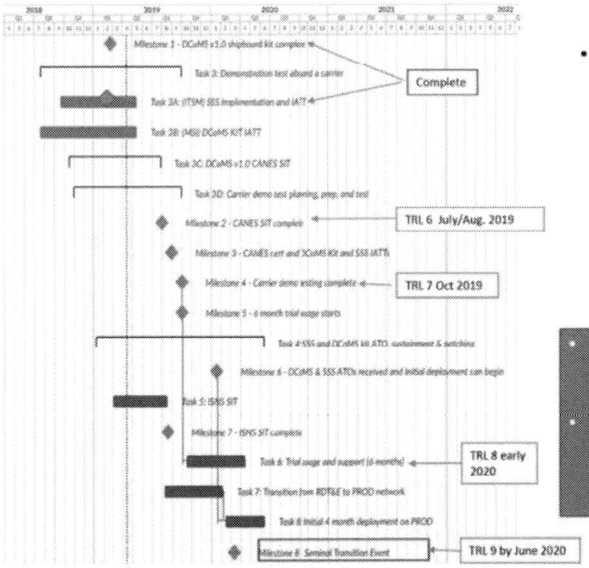

- **Seminal Transition Event**
  - **What is it:** Initial Deployment of 7 kits (1 kit each on 7 ships) aboard aircraft carriers.
  - **When it will occur:** March,2020
  - **Where it will occur:** Aboard in service aircraft carriers
  - **Who transitions the technology:** MSI with direct support from PEO Carriers.
  - **3rd Year transition:** Based on continuing deployment on aircraft carriers with additional platforms based on new funding.

- Funded RIF effort is on track to transition DCoMS to up to 7 carriers by mid-2020.
- Post-RIF – a) continue supporting 7 initial kits and b) DCoMS for remaining carriers and other surface ships including logistics support (need funding).

Distribution Statement A. Approved for Public Release

NAVSEA
NAVAL SEA SYSTEMS COMMAND

DCoMS
RIF-2017-000112

Mechanical Solutions, Inc.
Engineering Analysis, Test & Technology

Enables shore-site SMEs to remotely support more timely shipboard maintenance and troubleshooting activities without days of separation by travel.

Relevance / How Technology Addresses Requirement
- Maintenance costs reduced and mission readiness improved by amplifying shore-site SME and ISEA effectiveness.
- Optimizes maintenance activities over the platform life cycle.
- Companion Historical Archive Library (HAL) module improves sailor training.

Transition (Existing Military System, Program or Capability)
- Select RMC personnel and aircraft carriers (SBIR sponsor)
- Entire surface fleet plus aircraft maintenance (Fleet Readiness Centers)

TRL: Current – 6 Anticipated by June 2020– 9

Technical Approach/Qualifications

Project Objectives and Scope:
- Successfully complete a carrier demonstration test and a 6 month trial usage period.
- Transition from RDT&E to production NMCI network & complete an initial 4 month deployment

Key Personnel, Facilities / Equipment
- All in-place including an agreement with subcontractor NIWC (formerly SPAWAR) for implementing and hosting the shore-site infrastructure

Related Prior or Current Work (as of May 1, 2019)
- Completed – Development, ground-based demo test, & shipboard environmental test
- Funded/in-progress – Authority to Operate (ATO), shore-site network infrastructure & System Integration Test (SIT)

Programmatics and Contacts

Major Activities/Milestones:
1) Successful carrier demo test by October 2019;
2) Complete 6 month trial usage period by April 2020;
3) Complete an initial 4 month deployment by June 2020

Deliverables / Measures of Success:
- Up to 8 (7 deployed, 1 support) kits including training and usage support
- Effective maintenance related communications
- Maintain cybersecurity related certification

Potential Risks:
- Cybersecurity accreditation process timing and effort
- SIT test schedule (set by PMW 160)

Contacts:
Eric Olson DCoMS PI 973-326-9920 ext. 125 ejo@mechsol.com
Jerry Bobo Navy RIF PM 805-228-5745 jerry.bobo@navy.mil

Distribution Statement A. Approved for Public Release

NAVSEA
NAVAL SEA SYSTEMS COMMAND

DCoMS Development Support

MFPT

MSI would like to acknowledge the support from:
1. PEO Aircraft Carriers – Sponsor of both the SBIR and RIF Projects
2. The Navy Small Business Innovative Research (SBIR) Program
3. OSD and NAVSEA Rapid Innovation Fund (RIF) Program
4. PMS 312 In-service Aircraft Carriers – Cybersecuity and fleet coordination
5. Naval Surface Warfare Center (NSWC) Port Huereme Division – 1) SBIR and RIF Program Management by the NSWC Distance Support Advocate, and 2) Use of the Self Defense Test Ship (SDTS) for DCoMS prototype testing.
6. Regional Maintenance Center (RMC) Southwest - DCoMS System Requirements Input (Voice of the customer)
7. Fleet Readiness Center (FRC) Southwest - DCoMS System Requirements Input (Voice of the customer)
8. Fleet Integrated Sustainment Team (FIST) In Service Engineering Agent (ISEA)/Regional Maintenance Center (RMC) Lash-Up – Voice of the customer
9. Naval Information Warfare Center (NIWC, formerly SPWAR) IT Solutions Management (ITSM) – MSI subcontractor for implementing and hosting the DCoMS shore-site servers
10. PMW 160 Application Integration (AI) and the Development Integration Certification Environment (DICE, formerly the ITF) – Cybersecurity related certification and testing
11. PMW 130 Cybersecurity Design and Development (CDD) team – HBSS Policy Development and Test Support, as well as ACAS scan support
12. PMW 790 Ashore Application Integration Process (AAIP) – Cybersecurity certification support
13. NSWC, Dahlgren Division – EMI and HERO testing

NAVSEA RIF: RIF-2017-000112
Contract: N00024-18-C-4003
PI – Eric Olson, ejo@mechsol.com

Mechanical Solutions, Inc.
Engineering Analysis, Test & Technology

10

Distribution Statement A. Approved for Public Release

Approved for public release

U.S. ARMY COMBAT CAPABILITIES DEVELOPMENT COMMAND – ARMY RESEARCH LABORATORY

TOWARDS REPLACEMENT OF FAILED PARTS ON THE BATTLEFIELD VIA METAL CASTING IN 3D-PRINTED DESERT SAND MOLDS

Marc Pepi[a], Jerry Thiel[b], Nathaniel Bryant[b], Brandon McWilliams[a], Andelle Kudzal[a], Jennifer Sietins[a]
[a]CCDC – ARL
[b]University of Northern Iowa

Presented by: Marc Pepi
Materials Engineer
CCDC - ARL

DISTRIBUTION A: Approved for public release; distribution is unlimited

16 MAY 2019

Approved for public release

- **Outline**
  - Objective / Motivation
  - State of the Technology
  - Task 1: Mock Component Proof-of-Concept (Sand Analysis and 3D-Printing of Sand)
  - Task 2: Actual AAV Component Proof-of-Concept (Sand Analysis)
  - Task 3: Actual AAV Component Proof-of-Concept (3D-Printing of Sand)
  - Task 4: Actual AAV Component Proof-of-Concept (Concurrent AM- Builds)
  - Conclusions / Lessons Learned

Approved for public release

TOWARDS REPLACEMENT OF FAILED PARTS ON THE BATTLEFIELD VIA METAL CASTING IN 3D-PRINTED DESERT SAND MOLDS

Objective

The objective of this project was to show a proof-of-concept of producing replacement parts for long lead time DoD critical components in-theater utilizing a sand 3D-printer with actual desert sand to produce the sand molds from a CAD drawing of the actual part(s), and pouring molten scrap aluminum into the casting molds. Concurrently, metal AM parts were also built.

SECARMY Priorities
#1 Readiness
#2 Modernization
#3 Reform

CSA Priorities
#1 Army Readiness
#2 Future Army
#3 Take Care of Troops

This capability (making long lead time spare parts in-theater) will increase operational readiness, as well as reduce our logistical footprint

Approved for public release 3

Approved for public release

TOWARDS REPLACEMENT OF FAILED PARTS ON THE BATTLEFIELD VIA METAL CASTING IN 3D-PRINTED DESERT SAND MOLDS

Motivation

To provide a proof-of-concept demonstration using 3D-printed desert sand to form casting molds for potential use in resurrected Tactical Fabrication (TACFAB) system (or similar equipment).

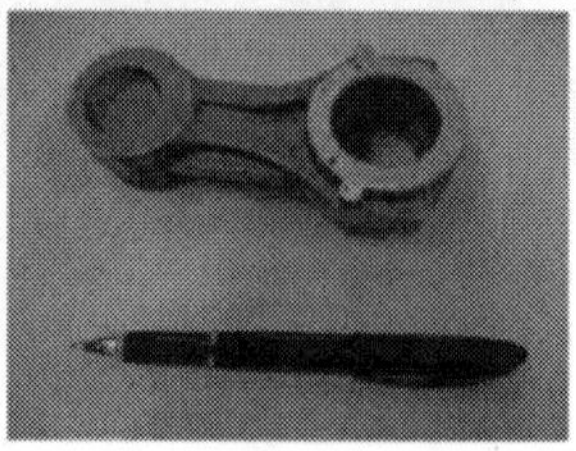

Approved for public release 4

TOWARDS REPLACEMENT OF FAILED PARTS ON THE BATTLEFIELD VIA METAL CASTING IN 3D-PRINTED DESERT SAND MOLDS

DEVCOM

State of the Technology

- **At the outset of this program, casting molds were being 3D-printed, but no one tried to use desert sand.**
- **University of Northern Iowa (UNI - through America Makes) 3D-printed casting molds with commercial grade sand [1]:**
  - Project, entitled Additive Manufacturing for Metal Casting "focused on faster, and more affordable casting for the automotive industry within the small and medium sized companies that make up the domestic metal casting supply chain".
- **Snell et. al. [2] determined:**
  - "Binder jet sand 3D printing could be used to manufacture metal components with similar properties to those of traditional sand casting processes".

[1] Additive Manufacturing for Metal Casting (AM4MC)", presentation by R. Lonardo and N. Pramanik for America Makes.
[2] Snelling, D., et. al., "The Effects of 3D Printed Molds on Metal Castings", Proceedings of the Solid Freeform Fabrication Symposium (2013), pp. 827-845.

TOWARDS REPLACEMENT OF FAILED PARTS ON THE BATTLEFIELD VIA METAL CASTING IN 3D-PRINTED DESERT SAND MOLDS

DEVCOM

State of the Technology (continued)

- **Work at Indiana University – Purdue University Indianapolis [3] concluded:**
  - "The surface finish of 3D printed sand molds was smoother than conventionally cast parts, and that closer dimensional tolerances could be achieved with 3D printed sand molds versus conventional sand casting"
- **Almaghariz, et. al. [4] claimed:**
  - "3D sand printing leads to significantly reduced lead times and labor costs as well as improved flexibility, reduced tooling and enhanced mold designs".
- **Connor [5] concluded:**
  - "3D printed sand molds were good for parts that were thin-walled".

[3] N. Hawaldar and J. Zhang, "Additive Manufacturing of Molds for Sand Casting", Poster at the Powdermet 2017 conference, Las Vegas, NV, 14 June 2017.
[4] E. Almaghariz et. al., "Quantifying the Role of Part Design Complexity in Using 3D Sand Printing for Molds and Cores", International Journal of Metalcasting, Volume 10, Issue 3, 2016.
[5] B. Connor, "Paradigm Shift", Defense AT&L Journal, Special Issue – Additive Manufacturing (3D Printing), November-December 2016.

Task 1: Mock Component Proof-of-Concept

.stl file of mock component (see photo) was furnished to UNI. They used metal casting software to determine optimum position of the gates/risers, etc.

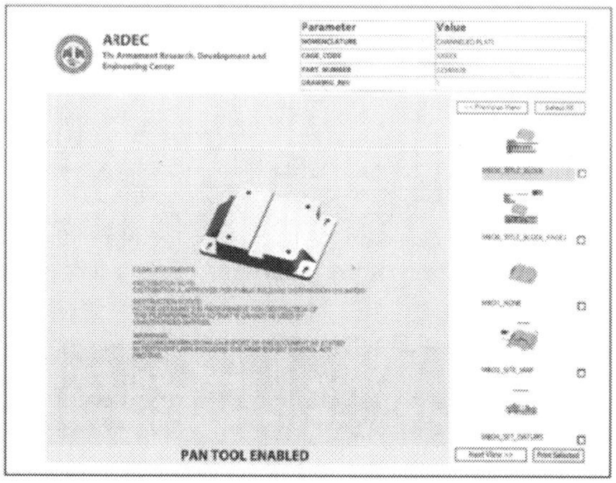

Task 1: Mock Component Proof-of-Concept (continued)

"Moon Dust" sand from Mojave Desert shipped to UNI for 3D-printing.

Desert sand chemical analysis
Mass%

SiO_2	Al_2O_3	Fe_2O_3	Cr_2O_3	TiO_2	CaO	MgO	Na_2O	K_2O
91.13	4.564	0.707	0.001	0.196	1.044	0.031	1.496	0.833

Mojave Desert sand used for 3D-printing casting molds.

Scanning electron micrograph (SEM) of the Mojave Desert sand (image courtesy of Clara Mock).

Task 1: Mock Component Proof-of-Concept (continued)

"Moon Dust" sand comparison.

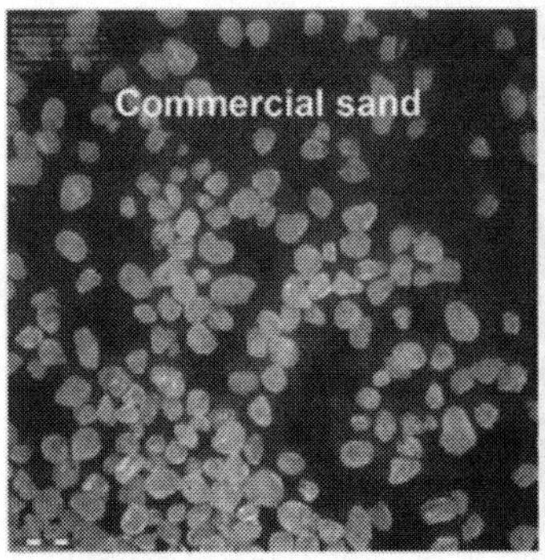

Optical macrograph of the commercial sand used with the ExOne S-Max 3D-printer. Micron marker is 500μm.

Optical macrograph of the Mojave Desert sand. Micron marker is 500μm.

Optical macrograph of beach sand from Whitsunday Islands, Australia. Compare to the commercial sand. Micron marker is 500μm.

Approved for public release

TOWARDS REPLACEMENT OF FAILED PARTS ON THE BATTLEFIELD VIA METAL CASTING IN 3D-PRINTED DESERT SAND MOLDS

Task 1: Mock Component Proof-of-Concept (continued)

3D-printing with "Moon Dust".

ExOne S-Max 3D-printer used for sand builds.

Printing head of the ExOne S-Max 3D-printer.

Task 1: Mock Component Proof-of-Concept (continued)

Actual part made from 3D-printed "Moon Dust".

Mock part in mold made of 3D-printed commercial sand, with gates and risers still attached.

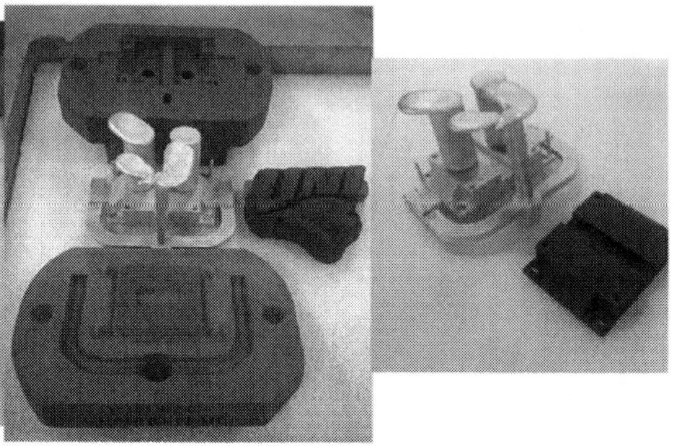

Mock parts made from indigenous sand molds.

Approved for public release

TOWARDS REPLACEMENT OF FAILED PARTS ON THE BATTLEFIELD VIA METAL CASTING IN 3D-PRINTED DESERT SAND MOLDS

DEVCOM ARMY RESEARCH LABORATORY

Task 1: Mock Component Proof-of-Concept (continued)

Computed Tomography (CT) of parts made from, 3D-printed "Moon Dust", and commercial sand.

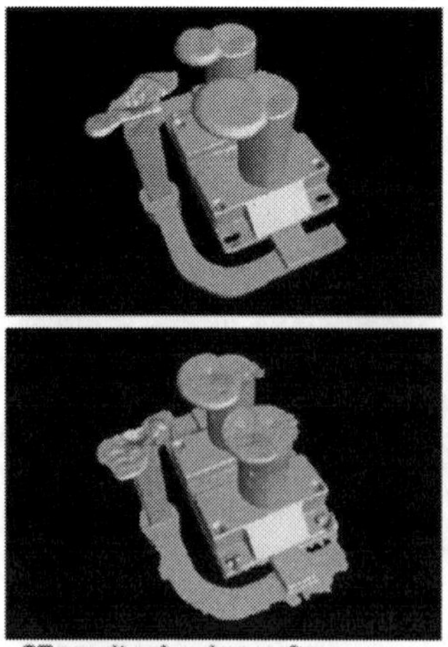

CT results showing surface roughness of parts made with commercial sand (top) and "Moon Dust" (bottom).

Cross-sectional defects of parts made with commercial sand (top) and "Moon Dust" (bottom).

Task 1: Mock Component Proof-of-Concept (continued)

Surface roughness and metallography of parts made from, 3D-printed "Moon Dust", and commercial sand.

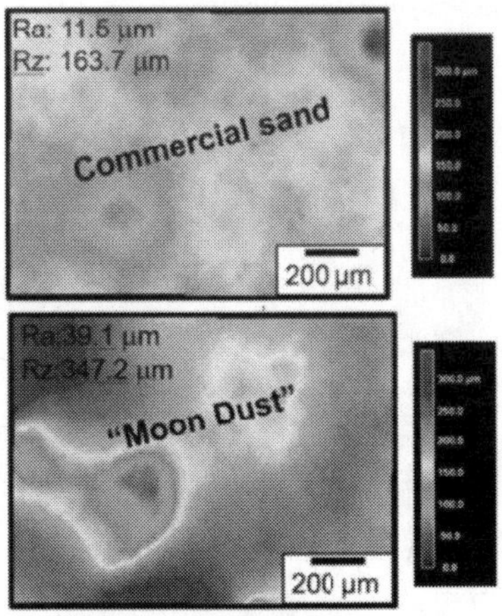

Surface roughness of parts made with commercial sand (top) and "Moon Dust" (bottom).

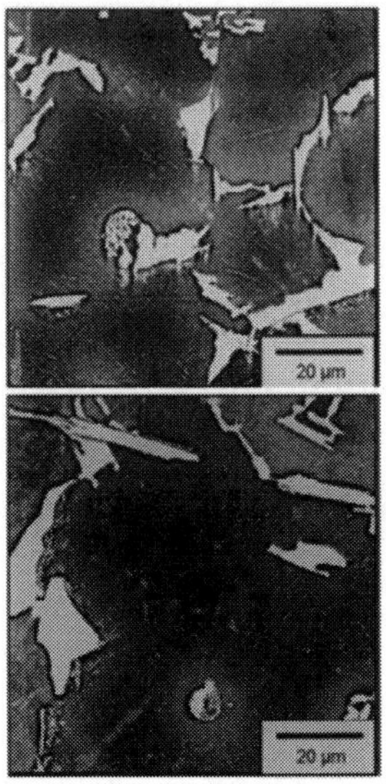

Microstructure of parts made with commercial sand (top) and "Moon Dust" (bottom).

TOWARDS REPLACEMENT OF FAILED PARTS ON THE BATTLEFIELD VIA METAL CASTING IN 3D-PRINTED DESERT SAND MOLDS

DEVCOM

Task 2: Actual AAV Component Proof-of-Concept

Long lead time item from engine of the USMC AAV.

CAD drawing showing coolant flange within AAV engine

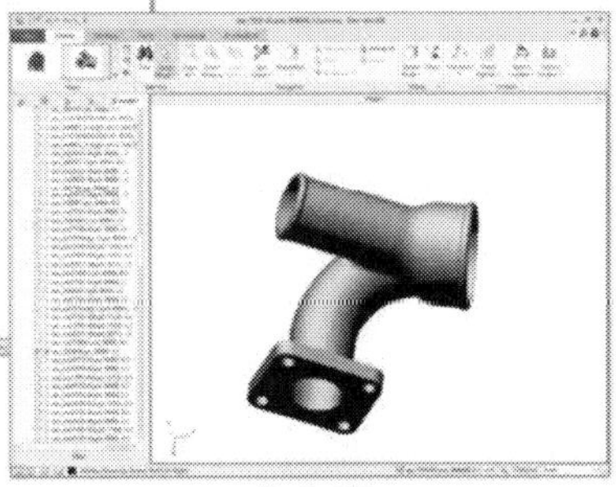

Coolant flange

Coolant flange within AAV engine

Task 2: Actual AAV Component Proof-of-Concept (Sand Analysis)

- **Challenge of Getting Usable Desert Sand:**
  - Scale-up proved challenging – only 35% of as-received sand was usable.
  - Initial stage – elimination of large particle (sieving, or "screening").
  - Baseline testing for sand pH and acid demand value (ADV) – needed for binder compatibility.
  - Second stage – too many fines? Lower screen wasn't eliminating the fine particles. UNI designed and built fluidized chamber. Angle of chamber varied for better results.

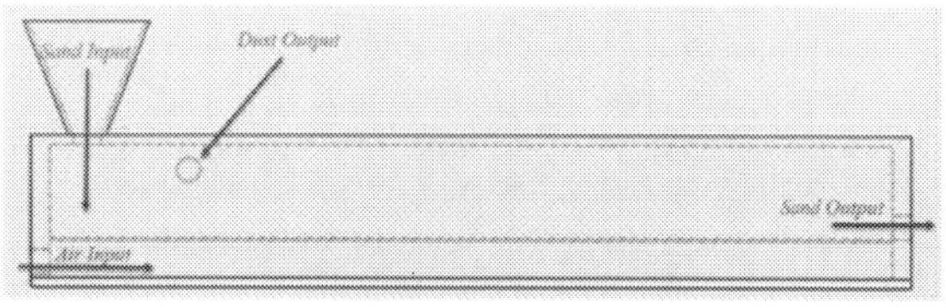

Approved for public release

TOWARDS REPLACEMENT OF FAILED PARTS ON THE BATTLEFIELD VIA METAL CASTING IN 3D-PRINTED DESERT SAND MOLDS

DEVCOM ARMY RESEARCH LABORATORY

Task 2: Actual AAV Component Proof-of-Concept (Sand Analysis)

- Final stage – "double-deck screening".
  - The 180-112 mesh screen combination retained most of the material
  - Retention of 70-mesh would have been ideal

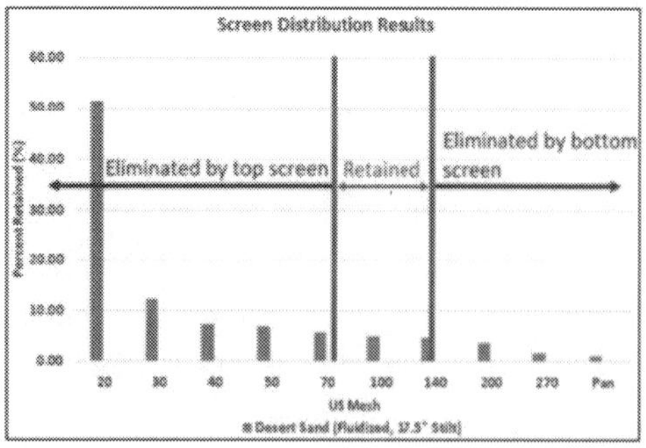

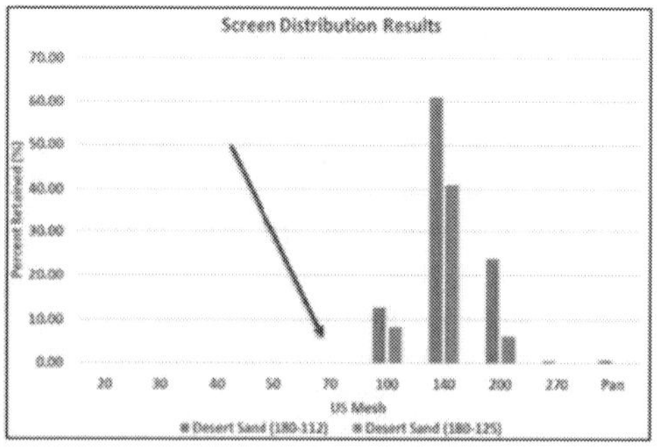

TOWARDS REPLACEMENT OF FAILED PARTS ON THE BATTLEFIELD VIA METAL CASTING IN 3D-PRINTED DESERT SAND MOLDS

Task 2: Actual AAV Component Proof-of-Concept (Sand Analysis)

- Final stage – "double-deck screening".
  - 300-125 mesh screen combination results:

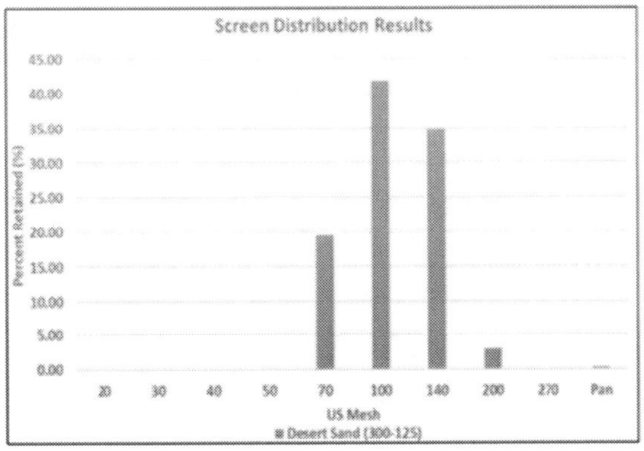

Approved for public release

TOWARDS REPLACEMENT OF FAILED PARTS ON THE BATTLEFIELD VIA METAL CASTING IN 3D-PRINTED DESERT SAND MOLDS

Task 2: Actual AAV Component Proof-of-Concept (Sand Analysis)

- **Flowability Trials**
  - This test allows estimation of overall aggregate "printability"

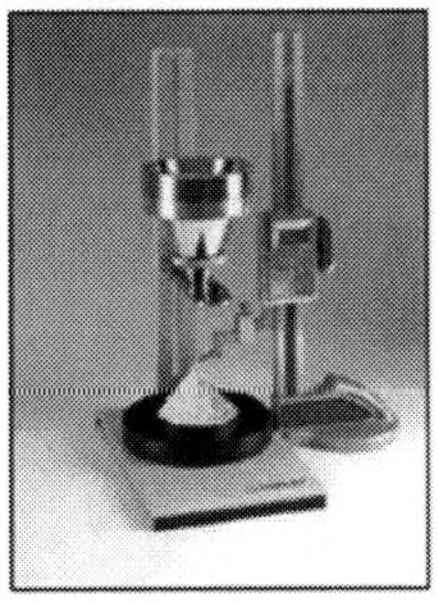

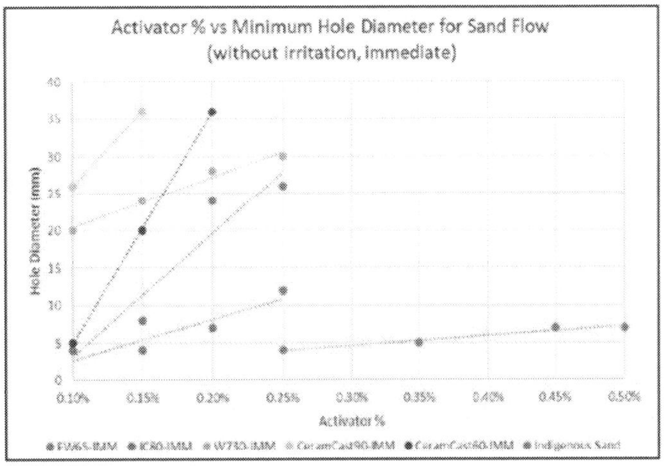

Task 2: Actual AAV Component Proof-of-Concept (Sand Analysis)

- **Flowability Trials**
  - Conclusion: it was much harder to retain the sand within the recoater than the commercial sands.

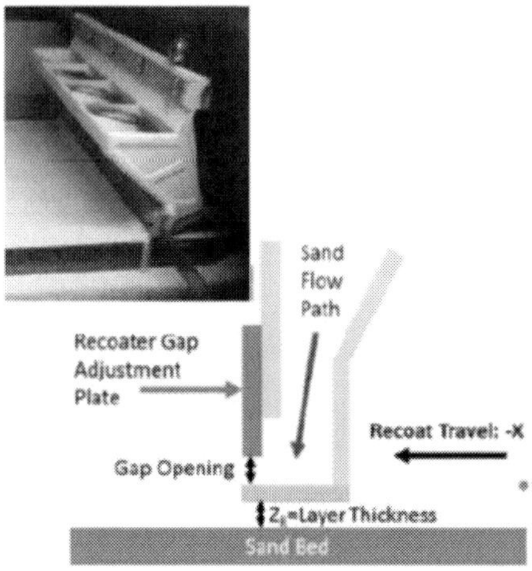

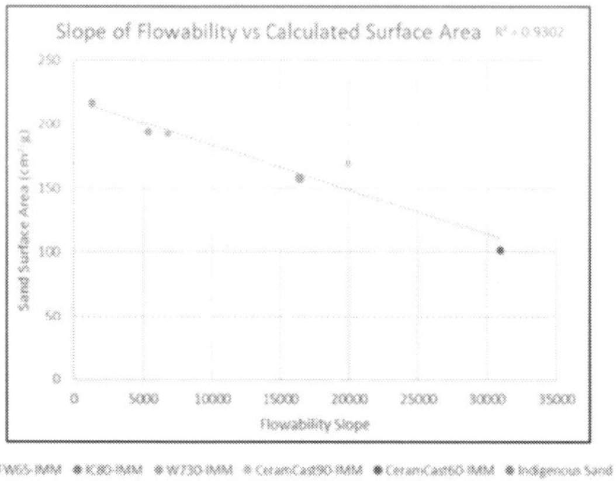

Approved for public release

TOWARDS REPLACEMENT OF FAILED PARTS ON THE BATTLEFIELD VIA METAL CASTING IN 3D-PRINTED DESERT SAND MOLDS

Task 2: Actual AAV Component Proof-of-Concept (Sand Analysis)

- **Bonding Strength**
  - Tensile strength of bonded 100% indigenous sand averaged 9.9 psi.
  - Blending sands improved strength, but still not acceptable.
  - Use of cement-based binder from Rutgers University improved strength

Sand Composition	TW40%	DA%	E3D%	Average Tensile Strength (psi)
100% Indigenous	0.5	5	2.5	9.9
75% Indigenous / 25% FW80	0.5	2	1.5	4.5
75% Indigenous / 25% FW80	0.5	3	2	10.8
75% Indigenous / 25% FW80	0.5	5	2.5	16.4
50% Indigenous / 50% FW80	0.5	5	2.5	37.3
9% N-400, 10.9% X11, Indigenous	N/A	N/A	N/A	64.9

TW40 - commercially available toluene sulfonic acid activator
DA - dry activator, a non-liquid sodium bisulfate powder activator
E3D - Enviroset 3D Jet resin, the furfuryl alcohol resin used to bind with the activator
N-400 - the amount of the powder used in the Rutgers cement binder formula
X11 - the amount of the liquid used in the Rutgers cement binder formula

Approved for public release

TOWARDS REPLACEMENT OF FAILED PARTS ON THE BATTLEFIELD VIA METAL CASTING IN 3D-PRINTED DESERT SAND MOLDS

DEVCOM ARMY RESEARCH LABORATORY

Task 3: Actual AAV Component Proof-of-Concept (3D-Printing of Sand)

- **Process Simulation Results**
  - Used MagmaSoft© software.

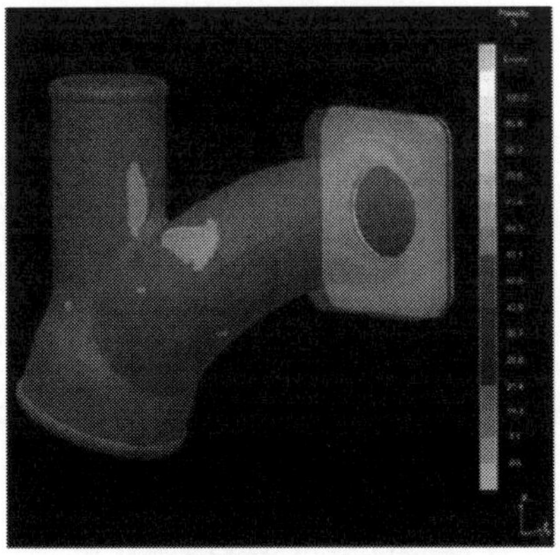

Software results looking at part as a liquid – last areas to freeze are shown in blue (shrinkage expected here)

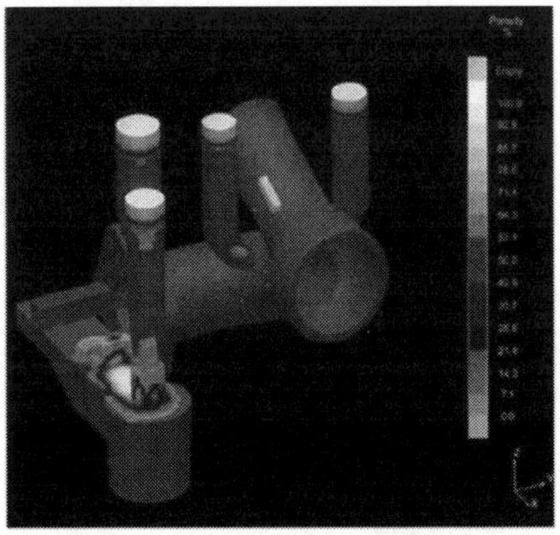

Optimized software results taking heat transfer and cooling rates into account – with proper positioning of gates and risers

Approved for public release

21

Approved for public release

TOWARDS REPLACEMENT OF FAILED PARTS ON THE BATTLEFIELD VIA METAL CASTING IN 3D-PRINTED DESERT SAND MOLDS

DEVCOM ARMY RESEARCH LABORATORY

Task 3: Actual AAV Component Proof-of-Concept (3D-Printing of Sand)

- **Cope & Drag Patterns / with Flask / Packed with Sand**

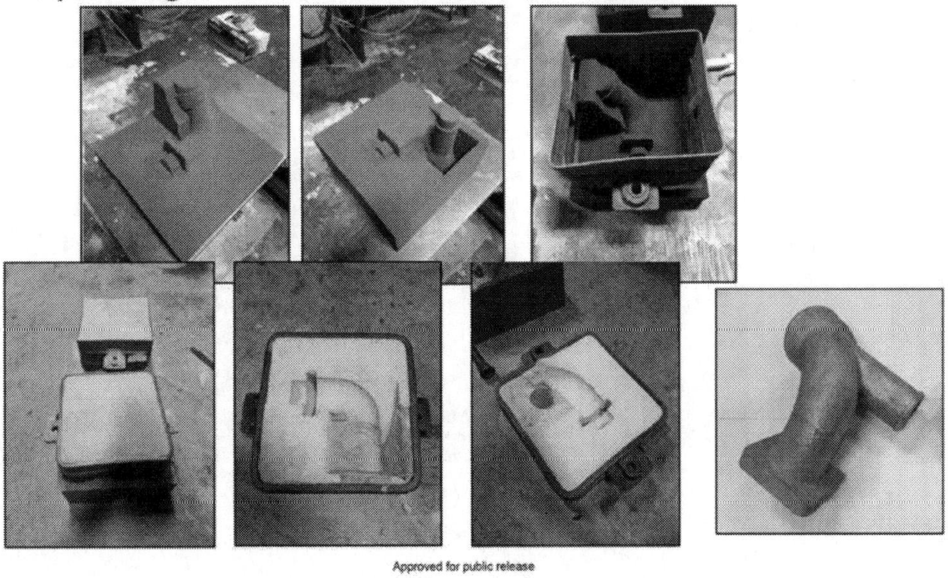

Approved for public release

22

Task 4: Actual AAV Component Proof-of-Concept (Concurrent AM- Builds)

- **Powder Analysis**
  - ALSi12 powder (used with 3DS ProX300)

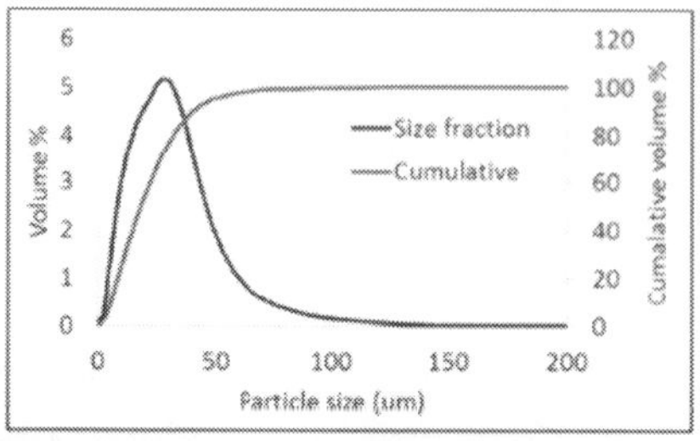

Chemical analysis of AlSi12 powder (wt %)

Element	Result
Carbon	0.003
Sulfur	<0.001
Oxygen	0.178
Nitrogen	<0.001
Iron	0.22
Silicon	12.0
Manganese	0.020
Chromium	0.016
Gallium	0.013
Nickel	0.021
Lead	0.028
Tin	0.013
Titanium	0.036
Zinc	0.063
Aluminum	remainder

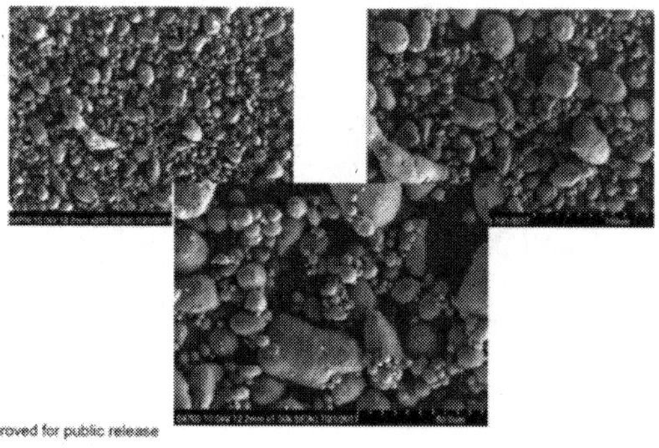

Task 4: Actual AAV Component Proof-of-Concept (Concurrent AM- Builds)

- **AM Builds – Round One**
  - Dimensional verification showed less than nominal wall thickness (attributed to compromised .stl file
  - CT showed witness line through part

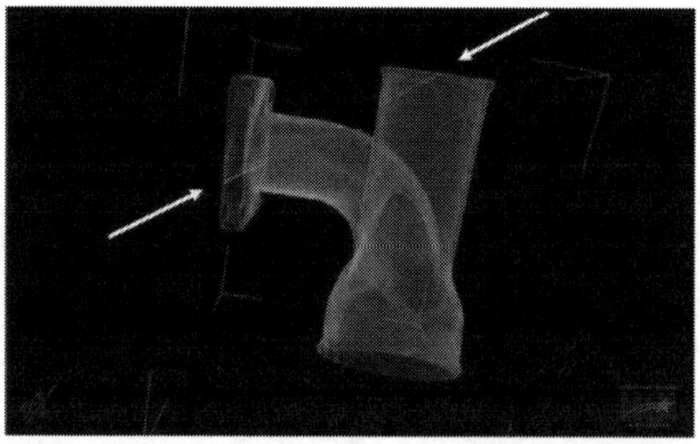

Task 4: Actual AAV Component Proof-of-Concept (Concurrent AM- Builds)

- **AM Builds – Round Two**
  - Dimensional verification showed all dimensions within specification
  - CT no visible deleterious internal defects (some porosity)

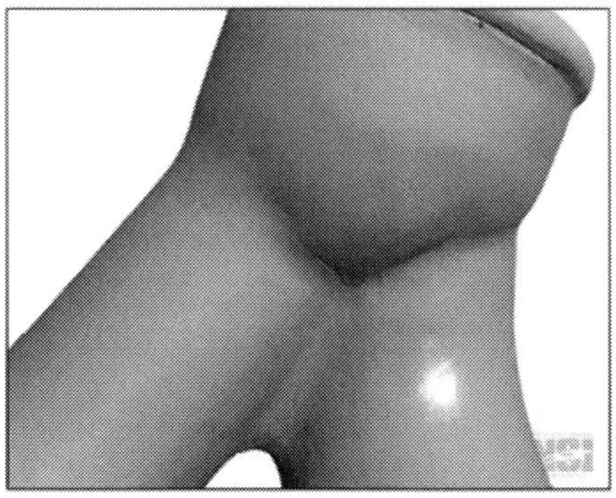

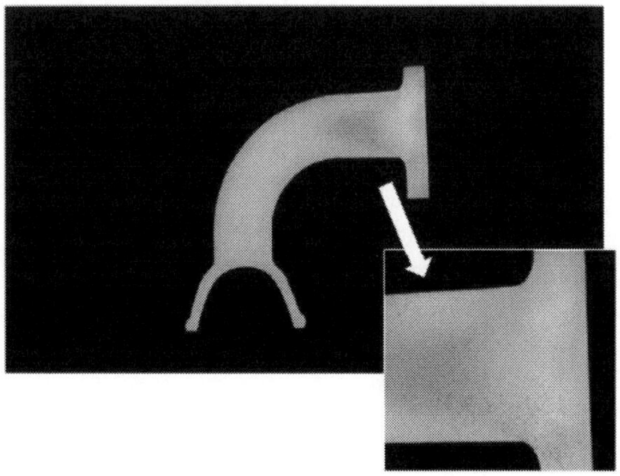

Approved for public release

TOWARDS REPLACEMENT OF FAILED PARTS ON THE BATTLEFIELD VIA METAL CASTING IN 3D-PRINTED DESERT SAND MOLDS

Task 4: Actual AAV Component Proof-of-Concept (Concurrent AM- Builds)

- **AM Builds – Round Two**
  - Density compared to theoretical bulk density of material
  - Average Vickers hardness (118 HV)
  - Tensile testing resulted in average YS of ~265 MPa, and average UTS of ~350 MPa

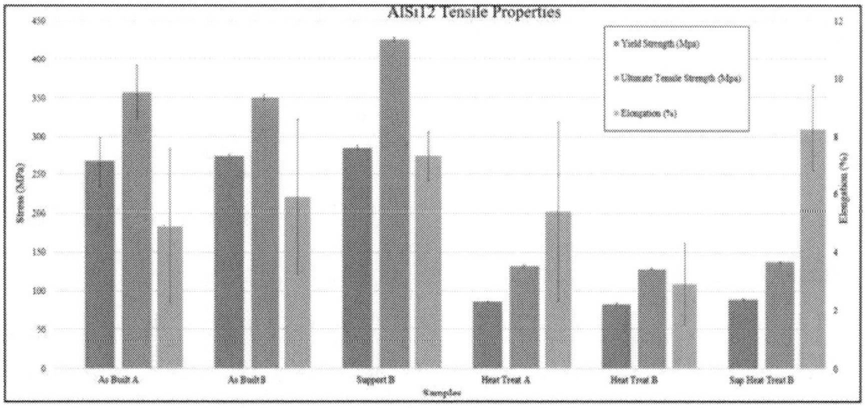

Task 4: Actual AAV Component Proof-of-Concept (Concurrent AM- Builds)

- **AM Builds – Round Two**
  - Tensile testing resulted in average YS of ~265 MPa, and average UTS of ~350 MPa

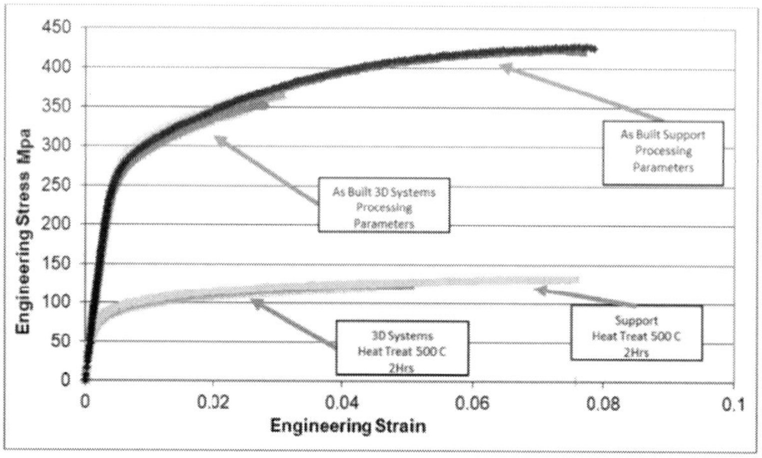

Approved for public release

TOWARDS REPLACEMENT OF FAILED PARTS ON THE BATTLEFIELD VIA METAL CASTING IN 3D-PRINTED DESERT SAND MOLDS

Task 4: Actual AAV Component Proof-of-Concept (Concurrent AM- Builds)

- **AM Builds – Round Two**
  - Metallography showed some internal porosity, and typical laser melt pool structure

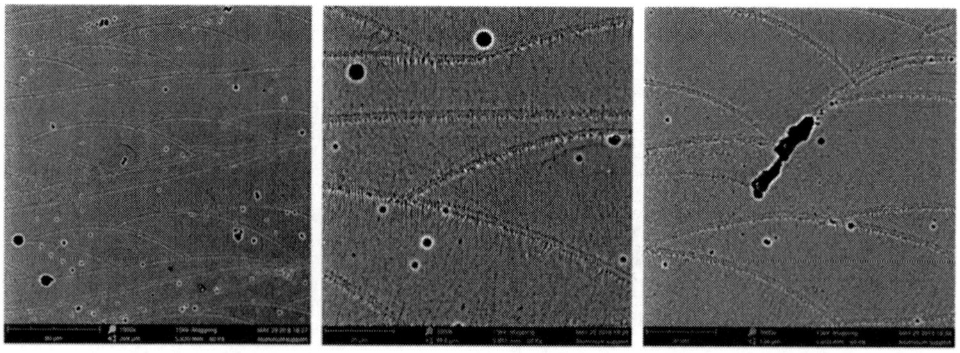

Task 4: Actual AAV Component Proof-of-Concept (Concurrent AM- Builds)

- **AM Builds – Round Two**
  - Attempting to have part tested in USMC AAV engine test stand

Approved for public release

TOWARDS REPLACEMENT OF FAILED PARTS ON THE BATTLEFIELD VIA METAL CASTING IN 3D-PRINTED DESERT SAND MOLDS

- **Conclusions / Lessons Learned**
  - Demonstration of 3D-printed desert sand molds for casting of small part was successful
  - Demonstration of 3D-printed desert sand molds for actual component (larger) proved difficult
    - Mojave desert sand particle distribution posed a problem
    - Desert sand proved "worst case", as clay contamination resulted in huge surface area – beach sand may have worked
    - Desert sand will require a modification to the 3D-printing process
    - Research on different binders may make desert sand more feasible
    - A new screening process would be needed to better remove aggregates and most fines
  - Demonstration of metal additive manufacturing of the actual component was successful
    - Need to ensure digital file is accurate prior to printing
    - Must have uninterrupted builds for better physical / mechanical properties

- **Acknowledgements:**
  - Tony Delgado and Eric Kirchner (DLA) for project funding, and continued suggestions.
  - Franklyn Kellogg and Clara Mock (SURVICE Engineering and ORAU, respectively) for technical reviews.
  - Richard Riman (Rutgers) for improved sand binder formulation.

MFPT 2019 Where Theory Meets Practice

MAY 14-16, 2019 | PHILADELPHIA, PA

Case Studies on Wear Debris Monitoring in Industrial Gearboxes

Mark Redding
President
mark.redding@poseidonsys.com
Phone: +1 (585) 633-8552

POSEIDON SYSTEMS

Why Wear Debris Monitoring?

- Gearbox Faults: **Despite significant advancements, gearboxes still:**
  - Experience catastrophic failures
  - Not reliably detected in earliest stages
  - Difficult to manage operation after fault detection
- Fault Detection/Severity: **Periodic borescopes and other widely used technologies cannot accurately detect all fault modes, failure rate, or severity**
- Offline Oil Sampling: **Particle counts and wear metal concentrations do not correlate with gearbox health**
  - Vary dramatically based on operating conditions
  - Cannot draw meaningful conclusions from periodic sampling
- Vibration Monitoring Limitations: **Certain failure modes aren't easily detected by vibration and damage can be difficult to quantify**
  - Coupled with wear debris monitoring can be very effective tool

Aftermath of Catastrophic Failure

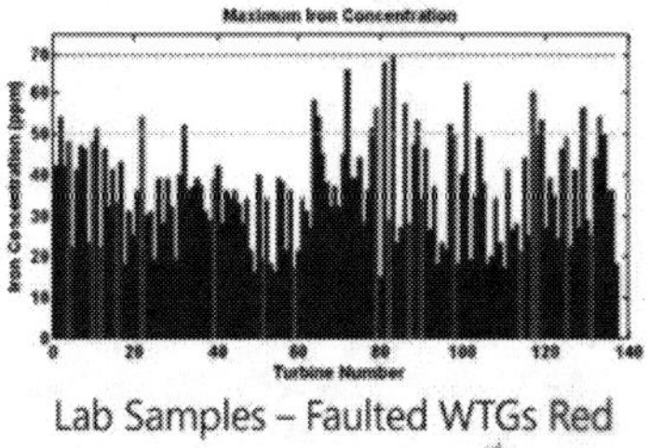

Lab Samples – Faulted WTGs Red

RCM Basis – Early Potential Fault Detection

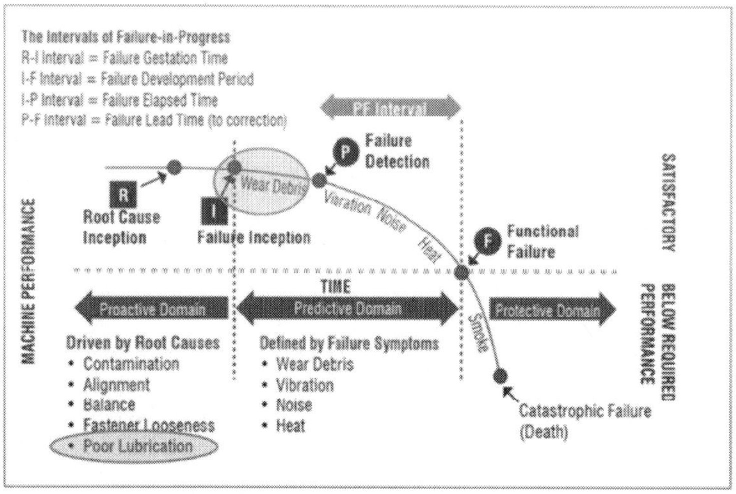

Preventative – Oil Condition Monitoring

Diagnostic – Wear Debris Monitoring

MFPT 2019 Where Theory Meets Practice

*Source: Machinery Lubrication

3

POSEIDON

Metallic Wear Debris Monitoring

Sense › Collect › Analyze

Inductive Coil Sensing Technology

- One or more coil windings around a non-conductive fluid passage (sensing bore)
- Coil is excited with a high frequency oscillating voltage
- Resulting magnetic field couples with metallic particle present within the field
- Particle passage results in a measurable impedance change in sensing coils
  - Detected as amplitude and phase shifts
- Magnitude indicates size of the particle
- Relative phase angle indicates particle type (ferrous or nonferrous)
- Width of disturbance represents the speed of the particle

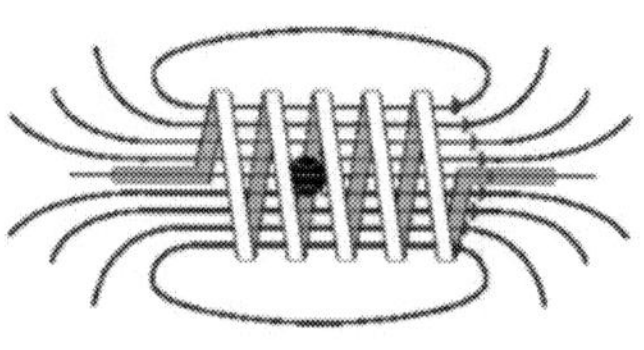

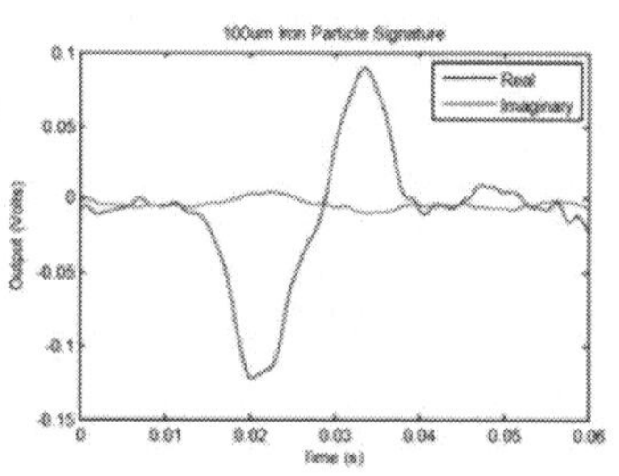

MFPT 2019 Where Theory Meets Practice

4

POSEIDON

Case Studies

Observations from over 2,500 continuously monitored gearboxes

- Fault Signature comparison for:
  - Gear tooth liberation
  - Bearing Spall
  - Slipping bearing
- Slowly progressing bearing spall
- Rapidly progressing bearing spall
- Particle size distribution observations
- Limitations of Offline Oil Sampling

Sun Pinion Liberated Gear Tooth

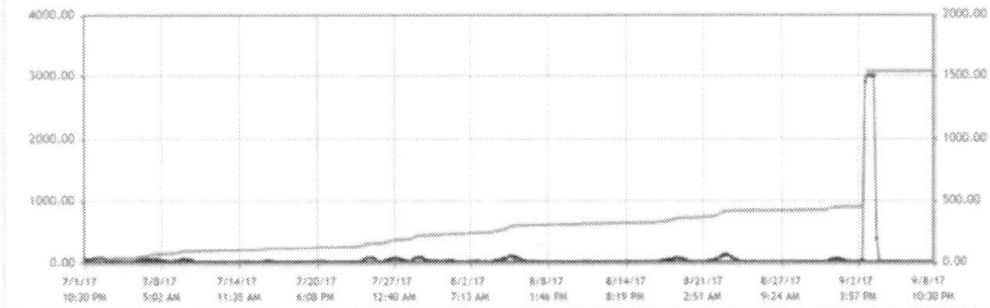

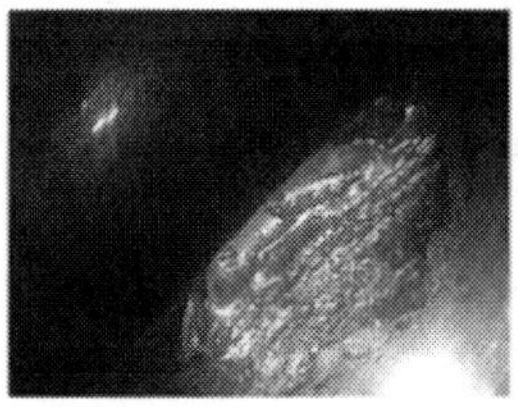

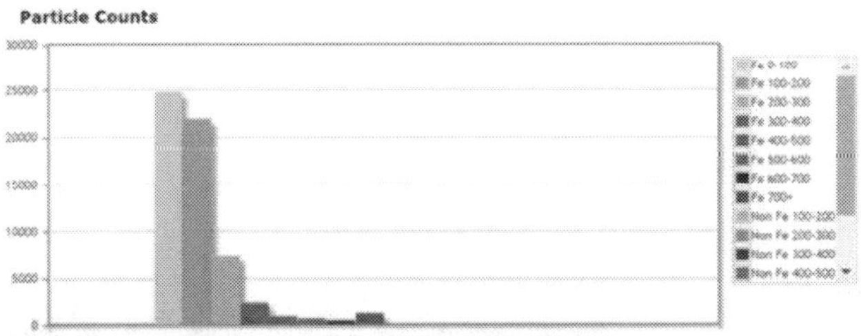

	Ferrous	Non Ferrous
0-100	24701	0
100-200	21889	0
200-300	7313	0
300-400	2267	0
400-500	851	0
500-600	634	0
600-700	398	0
700+	1175	0
Total Mass (mg)	3069.6	0

Export

Bearing Spall

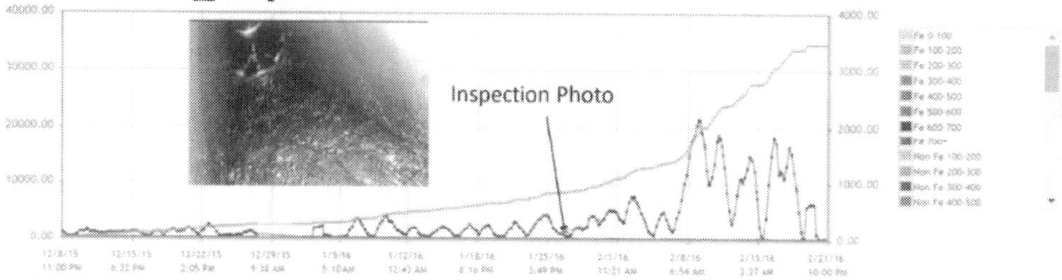

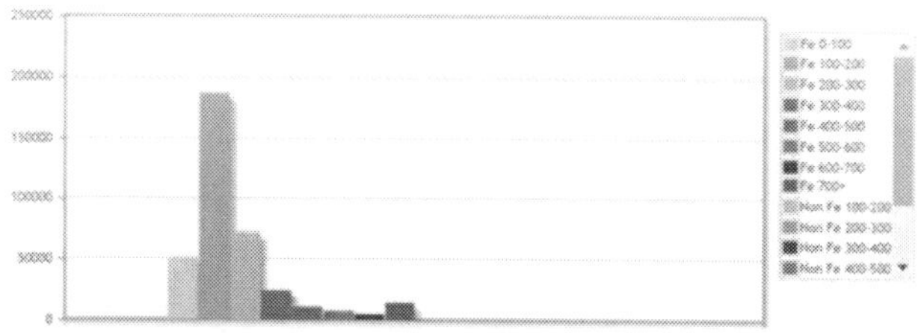

	Ferrous	Non Ferrous
0-100	51046	0
100-200	186497	2
200-300	71836	134
300-400	24246	0
400-500	10204	0
500-600	7409	0
600-700	4291	0
700+	13742	0
Total Mass (mg)	34567	4.3

MFPT 2019 Where Theory Meets Practice

7 POSEIDON

Slipping Bearing

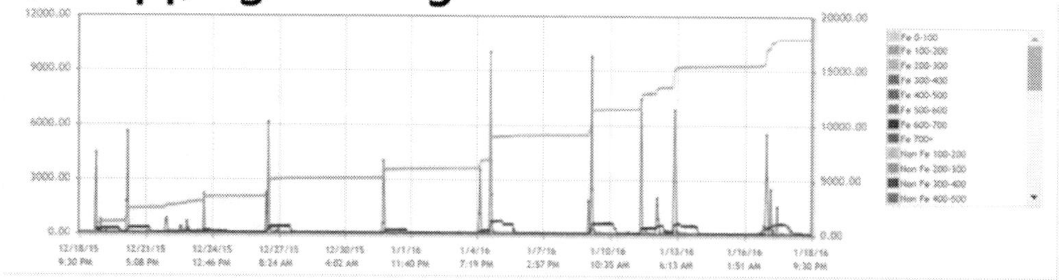

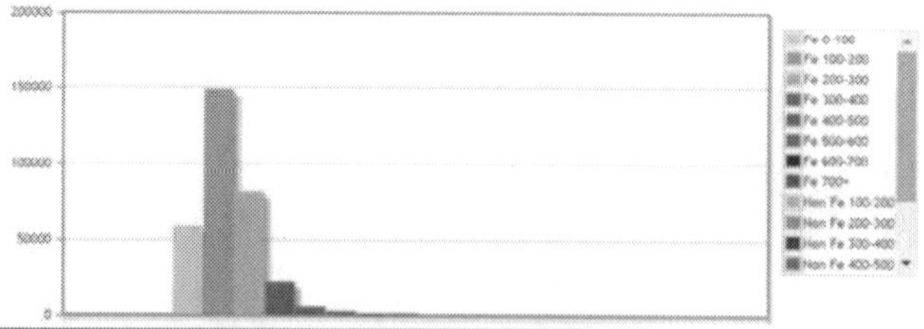

	Ferrous	Non Ferrous
0-100	58925	0
100-200	148149	44
200-300	80970	385
300-400	22301	7
400-500	5508	0
500-600	2449	0
600-700	965	0
700+	1210	0
Total Mass (mg)	10750	13.3

MFPT 2019 Where Theory Meets Practice

8 POSEIDON

Case #1 - Planetary Bearing – Life Extension Enabled by Condition Monitoring

- Technician identified metal flakes in filter canister during PM
- Borescope could not locate fault
- Deployed wear debris sensor to assess and monitor fault severity
- Observed abnormal wear levels shortly after deployment

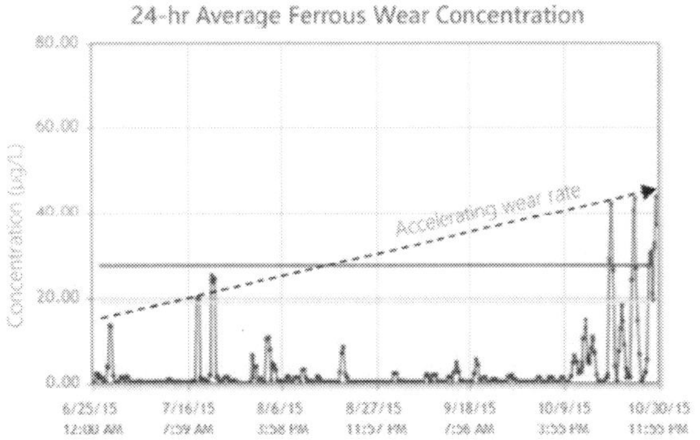

MFPT 2019 Where Theory Meets Practice

9 POSEIDON

Planetary Bearing – Life Extension

- Subsequent borescope isolates fault to planetary bearings
- Owner decided to run as long as possible and rely on sensor for shutdown notification
- Extreme wear levels observed in November 2017
- Shutdown for repair 2.5 years after fault detection

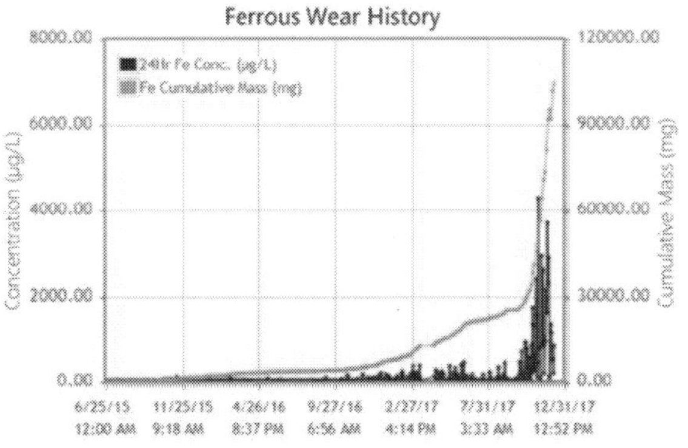

MFPT 2019 Where Theory Meets Practice

10 POSEIDON

Planetary Bearing – Derating

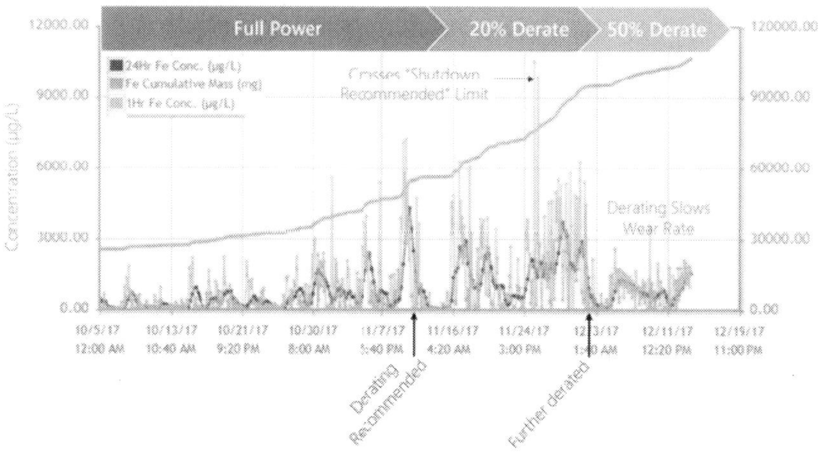

Wear debris monitor allows effectiveness of derating to be evaluated

Maximize production while managing risk of failure

MFPT 2019 Where Theory Meets Practice

11

POSEIDON

Planetary Bearing – Results

- Up-tower planetary bearing replacement performed catastrophic failure averted
- 2 of 3 rotor-side planet bearings found with through cracks on raceway
- Monitoring enabled:
    - 2.5 years safe operation at full power
    - Planned outage for repair
        - 15 days vs 45-60 days
        - 30+ days of additional power production
    - Up-tower repair instead of gearbox replacement
        - <$70k vs >$300k
    - Reduced number of required borescopes
- **Total savings: >$300k**

MFPT 2019 Where Theory Meets Practice

12

POSEIDON

Case #2 – Planetary Bearing – Rapid Developing Fault

- Gearbox operated for a year without any alarm exceedances and generating very little total mass
- Customer notified on 2/6/18 of an "event"
- Second, smaller event observed on 2/9
- Third, larger event on 2/13 – recommended customer perform borescope inspection

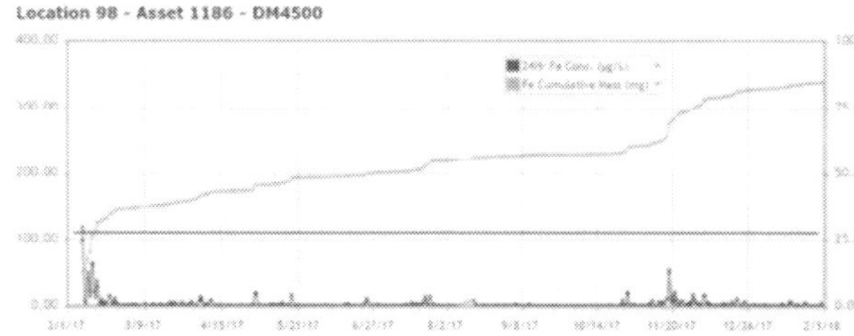

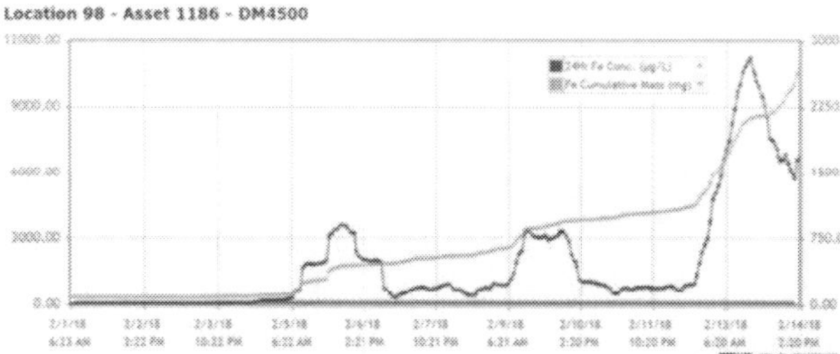

MFPT 2019 Where Theory Meets Practice

POSEIDON

Case #2 – Planetary Bearing – Rapid Developing Fault

- Borescope performed on 2/15 (wear stops while shutdown)
- Severe planet bearing damage confirmed
- Wear continues to accelerate after coming back online; derating recommended

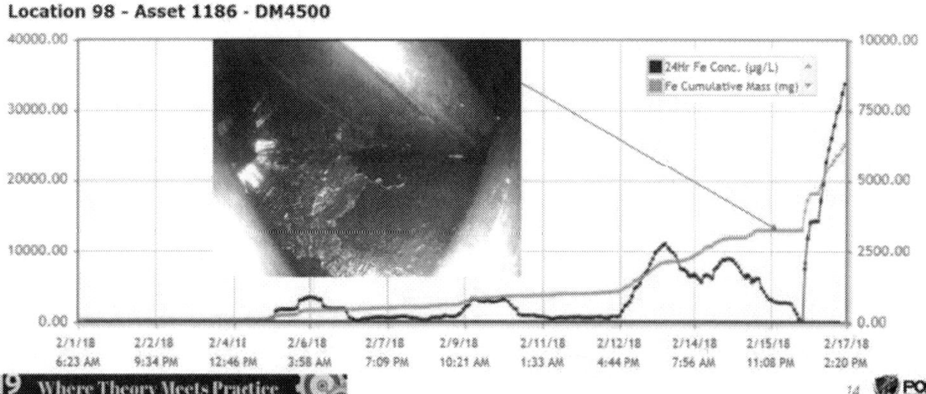

MFPT 2019 Where Theory Meets Practice

POSEIDON

Case #3 – Planetary Bearing – Derating Employed

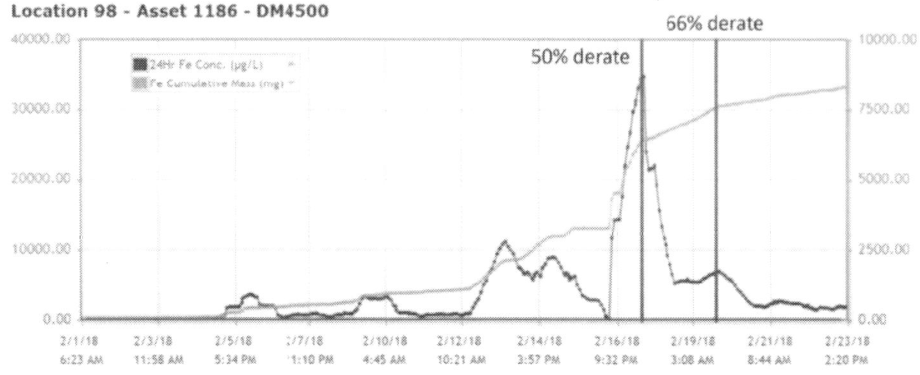

- 2/17 - 50% derating implemented. Wear rate slowed, but still 2500 mg/day
- 2/19 – 66% derating implemented. Wear rate drops to 500 mg/day (still well into alarm state)

Case #3 – Planetary Bearing – Life Extended

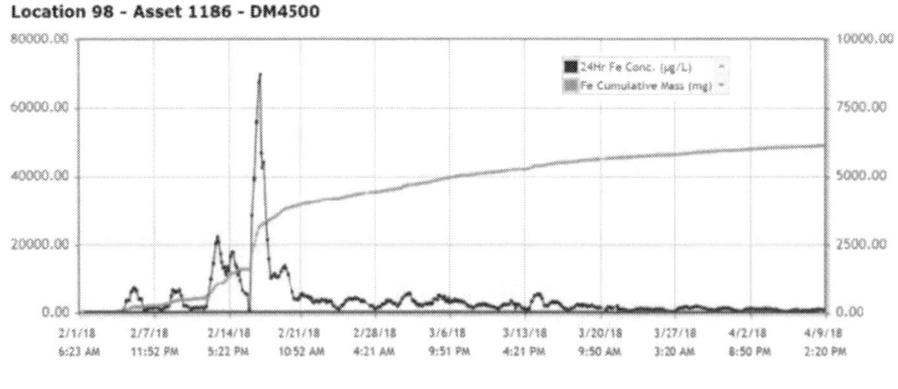

- Additional six weeks of safe derated operation while plans made for repair
- Catastrophic failure averted

Case #3 – Planetary Bearing – Life Extended

- Tear down photos showing extent of damage

Case #1 Planetary Bearing – Size Distribution

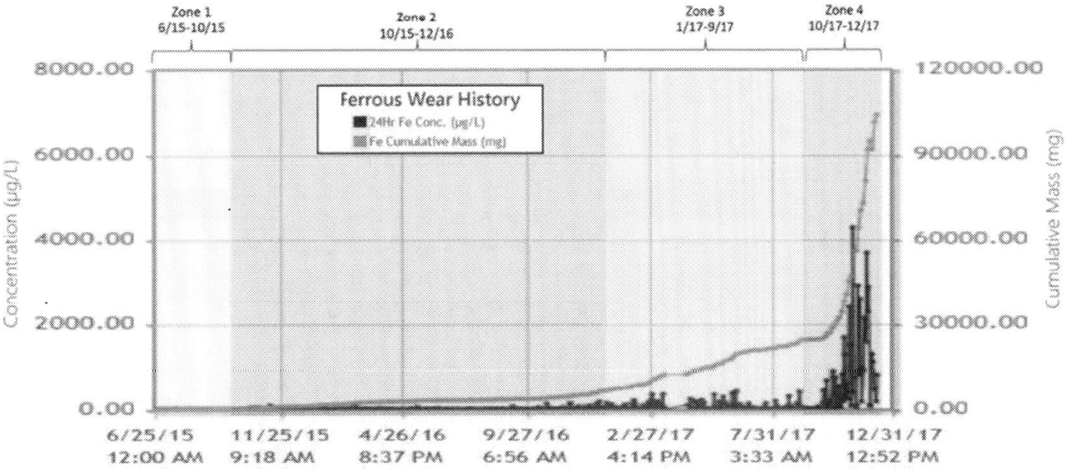

Case #1 Planetary Bearing – Size Distribution

Particle Count Totals

	Zone 1	Zone 2	Zone 3	Zone 4
Start Date	6/1/15	10/1/15	1/1/17	10/1/17
End Date	10/1/15	12/31/16	9/30/17	12/14/17
Fe Bin Size, µm (lower bound)	Particle Counts			
100	1,110	35,493	99,045	456,386
200	467	13,981	38,794	170,893
300	140	4,545	13,161	53,339
400	48	1,915	5,562	21,919
500	41	1,399	3,875	15,686
600	19	763	2,206	9,209
700	82	2,350	7,127	34,450
Avg mg/day	4	24	149	2,543

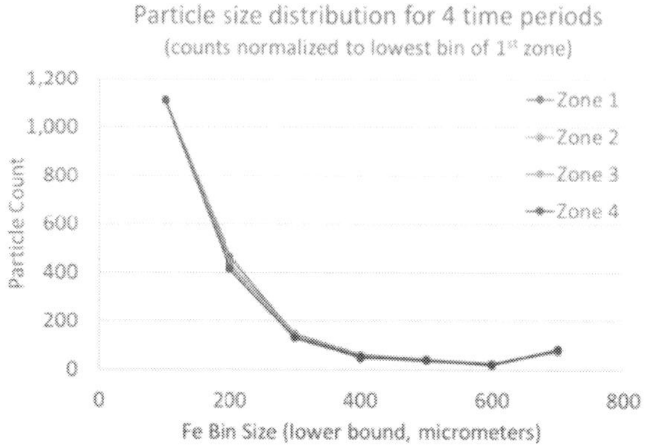

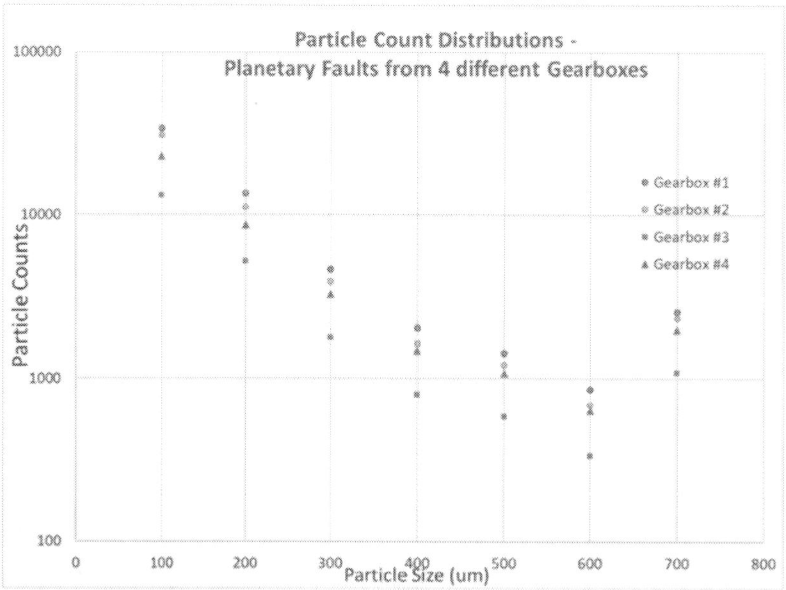

Periodic Oil Samples Not Enough

MFPT 2019 Where Theory Meets Practice

21 POSEIDON

Why Online vs Offline?

- Wear metal generation is not a continuous, predictable process
  - It is typically observed in bursts
- Samples drawn minutes apart can vary by orders of magnitude
- Impractical to analyze frequently enough to characterize gearbox health
- Very unlikely for a 4oz sample to contain significant wear metals, most filtered out immediately
  - For a faulted gearbox, **approximately one 100um-200um particle in every 700 oz of oil**
  - **Approximately 0.5% probability in typical oil sample**
- Online, inductive coil sensors are not skewed by:
  - Presence of nonmetallic particles
  - Oil changes
  - Filter media changes

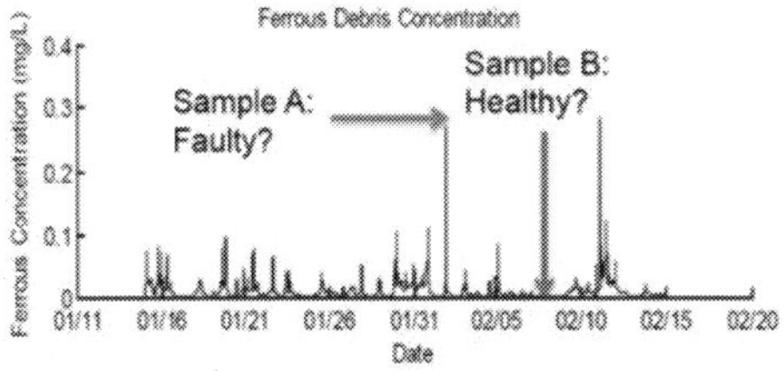

Wear metal concentrations vary dramatically based on operating conditions. Accurate conclusions cannot be drawn from periodic sampling.

MFPT 2019 Where Theory Meets Practice

Why Metallic Debris Monitor vs Optical?

- Most important in determining state of asset health is determining loss of metal, which is dominated by large particles
- Online optical devices only sample a small portion of the oil and have a maximum size limitation typically of 100-200 um.
- Particle size distribution follows a skewed normal distribution resulting in quantity of small particles (<10 um) being similar between small and large wear events

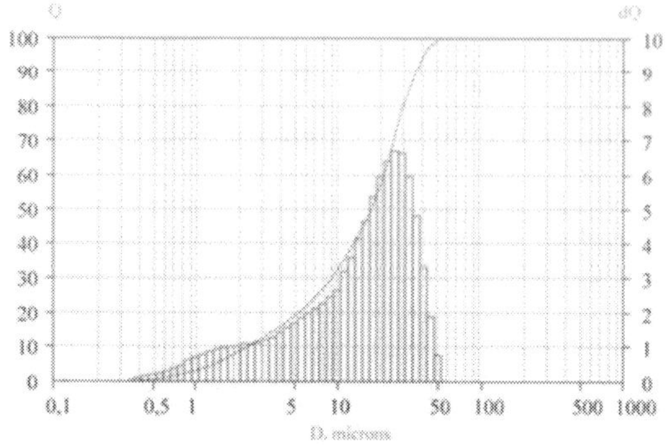

MFPT 2019 Where Theory Meets Practice

Lab Samples & ISO Codes Don't Predict Health

Viewing particle count data by ISO code or iron concentrations from lab samples is not a good indicator of gearbox health

- Case Study Across 137 Assets
  - 5 years of oil sample data
- Using Maximum Iron Concentration
  - 70ppm Alarm
    - 0.8% false alarms (1/119)
    - 100% missed detection (0/18)
    - 100% nuisance (1/1)
  - 50ppm Warning
    - 13% false alarms (16/119)
    - 72% missed detections (13/18)
    - 77% nuisance (17/22)

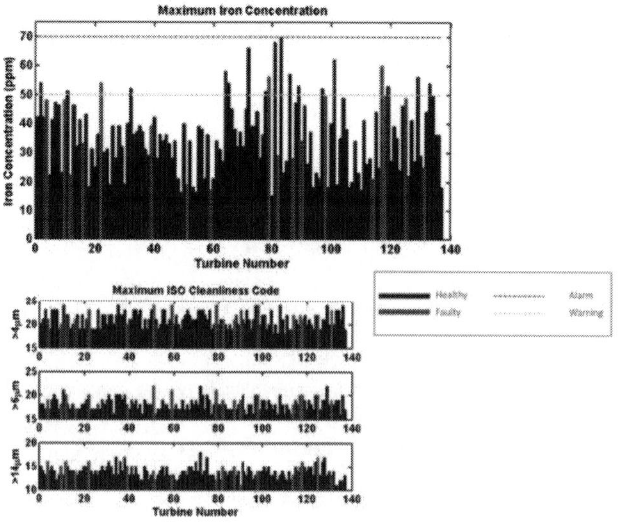

MFPT 2019 Where Theory Meets Practice

Online vs. Offline Example

Offline sampling unreliable for capturing fault events

Case Study – Oil samples did not indicate signs of fault, online sensor alerted years earlier

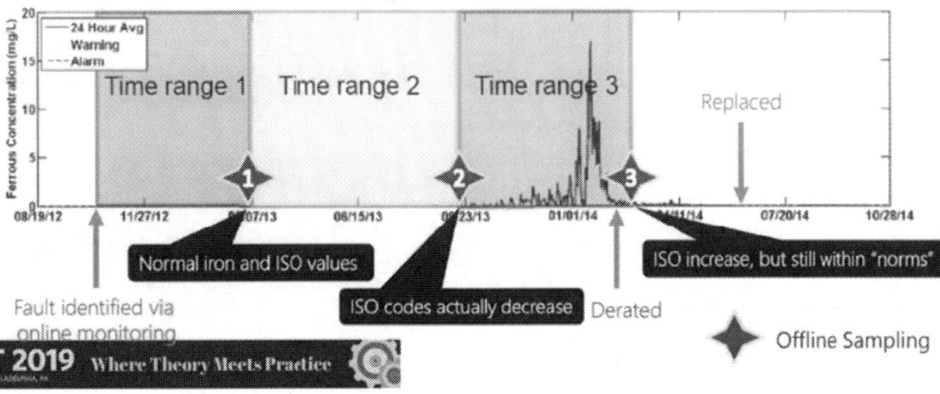

MFPT 2019 Where Theory Meets Practice

Time Range 1: **Offline Data**

Oil Sample 1 from 3/6/2013 (5 months after fault detected)

- Iron: 5 ppm
- ISO Code: 20/16/13
- Normal iron and ISO values

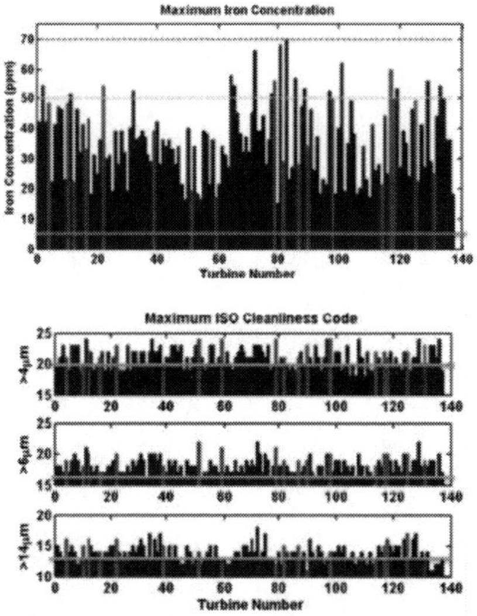

MFPT 2019 Where Theory Meets Practice

Time Range 2: **Offline Data**

Oil Sample 2 from 9/12/2013 (11 months after fault detected)

- Iron: 10 ppm
- ISO Code: 19/15/12
- Iron concentration increased, but still low
- ISO Codes actually decreased from previous sample

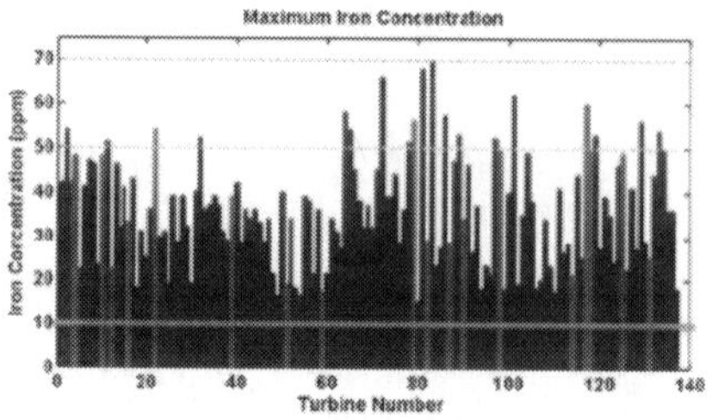

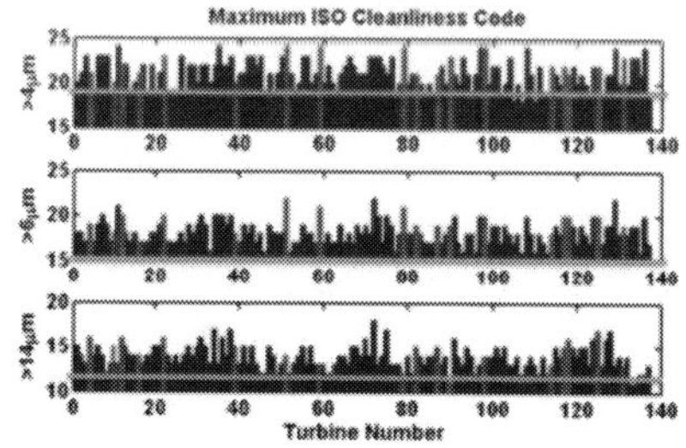

MFPT 2019 Where Theory Meets Practice

POSEIDON

Time Range 3: **Offline Data**

Oil Sample 3 from 2/03/2014 (16 months after fault detection)

- Iron: 15 ppm
- Iron Rate: 0.16 ppm/100hrs (~16% of limit)
- ISO Code: 22/18/14
- Iron concentration increase is consistent with aging oil
- ISO finally increase, but still within typical operator "norms"

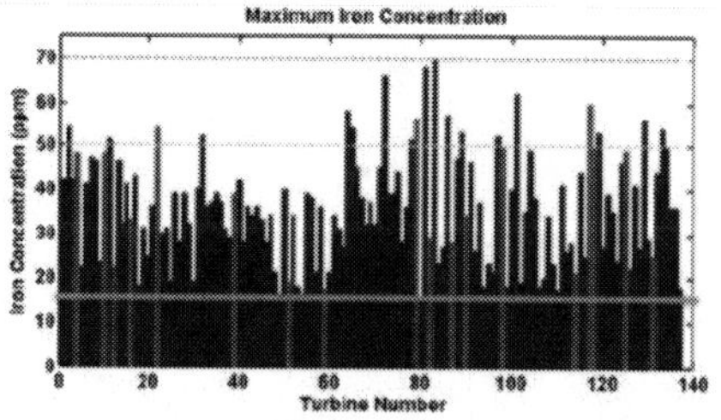

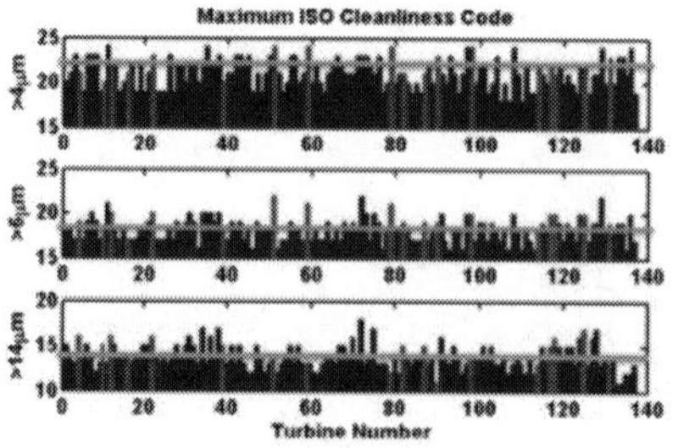

MFPT 2019 Where Theory Meets Practice

POSEIDON

Conclusions

- Wear debris monitoring can be extremely effective for providing:
  - Early fault detection
  - Quantification of damage (loss of metal)
  - Some degree of fault identification through particle size distribution and rate of generation
- Predicting RUL can be extremely challenging, BUT wear debris monitoring can extend useful life and prevent catastrophic failure
- Offline oil analysis is useful in determining ***oil health***, but not necessarily ***asset health***

Modern Laser Applications for Improving Machine Reliability

HAMAR LASER®

Ronaldo Sullivan

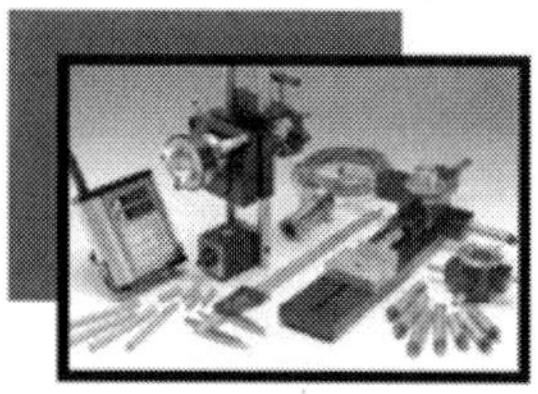

Prior Experience

HAMAR LASER®

- Brooklyn Tech
- Rensselaer Polytechnic Institute
- Carrier
- VibrAlign
- Commtest
- Wilcoxon Research
- Azima DLI

Carrier

VIBRALIGN

commtest
The Revolution

Wilcoxon
SENSING TECHNOLOGIES

AZIMA DLI

Align With The Best™

HAMAR LASER®

Hamar Laser Instruments

- Celebrated 50 years in 2017!

Align With The Best™

HAMAR LASER®

Introduction

- Incorporated in 1967 by Martin Hamar
- Part of laser team at Perkin Elmer - produced first laser alignment system
- 27 US and Foreign patents
- Only laser alignment system Made in the USA

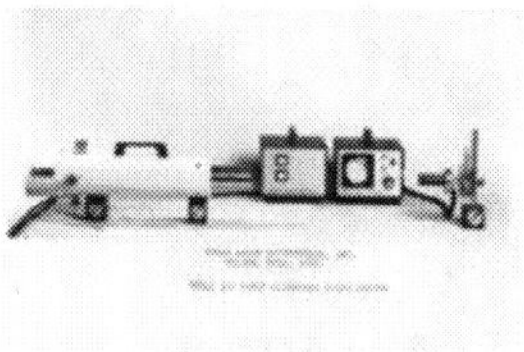

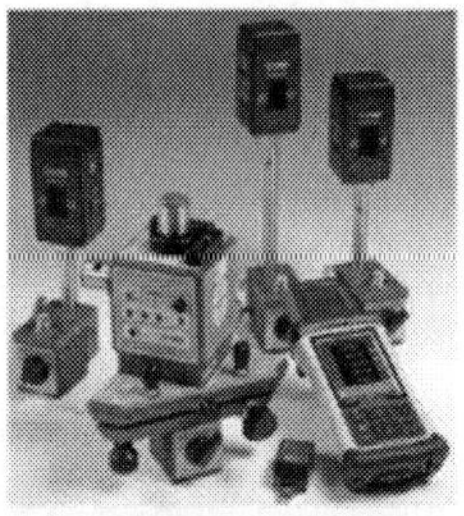

Align With The Best™

All Industrial Sectors

- Industry leader and innovator for over 52 years

Align With The Best™

Laser Alignment Systems
Geometrical Planes, Bore, Spindle, Shaft

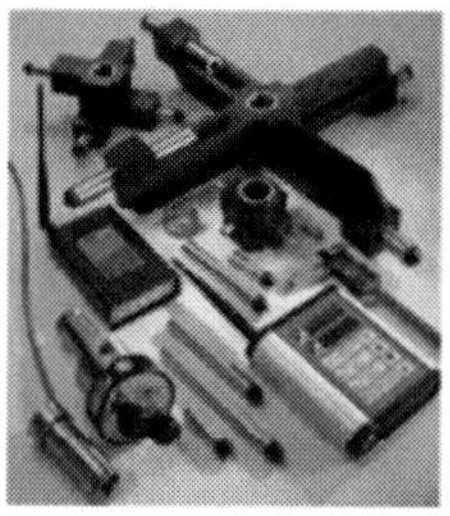

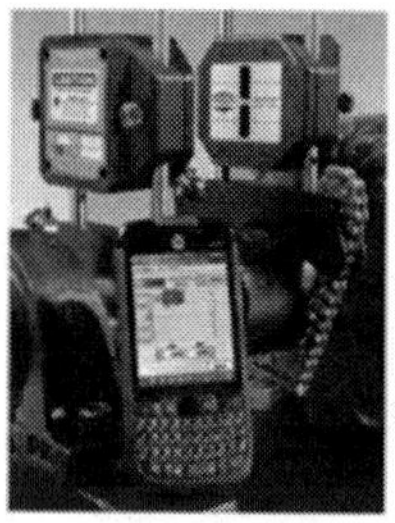

Align With The Best™

Main Advantages of Alignment Lasers

- Speed
  - 50%-70% faster than traditional methods
    - Squares, levels, straight edges, theodolites, interferometers, autocollimators
  - Much fewer setups
  - Live Data –auto-update when adjusting machine
  - Quick machine crash checks
- Accuracy – accuracies to 0.001 mm/M
  - Fewer accumulating errors
  - Common reference system
- Reduced machine down time
- Lower scrap rates
- Longer machine life
- Faster production – John Deere
- Deflection (load) monitoring

Align With The Best™

Resolution Versus Accuracy

- Resolution is the smallest measurement increment an instrument can display
- Accuracy relates to the degree of conformity to a standard of measurement
- Accuracy ≠ Resolution
- 0.01 mm indicator ≠ 0.01 mm accuracy
  - Mechanical sag – longer the arm the bigger the sag
  - Methods & technique
  - Interpretation (how many rotations?)
  - Indicator stickiness
  - Extremely small measuring "foot" (more subject to errors)

Align With The Best™

How Geometry Lasers Work

- 3 components of laser alignment systems:
  - Laser (straight line or plane)
  - Target using PSD
  - Readout and software

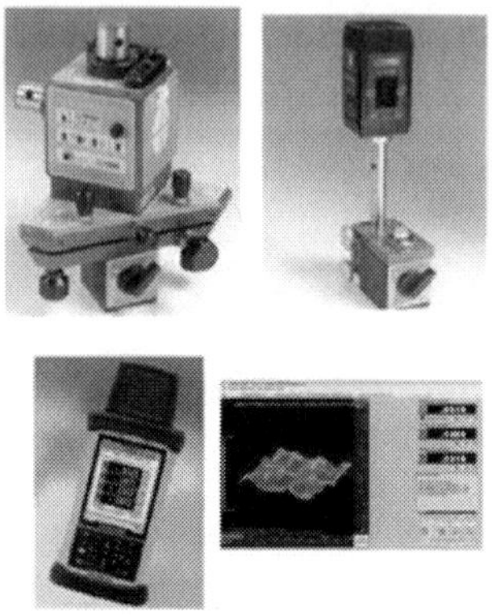

Align With The Best™

How Geometry Lasers Work

Flat Laser Planes

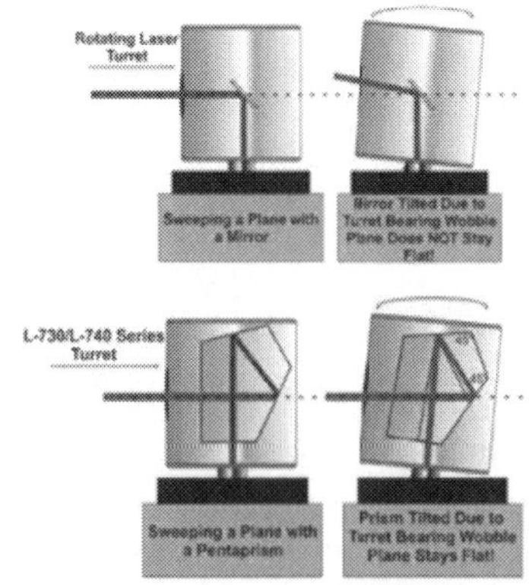

- Flattest laser planes use Penta Prisms
  - Prism has two reflecting surfaces.
  - Prism rotates about an axis perpendicular to the plane of this screen.
  - Two 45° angles bend input beam 90°
  - Any tilt in beam resulting from the first reflecting surface is cancelled by the second reflection.
  - Output laser beam remains square to the input laser beam.
- Penta prisms compensate for bearing wobble
- Mirrors do not

Align With The Best™

How Geometry Lasers Work
Position Sensing Detectors (PSDs)

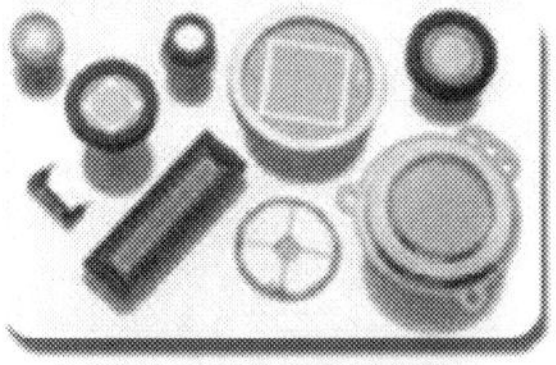

- Silicon photodiode that detects light
- Turns light into voltage, which is turned into a digital reading
- Finds center of energy of "fat" laser beam
- High resolution – as low as .00001" (.00025mm)
- Large measuring range – up to 33 mm

Align With The Best™

How Geometry Lasers Work
Readouts & Software

- Hand held readouts
- Wireless, multi-target displays
- Data taking & analysis software

Align With The Best™

Case History
Entrust Tool & Machine

- Builder of specialty machines
- L-742 Laser purchased to help align and install machines
- 30% time saving realized in the alignment and assembly process
- Customer able to align 2 axes for straightness, simultaneously
- Customer able to check parallelism from the same set up as straightness check
- 45% time saving realized in machine installation
- Laser replaced conventional tight wire methods, levels and granite squares

Case History
Lincoln Park Boring

- Problem: After milling, a long part had crown which caused Cincinnati Machine to frequently reject the part
- Solution: L-740 Laser used to measure part surface after rough machining
- Data gathered from L-740 used to create offset table that was input into the controller
- The finish cut was based upon laser data
- Customer able to either introduce or remove "crown" - both conditions are required for different parts
- Delayed the need to realign the machine which was very expensive
- Laser did the job 40% faster than Optodyne interferometer
- Part rejection for Cincinnati no longer a problem

Align With The Best™

Case History
Alcan Aluminum

- Very difficult alignment of ingot producing machine
- Laser was able to measure 20 points for every 1 point for optics
- Optics required 2 theodolites and 4 technicians
- Only needed one laser and 1 operator
- Aligned ingot forming machine to .14mm – the best it's ever been
- Resulted in total time savings of 2.5 days
- 2.5 days of production on this machine is worth $250,000
- Operators also report never seeing such straight ingots, which resulted in almost eliminating a second "shaving" operation

Align With The Best™

Bore Alignment Applications

- Cylinder QC checks
- Ball-screw bearing housings
- Diesel engine crankshaft bores
- Extruders
- Steam and gas turbines
- Stern tubes & bearings in ships
- Aircraft hinge lines
- Bar-turning machines
- Bar feeders

Align With The Best™

HAMAR LASER

Bore Alignment

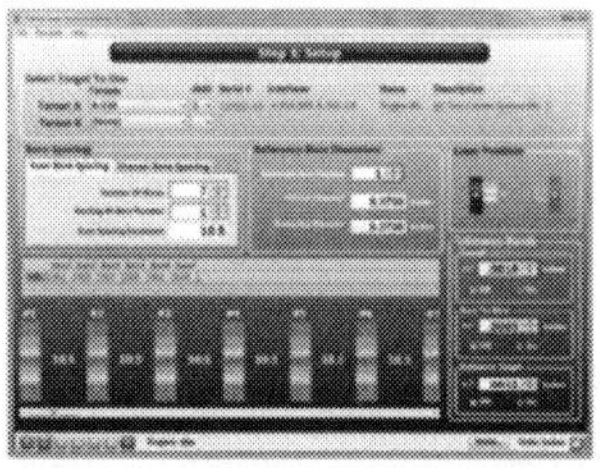

- Internal Mount for Stern Tube and Shaft Bearing Bores

Align With The Best™

HAMAR LASER

Bore Alignment

Self-Centering Bore Alignment Laser

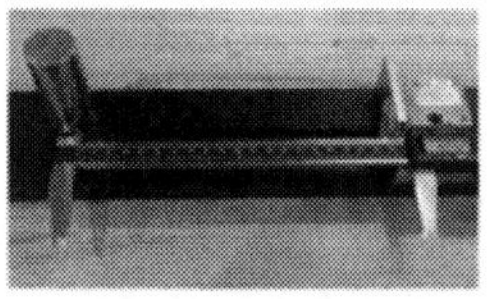

A-514 Adapter on A-514GS Leg Setting Gage

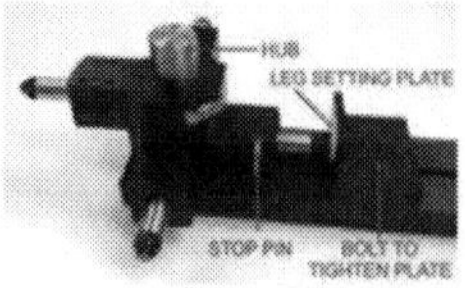

L-708/A-514 Mounted in a bore

Align With The Best™

Bore Alignment

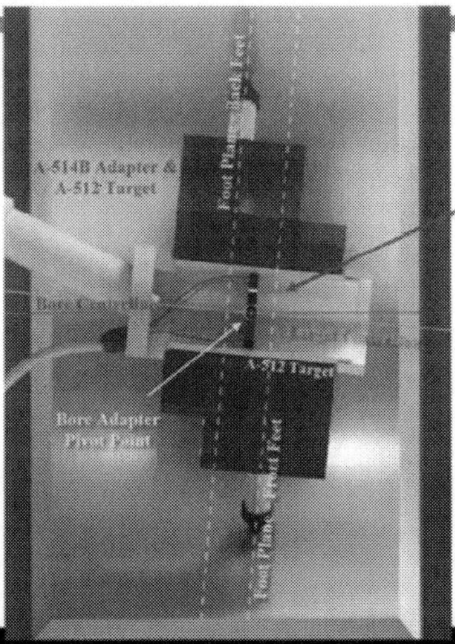

- **Self-Centering Target Principle**

Position Sensing Detector (PSD) is centered axially and radially in A-514 Bore Adapter and therefore the angle of A-514 centerline relative to laser beam and/or the bore centerline does not affect the centering in the bore.

Align With The Best™

Bore Alignment

Self-Centering Target in Different Bore ID's

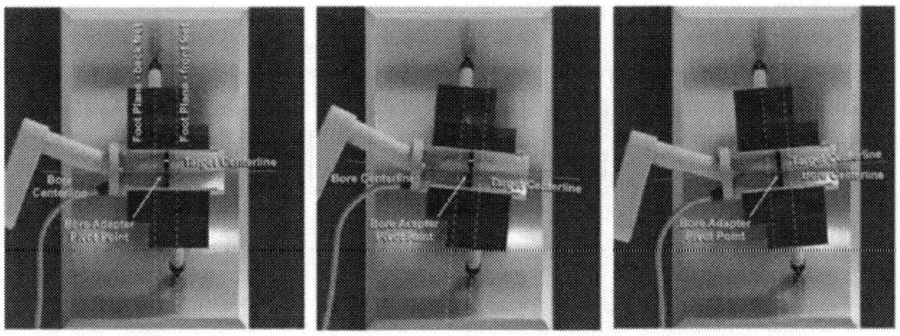

Align With The Best™

Impacts of Shaft Misalignment

- Machinery Availability
- Reliability
- Vibration
- Bearings
- Seals
- Energy
- Use less resources

Align With The Best™

Causes of Shaft Misalignment

- Studies in the past 10 years indicate that 50% of problems in machinery are caused by misalignment.

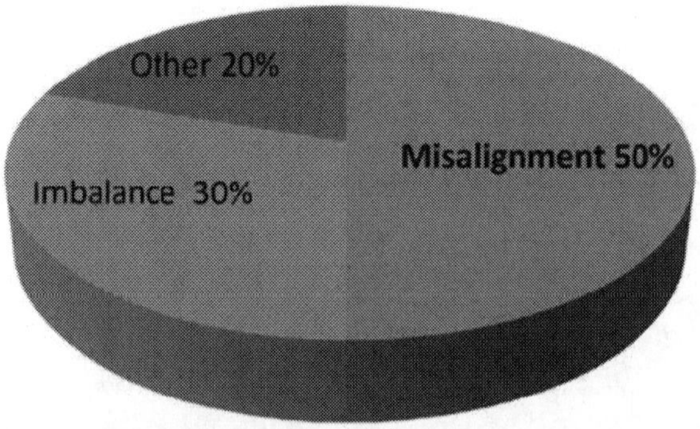

Align With The Best™

Impacts of Shaft Misalignment

- The expected life of a bearing is calculated using the following formula:

$$L_{10} = \left[\frac{\text{Bearing Constant}}{\text{Bearing Load}}\right]^3$$

Doubling the load will reduce bearing life 1/8 of its estimated life

Align With The Best™

Impacts of Misalignment

- Proper alignment can reduce energy consumption by up to 15% and sometimes more.

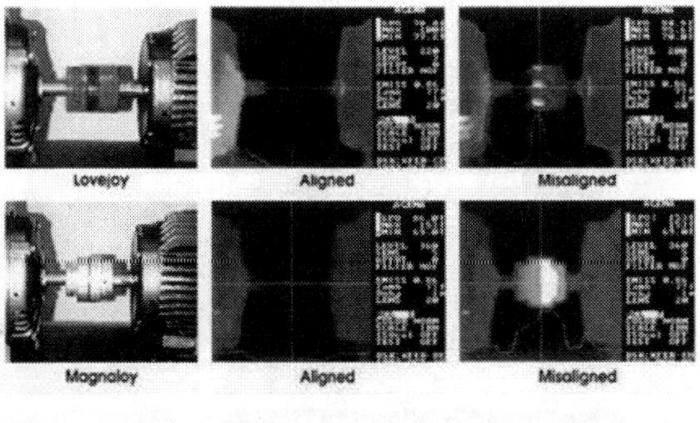

Align With The Best™

Shaft Alignment Defined

The objective of precision alignment is to get the rotational centerlines collinear when the machines are at operating temperatures and conditions.

Align With The Best™

Stationary Machine

The driven side is usually the considered the stationary machine.

Align With The Best™

Movable Machine

The driver is usually the movable machine.

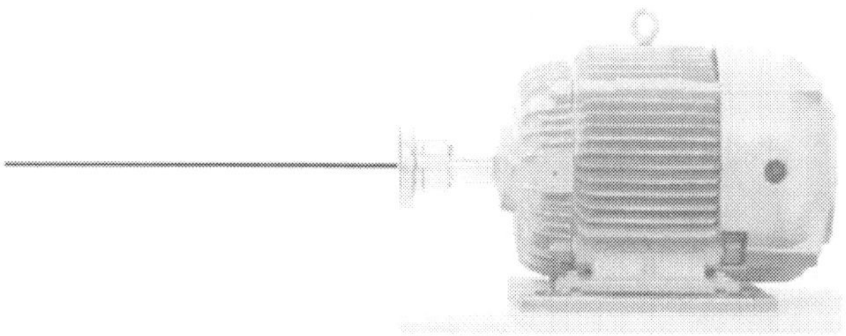

Align With The Best™

Align With The Best™

STEALTH Technology

Dual-Beam ™ Detector Technology
A patented technology, that measures offset and angle simultaneously with only one PSD, increasing accuracy by 50% (over 2-laser, 2-detector method)

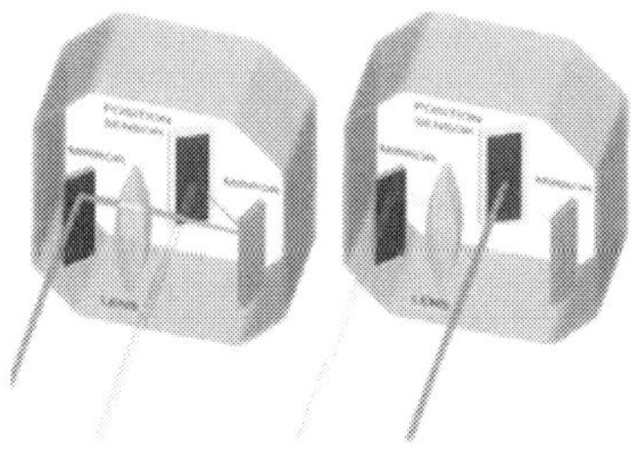

0.5 µm resolution

Error:
< 0.15% (offset)
<0.75% (angular)

Align With The Best™

Thank You!

- For More Information:
- Hamar Laser Instruments website: https://www.hamarlaser.com/
- Stealth website: https://www.stealthalignment.com/

Align With The Best™

MFPT 2019 Where Theory Meets Practice

MAY 14-16, 2019 | PHILADELPHIA, PA

RSL Fiber Systems, LLC

Distributed Temperature Sensing For Critical Electrical Systems Monitoring

MFPT Conference
King of Prussia, PA
May 14 – 16, 2019

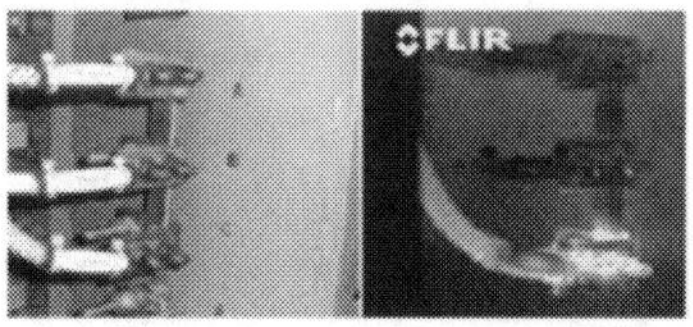

This presentation is the sole property of RSL Fiber Systems, LLC. It cannot be copied or duplicated in whole or in part without the written consent of RSL Fiber Systems, LLC.

RSL

MFPT 2019 Where Theory Meets Practice

MAY 14-16, 2019 | PHILADELPHIA, PA

DISTRIBUTED TEMPERATURE SENSING FOR CRITICAL ELECTRICAL SYSTEMS

MONITOR FOR:

- Loose Connections
- Real Time Thermal Rating (RTTR)
- Cable / Conductor Damage

TWO APPLICATIONS INVESTIGATED:

- U.S. Navy Medium Voltage Electrical Panels (4,160 VAC)
- Insulated Bus Pipe for U.S. Navy vessels

NOTE: The applications described in this presentation are for U.S. Navy programs however they are applicable to all critical electrical systems.

RSL

MFPT 2019 Where Theory Meets Practice
MAY 14-16, 2019 | PHILADELPHIA, PA

RAMAN BASED DISTRIBUTED TEMPERATURE SENSING

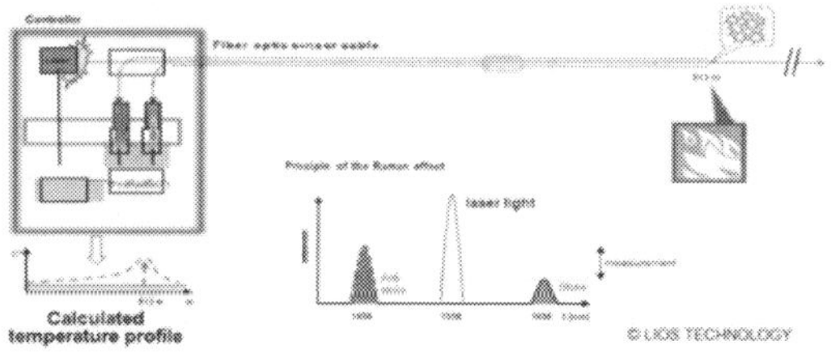

- Standard MM 50/125, 62.5/125 or SM fiber
- Entire fiber is the sensing unit – up to 1000 zones per fiber channel
- Up to 30 Km MM, 40 Km SM - real time monitoring through entire length
- Temperature rise of 0.5ºC
- Spatial Resolution down to 50 cm
- Temperature measurement is based on inelastic collisions of photons with molecules along the optical fiber. Temperature is calculated by comparing Stokes to Anti-stokes:
  - Stokes are quasi –temperature insensitive
  - Anti-stokes are temperature sensitive

MFPT 2019 Where Theory Meets Practice
MAY 14-16, 2019 | PHILADELPHIA, PA

OPERATION, DISPLAYS, DATA

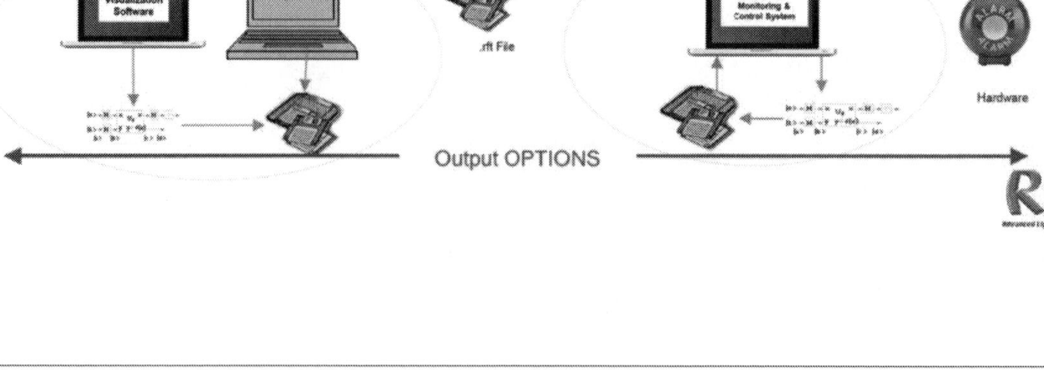

TEMPERATURE PROFILE

Identify temperature at exact locations...

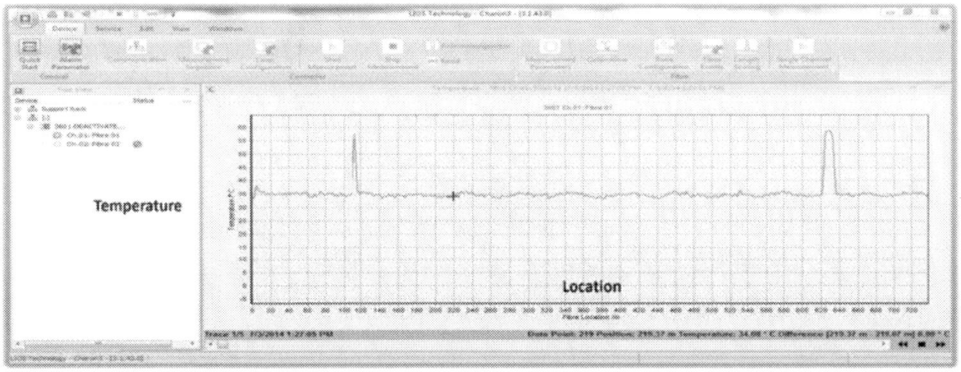

RSL

MFPT 2019 Where Theory Meets Practice
MAY 14-16, 2019 | PHILADELPHIA, PA

ZONES AND ALARMS PARAMETERS

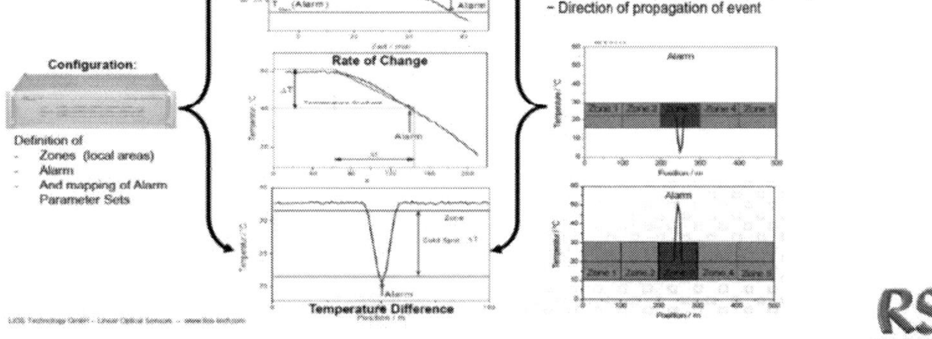

RSL

MFPT 2019 Where Theory Meets Practice
MAY 14-16, 2019 | PHILADELPHIA, PA

ELECTRICAL PANELS

Requirement:
- Annually inspect MV (4,160 VAC) and HV (13,800 VAC) electrical panels to identify loose connections.
- Temperature difference between phases indicates loosening connections.
- Inspection typically by IR Thermography.

Issue:
- Hazardous
- Labor Intensive
- Performed Annually – no real time monitoring.

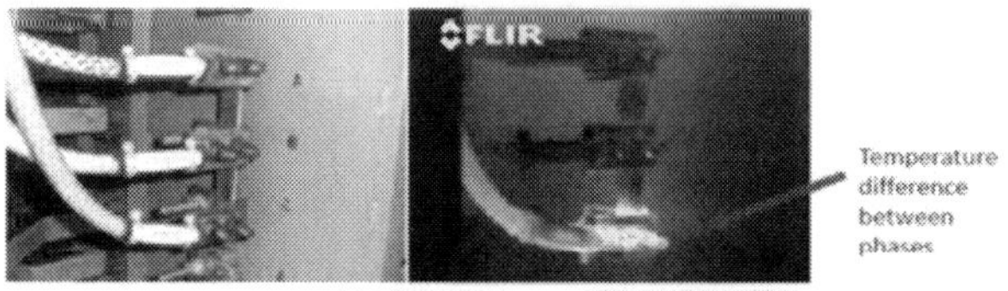

Load Center

Solution:
Fiber optic distributed temperature sensing system for real time monitoring.

RSL

MFPT 2019 Where Theory Meets Practice
MAY 14-16, 2019 | PHILADELPHIA, PA

DTS IN MV ELECTRICAL PANEL

__OBJECTIVE:__ Identify conditions that may indicate loosening of electrical connections by continuously monitoring the temperature of each bus bar and the temperature differences between the three phases.

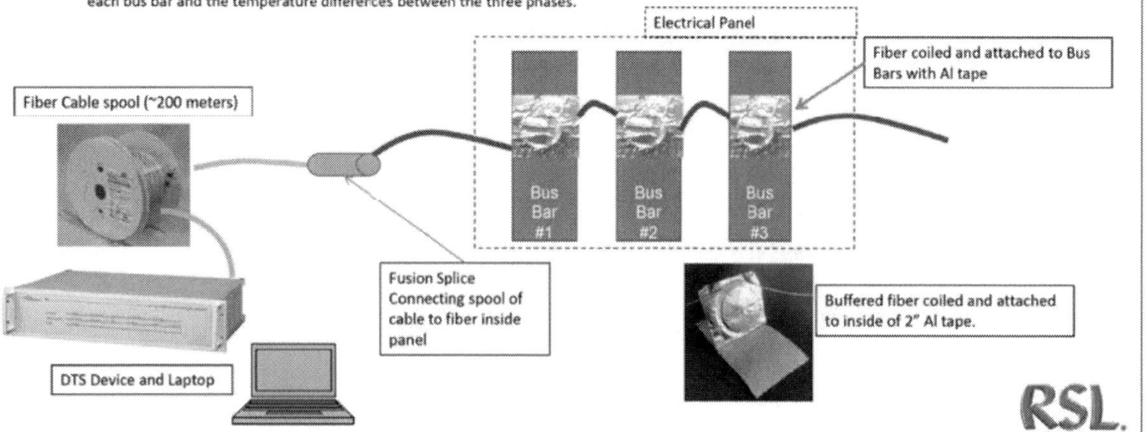

RSL

INSULATED BUS PIPE

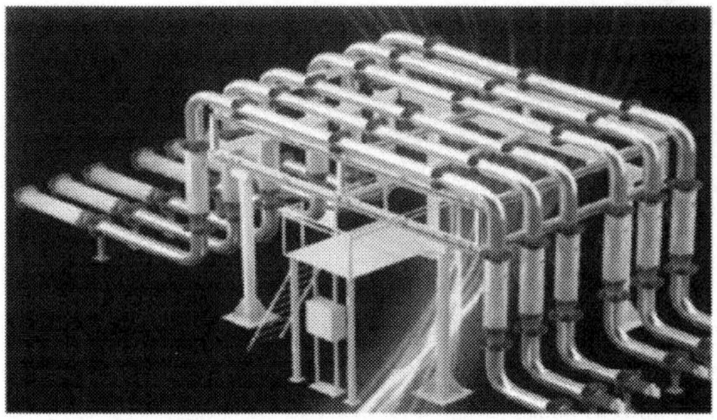

TEFELEN
Technology For Electrical Energy

RSL.
Advanced Lighting Technologies

MFPT 2019 Where Theory Meets Practice
MAY 14-16, 2019 | PHILADELPHIA, PA

INSULATED BUS PIPE

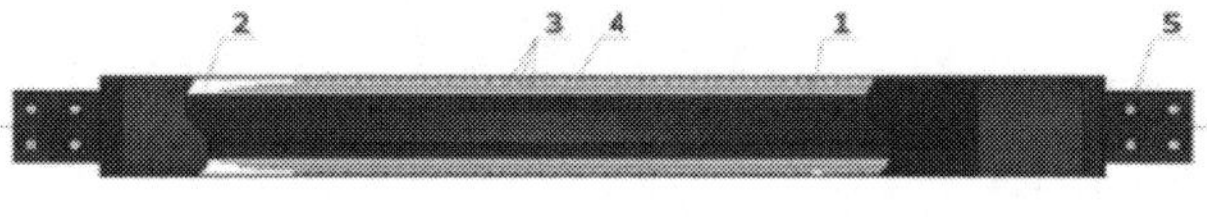

The Bus Pipe element's construction drawing

1. Conductor (Al, Cu)
2. Capacitive grading
3. Epoxy-impregnated paper wrapping
4. Grounding layer
5. End Connector

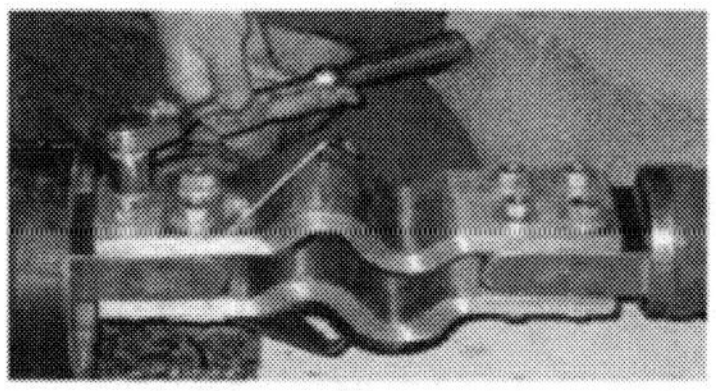

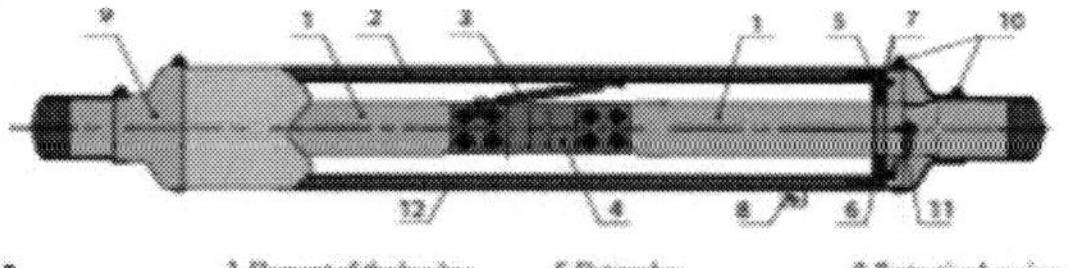

Figure 3. The bus bar elements connection

1. Element of the bus bar
2. Connection sleeve
3. Contact spring
4. Flexible
5. Flat washer
6. Sealing ring
7. Half flange
8. The connection sleeve's grounding
9. Protective housing
10. Clamp
11. The element's earthing point
12. Protective sleeve

TEFELEN
Technology For Electrical Energy

RSL.
Advanced Lighting Technologies

MFPT 2019 Where Theory Meets Practice

MAY 14-16, 2019 | PHILADELPHIA, PA

MODULAR SHIP CONSTRUCTION

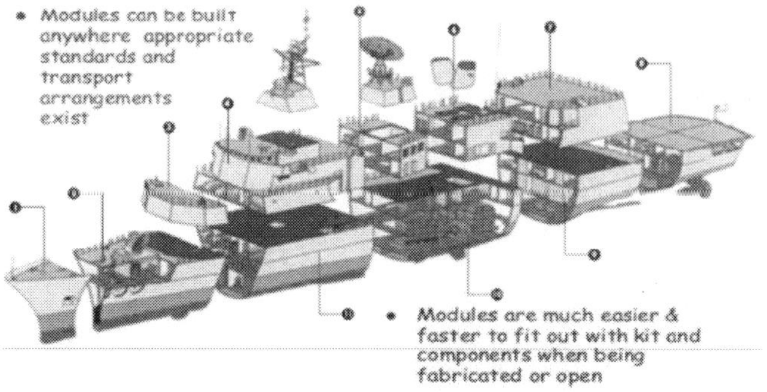

"Installing electrical power cables is one of the most labor intensive tasks of the entire shipbuilding process..."[1]

[1] David Ellis, Engineering Manager, Newport News Shipbuilding.

RSL

MFPT 2019 Where Theory Meets Practice

MAY 14-16, 2019 | PHILADELPHIA, PA

INSULATED BUS PIPE

Advantages:

- Can be pre-installed during construction of ships' sections.
- Cost savings, weight savings, space savings.

Challenges:

- Bolted connections must be inspected annually.
- Inspection would be very labor intensive and cost prohibitive.

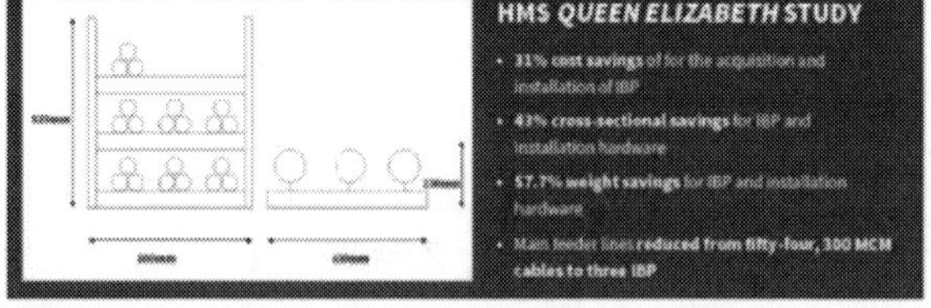

Solution:

Fiber optic distributed temperature sensing system for real time monitoring.

RSL

SENSING CABLE INSTALLATION

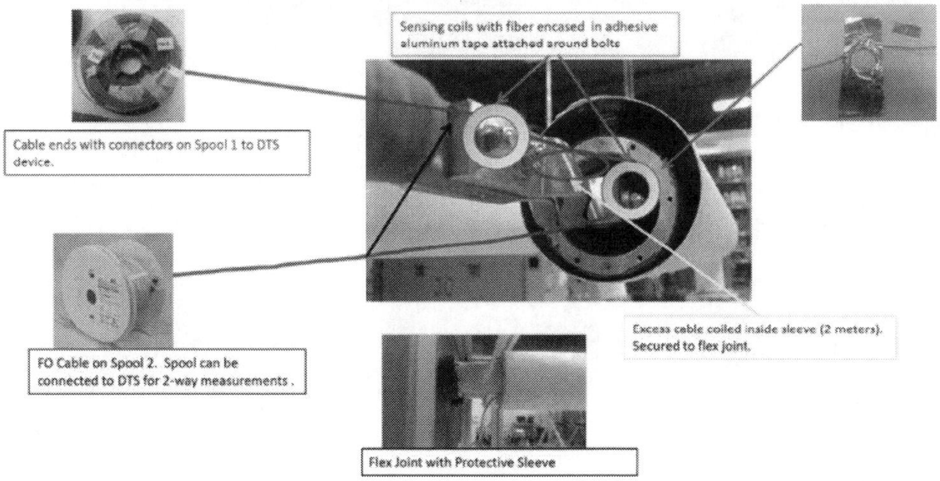

RSL

MFPT 2019 Where Theory Meets Practice
MAY 14-16, 2019 | PHILADELPHIA, PA

DTS ZONES SET-UP

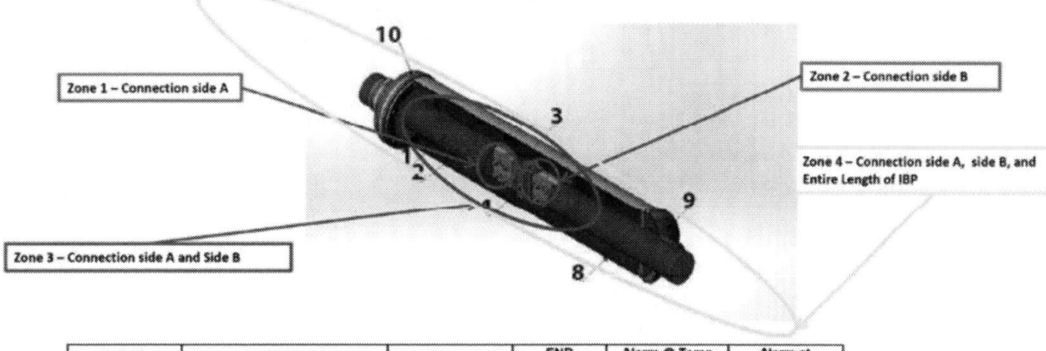

ZONE No.	ZONE NAME	START (meters)	END (meters)	Alarm @ Temp (°F/°C)	Alarm at Δ Temp (°F/°C)
1	Connector Side A	16	20	185 / 85	N. A.
2	Connector Side B	23	27	185 / 85	N. A.
3	Connector All	10	30	185 / 85	36 / 20
4	Full System Length	10	59	185 / 85	36 / 20

RSL

MFPT 2019 Where Theory Meets Practice
MAY 14-16, 2019 | PHILADELPHIA, PA

DTS SET-UP AND DISPLAY

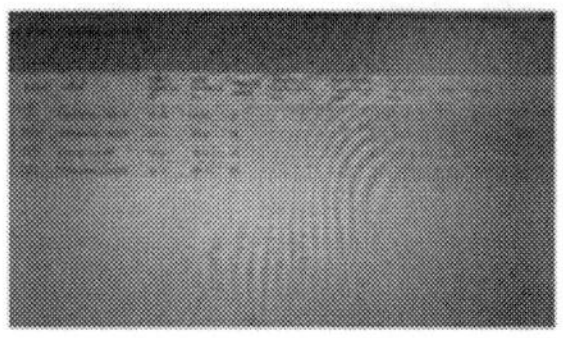

SENSING ZONES SET UP

BOTH SIDES OF CONNECTION HEATING UP
(NORMAL - NO ALARM TRIGGERED)

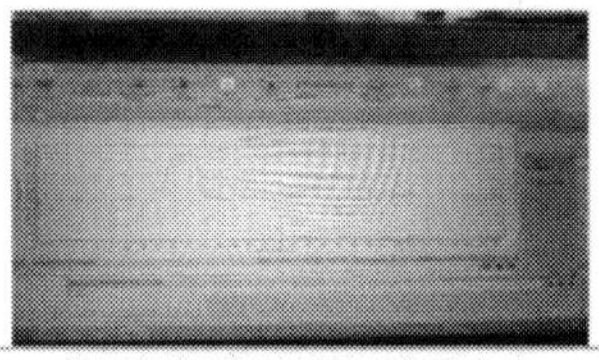

ONE SIDE OF CONNECTION HEATING UP
(ALARM TRIGGERED)

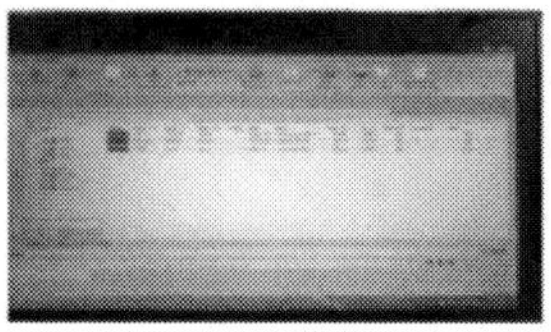

ALARM DISPLAY – SCREEN AND DTS DEVICE

RSL

MFPT 2019 Where Theory Meets Practice
MAY 14-16, 2019 | PHILADELPHIA, PA

DTS IMPLEMENTATION

Route Fiber Cable to Devices Monitored → Attach Cable and Fiber Coils to Devices Monitored → Configure Sensing Zones → Set Alarm Parameters → Integration Into control & monitoring system

<u>Real Time Monitoring</u> for early warning to loose connections and potentially dangerous electrical conditions

RSL

MFPT 2019 Where Theory Meets Practice
MAY 14-16, 2019 | PHILADELPHIA, PA

SUMMARY

A Raman based DTS system will:
- Monitor electrical connection points in real time;
- Accurately identify exact location of loosening connections;
- Detect current through conductors;
- Monitor Real Time Thermal Rating of conductors;
- Provide 100% assurance of critical electrical systems' reliability!

RSL.

MFPT 2019 Where Theory Meets Practice
MAY 14-16, 2019 | PHILADELPHIA, PA

Contact Information

RSL.

Giovanni Tomasi
CEO/CTO
(860) 282-4930 ext. 4929
(860) 305-5610
gptomasi@rslfibersystems.com
RSL Fiber Systems, LLC
473 Silver Lane
East Hartford, Connecticut 06118 USA
www.rslfibersystems.com

RSL.

MFPT 2019 Where Theory Meets Practice
MAY 14-16, 2019 | PHILADELPHIA, PA

Discriminative Dictionary Learning Based Sparse Classification: A Novel Framework for Intelligent Fault Identification of Planet Bearings

Yun Kong, Tianyang Wang, Fulei Chu

Department of Mechanical Engineering, Tsinghua University

Presenter: Dr. Yun KONG

PhD Student @ THU

Visiting Scholar @ NYU

Email: kongy15@mails.Tsinghua.edu.cn

Table of Contents

1. Introduction (1/2)

3

➢ Planet Bearing Fault Diagnostics

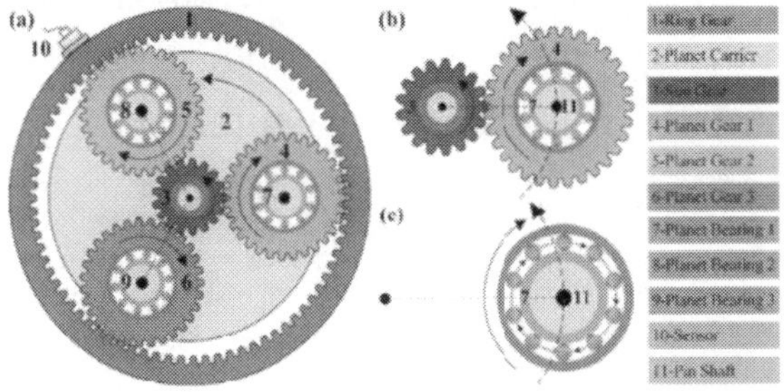

➢ Research Difficulties

✓ Complex Physical Configurations

✓ Complex Modulation Characteristics

✓ Strong Interferences from Gears

☐ Complex Modulation Sidebands

☐ Weak Fault-induced Impulse Features

➢ Vibration Signal Characteristics

(1) Time-varying vibration transmission paths

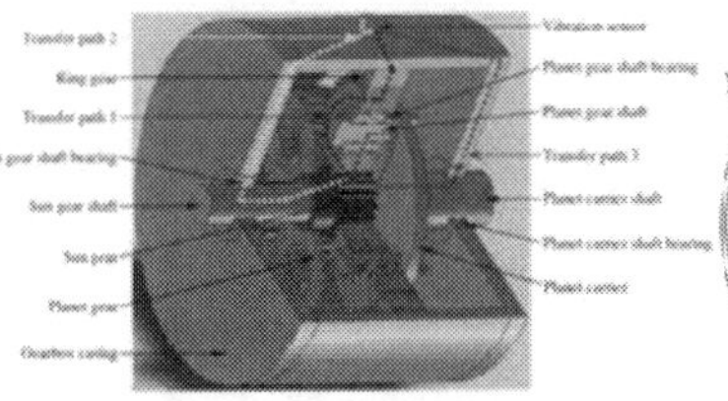

(2) Variations in impulse magnitude and impulse direction

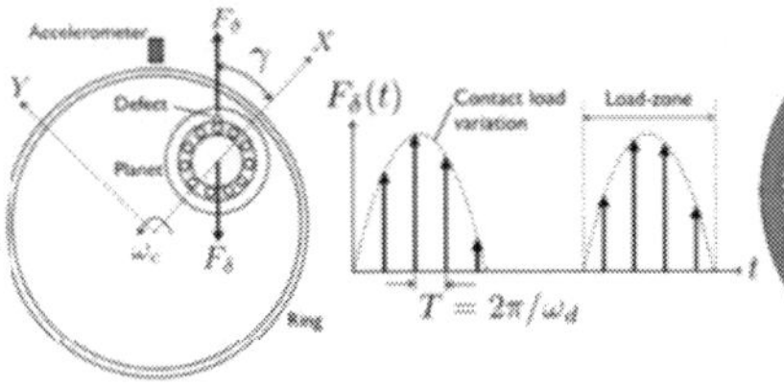

(3) Strong interferences from multiple meshing gears

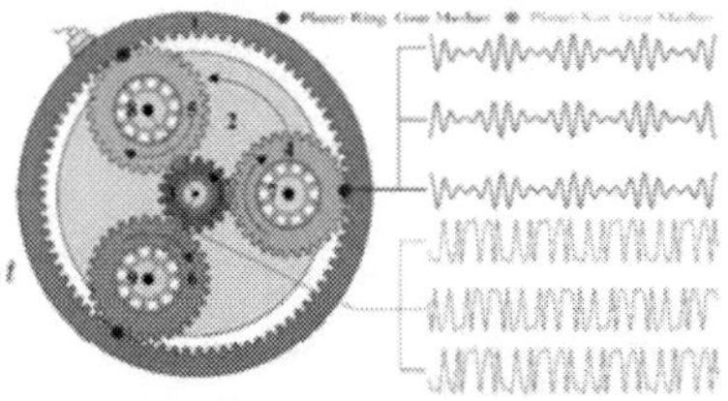

May 14-16, 2019 | PHILADELPHIA, PA, USA

MFPT 2019

1. Introduction (2/2)

4

Traditional Diagnosis Methods for Planet Bearing Fault

(1) Planetary gearbox dynamics considering the defective planet bearings

➢ Sharad Jain et al. 2011 and 2012

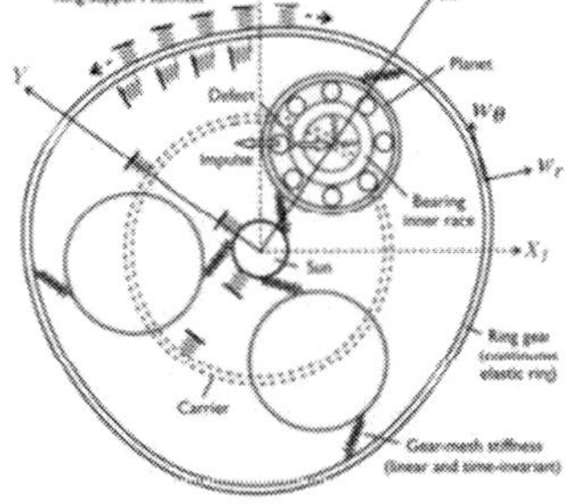

(2) Modulation sidebands identification using demodulations methods

➢ Zhipeng Feng: EMD (2016);

➢ Tianyang Wang: SKRgram (2017)

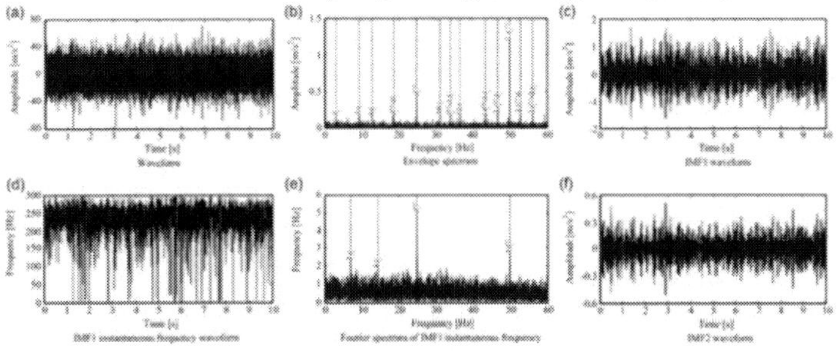

Frequency characteristics in demodulation spectrums are still not sufficient to reliably identify the planet bearing faults.

☐ **Our Proposal: Learn discriminative features from vibration data for intelligent identification of planet bearing faults .**

May 14-16, 2019 | PHILADELPHIA, PA, USA

MFPT 2019

2. Sparse Representation Classification (1/4)

2.1 Sparse Representation and Dictionary Learning

- Given the input signal $y \in \mathbf{R}^n$ and the representation dictionary $\boldsymbol{D} \in \mathbf{R}^{n \times K}$, sparse representation is to seek a sparse linear combination of few dictionary atoms for signal representation.

$$x^* = \arg\min_x \|y - \boldsymbol{D}x\|_2^2 \quad \text{s.t.} \quad \|x\|_0 \leq T$$

Reconstruction error **Sparsity constraint**

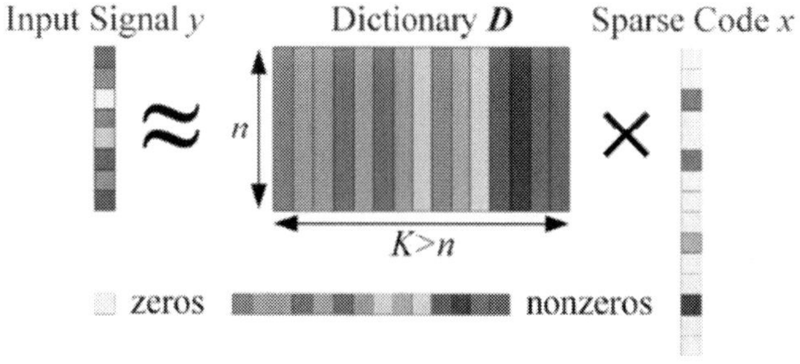

- Convex Relaxation
  L1-norm,etc
- Greedy algorithm
  - Matching Pursuit
  - OMP

2. Sparse Representation Classification (2/4)

2.1 Sparse Representation and Dictionary Learning

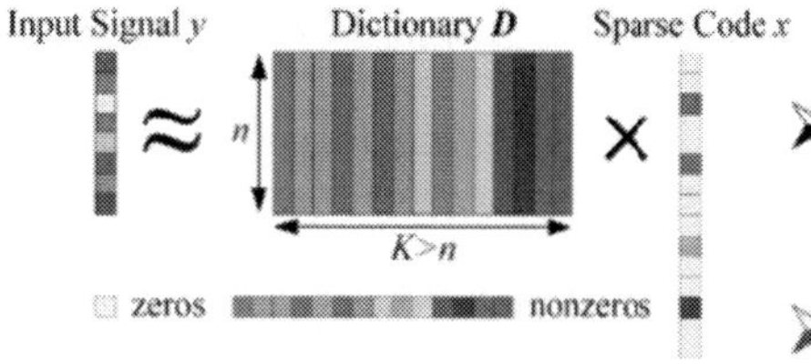

Dictionary Design

- Implemented by known transforms (e.g. FFT, wavelet transform, analytical, etc.)
- Dictionary Learning in a data-driven way.

➤ Optimization Problem for Dictionary Learning

$$\langle \boldsymbol{D}, \boldsymbol{X} \rangle = \arg\min_{\boldsymbol{D},\boldsymbol{X}} \|\boldsymbol{Y} - \boldsymbol{D}\boldsymbol{X}\|_F^2 = \sum_{i=1}^{N} \|y_i - \boldsymbol{D}x_i\|_2^2 \quad \text{s.t.} \ \forall i, \|x_i\|_0 \leq T$$

Knowns: Y

Unknowns: D and X

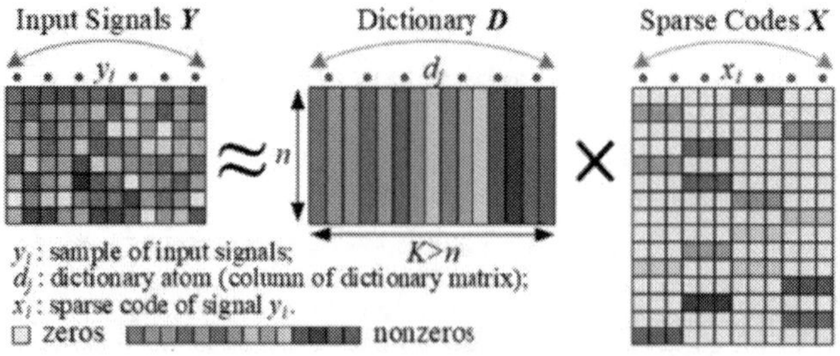

Dictionary Learning Algorithm

- K-SVD
- Method of Optimal Direction (MOD)

2. Sparse Representation Classification (3/4)

7

2.2 Traditional Sparse Representation-based Classification Methods

2.2.1 Reconstruction Error Based Sparse Representation Classification

Step 1: Learn the category-specific dictionary for each class *l*.

$$\langle \boldsymbol{D}_l, \boldsymbol{X}_l \rangle = \arg\min_{\boldsymbol{D}_l, \boldsymbol{X}_l} \|\boldsymbol{Y}_l - \boldsymbol{D}_l \boldsymbol{X}_l\|_F^2 = \sum_{i=1}^{N} \|y_i - \boldsymbol{D}_l x_i\|_2^2 \quad \text{s.t. } \forall i, \|x_i\|_0 \le T$$

Dictionary D_1 | Dictionary D_2 | ... | Dictionary D_3

Step 2: Calculate reconstruction errors of testing sample y with respect to dictionary D_l by sparse coding.

$$x^* = \arg\min_x \|y - \boldsymbol{D}_l x\|_2^2 \quad \text{s.t. } \|x\|_0 \le T$$

$$R(y, \boldsymbol{D}_l, x_l^*) = \|y - \boldsymbol{D}_l x_l^*\|_2^2, \quad l = 1,2,\ldots,L$$

Step 3: Classification according to the minimal reconstruction error.

$$label(y) = \arg\min_l R(y, \boldsymbol{D}_l, x_l^*), \quad l = 1,2,\ldots,L$$

Class 1

Class 2

Class 3

...

Class *L*

Representational Power of Dictionary Learning

2. Sparse Representation Classification (4/4)

8

2.2 Traditional Sparse Representation-based Classification Methods

2.2.2 Classifier Training Based Sparse Representation Classification

Basic Idea: Sparse Codes as Input Features of Classifier

2.2.2.1 Separately (Dictionary Learning + Classifier Training)

(1) $\langle \boldsymbol{D}, \boldsymbol{X} \rangle = \arg\min_{\boldsymbol{D},\boldsymbol{X}} \|\boldsymbol{Y} - \boldsymbol{D}\boldsymbol{X}\|_F^2 = \sum_{i=1}^{N} \|y_i - \boldsymbol{D}x_i\|_2^2 \quad \text{s.t. } \forall i, \|x_i\|_0 \le T$

(2) $\boldsymbol{W} = \arg\min_{\boldsymbol{W}} \sum_{i=1}^{N} C\{h_i, f(x_i, \boldsymbol{W})\} + \lambda \|\boldsymbol{W}\|_F^2$

$\boldsymbol{W}$ —— classifier model parameters;
$f(x, \boldsymbol{W})$ —— classifier output;
$\boldsymbol{H} = [h_1, h_2, \ldots, h_N]$ —— label information;
C —— classification loss function.

✓ Dictionary

Suboptimal !

✓ Classification

Class 1

Class 2

Class 3

...

Class *L*

2.2.2.2 Jointly (Dictionary Learning + Classifier Training)

$$\langle \boldsymbol{D}, \boldsymbol{W}, \boldsymbol{X} \rangle = \arg\min_{\boldsymbol{D},\boldsymbol{X},\boldsymbol{W}} \|\boldsymbol{Y} - \boldsymbol{D}\boldsymbol{X}\|_F^2 + \lambda_1 \sum_{i=1}^{N} C\{h_i, f(x_i, \boldsymbol{W})\} + \lambda_2 \|\boldsymbol{W}\|_F^2$$

$$\text{s.t. } \forall i, \|x_i\|_0 \le T$$

Not consider the structure information of sparse codes !!!

3. Discriminative Dictionary Learning Based Sparse Classification Framework (1/5)

9

3.1 Discriminative Dictionary Learning (DDL)

➢ **Discriminative Sparse Codes**

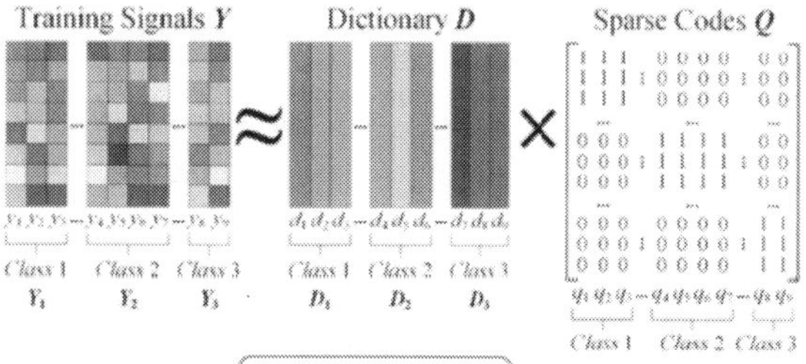

Advantages

I. Explicit correspondences between dictionary atoms and label information;

II. Leverage the structured sparse codes.

Superior Classification Performance !

➢ **Optimization Objective for DDL**

Reconstruction error

$$\langle D, W, A, X\rangle = \arg\min_{D,W,A,X} \|Y - DX\|_F^2 + \alpha\|Q - AX\|_F^2 + \beta\|H - WX\|_F^2 \quad s.t.\ \forall i, \|x_i\|_0 \le T.$$

Discriminative sparse code error **Classification error**

Advantages

I. Introduce the discriminative sparse code error term to enhance discriminability;

II. Leverage the label information;

III. Learn the discriminative dictionary and classifier model simultaneously.

3. Discriminative Dictionary Learning Based Sparse Classification Framework (2/5)

10

3.1 Discriminative Dictionary Learning (DDL)

➢ **Optimization Objective for DDL**

$$\langle D, W, A, X\rangle = \arg\min_{D,W,A,X} \|Y - DX\|_F^2 + \alpha\|Q - AX\|_F^2 + \beta\|H - WX\|_F^2,$$

$$s.t.\ \forall i, \|x_i\|_0 \le T.$$

$$Y_{new} = (Y^T, \sqrt{\alpha}Q^T, \sqrt{\beta}H^T)^T$$

$$D_{new} = (D^T, \sqrt{\alpha}A^T, \sqrt{\beta}W^T)^T$$

➢ **Optimization Algorithm**

Standard Dictionary Learning Problem

$$\langle D, W, A, X\rangle = \arg\min_{D,W,A,X} \left\| \begin{pmatrix} Y \\ \sqrt{\alpha}Q \\ \sqrt{\beta}H \end{pmatrix} - \begin{pmatrix} D \\ \sqrt{\alpha}A \\ \sqrt{\beta}W \end{pmatrix} X \right\|_F^2, \quad s.t.\ \forall i, \|x_i\|_0 \le T$$

Solution: K-SVD

$$\langle D_{new}^*, X^*\rangle = \arg\min_{D_{new}, X} \|Y_{new} - D_{new}X\|_F^2, \quad s.t.\ \forall i, \|x_i\|_0 \le T$$

Learn Discriminative Dictionary and Optimal Classifier Model

$$\hat{D} = \left[\frac{d_1}{\|d_1\|_2}, \frac{d_2}{\|d_2\|_2}, \ldots, \frac{d_K}{\|d_K\|_2}\right], \hat{W} = \left[\frac{w_1}{\|d_1\|_2}, \frac{w_2}{\|d_2\|_2}, \ldots, \frac{w_K}{\|d_K\|_2}\right]$$

3. Discriminative Dictionary Learning Based Sparse Classification Framework (3/5)

3.2 Sparse Classification Approach

I. Sparse Coding of Testing Signals over the learnt discriminative dictionary *D**

$$\hat{x}_i^* = \arg\min_{x_i} \left\| y_i - \hat{\boldsymbol{D}}\hat{x}_i \right\|_2^2, \quad \text{s.t. } \forall i, \left\| \hat{x}_i \right\|_0 \le T$$

II. Classification using maximum absolute value of Classifier Prediction using the learnt classifier model W^*

$$\boldsymbol{l} = \hat{\boldsymbol{W}} x_i^* = [l_1, \ldots, l_L]^T,$$

$$label(y_i) = \arg\max_j \ \text{abs}\,(l_j)$$

Sparse Coding **Classifier Prediction** **Max Pooling Layer**

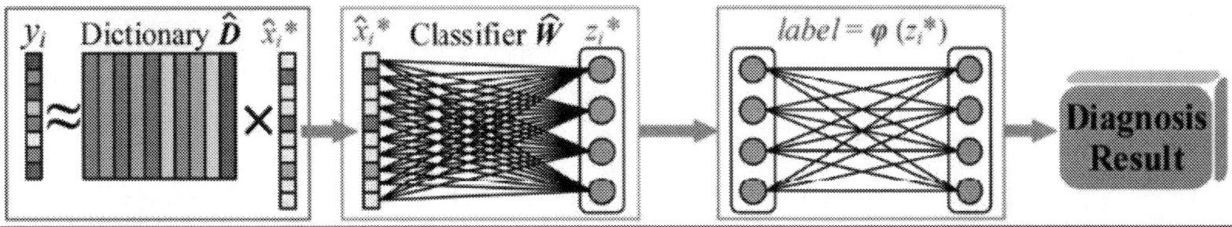

3. Discriminative Dictionary Learning Based Sparse Classification Framework (4/5)

3.3 Intelligent Fault Identification for Planet Bearings by DDL-SC

STEP 1 Mechanical Vibration Signal Acquisition

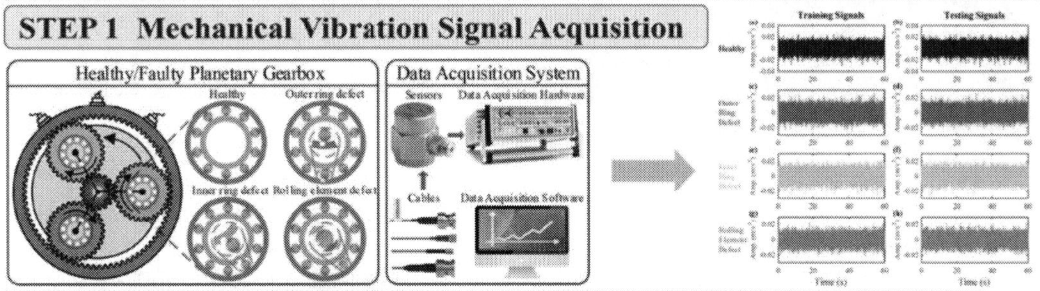

STEP 2 Overlapping Segmentation for 1-D Mechanical Vibration Data

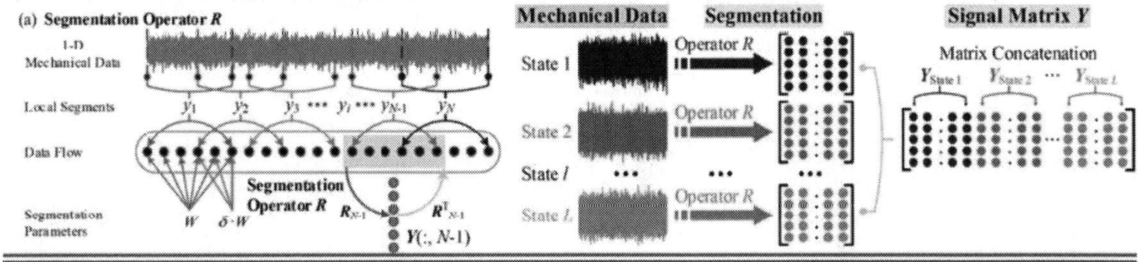

3. Discriminative Dictionary Learning Based Sparse Classification Framework (5/5)

3.3 Intelligent Fault Identification for Planet Bearings by DDL-SC

STEP 3 Discriminative Dictionary Learning

$$\langle D, W, A, X \rangle = \arg\min_{D,W,A,X} \|Y - DX\|_F^2 + \alpha\|Q - AX\|_F^2 + \beta\|H - WX\|_F^2,$$

$$s.t. \forall i, \|x_i\|_0 \le T.$$

Reconstructive Power **Discriminative Power** **Classifier Learning**

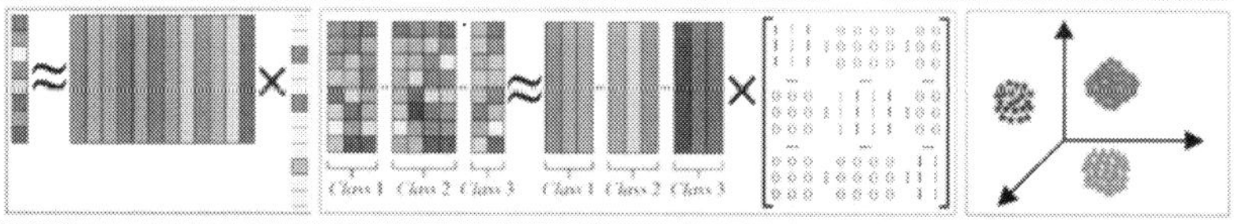

STEP 4 Intelligent Fault Identification for Planet Bearings

Sparse Coding **Classifier Prediction** **Max Pooling Layer**

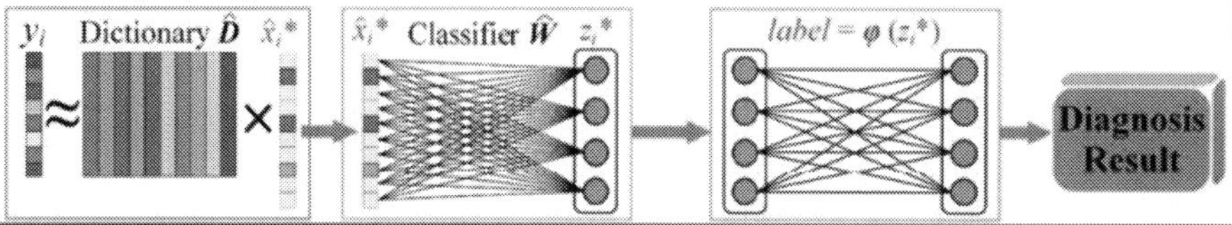

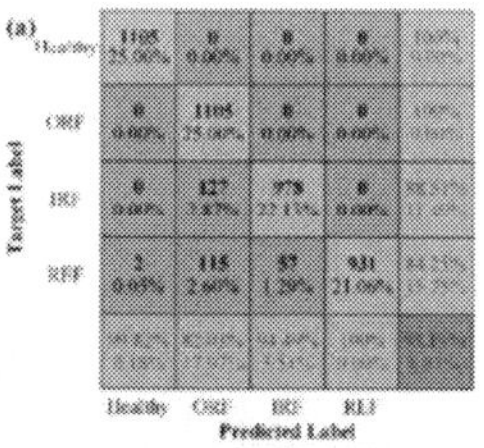

4. Experiment Validations (1/6)

➢ Planetary Gearbox Test Rig

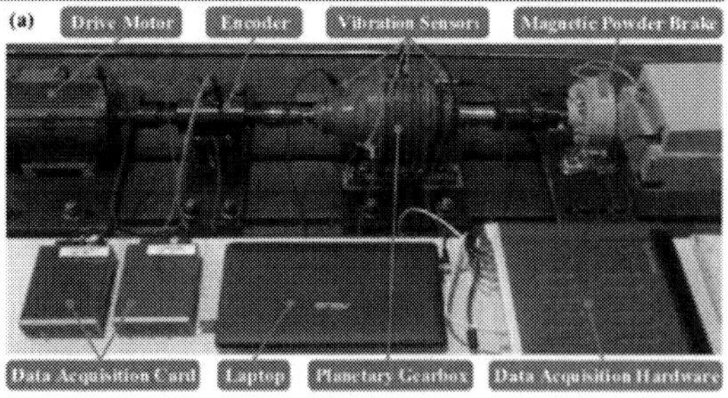

➢ Experiment Descriptions

Gear	Tooth number	Planet bearing	Size (mm)
Carrier	--	Roller diameter	9
Sun gear	13	Diameter of pitch circle	36
Ring gear	92	Number of rollers	10
Planet gear	38(3)	Contact angle (°)	0

Components	Length (mm)	Width (mm)	Depth (mm)
Outer ring	14.90	0.44	2.20
Inner ring	10.18	0.40	1.30
Rolling element	14.90	0.28	1.10

Single Localized Defect Size

(1) Constant speed: $f_{sun} = 24.97$ Hz;
(2) Sampling Frequency: 20480Hz;
(3) Measurement Time: 60 seconds. (training and testing)

➢ Planet Bearing Health States

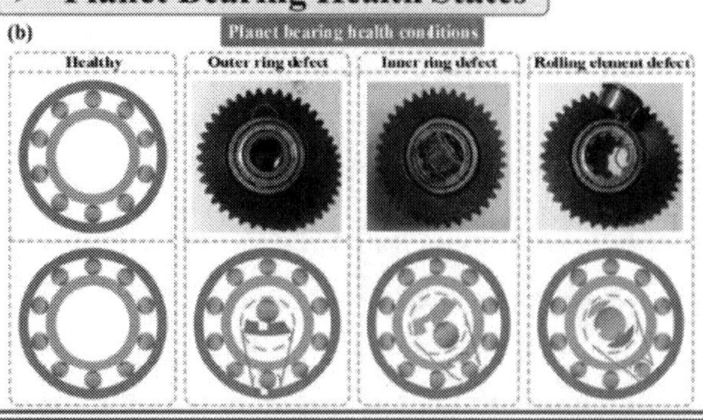

➢ Measured Vibration Signals

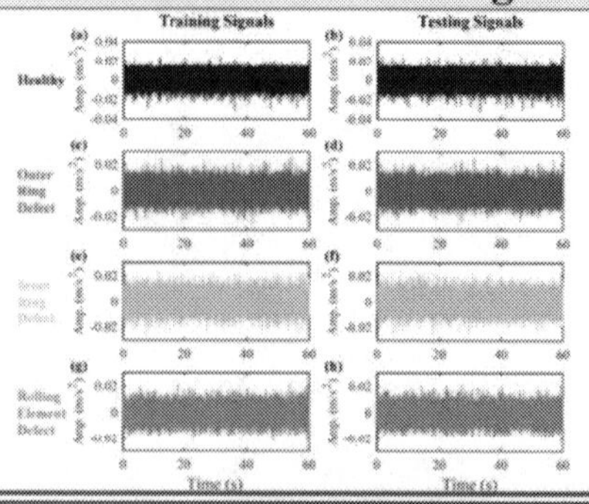

4. Experiment Validations (2/6)

4.1 Simple Case: Three Classes of Planet Bearing Health States

◆ Algorithm Parameter Settings

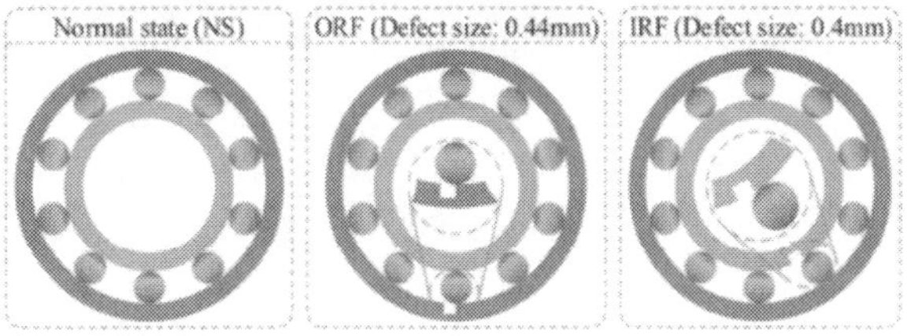

- Window size W is 1750;
- Overlapping rate δ is 0.93;
- Dictionary size K is 1430;
- Regularization parameters α=0.01 and β=0.001;
- Sparsity threshold T = 4 and maximum iterations $J_{\max}$ = 25.

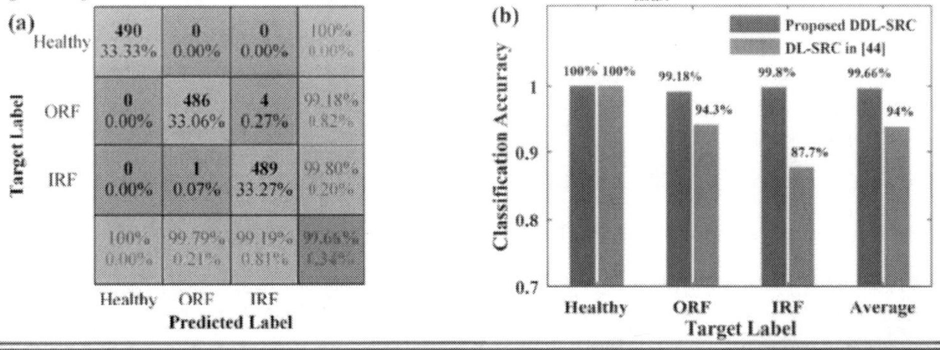

4. Experiment Validations (3/6)

4.2 General Case: Four Classes of Planet Bearing Health States

◆ Algorithm Parameter Settings

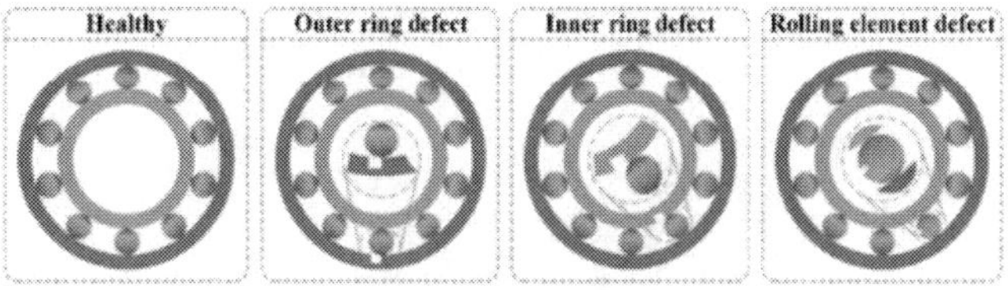

- Window size W is 1800;
- Overlapping rate δ is 0.97;
- Dictionary size K is 4350;
- Regularization parameters α=0.006 and β=0.001;
- Sparsity threshold T = 4 and maximum iterations $J_{\max}$ = 35.

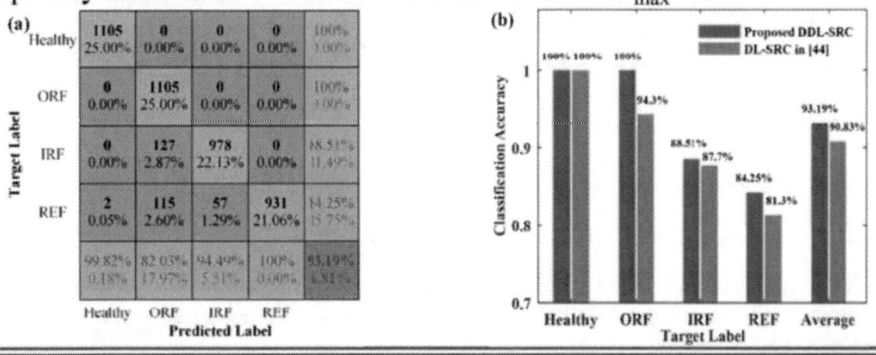

4. Experiment Validations (4/6)

4.3 Parameter Analysis

(1) Dictionary size K

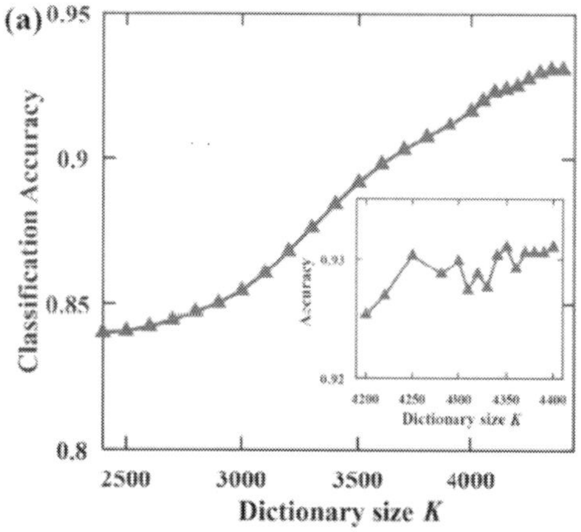

(2) Window Size W and Overlapping Rate δ

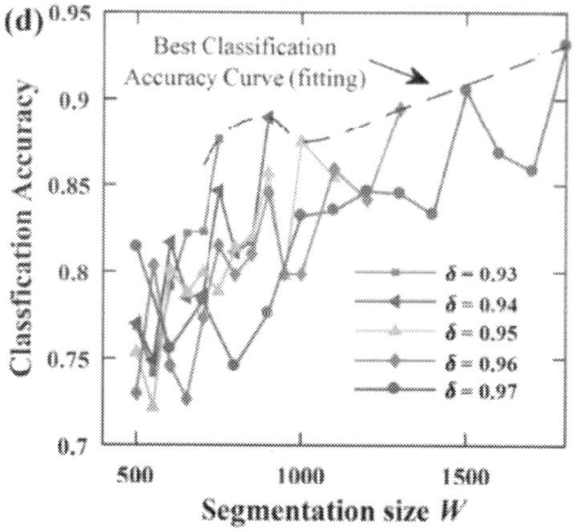

- Higher dimension of feature space;
- Representation and discriminability;
- $K\uparrow$, classification accuracy $\uparrow$.

- Overlapping segmentation strategy;
- $W\uparrow$, classification accuracy $\uparrow$;
- $\delta\uparrow$, best classification accuracy $\uparrow$.

4. Experiment Validations (5/6)

4.3 Parameter Analysis

$$\langle D,W,A,X\rangle = \arg\min_{D,W,A,X} \|Y-DX\|_F^2 + \alpha\|Q-AX\|_F^2 + \beta\|H-WX\|_F^2 \quad s.t.\ \forall i, \|x_i\|_0 \le T.$$

(3) Regularization parameters α and β

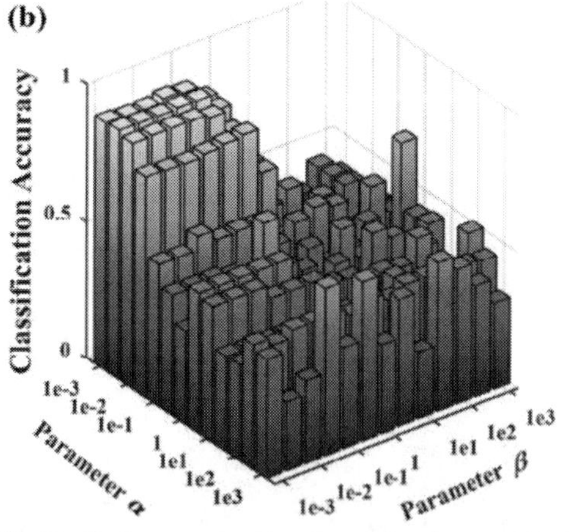

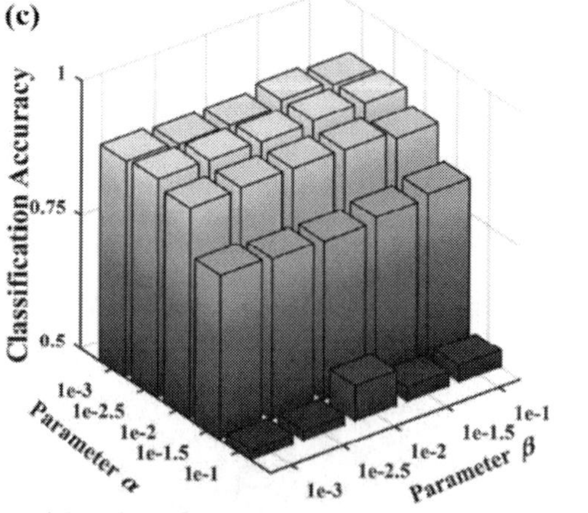

- Relative contributions in the optimization objective for DDL;
- (α, β) with relatively small values, classification accuracy $\uparrow$;
- (α, β) with the order of magnitude of 10^{-3}, superior classification accuracy.

4. Experiment Validations (6/6)

19

4.3 Parameter Analysis

(4) Sparsity Threshold T and Maximum Iterations J_{max}

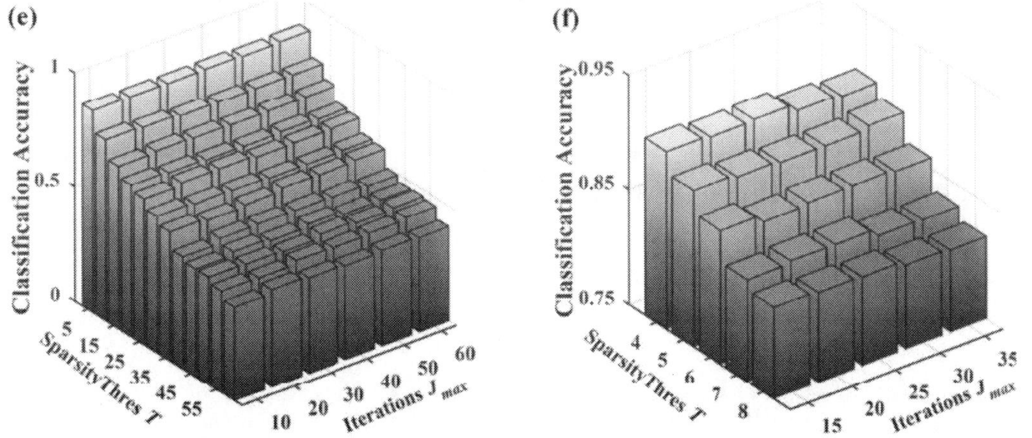

- ☐ OMP for sparse coding and K-SVD for DDL;
- ➢ sparsity thresholds $T \downarrow$, classification accuracy $\uparrow$;
- ➢ Maximum iterations $J_{max} \uparrow$, classification accuracy is promoted little $\leftrightarrow$.

5. Conclusions

20

A novel discriminative dictionary learning based sparse representation classification (DDL-SRC) framework is proposed for intelligent fault identification of planet bearings.

- ➢ The DDL-SRC framework could learn a discriminative dictionary and an optimal multiclass linear classifier simultaneously for superior classification performance.
- ➢ The DDL-SRC framework is free of feature design and selection. It can adaptively learn adaptive sparse features from raw vibration data for data-driven machine fault identification.
- ➢ The DDL-SRC framework outperforms the existing state-of-the-art SRC method and achieves classification accuracy for intelligent fault identification of planet bearings.

Thank you for your attention!

Intelligent Technology for Maintenance Free Measurement

Micro Motion

EMERSON. Process Management

Digital Transformation

- Intelligent field measurements are shown to improve the bottom line by reducing downtime, improving fuel management and eliminating need for costly calibration procedures. The presentation will cover the digital transformation of measurement technology and smart device technology that together enable operational performance improvements

Micro Motion

EMERSON. Process Management

Digital Transformation Defined

"Getting the right data in the right format to the right people so they can take action at the right time."

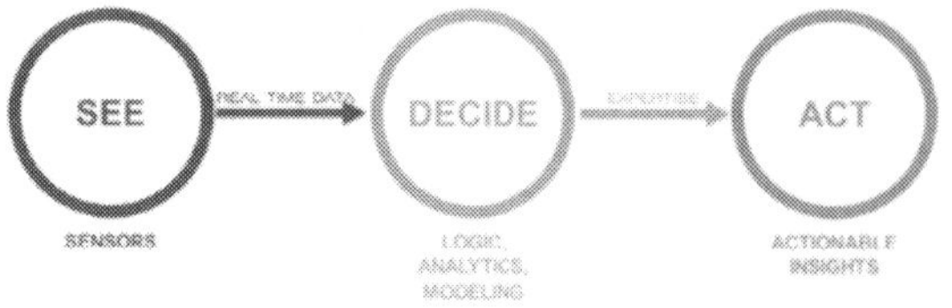

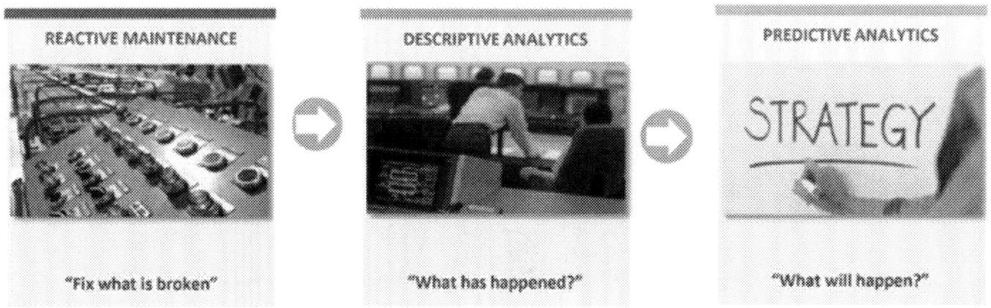

Micro Motion

EMERSON.
Process Management

Flowmeter Selection Process

- **Tendency to stay with what's worked in past**
- **Gathering information often difficult**
- **Not an Exact Science**
  - **Usually more than one answer**
- **Potential LARGE gains**
  - **Reduced CAPEX**
  - **Reduced OPEX**

Micro Motion

N.
Process Management

Flow Technology Selection Process

Application/Purpose of Measurement

Actions: Review purpose of measurement, key performance criteria, environmental considerations, approvals

Purpose of Measurement

Performance

Actions: Complete Spec Sheet, Size Flowmeters for application, complete Installed Performance analysis

Installed vs Reference Performance

Cost/Economics

Actions: Consider schedule, price, cost to Install, operating costs (Reliability, Maintenance, Energy, etc.)

Price/Installed Cost/Life-Cycle Cost

Select Flowmeter

Micro Motion

EMERSON. Process Management

Precision Flow Market

$3.7 Billion Market

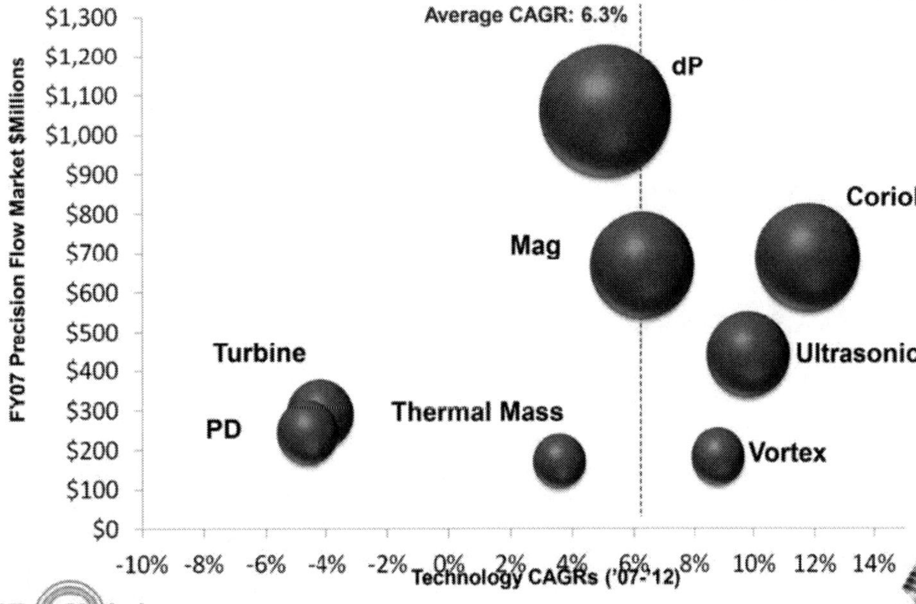

Micro Motion

EMERSON. Process Management

Why Coriolis – Direct Mass Measurement

- Coriolis meters have been the fastest growing flow measurement technologies over the past decade.
- Growth has been driven by a unique ability to measure mass directly
  - Better control of Chemical reactions (based on molecular ratios)
  - Better Mass balance
  - Reduced Process Variability especially in compressible fluids (Gas)

The same amount of fluid (342 lb) looks like 2.7% difference with volumetric measurement (42.0 vs. 40.9 gallons)

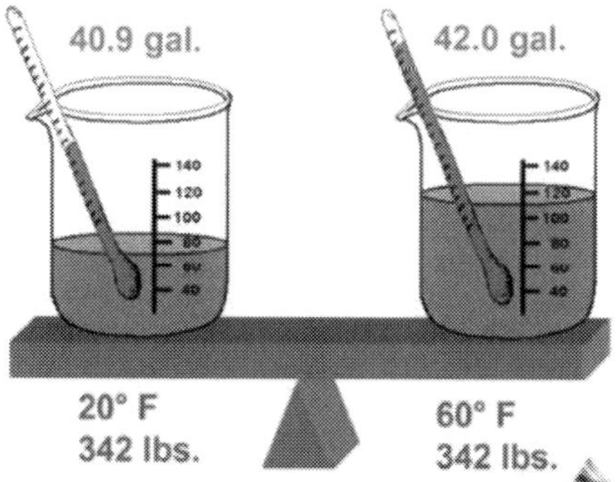

Micro Motion

EMERSON. Process Management

Theory of Operation

- The Coriolis effect is an inertia force.
- In 1835, Gustave-Gaspard de Coriolis showed that this inertia force must be taken into consideration if the simple Newton's Law of Motion of bodies are to be used in a rotating frame of reference.

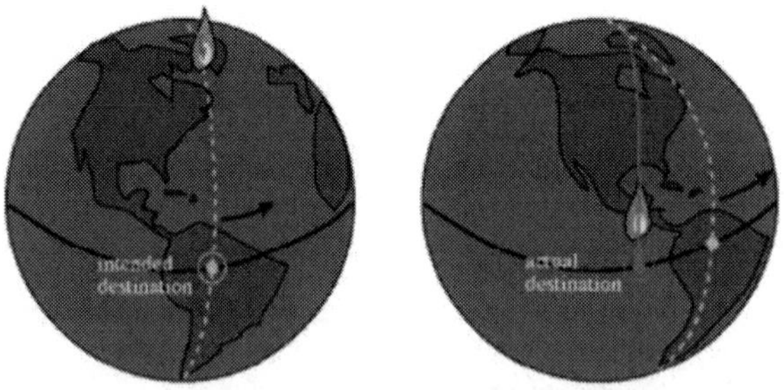

Gasparde de Coriolis

Coriolis Effect: The original path is deflected westward by the rotation of the planet

Micro Motion

EMERSON. Process Management

Theory of Operation – Mass Flow

- In a Coriolis meter, the inertial force is provided by vibrating the flow tubes. The tube twist or angle of deflection from the vibration plane is measured and converted into a mass flow measurement.
- Most implementations use a pair of counter-vibrating tubes to cancel the effect of external vibrations, but many geometries exist.

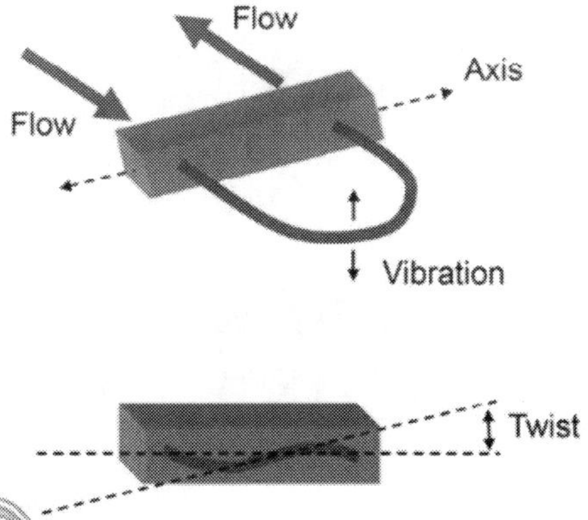

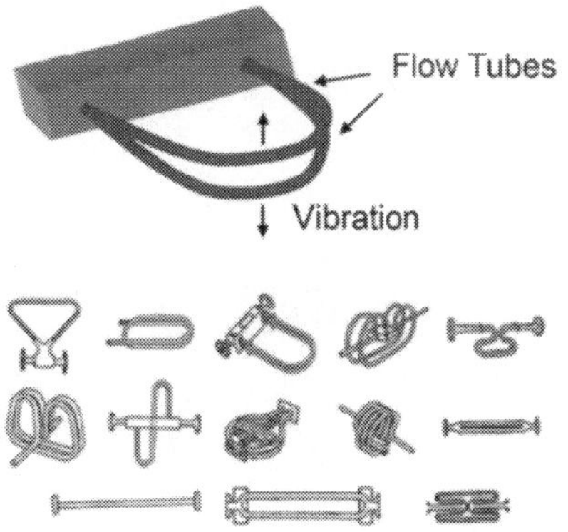

Micro Motion

EMERSON.
Process Management

Theory of Operations – Mass Flow

- Process fluid enters the sensor and flow is split with half the flow through each tube. The sensor flow tubes are vibrated in opposition to each other by energizing a drive coil. Tubes are oscillated at their natural frequency.
- Magnet and coil assemblies, called pick-offs, are mounted on the flow tubes. As each coil moves through the uniform magnetic field of the adjacent magnet it creates a voltage in the form of a sine wave.

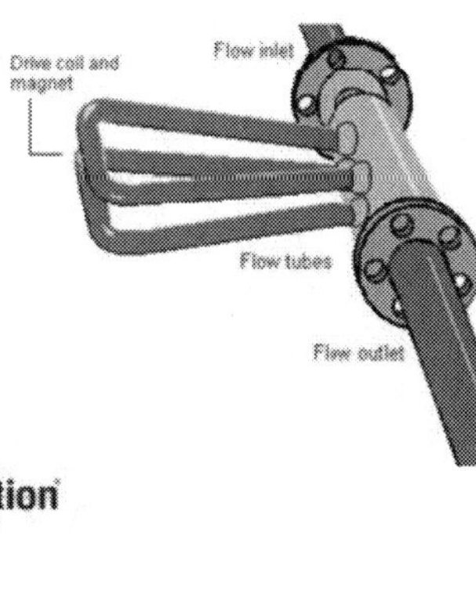

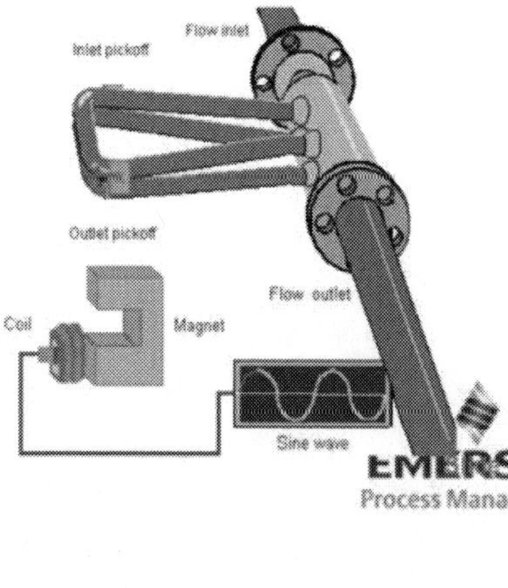

Micro Motion

EMERSON.
Process Management

Theory of Operations – Mass Flow

- During a no flow condition, there is no Coriolis effect and the sine waves are in phase with each other.
- When fluid is moving through the sensor's tubes, Coriolis forces are induced causing the flow tubes to twist in opposition to each other. The time difference between the sine waves is measured and is called Delta-T which is directly proportional to the mass flow rate.

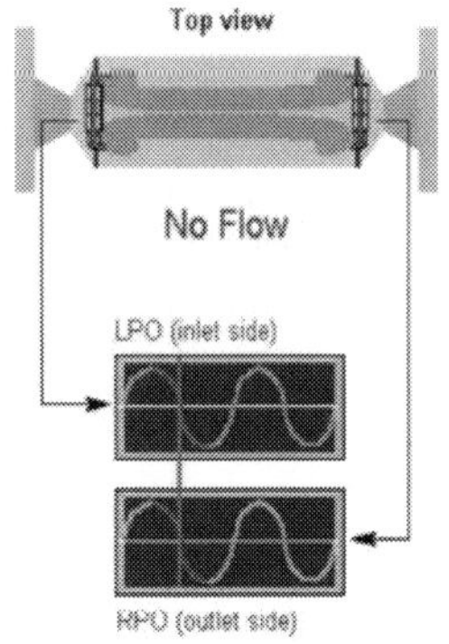

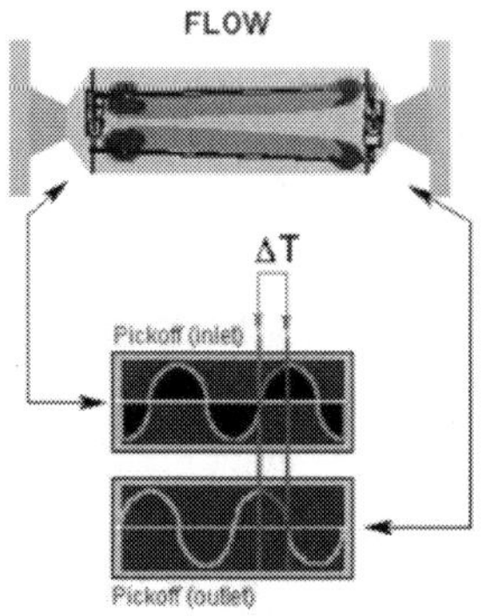

Micro Motio

EMERSON. Process Management

Theory of Operations – Mass Flow

- We also measure temperature – why?
- Compensates for the effect of temperature on tube rigidity (% change in rigidity per 100°C).

→ Three wire platinum RTD measures process temperature

- Accurate to +/- 1.0°C
- Available as additional process variable

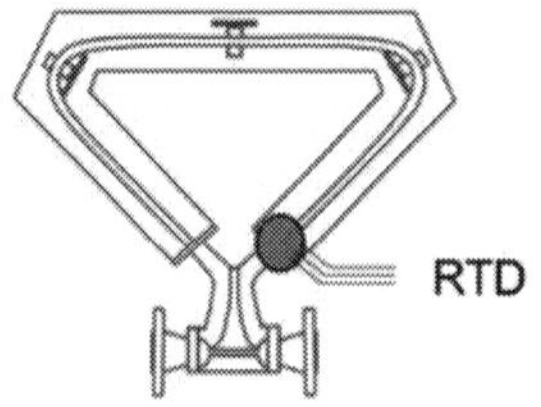

Micro Motion

Micro Motion Confidential
Page 12

EMERSON. Process Management

Theory of Operation - Density

- Density measurement is based on the natural frequency of the system including the flow tubes and the process fluid.
  - As the mass *increases*, the natural frequency of the system *decreases*.
  - As the mass *decreases*, the natural frequency of the system *increases*.

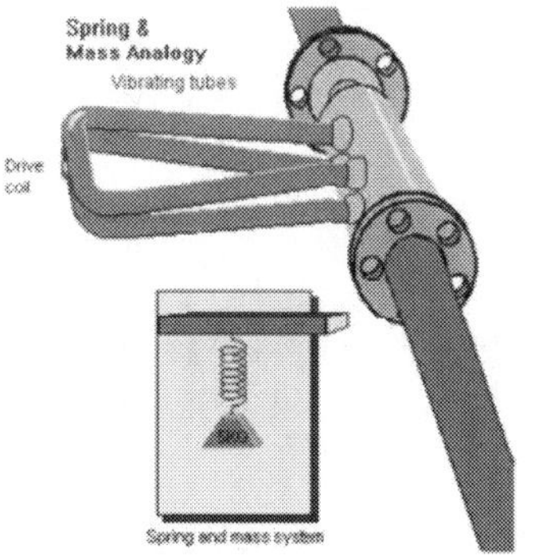

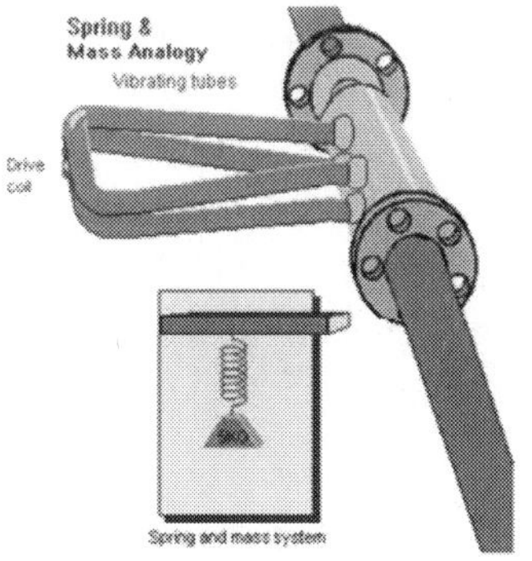

Micro Motion

EMERSON.
Process Management

Theory of Operation - Density

- The density of the process fluid can be derived from the frequency of oscillation of the sensor. This frequency signal is taken from one or both pickoff coils.
- The volume of the fluid contained in the flow tubes remains constant, so the only way mass can change is if density also changes. Because of this relationship between mass and density, the natural frequency of the flow tubes indicates not only the mass of the fluid contained, but also the density.

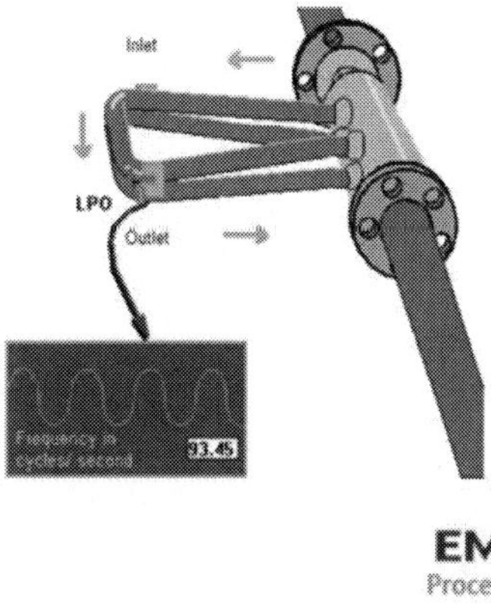

Micro Motion

EMERSON.
Process Management

Theory of Operation - Density

- Density calibration is performed at the factory on air and water.
  - Tube period of air (K1) 10484
  - Tube period of water (K2) 10966
  - Density of air (D1) 0.0010
  - Density of water (D2) 0.9982
  - Temperature coefficient 4.39
- The transmitter automatically performs a calculation based upon the data points stored in its memory during calibration.
- Field calibrations can also be performed using air, water, or alternate fluids depending on the density span desired.

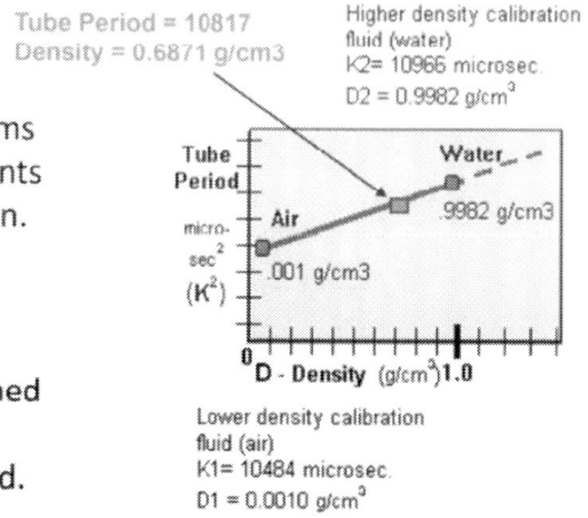

Micro Motion

EMERSON. Process Management

In-Situ Meter Verification

- Calibration
  - Micro Motion Super Service Centers only reliable facility for calibrating (or re-calibrating) meters!
  - Typically, customers will request factory cal for master meters or cal carts
- Validation
  - Primary flow standards such as provers, flow labs, and master meters
  - Required for regulatory, agency, or contractual agreements (AGA, API, OLNL, Country Specific Custody Transfer requirements)
- Verification
  - Most customers need to verify – not validate or calibrate
  - Secondary comparison such as cal carts, catch & weigh, etc.
  - Driven by Standard Operating Procedures and Quality Assurance
  - Troubleshooting

Micro Motion

EMERSON. Process Management

Smart Meter Verification

Micro Motion

EMERSON. Process Management

Easily Verify Meter Performance and Health

- **"Can I tell if my Coriolis meter has changed?"**
- **"Can I detect when the meter is being coated?"**
- **"How often do I need to recalibrate my meter?"**

Characteristics that alter meter performance

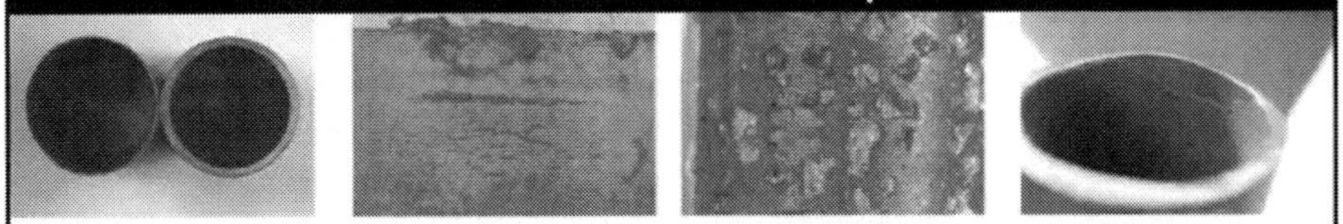

Erosion **Cracking** **Pitting** **Coating**

"We calibrate manually by taking meters out of the process, and it is a pain. It is an all-day deal and we pay $2k to 3k per meter per year for this." --- Oil & Gas Customer

Micro Motion

EMERSON. Process Management

Using Verification in the Field

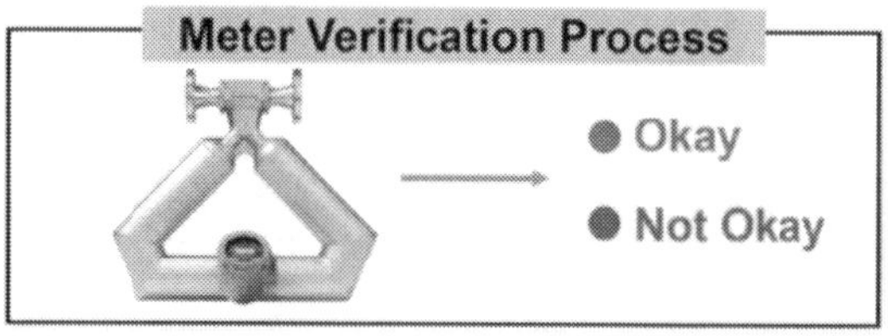

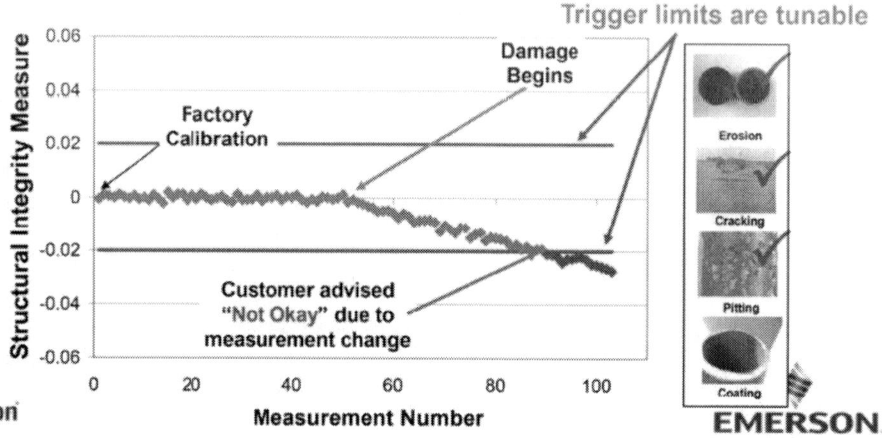

Micro Motion

EMERSON. Process Management

Sample of Verification Report

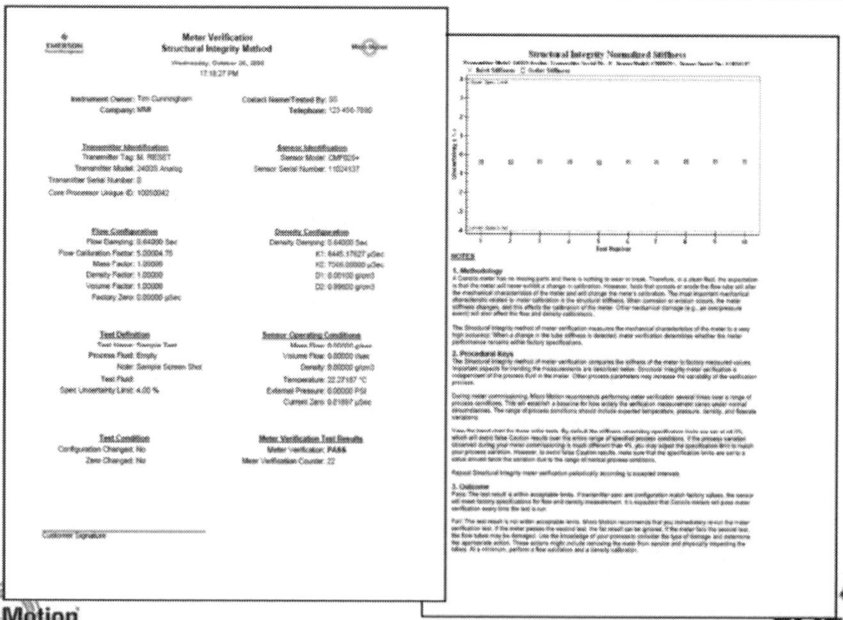
Meter Verification

Micro Motion

EMERSON. Process Management

Why Coriolis – Summary

- Multivariable Measurements
  - Mass Flow, Volume Flow, Density, Temperature, % Solids, Concentration
  - Measures liquids, Gases, and Slurries
- Insensitive to fluid properties
  - Pressure, Temperature, Density, or Viscosity
  - Calibration on water transfers to all other fluids including gases
- Easy to Install and Maintain
  - Measures independent of flow profile, no Straight run required
  - No Moving Parts
  - Bi-directional
- High Accuracy
  - +/- 0.05% of rate Liquid Mass Flow
  - +/- 0.10% of rate Liquid Volume Flow
  - +/- 0.35% of rate Gas Flow
  - +/- 0.0002 g/cc Liquid Density
  - High turndown, >100:1 possible
  - Batch accuracies to 0.10%

Micro Motion

EMERSON.
Process Management

Magnetic flow meter

From Wikipedia, the free encyclopedia

The third most common flowmeter behind differential pressure and positive displacement flow meters, is the **magnetic flow meter**, also technically an *electromagnetic flow meter* or more commonly just called a *mag meter*. A magnetic field is applied to the metering tube, which results in a potential difference proportional to the flow velocity perpendicular to the flux lines. The physical principle at work is electromagnetic induction. The magnetic flow meter requires a conducting fluid, for example, water that contains ions, and an electrical insulating pipe surface, for example, a rubber-lined steel tube.

Usually electrochemical and other effects at the electrodes make the potential difference drift up and down, making it hard to determine the fluid flow induced potential difference. To mitigate this, the magnetic field is constantly reversed, cancelling out the static potential difference. This however impedes the use of permanent magnets for magnetic flowmeters.

EMERSON.
Process Management

Vortex - Theory of Operation

- Based on the von Karman Effect
  - Fluid alternately separates from each side of the shedder bar face
  - Vortices form behind the face and cause alternating differential pressures (DP) around the back of the shedder bar
  - The frequency of the alternating vortex development is linearly

Micro Motion

EMERSON. Process Management

Questions ???

Micro Motion

EMERSON. Process Management

NAVAL SURFACE WARFARE CENTER
PHILADELPHIA DIVISION

Navy Arc Fault Detection and Prevention Systems

Presented at the Society for Machinery Failure Prevention Technology

MFPT 2019

King of Prussia, PA

May 14, 2019

Brandon Whaley

Naval Surface Warfare Center – Philadelphia Division

Code 445 – CVN and Submarine Power Systems

Branch Head Stephen Carroll

- Submarine switchboard fires in the late 70's: John's Hopkins University Applied Physics Lab contracted for research and development
- First systems fielded early 1980's
- Two functions of the system:
  - Arc Fault Detection: Reduction in catastrophic damage to switchgear
  - Continuous Thermal Monitoring: Provides an alert for the crew that a condition exists within an enclosure which may lead to an arc fault

DISTRIBUTION STATEMENT A. Approved for public release. Distribution is unlimited.

System Variants

- Three main variants:
- 1st generation
  - Sensors: Photo, Pressure, Ionization
  - JHU/APL designed, Navy-owned
- 2nd generation
  - Sensors: Photo, Ionization
  - JHU/APL designed, Navy-owned
- 3rd generation
  - Sensors: Photo, Pressure, Ionization
  - Developed from JHU/APL design, commercial-owned

DISTRIBUTION STATEMENT A. Approved for public release. Distribution is unlimited.

Gen 1 & Gen 2

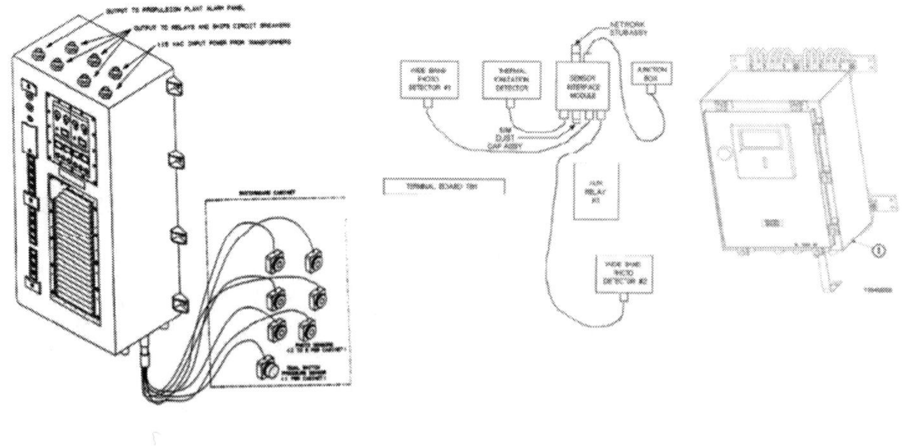

DISTRIBUTION STATEMENT A. Approved for public release. Distribution is unlimited.

Sensors

- Photo:
  - Narrow vs Wide band
  - UV-enhanced photodiode
  - <211 LUX
- Pressure:
  - Dual-switch
  - 0.2" WC
- Ionization:
  - Sealed-source (0.9µCi Americium-241)
  - UL268 0.5 – 4.0% obscuration

DISTRIBUTION STATEMENT A. Approved for public release. Distribution is unlimited.

Arc Spectra Measurements

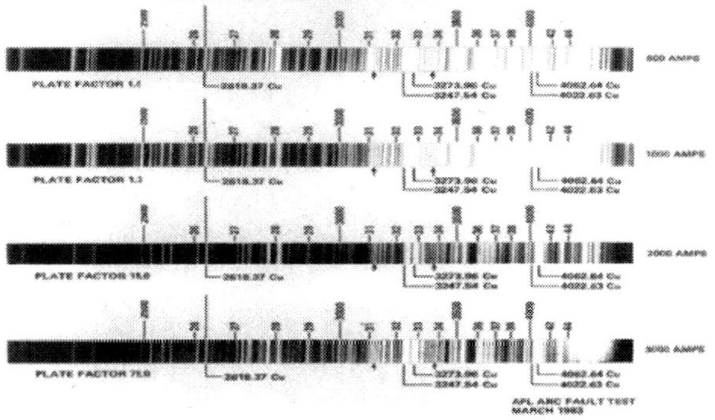

DISTRIBUTION STATEMENT A. Approved for public release. Distribution is unlimited.

NAVSEA WARFARE CENTERS Philadelphia

Control Unit

- 1st Gen is payload-specific, $2^{nd}/3^{rd}$ are configurable
- Receives voltage signal directly from sensors or over serial (RS-485) variant-specific
- Validates arc condition
- Outputs to isolation breakers

5/20/2019 DISTRIBUTION STATEMENT A. Approved for public release. Distribution is unlimited. 7

NAVSEA WARFARE CENTERS Philadelphia

Thermal Imaging

- Catch connection issues prior to failure
- Yearly / triennial platform dependent
- 450VAC and below only
- Thermal windows 450VAC+
- Certification Program

5/20/2019 DISTRIBUTION STATEMENT A. Approved for public release. Distribution is unlimited. 8

BEARING FAULTS DETECTION AND IDENTIFICATION USING RELATIONAL DATA CLUSTERING WITH COMPOSITE DIFFERENTIAL EVOLUTION OPTIMIZATION

Issam Abu-Mahfouz and Amit Banerjee
Penn State University Harrisburg
777 West Harrisburg Pike
Middletown, PA 17111
USA

Abstract: Bearing faults in machinery are among the most critical faults that require attention by maintenance personnel at early stages of fault initiation. In many cases it is difficult to directly and accurately identify the fault type and its extent under varying operating conditions. This work demonstrates a novel procedure for bearing fault detection and identification in an experimental set-up. Three seeded faults, in the rotating machinery supported by the test ball bearing, include inner race fault, outer race fault and one roller fault. The rotor is run at different speeds and with small level of rotating mass unbalance. Accelerometer based vibration signals are analyzed for the different bearing faults' signatures using statistical features, frequency spectra and wavelet coefficients. The composite differential evolution technique is proposed for parameter estimation when the system response is known *a-priori*. The algorithm is compared to five other differential evolution algorithms using conventional crossover and mutation operators. The objective is to correlate bearing faults to the extracted vibration features. The results of this analysis will be extended for applications in real time bearing condition monitoring system.

Key words: Bearing fault; Condition monitoring; diagnostics; differential evolution; evolutionary algorithms; parameter identification; vibration; wavelets.

1. INTRODUCTION

Vibration monitoring in combination with signal processing techniques are effective indicators of wear and damage for bearings used in rotating machinery [1]. The capability of successfully and accurately detecting the faults and their types at early stages is of great importance for the safety and economical aspects in any industrial setup. Therefore, predictive condition monitoring procedures are required to extract the features related to the faulty element bearing. Several well-established methods have been used for bearing fault diagnosis such as spectral analysis, envelop detection, autoregressive modeling, cyclostationary analysis, hidden Markov models (HMM), Helbert-Huang transform, and wavelet decomposition. Recent research in the field for machine condition and predictive health monitoring have concentrated on the development of advanced signal processing and machine learning techniques using artificial Intelligence (AI) and fuzzy logic algorithms. See for example [2-5] as a sample. El-Thalji and Jantunen [6] provided a critical review of the Predictive Health Monitoring methods of the entire defect evolution process i.e. over the whole life time and suggests enhancements for rolling element bearing monitoring.

The purpose of this study is to explore the potentials and capabilities of an artificial intelligence algorithm, namely the differential evolution (DE) in developing machine condition monitoring (MCM) methods for ball bearings supporting a rotor-disk system under periodic unbalance excitation. Evolutionary algorithms (EA) are metaheuristic optimization algorithms that uses mechanisms inspired by processes of biological evolution such as reproduction, mutation, recombination and survival of the fittest. They are used to successfully solve problems that cannot be efficiently solved by classical mathematical polynomial methods. Differential Evolution (DE) is a stochastic population-based evolutionary algorithm (EA) in which current information at each iteration is used to guide the search process. In this work, the vibration signals obtained from an experimental setup with faulty bearings, as described in the following section, are processed to extract features that form the search landscape for the DE technique.

2. EXPERIMENTAL SET-UP AND SIGNAL PROCESSING

A rotor dynamic test rig is used for collecting vibration signals from bearings with seeded artificial faults. The schematic of the set-up is shown in Fig. 1. The bearing used for this test is a SKF single row radial ball bearing (1.25 inch I.D., 2.5 inch O.D. and 0.625 inch width dynamic load capacity of 2160 lbf.) For the normal (no fault) bearing case, balls and races are inspected for defects prior to the test runs. In addition, three more bearings were artificially introduced with defect points using a small carbide grinding bit such that each single bearing will have only one fault. The defects are; inner race, outer race and single rolling ball fault. The vibration signals are measured by a PCB accelerometer (model PCB 302A, with sensitivity 10mV/g, and frequency range of 0.7-10000 Hz, and amplitude range of + - 500 g pk). The accelerometer is mounted on the outboard bearing block (Fig. 1). The vibration signals are sampled at 10 KHz and then recorded via a National Instruments USB-6211 data acquisition device. Bearings conditions are evaluated at 10 operating speeds from 500 rpm to 1400 rpm with increments of 100rpm. A small rotating unbalance is added to the rotor disk to create periodic but speed dependent load on the bearings. The raw time vibration signals are analyzed using the Fast Fourier Transform (FFT) and the continuous wavelet transform (CWT) techniques. Figures 2 and 3 show sample time signatures and their corresponding FFT spectra plots for two fault conditions at different running speeds. The time signals are presented by plotting the output of the vibration sensor versus the time sample points. Therefore, at a sampling rate of 10,000 s/sec the segments in Figure 2 represent a duration of 0.5 second. Statistical parameters such as the standard deviation and skewness are also computed. Sample statistical features are shown in Figures 4 and 5 for all four bearing cases.

Wavelet transform is being increasingly used as a signal processing technique [7]. The continuous wavelet transform (CWT) of a signal $x(t)$ is defined as the convolution integral of $x(t)$ with a translated and dilated versions of a mother wavelet function. For this work and after experimenting with several wavelets, it has been found that the Mexican hat wavelet gave acceptable results. Figures 6 and 7show sample wavelet scalograms, over 500 data points, for the two bearing fault cases presented in Figures 2 and 3 respectively. The scalogram is a two-dimensional plot of scales (inverse of frequencies) versus time (or in this case sample points).

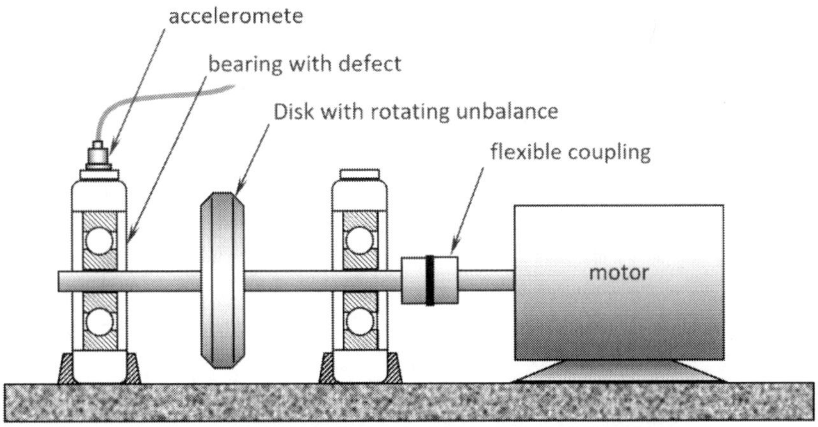

Figure 1. Experiment setup for ball bearing fault.

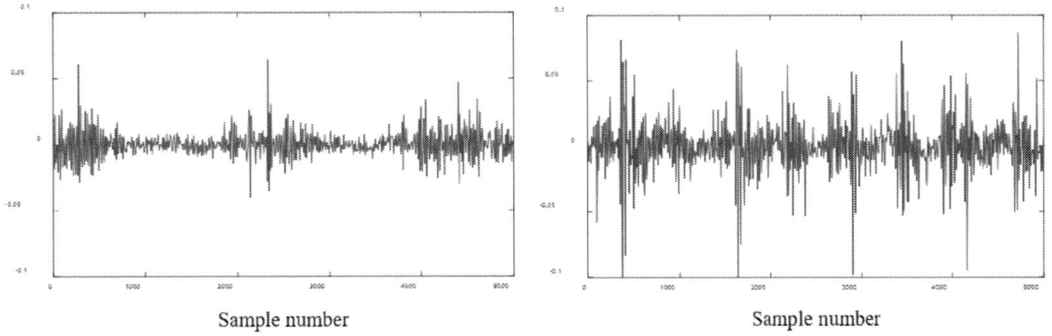

Figure 2. Vibration signals for (a) 500 rpm _ ball fault, and (b) 800 rpm _ inner race fault.

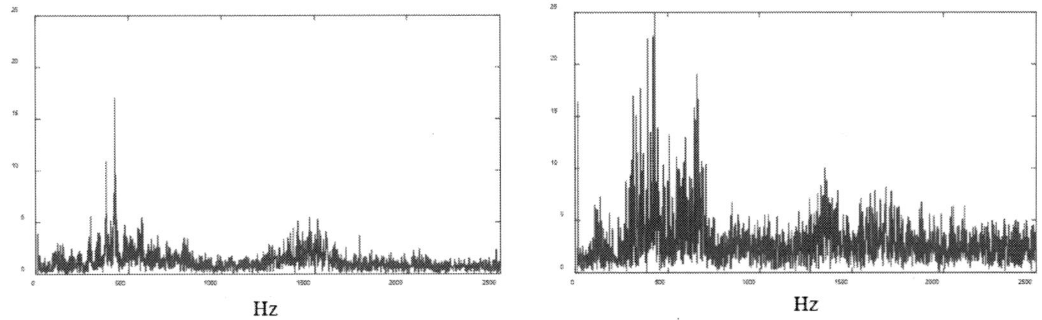

Figure 3. FFT Spectra for (a) 500 rpm _ ball fault, and (b) 800 rpm _ inner race fault.

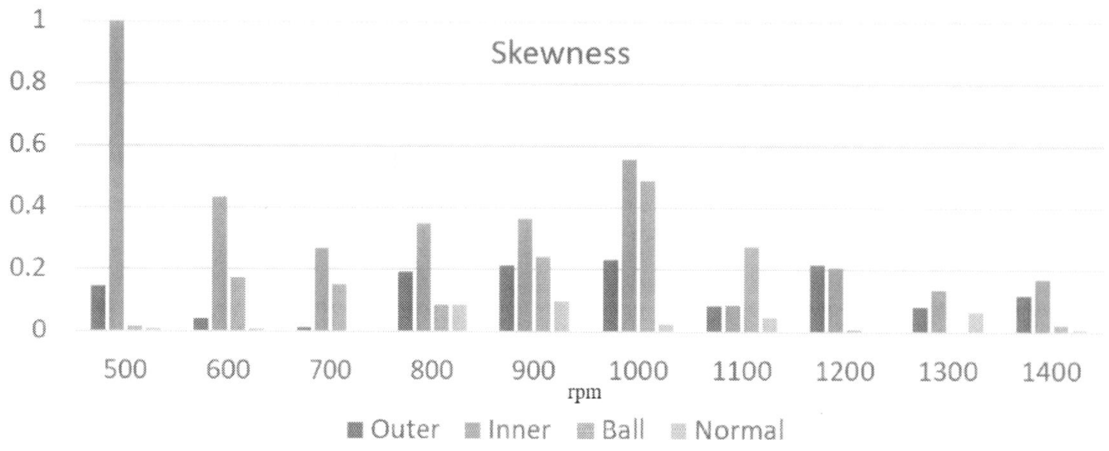

Figure 4. Normalized skewness at different speeds in rpm.

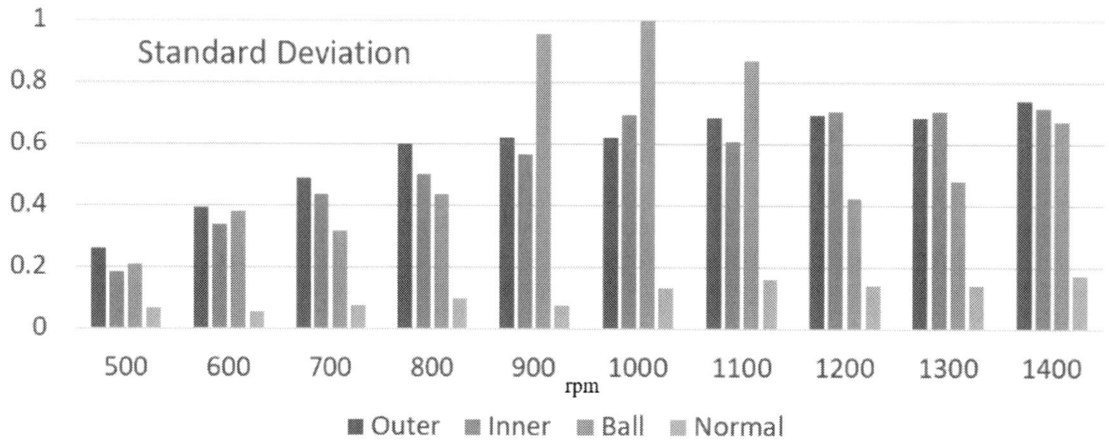

Figure 5. Normalized standard deviation at different speeds in rpm.

Small scaling parameter values result in narrow windows and serve for precisely localized registration of high frequency phenomena. On the other hand, large scaling parameter values result in wide windows and serve for the registration of slow phenomena. This way, information in the time and frequency domains are represented simultaneously. The Mexican hat wavelet, as selected for this investigation, is applied to segments of 5000 data points each. The 64 wavelet scales, for each bearing fault type and at each rotating speed, are averaged over these 5000 points to represent an additional feature vector.

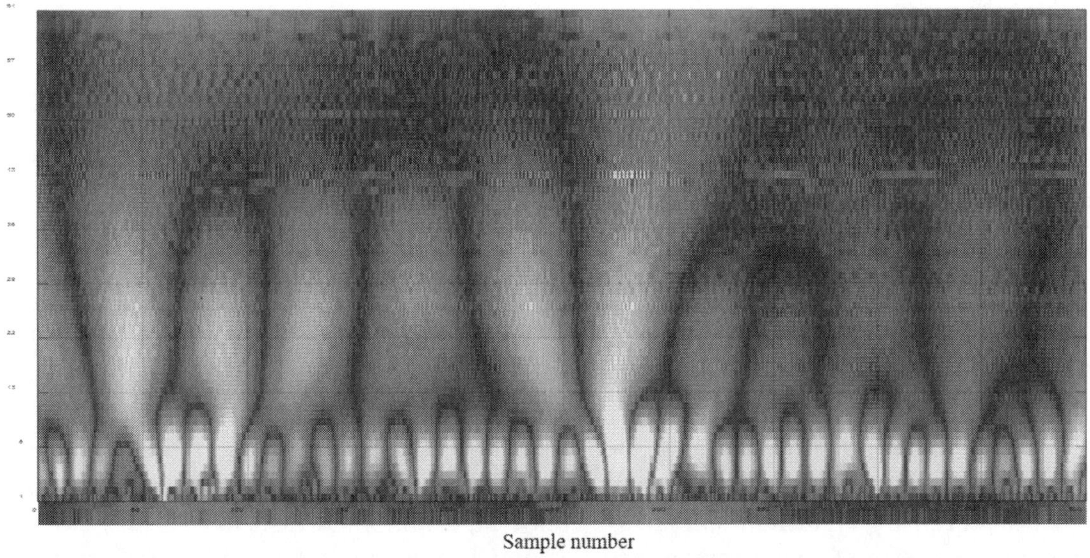

Figure 6. Wavelet scalogram for a bearing with ball fault running at 500 rpm.

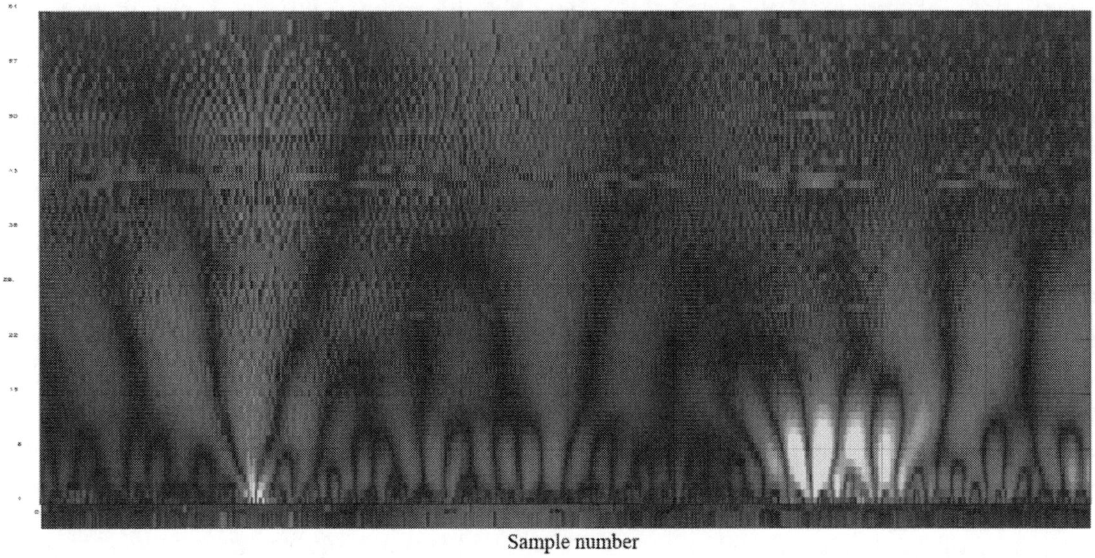

Figure 6. Wavelet scalogram for a bearing with inner race fault running at 800 rpm.

In this paper, we formulate the inverse problem of parameter estimation where the system response is known. The parameters are then estimated by solving a minimization problem in the response space formulated as,

$$f = \min\left[\sum\left\{x(\Theta) - x(\hat{\Theta})\right\}^2\right] \tag{1}$$

where $x(\Theta)$ is the known set of response for an unknown parameter set Θ and $x(\hat{\Theta})$ is the response for the estimated parameter set $\hat{\Theta}$. The known response is in the form of a finite set of either the standard deviation and skewness of the vibration signal, or the normalized averaged FFT bands or the averaged wavelet coefficients. The minimization functional f used by the differential evolution algorithm is the average sum of differences between the measured response and estimated response over the entire response spectrum. The algorithm will guide the search for parameters based on minimizing this functional. The composite differential evolution algorithm is used for optimizing the estimated parameters in this paper and algorithmic performance is compared to basic versions of the differential evolution algorithm.

3. Composite Differential Evolution (DE)

A potential candidate solution for the parameter estimation problem is encoded in 6-bits – the first four encoding for speed (rpm) and the other two bits encoding for the four defect types. A population of size N is used and evolved using the differential evolution (DE) algorithm till convergence. In this paper, we define convergence when there is no change (within a threshold) in the population over two consecutive generations. The DE algorithm is an evolutionary algorithm similar to genetic algorithms but instead of using crossover and mutation to evolve new population of candidates, uses a simple differential operator to create new candidate solutions and employs a one-to-one competition scheme to greedily select new candidates [9]. Similar to genetic algorithms, DE is a population-based evolution technique where the starting population is randomly populated by potential candidate solutions. The next generation is evolved by first creating a trial vector for each target vector (from the current population) using a trial vector generation strategy, followed by a crossover operation which creates an offspring vector for every target vector that has aspects of both the target and the trial vector. The offspring vector replaces the target vector in the new generation if it is better (fitter) than the target vector. The performance of the differential evolution algorithm depends on the strategy used to generate trial vectors, and how well the control parameters are tuned to solve the specific problem. To overcome these issues, a variant of the differential evolution algorithm called the composite differential evolution (CoDE) algorithm was proposed [10]. It combines three most popular trial vector generation strategies with three control parameter settings which are used in a random way to generate trial vectors. It has a simple structure and is easy to implement. The three trial vector generation strategies used in this paper are:

1) rand/1
   $$\upsilon_i = x_{r1} + F.[x_{r2} - x_{r3}]$$
2) best/1
   $$\upsilon_i = x_{best} + F.[x_{r1} - x_{r2}] \quad (2)$$
3) current-to-best/2
   $$\upsilon_i = x_i + F.[x_{best} - x_i] + F.[x_{r1} - x_{r2}],$$

where υ_i is the trial vector and x_i is the parent vector. The best candidate solution in a generation is labeled x_{best} and x_{r1} and x_{r2} are two randomly selected vectors from the current generation. Empirical studies have shown that rand/1 maintains a good diversity, while current-to-best/2/bin shows good convergence characteristics. The DE crossover operator implements a discrete (bit-wise) recombination of the trial vector, υ_i, and the parent vector x_i to produce offspring x'_i as follows:

$$x'_{ij}(g) = \begin{cases} \upsilon_{ij} & \text{if } \eta < c_r \text{ and } j \in \Im \\ x_{ij} & \text{otherwise} \end{cases} \tag{3}$$

where x_{ij} refers to the jth bit of the vector x_i, c_r is the crossover probability [0-1] and $\Im$ is the set of bit indices that will undergo crossover The binomial crossover (/bin) is the most widely used method to determine the set $\Im$ where crossover points are randomly selected from a et of possible crossover points, $\{1, 2, .. p\}$ where p is the total number of bits in genotype definition. The larger the crossover probability c_r, the more crossover points will be selected. This also means that more elements of the trial vector will be used to produce the offspring with fewer bits coming from the parent vector. The binomial crossover operator is used with the rand/1 and the best/1 trial vector generation strategies but not with the current-to-best/2.

The scaling factor F controls the amplification of the differential variations and controls what is akin to a mutation operator in genetic algorithms. The smaller the value of F, the smaller the mutation step sizes and the longer it will make the algorithm to converge. On the other hand, larger values help in exploration but may cause the algorithm to miss a good optima. Therefore, the value of F should be small enough to allow differentials to exploit tight spaces in the search space, and large enough to maintain diversity (exploration). In CoDE, three control parameter settings are used because of the drawback in using fixed values of c_r and F throughout the algorithm. In this paper, we use:

1) $F = 1.0$, $c_r = 0.1$
2) $F = 1.0$, $c_r = 0.9$ (4)
3) $F = 0.8$, $c_r = 0.2$

These settings are the same as those used in (Refer to CoDE paper). At each generation, all three trial vector strategies are used to create new trial vectors followed by offspring vectors with a control parameter setting chosen randomly from the three control parameter settings. Therefore, three offspring vectors are generated for each target vector. The best offspring vector then enters the next generation if it is better than its target vector. In this paper we compare the performance of CoDE to that of DE/rand/1/bin and DE/current-to-best/2 with all three control parameter settings used one at a time. The algorithms will be referred to with a number prefix which will correspond to the control parameter setting.

4. RESULTS

All the algorithms are run independently for 5 times for population size of $N = 10$, and number of generations till convergence and the quality of the best solution for the median run are compared for the different algorithms. The CPU run time on MATLAB 2018 running on dual core $i7$ processor is also provided in Tables 1-3 for three different feature sets. The feature sets used are (1) statistical feature set with two features – standard deviation and skewness in Table 1, (2) 16 PSD (FFT) bands in Table 2, and (3) 64 wavelet coefficients in Table 3. As can be seen CoDE outperforms the other algorithms in both convergence characteristics and quality of best solution but takes significantly more run time than the basic versions of the DE algorithm. This is because of multiple trail vectors generated by for every target vector, followed by a comparison phase.

Table 1: Comparative Performance for the DE algorithms for Statistical Feature Set.

	Convergence (# generations)	% Error	CPU Runtime (ms)
CoDE	8	1.924	62
DE/rand/1/bin/1	16	2.996	27
DE/rand/1/bin/2	15	3.412	32
DE/best/1/bin/3	18	2.946	29
DE/current-to-best/2/1	14	3.218	33
DE/current-to-best/2/2	14	3.184	34

Table 2: Comparative Performance for the DE algorithms for FFT Coefficients Feature Set.

	Convergence (# generations)	% Error	CPU Runtime (ms)
CoDE	11	1.247	132
DE/rand/1/bin/1	18	1.895	58
DE/rand/1/bin/2	18	2.887	75
DE/best/1/bin/3	21	1.924	72
DE/current-to-best/2/1	15	2.560	79
DE/current-to-best/2/2	17	2.677	82

Table 3: Comparative Performance for the DE algorithms for Wavelet Coefficients Feature Set.

	Convergence (# generations)	% Error	CPU Runtime (ms)
CoDE	17	4.284	185
DE/rand/1/bin/1	23	5.778	120
DE/rand/1/bin/2	24	6.820	127
DE/best/1/bin/3	20	4.818	121
DE/current-to-best/2/1	19	5.320	125
DE/current-to-best/2/2	21	5.320	125

Among the feature sets tested, the FFT coefficients feature set (16 features) provided the more accurate results than the Wavelet coefficients and the statistical feature set. The statistical feature set (2 features) is the smallest feature set but encapsulates enough information to provide errors < 5% across all algorithms. The statistical feature set also takes the least amount of CPU runtime.

5. Conclusion

In this paper, a framework for parameter estimation using differential evolution is created. The framework is based on features extracted from the rotor system vibration response with the purpose of bearing fault identification. It is scalable and can be applied to a larger sample size with very little change to the methodology presented in this paper. The composite differential evolution algorithm, an adaptive variant of the basic differential evolution algorithm is shown to outperform the basic other variants of the algorithm on the parameter identification problem.

References

[1] I. Howard. A Review of Rolling Element Bearing Vibration: Detection, Diagnosis and Prognosis: Melbourne, Australia, DSTO Aeronautical and maritime Research Laboratory, 1994.

[2] T. I. Liu; J. H. Singonahalli; and N. R. Iyer. "Detection of roller bearing defects using expert system and fuzzy logic," Sep., Mech. Syst. Signal Process. Vol. 10, pp. 595–614, 1996.

[3] M. Schwabacher and K. Goebel. "A survey of artificial intelligence for prognostics," NASA Ames Research Center, pp. 107–114, 2006.

[4] S. Tyagi. "Acomparative study of SVM classifiers and artificial neural networks application for rolling element bearing fault diagnosis using wavelet transform preprocessing," World Acad. Sci. Eng. Technol., Vol. 43, pp. 309–317, 2008.

[5] X. Li; A. Zheng; X. Zhang; C. Li and L. Zhang. "Rolling element bearing fault detection using support vector machine with improved ant colony optimization," Oct., Measurement, Vol. 46, pp. 2726–2734, 2013.

[6] I. El-Thalji and E. Jantunen. "A summary of fault modelling and predictive health monitoring of rolling element bearings," Mechanical Systems and Signal Processing, Vol. 60-61, pp. 252–272, 2015.

[7] R. Yana; R. X. Gaob and X. Chenc. "Wavelets for fault diagnosis of rotary machines: A review with applications," Signal Processing, Vol. 96, pp. 1–15, 2014.

[8] R. Storn and K. Price. "Differential evolution – A simple and efficient heuristic for global optimization over continuous spaces," Journal of Global Optimization, Vol. 11, pp. 341-59, 1997.

[10] Y. Wang, Y.; Z. Cai and Q. Zhang. "Differential evolution with composite trial vector generation strategies and control parameters," IEEE Transactions on Evolutionary Computation, Vol. 15, No. 1, pp. 55-66, 2011.

CONDITION MONITORING OF A CYCLOID GEARBOX

Eric Bechhoefer
GPMS Inc.
40 Ridge Rd
Cornwall, VT 05753
eric@gpms-vt.com

Abstract: A cycloid drive for gearboxes allows for high reduction ratio and zero or very low backlash. The cycloid gear design is based on compression, whereas most gear interactions are based on shear. Further, the contract of a cycloid gear is typically subject to rolling forces vs. sliding, which are seen in traditional gearboxes. These features of a cycloid gearbox allow for high shock load capacity, high torsional stiffness, and quiet operation. This paper details the modeling required for correct configuration to perform analysis on the cycloid gearbox and then is demonstrated on a 51:1 ratio, run to failure test. This paper documents the sensitivity of standard condition indicators for gear/bearing during the run to failure test.

Keywords: Cycloid; Smart Sensor; vibration diagnostics; Resonance; Spectral Estimation; Model Bases Dynamics;

Introduction: There are many non-standard, condition monitoring application which use relatively unusual gearbox design. While most reduction (example: epicyclical) gearboxes have well-understood dynamics, other, such as a Variator (continuously variable transmission) or cycloidal gearbox, have not been reported on. This paper covers the dynamics, configuration and some test observation of work done on a cycloidal gearbox.

The interest in cycloidal gearbox is derived in that they are used in many applications were a low-cost drive motors is needed. For example, many conveyer belt systems (sorting, moving bulk media, slew drives) use cycloid drives. When the drive itself is a relatively low-cost asset, the process, which they support, can have significant economic impacts if they fail. For example, one of the larger courier delivery services has a distribution center with 3000 cycloid drives to move packages. The loss of a drive unit halts sorting, which could impact revenues of up to $200,000 per day.

The cycloid drive for gearboxes allow for high reduction ratio and zero or very low backlash. The cycloid gear design is based on compression vs. shear forces, where the contact is typically rolling vs. sliding. These features allow for high shock load capacity, high torsional stiffness, and quiet operation. Single stage ratios more than 200:1 are possible.

The gearbox chosen for the test was an integrated induction motor and gearbox. This gearbox is rated for 0.75 kW, approximately 1 Hp drive. For 60 Hz operations, using a four-pole motor, the drive unit has a 100% input shaft rate of approximately 1795 rpm. The gearbox a 51:1 reduction ratio. Figure 1 shows the typical arrangement of a cycloid gearbox.

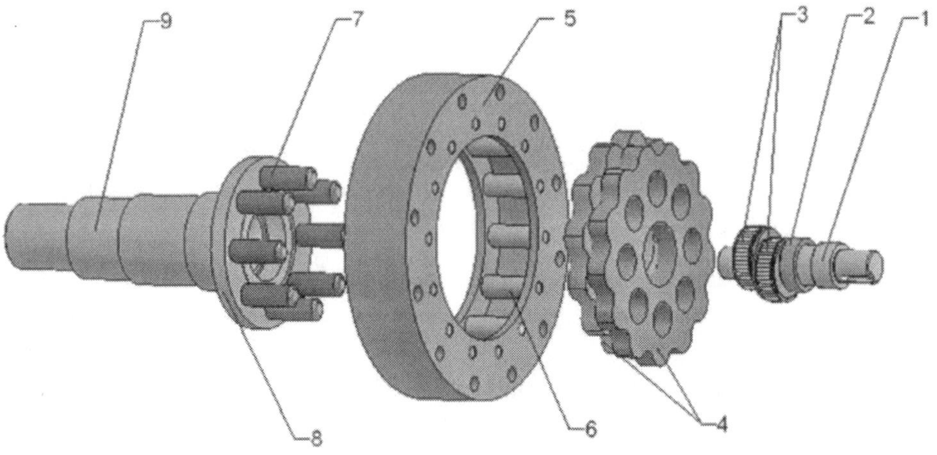

Figure 1. Typical Cycloid Gears (*copied from http://mechanicaldesign.asmedigitalcollection.asme.org/data/journals/jmdedb/27951/085001_1_1.jpeg)*

The main components of the gearbox are: 1) inputs shaft, 2) support bearing, 3) eccentric bearings, 4) the cycloid gears, 5) the pin teeth-case, 6) the pins, 7) output rollers, 8) output shaft, 9) support bearing.

The ratio for the gearbox is given as:

$$\text{ratio} = \frac{(n_{\text{teeth}}-1)\times n_{\text{pins}}}{n_{\text{teeth}}-n_{\text{pins}}} \quad \text{(eq 1)}$$

The test gearbox has dual disc (4) with 26 teeth and 51 pins.

Equations of Motion and Configuration: Configuration is driven by the equations of motions for the monitoring components. This consists of describing synchronous motion analysis of the shafts and gears, and the asynchronous motion of the bearings.

Bearing Motion: The simple input/output gearbox design uses three bearings on the input shaft: bearing D (the eccentric bearing) and input shaft bearing C. The output shaft was supported by two bearings: bearing B and bearing A.

The bearing rate fault frequencies are determined by the shaft rate, and:

- the number of rolling elements (*b*),
- the roller element diameter (*d*),
- the bearing pitch diameters (*e*), and
- the bearing contact angle (α).

The fault features are related to damage accumulated on the bearing itself.

There are typically six fault features calculated for the bearing associated with bearing elements: cage, ball, inner race, outer race. For mechanical looseness, the bearing may also generate signatures associated with whip/whorl (in the base spectrum) or a 1/revolution impact (tick) in the heterodyne analysis. The bearing feature rates are calculated as::

$$cage = 0.5(1 - d/e * \cos(\alpha)) \quad \text{(eq 2)}$$

$$ball = e/d\,(1 - (d/e)^2 * \cos(\alpha)^2) \quad \text{(eq 3)}$$

$$inner\ race = b/2\,(1 + (d/e) * \cos(\alpha)) \qquad \text{(eq 4)}$$
$$outer\ race = b/2\,(1 - (d/e) * \cos(\alpha)) \qquad \text{(eq 5)}$$

Because the outer rate of the eccentric bearing is in contact with the cycloid gear and the input shaft, the total rate seen by the bearing is the input shaft + output shaft. The eccentric bearing analysis was assigned to the input shaft. To capture the effect of the relative motion, the bearing rates were multiplied by 1 + 1/51 = 1.0196. This speed correction is needed in order to determine the correct bearing rate fault features.

Shaft/Gear Motion: Shaft and gear analyses are based on the time synchronous average, which requires an accurate rate from the tachometer. The tachometer is used to resample the vibration data and correct for any changes in shaft rate. Gear analysis, and more importantly gear mesh tones, is a function of the shaft rate and the number of teeth on the gear. A conventional input shaft with a 29.23 Hz rate with 26 teeth would have a gear mesh frequency of 29.23 x 26 = 759.96 Hz. However, the cycloid gear has a relative motion to the shaft, driven by the eccentric gear and the output shaft. The relative motion of the cycloid to the ring gear means that, for each revolution, there is one extra gear mesh. Hence, the actual gear mesh frequency would be 789.19 Hz. Figure 2 shows the corresponding TSA order of 27 (26 teeth + 1) and the accompanying raw spectrum (in Hz).

For this reason, gear analysis is based on 27 teeth, not 26 teeth.

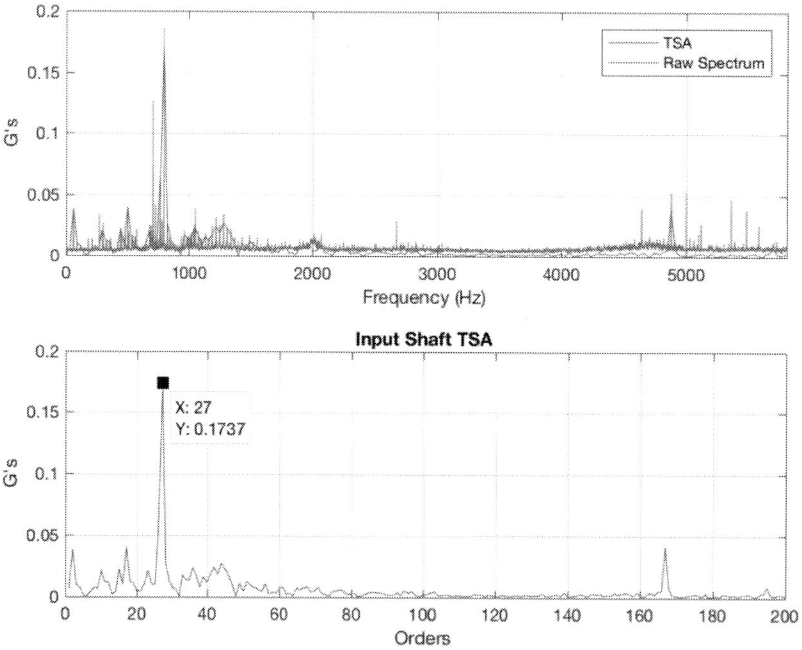

Figure 2, Comparison of the Raw to TSA Spectrum for the input shaft.

The ring gear analysis would usually be associated with the number of ring gears teeth, but because there are two cycloid gears (of 26 teeth), the measured mesh is 51 x 2 or 102 mesh. The TSA spectrum and raw spectrum then show frequencies at 29.23/51 x 102 = 58.46 Hz. Due to the modulation of two disc, there are sidebands at 102 +/-51 = 51 and 153, or 29.22 and 87.69 Hz, as seen in Figure 3.

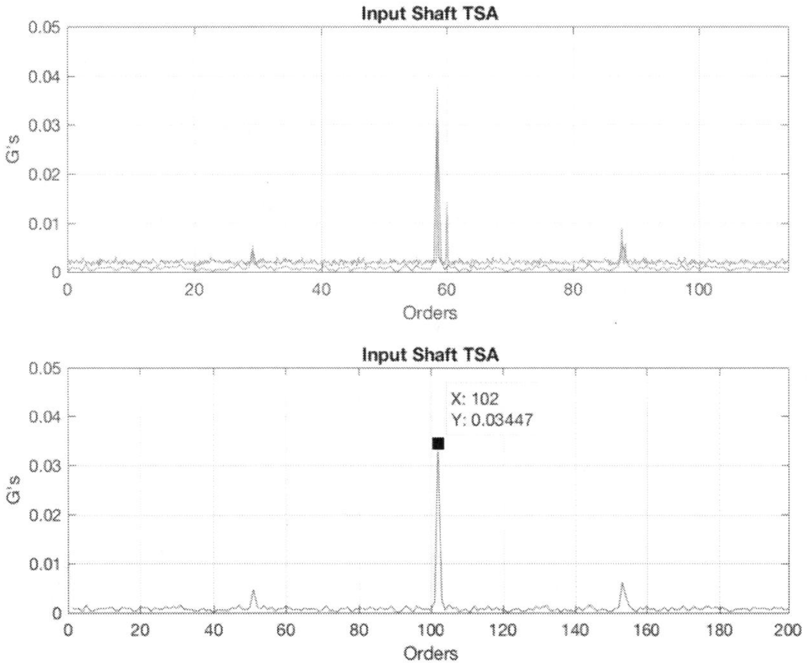

Figure 3, Comparison of the Raw to TSA Spectrum for the output shaft.

Window Selection for the Envelope Analysis: The envelope analysis is based on demodulation of high-frequency resonance, which is based on impact. Poor selection of a window results in poor envelope/bearing analysis. Initially, with no faulted bearings to test, a "rap" test is conducted to excite a modal response. This is sampled, and the spectrum calculated, (see Figure 4 below).

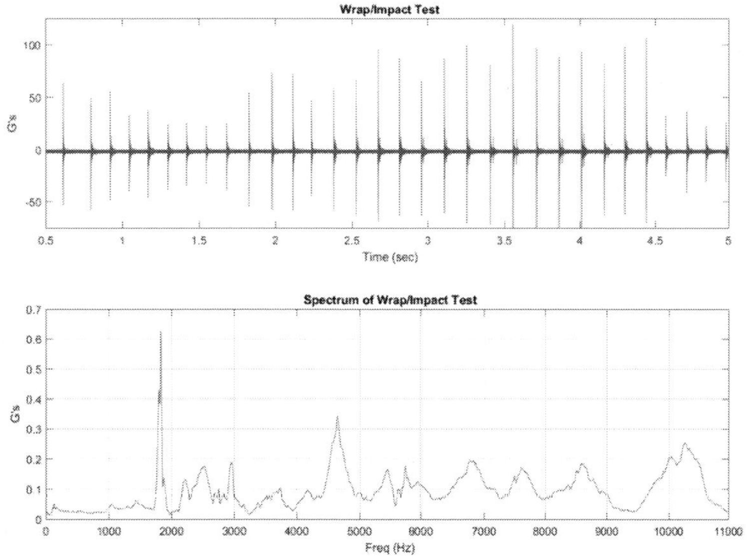

Figure 4. Impact and resulting modal response.

Window selection is based on a sample rate of the sensor. The sample rate also affects the length of the TSA:

$$TSA\ length = 2^{ceil(log_2(Sample\ Rates/Shaft\ Rate))} \quad \text{(eq 6)}$$

Given the low output shaft rate of approximately 0.57 Hz, the measured acceleration will be low. For this reason, the acquisition length must be adequately long to capture perhaps 20 revolutions. Hence, a high sample rate taken over an extended period results and a very large data set, which takes more time to process and to download raw data (if needed).

For this reason, the sample rate of the output shaft was taken at 2930 sps, for 60 seconds. Note: at 0.57 Hz, this is only 34 revolutions. The TSA length is then 8192. For the input shaft, which is close to 30 Hz, only eight seconds of data was taken at 23438. This allows a Nyquist frequency of 1465 Hz for the output shaft, and 11719 for input shaft. From the model response in Figure 4, the window for output shaft analysis was taken at 300 to 1300 Hz, which covers the small resonant mode at 1000 Hz. For the input shaft, the window was taken from 9000 to 11000 Hz, covering the modal response at 10000 Hz (Figure 5).

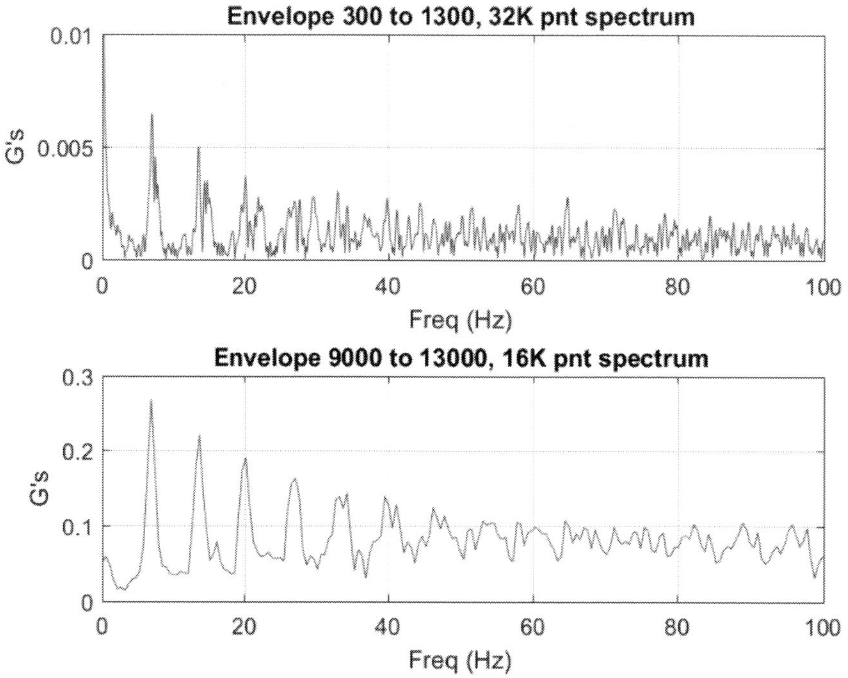

Figure 5. Envelope Response from Impact

Gear Feature Analysis: The following CIs were used on the cycloidal gear:

Residual RMS, Residual Kurtosis, Residual Crest Factor, Energy Ratio, Energy Operator Kurtosis, Energy Operator Crest Factor, Figure of Merit 0, Side Band Lifting Factor, Side Band Analysis, Narrow Band Kurtosis, Narrow Band Crest Factor, Amplitude Modulation RMS, Amplitude Modulation Kurtosis, Frequency Modulation RMS, Frequency Modulation Kurtosis, Gear Mesh Energy [2].

Health Indicator Paradigm: In order to reduce maintainer workload, and improve fault detection, the concept of the health indicator was introduced in [2]. In general, the Health Indicator (HI) is a function of distributions of *n* CIs, where the HI is scaled such that:

- HI ranges from 0 to a positive value, where the probability of exceeding an HI of 0.35 is the PFA (probability of false alarm),
- A warning (yellow) alert is generated when the HI is greater than or equal to 0.75. Maintenance should be planned by estimating the RUL until the HI is 1.0.
- An alarm (red) alert is generated when the HI is greater than or equal to 1.0. Continued operations could cause collateral damage.

Note that this nomenclature does not define a probability of failure for the component, or that the component fails when the HI is 1.0. Instead, it suggests a change in operator behavior to a proactive maintenance policy: perform maintenance before the generations of collateral or cascading faults. For example, by performing maintenance on a bearing prior the bearing shedding extensive material, costly gearbox replacement can be avoided.

Hence, from a maintainer perspective, this is a stoplight-based threshold setting/alerting system: when a component is yellow, plan maintenance, and when the component turns red, do maintenance.

Controlling for the Correlation Between CIs: All CIs have a probability distribution (PDF). Any operation on the CI to form a health index (HI) is then a function of distributions. The HI function is defined as The norm of *n* CIs (energy):

$$HI = {}^{0.35}/_{crit}\sqrt{\mathbf{Y}^T\mathbf{Y}} \qquad \text{(eq 7)}$$

where **Y** is the whitened, normalized array of CIs, and *crit,* is the critical value.

The function is valid if and only if the distribution (e.g., CIs) are independent and identical (e.g., IID). As an example, for Gaussian distribution, subtracting the mean and dividing by the standard deviation will give identical Z distributions. The issue of ensuring independence is much more difficult. In general, the correlation between CIs is non-zero. For instance, Table 1 shows the correlation coefficients for 6 CIs used for gear fault analysis: most are statically significant.

Table 1: Correlation Coefficients for the Six CIs Used in the Study

ρ_{ij}	CI 1	CI 2	CI 3	CI 4	CI 5	CI 6
CI 1	1	0.84	0.79	0.66	-0.47	0.74
CI 2		1	0.46	0.27	-0.59	0.36
CI 3			1	0.96	-0.03	0.97
CI 4				1	0.11	0.98
CI 5					1	0.05
CI 6						1

This correlation between CIs implies that for a given function of distributions to have a threshold that operationally meets the design PFA, the CIs must be whitened (e.g., de-correlated). A whitening solution can be found using Cholesky decomposition. The Cholesky decomposition of Hermitian, positive definite matrix results in $\mathbf{A} = \mathbf{LL}^*$, where **L** is a lower triangular, and $\mathbf{L}^*$ is its conjugate transpose. By definition, the inverse covariance is positive definite Hermitian. It then follows that if:

$$\mathbf{LL}^* = \mathbf{\Sigma}^{-1}, \text{ then } \mathbf{Y} = \mathbf{L} \text{ x } \mathbf{CI}^T \qquad \text{(eq 8)}$$

The vector **CI** is the correlated CIs used for the HI calculation, and **Y** is 1 to *n* independent CI with unit variance (one CI representing the trivial case). The Cholesky decomposition, in effect, creates the square root of the inverse covariance. This, in turn, is analogous to dividing the CI by its standard deviation (the trivial case of one CI). In turn, it can be shown that $\mathbf{Y} = \mathbf{L} \times \mathbf{CI}^T$ creates the necessary independent and identical distributions required to calculate the critical values for a function of distributions.

Finding the Critical Value: The critical value is taken from the inverse cumulative distribution function for the HI. The CIs used have Rayleigh like PDFs (e.g., heavily tailed). For magnitude based CIs, it can be shown that for the nominal case, the CI PDF is Rayleigh. For Gear CIs, and Bearing CIs (where magnitudes which are biased by RMS), a transform is used by make the CI more Rayleigh like. The transform "left shifts" the CI. For example, a shift such that the .05 CDF (cumulative distribution function) is assigned to 0.0.

Consequently, the HI function is based using the Rayleigh distribution. The PDF for the Rayleigh distribution uses a single parameter, β, defining the mean $\mu = \beta^*(\pi/2)^{0.5}$, and variance $\sigma^2 = (2 - \pi/2) * \beta^2$. The PDF of the Rayleigh is: $x/\beta^2 \exp(x/2\beta^2)$. Note that when applying these equations to the whitening process, the value for β for each CI will then be: $\sigma^2 = 1$, such that: $\beta = \sigma^2 / (2 - \pi/2)^{0.5} = 1.5264$.

For the HI equation in (7), the normalized energy of the CIs, it can be shown that the function defines a Nakagami PDF [2]. The statistics for the Nakagami are $\eta = n$, and $\omega = 1/(2-\pi/2)^*2^*n$, where n is the number IID CIs used in the HI calculation.

Test Stand Results: Using a nominal gearbox, run at approximately 50% load for 45 hours. Acquisitions were taken every 5 minutes, Figure 6.

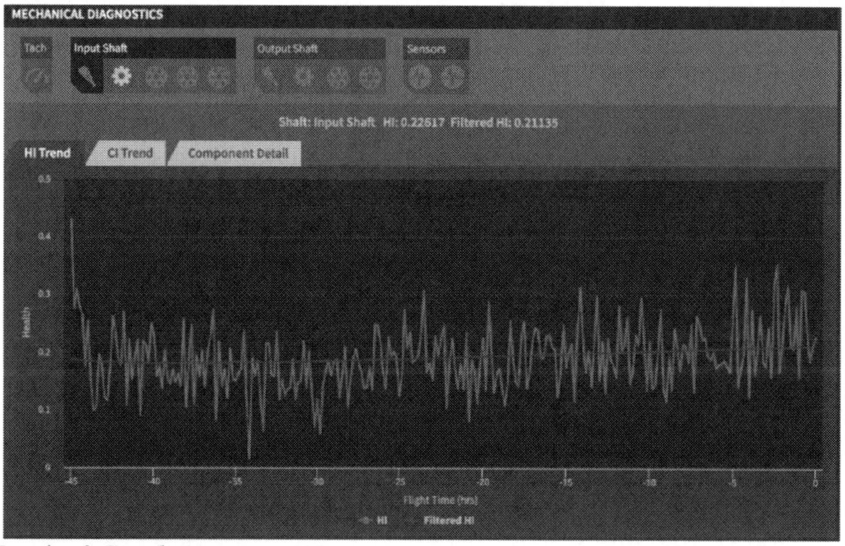

Figure 6. Nominal Gearbox

Initiation of Run to Failure: After the initial test run to set threshold, the gearbox was run at 150% for one hour, then at 100% until failure. Note the resulting substantial step change. The step-change is seen on most components. It is likely that the eccentric bearing was damaged during the overload (Figure 7, 8).

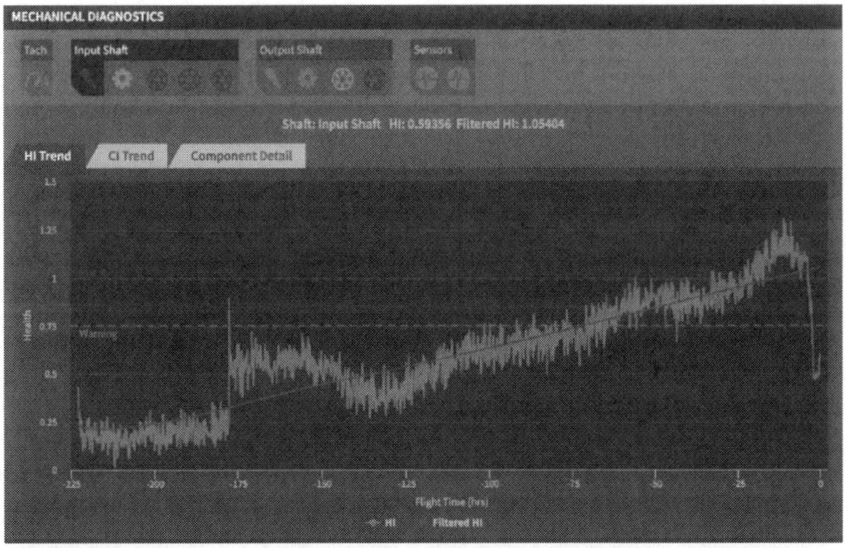

Figure 7, Input Shaft driven by 1/Rev. Step change occurs during overload.

While the HI for both bearing C and D increase drastically (input shaft), it seems this is more a function of the generalized wear in the gearbox. This wear caused a low-frequency resonance within the gearbox, confounding the gear analysis for bearings A and B (Figure 9).

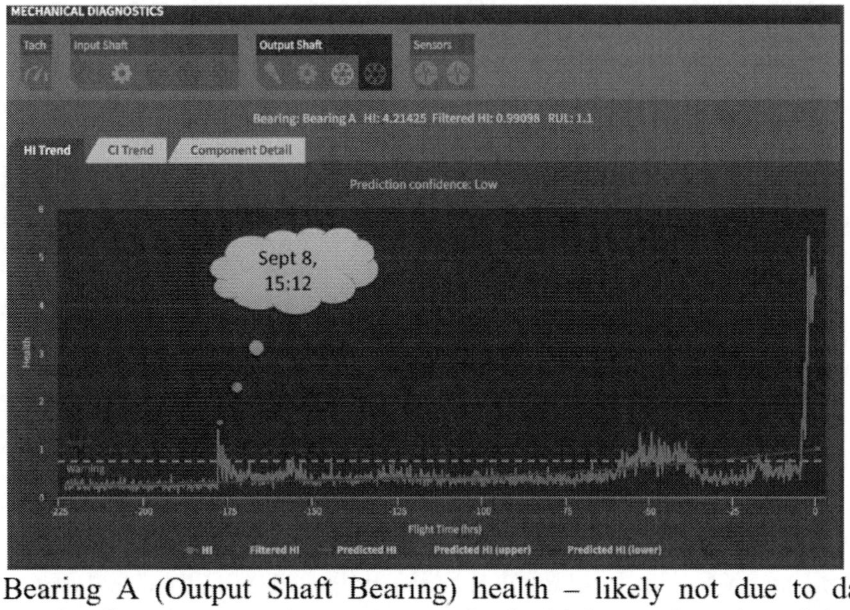

Figure 8, Bearing A (Output Shaft Bearing) health – likely not due to damage but measuring a noise floor increase due to damage in the high-speed section of the gearbox.

General Comments on the Run to Failure Test: The large number of failed components, as indicated by most of the components being in Alarm, suggests a high level of generalized wear. The high-speed input shaft shows an increase in SO1 (imbalance). It is suspected that there was damage to bearing C or D cage (output shaft bearings), allowing the shaft to no longer be centered. Again, this may have occurred during the overload (Figure 6).

The evidence to suggest that neither bearing A, nor B (output shaft) were damaged is that all four bearing components (Cage, Ball, Inner Race, Outer Race) are correlated – no individual component indicating damage. This supports the hypothesis that the noise floor for the envelope analysis was increased due to damage in some other part of the gearbox.

The Eccentric bearing showed bearing wear due to mechanical looseness (as evident by the 1/Rev "Tick" at 29.75 Hz and a Ball failure, approximately 24 hours before failure (Figure 9)

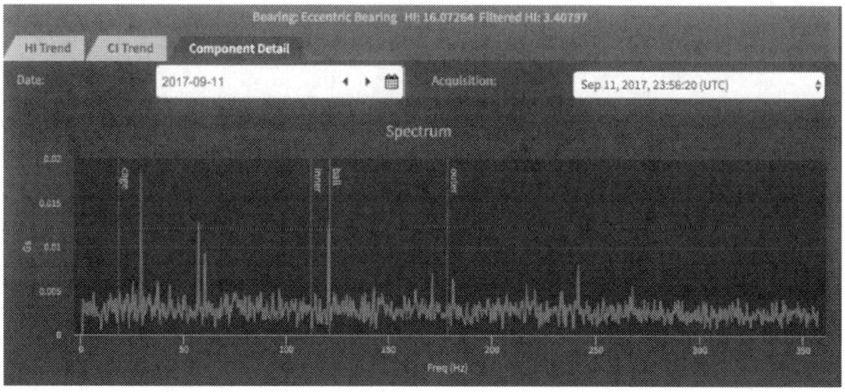

Figure 9, Eccentric bearing Envelope analysis, one day before failure. Note 1/Rev (29 Hz) tone indicating mechanical looseness.

The unfiltered bearing health eventual spiked at a level of 100, one hour before failure. It is not unusual to see a dip in bearing fault feature energy with severe faults. This is due to the ball elements no longer rolling, but sliding – hence the slip is increased, and the frequency of the bearing fault feature drops, exceeding the search window of the bearing energy algorithm (Figure 10).

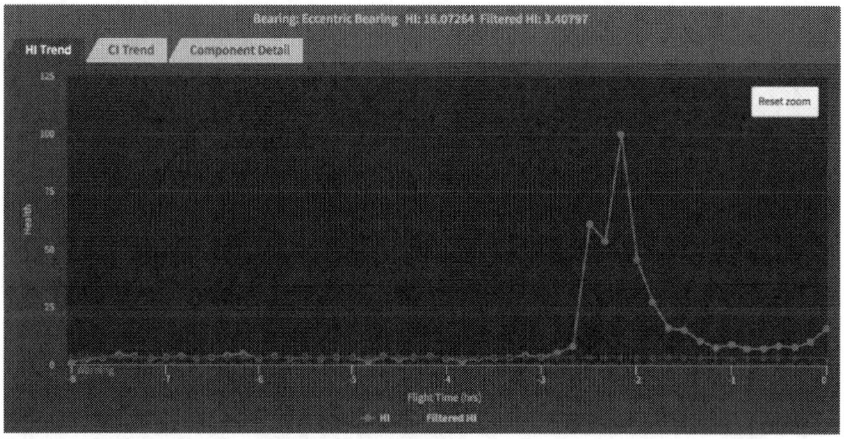

Figure 10. Eccentric bearing health spikes at 100.

The cycloid gear shows in alarm level gear mesh, from 50 to 20 hours before failure. Gear mesh energy is not usually a consistent indicator (Figure 11).

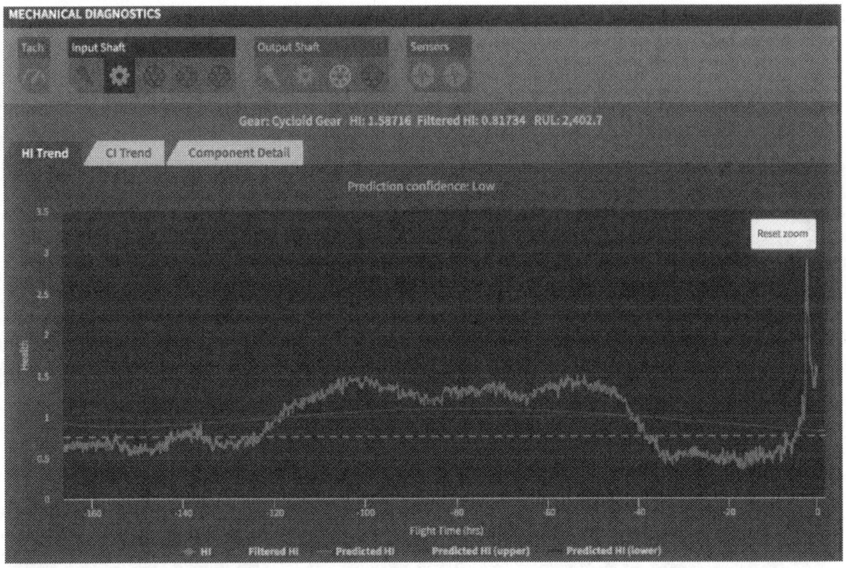

Figure 11. Cycloid Gear Health.

Note that from 20 hours before failure, Residual RMS, Energy Ratio, and FM0 is sensitive to the impending fault. From this, it was learned that the best five indicators for the cycloid gear health: Residual RMS, Energy Ratio, FM0, AM Kurtosis, and Gear Mesh. This suggests that during the last 10 to 20 hours of the run, the cycloid gear experienced a second failure mode that was detected by the more traditional gear faults (Figure 12).

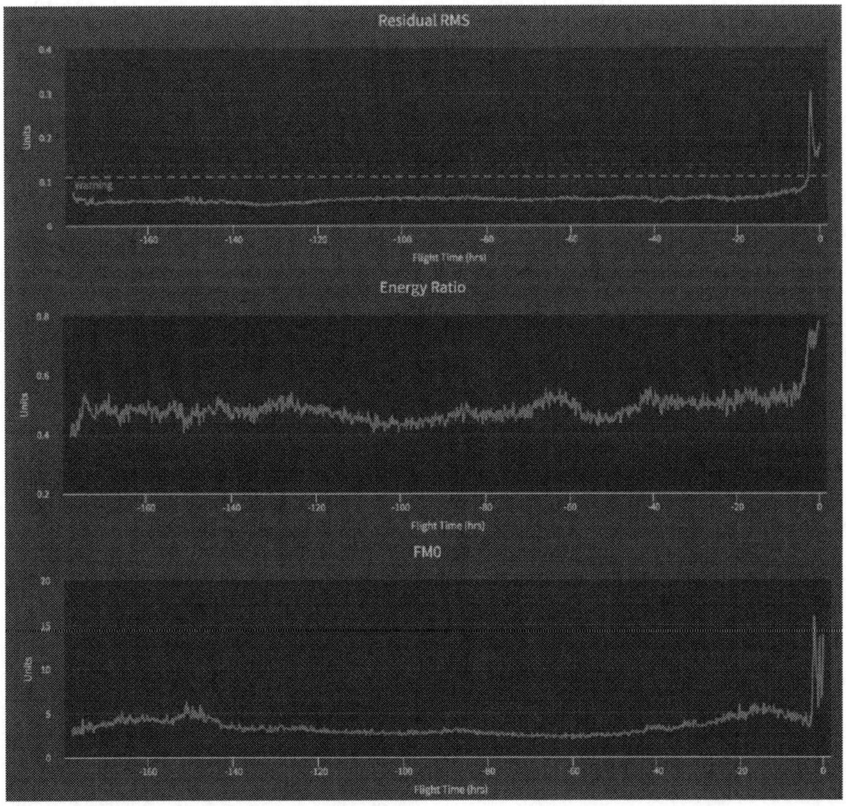

Figure 12, Cycloid gear condition indicators

The ring gear showed damage as well. Because the gear health is based on the filtered HI, it does not reflect the step change HI of 2 in the raw HI data just prior to failure (Figure 13)

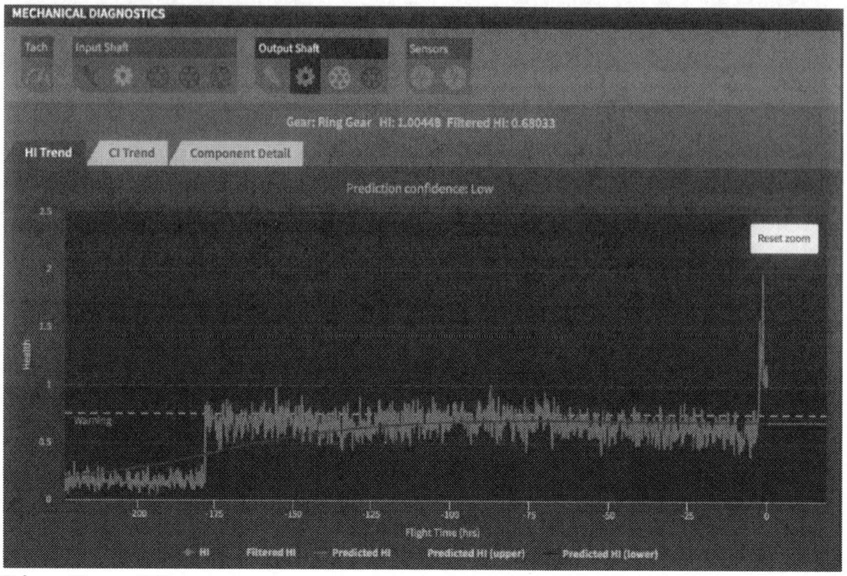

Figure 13: Ring Gear HI.

The gear CIs that were sensitive to this fault was found to be: Residual RMS, Energy Ratio, FM0, AM Kurtosis, and Gear Mesh. These condition indicators will be used in the future.

Conclusion: The cycloid gearbox has unique dynamics, which require using the correct ratios for the TSA and bearing rate calculation. The bearing analysis for the Cycloid gearbox is relatively standard. The eccentric bearing rate was multiplied by a correction factor to account for the rotating frame of the outer race. The eccentric bearing is the weak point of the gearbox. During the run to failure test, the eccentric bearing roller elements were found to be faulted. Towards the end of the run to failure test, the high level of damage (resonance energy) associated with the eccentric gear raised the noise floor of the envelope analysis, which contributed to high HI level of all the bearings in the gearbox.

The cycloid gear itself posed a challenge in that the apparent gear mesh frequency is based on the gear eccentric behavior and not on the gear mesh frequency alone. The observed gear mesh frequency is the gear tooth + 1 vs. gear tooth, multiplied by the shaft rate.

Because the gears are under compression vs. shear, care must be used in the CI selection used for the health indicator, as many of the traditional analyses were not sensitive. It was found that residual RMS, amplitude modulation RMS and gear mesh energy for gear tooth + 1 the shaft order worked well.

References:

[1] Zhu, J., Nostrand, T., Spiegel, C., Morton, B., "Condition Indicators for Condition Monitoring Systems," Annual Conference of the Prognostics and Health Management Society, 2014

[2] Bechhoefer, E., He, D., Dempsey, P., "Gear Health Threshold Setting Based On a Probability of False Alarm," Annual Conference of the Prognostics and Health Management Society, 2011.

Bibliography:

[1] Eric Bechhoefer is the president of GPMS, Inc., a company focused on the development of low-cost condition monitoring systems. Dr. Bechhoefer is the author of over 100+ juried papers on condition monitoring and prognostics health management and holds 23 patents and five patent applications in the field of CBM.

THE CASE OF THE MISSING BEARING FAULT FREQUENCY

Eric Bechhoefer
GPMS Inc.
40 Ridge Rd
Cornwall, VT 05753
Telephone: (937) 256-2285
eric@gpms-vt.com

Abstract: During testing on a small gas turbine, frequencies related to the power turbine shaft and a bearing were observed. The frequencies of the fault were not associated with any known bearing on the power turbine shaft. This paper is an investigation of why the observed bearing fault frequency of 6% higher than anticipated. It will be shown that because the faulted bearing was a worn thrust bearing, the contact angle and pitch diameter of the roller element had changed. This is an infrequently observed phenomenon, which can lead to missed fault detection on a critical component. A mitigation strategy for this type of failure is discussed.

Keywords: Spectral Kurtosis; vibration diagnostics; resonance; performance testing; hardware testing

Introduction: Using two MEMS accelerometer-based smart sensors along with tachometer inputs to supply rotational speed, a blind, seeded fault test was conducted to generate condition indicators for 10 shafts, 12 gears and 26 bearings for the Rolls Royce M250-C47 gearbox. The seeded fault was an unknown bearing installed and run on Day 1, while a nominal bearing replaced the faulted bearing for a run on Day 2. The data acquisition system collected 31 acquisitions on day 1, and 32 acquisitions on day 2. Acquisitions included both processed (condition indicator data) and raw vibration and tachometer data.

The spectral response from the two sensors indicated content up to the bandwidth of the sensors (32 KHz). This far exceeded the estimated modal response of the gearbox (by rap test). Validation of the configuration was performed with known frequencies of interest (corresponding to shaft rates and gear mesh rates) to ensure Condition Indicator (CI) values matched spectral data.

Based on the observed spectral content, the envelope configuration was set from 18 to 25 KHz. A consistent frequency of 4029 Hz was then seen on day 1 and not on day 2. It is hypothesized that this frequency was associated with the power turbine shaft, and further, that the fault feature was from an ball element fault. The 4029 Hz frequency would correspond to an outer race rate of 7.58, which did not match any know bearing fault rate on the engine – the closest being an roller element rate of 7.05. The issue is: why is there such a large disparity between the measures and published bearing data?

Review of Bearing Envelope Analysis: Rotating equipment in general and equipment such as helicopters are dependent on a transmission to condition the power for useful work. In the case of helicopters, the power of a high speed, low torque power shaft is converted into a low speed, high torque main rotor shaft, which delivers lift. Integral to the reliable operation of this transmission, are bearings. Reliability and availability of the helicopters are improved if monitoring techniques are developed, which can detect when a degrading or faulty bearing requires maintenance. This is the essence of condition monitoring for bearings.

Because the vibration signals of a faulty bearing are small compared to shaft order and gear mesh frequency, detection of faults at the bearing rate frequencies using Fourier analysis is difficult. Fault detection at the baseband frequencies of the bearing rate is "Stage 1" fault detection. Bearing faults detected using these types of analysis are Late Stage; the bearing can be close to catastrophic failure. At the very least, a bearing in this state is generating metal, which can cause damage to other components within the gearbox.

Ultrasonic emission can detect bearing inner and outer race roughness (a "Stage 3" fault), but the remaining useful life of a bearing at this stage is relatively long compared to the overall life of the bearing. Bearing envelope analysis (BEA) can typically detect bearing faults 100s if not 1000s of hours before when it is appropriate to do maintenance. It is for this reason that many conditions monitoring systems manufacturers are using envelope analysis techniques.

Bearing Envelope Analysis: BEA is based on demodulation of high-frequency resonance associated with bearing element impacts. For rolling element bearings, when the rolling elements strike a local fault on the inner or outer race, or a fault on a rolling element strikes the inner or outer race, an impact is produced. These impacts modulate a signal at the associated bearing pass frequencies, such as associated with the Cage (FTF, fundamental train frequency),

$$FTF = \frac{S}{2}\left(1 - {}^{Bd}/_{D}\, cos(\phi)\right) \qquad \text{(eq 1)}$$

or the Ball Pass Frequency Inner Race (BPFI),

$$BPFI = \frac{Nb \times S}{2}\left(1 + {}^{Bd}/_{D}\, cos(\phi)\right) \qquad \text{(eq 2)}$$

or the Ball Pass Frequency Outer Race (BPFO),

$$BPFO = \frac{Nb \times S}{2}\left(1 - {}^{Bd}/_{D}\, cos(\phi)\right) \qquad \text{(eq 3)}$$

and the Ball Pass Spin Frequency (BSF),

$$BSF = \frac{Pd \times S}{Bd}\left(1 - \left({}^{Bd}/_{D}\right)^{2} cos(\phi)^{2}\right) \qquad \text{(eq 4)}$$

Where:

S is the shaft frequency,
Bd is the ball or roller element diameter
Nd is the number of balls or rollers
D is the pitch diameter and
Φ is the contact angle.

Note that the BSF is usually constructed as half of the rate (divided by two), but because a spall or damage to the roller hits both the outer and inner trace, the observed frequency is 2x, or as given in eq 2.

To illustrate BEA, Figure (1) is an example of an outer race fault, where the BPFO is approximately 80 Hz. Note that the modulation rate, T1, is approximately .0125 seconds (e.g., 1/80 Hz). The time T2, the period of the resonance, is approximately 1.12e-4 seconds, or about 9000 Hz. Note that the time domain representation is the superposition of many resonances of the bearing itself.

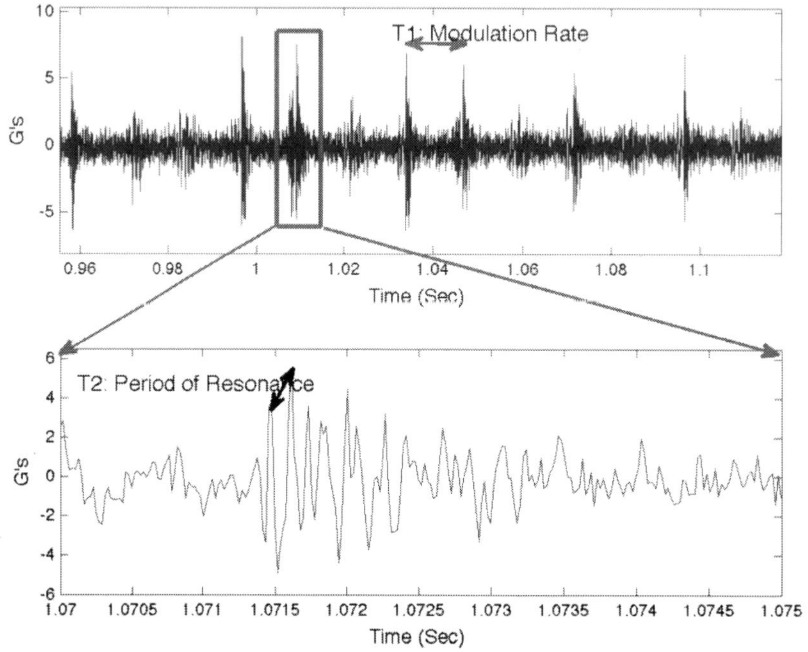

Figure 1 Example Outer Race Fault

Mathematically, the modulation is described as:

$$\cos(a) \text{ x } \cos(b) = \tfrac{1}{2} [\ \cos(a - b) \text{ x } \cos(a + b)] \qquad (\text{eq } 5)$$

This is an amplitude modulation of the bearing rate (a) with the high-frequency carrier signal (resonant frequency (b)). This causes sidebands in the spectrum surrounding the resonant frequency. It is sometimes difficult to distinguish the exact frequency of the resonance. It is usually not known *a priori* and cannot be determined easily without a faulted component. However, demodulation techniques typically do not need to know the exact frequency. One method for the BEA involves multiplying the vibration signal by a resonant frequency (example, 9 kHz). This is then low pass filtered to remove the high-frequency image, decimated, and the spectral power density is estimated (Eq 6)

$$\cos(b) \text{ x } \cos(a + b) = \tfrac{1}{2}[\cos(a+b-b)+\cos(a+b+b)] \text{ -> } H(\omega\Downarrow)\text{ -> }\cos(a) \qquad (\text{eq } 6)$$

An alternate way to generate the envelope is to take the FFT (fast Fourier transform) of the time domain signal and calculate the frequency of each bin (e.g., sample rate/length of time series). Calculate the "window" indexes (e.g., the range of frequencies in which the bearing is resonating, say 8 to 9 kHz). Copy the Fourier coefficients from the desired window (e.g., 8kHz to 9 kHz) into the baseband (e.g., 0 to 1 kHz) indexes (this is a Heterodyne operation). Then copy zeros into the remaining Fourier coefficients (Fourier filtering and Hilbert transform, recalling that the Hilbert transform is defined at the real frequency of the Fourier domain, see Ref 1). Finally, take the absolute value of inverse FFT Fourier coefficients. A Matlab® function is given as:

```
function [env,dty] = envelope( data, dt, lowf, highf)
% [env,dty] = envelope1(data,dt,nfilt,lowf,highf);
%Inputs:
% data        :data vector, time domain
% dt          :sampling time interval
% lowf        :low frequency limit of the bandpass filter
% highf       :high frequency limit of the bandpass filter
%Outputs:
% env         :Envelope of data
% dty         :decimated sample rate

n = length(data);
dfq = 1/dt/n;
idxLow = floor(lowf/dfq);        % get the bin for the window low
idxHi  = ceil(highf/dfq);        % get the bin for the window high
D = fft(data);
idx = idxHi-idxLow + 1;

D(1:idx)  = 2*D(idxLow:idxHi);  % this is the definition of the
D(idx+1:end)  = 0;               % Hilbert Transform
data = abs(ifft(D));
bw = highf - lowf;
r = fix(1/(bw*2*dt));           % Decimating after low pass
filtering
env =  data(1:r:n);              % crop first nfilt elements
dty = dt*r;                      % New sample rate after decimation
```

The bearing components have many vibration modes, which will correspondingly generate resonance at various frequencies throughout the spectrum. The selection of the frequency range used to demodulate the bearing rate signal (e.g., the window center frequency) should take into account some issues: First, the gearbox spectrum contains many high-energy frequencies from shaft and gear harmonics, which would mask analysis at lower bearing frequencies. Second, there are a number of accelerometers with natural resonance at frequencies that are similar to the bearing modes. Using a higher frequency window close to the accelerometer resonance can amplify the bearing fault signal, increasing the probability of fault detection.

BEA should be performed at frequencies higher than the shaft and gear mesh frequencies. This ensures that the demodulated bearing frequencies are not masked by the other rotating sources, such as shaft and gear mesh, which are present at FTF, BPFO, BPFI, and BSF frequencies. Typical shaft order amplitudes of 0.1 G's and gear mesh amplitudes of 10s of G's are typical. Damaged bearing amplitudes are 0.001 G's.

A technique used to identify an appropriate envelope window is Spectral Kurtosis (ref 2), where spectral kurtosis (SK) is used to determine the best window to determine the condition of a test gearbox with seeded faults. Kurtosis is a non-dimensional quantity that measures the relative "peakedness" of a distribution relative to the Gaussian distribution. Spectral kurtosis (SK) is a statistical parameter indicating how the impulsiveness of a signal varies with frequency. As noted, faults associated with rolling element bearings give rise to short impulse. The SK will be significant in frequency bands where the fault signal is dominant and small where the spectrum is dominated by stationary signals. Antoni developed the kurtogram, which is a map indicating the optimum center frequency and bandwidth combination.

In this kurtogram example, filtering was limited between 1 kHz and 24.4 kHz (Nyquist). The lower limit was chosen because typically, below 1 kHz, there are significant gear mesh frequencies. The kurtogram frequency space spans eight octaves, such that frequency band is approximately halved with each increase in octave.

TABLE I. KURTOGRAM FREQUENCY MAP

Octave	Bandwidth	Number of Bands	PSD Window
1	12.2 kHz	2	32768
2	8.1 kHz	3	16384
3	6.1 kHz	4	8192
4	4.1 kHz	6	4096
5	3.05 kHz	8	4096
6	2.03 kHz	12	2048
7	1.5 kHz	16	1024
8	0.76 kHz	32	512

Figure 2 shows the raw spectrum and the Kurtogram. The largest kurtosis value was in Octave 5, or a window of 1 to 4 kHz. This is compared to the second example window in Octave 8, from 2.51 kHz to 3.266 kHz.

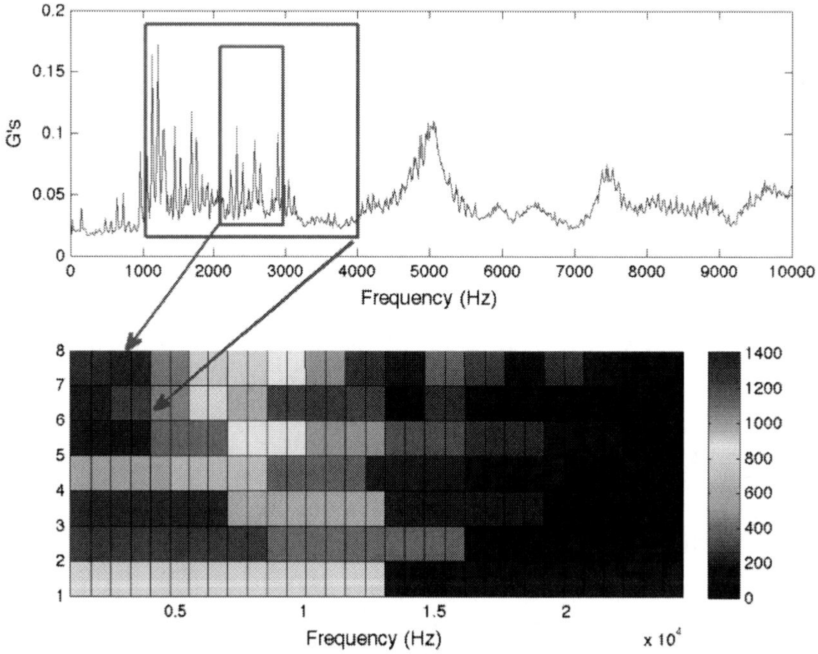

Figure 2 Kurtogram example

The envelope analysis spectrum for Window 1 and Window 2 is plotted in Figure 3. Note that a 3rd envelope analysis is also demonstrated for a poor window: 15 to 18 kHz (band 5, Octave 4). The SNR for the Window 1 was 23 dB, Window 2: 15 dB, and Window 3: 10 dB. There have been many reported cases where poor window selection has resulted in a missed detection of a faulted bearing.

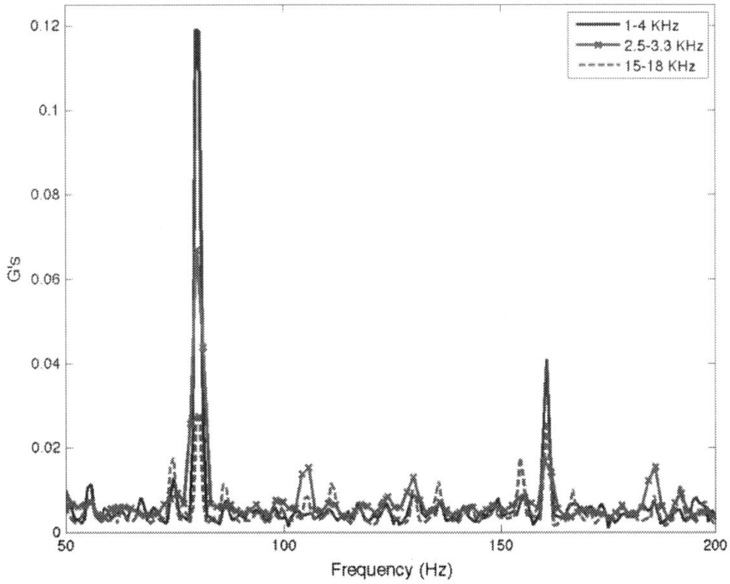

Figure 3 Example of the effect of the Envelope Window on the Spectra

The Test Article: The test article was a small, gas turbine engine used to power a helicopter. The power turbine shaft rate was: 32,184 rpm or 536.4 Hz at 100%. This type of engine typically runs at 55% for idle, and at 100% for flight. Power is delivered from the turbine shaft via a gearbox to the power output drive via a gearbox integrated into the engine. Four bearings support the power turbine shaft. The bearing of interest has the following rates: 231, 3779, 5782, and 4409 Hz for the *FTF, BSF, BFPI,* and *BPFO.*

The envelope analysis frequencies were chosen as 18 to 25 kHz, while the spectral density window as 8192. For a typical run, the envelope response was seen as in Figure 4. Note that there is a great deal spectral content, with energy associated with both the shaft (at 536 Hz) and cage at 230 Hz. However, the other spectral content at 3493, 4029 and 4565 Hz does not match any of the calculated bearing fault frequency. Curiously, it was noted that that 3493 and 4565 is -/+ 536 Hz, which was the shaft rate. This seems to indicate sidebands of the bearing fault modulated by the shaft, but 4029 Hz is almost 6.5% higher than the ball fault frequency, as calculated in eq 2 (see Figure 4).

Further investigation showed that these frequencies were associated with the power turbine, that they were present only when the engine was above 65% torque. The frequency of 4029 Hz was too high a frequency to be associated with any other bearing other than the bearing of interest. The problem was, the fault frequency did not match the published bearing data. What is the cause of this discrepancy? One clue is that this is a thrust bearing.

From (ref 3) it is seen that for rigidly mounted bearings incapable of radial deformation, the contract angel β due to a thrust load can be written as:

$$\beta = cos^{-1}\left({}^{D - P_d/2}/_{D + \delta} \right) \quad \text{(eq 7)}$$

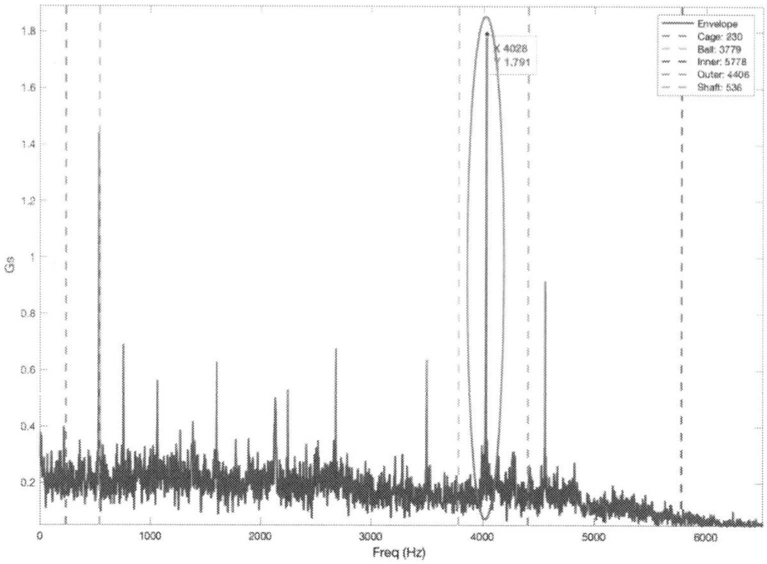

Figure 4 Envelope Spectrum of a Turbine Engine

Where D is the pitch diameter, P_d is the diametral clearance, and δ is total elastic deformation or wear. From (ref 4) it is given that:

$$\cos(\beta_f) = {D - P_d/2}/{D} \qquad \text{(eq 8)}$$

which is the manufactures contract angle β_f of 14.7 degrees. This allows one to calculate P_d as 0.0183. Rearranging terms gives:

$$\delta = cos^{-1}\left({cos(\beta_f)}/{cos(\beta)}\right) \qquad \text{(eq 9)}$$

The thrust loads axial deflection, δ_t, give in in 3, is then:

$$\delta_t = (D + \delta)sin(\beta) - Dsin(\beta_f) \qquad \text{(eq 10)}$$

and finally:

$$\delta_t = {D\ sin(\beta - \beta_f)}/{cos(\beta)} \qquad \text{(eq 11)}$$

Given that this is a thrust bearing, one can hypothesize that the turbine axial load is causing a displacement δ_t, which is both changing the contact angle of the bearing from β_f to β and the pitch diameter from D to $D+\delta_t$. This change in contact angle and apparent pitch diameter then *increases* the observed frequency. In our case, from 3779 Hz to 4028 Hz, as per Figure 5.

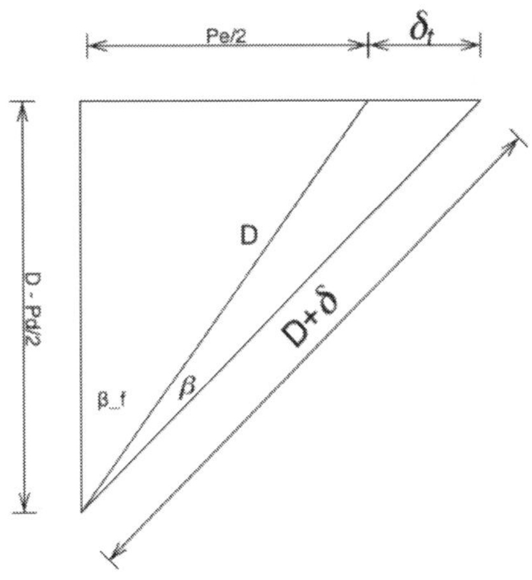

Figure 5 Change in Pitch Diameter and Contact Angle due to thrust load

If the contact angle changed from 14.7 to 18.16 degrees, the δ_t becomes 0.128. This intern suggests the apparent pitch diameter under load is: 2.129. The calculated *BSF* becomes the desired 4029 Hz.

A Mitigation Strategy: These phenomena will only be observed on thrust bearings, and generally only under load. Further, it is suspected that due to the age of the test article, seen only on system that are worn or towards end of life. While a rare event, a mitigation strategy is needed to allow for fault identification.

Typically, diagnostic systems, which rely on RMS or other, gross time domain identification, will not work on these complex systems. For example, the peak-to-peak measured G on the test article was close to 250 Gs. This is because there are a number of gear mesh interactions at different rates, and in time domain, the measured signal is the superposition of all vibration sources. While each gear mesh may be 5 or 10 Gs, there are 12 gears in the gearbox, and the measured, time domain peak to peak is the superposition of all signals.

Instead, the mechanization for diagnostics in complex gearboxes is based on configuration for a known fault frequency. Because bearings have non-Herztian contact, there tends to be some slip. As such, the calculated frequencies (eq 1, 2, 3 and 4) tend to be 0.5 to 1 percent low. Hence, a system does a local, directed search around the anticipated fault frequencies and picks the peak energy in search window. This becomes the condition indicator.

It is recommended that for thrust bearings, the search window be set from 1% low to 7% high for the anticipated frequency bear fault frequency. For example, in case the anticipated frequency was 3778 Hz. The search window for the fault feature would then be 3740 Hz to 4042 Hz. Were these a non-thrust bearing application, the window would be: 3740 to 3816 Hz.

Note: On a separate, operational turbine engine installation, the same thrust bearing phenomena was seen, albeit with a change of frequency of just 2.36% change. This newer engine had less wear. As can been seen in Figure 6, were the ball rate is calculated at 3775 Hz, the observed frequency was 3864 Hz. Note the side bands at both 3328 Hz and 4399 Hz, or +/- the shaft rate of 535, again suggested a ball fault modulated by shaft.

Conclusion: During a test of a light turbine engine, indications of a bearing fault were found in the envelope spectrum. However, the fault frequency was not associated with any know bearing. The fault frequency had side bands at the power shaft rate (536 Hz), which indicated an inner race fault, or roller/ball element fault: side bands are commonly observed when the shaft rate modulates the bearing fault frequency. The observed fault frequency was 6% higher than would be expected for a roller/ball fault, whereas the 8% lower for the next know frequency (outer race frequency).

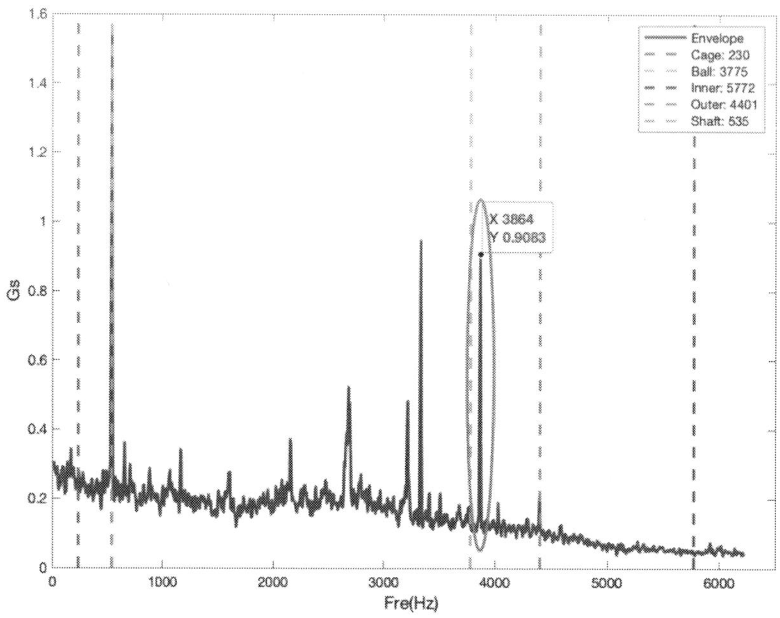

Figure 6 Ball Fault on Turbine, 2.6% error

The fault was observed when the engine output was greater than 65% rated power. Further, it was observed that the bearing with fault frequencies that was closes to the observed feature was a thrust bearing. Thrust bearing under axial loads can change their contact angle. A change in the contact angle additionally changes the axial displacement of the bearing. This change in axial displacement increases the observed pitch diameter, which increases the observed fault frequency. As mitigation, it is recommended that for thrust bearings, the frequency bound should be set from one percent below to seven percent above the specified fault frequency.

References:

[1] Randal, R. B., *Vibration-based Condition Monitoring*, John Wiley & Sons, 2011.

[2] Antoni, J., Randall, RB., "The spectral Kurtosis: application to the vibratory surveillance and diagnostics of rotating machines," Mechanical systems and signal processing 20(2), 2006

[3] Hamrock, B., Dowson, D., "Ball Bearing Mechanics," NASA Technical Memorandum 81691, 1981

[4] Hamrock, B., Dowson, D., "Introduction to Ball Bearings," NASA Technical Memorandum 81690, 1981

Bibliography:

[1] Eric Bechhoefer is the president of GPMS, Inc., a company focused on the development of low-cost condition monitoring systems. Dr. Bechhoefer is the author of over 120+ juried papers on condition monitoring and prognostics health management and holds 23 patents and 5 patent applications in the field of CBM.

SELECTION OF THERMAL ISOALTOR FOR A PROPYLENE TRANSFER PUMP – A CASE STUDY

Mantosh Bhattacharya
Petrofac International ,
PO Box 23467 Sharjah , UAE
Tel 00971501469853
Mantosh.b@petrofac.com

Abstract –
In a Petrochemical Complex, liquid Propylene as a product is transferred by a transfer pump from storage tanks for ship loading / other required facilities. Liquid Propylene has a boiling point of minus 47.7 degree C, hence operating fluid must be kept below minus 47.7 degree C during complete pumping operation and storage.

To protect concrete from such a cold temperature exposure a thermal isolator was selected. In addition, isolator was supposed to tolerate pump dead weight, provide proper anchorage during dynamic motion, shall not fail even at high stresses generated during a trip due high vibration or a short circuit. Most importantly, it was supposed to isolate high vibration and dynamic forces thus ensuring smooth operation of other pumps of similar type. The case study provides outline how such objectives were achieved by in-house detailed engineering and calculations.

Key words - **API** VS6 pump, unbalance force, transmissibility, isolation, cold transfer

Introduction – In a Petrochemical Complex, Propylene as a product, is transferred by a transfer pump from storage tanks. Liquid Propylene has a boiling point of minus 47.7 deg C, hence operating fluid must be kept below minus 47.7 deg C.

A VS6 (a double casing diffuser type API 610 nomenclature) Single suction pump is normally selected for the purpose. VS6 type pump can create its own NPSHa using its can length. VS6 type pumps are quite different from other pumps because they have more flexible motor and pump discharge casings than comparable horizontal pumps and a more flexible attachment of these casings to the foundation. This is in addition to having very long line-shafting that connects the motor to the belowground liquid-end pump bowl assembly. Because of this unique construction, vertical turbine pumps (VS6) are more prone to structural and shaft vibration problems This type of pumps have a major usage in hydrocarbon industries.

In such a low temperature operation, "cold flow "from pump base plate to concrete can cause cracks and damage to foundation. Use of Epoxy grout between base frame and concrete foundation becomes a borderline case. The reason was high ambient temperature and very low operating fluid temperature. Subject pump which was transfer liquid propylene was to be fixed on concrete foundation, wet portion of pump in pit below the ground level. Pit wall opening was increased to take care of insulation requirement of can to prevent cold flow / increasing flowing fluid temperature. Refer fig 1

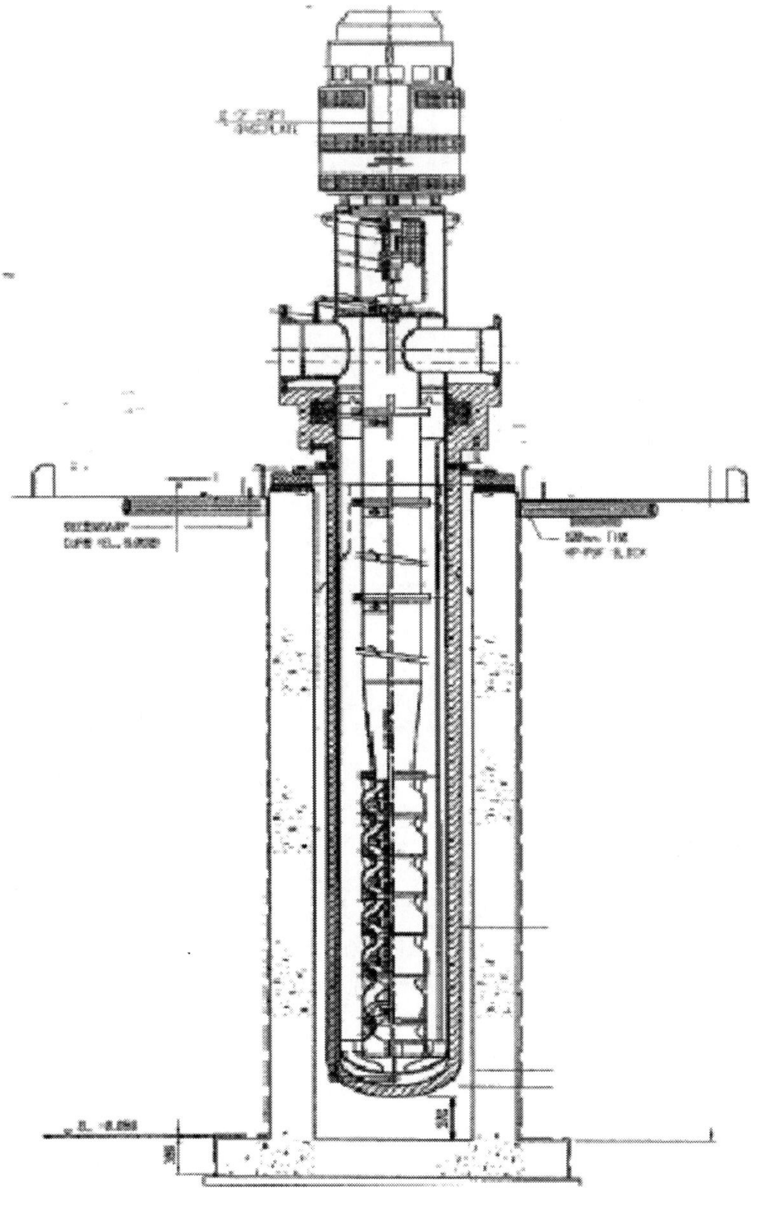

Fig 1 -Installation arrangement of pump under case study

Since the existing base plate of pump was undersized with respect to new pit wall square opening, an additional sole plate of 1¾ inch thick of SS 316L was considered. Sole plates were stiffened at 8 places for better stability in case of swaying of pump during operation. Number of Anchor bolts increased from four to eight to ensure better dynamic stability utilizing added sole plate area of coverage on foundation. Refer Fig 2 for sole plate details.

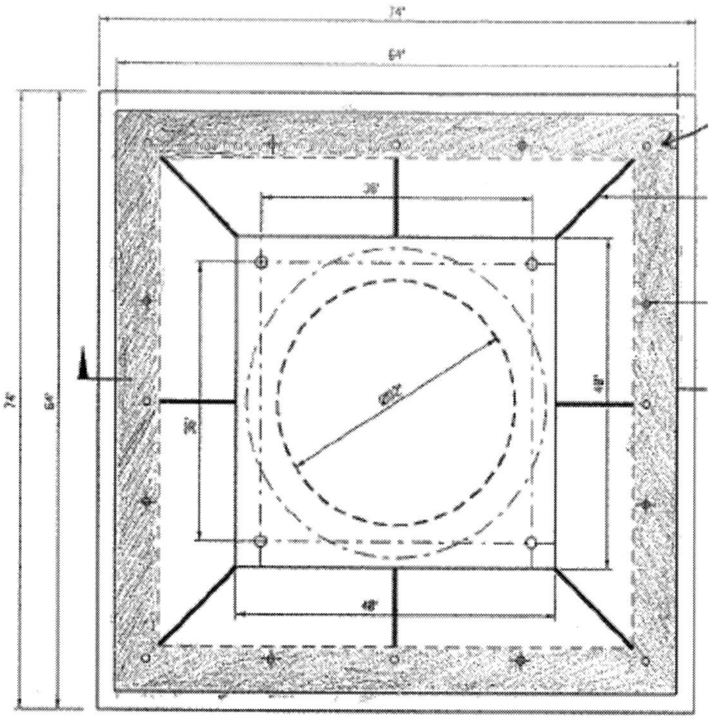

Fig 2 –Sole plate [shaded area of sole plate shown where Isolator was to be placed]

The challenge was to select a thermal isolator between metallic base frame of pump and concrete. The isolator was to withstand all dynamic forces acting during pump operation even considering worst operating scenarios as well which means it shall act as vibration isolator as well to protect concrete and nearby pump installation. An isolator is a resilient support which decouples an object from steady state or forced vibration. To reduce the transmitted vibration, isolators in the form of springs are used. Common springs used are pneumatic, steel coil, rubber (elastomeric) and other pad materials.

To avoid major modification in already laid piping at site, the target of vibration isolator thickness was kept 2 inches or below. The target vibration isolation was kept more than 80%. The anchor bolts were also to be isolated with sleeves made of same material to eliminate any metal to metal contact.

Subject Pump data - The subject VS6 type pump operating details are as follows –

Rated Flow – 3965 US GPM, Pump RPM- 1800, Power -500HP, Minimum Continuous Flow – 2500 USGPM. Impeller dia – 13.4inch No of stages- 6, Discharge pressure Pd =153.6 psig , suction pressure – Ps – 3.7 psig ,motor and pump directly coupled and total wt of system- 20000 lb. Pumping fluid temperature – minus 54 deg F. Balance quality of machine - Gr 2.5 ISO 1940, Concrete- C 40 grade concrete . Vibration acceptance criteria – ISO 10816-3

The scheme was to arrange the thermal isolator in such a way that there is no cold transfer from metal parts to concrete. To have this arrangement from drawing board to site execution various vendors were contacted and inhouse assessment of the proven features and design were carried out. Then the chosen type of thermal isolator was to be reviewed in terms of vibration isolation also.

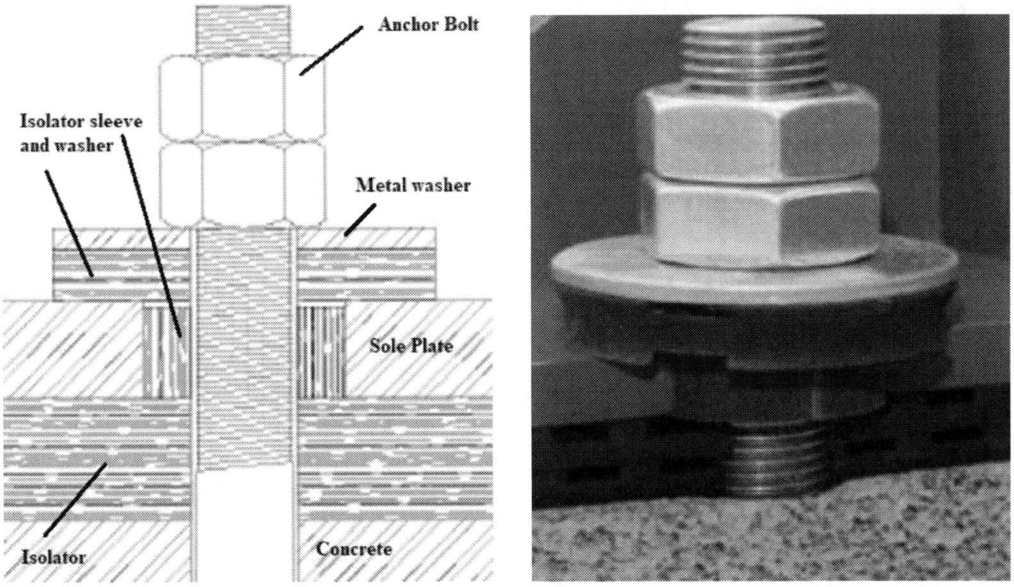

Fig 3 Cross sectional arrangement showing Isolator Pad, Isolator sleeve and washer

First objective was to calculate major forces those shall act on machine foundation and second objective was to select a suitable thermal isolator which could withstand pump dead weight, provide proper anchorage during dynamic motion, shall not fail even at high stresses generated during a trip due high vibration or a short circuit. Most importantly, it was supposed to isolate vibration and dynamic forces and cold flow to concrete thus ensuring smooth operation of other pumps of close vicinity.

A. Downward Axial thrust & Dead Load –

Downward axial thrust was calculated by vector summation of downward forces due to discharge pressure and upward forces due to suction pressure as shown below with 6 number in line impeller arrangement.

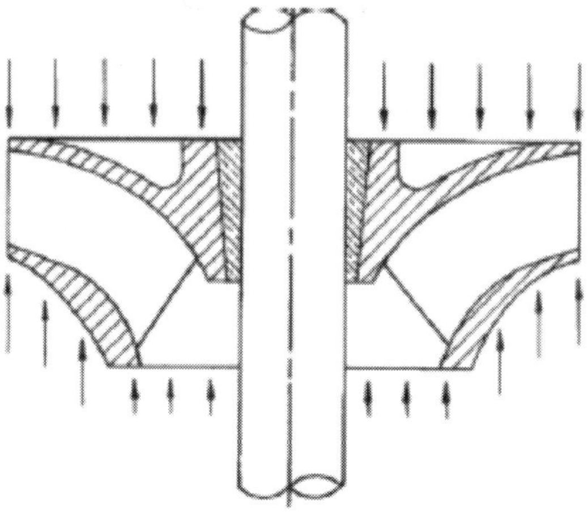

Fig -4 – Direction of forces acting on individual impeller in VS6 type pump.

Note - Since all data supplied by Isolator Supplier was in Imperial Units, all results are shown in Imperial units.

1a. Hydraulic Axial thrust (downward) calculated using discharge pressure, shroud area, suction pressure and intake area and number of stages as 114400 lbf
1b. Hydraulic Axial thrust (downward) calculated using Darfur and Nelson equations as **8190** lbf

Higher among 1a and 1b was considered -
2. Dead wt of pump & motor including piping volume was calculated as **24300** lbf

Note -Shaft thrust (upward) & sleeve thrust (upward) considered as ZERO effect on compressing force on isolator.

Hence Total Thrust was calculated as @ 114400+24300 = **138700-pound force**

Axial stress on bolts = [(Maximum bowl pressure *discharge flow area) + (bearing load) + bolt preload * no. of bolts] / (no. of bolts *effective cross-sectional area of bolt in perpendicular plane of snub tightening)

B. Forces generated when vibration reaches trip value.

An unbalance will cause vibration in rotor bearing system which shall be transferred by antifriction bearing to pump casing and subsequently to foundation bolt and isolator pads. To simulate worst case scenario, damping is not considered at all.
The radial component 1x vibration will act as force of shear to foundation bolts and isolator. Hence isolator as well should strong enough to avoid damage by such forces.
Total line shaft weight -231 lb.
Impeller assly weight – 272+85 = 357 lbs.
Coupling weight – 100 lbs.
Motor rotor weight – 7787/3 = 2629 lbs.

1/3 rd of Motor wt. is taken because coupled with a flexible coupling motor unbalance does not really affect the pump

Total rotor wt – **3317** lbs

RPM – 1800, 30 Hz

Two different scenarios were selected for disturbing forces –

Forces caused by unbalance in rotating parts –

F = 22.5*6.015*(Balance grade as per ISO) * Rotor weight / RPM. Normal balance grade is 6.3 however Balance grade 16 was selected as conservative approach.

With balance grade 16, F= 3600 lbf

Forces generated at the event of trip -

A higher than trip value (10 mm/ sec RMS) of vibration was set as. 15mm/ sec RMS (0.6 inch / sec. RMS) to analyze the integrity of thermal isolator. This was because rotor system becomes less stiff due to introduction of isolator and thus reacts more to unbalance. We assume that there is only one peak in complete spectrum and that is at 1X. (Note -For simplicity of calculation, true sine waves are considered.)

Referring the conversion table of Vibration Severity Chart as per ISO 10816-3, the calculated peak to peak displacement in inches i.e.

D = (19100 / 1800) * 0.6*2√2 = 9.55 mils pk-pk as per chart below -

CONVERSION FORMULAS FOR VIBRATION UNITS

$$D = \left(\frac{19100}{F}\right) V$$

$$D = \left(\frac{8383}{F}\right)^2 A$$

$$V = \left(\frac{F}{19100}\right) D$$

$$V = \left(\frac{3687}{F}\right) A$$

$$A = \left(\frac{F}{8383}\right)^2 D$$

$$A = \left(\frac{F}{3687}\right) V$$

$$G's = 0.0142 \left(\frac{F}{1000}\right)^2$$

$$OZ\text{-}IN = D \left(\frac{W}{125}\right)$$

$$OZ\text{-}IN = G \frac{(0.563W)}{\left(\frac{RPM}{1000}\right)^2}$$

$$\#FORCE = 1.79\ OZ\text{-}IN \left(\frac{RPM}{1000}\right)^2$$

$$C.F. = (W)(.00034\ F^2 R)$$

D = PEAK TO PEAK DISPLACEMENT - (mils)
1 mil = 0.001 inch
V = PEAK VELOCITY -inches/second
A = PEAK ACCELERATION -G's
1 G = ACCELERATION DUE TO GRAVITY = 386.087 inches/second
F = FREQUENCY - CYCLES PER MINUTE

OZ-IN = THE UNBALANCE IN OUNCE-INCHES
(1 OZ AT 1 IN RADIUS = 1 OZ-IN; 1 OZ AT 10 IN RADIUS = 10 OZ-IN)
W = WEIGHT OF ROTATING ELEMENT IN POUNDS
C.F. = CENTRIFUGAL FORCE
R = RADIUS IN FEET

Fig -5 Vibration Conversion Chart

Equivalent oz-inch = 9.55 * (3317 / 125) =253.41 oz-inch

Centrifugal Force due to Rotation= 3317 * 0.00034*1.1 *30*30 = 1030.002 lbf

Unbalance Force = 1.79*253.41*1.8*1.8 = **1469.67** lbf

Total force = @ **2500 lbf**

ISOLATOR PAD SELECTION BASIS

Type of Isolator was selected keeping very important things in mind –

1. Selected Isolator dynamic natural frequency was to be less than half of lowest disturbing frequency.
2. It shall withstand all dynamic and static forces encountered during pump operation for a prolonged duration as calculated above.
3. It should not de rate by weathering and cold. It maintains its property of isolation in all ranges of ambient temperature required in Pump data sheet.
4. It should comply to ASTM2000 D-04

Size selected for isolator pad which shall be placed underneath sole plate – 64"x50"x2 "thick as shown in shaded area in Fig.2

Selection Procedure – To start with, all information was passed on to prospective suppliers of Vibration Isolators with a drawing (as shown in fig 1 and 2). Response were solicited with all back up calculations with temp.-stiffness curve, dynamic spring rate, transmissibility curve, mech/ physical properties.

Vibration Isolation analysis -

Total wt of unit including piping load, liquid wt etc. = 24249 lbf

RPM – 1800, Lowest Forcing / disturbing frequency Ff= 1800 / 60 = 30 Hz (1800/60)

The dynamic natural frequency of selected type of isolator pads Fn= 9 Hz.

$Ff / Fn = \sqrt{\{(1 + T) / T\}}$

T is Transmissibility.

If the simple system is subjected to Forced Vibration at frequency f, and sinusoidal foundation motion at amplitude x, the absolute value of the mass response amplitude y expressed as a ratio $|y/x|$, also known as Transmissibility T. In simple words, Transmissibility is a measurement used in the classification of materials for vibration management characteristics. It is a ratio of the vibrational force being measured in a system to the vibrational force entering a system. If a vibration isolator pad has a transmissibility of 11.64%, it means that 11.64% of the vibrating force is being transmitted through the pad.

Isolation = 100- Transmissibility. considering the system as linear system.

Isolation = 89.36 % from transmissibility chart which served the purpose of selection.

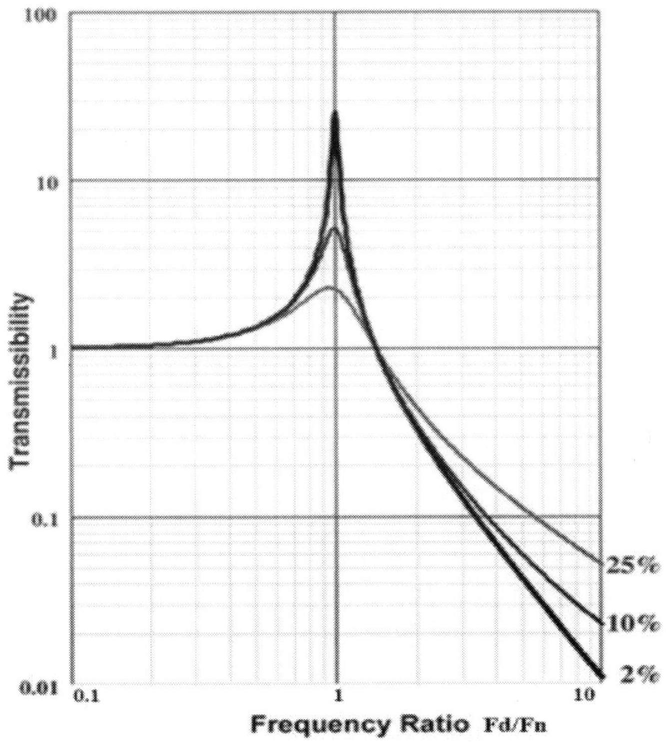

Fig 6-Transmissibility curve

Compressive stress on isolator pad (having full coverage as shown in Fig
Total Static Load / Isolator Coverage area = 138649 lbs. / 1392 Sq inches
= @99 psi < design value of 120 psi.

The isolator performance was also cross checked with Load Deflection graph which was available in Isolator pad Technical Information sheet. The relation between natural frequency and static deflection of a linear, single-degree-of freedom system is shown below.

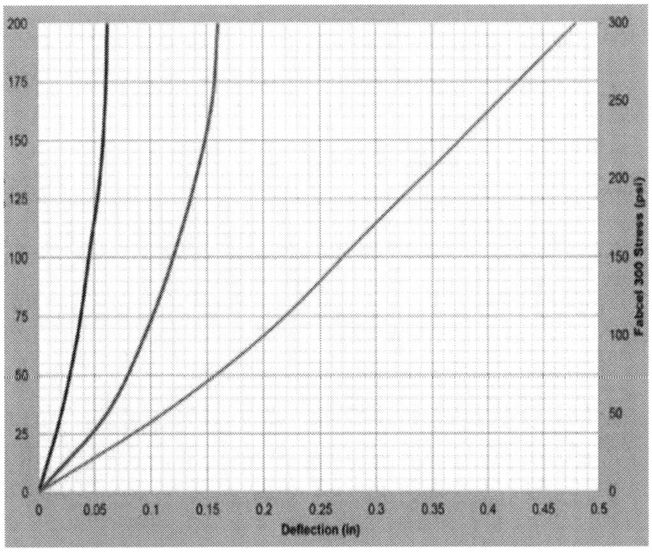

Fig 7- Load deflection diagram

Thermal isolation

Heat transfer coefficient of Isolator material k = 0.274 W / m deg C = 0.23 kcal / m deg C. Note -The thermal conductivity calculation done in house excel sheet using SI units.

Based on heat transfer coefficient of Isolator and pump constructional details, a cold flow calculation was performed. Results showed that most of the cold shall pass towards open face of sole plate. The "Cold "transfer calculation showed that a mere 1.2 mm thick isolator shall start resisting cold flow.

Heat transfer calculation table

Minimum allowable temperature of concrete	0 deg C
Pumping temperature	- 47 dig
Can inside radius r1	400 mm
Can outside radius r2	406.4 mm
Can insulated radius r3	440 mm
K conductivity of can W/m deg C	19
Total Cold Flow	-353479 W.
Actual band width of pump CAN transferring the Cold	0.035 m
Actual cold flow	-12371.74 W
KW	-12.371
Kcal / hr.	-10639.89

A – area of sole plate in contact with isolation pad -1.029m2.
Q/A = T1-T2 (t/k), t = @1.025 mm. The actual pad thickness was 50 mm so it will only permit 195kcal / hr. of coldness into concrete.

Forces generated due to high Vibration

When high vibration is occurring, the radial forces is transmitted to pump body and base plate which in turn transmitted to foundation bolts in a particular side.

This acts as Shear force applied in isolator sleeve / washer.

Shear stress on these bolts as calculated was far less than design value suggested in supplier catalog.

FEA analysis by Pump Supplier for Wet structural natural frequency –

Some iteration was to be done in finalizing the level of constraints for isolators in FE modeling of system.

The analysis conducted by Pump Supplier showed first six structural natural frequencies are well separated for 1x synchronous speed excitation. FEA Model and sample Operational Defection shape shown in fig.

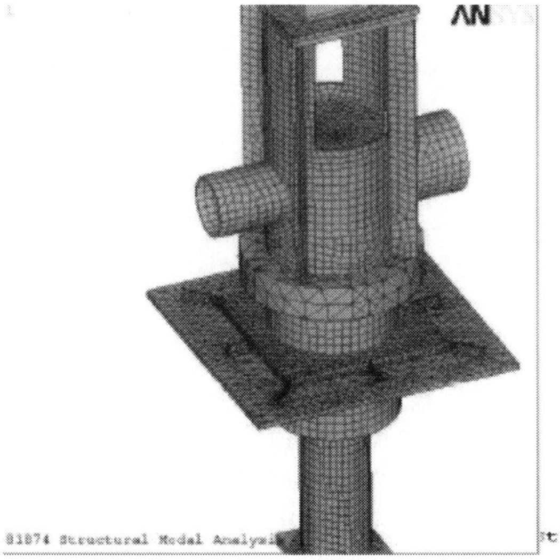

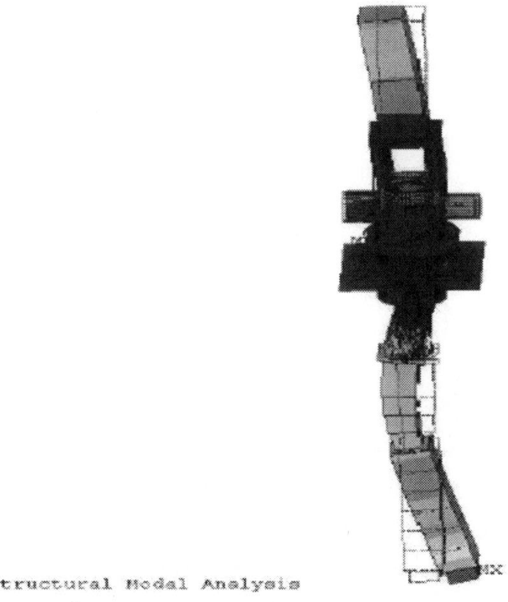

FEA Model **3rd Below Grade Mode 23 Hz.**

Fig -8 – Structural Natural Frequency analysis by FE software.

Besides all above studies for isolator selection, utmost care was taken to install the pump with Isolator pads in presence of Pump installation expert, Isolator application engineer and E&C Rotating equipment engineer.

The pump was commissioned and was put to operation in year 2007. Since then, it has been in operation without any reported trouble.

References –

1. API 610 10th edition was referred during study.
2. ISO 10816- 3, Evaluation of machine vibration by measurements on non-rotating parts
3. ISO 1940 - Balance quality requirements for rigid rotors
4. An End-User's Guide to Centrifugal Pump Rotor-dynamics, - William Marcher 32nd international Pump symposium TAMU Texas
5. Various Technical Communications with Isolator Supplier.
6. Effect of mechanical unbalance on vibration, forces and reliability of a single stage centrifugal pump – Lev Nelik & Charles Jackson – Proceedings of 12th international pump symposium – TAMU, Texas
7. Special vibration considerations for Vertical turbine pumps – Allan R. Budris – Water World
8. The perils of installing VFD drives on Vertical turbine pumps – Nelson .L.Baxter
9. ISO 10816-7:2009- Evaluation of vibration on rotodynamic pumps for industrial applications

DEVICES FOR INSITU TUNING OF MACHINERY STRUCTURES TO MINIGATE RESONANCE

Mantosh Bhattacharya
Petrofac International ,
PO Box 23467 Sharjah , UAE
Tel 00971501469853
Mantosh.b@petrofac.com

Abstract –

For on-shore and grout less design applications, the support structure for a rotating machine consists of a base-frame (base plate) normally known as skid. Most of the time these structures show up with high vibration on machine bearing cap due to structural resonance, unidentified cracks. Detection and mitigation of such anomaly is time consuming job. The paper proposes a blueprint of active and passive resonance control at location without removal of base frame from foundation / main support and any structural modification which requires any hot work.

Introduction -

Structural resonance is commonly identified as symptom of excessive vibrations of non-rotating components, usually machinery parts or supporting structures. This is the more common resonant condition usually occurs at or near the rotating speed of the machine. Even slight vibratory forces due to residual unbalance and misalignment effects of the machine can excite to resonate the base structure, resulting in severe vibration.

During resonant condition, the support structure normally demonstrates vibrations in rigid body modes and aggravating the shaft vibration conditions. This issue occurs normally with rigid rotors with antifriction bearings or with machines where ratio of pedestal dynamic stiffness and bearing stiffness are less than 3 such as electric motors, generators, steam and gas turbine casing structures.

Sometimes, the evaluation of resonant frequencies for equipment base frame is not given its due importance by the OEM particularly for medium sized rotating machines. To identify such condition at site and rectification is cumbersome due to the complexity of the components in structural base frame.

Descriptions of structural resonance and methods to detect -

Sometimes in spite of good quality of balance, correct alignment, rotor exhibits excessive vibration at 1x harmonics. Often it is assumed that simply balancing the rotor with a stricter grade will make the vibration problem go away. However, upon further investigation it is not uncommon to find that a structural resonance is the root cause of excessive vibration, and balancing is not a viable or long-term solution to the problem.

The large amplitudes of vibration are seen when a machine's running speed is at or near a natural frequency even if the vibration amplitude due to unbalance or misalignment or other higher harmonics such as vane pass frequency are low This problem is particularly common in variable speed equipment or for those machines having specially fabricated base plate with non- grouted design option.

Various field tests are performed to verify that "resonance is really in fact the cause" of excessive vibration in a system. Two of the most common tests are a modal impact test and a start-up or coast-down data collection with a cascade plot. The cascade plot below shows an example of typical vibration exhibited during start-up of a machine with a structural natural frequency in its operating speed range as shown in fig 1.

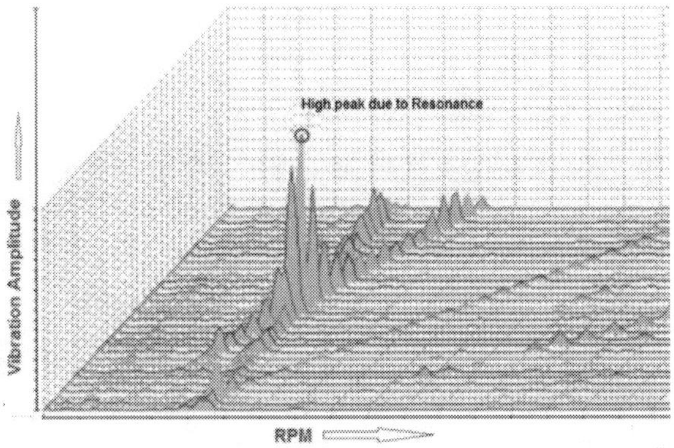

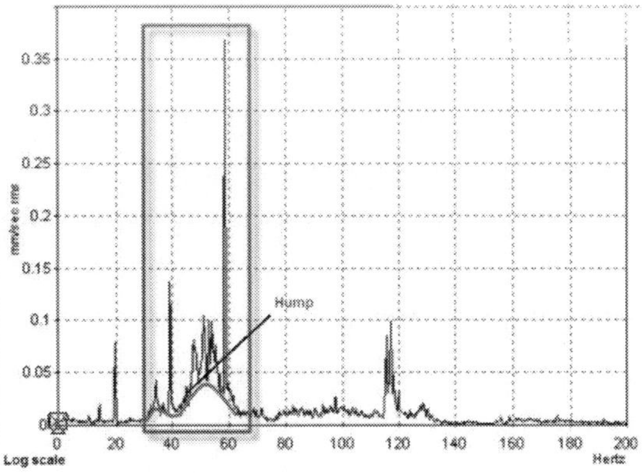

Fig 1 – Occurrence of resonance as detected in cascade plot and FFT spectra

Resonance is characterized by a large increase in vibration at a particular frequency with humps but generally lower amplitudes at all other operating speeds / frequencies as shown in below FFT spectra of a machine.

Sometimes, bump test combined with ODS (operational deflection shape) analysis also used as verification of existence of structural resonance. It is easy to detect pure horizontal or vertical movement of rotor / machine but without visual aid such as wmv or animation file, it is not possible to find location of deflection on structure.

The normal modes of structural and machinery movement due a resonant condition is shown in simplified manner.

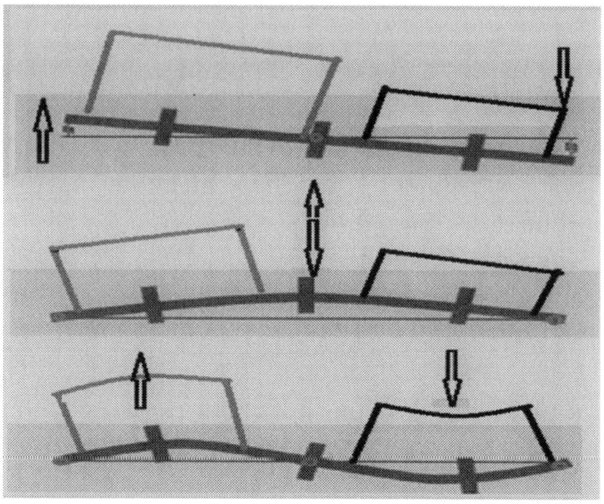

Fig 2 various modes of structural vibration

Any rigid-body displacement of a structural element can be resolved into these six independent displacements as shown in fig. Hence, the rigid block has six degrees of freedom and six natural frequencies. Of six types of motion, translation along the Z axis and rotation about the Z axis can occur independently of any other motion. However, translation about the X axis (or Y axis) and rotation about the Y axis (or X axis) are coupled motions. Therefore, in the analysis of a block, we have to concern ourselves with four types of motions. Two motions are independent and two are coupled. For determination of the natural frequencies, in coupled modes, the natural frequencies of the system in pure translation and pure rocking need to be determined

The results with wmv or animation file of modal analysis or ODS can provide a clear picture of coupled rigid body modes as example below.

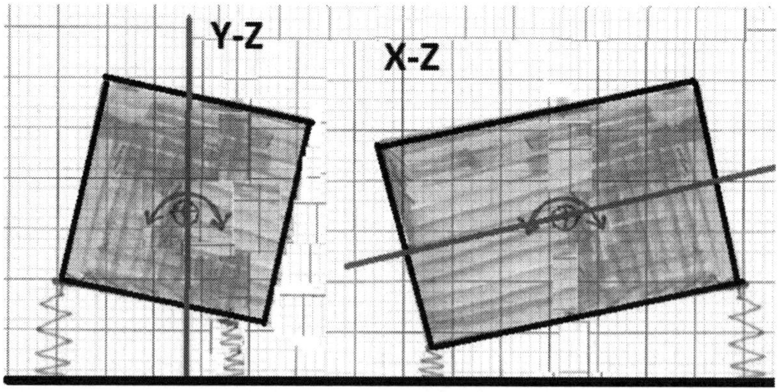

Fig3 - Coupled rigid body modes - most common due to weak structural members

Practiced Solutions -

As we've seen, resonance occurs when the natural frequency of a system coincides with any expected forced vibration frequencies (such as unbalance) which can lead to severe levels of vibration. If it is determined that resonance is in fact the cause of excessive vibration, what can be done to stop or minimize the effect of a resonant condition?

The natural frequency of a system is dependent upon two main factors; stiffness, and mass. If the natural frequency is ω, $\omega = \sqrt{(k/m)}$.

Where k is the stiffness and m are the mass of base frame. Therefore, in order to change the natural frequency, we need to change either k or m or both.

•Increasing mass or decreasing stiffness will reduce the natural frequency

•Decreasing mass or increasing stiffness will increase the natural frequency

Typically, the objective is to increase the natural frequency such that it is above any expected vibration frequencies. If the natural frequency is above or significantly far away from any expected vibration frequencies the resonance will likely no be excited. This theory forms the basis for any structural redesigns implemented to avoid resonance.

In practice, the following rules also can be used to shift a natural frequency and minimize the vibration response of a system;

- Increasing damping reduces the peak response but widens the response range
- Decreasing damping increases the peak response but narrows the response range
- Reducing forcing amplitudes reduces response at resonance

Tuning of the base frame are mostly done by increasing stiffness k by stiffing the base frame to shift its resonant frequency. This is done by adding structural members or by changing the shape / size of structural member to decrease the deflection / twist in particular direction.

Complications with present day analytical method -

Detection of resonance need special equipment and multichannel vibration analyser with trained personnel to conduct such tests.

The natural frequencies of a structure are related to its design and are greatly affected by the mass and stiffness of the structure. If we made the tines of a tuning fork stiffer, the note made by the tuning fork would increase in frequency. And the same is true for the structure supporting our machine; which is normally what is resonating. If we know that we have a resonance problem we can consider either adding mass to reduce the natural frequency or increasing stiffness to increase the natural frequency. The aim is to change the natural frequency so that it is no longer excited by the machine.

This is a very sophisticated process and there are a number of issues that must be considered. In essence this is what we are trying to do – modify the structure so that the vibration generated by the machine is no longer amplified and therefore harming the machine or the structure

If stiffener / mass is arbitrarily added to the structure, it could easily shift the natural frequency into another range of operation thus making the vibration significantly worse. Calculation of adding mass to the system and shift the natural frequency down to a frequency where the forces are lower is a cumbersome exercise as addition of mass increases the stiffness of that local area.

Hence a validated finite element analysis (FEA) model of a structure is constructed to determine the optimal design changes in order to fix and avoid resonant conditions. Validation of model is done first by free -free resonant test of structure, putting the data to FE model and then correcting the FE model which is a complicated work. This allows to test various different potential design changes in a computer simulation first before recommending any structural changes.

Based on criticality of machine, combination of frequency sweep test, bump test and then creating an animation file of operating deflection shape the resonant frequency and nodal points of deflection can be found. Based on location of nodal and anti-nodal points structural modification can be carried out.

To carry out structural modification, the machine needs to be removed from base frame and the base frame need to be sent to work shop / fabrication shop to undertake controlled welding for adding structural elements to increase stiffness. All these activities may render the unit out of operation for 2-3 months.

Proposed solutions -

If the particular structural member is subjected to lateral compression or tension, its bending stiffness and torsional stiffness increases and this is the basis of proposed solution detailed in forthcoming paragraphs. As we know, structural member undergoes a temporary deflection during a coupled rigid body mode and if that defection is arrested then we increase the directional stiffness of subject structural element.

Two types of controls to restrain the temporary defection are being proposed based on location, criticality and sensitivity of machinery installation.

For **passive control,** some devices shop fabricated devices can used which can mitigate the structural resonance problem. The proposed methods shall try to cover the possible discrepancy on-site grouting issues where a support structure may undergo push pull action. The push-pull phenomena aggravate machinery vibration. The device proposed can provide additional directional stiffness.

The proposal is to design base frame for a horizontal centrifugal machine where the effective stiffness can be increased in a particular direction based on readings taken at installation site itself. The design incorporates devices on the peripheral structural parts through which adequately pre-sized structural element is to be introduced, thus increasing the stiffness and restricting coupled modes of concern such as rolling or pitching which is mainly caused by structural resonance modes of base frame.

To facilitate a base frame tuneable at site, an overall drawing of a proposed base frame is shown in figure 10. To increase the stiffness of cross (traverse) members, additional devise (hollow cylindrical elements) are inserted across the two longitudinal main members.

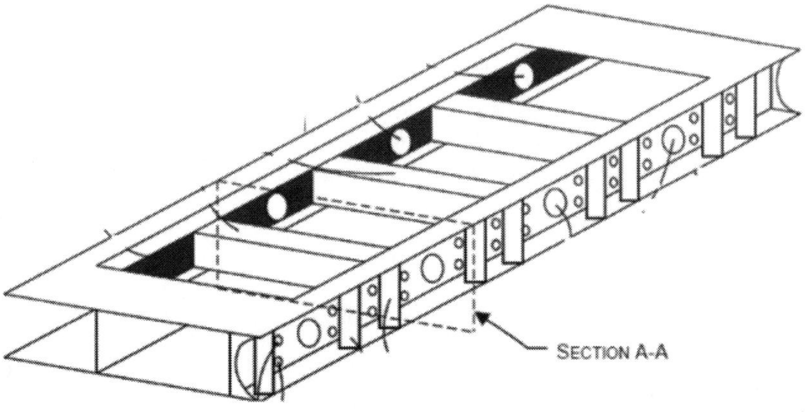

Fig 4 - Base frame with open channel type design

To overcome the rolling mode of base frame and machines during operation, directional stiffness shall be increased by inserting hollow stiff cylinders and then tightening with spring washers as shown in picture. To prevent the cylinder locking device getting loose, tab type washers or locknut can be used.

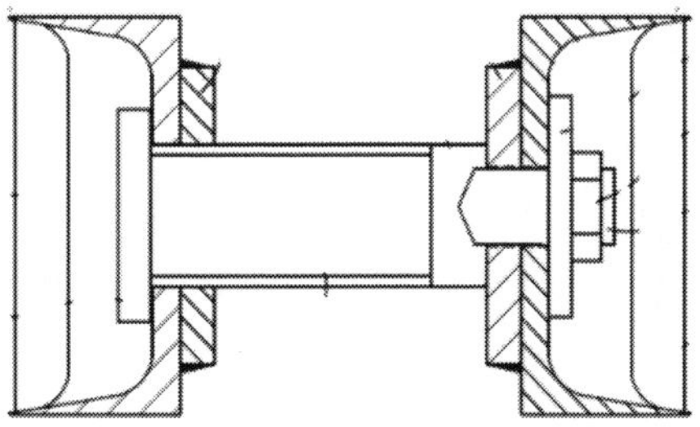

Fig 5 cylindrical Stiffening device attached across channels section A-A

To overcome the pitching motion of base frame and machines during operation, directional bending stiffness shall be increased by a fastening a plate along with the web portion of open sections. This is shown in figure 13.

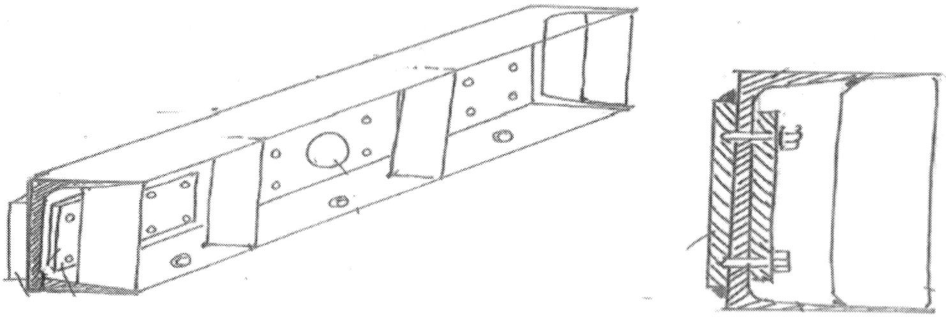

Fig 6 - Stiffening of longitudinal beams

To prevent the rolling and pitching movement of stool mounted motor, a turn buckle type arrangement can be used with circular slot having angular freedom of movement as shown in figure 14.

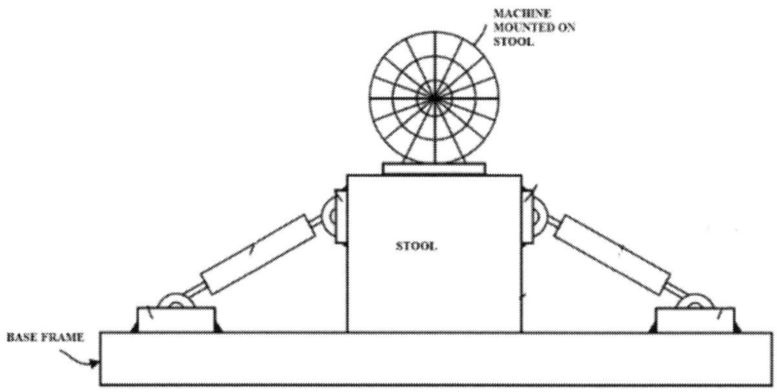

Fig 7- Stool holding device using turn buckle arrangement to prevent swaying

With some fine tuning such as adjusting torque values of bolt, spring washers changing the thickness to higher and lower range as part of supplied device, the machine vibrations resulting due to structural resonance can be brought down to acceptable values.

Active control -

An active control of coupled modes of vibration (rocking and pitching) can be achieved with instrumentation using strain sensors, servo valves and hydraulics by increasing the directional stiffness.

For proposed active control shown in figure, miniature opposed single acting hydraulic cylinders are inserted into hollow part of base frame area just besides the welded cross member.

As we know, structural member undergoes a temporary deflection during a coupled rigid body mode. This deflection shall cause strain and change in force acting on particular direction.

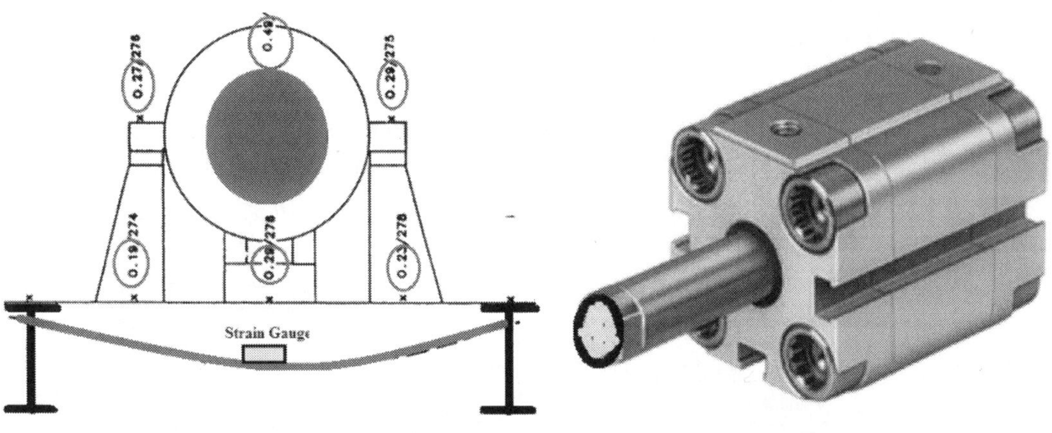

Fig 8 -Left- location of a strain gauge, Right - Miniature hydraulic cylinder

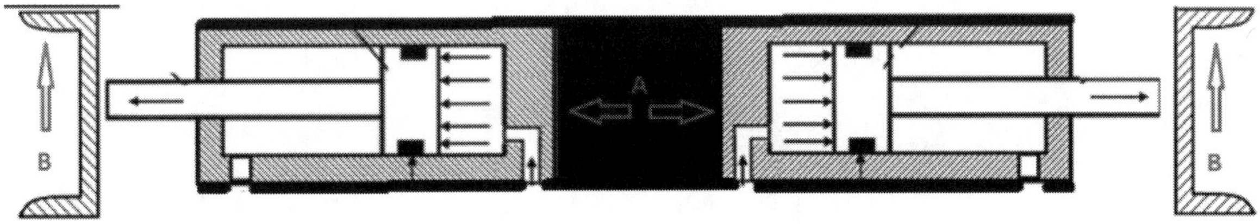

Fig 9 Hydraulic device arrangement with two miniature opposed hydraulic cylinder

So, if the bonded strain gauge tape (deflection / twist sensor) or strain rosette is used on subject structural element then they can give exact position of equipment in terms of deflection and coupled modes of vibration. If the signal output is in form of x-y milliamps, it can control any device based on closed loop feedback control.

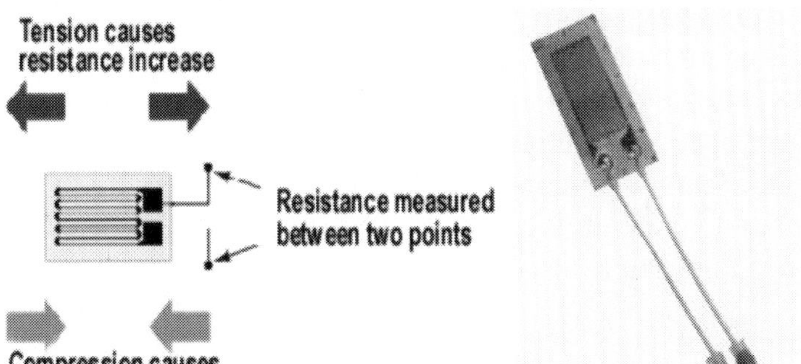

Fig 10 Pre-wired strain gauge

The proposed device uses strain gauge / strain rosette pasted on structural member. A structural member bends by an applied moment or load exerted perpendicular to that member. One side of beam shall be in tension and other side at compression (shape attained by member as convex - concave). A feedback device wired with a signal conditioner shall control directional control valve to create required push action by opposed single cylinder with extended pistons which shall work as additional traverse member. In this proposed design, multiple miniature opposed single acting hydraulic cylinder shall be used. Adjustable force ranging from "negligible" to maximum cylinder capacity, by adjusting the input pressure can be achieved.

If the deflection / strain is found in longitudinal structural elements such miniature single direction hydraulic device can create required vertical stiffness. The strain gauge tape and location of hydraulic device should be in middle of foundation bolt location to have maximum effect.

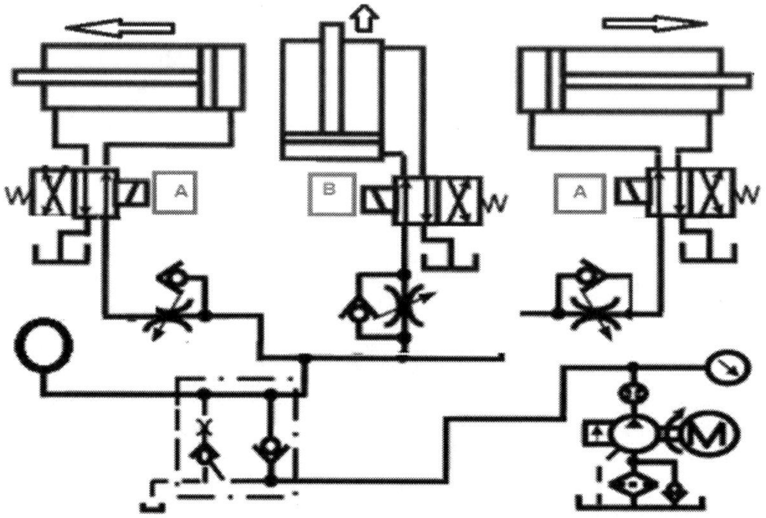

Fig 11 - Hydraulic circuit

Solenoids A shall be connected to banded strain gauges on traverse structural elements to pick command to extend pistons in horizontal directions. Solenoid B shall be connected to force sensor or banded strain gauges on vertical direction of structural elements to pick command to extend pistons in vertical directions to prevent spring action.

Three traverse structural elements and two longitudinal structural elements with hydraulic actuated devices can control any unwanted coupled modes by shifting local stiffness to higher values than operating frequency of machine. Hence when the state of no change in resistance in bonded strain gauge is registered after action by hydraulic cylinders, then system stays put and no action is required. Thus, system becomes completely adaptive to coupled mode due to possible structural resonance.

For controlling rocking mode of a motor or slow operating equipment stool, two hydraulic actuators with hinge arrangement can control the mode as shown below -

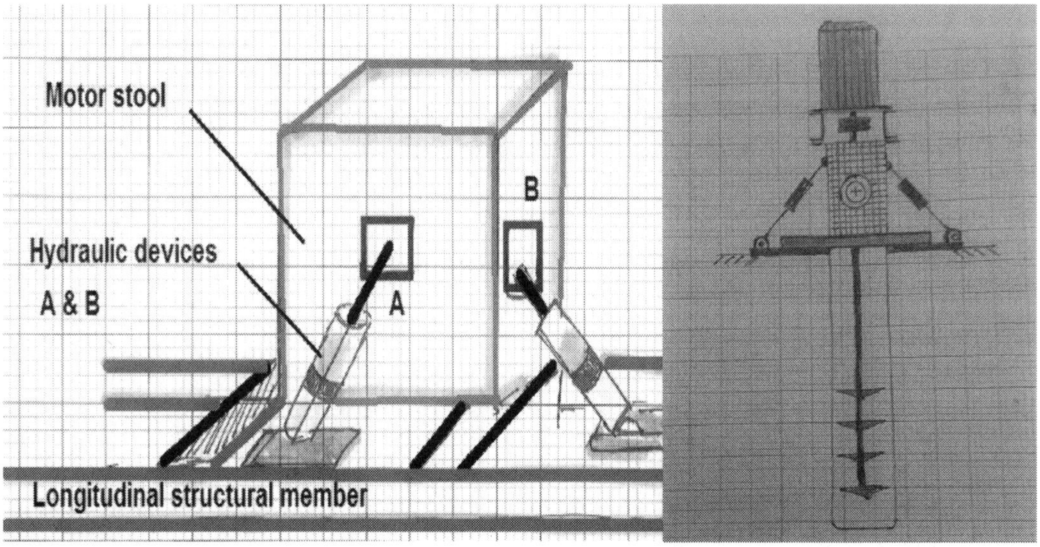

Fig 12 Motor Stool (left), vertical pump (right) rocking and rolling mode control

Conclusion –

The proposed devices can be very helpful on offshore installations or a live gas plant where obtaining with hot work permit is a long exercise. With passive tuneable device, the structural related vibration issues can be greatly mitigated. Even some the cases it is not even necessary to undertake modal analysis.

Active structural resonance control can be a very useful tool to find the location reinforcement required to mitigate structural resonance by adding stiffness to that particular section. For variable speed driven machines, the concept and blue print of active vibration control can be very useful.

Reference and Bibliography –

1. Baffling vibrations- Mantosh Bhattacharya, Hydrocarbon Engineering 2015
2. Experimental Modal Analysis A Simple Non-Mathematical Presentation-Peter Avitabile, University of Massachusetts Lowell, Lowell, Massachusetts. - S&V magazine
3. Introduction to operating deflection shapes -Brian J. Schwarz & Mark H. Richardson-CSI Reliability Week, Orlando, FL-2009
4. Shock and Vibration hand book – Harris
5. Strength of Materials - M. Rajaratnam; Dhanpat Rai Publications; Edition 12; 2012
6. Handbook of structural steelwork – TATA steel
7. Design, Analysis and Optimization of Skid Base Frame -Pandhare , Chaskar ,Patil et al.
8. Force and Displacement Measurement - Prof. R.G. Longoria-ME 144L - Department of Mechanical Engineering, University of Texas at Austin

9. Use of Strain Gages to Determine the Strain in Cantilever Beams-lab report by Derek Nichols
10. Strain Measurement in a Simply Supported Beam-Dr. Bridget M. Smyser, Northeastern University-ASEE conference Seattle, 2015
11. Various literatures on Strain gauge by manufacturers
12. Fast Estimation of Strains for Cross-Beams Six-Axis Force/Torque Sensors by Mechanical Modelling -Junqing Ma and Aiguo Song Jiangsu Key Lab of Remote Measurement and Control, School of Instrument Science and Engineering, Southeast University, Nanjing 210096, China-Sensors 2013,

DELIBERATIONS ON FOUNDATION DESIGN METHODS FOR ROTATING MACHINERY.

Mantosh Bhattacharya
Petrofac International ,
PO Box 23467 Sharjah , UAE
Tel 00971501469853
Mantosh.b@petrofac.com

Abstract –

For on-shore applications, the support structure for a rotating machine consists of a block foundation and base-frame (base plate) normally known as skid. Adequate dynamic analysis of support structures for rotating equipment is necessary to safeguard other machinery in the vicinity of the subject rotating machine, as well as ensure good conditions for the operation of the supported machine.

The present method tries to achieve an under-tuned support structure which considers the machine mass and block foundation mass as one entity. The design procedure assumes that the assume that all vibrations are transferred to machinery foundation irrespective to the type of machine. This paper endeavours to highlight that how and why current design procedures for support structures of rotating machinery should consider the stiffness ratio of the bearing support (casing) and the rotor - bearing system, With the advent of foundation damping methods, the size of foundation can be lowered with a correct deliberation on subject.

Introduction

Heavy duty rotating machinery train in on shore application with high speed rotating masses requires a support system that can tackle dynamic forces minimize vibrations and transfer minimum vibration to surrounding soil. Excessive vibrations are detrimental to machinery, its support system, and pose hazard for operating personnel. The present practice of design of rotating machines includes gas turbines, steam turbines, turbo-expanders, compressors, fans, electric motors, and high energy centrifugal pumps etc. Normally block type foundations are used to support this type of machinery for on shore applications. In block-type foundation, dynamic machines are preferably located close to grade to minimize the elevation difference between the machine dynamic forces and the centre of gravity of the machine-foundation system. Block foundations are generally designed as rigid structures with pedestal and thick footing or mat. The loads on machine foundations are static and dynamic. Static loads are principally a function of the weights of the machine and all its auxiliary equipment along the weight of foundation itself. Dynamic loads, which occur during the operation of the machine, result from forces generated by unbalance, inertia of moving parts, or both, and by the flow of fluid and gases for some machines. The magnitude of these dynamic loads primarily depends upon the machine's operating speed and the type, size, weight, and arrangement (position) of moving parts within the casing. The basic goal in the design of a machine foundation is to limit its motion to amplitudes so subject machine ensures satisfactory operation of the machine and other machines in immediate vicinity the dynamic response of a rigid block foundation depends on the transmitted dynamic load from rotating machinery, foundation's mass, it's dimensions, and soil characteristics.

Present practice of foundation design: civil & structural engineering perspective

Machinery Foundations are subjected to free damped vibrations and forced vibrations. Forced vibrations occur with continuous external forces on machine foundation. When a damped system is subjected to external dynamic force arising from the operation of machine F (t), the equation of motion can be written as-

$$M(d^2 z/ dt^{2)} + C (dz/dt) +Kz = F(t)m$$

Where M is mass, C is damping and k is stiffness of foundation. z is the displacement of foundation at particular direction. Normally machine foundation has 6 degree of freedom (known as rigid body modes) and above equations need to be solved separately for all rigid body modes.

Accordingly, we can say that for successful operation two aspects become critical for foundation design: 1) The machine itself should run smoothly (round the clock); 2) The foundation supporting the equipment is capable of sustaining the various loads coming from the machine under operation (dynamic loads) as well as those that could develop during transient operations thermal bows, electrical faults, short circuits etc. The analysis is hence considered a very complex problem because of the interaction of the structure, the subsurface soil, and the vibrating machine. Two different models for machine mass location are considered. First one by applying the seismic force at the machine anchorage locations, and the second model by applying the seismic force at the centre of gravity of the machine.

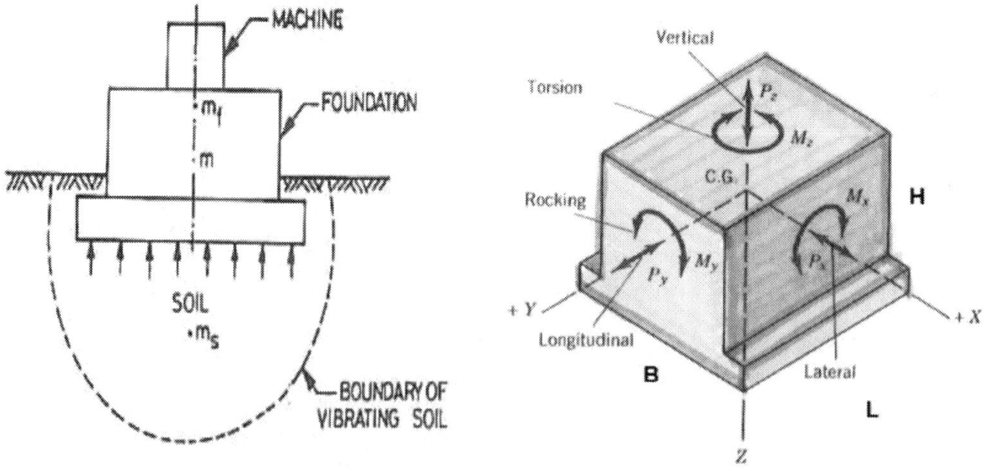

Fig 1 Block type Foundation layout and rigid body modes

The dynamic forces those are produced by the operating machine are transferred to the machine foundation. The major contributor to the dynamic forces is unbalance forces during machinery operation caused by residual unbalance (eccentricity of mass distribution) of rotor. Other limiting dynamic criterion is to avoid event of resonance and excessive transmissibility to the supporting soil or structure. If there is excessive vibration transmitted to adjoining soil by machinery foundation, then soil also vibrates randomly and continuous compacting takes place. This can cause differential settlement of foundation and eventual machinery failure.

To avoid resonant conditions, tuning of foundation based on operating frequencies is required. The value of r (frequency ratio) is considered based on required separation margin i.e. resonant frequency of foundation and affected soil as shown in figure2. Figure 2 below shows the effect of amplification of high tuned and low tuned foundation. It is clear from figure 2 that low tuned foundation should be preferred for high speed machines for better attenuation of resonance.

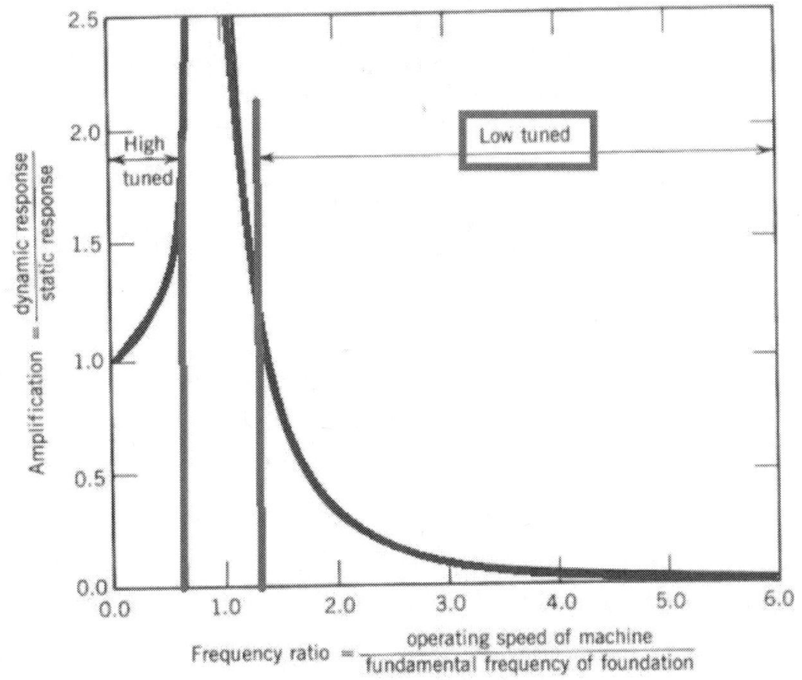

Fig 2 tuning of foundation

Above evaluation becomes more complicated of equipment operating with variable speed drives. For that reason, the natural frequency of foundation should be kept away from complete operating frequency band.

Another key ingredient to a successful design is the careful engineering analysis of the soil-foundation response to dynamic loads from the machine operation. The foundation's response to dynamic loads can be significantly influenced by the soil on which it is constructed.

Effect of Embedment of Foundation

All machine foundations are invariably embedded partly in to the ground. There are two different school of thoughts - a. effect of embedment causes increase in natural frequency and b. it causes reduction in amplitudes. By and large, it has been generally agreed that embedment tends to reduce the dynamic amplitudes. The reduction in the amplitudes could either be on account of change in stiffness, change in damping, change in soil mass participation or their combination.

For the soil to behave as an elastic material, it is necessary that the total pressure (static + dynamic) exerted by the foundation on the soil remains within elastic limits. A reasonable margin therefore should be kept while assigning bearing capacity to the soil intended to be used for machine foundation application. The dynamic pressure produced by machines not only' affects the foundation directly under the machine but to other foundations too, which are away from machine as the energy gets transmitted through soil in all directions. It is therefore desirable to keep intended margins for even static foundations i.e. foundations for static equipment including foundations of the building housing machines etc.PIP recommendation states bearing pressure should be 75% for combined static and dynamic loads which means 50% for static case only.

Static bearing capacity: proportion of footing area for 50% of allowable soil pressure, which means that the actual soil pressure should be less than 50% of static bearing capacity. The actual soil pressure equals to the weight of machine and foundation divided by the base area of footing as shown -

Actual soil pressure = $W_{m/c} + W_{found.}$ / Area of foundation

However, practised guidelines for permissible soil pressures for machine foundations and buildings/structures housing machine foundations are:

~ For low rpm machines, no reduction of soil stress is needed i.e. one can go up to 100% of the bearing capacity.

~ For medium rpm machines, reduction factor should be 10% i.e. permissible bearing pressure should be limited to 90% of the allowable bearing capacity.

~ For high rpm machineries, reduction factor should be 20% i.e. permissible bearing pressure should be limited to 80% of the allowable bearing capacity.

To determine soil property, for a construction site a site, dynamic soil investigation is carried out using wave propagation test or by vertical vibration resonance test, cyclic plate load test. The details of these tests are not the base scope of this paper.

The amplitude of vibration of a machine at its operating frequency is the most important parameter to be determined in designing a machine foundation, in addition to the natural frequency of a machine foundation soil system.

During design of block type foundation following loads and parameters are considered -

Gravity (dead) loads due to the weight of the machine, auxiliary equipment, pipe, valves, and deadweight of the foundation structure are to be considered in design. However, it is common to assume the machine frame is rigid and that its weight is uniformly distributed between support points

Static operating loads include the weight of gas or liquid in the machinery equipment during normal operation and forces, such as the drive torque developed by machines at the connection between the drive mechanism and driven machinery. The torque load is generally resolved into a vertical force couple by dividing it by the centre-to-centre distance between longitudinal soleplates or anchor point. A wider sole plate on a levelled foundation gives a better stability. Start-up torque and emergency drive torque such as short circuit torque, phase fault, auto-start obtained from the motor manufacturer.

Dynamic loads due to unbalanced masses-Dynamic forces transmitted to the bearing pedestals and then to casing under the following conditions— a) Under design unbalance levels over operating speed range; b) At highest vibration when passing through critical speeds; at or near a critical speed, the force from rotating unbalance is highly amplified. c) At a vibration level where the machine is just short of tripping on high vibration; and d) Under the maximum level of upset condition / transient condition such as passing through critical speed the machine is designed to survive. The forcing function F(t) equation shown below is a generic function to determine the dynamic force applied on the foundation at any operating frequency **Ω** corresponding to each of the rotating machine.

$$F(t) = M_i.(G.\Omega^2 / \omega) \times \text{Sin}\,(\Omega t + a_i)$$

Where M_i is mass of rating part i , G is balance quality, G = e.ω , ω is designed rotation speed in radian / sec , Ω is rotational speed of machine for which unbalance force is calculated.

Most of the rotors are balanced to an initial balance quality either in accordance with the. ISO 1940 - balance quality of rigid rotors in terms of a constant emωo. The normal balance quality Q for parts of process-plant machinery is 6.3 mm/s-(Balance quality 6.3). The rotor intended for higher speeds are balanced to balance quality grade G2.5 or less as per ISO 1940 table.

Machine unbalance equivalent to trip vibration level and effective bearing stiffness should be considered. Since a rotor is often set to trip on high vibration, it can be expected to operate continuously at any vibration level just below the trip limit and hence foundation must withstand this type of abnormal situation which may occur.

Transients - emergency torque due to entire short – circuit faults which can occur. A line to line short circuit at the driver terminals causes the most severe loading of foundation. Load due to bowed rotor-A bowed rotor can impose large dynamic forces on the machinery foundation. This is very difficult to quantify during foundation design stage.

Loads from multiple rotating machines—if a foundation supports multiple rotating machines, such as speed increaser or reducer gear drives, the foundation designer computes unbalanced force based on the mass, unbalance, and operating speed of each rotating component. The response to each rotating mass is then combined to determine the total response-While we state foundation as rigid, that means elastic deformation of thick foundation shall be negligible. But the soil on which machinery block foundation is placed with some embedment is not so rigid. Hence elastic properties of soil -foundation interface must be addressed.

During dynamic analysis of machine foundation, following design parameters are taken for soil and machinery operating conditions -

Soil – These include soil density, Poisson ratio, shear wave velocity / dynamic shear modulus, coefficient of friction for sliding, damping ratio, allowable soil bearing capacity.

Machinery operating and design parameters- These include Machine and Foundation Dimensions, Operating Frequency of Rotating Parts, Coordinates of CGs, Coordinates for Point of Application of Forces, Eccentricity of Unbalanced Forces for Rotating Parts, Critical Speeds, Rotor Weights, complete machinery weight

Foundation- These include grout type and extent in plan, levels, base plate and anchor bolt details / setting details with uprooting force.

As a rule of thumb, dynamic analysis of foundation is carried out for equipment heavier than 22kN with frequency checks, amplitude and velocity checks are carried out with time-based plots.

Torsional and rocking modes are controlled by adequately sizing length, breadth and height of foundation. Typically, stability checks are done with static loads

Methods of Shear, compression waves absorption / dampening

A foundation on absorbers is usually made of two parts: a lower slab or a sole plate on which the absorbers are placed and an upper foundation block resting on the absorbers. The machine is anchored to the upper foundation block. A schematic sketch of a machine foundation on absorbers is shown in Fig 8. V_c is compression wave velocity and V_s is shear wave velocity.

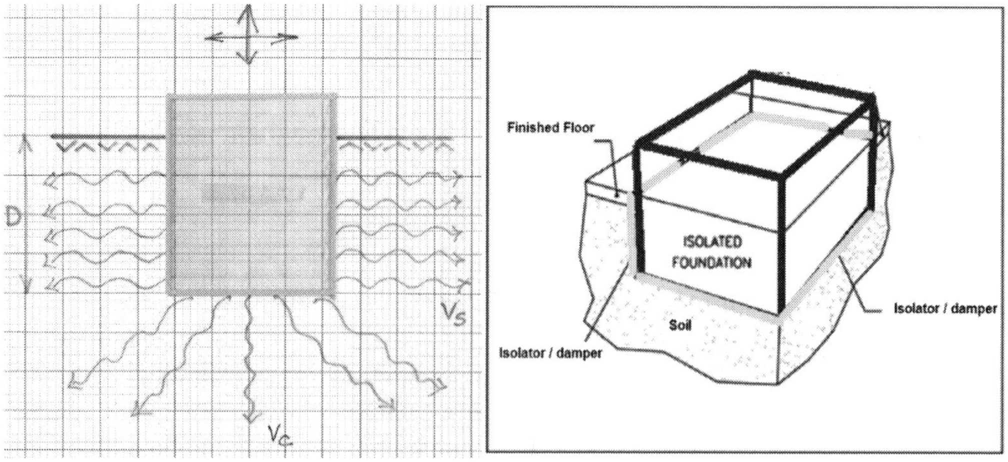

Fig 3 -Shear & compression waves and dampening devices (shown in yellow)

Materials capable of undergoing elastic deformation can be used as vibration absorbers. Commonly used vibration absorbers are; steel or metal springs, cork pads, rubber pads, timber pads, neoprene pads and pneumatic absorbers. However, it is important to consider that waves should not get reflected back to origin i.e. machinery foundation.

A correct isolation between machine foundation and the surrounding area will result in trouble free operation to machine and neighbouring machines, negligible possible compacting during operation.

The isolation / damping material absorbs machine-induced energy, limits the transmission of higher frequency disturbances and provides isolation from ambient and induced shock and vibration

A foundation block or Vibration isolation mountings for a high dynamic machine is required in order to reduce the transmission of vibration and shock to nearby machines / building structures. To control the source of vibration disturbance through the use of resilient insulating materials is known as active vibration isolation.

When it is not possible to prevent or sufficiently lower the transmission of shock and vibration from the source, a resiliently supported vibration insulating foundation block can be used for the passive vibration isolation of critical equipment in vicinity.

Open trenches and (gas cushions type panels) are the only wave barriers that have been extensively studied and used since long time. However, many designers have successfully tested elastomer-based isolation / damping devices.

A seismic isolation pad produced by utilizing the scrap tire rubber which contains interleaved steel reinforcing cords has been proposed. The steel cords are expected to function similar to the steel plates used in conventional laminated rubber bearings.

A new type of seismic isolation system has been introduced using the scrap tire rubber as easily available material at low or negligible cost. It can be found that the research on scrap tire pad has been initiated by Turer et al. In the previous studies, the researchers produced the specimen samples and tested in vertical compression and in horizontal shear. These bonded samples would be produced by stacking one on top of another with the application of adhesive. The car tire usually contents synthetic fibre as reinforcing cords and are very stable in such kind of forces.

Observations on current methodology of design -Machinery dynamics perspective

Following are major observations in present theory of foundation design –

The theory does not identify the type of rotors and bearings being used in rotor bearing system of subject rotating machine. It assumes that all vibratory forces generated during machine operation are completely transferred to machine foundation. This can only happen when rolling element bearings are used where the structural damping due to metal to metal contact is very less. Machinery having rigid rotors uses vibration measurement at bearing cap which is a part of casing because vibration magnitudes are indicative of excitation forces generated when rolling element bearings are used. Medium energy pumps, Fans and blowers, LV electric motors and aero-derivative gas turbines measure vibration on bearing cap / housing in terms of mm/ sec RMS as per ISO and API guidelines.

Machinery having flexible rotors (such as high energy pumps, centrifugal compressors), vibration measured only on bearing cap is not be good indication of machinery health as flexible rotor may experience high displacement while due to fluid film damping and stiffness of bearing pedestal, vibration measured at bearing cap is quite low. As experimental verification, shaft displacement measured (overall vibration up with FFT range of 1000 Hz) on an operating centrifugal compressor as 25 micron pk pk (meeting API 617 criteria) showed 0.3 mm/sec rms as bearing housing vibration. For easy understanding, we considered this phenomenon occurring at a single frequency (1 X only) and converted displacement into velocity mm/sec rms using Sinusoidal Waveform Vibration Calculator. This showed the vibration at bearing cap as 9.2 mm/ sec rms, which may be termed as faulty machine as per table A.1 of ISO 10816-3.

Frequency					
					166.666 Hz
					10000.0 cpm
Displacement	0.348	mil rms	8.839	μm rms	
	0.492	mil pk	12.500	μm pk	@ 0.0 °
	0.984	mil pk-pk	25.000	μm pk-pk	
Velocity	0.364	ips rms	9.256	mm/s rms	
	0.515	ips pk	13.090	mm/s pk	@ 270.0 °
	1.031	ips pk-pk	26.180	mm/s pk-pk	

Fig 4 a – snap shot of vibration conversion calculator. Displacement to velocity

When we converted 0.3 mm/sec RMS to displacement assuming same "one" frequency of interest, we found equivalent displacement is quite low as shown below –

Frequency					
					166.666 Hz
					10000.0 cpm
Displacement	0.011	mil rms	0.286	μm rms	
	0.016	mil pk	0.405	μm pk	@ 0.0 °
	0.032	mil pk-pk	0.810	μm pk-pk	
Velocity	0.012	ips rms	0.300	mm/s rms	
	0.017	ips pk	0.424	mm/s pk	@ 270.0 °
	0.033	ips pk-pk	0.849	mm/s pk-pk	

Fig 4 b – snap shot of vibration conversion calculator. velocity to Displacement

This observation clearly proves that amount of vibration occurring at rotor is not fully transmitted to casing for every rotating machine even though the readings may close to each other as per graph below –

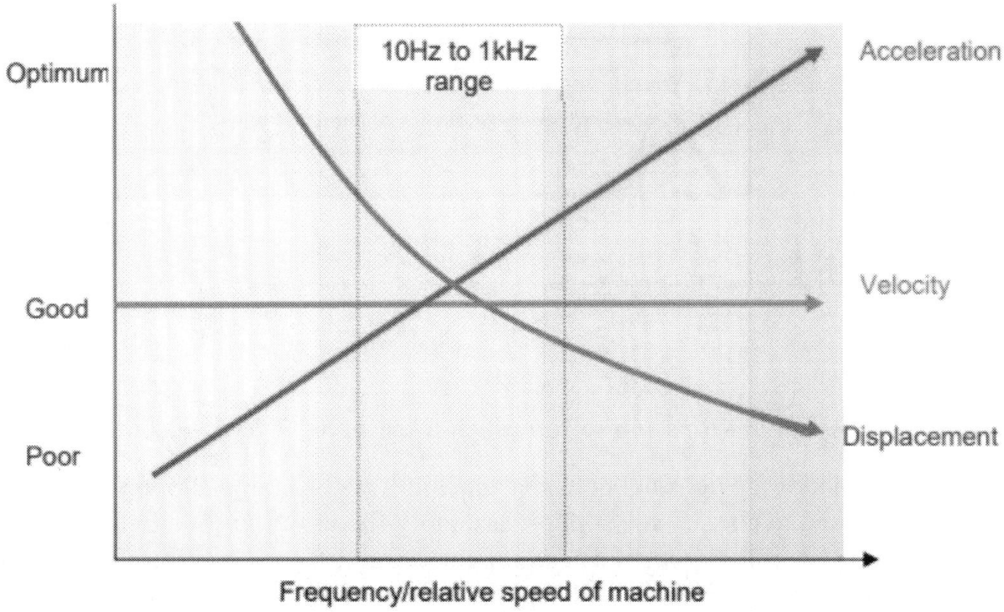

Fig 5 – Range of various measurement accuracy of vibration and trade-off

In case of rotors using hydrodynamic journal fluid film bearings, a substantial amount of vibration is attenuated due to dynamic stiffness of rotor bearing system as shown -

K Dynamic Stiffness= $(K-M\Omega 2) +j(D\omega - \lambda D\,\Omega)$

Where K is the spring stiffness, M is the rotor modal mass, D is the rotor/bearing viscous modal damping, and ω is a term that represents the frequency of the vibration precession (or orbiting), which may be different from the rotation speed, Ω. λ is the Fluid Circumferential Average Velocity Ratio of the fluid in the bearing or seal.

As known, Vibration Displacement =

Summation of Forces Acting on the Rotor System divided by the System Dynamic Stiffness

In the case of rotor with fluid film bearing, vibration energy is absorbed and dissipated via the direct damping term, $D\omega$, which is the rotor pushing on the fluid, as a shock absorber effect. In high speed turbo machines, adequate damping occurs within the bearings as they are rigidly mounted to the casing and foundation and the foundation spring rigidly connected to ground.

Due to this viscous damping by lube oil and viscous shear, energy is dissipated per cycle in form of heat and hence oil temperature is controlled by air cooler / heat exchanger in a rotating machinery employing hydrodynamic bearings.

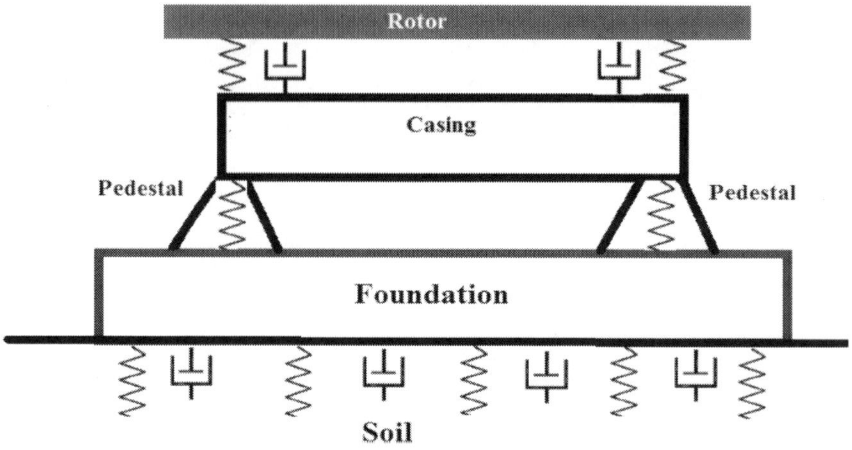

Fig 6 -Transmission of vibration to foundation -rotor with hydrodynamic bearings

The design methodology does not consider machines according to support flexibility Two conditions are used to classify the support assembly flexibility in specified directions: – rigid supports; – flexible supports. These support conditions are determined by the relationship between the machine and support / pedestal flexibilities/ stiffness.

With current method of foundation design, it can be seen below that maximum amplitude are found much below than as required or prescribed by OEM / text book requirement (20 am vs 1.6 am) which means foundation is overdesigned in terms of dynamics.

Type	Permissible amplitudes (m)
Low-speed machinery (500 rpm)	0.0002 to 0.00025
Hammer foundations	0.001 to 0.0012
High -speed machinery: a. (3000 rpm) b. (1500 rpm)	 0.00002 to 0.00003 0.00004 to 0.00006

Fig 7 - Generic permissible amplitudes of foundation vibration

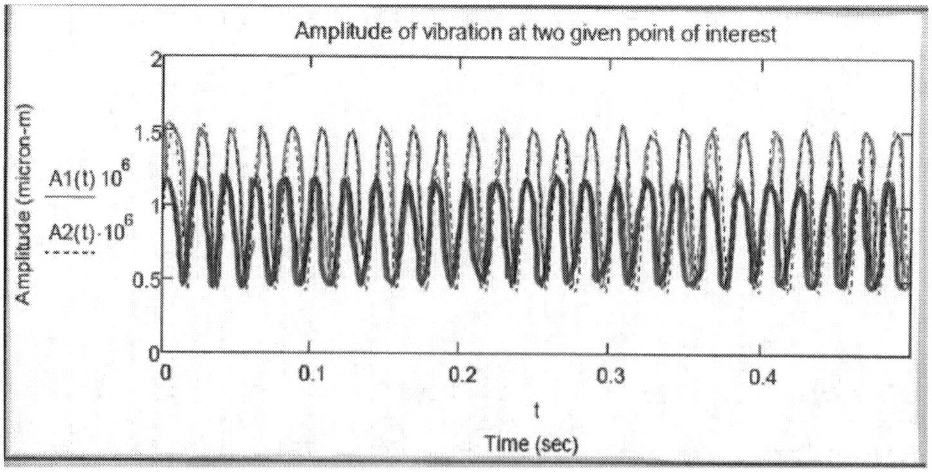

Fig 8 Foundation analysis Output file for actual amplitude

Proposed approach for foundation design - machinery dynamics -

In rotor-dynamic point of view the foundation design should consider classification of machine in terms of type of bearings, actual balance grade, dynamic stiffness ratio of pedestal and bearings and type of device used to measure vibration.

It is seen that fluid film bearings dampen the rotor vibration and a low fraction of vibration is transmitted to foundation which is based on stiffness ratio of bearing pedestal and rotor bearing. With high thickness of casing of heavy-duty rotating equipment, the stiffness is quite high compared to rotor and hence negligible amount of vibration is transmitted to foundation. Fig 8 illustrates the visual comparison of bearing stiffness vs casing stiffness.

Fig 9 - High speed turbo-machine Rotor (red) with highly stiff casing (in blue)

If the support stiffness is 3.5 times higher than bearing stiffness at particular operating frequency, support is considered adequately rigid as suggested by API. With conservative value of estimated amplification factor, unbalance response of rotor at first critical speed and vibration values at centre span can be calculated. This value can be used as eccentricity to calculate total force exerted on bearing pedestal / casing even in case of machine having rotor sag or temporary rotor bow. When machine and foundation is considered as one entity, vibration measured and transmitted shall be quite lower than required.

According to API standards the permitted initial mass eccentricity is tighter than ISO balance quality grade G2.5, which is normally applied to high speed equipment under ISO 1940. With API consideration, actual grade is ISO grade G0.67 equivalent. Rotating machine manufacturers often do report the unbalance that remains after balancing to demonstrate that their internal Quality plan and API criteria are met. For machines with high casing stiffness and using active magnetic bearings it can be considered that very low vibration is transmitted due to magnetic levitation of rotor.

Another point is to consider the method of measuring vibration of various types of machines based on the pedestal / casing stiffness and bearing stiffness. Some of high-speed machines use relative shaft displacement methods since long time to measure onerous data for machinery protection and monitoring. As per ISO TR 19201:2013(Titled

-Mechanical vibrations – methodology for selecting appropriate machinery vibration standards), this choice is based on ratio of dynamic stiffness of pedestal (bearing housing connected with casing) and dynamic stiffness of bearing. The dynamic stiffness related to damping and mass effect of components.

For massive casings rigidly connected to foundation designed for static loads, the overall vibration amplitude is miniscule and can be considered as negligible as they normally use relative shaft vibration measurement (zone in yellow) based on following chart from ISO TR 19201:2013.

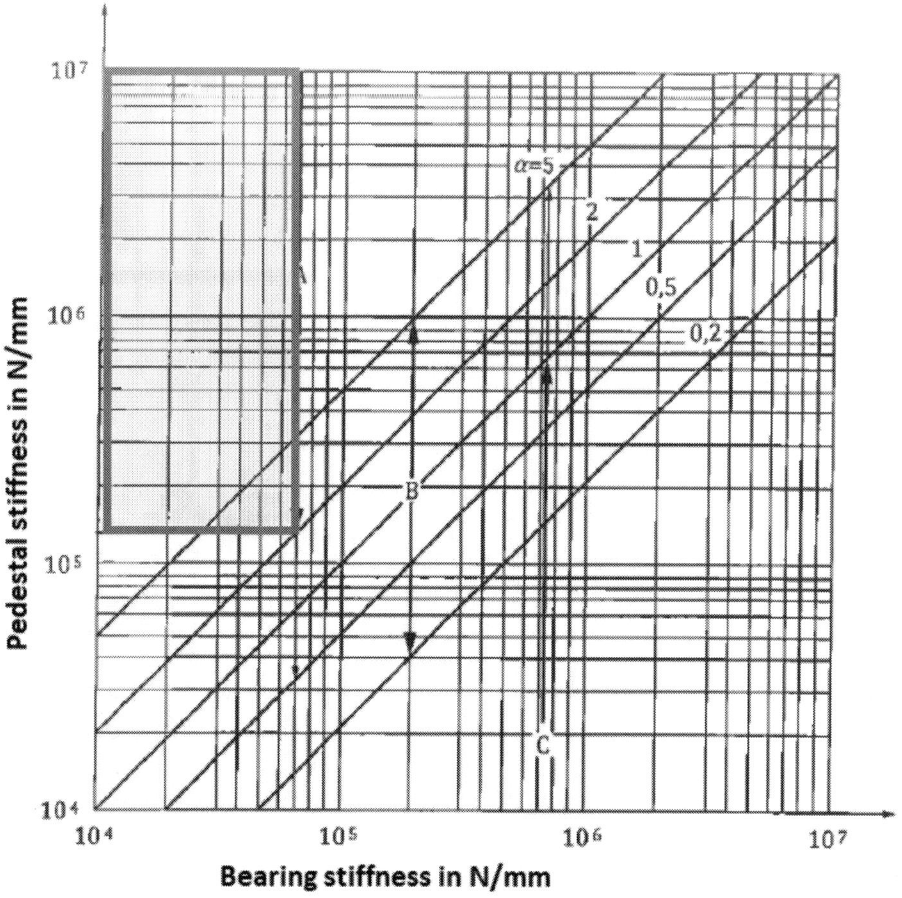

Fig 10 -Zone of measuring relative shaft vibration based on stiffness ratio

It is worth to note that machines equipped with active magnetic bearing create adequate stiffness and damping by magnetic flux and enable rotor to levitate. Hence very low vibration transmission occurs from rotor to stator. This means the foundation also is subjected to very low vibration amplitude.

Effect of transients – Foundation Design against short circuit - Now a days the protection system such as IGBT type transistors, rotor flux monitoring system can dictate and isolate the short circuit from source and in motor itself. The instrument can monitor and evaluate up to four machines at the same time with stored the base line measurements, so that when a rotor shorted turn is detected, an alert relay is activated.

Conclusion with proposed methodology

The rotating machinery should be classified as machine with antifriction bearings and / or machines with lower ratio (<3.5) of dynamic stiffness as listed in ISO TR 19201:2013 to conduct dynamic analysis of foundations. They include - OH2 pumps, low energy pumps as defined in API 610, centrifugal fans, electric motors, screw compressors, steam and industrial gas turbines turbine. For high speed turbomachines with higher ratio of dynamic stiffness, relevant sections of ISO 7919 with a conjunction of ISO 19201 to be followed. Below table Table A.1 of ISO 10816-3. shows shaft displacement and bearing cap vibration velocity both are to be considered together as acceptance criteria as shown

Support class	Zone boundary	R.m.s. displacement µm	R.m.s. velocity mm/s
Rigid	A/B	29	2,3
	B/C	57	4,5
	C/D	90	7,1
Flexible	A/B	45	3,5
	B/C	90	7,1
	C/D	140	11,0

However , in both type of machines, it is imperative to ensure that bearing housing natural frequencies do not coincides with any operating frequency and its harmonics .

For turbo-machines, dynamic analysis of foundation is not needed to be carried out if –

- Machine incorporates a "between bearing design" with operating frequency equal or higher than 100 Hz or machines with active magnetic bearings or hydrodynamic fluid film and
- Rotor dynamics analysis has considered absolute vibration and relative vibration as close or same and
- Machinery protection probes are installed on casing and rotor meets ISO 1940 balance quality as 1 or lower (API residual unbalance standard) and
- No vibration trip multiplier is incorparted in set up.

In that case, only static analysis of foundation can be done with following guidelines –

1. The bottom of block foundation should be above water table. It should not be resting on back filled soil or soil prone to excessive settlement.

2. The mass of rigid foundation shall be 2- 3 times the mass of supported machine (for centrifugal machines with between bearing, fluid film design. In addition, foundation mass shall not be less than 10 times mass of rotor.

3. The top of block is to be kept (0.3 m) above finished floor or high point of paving to prevent damage from surface water run-off.

4. The vertical thickness of block should not be less than .600 mm. However, if the installation is under the shade, same can be optimized to 300 mm. Also, thickness of block should be enough to accommodate anchor bolt embedment

5. The foundation should be wide enough to increase damping in the rocking mode. The width should be at least 1.5 to 2 times height of machine centreline from machine base

6. The combined centre of gravity of machine and foundation should coincide with the centre of gravity of the foundation.

7. Static bearing capacity is proportion of footing area for 50% of allowable soil pressure, which means that the actual soil pressure should be less than 50% of static bearing capacity.

8. Static settlement must be uniform; centre of gravity of footing and machine load should be within 5% of each linear dimension from the foundation centre.

9. End user can insist machinery OEM that rotor should have lower amplification factor i.e. less than 4. Lower amplification factors mean high damping and that means lower force on bearing during resonance condition.

References & Bibliography –

1. Foundations for dynamic equipment- ACI 351.3R-04
2. Design of structures and foundations for vibrating-machines-by Neil & Arya
3. Foundation for Industrial Machines- K.G. Bhatia
4. Foundation Analysis and Design- Bowles
5. ISO 10816-1 Mechanical vibration -- Evaluation of machine vibration by measurements on non-rotating parts -- Part 1: General guidelines
6. ISO 10816-3 Mechanical vibration -- Evaluation of machine vibration by measurements on non-rotating parts -- Part 3: Industrial machines with nominal power above 15 kW and nominal speeds between 120 r/min and 15 000 r/min when measured in situ
7. ISO/TR 19201:2013 Mechanical vibration -- Methodology for selecting appropriate machinery vibration standards
8. ISO 1940 Mechanical vibration -- Balance quality requirements for rotors in a constant (rigid) state
9. Structures subjected to start-up and shutdown of rotating machinery- A A Elvin; N G Elvin- South African Institution of Civil Engineering
10. Vibration absorption and isolation in dynamically loaded foundation - T. Öztürk and Z. Öztürk- The 14th World Conference on Earthquake Engineering October 12-17, 2008, Beijing, China
11. Advanced Foundation Engineering - Prof. T.G. Sitharam -Indian Institute of Science, Bangalore
12. Vibrations of machine foundations and surrounding soil - H. van Koten, and P.C.J. Hoogenboom, the Netherlands
13. The Dynamic Principles of Machine Foundations and Ground -J. H. A. Crockett, B.Sc. (Eng.), and R. E. R. Hammond
14. Handbook of Machine Foundations - Srinivasulu & Vaidyanathan
15. Various OEM catalogues for foundation isolation
16. Various OEM cataloges for partial discharge and Continuous On-line Rotor Flux Monitoring system .

ANALYZING THE CENTER OF MASS OF ATHLETES PERFORMING BOTH AN OPEN AND CLOSED SKILL EXERCISE

Amanda Delaney, Kimberly Bigelow
University of Dayton
Department of Mechanical Engineering
300 College Park
Dayton, OH 45469-0238
Telephone: 937-229-5374
Amanda Delaney delaneya2@udayton.edu

Mark Derriso
Wright Patterson Air Force Base
1253 Chestnut Street
Wright-Patterson AFB, OH 45433

Christine Kabban-Schubert
Air Force Institute of Technology
2950 Hobson Way
Wright-Patterson AFB, OH 45433

Ed Downs
ProTERF
6808 SW 81 Street
Miami, FL 33143

Abstract: In looking into the components of human monitoring systems, there are three main elements that comprise a system: sensors; data acquisition and communication; and data processing and analytics. Sensors that function with the purpose of sensing body movements or collecting specific physiological or biological parameters of an individual are typically known as wearables. Wearables used for capturing body movements are primarily inertial measurement units (IMUs) which utilize sensor fusion to combine the technology of an accelerometer, gyroscope, and magnetometer. Data obtained from these wearables may provide important insight into subtle differences in body movements that influence performance outcomes. The long-term goal of our work is to develop approaches that enable the prediction of an individual's performance in an open-skilled environment. The specific aim of this research was to determine how data from a full body IMU-based system could be used in detecting subtle movement differences in the execution of a pre-planned agility test versus a reactive agility test.
Ten healthy young adult males who were regularly physically active and had played on a

sports team within the past four years participated in this study. An Xsens Awinda 17 sensor suit was used to capture body movement data during the two agility tests. In both the pre-planned and reactive agility test, study participants stood facing six programmable illuminating lights, arranged in a 3 meter arc, 1.5 meters apart from each other, with the center being the starting position. For the pre-planned test the participant was informed which three lights, in order, that they would run to, with the requirement that they return to the center before advancing to the next light. For the reactive agility test, the participant was informed that all 6 lights would turn on, but the light they must run to and turn off would not have a color pair (ex. 3 red, 2 blue, 1 green). Participants were also asked to remember the color sequence of the three lights they turned off for an additional stressor.

Throughout the tests, each participant's center of mass (CoM) was tracked and compared to the direct path to the light with deviations calculated using standard sums of squares error (SSE). Study participants were broken into two groups for comparison: faster individuals and slower individuals based off of their completion time on the pre-planned agility test. Results from the pre-planned agility assessment indicated that the mean SSE for the CoM deviations to the light averaged 1.66 ± 1.43 meters for the faster performers and then 2.96 ± 2.05 meters for the slower performers. For the reactive agility tasks, the CoM deviations were higher with an average SSE of 5.82 ± 4.16 meters for the faster performers and 6.97 ± 7.60 meters for the slower performers. This suggests that the slower individuals were not just slow, but were likely slow because they were inefficient. Additionally, these findings indicate the notable increase in inefficiency generated by the uncertain scenario of the reactive agility test which corresponds to the 84% time increase from the pre-planned test.

Key Words: Human monitoring systems, Reactive Agility, Pre-planned Agility

Introduction: Over the years, training regimens continue to adapt to include new innovative training drills in order to enhance an athlete's skills. While revisions continue to shape the training realm, the essential goals remain the same: to assess athletes and improve their agility. With the evolution of training drills, the term agility has been modified from its previous meaning of solely the ability to move both quickly and effectively to now include an individual's ability to react to an uncertain stimulus [1]. To prevent confusion, the field has introduced the word reactive agility to coincide with the new definition. In adding the word reactive, the term more holistically describes the type of agility required in an open skilled sport. An open skilled sport is considered to be any sport in which the environment is unstable and requires the athlete to adapt to sudden changes (ex. football, basketball, soccer, etc.) [2]. A prime example of this is when a substitution is made and the defender has to change his or her approach to defending, or when a team is running a new play unknown to the opposing team, or simply during play where having a faster reaction and decision making time allows an athlete to beat their opponent. Reactive agility is more representative of what is seen during game play.

One of the drills innovated to incorporate this new inclusion of a response to a stimulus, which is essentially creating some sense of uncertainty, is known as the Reactive Agility

Test (RAT) [3,4,5]. The RAT has many different test setups. One of the setups includes a "Y" shaped path where two lights or lighted gates are placed at the ends of the fork and the player starts at the bottom of the "Y" [3]. Once a player runs to the fork a sensor trips one of the lights or gates to turn on and the player is to run to the illumination. The RAT setup has been used for both performing a pre-planned drill as well as the reactive agility drill. For the pre-planned test, the player is told which light/gate will be the one that illuminates before the drill starts, erasing the uncertainty of the drill. Another test variation includes a tester who initiates a movement to the left or right where the athlete is required to mimic the movements [4]. This version of the test is easy to run and includes the sense of an actual opponent. Timing equipment and cameras capture the athlete's movements for further analysis. Instead of using a person as a tester, the test can also be virtual via a video screen and have the option of a pre-planned or reactive test as well [5]. Research comparing a pre-planned version to the reactive agility version has found that the reactive version yields slower reaction and completion times [3,4,5]. This increase in time lends itself to correspond to the fact that the athlete has to use his/her cognitive skills, such as decision making and reaction time, to overcome the uncertainty that was added. Cognitive flexibility, while not often assessed, is critical for making the correct adaptations to an environmental stimuli [6].

When looking to sense and measure the variables that allow a person to succeed, human monitoring systems become crucial. These monitoring systems are usually comprised of three main elements which include sensors, data acquisition and communication, and data processing and analytics [7]. For the purposes of monitoring human performance, wearable technology has emerged to sense anywhere from physiological data to biomechanical data. Wearables specific to biomechanical data, capture body movements through incorporating Inertial Measurement Units (IMUs) with algorithms to process signals and collect data. The IMUs work by combining the functions of an accelerometer, gyroscope, and magnetometer for the purpose of body movement sensing. These wearables can allow for further analysis of both an athlete's movements and efficiency during a training drill, especially as they can provide insight on subtle differences that may not be captured without them.

Wearables enable agility to be examined in a more objective way, where slight movements and form are able to be detected, measured, and quantified. While movements of anywhere on the body may be of interest, efficiency of movement may best be represented by examination of center of mass movements (CoM). In looking into an athlete's biomechanics, one's Center of Mass (CoM) is defined as the location where his/her mass is distributed equally [8]. This location is commonly used to determine the stability of a person during a static or dynamic movement. Essentially we must keep the displacement and velocity of our CoM within our base of support or else we will become unstable and fall [9]. In looking at CoM in the athletic realm, it is used as a marker for measuring postural control and balance. This is not only important for enhancing performance and efficiency, but also for preventing injuries. In golfers, the CoM can be tracked throughout a swing to correct form [10]. For hurdlers and runner, studying CoM trajectories and deviations can assist a trainer in making the corrections that maximize efficiency [11-13]. While not often used in agility-related research, studying the CoM may add significant value.

As part of a larger research study, the long term goal of this work is to develop a method for predicting which individuals will prove to be high performers in an open skilled environment.
However, the specific aim for this research was to analyze the CoM movements between the pre-planned and reactive agility drill. It was hypothesized that there would be a greater deviation from an efficient path found in the reactive agility drill as a result of the added uncertainty. A secondary aim was to compare the CoM deviation in a group of poorer performers to better performers based on their pre-planned times to assess any difference between those who were faster and those who were slower. It was hypothesized that there would be a noticeable difference in CoM efficiency between those who were slower in comparison to those who were faster as this may account for some of the time difference.

Methods:

Participants:
Twenty healthy, athletic male participants were recruited from the University of Dayton's campus. All participants met the inclusion (>150 minutes of exercise per week which includes working out, running, lifting weights, or biking, and currently participating on a sports team of any level or have participated on a sports team of club level or higher within the past 4 years) and exclusion criteria (have not had a serious injury within the past 6 months, have not had a concussion within the past year, is not colorblind, has no heart conditions, does not have moderate to severe asthma, and has no current pain with physical activity). Colorblindness was an exclusion criteria as differentiating between light colors was important for the reactive agility drill. The population was as follows: ages 19 – 23 years (21 ± 1.214 years), in weight from 58.97 - 104.33 kg (79.424 ± 10.674 kg), and in height from 169 - 197 cm (180 ± 8.67 cm). A Physical Activity Readiness Questionnaire (PAR-Q) was administered after written informed consent to further clear all participants for the study [14]. The University of Dayton Institutional Review Board approved the study protocol.

Protocol:
As part of a larger research study, participants performed their pre-planned and reactive agility drills on the second (final) day of testing. Both testing days required an 8 minute warm up before advancing to the physical tests. The pre-planned and reactive agility tests followed six physical assessments and were the last two drills of the day for all participants. All participants were wearing a Zephyr bioharness band and a Xsens Awinda 17 sensor suit to record physiological and biomechanical data respectively.

For this research, the RAT test setup was amended even further to encompass a more challenging situation. The test setup mimicked a 3 meter arc with 6 Fitlights equally spaced among the arc. The center point of the arc was deemed as the starting position for all of the athletes. An example of this can be seen below in Figure 1. This newer setup increased the variability in comparison to the RAT test as there were more directional changes. Rather than just running in the direction of the illuminated light and being done, the drill required them to return to start and then run to two more lights. Fitlights allowed for researchers to program certain sequences and utilize different colors.

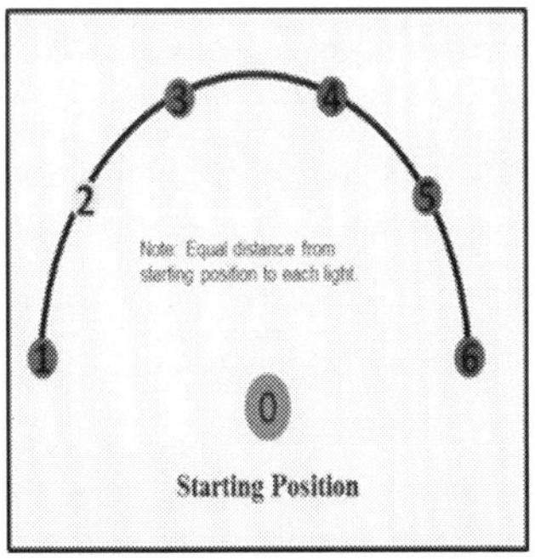

Figure 1: Test Set-up

For the pre-planned test, all participants were instructed to run to lights 1, then 3, and then 6, returning to home in between each light. This counted as one trial and three trials were completed for each subject with a 60 second break required in between. Before advancing to the reactive agility test, a 3 minute break was required in attempt to mitigate fatigue.

For the reactive agility drill, the sequence of lights was unknown to the athlete and all 6 Fitlights turned on in sync. With the concurrent lights, the sequence was programmed to utilize different colors for creating the element of visual scanning. Furthermore, all participants were instructed to run to whichever light had no color pairs, essentially the odd color light (example: 2 red, 3 green, and 1 blue). Coinciding with the pre-planned drill, there were a total of 3 lights to turn off in one trial and the participant was instructed to return to home in between each light. To add an additional stressor to the reactive agility drill, the participants were told to remember the three color sequence in order of the lights they turned off. Three trials of the reactive agility drill were ran with the same 60 second break between each trial.

Time to completion was recorded for both the pre-planned and reactive agility drill via a stopwatch. The start for each test was initiated by the researchers saying "Go" and was stopped when the athlete returned back to the starting position after the final light.

<u>Analysis:</u>
MVN 4.4 software was used in compatibility with the Xsens Awinda suit. All body movements were collected at an update frequency of 60Hz. Out of the 20 subjects, 10 subjects were used for analysis as a result of equipment malfunction or incomplete data. These 10 subjects were then divided into two groups ("above average performers" and "below average performers") based off of their times on the pre-planned agility drills. The best trial out of the 3 was chosen for plotting and further analysis.

For determining CoM efficiency based off of the path to the light, the Xsens system provided the raw CoM positioning data in three-dimensional space over time. These values were exported into Matlab. Plots of the CoM in the forward-backward direction versus the CoM side to side direction throughout the drill were plotted in Matlab for both the pre-planned and reactive agility drills. This was done to capture the athlete's path to the three lights as they were diagonal curves. On the same plot, three straight line paths were plotted to represent the direct path to all three lights from the start, figure 2. To calculate the efficiency of staying on the linear path to the light, Sum of Squared Errors (SSE) was used. SSE was calculated by taking points on the curve (only on the path to the light) and subtracting them from points on the straight line path to the light, summing the differences for a final error. This error is essentially the area between the two paths to determine the efficiency of the athlete's movements.

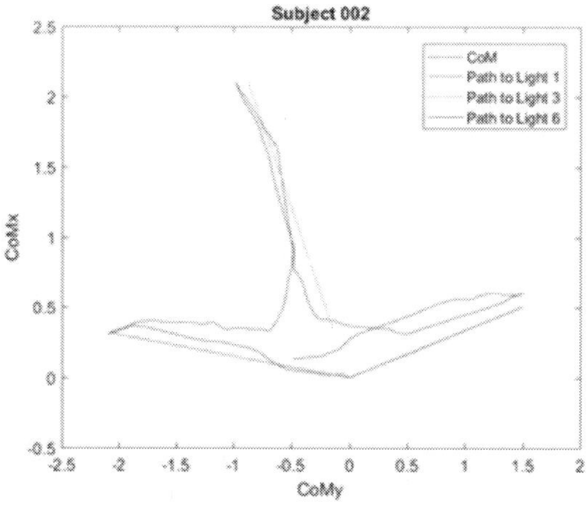

Figure 2: Example of a CoM trajectory for the Pre-planned Agility Drill

Results: Results for the ten athletes' best trial times for both the pre-planned and reactive agility drills can be seen in Table 1 below.

Table 1: Time to Completion in seconds

	Better Performers					Poorer Performers				
Pre-planned (s)	5.13	5.82	5.57	5.56	5.36	7.31	6.55	6.63	8.23	6.78
Reactive (s)	10.35	11.25	11.38	12.26	12.08	13.40	12.56	11.90	13.63	12.29

Pre-planned agility CoM plots portrayed curves that did not generally return back to (0,0) as only one foot was needed to touch start. For the reactive agility test, all subjects fully returned to the home location in order to visually assess the next light sequence. The SSE used to calculate CoM deviations from the path for the 5 good and 5 poor performers can be seen below.

Table 2: SSE for all three lights during the Pre-planned Agility Drill

	Better Performers					Poorer Performers				
Light 1	0.283	0.069	0.207	0.854	0.158	0.048	0.509	0.114	1.002	0.853
Light 2	1.943	0.682	2.327	1.304	9.372	1.170	12.95	6.845	7.097	1.617
Light 3	1.092	0.084	1.956	0.918	3.747	0.293	3.135	8.151	0.293	0.314
Average	1.106	0.278	1.497	1.025	4.426	0.504	5.530	5.037	2.797	0.928
Overall Average ± Standard Deviation				**1.66 ± 1.4349**		**2.959 ± 2.0545**				

Deviations from the path during the reactive agility drill can be seen in Table 3 below.

Table 3: SSE for all three lights during the Reactive Agility Drill

	Better Performers					Poorer Performers				
Light 1	2.163	0.062	1.407	1.884	3.398	1.087	2.381	12.711	0.911	1.802
Light 2	3.275	1.431	14.915	6.800	6.112	5.978	0.696	12.109	42.126	1.904
Light 3	10.134	1.539	2.621	29.946	1.537	0.916	0.313	1.755	15.420	4.495
Average	5.191	1.011	6.314	12.877	3.682	2.660	1.130	8.858	19.486	2.733
Overall Average ± Standard Deviation				**5.815 ± 4.1577**		**6.974 ± 7.5967**				

Discussion: The objective of this study was to analyze CoM as a biomechanical predictor of efficiency in an athlete's performance in a pre-planned versus reactive agility test; while also acknowledging that with an increase in variability, comes a decrease in efficiency. While the RAT is becoming more popular in literature, it has largely been analyzed based on time differences and has not seen enough analysis based off the biomechanical differences in comparison to the pre-planned version. Confirming CoM as a biomechanical marker to be used for efficiency analysis will allow trainers to make the proper form corrections on their open skilled athletes. This is where human monitoring systems such as wearables come into play to detect the biomechanical markers that trainers cannot see.

With time being a commonly used variable for quantifying performance, especially that of speed, it was used to break up the study population into two comparison groups: above average and below average performers. Specifically time to completion of the pre-planned agility drill was used as it did not include the variability that the reactive agility drill saw. In looking at the completion times of the pre-planned and reactive agility drill, there was an average increase in time of 84% seen for the reactive agility drill. This was expected due to the cognitive demand and uncertainty of the reactive agility drill. However, it is unclear if this change is due to the increase in processing time, inefficient movements, or a combination of factors.

In comparing the CoM deviations from the pre-planned to the reactive agility drill, the above average performers saw a 250% increase and the below average performers saw a 136% increase. While it is somewhat surprising that the above average performers were more affected by the open task, they still, on average, were more efficient than the below average performers. This may be due to a different technique approach where some athletes are programmed to run in a more curved fashion, a common defensive move, but still maintained a fast speed while doing so. Additionally the above average performers may not perform well under the cognitive load and uncertainty, whereas some below average performers were able to use their cognitive skills to make up for their lack of speed.

It is likely that the inefficiency in movement added additional time to complete the task; though this alone did not likely cause slower times. Additional information which could be captured with additional sensors, such as processing and reaction time can provide further insight to why an athlete performed differently in addition to their efficiency to the light. Furthermore, analyzing reactive agility requires more than just one factor to be examined and more wearables besides just ones that capture body biomechanics to effectively conclude results. This research attempts to dive into one of the factors of reactive agility for the benefit of future research to continue analysis on the drill.

The method for solving for the CoM deviations was not the most robust, so even though these findings follow expectations, the accuracy of the results may be questionable due to error and uncertainty within the wearable sensors. The use of the Xsens suit to predict the CoM location is dependent on accurate body dimensions, sensor placements, and the system program. Noticeable drifting or slight avatar deviations from the person warranted us to remove some subjects from further analysis, while those kept in for this analysis did not exhibit any evidence of problem, there could be some inaccuracies as well. This acknowledges some of the issues with relying on human monitoring systems and in fact the monitoring system has since been evolved to create a more robust system.

Our results do match the expectations in findings between the two groups as well as the two drills. With time to completion being an important variable in assessing an athlete's performance, decreasing the CoM deviations may present as a method for augmenting future performances. Therefore, using this biomechanical marker can provide insight into an athlete's performance and can allow for trainers to correct form and improve performances. Researchers should continue to analyze the impact of CoM in dynamic drills and the impact of efficient movements on performance.

Conclusion: Overall, results supported both of our hypotheses: higher CoM deviations were observed in the reactive agility drill and in those who were deemed as poorer performers. These results suggest that there is in fact a difference between the pre-planned and reactive agility drill besides solely the time to completion. With CoM currently being used for golfers, runners, and hurdlers, more research should be conducted to fully understand the impact it could have in augmenting athletic performance. As poorer performers generally had larger CoM deviations, fixing their form may enable faster times. However, researchers should aim to ensure their human monitoring systems are providing accurate results before making clear conclusions.

Acknowledgments: The Authors would like to thank ORISE for funding of the analysis of this project.

References:

[1] Sheppard, J. M., & Young, W. B. (2006). Agility literature review: Classifications, training and testing. *Journal of sports sciences*, *24*(9), 919-932.
[2] Arvinen-Barrow, M., Weigand, D. A., Thomas, S., Hemmings, B., & Walley, M. (2007). Elite and novice athletes' imagery use in open and closed sports. *Journal of Applied Sport Psychology*, *19*(1), 93-104.
[3] Young, W. B., & Willey, B. (2010). Analysis of a reactive agility field test. *Journal of Science and Medicine in Sport*, *13*(3), 376-378.
[4] Sheppard, J. M., Young, W. B., Doyle, T. L. A., Sheppard, T. A., & Newton, R. U. (2006). An evaluation of a new test of reactive agility and its relationship to sprint speed and change of direction speed. *Journal of science and medicine in sport*, *9*(4), 342-349.
[5]Young, W., & Rogers, N. (2014). Effects of small-sided game and change-of-direction training on reactive agility and change-of-direction speed. *Journal of sports sciences*, *32*(4), 307-314.
[6]Scott, W. A. (1962). Cognitive complexity and cognitive flexibility. *Sociometry*, 405-414.
[7]Mukhopadhyay, S.C. (2015). Wearable sensors for human activity monitoring: A review: *IEEE Sensors Journal*, 15(3), 1321-1330. Chicago.
[8]Boundless. "Center of Mass." *Center of Mass | Boundless Physics*, courses.lumenlearning.com/boundless-physics/chapter/center-of-mass/.
[9]Yamaguchi, T., Higuchi, H., Onodera, H., Hokkirigawa, K., & Masani, K. (2016). Misalignment of the desired and measured center of pressure describes falls caused by slip during turning. *PloS one*, *11*(5), e0155418.
[10]Najafi, B., Lee-Eng, J., Wrobel, J. S., & Goebel, R. (2015). Estimation of center of mass trajectory using wearable sensors during golf swing. *Journal of sports science & medicine*, *14*(2), 354.
[11]Krzeszowski, T., Przednowek, K., Iskra, J., & Wiktorowicz, K. (2015). Monocular Tracking of Human Motion in Evaluation of Hurdle Clearance. 16-29. 10.1007/978-3-319-25249-0_2.
[12]Bollschweiler, L. R. (2007). A biomechanical analysis of male and female intermediate hurdlers and steeplechasers.

[13]Gindre, C., Lussiana, T., Hébert-Losier, K., & Mourot, L. (2016). Aerial and terrestrial patterns: a novel approach to analyzing human running. *International journal of sports medicine*, *37*(1), 25-29.
[14] Bredin, S. S., Gledhill, N., Jamnik, V. K., & Warburton, D. E. (2013). PAR-Q+ and ePARmed-X+: new risk stratification and physical activity clearance strategy for physicians and patients alike. *Canadian family physician*, *59*(3), 273-277.

Full Circle Mechanical Dynamic Characterization Including Experimental Modal Analysis and Finite Element Analysis

Jason Cook, Thomas Hazelwood, Clay Jordan, and Blake Van Hoy

Oak Ridge National Laboratory
Oak Ridge, TN, USA
865-574-3010
hazelwoodtj@ornl.gov

Abstract: During operation, it was observed that a specific mechanical system experienced undesirable vibration and it became necessary to understand and mitigate this phenomenon. This document investigates the tools, methodology, and results of the dynamic characterization of the system. The characterization makes use of the experimental modal analysis (EMA) methods of single input multiple output (SIMO) and single input single output (SISO). The validity of the theory of reciprocity is confirmed to minimize measurement error, cost, and time of repeat testing. Finite element analysis (FEA) is used in choosing transducer and modal impact locations to adequately characterize the system. Single degree of freedom (SDOF) and multiple degree of freedom (MDOF) curve fitting is used to fully characterize the system's mode shapes and natural frequencies. The EMA characterization results are used to modify and validate the FEA model so that FEA can be used to model potential structural modifications to the system to mitigate the undesirable vibration. Structural modifications are chosen, implemented, and their effectiveness is quantified using EMA. A qualitative evaluation of the methodology of FEA validation by EMA and tuning of the model to match the experimental results is discussed.

Keywords: Case Study, Non-destructive Testing, Signal Analysis

INTRODUCTION

Undesirable vibration occurred during operation of a mechanical system and investigation was necessary to understand the event. A combination of experimental modal analysis (EMA) and finite element analysis (FEA) was successfully used in a feedback loop to improve the efficacy of the investigation. FEA provided insight into preferred transducer and modal impact locations needed to utilize single input multiple output (SIMO) and single input single output (SISO) EMA methods, and the results of those tests were then used to refine the FEA model as needed. Modifications were added to the mechanical system to alter its natural frequencies. EMA and FEA was performed on these modifications and compared to ensure similar trends in frequency shifts.

INITIAL EXPERIMENTAL MODAL ANALYSIS

Experimental modal analysis was performed using a modal impact hammer, accelerometers, and data acquisition equipment. For further information on performing

This manuscript has been authored by UT-Battelle, LLC, under contract DE-AC05-00OR22725 with the US Department of Energy (DOE). The US government retains and the publisher, by accepting the article for publication, acknowledges that the US government retains a nonexclusive, paid-up, irrevocable, worldwide license to publish or reproduce the published form of this manuscript, or allow others to do so, for US government purposes. DOE will provide public access to these results of federally sponsored research in accordance with the DOE Public Access Plan (http://energy.gov/downloads/doe-public-access-plan).

EMA, refer to the Shock and Vibration Handbook [1]. Single input multiple output testing was performed first, making use of transducers already in place in a single axial line along the cylinder measuring radially. The primary function of these transducers was system condition monitoring during normal operation, but, in this case, they were used in conjunction with a modal hammer while the system was not operational. The frequency response functions (FRF) obtained from the SIMO testing were preliminarily curve fit using a single degree of freedom (SDOF) curve fitting software to produce representations of the system's mode shapes. The investigation of these mode shapes indicated that there were not enough measurements made to characterize the system because there were shapes more complicated than the standard rigid body modes and bending modes. A finite element analysis can be used to determine the resonant frequency and mode shapes of a system and provide insight to measurement locations.

FEA DESCRIPTION

ANSYS mechanical was used to conduct the modal analysis of the mechanical system [2]. A modal analysis calculates the resonant frequencies of a structure based on its mass and stiffness assuming no excitation. The advantage of a simulation is the ability to make changes to the structure and quickly determine how these changes affect its resonant frequencies. Also, the simulation produces mode shapes for each of the resonant frequencies which aids in determining whether they are rigid bodies, bending, or shell modes.

A modal analysis of the mechanical system was created to both determine optimal locations for modal measurements and provide insight into how slight modifications to the test stand may affect its resonant frequencies. The solid models used to create the drawings for fabrication were also used to develop the finite element analysis model. These models were simplified for use in the FEA model to increase computational efficiency while not reducing accuracy. These simplifications include filling minor holes in components and removing noncritical radii and chamfers. Some components were excluded such as minor tubing and small fastener hardware that were not considered structurally significant. The structure consists of primarily isotropic metallic materials.

There are several bolted sections of the mechanical system. They are connected through bolts of several sizes and can be torqued to various values. These torque values were used to adjust the model to match measured data. Connections between components consisted of either frictional or bonded contact. Bonded contact was used between small internal components and for bolts that were threaded into attaching components. Frictional contact was used between major bolted connections and could also be adjusted to match measured results. The model used for the modal analysis is shown in Figure 1.

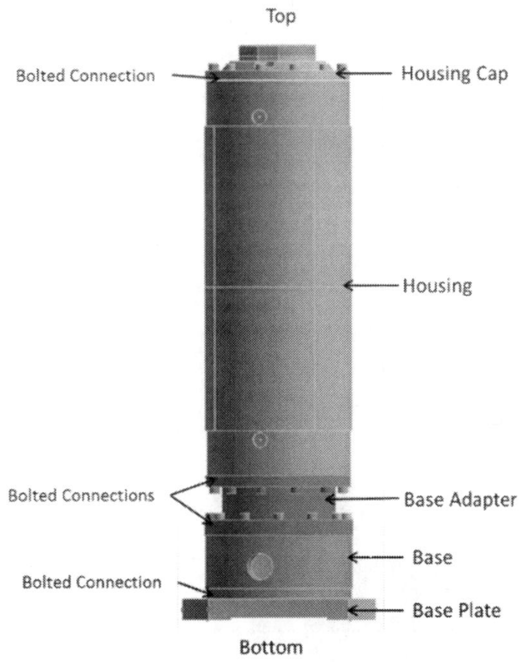

Figure 1: Mechanical System Model Description

Three orthogonal springs were added to each location of the base plate where it connected to the ground. Each of the three springs were aligned with the global X, Y, Z coordinate axes. The spring stiffness was adjusted to aid in aligning the model's resonant frequencies with the measured data. A coarse mesh was applied to the structure consisting of 2nd order tetrahedral elements. Figure 2 shows the spring locations anchoring the mechanical system to ground and analysis mesh.

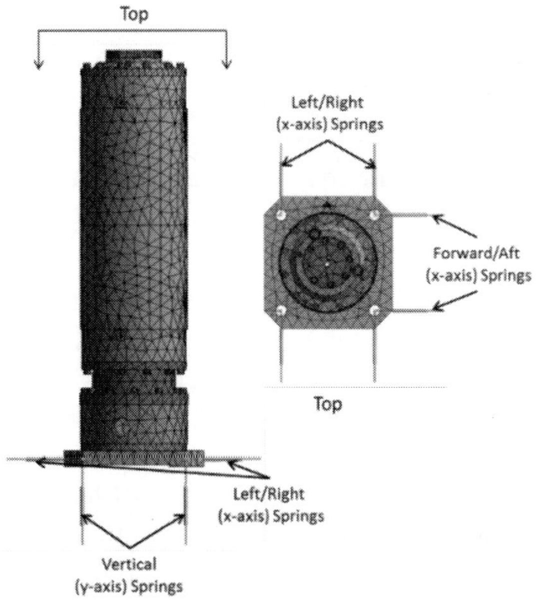

Figure 2: Model mesh and spring to ground boundary conditions

MODIFIED TRANSDUCER PLACEMENT AND IMPACT LOCATIONS

Initial stiffness values were assigned to the ground springs and bolts for the first modal analysis. The intent of this analysis was to determine accelerometer locations on the housing that would be suitable for determining mode shapes. The key at locating these accelerometers was to determine the antinodes of the structure that captured the important mode shapes. The initial FEA model exposed several shell modes that occurred in the frequency range of interest and identified more measurement locations than used for the initial SIMO EMA are needed to differentiate the shell modes from bending modes. Important mode shapes include; 1st bending, shell, and 2nd bending. Figure 3 shows the suggested minimum accelerometer locations determined by the initial modal analysis and the actual number of locations used in the roving hammer tests. Further points were added during the roving hammer empirical modal tests to increase mode shape fidelity.

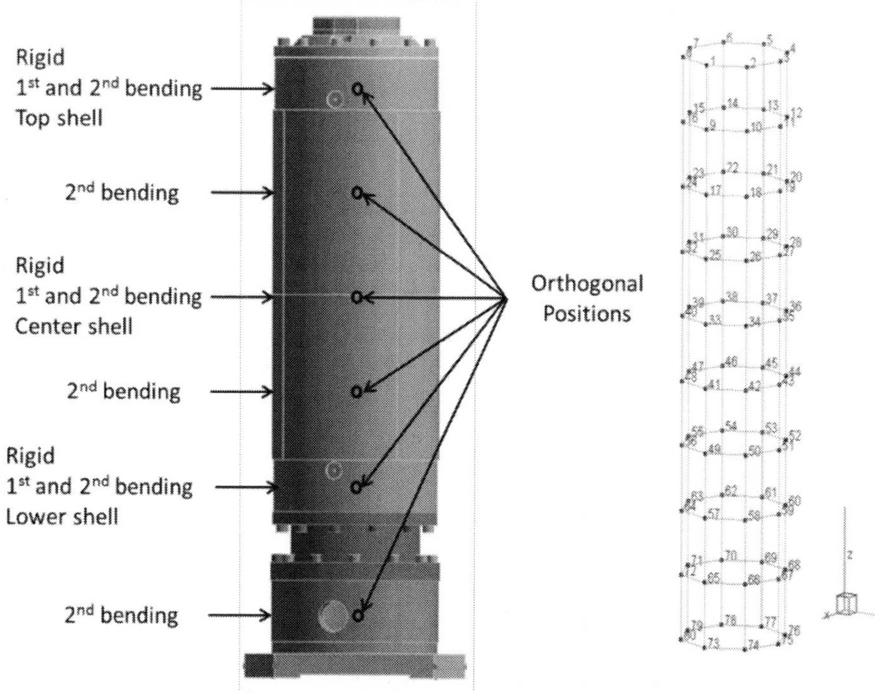

Figure 3: : Suggested accelerometer locations (left) and actual location for the roving hammer empirical modal tests (right)

ROVING HAMMER EMPIRICAL MODAL ANALYSIS

In order to accurately differentiate between and clearly display the mode shapes within the frequency range of interest, ten concentric circles of eight equally spaced radial measurement locations were excited using the SIMO roving hammer technique. Before measuring all eighty points, the measured response locations from the initial SIMO testing were impacted and the response was measured at one of the initial impact locations. The frequency response functions generated from the initial SIMO measurement closely matches that of the SISO roving hammer measurement, thus, the theory of reciprocity is validated for use with this measurement. Figure 4 shows the normalized FRF of the standard mechanical system. This significantly sped up the measurement process as one modal impact hammer and one fixed location accelerometer were used to collect the measurements rather than moving a column of accelerometers for each measurement. Each of the eighty points were impacted radially three times to collect an average measurement at each location. A multiple degree of freedom (MDOF) curve fitting software generated the mode shapes within the frequency range of interest of the mechanical system. The resonant frequencies and their respective mode shapes were then compared with the FEA results.

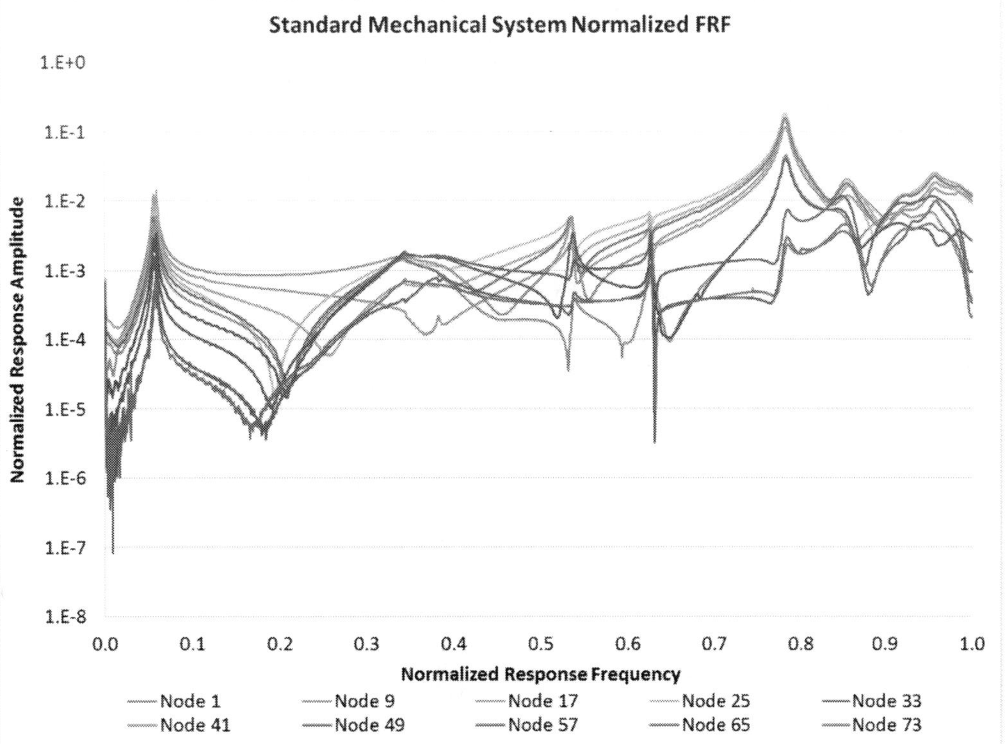

Figure 4: Normalized FRF of the standard mechanical system.

FEA MODIFICATION AND VALIDATION

Several parameters in the FEA model were adjusted to obtain a reasonable comparison with the measured data. The springs at the base of the model were adjusted to match the stiffness of the floor beneath the physical mechanical system. These primarily adjusted the rigid body modes. The bolt torque and friction coefficients at the major joints were adjusted to match the upper and lower shell and bending frequencies more accurately. After several iterations of adjusting bolt torque and fiction coefficient, it was determined that removal of the bolt torque and bonding the major joints of the mechanical system matched the roving hammer results best.

Once the model predicted the major modes of the mechanical structure, a total of 25 modes were calculated and the resulting frequencies were normalized across the normalized range of interest from 0 and 1. Table 1 lists the resulting modes and a description of the mode shape.

Table 1 Modes of interest of the mechanical system.

Mode	Model Frequency	Measured Frequency	Shape Description
1	0.040	0.055	Rigid
2	0.040	0.059	Rigid
8	0.305	0.346	1st Bending
10	0.470	0.417	Upper End Shell
12	0.518	0.536	Lower End Shell
19	0.800	0.835	Central Shell
21	0.835	0.855	2nd Bending

Table 2 shows a representation of the major mode shapes from the model analysis and the roving hammer test. Note that the deformed shapes are exaggerations of the actual mode shapes.

Table 2: Major mode shapes of the test stand.

Mode Shape Description	Normalized Frequencies	Model Mode Shapes	Measured Mode Shapes
Rigid	0.040 - Modeled 0.055 - Measured		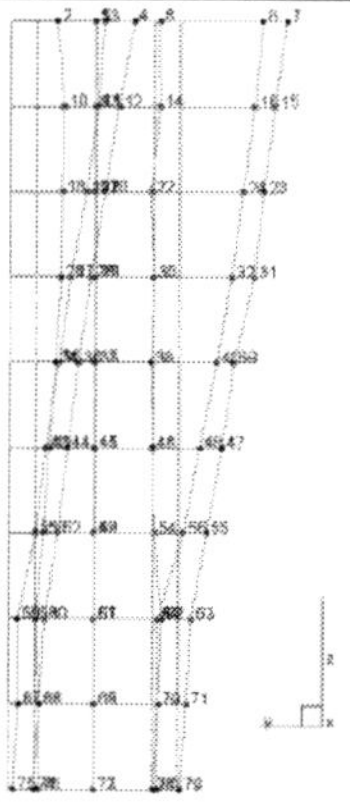

1st Bending	0.309 – Modeled 0.346 – Measured		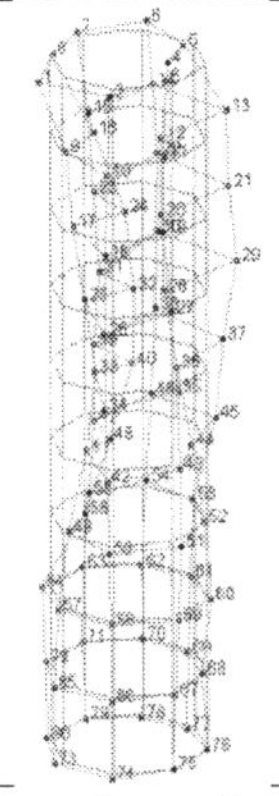
Upper Shell	0.475 – Modeled 0.417 – Measured	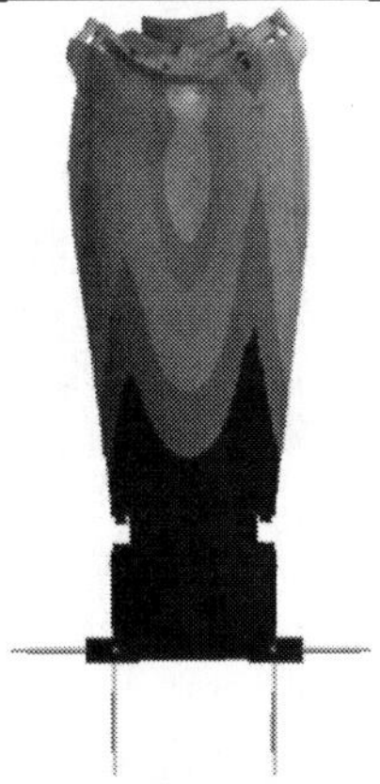	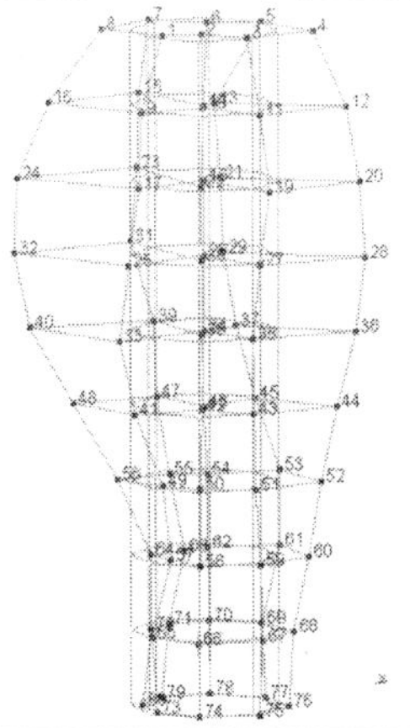
Lower Shell	0.518 – Modeled 0.536 – Measured		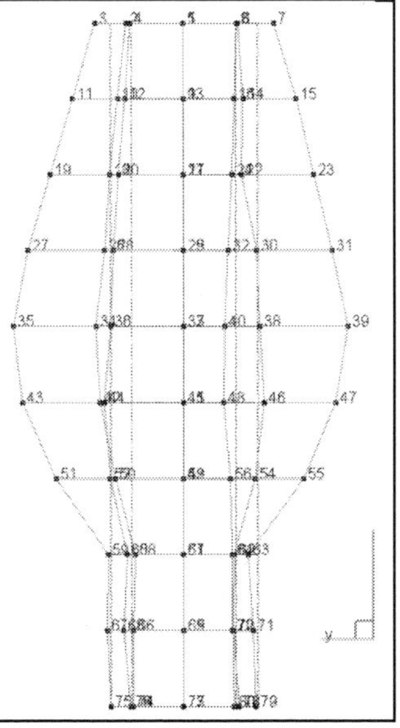

Central Shell	0.798 – Modeled 0.783 – Measured	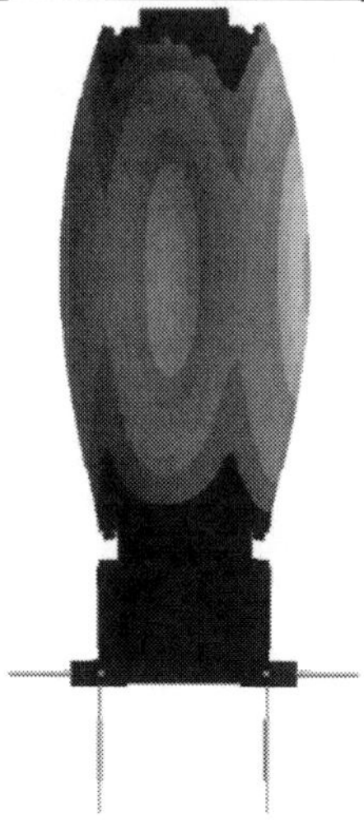	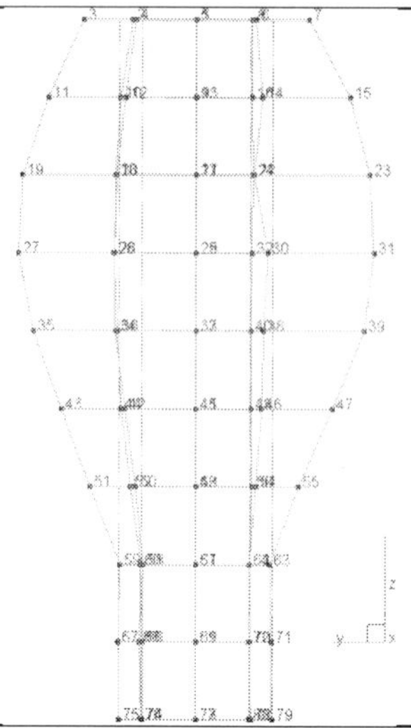
2nd Bending	0.838 – Modeled 0.855 – Measured	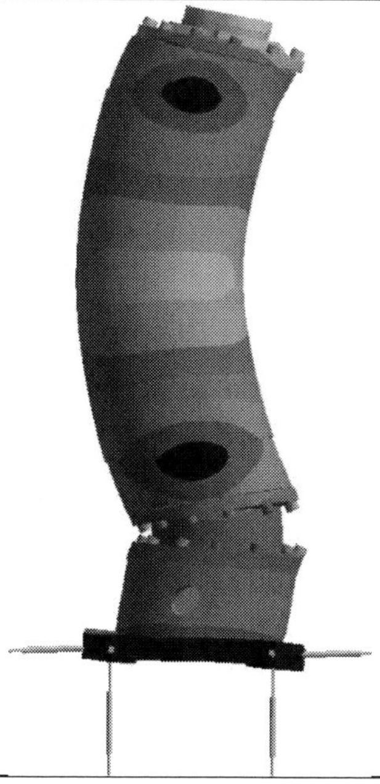	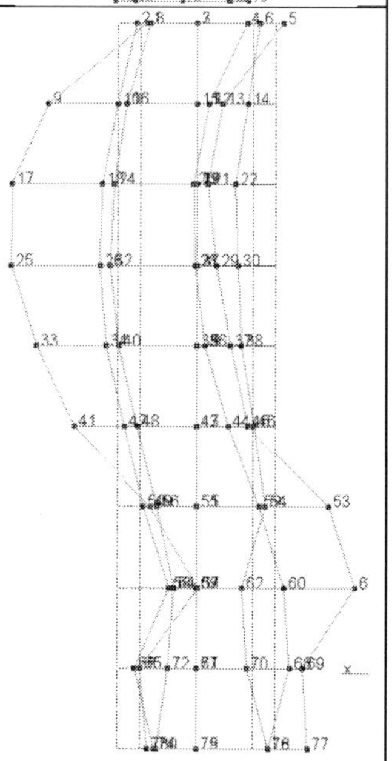

SYSTEM MODIFICATION

The modal analysis adds value to the modal testing results by aiding in determining mode shapes and performing quick parametric studies that show how the resonant frequencies of the structure shift due to geometry modifications. These geometry modifications include lengthening sections of the mechanical system, adding mass, or adding stiffeners.

A series of changes were implemented to the mechanical system both in the FEA model and the physical system. The first was an addition of a steel strap located near the center of the system. The intention of the strap was to shift the frequencies of the central shell modes. The second was an extension of the base component which was intended to shift the first bending frequency. Refer to Figure 1 for the location of the base component. Modal analysis was performed for each change and compared back to the standard system analysis and the EMA results.

The EMA roving hammer results of the mechanical system, shown in Figure 5, with a lengthened base primarily shifted the first bending mode down by 0.1 normalized frequency units, but the second bending mode shifted slightly up and its damping decreased. The FEA results showed a similar shift in the first bending mode and shows similar trends in behavior for the higher frequency modes. Table 3 lists the normalized 1st bending, center shell, and 2nd bending frequencies for the standard mechanical and lengthened base FEA models.

The EMA roving hammer results of the mechanical system, shown in Figure 5, with a steel strap near centrally located on the housing primarily showed that the modification was not as effective. The central shell shifted down slightly by about 0.02 normalized frequency units rather than up and the mode was split into two modes due to non-symmetric stiffness. The shift down in frequency is a result of the straps preloading the housing reducing its circumferential strength. The FEA results show a very little change in frequencies as indicated by the EMA tests. Table 3 lists the normalized 1st bending, center shell, and 2nd bending frequencies for the standard mechanical and steel strap FEA models.

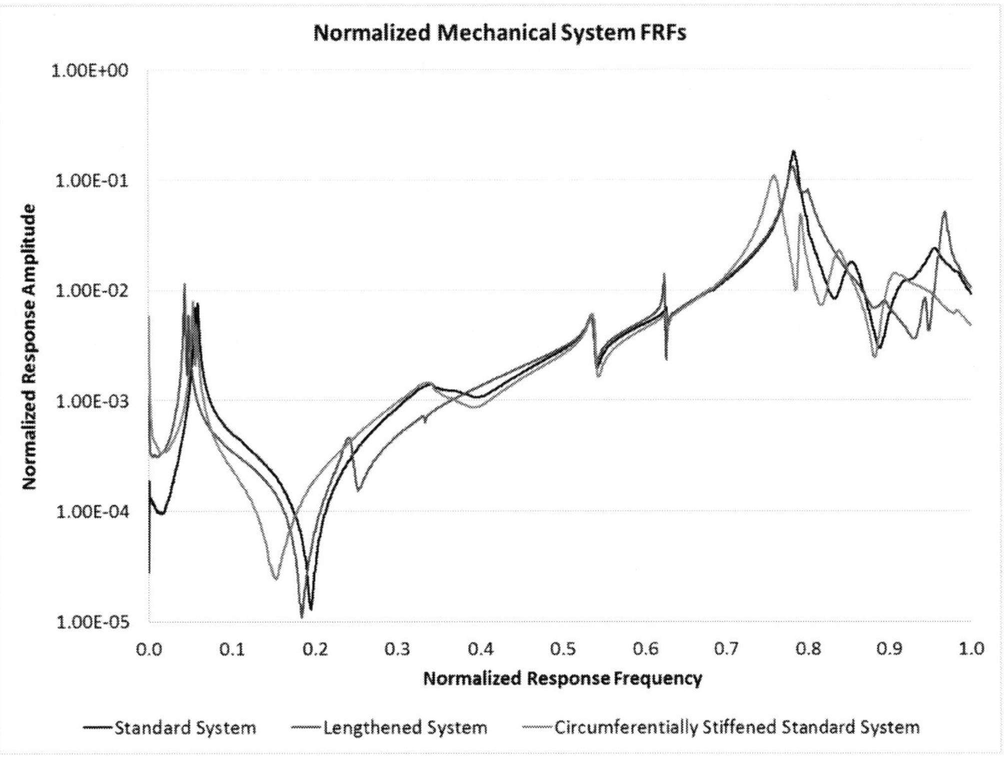

Figure 5: Normalized FRFs of the standard system and modified system

Table 3: FEA normalized frequencies for the standard system and the two variations.

Mode Shape	Standard System	Lengthened System	Circumferential Strap
1st Bending	0.309	0.247	0.312
Central Shell	0.798	0.798	0.809
2nd Bending	0.838	0.782	0.792

CONCLUSION

This paper has presented a successful implementation of a feedback loop using EMA and FEA in concert to increase the effectiveness of an investigation into aberrant performance in a mechanical system. FEA provided recommended transducer and modal impact locations for SIMO and SISO EMA methodologies. The combination of analytical and experimental approaches resulted in improved understanding of the mechanical system and allowed appropriate actions to be taken to address the undesirable vibrations.

FUTURE WORK

Recommended future work focuses on improving the fidelity of both the analytical and experimental analyses. An FEA harmonic analysis on the mechanical system would add information about the severity of each resonant frequency. An FRF over the entire frequency domain could be created from the harmonic analysis and compared with the measured data. The advantage of the harmonic analysis gives the ability to inflict a sinusoidal force to the mechanical system and predict the response at specific locations. The location of the force and locations of response could be the same as the location of the strike of the modal hammer and the locations of the accelerometers.

EMA testing of the axial and torsional modes would increase the fidelity of the analysis by characterizing their effect on the mechanical system and identifying their presence in the radially excited data already collected.

Other structural modifications could be considered and tested to increase the vibration mitigation of the desired modes.

REFERENCE

[1] *Shock and Vibration Handbook*, 4th Edition. Editors Cyril M. Harris, Charles E. Crede, McGraw-Hill Professional Publishing, 1995, chapter 21.

[2] ANSYS Mechanical Version 19.

BATHTUB, FAILURE DISTRIBUTION, MTBF, MTTF, AND MORE: THEY ARE RELATED

Thomas Heiser and James P. Hofmeister
Ridgetop Group, Inc.
3580 West Ina Road
Tucson, AZ 85741
520-742-3300
theiser@ridgetopgroup.com

Abstract: This paper describes the relationship between a bathtub curve, failure distributions, and commonly used metrics: mean-time-before failure (MTBF), mean-time-between failure (MTBF), and mean-time-to failure (MTTF) – metrics that are often misunderstood and misused. A bathtub curve is a statistical depiction of the failure rate over the lifetime of a population of products and is related to a failure-distribution curve: they can be combined to form a continuous curve. A bathtub curve graphically relates three types of failure: early, random, and wear out. Manufacturing and material defects typically result in early, rapid failure of products: that region of a bathtub curve is called infant mortality. After the infant-mortality region, there is a constant-failure region during which a low number of failures occur that are often referred to as random failures that are characterized as having a mean-time-between failures (MTBF). In-use products are subjected to stresses and strains that are cyclic in nature, plastic work, and which eventually causes irreversible damage and the onset of degraded operation. Damage accumulates until the product is no longer capable of operating within specifications and is said to have functionally failed – it has worn out. Such functional failures are characterized by a failure distribution having two common metrics: mean-time-before failure (MTBF) and mean-time-to failure (MTTF). Associated with bathtub and failure distribution curves are other metrics, including the following: failure rate, prognostic trigger point, prognostic distance (PD), failures-in-time (FIT), and useful life. Those metrics and how they are related are the focus of this paper.

Keywords: Bathtub; degradation; failure; metrics; prognostics; trigger; useful life

1. Introduction

There is a relationship between a bathtub curve failure distributions and commonly used metrics: mean-time-before failure (MTBF), mean-time-between failure (MTBF), and mean-time-to failure (MTTF). These metrics are often misunderstood and misused. A bathtub curve, illustrated in Figure 1, is a statistical depiction of the failure rate over the lifetime of a population of products; and it is related to a failure-distribution curve (see Figure 2). They can be combined to form a continuous curve [1]-[4].

Prognostics

Prognostics is an ability to accurately detect and report future failures in systems. The purpose of prognostics is to detect degradation and create prognostic information such as estimates of state-of-health (SoH) and remaining-useful-life (RUL) [5]-[7].

Bathtub Curve

A bathtub curve graphically relates three types of failure: early, random (constant failure), and wear out. Manufacturing and material defects typically result in early, rapid failure of objects: that region of a bathtub curve is called infant mortality. After the infant-mortality region, there is a constant-failure region during which a low number of failures occur that are often referred to as random failures. These failures are characterized as having a mean-time-between failures (MTBF). In-use products are subjected to stresses and strains that are cyclic in nature: plastic work. Plastic work eventually causes irreversible damage and the onset of degraded operation. Damage accumulates until the product is no longer capable of operating within specifications and is said to have functionally failed – it has worn out. The random, or constant, failure region is sometimes referred to as the useful life region [7],[8].

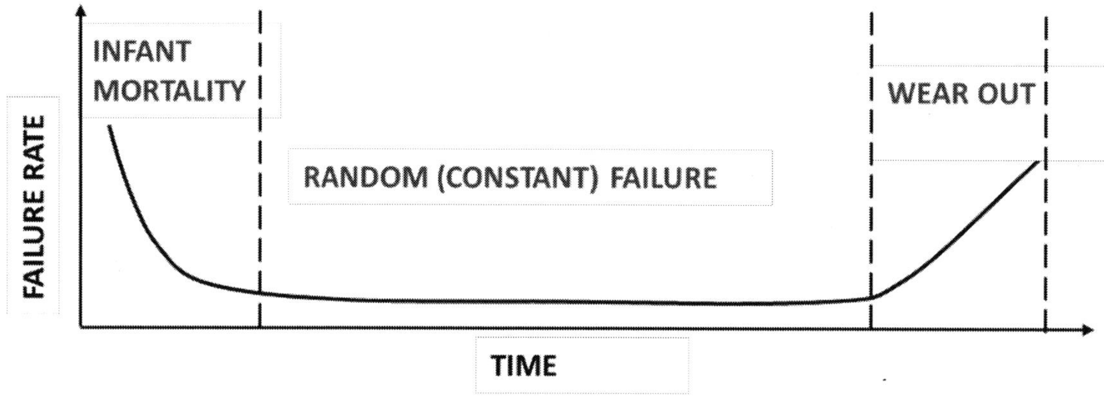

Figure 1: Example of a bathtub curve

Failure Distribution Curve

Products fail at different rates, creating a failure distribution (see Figure 2) that may or may not be a normal distribution.

Referring to Figure 3, mean-time-**between** failure (MTBF) is defined as the time between low-rate failures. Such failures are said to be random failures occurring at constant rate of failure: products having an exponential distribution. MTBF is also defined as mean-time-**before** failure (MTBF). More confusing, MTBF (between) has been used in the context of repairable products while MTBF (before) has been used in the context of non-repairable products. Even more confusing: EQ. (1) is used to calculate mean time to failure (MTTF) and EQ. (2) is used to calculate MTBF (before) – the calculations are identical [3], [4], [9]-[12].

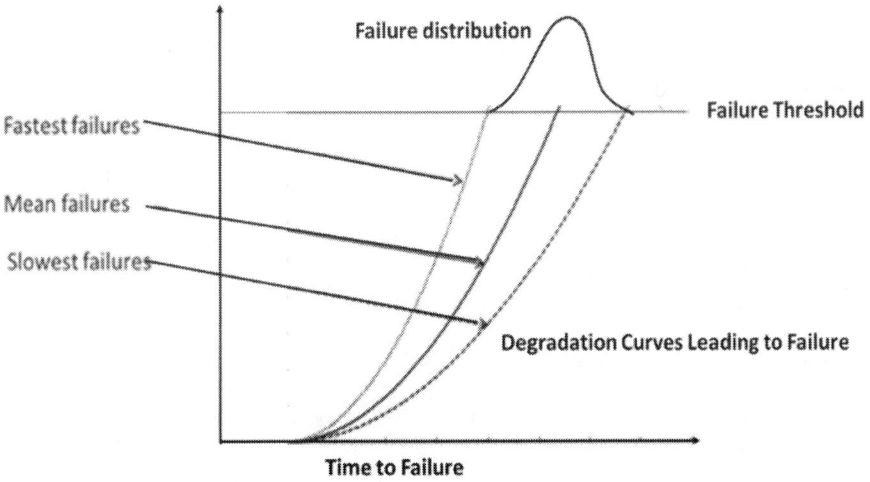

Figure 2: Example of a failure distribution curve due to degradation failures (wear out)

MTBF and MTTF: Relationship to a Bathtub and a Failure-Distribution Curve

$$MTTF = (Total\ Time)/(Number\ of\ Failures) \tag{1}$$

$$MTBF = (Total\ Time)/(Number\ of\ Failures) \tag{2}$$

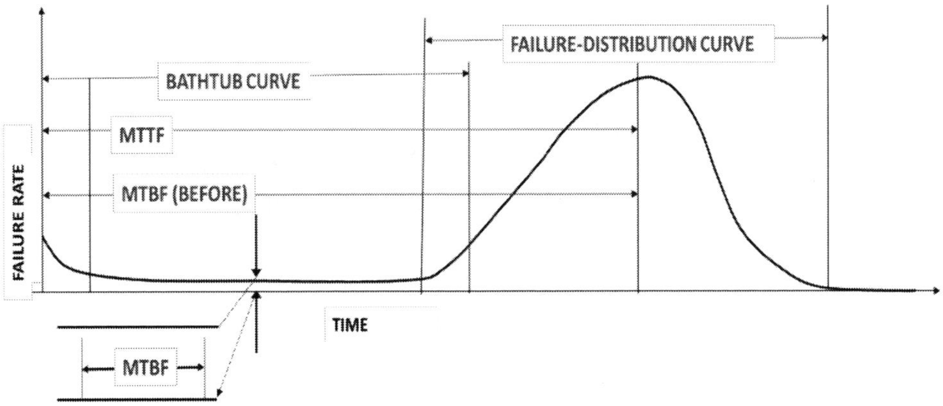

Figure 3: Combined bathtub and failure-distribution curve showing MTBF (between)

2. Bathtub: Other Terms

Other metrics related to failures include prognostic trigger point, prognostic distance (PD), failures-in-time (FIT), and useful life. Related to useful life are the following: (1) remaining useful life (RUL), and (2) end of life (EOL).

Prognostic Trigger, PD, RUL, and EOL

In the early days of IC devices, 1993 for example, it was believed that those devices did not experience wear out: they failed randomly. IC devices were subjected to 'burn in' testing where high stresses were applied to cause devices with manufacturing and/or material defects to fail: the remaining devices would then fail randomly at a low rate of failure,

$$failure\ rate = 1 - \exp[-\lambda t^{\beta}] \qquad (3)$$

EQ. (3) is related to a bathtub curve as follows (see Figure 4):

For $\beta < 1$, the failure rate decreases as time increases – infant mortality
For $\beta = 1$, the failure rate is constant – useful life
For $\beta > 1$, the failure rate increases – end of life

One method of prognostic-enabling an IC product, such as a microprocessor, is to include a device that fails earlier than any other device: much like a canary used in mines to detect lethal gases such as methane. Referring to Figure 4, that earlier time of failure is called a *prognostic trigger point* and *prognostic distance (PD)* is defined as

$$PD = useful\ life - (trigger\ time - begin\ time\ of\ useful\ life) \qquad (4)$$

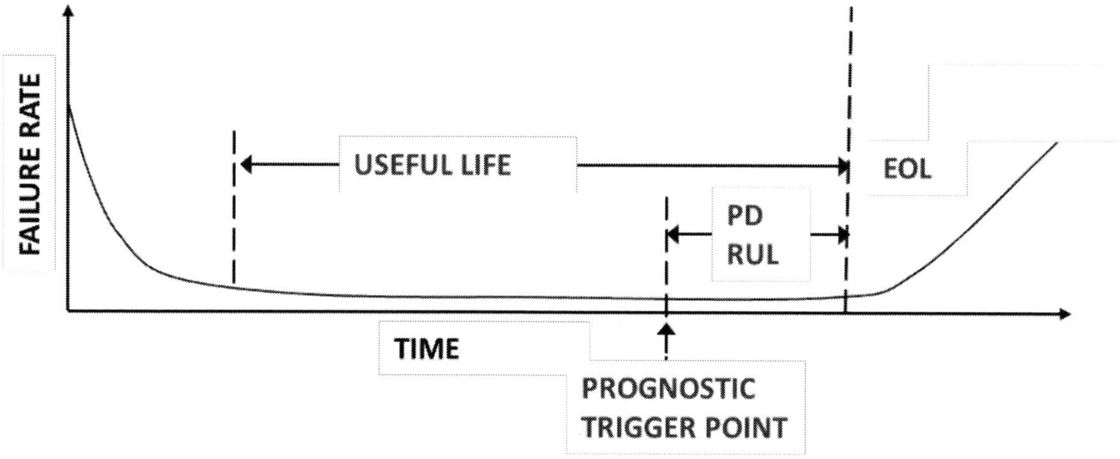

Figure 4: Bathtub showing other metrics

A canary is used as an early warning (alert) that a product needs to be replaced because it is about to enter the (EOL) region of operation. Such use is predicated on a belief that careful design, high-quality manufacturing, and thorough understanding of physics of failure would lead to canary failures that could be used to accurately calculate PD and RUL estimates, where RUL is the amount of time between the trigger point and the beginning of the EOL region [6], [7].

FIT

FIT values are used as a measure of the reliability of a product [13],

$$\text{Failure Rate, Lifetime} = \lambda = \frac{(\text{number of failures})}{(\text{number of tested parts})(\text{hours of test})\,AF}\,10^9\ \text{FIT} \quad (5)$$

$$\text{FIT} = 1/\lambda \quad (6)$$

Suppose a particular system has a product that is tested using a regime that calls for 40 units to be tested for 2,500 hours and the AF value for the required test is 10,000 hours using a test cycle of 1 hour. During the test, only 1 of the 40 fails: the failure rate would be calculated as

$$\lambda = \frac{(1)}{(40)(2{,}500\ \text{cycles})(10{,}000\ \text{hours/cycle})}\,10^9 = 1\ \text{FIT}$$

This does not help you because you typically only know the FIT number. Suppose the product specifications list FIT number of 50,

$$\lambda = 1/(50\ FIT) = 10^9/50 = 20{,}000{,}000 = 20\ \textit{million hours}!$$

Because the calculated rate of failure is so large, you decide to calculate MTBF using EQ. (2),

$$MTBF = (Total\ Time)/(Number\ of\ Failures)$$

But this does not help because you do not know the total time, so you turn to EQ. (7),

$$Total\ Time = (\#\ of\ tested\ units) * (test\ time) \quad (7)$$

which is no help because you don't know the number of tested units nor the test time. So, you find another expression for MTBF:

$$MTBF = 10^9/FIT \quad (8)$$

which, for FIT = 50 results in an MTBF (life time) of 20,000,000 hours: over 2,000 years! We highly recommend for prognostics never to use MTBF and MTTF metrics related to a bathtub curve. Similarly, never use FIT values for prognostics.

3. Failure Distribution: Other Terms

Referring back to Figure 3, you see that MTBF (before) is related to failure distribution and is interchangeable with a more modern use of MTTF. Further research leads to the following discoveries:

- MTTF is defined as 'mean time to failure' for a non-repairable product

- A relationship of MTBF to MTTF is defined as follows [13]:

$$MTBF = MTTF + MTTR \tag{9}$$

  Where MTTR is defined as 'mean time to repair'

- MTTR is usually not specified

More importantly, from EQ. (3), an MTTF and/or MTBF value only means there is 63% probability the system will fail prior in time to that value: not very useful in estimating when a particular instantiation of product is going to fail.

Failure Distribution: Degradation Signatures and Time-to-Failure

Referring to Figure 2 and Figure 5, let the fastest failure curve be replaced by a degradation signature, CBD1, let the slowest failure curve be replaced by another degradation signature, CBD2, and let each degradation signature begin at and reach failure at different times. Then we define TTF (time-to-failure),

$$TTF = (functional\,failure) - (onset\ of\ degradation) \tag{10}$$

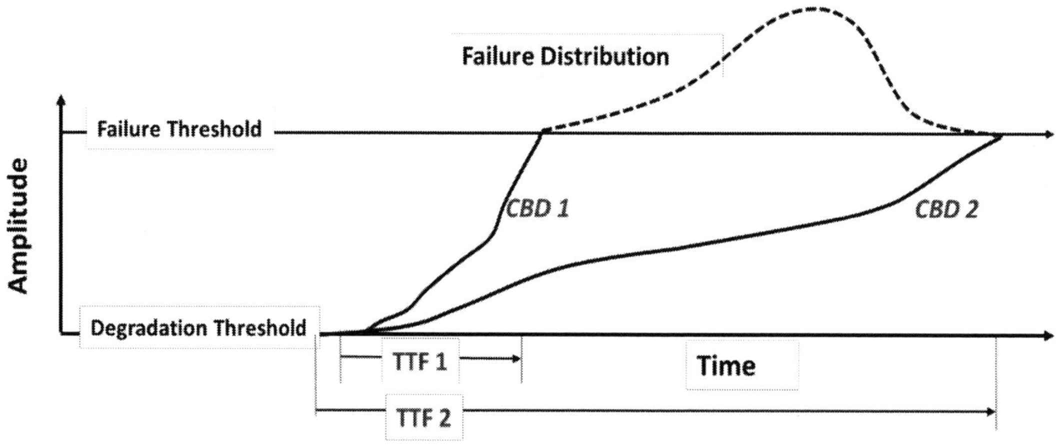

Figure 5: Degradation signatures: TTF relationship to failure distribution

Degradation signature, TTF, sampled data, and RUL

Referring to Figure 6, suppose you have a PHM system with a set of algorithms for conditioning, signature processing, and prediction estimating that accurately detects the onset of degradation, processes input signature data to accurately estimate when functional failure is likely to occur. Then from EQ. (10) for every sampled data at time S,

$$RUL = TTF - \text{(sample time)} \tag{11}$$

To produce RUL estimates, you need to use a PHM system that uses condition-based data (CBD) to produce degradation signatures that are processed to produce accurate estimates of TTF. Then for each sampled data point, that PHM system calculates accurate RUL estimates.

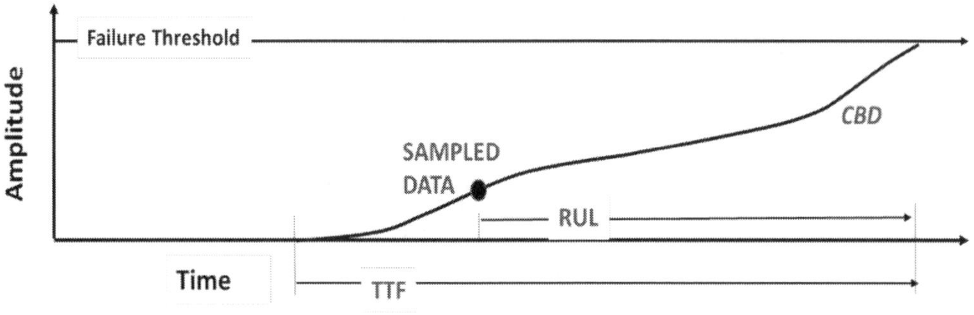

Figure 6: Degradation signature, TTF, sampled data, and RUL

Degradation and Prognostic Information

You cannot use metrics related to a bathtub curve or a failure distribution curve to produce accurate estimates of RUL for a particular product. Instead, you must process CBD, detect the onset of degradation, transform CBD points into degradation signature data, project that signature and estimate a time when the projected signature is likely to reach a defined threshold level for functional failure. You need to use algorithms that (1) mitigate time of functional-failure variability not caused by degradation and (2) minimizes variability caused by degradation.

A design approach is to treat data points as particles having inertia and momentum: particles do not exhibit rapid changes, instead they tend to maintain velocity and direction. A satisfactory design objective is to employ a dampening factor to changes in amplitude and velocity. Another design approach is to develop a random-walk solution for particles that progress from a zero-degradation state (lower-left corner) to a maximum-degradation state (upper-right corner). For example, (1) use the previous data point; (2) calculate the predicted location of the next data point; (3) adapt a signature model to an adjusted location between the predicted and the actual location of the next data point; (4) use dampening factors and coefficients to adjust the signature model; (5) use the adapted model to estimate when functional failure is likely to occur; and (6) calculate prognostic information such as RUL and prognostic horizon (PH), the relative time between the onset of degradation and the time of functional failure. Use EQ. (10) and EQ. (11) to calculate RUL, then calculate PH using EQ. (12),

$$PH = RUL + (sample\ time) - (onset\_of\_degradation\ time) \tag{12}$$

From the definition of PH and TTF (see Figure 6), PH and TFF are the same. The authors prefer using PH because TTF is sometimes referenced to a time before the onset of degradation: for example, the time when a product is first used or a value of time reference to a calendar time (yy/mm/dd hh: mm: ss).

Predefined RUL Value

Prior to the onset of degradation, you cannot make any accurate estimate of RUL: you must use a predefined value. For example, use a customer-specified value or a vendor-specified value, or a value derived from evaluating historical or experimental data. That initial RUL value is likely to be higher or lower than actual. Figure 7 is an example of a solution applied to an initial-estimate error [14].

In addition to a high-value initial-estimate error, a reliability engineer might specify an initial RUL value that is low compared to that for a prognostic target. Suppose for a prognostic target a reliability engineer specified 100 days as the estimated TTF after the onset of degradation and it actually takes 200 days to become functionally failed: a random-walk with Kalman-like solution will result in curves that resembles those shown in Figure 8 [14].

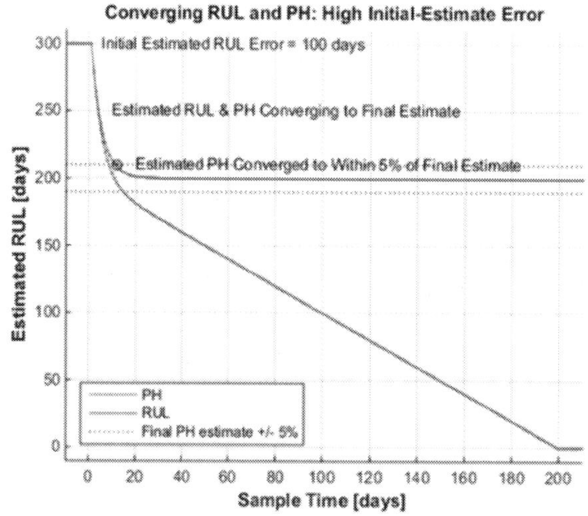

Figure 7: RUL and PH estimates when initial RUL is higher than actual

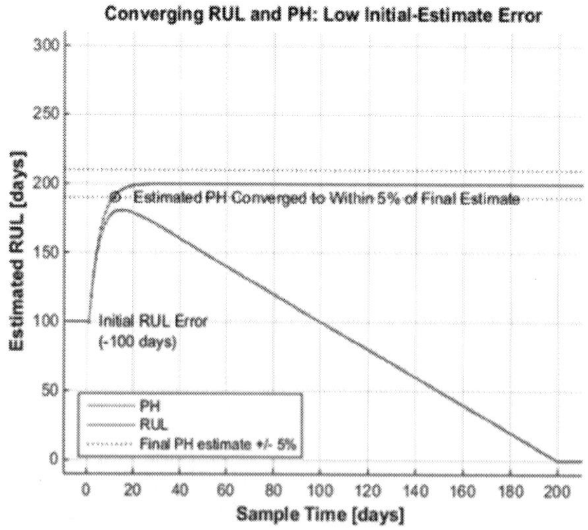

Figure 8: RUL and PH estimates when initial RUL is lower than actual

4. Conclusion

This paper described problems in relating metrics related to bathtub and failure distributions: two MTBF metric and an MTTF metric. Those metrics have multiple definitions that are confusing and, more importantly, are not suitable for use in prognostics for products. Other metrics related to a bathtub curve (prognostic trigger point, RUL, and PD, and FIT) are also not suitable for use in prognostics.
Metrics related to degradation were introduced and shown to be useful in prognostics: TTF, sample time, RUL, and PH. In the absence of detected degradation, TTF cannot be calculated: instead, a predefined value must be used, which results in an initial RUL value that is higher or lower in value than actual.
To produce accurate RUL estimates, it is necessary to use prediction algorithms that do not rely on statistical-based and reliability-based metrics such MTBF, MTTF, and FIT.

Acknowledgments

The authors thank Naval Air, Naval Sea, U.S. Army, U.S. Air Force, and NASA research centers for their support and funding of multiple projects that led to the solutions and the results described and shown in this paper.

References

[1] NIST (2018), "Assessing Product Reliability," Engineering Statistics Handbook, 8.1, 8.3, and 8.4. [Online], accessed 2018, https://www.itl.nist.gov/div898/handbook.
[2] Seastrunk, B. (2016), "Reliability, Warranty, and Why Nothing Lasts as Long as it Used To," Blog post. [Online], accessed 2018, http://opinionbypen.com/reliability-warranty-and-why-nothing-lasts-as-long-as-it-used-to/
[3] Speaks, S. (2005), Reliability and MTBF Overview. VICORP. Andover, MD. [Online], accessed 2018, www.vicorpower.com/documents/quality/Rel_MTBF.PDF.
[4] Wilkins, D.J. (2002), "The Bathtub Curve and Product Failure Behavior, Part Two – Normal Life and Wear-Out," HotWire Issue 22, 2002. [Online], accessed 2018, https://www.microsemi.com/document-portal/doc_view/124041-calculating-reliability-using-fit-mttf-arrhenius-htol-model.
[5] Kumar, S. and Pecht, M. (2010) "Modeling Approaches for Prognostics and Health Management of Electronics," *International Journal of Performability Engineering*, Vol. 6, No. 5, pp. 467-476.
[6] Pecht, M. (2008) *Prognostics and Health Management of Electronics*. Hoboken, NJ: John Wiley & Sons.
[7] O'Connor, P. and Kleyner, A. (2012) *Practical Reliability Engineering*. Chichester, UK: John Wiley & Sons.
[8] Hofmeister, James P.; Goodman, Douglas, L.; and Szidarovszky, Ferenc (2018); "Transforming Condition-based Data Signatures into Functional Failure Signatures," IEEE 2018 Aerospace Conference, Big Sky, MT, 3-9 March, 2018.
[9] Abernethy, R.B. (2006), The New Weibull Handbook 5th edition, Berringer and Associates, 2006.
[10] RAC (2005), Reliability Toolkit: Commercial Practices Editions, Reliability Analysis Center, Rome Laboratory [Online]. accessed 2018,

[11] https://www.dsiac.org/sites/default/files/journals/2Q2005.pdf.
[12] Seymour, B. (1993), "MTTF, FAILRATE, RELIABILITY AND LIFE TESTING," Application Bulletin, AB-059, Burr-Brown Corporation, Tucson, AZ
[13] Weibull (2008), Reliability Engineering, "MTTF, MTBF, Mean Time Between Replacements and MTBF with Scheduled Replacements," HotWire Issue 94, 2008. [Online], accessed 2018, https://www.weibull.com/hotwire/issue94.
[14] Ellerman, P. (2012), "Calculating Reliability using FIT & MTTF: Arrhenius HTOL Model," MicroNote™ 1002, Rev 0, MicroSemi Corp., 1/9/2012.
[15] Goodman, Douglas L.; Hofmeister, James P.; and Szidarovszky, Ferenc; 2019 (Pending), *Prognostics and Health Management: A Practical Approach Using Conditioned-Based Data,* 1st Edition, John Wiley & Sons, Ltd, The Atrium, Southern Gate, Chichester, West Sussex, PO19 8SQ, United Kingdom, *ISBN: 9781119356653,* 2019, pp. 44-46.

Electrically Improving a MEMS Sensor for Rolling Stock

James P. Hofmeister, Wyatt Pena, Erik Sandberg, and Robert S. Wagoner
Ridgetop Group, Inc.
3580 West Ina Road
Tucson, AZ 85741
520-742-3300
jhofmeister@ridgetopgroup.com

Abstract: This paper describes work related to improving the electrical performance of an accelerometer-based sensor, RotoSense™, used for monitoring rolling stock: the locomotives and cars used in trains. At the 2018 MFPT conference, a paper, "Improved RotoSense™ for Rolling Stock: Locomotives and Cars," focused on physical improvements to the shaft-mounted, wireless sensor, although there were improvements in signal performance. This paper describes subsequent improvements to that sensor, with focus on signal quality and battery life. The original version of the sensor described in this paper is the first and, still, only known to survive, intact, three days of testing at the National Test Track Center in Pueblo, Colorado, including a 10-hour, non-stop, 400-mile test run. The rationale, the methods, and the results of those electrical improvements are the focus of this paper.
*RotoSense is a trademark of Ridgetop Group, Inc.

Keywords: Accelerometer; gear; MEMS; rolling stock; RotoSense, signal quality; train, wheel hub

1. Introduction

Equipment such as robots and gear boxes that incorporate rotating shafts often need to monitor rotational vibration and shaft speed, as part of broader condition-based maintenance (CBM) systems. Fault detection equipment on drive systems typically use accelerometers mounted on transmission housings to capture, measure, and process vibration signals. The usefulness and flexibility of such detection equipment for applications involving rotating shafts, including pinion and planetary gears, have been limited by cabling, slip-ring approaches, and multiple sensors to obtain monitoring information. For more complex systems, especially those with poor signal transmission paths, a shaft-mounted, wireless solution based on a micro-electro-mechanical system (MEMS) is needed (see Figure 1) [1] [2] [3] [4].

A prototype solution was designed, developed, and imbedded on a rotating shaft used in helicopter transmissions to capture and measure vibration signals, and then a MEMS version mounted on the wheel hubs of rolling stock was developed.

This paper focuses on the methodology and results in ruggedization, improved signal quality, and increased battery life of those solutions.

2. Theoretical Background

MEMS-based accelerometers that measure acceleration and force, such as produced by vibration and shock, were used in prototype solutions for the following applications: (1) mounting inside helicopter transmissions and (2) mounting on the hubs of rolling stock of trains to detect features of railroad tracks.

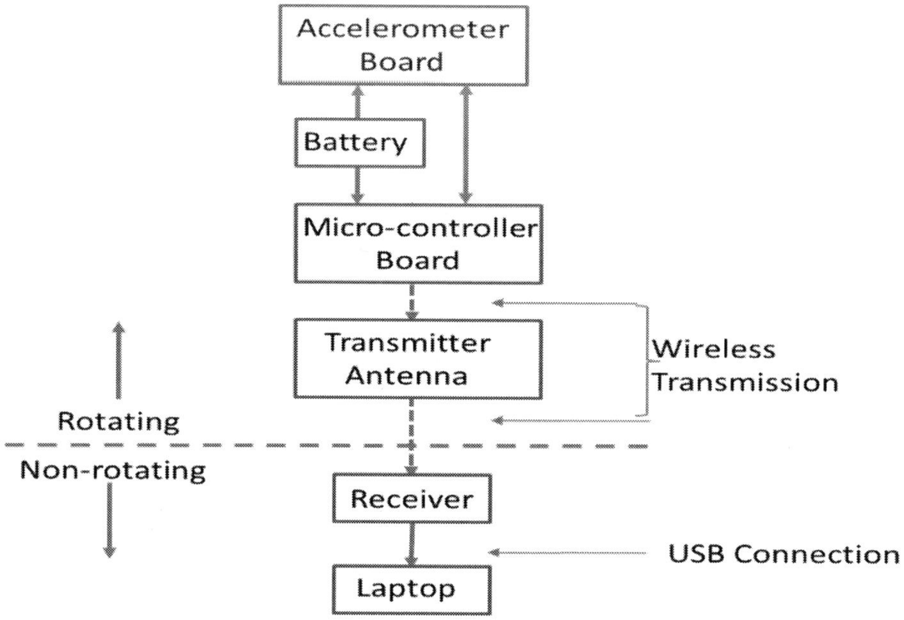

Figure 1: Simple MEMS Block Diagram. [2]

Helicopter Transmission: Pinion Gear

A specific application was the spiral-bevel pinion of an OH-58C transmission (see Figure 2). For that application, MEMS Sensing Package was designed and developed to contain a microcontroller board, an accelerometer board, and a battery pack mounted in a cylindrical canister (Figure 3).

A NASA Glenn Research Center, Small Business Innovation Research (SBIR) award led to an experiment in which a tooth on the spiral bevel gear was pre-notched and then the transmission was run to tooth failure. The feasibility of shafted-mounted solutions was proven comparable to traditional stationary, housing mounted accelerometer solutions.

Wheel Hubs of Rolling Stock

Figure 4 (left) shows the physical adaptation of the MEMS-based sensor for mounting on wheel hubs of the shafts of the trucks of train locomotives and cars. This version of the sensor was designed and developed to prove feasibility for using such sensors to locate and identify anomalies related to railroad tracks. The adapted system for rolling stock applications comprises a MEMs sensor and a gateway to collect data and write data to disk storage as files. The MEMS sensors were mounted on wheels of a train to produce shock data during test runs of a train over a High Tonnage Loop (HTL) test track (TT) to process the data, identify high-force events (HFEs), and locate the position on the HTL TT where HFEs occurred. The purpose was to demonstrate/validate an ability to support focused inspection of tracks to identify and locate anomalies requiring monitoring and service [2].

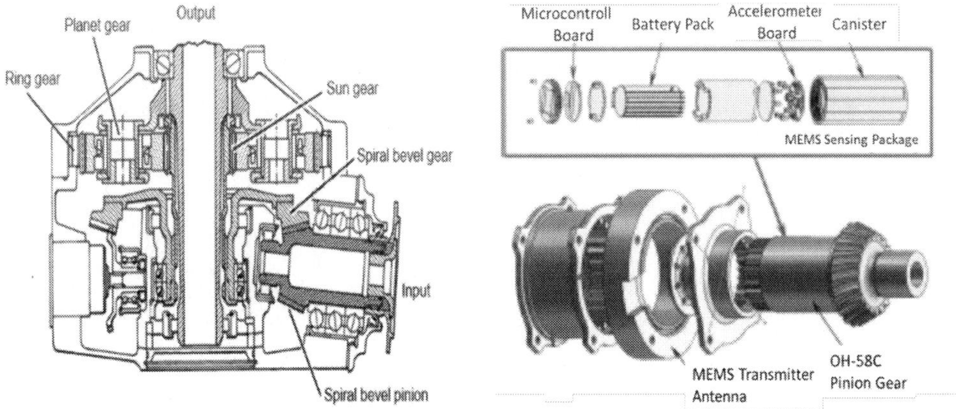

Figure 2: OH-58C Transmission (Left) and RotoSense on the Pinion Gear (Right).

Figure 3: Canister, Boards, End Cap (Left) and Boards w/Batteries and Housing (Right).

Initial Results: Shaft-mounted, beveled-pinion gear

A RotoSense configuration of triple shaft-mounted sensors were installed on pre-notched OH-58C spiral-bevel pinion gears and endurance tests at NASA's Glenn Research Center were performed and run to tooth fracture failure: the notch was extended at run time = 51.9 hours and widened at run time = 106 hours (Figure 4, right).

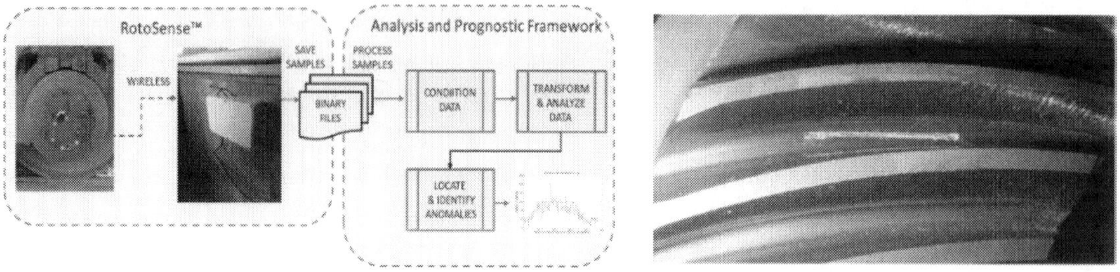

Figure 4: Rolling-stock Sensor System (Left) and Final Gear-Tooth Notching (Right).

The sensors performed well, lasted the entire test, and all MEMS accelerometers gave an indication of failure at the end of the test. The MEMS systems performed as well, if not better than, the existing stationary accelerometers mounted on the gear box housing with regards to gear tooth fault detection (Figure 5, left) and time-synchronous-average (TSA) signals (Figure 5, right) [1].

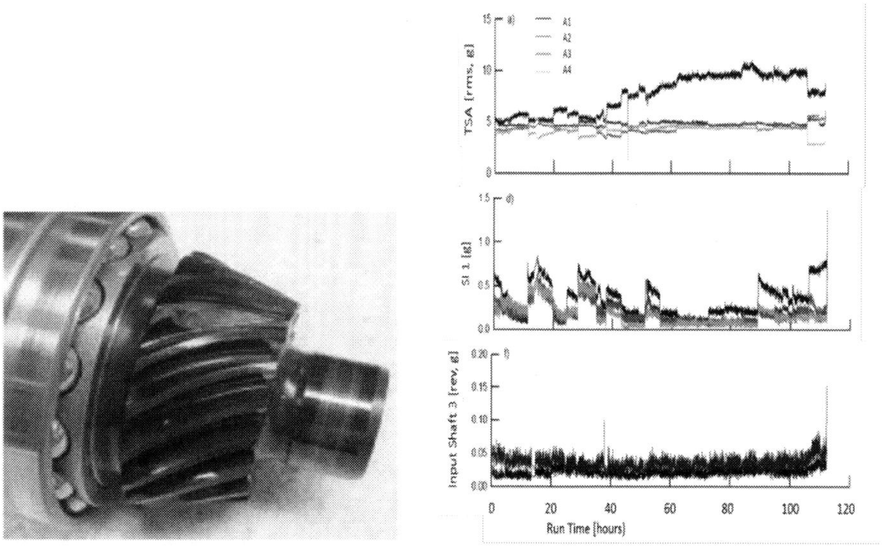

Figure 5: Fractured-Tooth Failure (Left) and Data (Right).

For both the MEMS sensors and stationary sensors, the fault detection time was not much sooner than the actual tooth fracture time. The MEMS sensor spectrum data showed large first-order shaft frequency sidebands due to the measurement rotating frame of reference. The method of constructing a pseudo tach signal from periodic characteristics of the vibration data was successful in deriving TSA signals without an actual tachometer: the method proved to be an effective way to improve fault detection for the MEMS [1].

Initial Results: Hub-mounted, freight car of a train

The configuration for train applications includes RotoSense, a MEMS sensor, supporting firmware and software to support collecting, wireless transmitting of data to a gateway, and saving data in binary files. The assemblies are mounted to the wheel hubs to rotate with the axle so any anomalies in the wheels or track can be detected (see Figure 8).

MEMS Configuration

The MEMS is configured as a three-axis accelerometer with 57mV/g sensitivity with a 158 Hz sampling rate: sensor was mounted concentrically on each end of a freight car axle and also on a locomotive axle of a train, as shown in Figure 6.

Figure 6: RotoSense – Wheel-Hub Mounted.

Test Train, Track, and Train Movement

Test Train: A test train at the National Test Track Center at Pueblo, Colorado, comprised three (3) locomotives and 110 freight cars and was run on a high-tonnage loop (HTL) test track (TT) used for research under heavy axle-loads to test track-component reliability, wear, and fatigue.

Test Track: The HTL track length is 2.7 miles divided into test sections that generally correspond to tangents, spirals, curves, and turnouts that are populated with features and test sections as seen in Figure 7. Table 1 lists the TT features and Figure 8 and Figure 9 show some of features listed in Table 1.

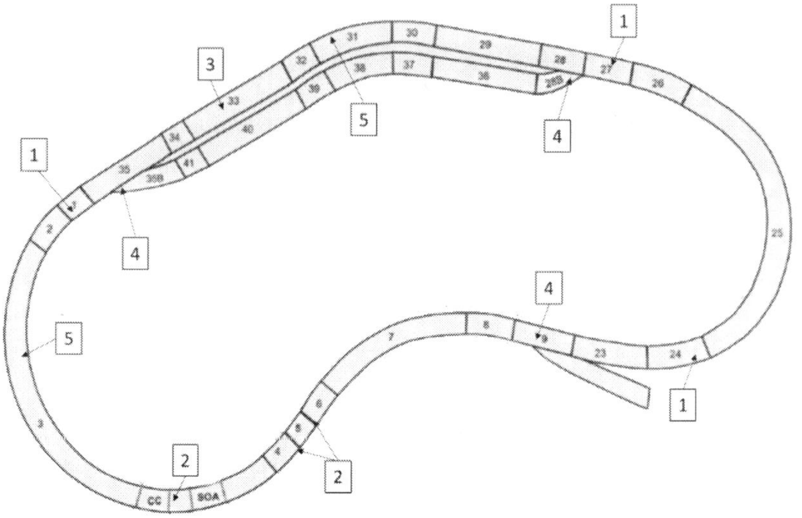

Figure 7: Layout of the Heavy Tonnage Loop Test Track.

Table 1: Key Physical Features of the HTL TT Shown in Figure 11.

1. Lubricator	2. Steel Bridge	3. Crib Ties and Fasteners
4. 405, 407 or 408 turnout and frog or switch	5. Thermite or Overlay Welds	6. Concrete Bridge

Figure 8: Turnout and Frog, Switch left [A] and Switch Right [B]; Steel Bridge [C].

Figure 9: Concrete Bridges [A] & Bot of [B]; Crib Ties @ Top of [B].

Train Movement: The train was auto-controlled to run 15 laps per hour: 4 minutes per lap and 38,640 samples per lap. Four test runs were started on four days (May 11 – May 14 in 2015). Table 2 summarizes the May 14 test run from 2000 to 0632 the next morning: 10 hours, 32 minutes and over 4 million sets of six-byte data. The train started to move about 30 minutes after the sensors were turned on: the train was moved to the test track and two laps of test conditioning were run. After that, the train was kept at a constant speed of 15 laps per hour (4-minute laps).

Data Collection and Analysis

Test Data: Data was collected and buffered for each axis at a sampling rate of 158 Hz and the buffered data was transmitted to a collection hub and saved in output files about once every 1.11 seconds. The data was analyzed, nominal values determined for zero-force conditions, and transformed into +/- values with respect to zero-force.

Data Analysis: Data analysis was hampered by the loss of Global Positioning Data (GPS) caused by a broken antenna. Consequently, raw data (Figure 10, left) was analyzed by binning the data in terms of magnitude (xy-plane, the z-plane, and both planes) and relative laps (Figure 10, right), examining the peaks, and comparing those peaks to features of the test track: pattern matching.

Table 2: Summary of the Test Run Started on May 14, 2015.

Description			**Comments**
Train	3 locomotives 110 cars		6,780' long (1.3 miles). Hopper car lengths, coupler to coupler, range from ~58.5 to 60.5 feet: used 60-foot length.
Build	92 minutes		2000 start; 2132 completed build
Run	540 minutes	132 laps	2132: started test conditioning run (TCR)
	17 minutes	2 laps	2149: completed TCR
	518 minutes	129 laps	2149 – 0627: testing
	5 minutes	1 lap	0632: end of test
Wheel 1	20645 files		
Wheel 4	20565 files		

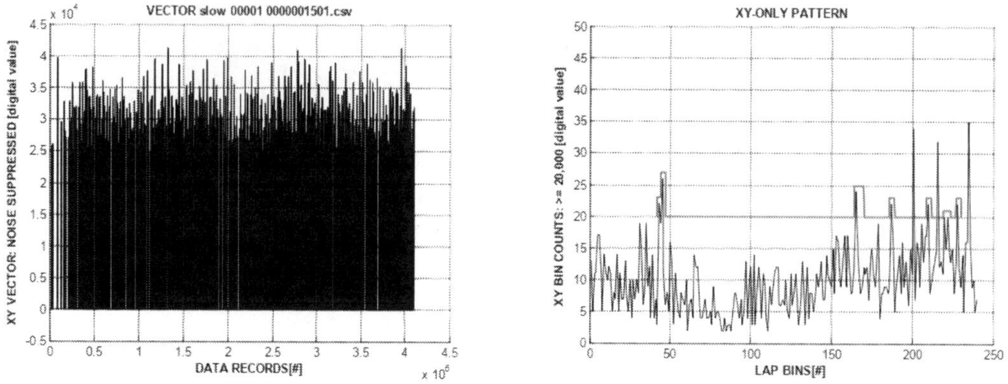

Figure 10: XY-plane Data, Raw (Left) and Binned by Bin Number, Right).

We successfully proved that, even without GPS, we were able to synchronize the data to the start of the test track and then locate those features that could be located: as seen indicated in Figure 11 and Table 3.

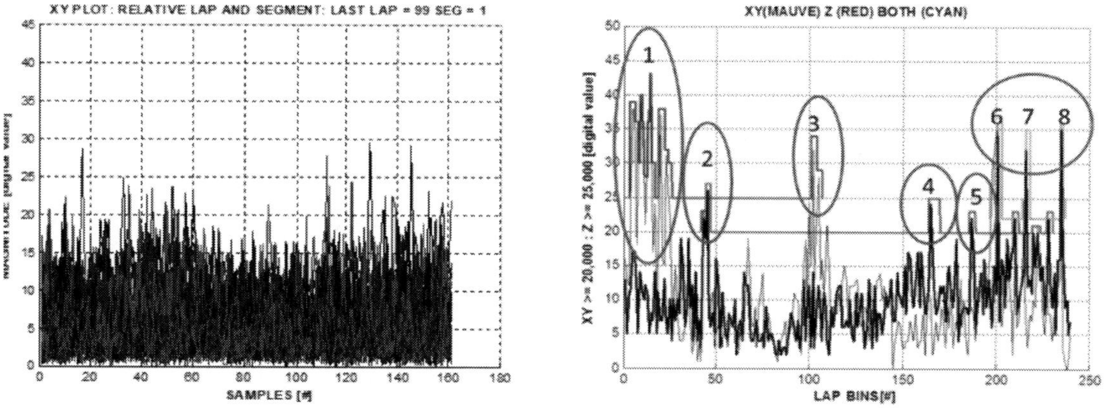

Figure 11: Data & Identified Features (1) Through (8) of Table 3.

Table 3: TT Description and Detection Evaluation.

Track Sections	Feature Sections	TT ID	Track Feature	Detection Evaluation	
				XY-vector	Z-vector
1 – 3		S1	Lubricator	ND	ND
4 – 5		S2			
6 - 62	5-26	S3	Repair/overlay welds		Yes (1)
	30-40	S3	Concrete bridge	Maybe	maybe
	42-46	S3	Concrete bridge	Maybe	Yes (2)
63 – 66		S4	Steel bridges	ND	ND
67 – 69		S5	Bridge deflection	ND	ND
70 – 73		S6	Steel bridges	ND	ND
74 – 92		S7	Rail performance	ND	ND

93 – 97		S8	Fiber optic cable	ND	ND
98 – 108		S9	405 turnout/frog		Yes (3)
109 - 117		S23	405 turnout/frog		Yes (3)
118 – 125		S24	Lubricator	ND	ND
126 – 163		S25	TPO, Tie and fastener, performance	Yes (4)	No
164 – 170		S26			
171 – 175		S27	Lubricator	ND	ND
176 – 180		S28	Turn out, steering switch, foundation	Yes (5)	
181 – 193		S29	LTM Tests	ND	ND
194 – 198		S30			
199 – 208		S31	FRA: Rail-seat deterioration, Thermite welds	Yes (6)	Yes (6)
209 – 212		S32			
213 – 225		S33	Crib ties	Yes (7)	Yes (7)
226 – 229		S34			
230 - 240		S35	407 turnout/frog	Yes (8)	Yes (8)

Note: ND means Not Detectable.

3. Methodology: Ruggedization, Signal Quality, and Battery Life

Subsequently, a manufacturer of rolling stock obtained sensor units and software and performed additional testing and evaluation. The units were found deficient because of the following: (1) physical failure at high-force testing at 40 g vibration and shock leading to board flexing and subsequent solder joint failures and battery displacement; (2) inaccurate vibrational readings caused by flexing of the printed circuit boards.

Operational concerns were reported in the following areas: (1) communication setup between the Sentinel Gateway and Laptop; (2) communication setup between RotoSense sensor and Sentinel Gateway; (3) sending commands to the RotoSense sensor; and (4) setting a synchronized time on a RotoSense sensor.

An improved MEMS-based sensor (Figure 12) was developed using the following approaches: (1) employ potting; (2) reposition the PCB board and the battery; (3) improve the quality of the built assembly to ensure an ability to withstand stresses and flexing due to high-g vibration and shock; and (4) operational design changes to reduce power consumption and thereby improve battery life.

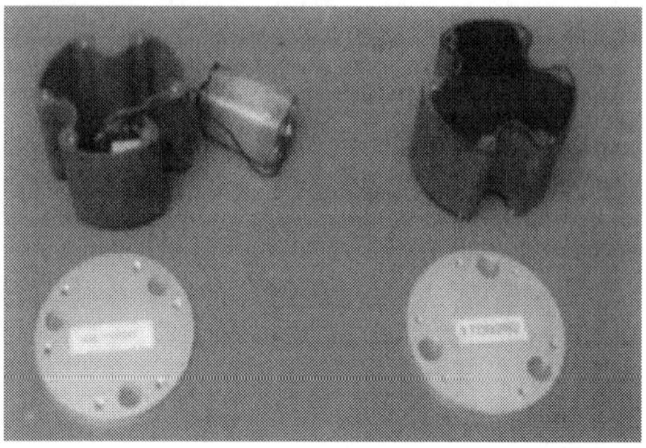

Figure 12: Original (Left) and Improved (Right) Sensors

Quality improvements included inspections and procedures regarding use, assembly, and testing. Improvements included hardware, software solutions, and documentation.

Final testing of the improved MEMS-based sensor, was performed at the National Technical Systems (NTS) test facilities in Tempe, Arizona and at Ridgetop Group laboratories in Tucson, Arizona.

Epoxy-based Potting

An epoxy-based potting was employed: (1) prevented battery displacement; (2) PCB protection from internal vibration and shock forces; (3) increased accuracy in sensor readings; and reduction of additional internal forces. Experiments were performed on sensor units with no potting, partial potting, and full potting. Partial and full potting addressed battery displacement, but partial potting did not fully address board flexing: full potting yielded the best results: zero defects related to shock and vibration occurred during initial and final testing conducted at Ridgetop Group and at NTS.

Component Repositioning

Experiments were performed using variations of component placement: the highest quality signals were obtained using the component placement shown in Figure 13.

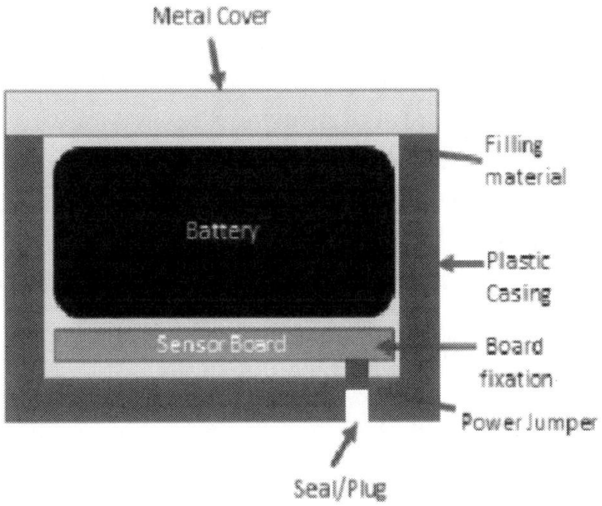

Figure 13: Component Placement Diagram.

Software and Firmware Improvement

Supporting software updates include an updated file structure for more flexible control of future parameter changes. Firmware updates include the following:

1. Updated sensor-gateway communications
2. Updated gateway firmware to reduce timeouts
3. Sensor firmware, functional improvements:
   - Ability to change sensor node address
   - Ability to change the computer IP address to which the data is transmitted
   - Set time on sensors
   - Read sensor temperature
   - Read radio frequency
   - Improved the wireless speed capabilities from 5.5KB/S to 11KB/S.

Documentation Improvement

The documentation was improved as follows:

1. Improved description of software commands
2. How to change sampling rates
3. How to change node addresses
4. Improved description of use and operation:
   - Procedure to turn off the firewall
   - Memory map functions
   - Operating as an administrator (Admin)

Operational Changes to Reduce Power Consumption

Power consumption was reduced to increase battery life as follows:

1. Development of software and firmware to dynamically support the following usage modes:
   - Burst: high-sample rates – 1 kHz to 100 kHz
   - Streaming: actively sampling and transmitting data at < 1 kHz
   - Standby (default): actively waiting for a command – default mode
   - Sleep: periodic sampling of accelerometer with no data storage or transmission when acceleration below defined threshold
   - Deep sleep: sensor is put into a static state a 'no vibration' period of 3 to 12 minutes – dependent on which add-on version is obtained
2. Redesign and reimplement sub-circuit stages, such the oscillator, voltage regulation, and dividers, to reduce power consumption.
3. Development of a deep Sleep operating mode: an electro-mechanical latch-circuit was implemented to allow the sensor to shut-off after being in a static sate for an adjustable time period of 3-12 minutes.

4. Data Collection and Analyses at NTS and Ridgetop Group

Test plans were updated from one to three types of vibration test: (1) fixed frequencies, (2) sweep frequency, and (3) pulse (shock). Testing included tests performed in the Ridgetop lab and external testing at NTS in June of 2017.

NTS Testing

Fixed Vibration Frequencies: Three vibration frequencies at seven levels of force defined and tested: Table 4. The sequences were run at four sampling rates: 160 s/s, 250 s/s, 500 s/s, and 1000 s/s.

Table 4. Fixed Vibration Frequency.

Vibration Frequency: Force
Dwell @ 50 Hz: 5g, 10g, 20, 30g, 50g, 70g, 100g
Dwell @ 100 Hz: 5g, 10g, 20, 30g, 50g, 70g, 100g
Dwell @ 250 Hz: 5g, 10g, 20, 30g, 50g, 70g, 100g

Sweep Vibration Frequency: The sweep frequency as defined as 10 Hz to 500 Hz at two different levels of force: 10g and 40g. The sequences were run at the defined four sampling rates.

Pulse Vibration (Shock): The pulse was defined as a six-millisecond, positive half-sine wave repeated five times at the six different magnitudes of force used for fixed-frequency testing. The sequence was repeated for each of four different sampling rates.

5. Discussion and Results

Ruggedization

No physical damage or anomalies were found during and after the testing regime at NTS. At no time was there any loss of signal.

Improved Signal Quality

After potting and repositioning of components, the signal quality is significantly improved as seen in Figure 14 through Figure 17.

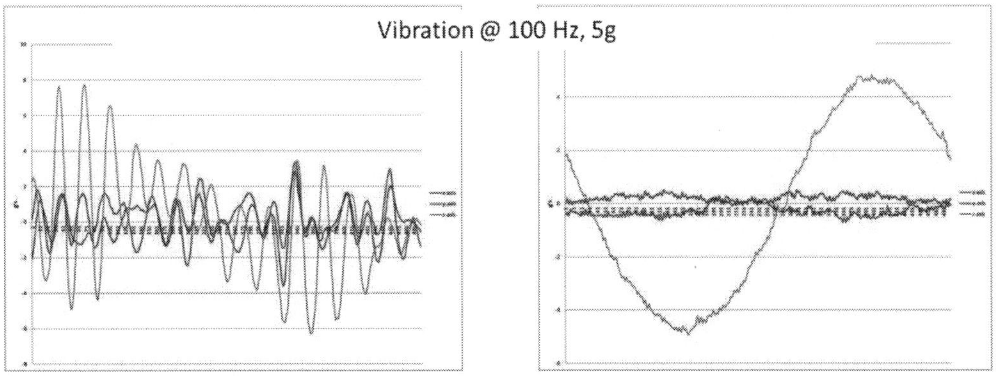

Figure 14: Vibration @ 100 Hz, 5g – Before (Left) and After (Right).

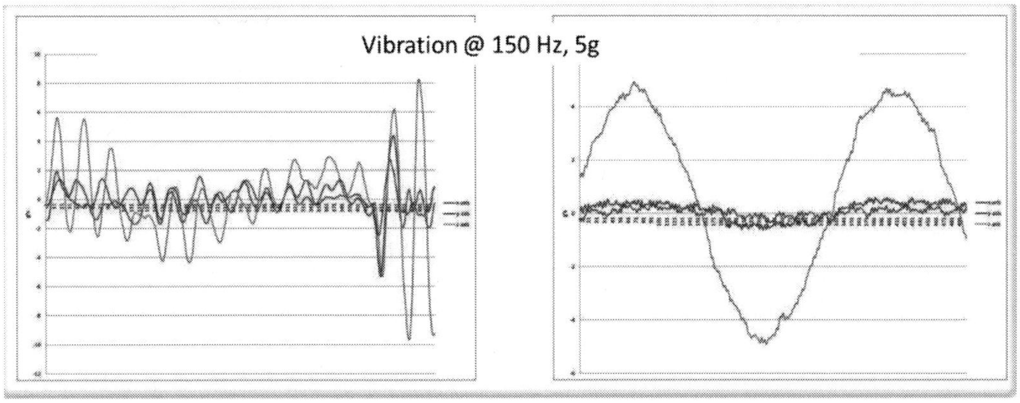

Figure 15: Vibration @ 150 Hz, 5g – Before (Left) and After (Right).

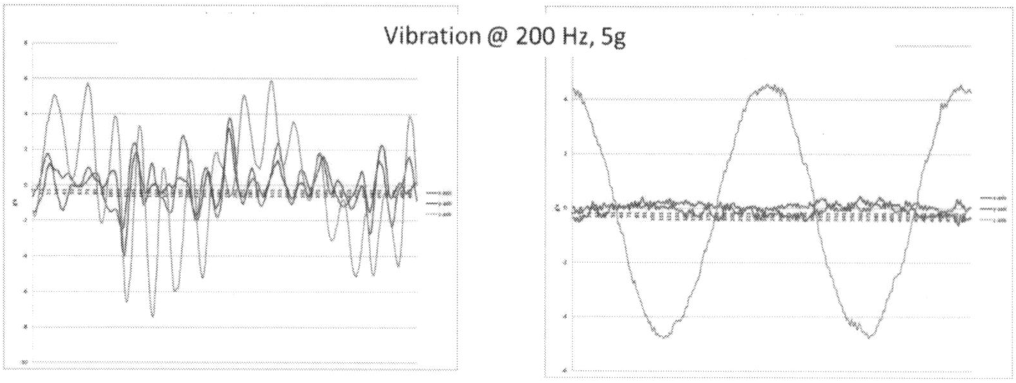

Figure 16: Vibration @ 200 Hz, 5g – Before (Left) and After (Right).

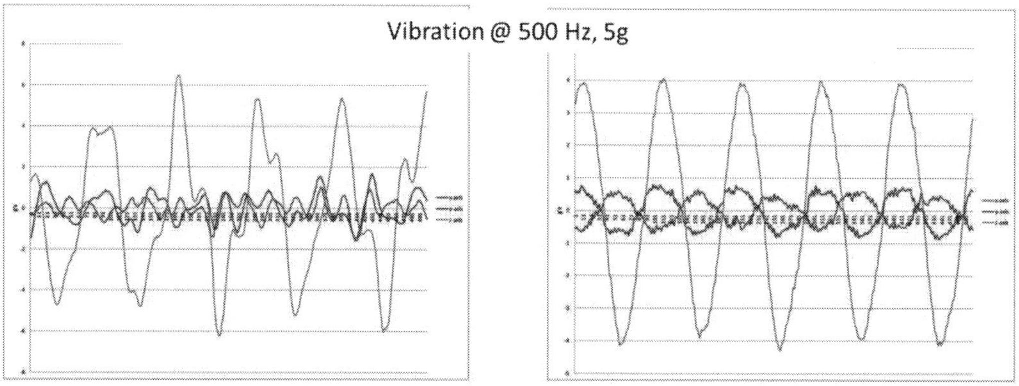

Figure 17: Vibration @ 500 Hz, 5g – Before (Left) and After (Right).

Reduced Harmonic Distortion and Noise, More Accurate Amplitude Measurements: The output signals from the MEMS-based sensor after repositioning and potting of the boards and battery pack no longer exhibit extreme harmonic distortion and noise; the amplitudes more accurately reflect the magnitude of the vibration force; and the bandwidth has been improved from 4-6 Kb/s to 20-22 Kb/s.

Further improvements in signal quality have been made: primarily through improvements in configuring the ground planes, signal paths, and voltage references.

Improved Gateway: A new version of the Gateway for helicopter use is significantly more rugged: (1) it uses super strong Vicor power supply, that went through extensive power cycle testing on the Apache helicopter; and (2) all internal components are compact, soldered to a carrier board, and/or immobilized.

The new Gateway is an Edge Computing, IoT device that taps and processes data where it is generated: increased efficiency, faster transmission, and reduced resources.

NTS - Fixed Frequency:

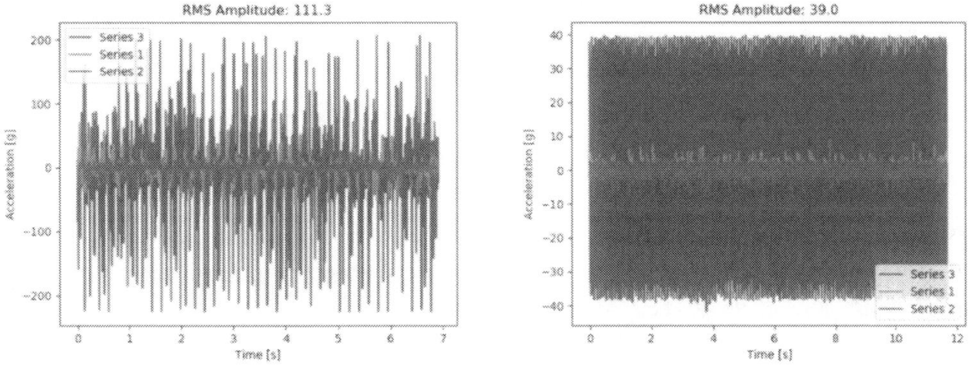

Figure 18: NTS Fixed Frequency Comparison.

NTS Sweep Test - 1 Hz/s Interval, 10 Hz - 250 Hz Frequency Sweep @ 40 g

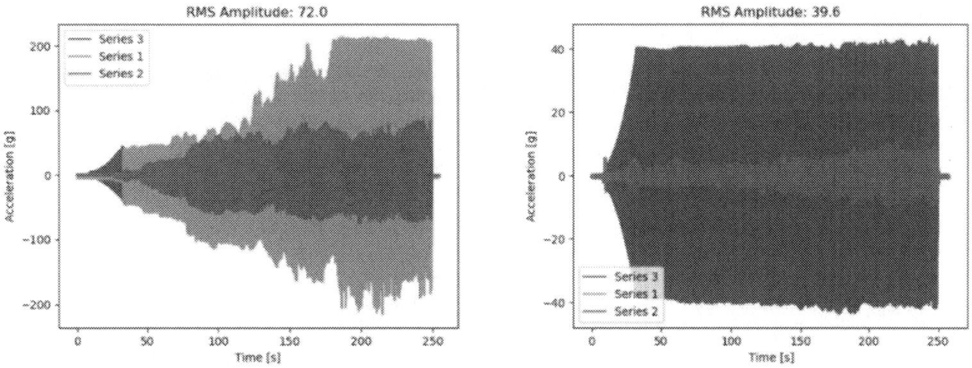

Figure 19: Sweep Frequency Comparison – Non-Potted (Left), Potted (Right).

NTS Shock Test

The non-potted version of the sensor failed during the fixed-frequency and sweep-frequency tests at NTS. The ruggedized (potted) and electrically-improved version of the sensor performed exceedingly well.

Decreased Power – Increased Battery Life

The operational power consumption has been reduced from the levels at the time of the test runs at the NTTC (2015) to those listed in Table 5.

NTS Verification

Ridgetop Group returned to NTS in December of 2017 with a revised test plan that was tailored to the IEC-61373 certification standard. The test conditions were for the body-mounted, bogie-mounted, and axle-mounted categories.

The units under test all survived the testing regime, and provided further repeatability and validation of what was observed in June 2017 with the potted RotoSense design.

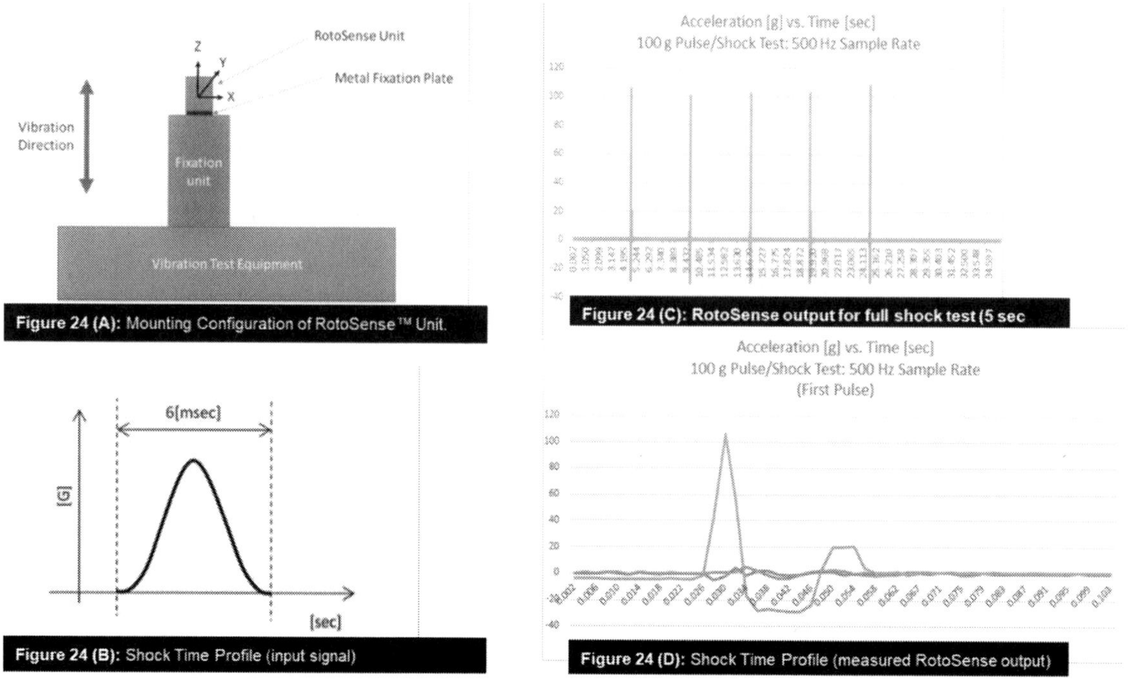

Figure 20: NTS Results - Shock Test

Table 5: Power Usage (9600 mAh Battery):

Usage Mode	2015	2018
OFF	0	0
Deep Sleep	N/A	< 0.001 mA
Sleep	N/A	< 4.0 mA
Standby	45 – 55 mA	< 22 mA
Streaming	40 – 60 mA	< 25 mA
Burst	45 – 75 mA	< 70 mA sampling < 25 mA data transfer

6. Conclusions

Prior to the 2018 conference, two versions of MEMS-bases sensors were designed and developed: one version for use in gear boxes of helicopters and the other version for hub mounting on axles of rolling stock. Each version of the sensors, their application, and their testing and results were described and the results evaluated: both versions functioned and operated successfully.

Subsequently, it was discovered that the original RotoSense design for rolling stock was not rugged enough to sustain a customer requirement of applied vibration at 40 g. After conducting a root cause analysis and research, that version of the sensor was ruggedized by repositioning and potting of the components: the boards and the battery pack. The ruggedized version of the sensor was tested at the National Technical Systems at various regimes: fixed frequency, sweep frequency, and pulse mode vibration at various frequencies and levels of force - no physical failure or anomalies occurred.

Improvement in signal quality (level of harmonic distortion) and the accuracy of measured force were verified by comparing signals from non-ruggedized sensors to those from ruggedized sensors. In addition to ruggedization, improvements were made to the hardware to increase sensor bandwidth and to improve the Gateway with respect to ruggedness, to IoT compatibility, and to communication speeds. Hardware design changes in grounding and signal paths were made to further reduce noise.

The software, firmware, and documentation were updated to increase communication, to reduce timeouts, and to increase functionality and usability.

We achieved a significant decrease in power consumption and a corresponding increase in average battery life. We anticipate further improvements in the future, especially with regards to further reducing power consumption; and we are investigating the feasibility of and approaches to energy harvesting and rechargeable methods to achieve further improvements in extending battery life: enabling design changes have already been developed. In summary:

1. All tested RotoSense units showed high correlation with the input test conditions on the T-1000 shaker.
2. The new RotoSense assembly process demonstrated that it could sustain high [g] operating conditions.
3. The signal to noise ratio was drastically improved, especially when compared to the original assembly.
4. The second trip to NTS provided further validation that the sensor assembly could operate at test conditions similar to that of the IEC-61373 certification standard.
5. The second trip to NTS also validated that the firmware and software enhancements increased data transmission bandwidth and reliability while maintaining high input/output data correlation.

We have designed and developed a third version of RotoSense for embedded planetary gearbox applications. It is the rectangular form factor that was also evaluated with the pinion gear sensor during the NASA project that was embedded in the top half of the gearbox. This particular design does have a 3-axial d-accelerometer instead of a 3-sensor configuration.

Plans are to apply the potting approach to increase the ruggedization of the helicopter version of the sensor. That version has already been improved by increasing the sensor bandwidth, making the Gateway more rugged, and improving Gateway communications.

Acknowledgments

The authors thank Naval Air, Naval Sea, U.S. Army, U.S. Air Force, and NASA research centers for their support and funding of multiple projects that led to the design and development of the prototype solutions and the results described and shown in this paper.

References

[1] Lewicki, D.G., Lambert, N.A., and Wagoner, R.S., (2015) "Evaluation of MEMS-Based Wireless Accelerometer Sensors in Detecting Gear Tooth Faults in Helicopter Transmissions," NASA/TM-2015-218722.

[2] Hofmeister, J.P., Goodman, D., and Wagoner, R. (2016), "Accurate Vibration and Speed Measurement on Rotating Shafts Using MEMS and IoT Single Wireless Triaxle Sensor," Conference of the Society for Machine Failure Prevention Technology, Dayton, OH, 24-26 May, 2016.

[3] Hofmeister, J.P., Pena, W., Wagoner, R., and Nielsen, M., (2017) "Improved RotoSense™ for Rolling Stock: Locomotives and Cars," Conference of the Society for Machine Failure Prevention Technology, Virginia Beach, VA, 16-18 May, 2017.

[4] Lindstrom, M., (2018) "What's Wrong with My Piezoelectric Accelerometer?" PCB-TN-30-0418, MTS Systems Corporation, PCB Piezotronics, Depew, NY, April 2018.

Evaluating The Accuracy Of Prognostic Estimates:

Remaining Useful Life, State Of Health, Prognostic Horizon

James P. Hofmeister, Erik Sandberg, and Robert Wagoner
Ridgetop Group, Inc.
3580 West Ina Road
Tucson, AZ 86741
520-742-3300
jhofmeister@ridgetopgroup.com

Abstract: This paper describes how to evaluate the accuracy of prognostic estimates for remaining-useful life (RUL), state of health (SoH), and prognostic horizon (PH). The primary goal of prognostics is to accurately predict future failure in systems. Condition-based-maintenance (CBM) systems use condition-based data (CBD) to detect degradation and project that degradation to a failing level at a future time. Prognostic estimates are produced by a Prognostic Health Monitoring (PHM) system. The process comprises three stages: sensing, data processing, and prediction. The time when data is sampled, the amplitude of features extracted from that data, the rate at which the amplitudes change, and a model for that data are used to estimate that future time of failure and to calculate prognostic estimates (RUL, SoH, and PH): collectively known as prognostic information. Evaluation of the accuracy of prognostic information is critical and necessary to install confidence in those estimates. Evaluation begins with knowing that ideal, zero-error estimates are not achievable in practice because of, for example, initial-estimate errors, offset errors due to sampling, noise, and data nonlinearity. Performance metrics to evaluate accuracy of RUL, SoH, and PH estimates are developed and tabulated: convergence efficiency, prognostic distance (PD), and PH accuracy (PHα). Other prognostic terms, names, and descriptions are also provided and examples of evaluation methods and how to improve prognostic accuracy are included.

Key words: CBM; convergence; estimates; PD; PH; prognostic; RUL; SoH.

1. Introduction

Prognostics has the primary goal of accurately predicting future failure in systems. Condition-based-maintenance (CBM) systems use condition-based data (CBD) to detect degradation and project that degradation to a failing level at a future time. The time when data is sampled, the amplitude of features extracted from that data, the rate at which the amplitudes change, and a model for that data are used to estimate a future time of failure and to calculate prognostic estimates (RUL, SoH, and PH): collectively known as prognostic information. You need to evaluate and ensure that prognostic information produced by a PHM system meets accuracy requirements.

PHM System

The job of a PHM system is to sense a node and isolate one or more features to be data processed as input to prediction algorithms that produce accurate prognostic information for subsequent health/fault management. You ensure the accuracy of prognostic information by changing operating parameters of a PHM system. Figure 1 is a block diagram of an exemplary PHM system comprising three (3) stages: a sensing stage, a data processing stage, and prognosing stage.

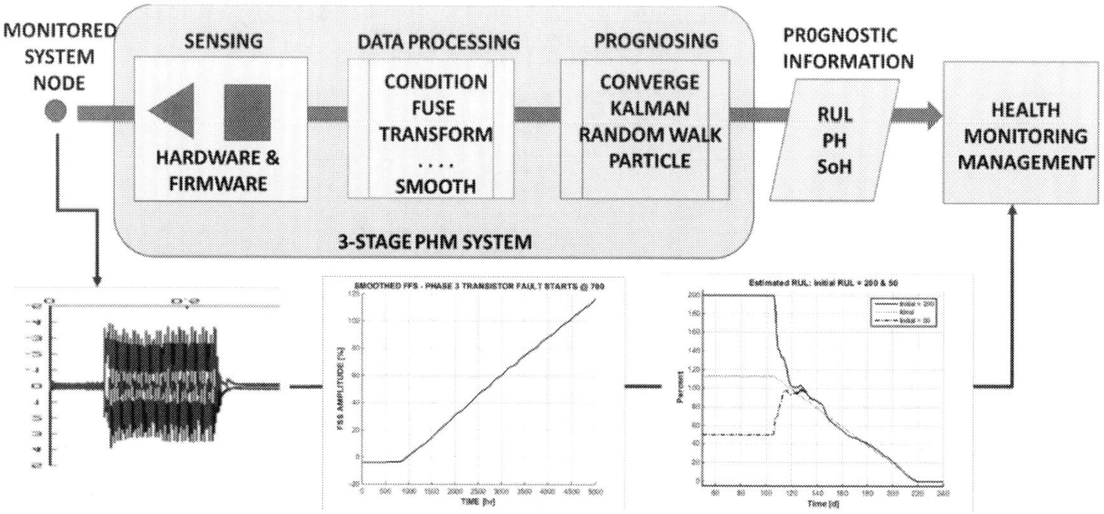

Figure 1. Diagram of a 3-Stage PHM System.

The sensing stage, attached to a node in a system being monitored for health, is comprised of one or more sensors that monitor and collect condition-based data (CBD): CBD examples include, vibration, shock, temperature (heat), light, and electrical (voltage and current). The output of the sensing stage is usually a filtered and/or windowed CBD. Sometimes, a sensing stage outputs reconditioned CBD as feature data (FD).

The data-processing stage conditions the output of the sensing stage to produce a linearized signature called Fault to Failure Progression (FFP) signature through the use of computational routines, functional routines, and program modules. Those routines and program modules are reusable and are selected and used as necessary for application-specific solutions.

The prognosing stage comprises a set of prediction algorithms that processes Functional Failure Signature (FFS) data to produce accurate prognostic information. The one in the example PHM system uses fast-convergence algorithms and Kalman-like filtering. In Kalman-like filtering, each data point is compared to a model, the model is adapted to the data, and then used to predict a future time of failure. The prediction algorithms also treat data as particles having momentum (they tend to keep moving in the same direction) and inertia (abrupt changes are not related to degradation). A random-walk type of solution is also used wherein data moves from the lower-left of a data space (zero degradation) to the upper-right of a data space (functional failure). The prediction algorithms support self-healing wherein degradation is reduced.

The purpose of evaluating the accuracy of prognostic information is to ensure accuracy requirements are met: if not, adjustments to and/or modification of the PHM system tuning become necessary. The exemplary PHM system has parameterized the selection, functionality, and operational values and thresholds to support tuning for each monitored node. For high accuracy, a 3-stage PHM systems needs the following:

- One or more sensors to monitor nodes to capture CBD, with or without preliminary data conditioning.
- Data processing routines, algorithms, and programs to filter, cancel, or otherwise mitigate noise and to transform data from CBD that is typically curvilinear and very noisy to signature data that is linearized and relatively noiseless;
- A very fast and accurate set of prediction algorithms – the exemplary PHM system uses a set of prediction algorithms in a program called Adaptive RUL Estimator, Advanced Version* (ARULEAV). [* trademarks of Ridgetop Group, Inc.].

Features and Noise. Degradation occurs when, because of continuing or cyclic stresses and strains (plastic work), damage accumulates and one or more features change accordingly. Such features are often referred to as condition indicators or leading indicators of failure.

In addition to features, system nodes exhibit noise, including thermal noise, harmonic distortion, switching nose (glitches and spikes) that are often found to be damped-ringing responses, and so on. Noise is defined as anything that is measurable and which is not a feature of interest. For example, referring to Figure 2, the CBD is analog voltage comprising at least eight forms of features and noise com: (1) DC voltage; (2) voltage droop; (3) ripple voltage; (4) thermal noise; (5) damped ringing response; (6) harmonic distortion; (7) switching spikes and glitches; (8) periodicity of features and noise; and so on.

$$V = CBD_1 + CBD_2 + \cdots CBD_n + N_1 + N_2 + N_m \quad (1)$$

In EQ. (1), CBD_{n} refers to one or more features of interest and $N_{\mathrm{n}}$is everything else that is measurable and collectible that is not of interest for a particular solution.

Features and Noise Within a Feature. In practice, a given feature may comprise other features. For example, the damped-ringing response shown in Figure 2 and Figure 3 comprises other features (and noise); primary features are the amplitude (A_R), the period of the resonant frequency (T_R), and the time constant, often referred to as lifetime, (λ) of the exponential decay.

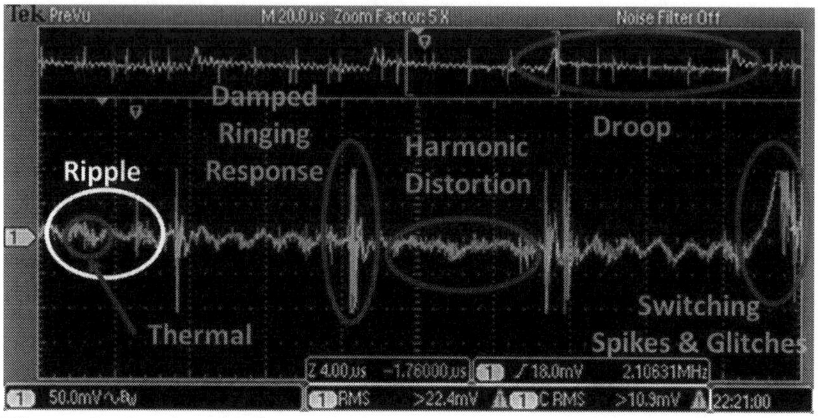

Figure 2. Analog Voltage Present at a Particular Node in a System

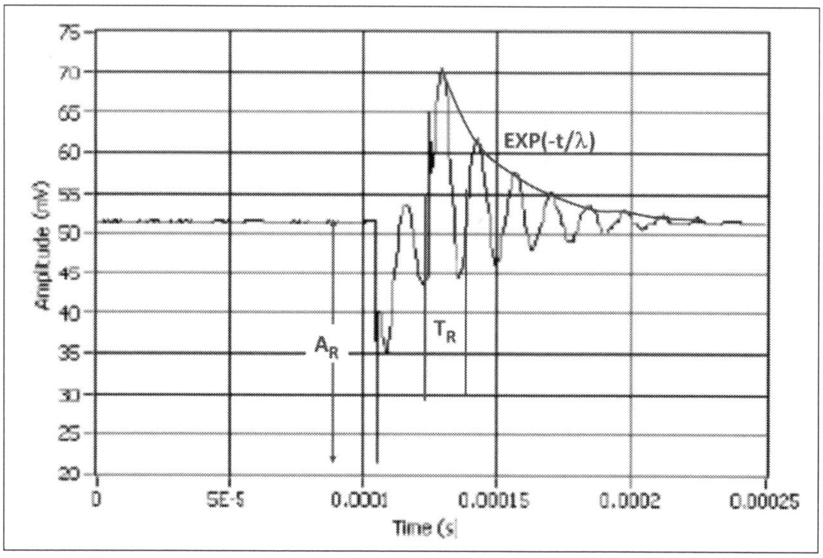

Figure 3. Damped-Ringing Response Showing A_R, T_R, and λ Features.

Prognostic Estimates and Terminology: Prior to the Onset of Damage

Figure 4 plots an ideal RUL transfer curve: when a product is first used, it has an expected lifetime (λ) that is exactly equal to the time when it fails (EOL). However, ideal RUL estimates are unrealistic and are not going to happen. If three identical products are manufactured the same day with a first product left unused, a second product put into heavy use on the 1st[th] day and ***fails*** 100 days later, and the third product put into normal use on the 101st day, what is the expected RUL of that third product? An ideal RUL transfer curve would result in a estimated RUL of 100 days. But wait, how does ***normal use*** versus ***heavy use*** change the RUL estimate? That answer is not only unknown, it cannot be known.

Traditional Approaches to Prognostics. A primary problem with traditional model-driven, data-driven, and hybrid-driven approaches to prognostic estimation is they do not and cannot take into account all of the stress and strains, plastic work, extreme events, and so on that result in the onset of damage and henceforth an ongoing accumulation of damage (degradation) leading to failure.

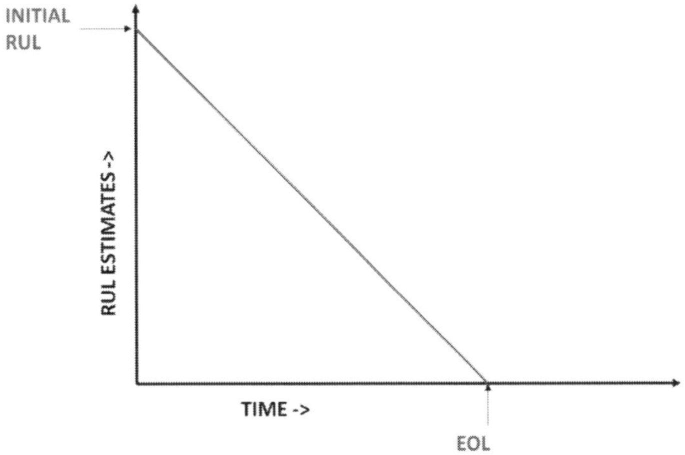

Figure 4. Ideal RUL Transfer Curve: RUL Estimate Linear with Sampling Time.

Failure versus Functional Failure. Failure, as it relates to prognostics, is often ill-defined. For example, does failure occur when a drive shaft shears? Or does it occur when friction between the shaft and a roller bearing reaches a level where torque on the shaft exceeds design specifications? Or does failure occur when the shaft becomes warped and the unit begins to vibrate?

We define and use ***functional failure*** rather than simply ***failure*** or ***physical failure***. Functional failure is defined to occur when a device, component, subassembly, … system is no longer capable of operating within specifications: functional failure, therefore, occurs before physical failure or operational failure.

2. Prognostic Information: Terms, Names, and Description

Figure 5 is a plot of a more realistic and practical RUL transfer curve: until damage occurs that causes a measurable change in a signal, it is not reasonable to provide RUL estimates that become smaller as time passes.

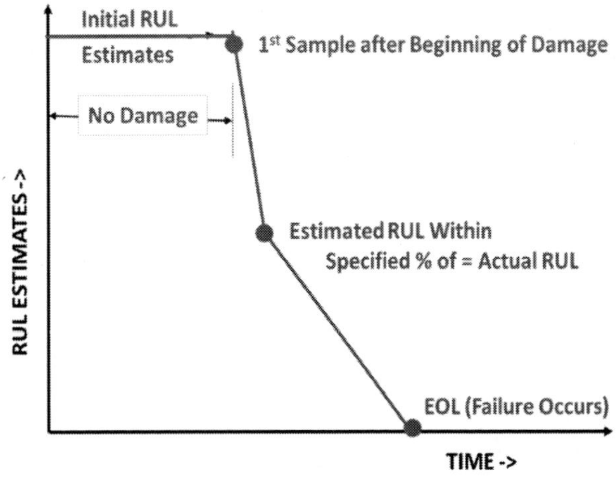

Figure 5. A More Realistic RUL Transfer Curve.

Once damage is detected, then RUL estimates can be made using knowledge of the following: physics-of-failure (PoF), association of a change in a measured signal to such PoF, and a model of how the signal changes as damage accumulates. Within some number of samples, prognostic algorithms should produce estimates that converge from an initial-estimate error to estimates within a specified % accuracy.

The accuracy of prognostic information, estimates of RUL, PH, and SoH, cannot be 100%, if for no other reason than it is not possible to know when a specific prognostic target is going to fail, and this is especially true in the absence of any detectable degradation. However, for a single failure mode, a PHM system should produce increasingly accurate estimates as degradation progresses toward EOL.

Terminology and Performance Metrics

Figure 6 illustrates the basic terminology that is associated with RUL and Table 1 lists prognostic terms, names, and descriptions (not all of which are shown in Figure 6).

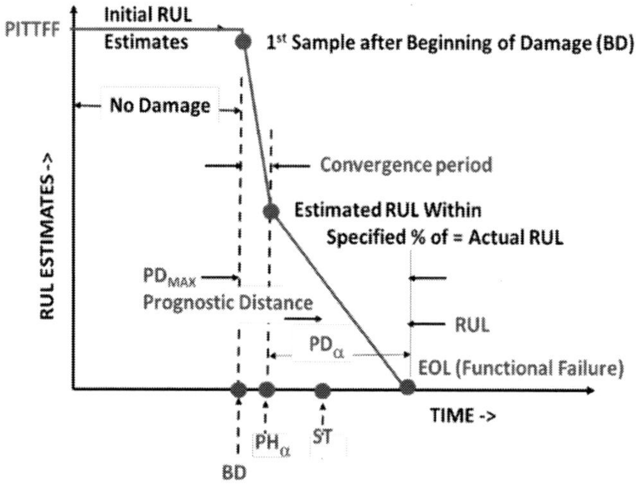

Figure 6. RUL Transfer Curve: Basic Terminology.

Table 1: List of Prognostic Terms, Names, and Description

Term	Name	Description/Explanation
α	Alpha (accuracy)	Used with **PD**, **PH**, and **χ** ▪ When estimates are within **α** of actual
BD_0	True BD	Actual time when degradation begins
BD ($BD_{DETECTED}$)	Estimated beginning of degradation	Estimated because of ▪ Measurement uncertainty, ▪ Noise (margin) ▪ Sampling period
BD_{OE}	Offset error for BD ▪ $BD_{OE} = BD_{DETECTED} - BD_0$	The error (time) between actual and estimated beginning of degradation

χ_α	Convergence figure of merit: ▪ $\chi_\alpha = PD_\alpha / PD_{MAX}$ Defined by Ridgetop Group, Inc.	Figure of merit value when **RUL** or **PH** within α of actual
Convergence	Convergence of estimates to actual	Transition from high-error values of **RUL** to within a specified α accuracy
EOL_0	True EOL	Actual time when end of life (functional failure) occurs
EOL ($EOL_{DETECTED}$)	Estimated end of life	Estimated because of ▪ Measurement uncertainty, ▪ Noise (in FF threshold) ▪ Sampling period
EOL_{OE}	Offset error for EOL ▪ $EOL_{OE} = EOL_{DETECTED} - EOL_0$	Time between actual end of life and detect time of functional error
FF	Functional Failure	State in which a prognostic target no longer operates within specifications
PD	Estimated prognostic distance ▪ $PD = EOL - ST = RUL$	Estimated time to functional failure
PD_0	Maximum prognostic distance ▪ $PD_0 = EOL_0 - BD_0$	Time between the actual beginning of degradation and the time when **FF** occurs
$PD_\propto$	Prognostic distance at $PH_\propto$	Estimated prognostic distance when RUL estimates are within α accuracy
PD_{MAX}	Estimated maximum prognostic distance ▪ $PD_{MAX} = EOL - BD$ For relative time (BD = 0) ▪ $PD_{MAX} = EOL$	Used in lieu of **PD_0** because **BD** and **EOL** are estimated
PH	Prognostic horizon, For relative time (BD = 0) ▪ $PH = ST + RUL$	The estimated time when **FF** occurs
PH_α	All subsequent PH estimates within ▪ $\alpha \leq abs((PD_{MAX} - PH)/PD_{MAX})100)$	When the percent difference between **PD_{MAX}** and **PH** is within **α** percent
PITTFF	Prognostic Initial Time-to-Functional-Failure	A specified value to be used for estimated RUL in the absence of detected degradation
RUL	Remaining useful life ▪ $RUL = EOL - ST$	The distance between **EOL** (estimated) and the sample time
RUL_α	RUL for PH_α	The value of **RUL** when **PH_α** is achieved
ST	Sample time	Time recorded for when the data was taken: includes count, record, ...

Ideal Prognostic Estimates

Given an initial estimate errors of ± 50% and assuming prediction algorithms that can converge to within α in less than 10 data points, Figure 7 illustrates ideal RUL and PH estimates and Figure 8 illustrates ideal SoH estimates.

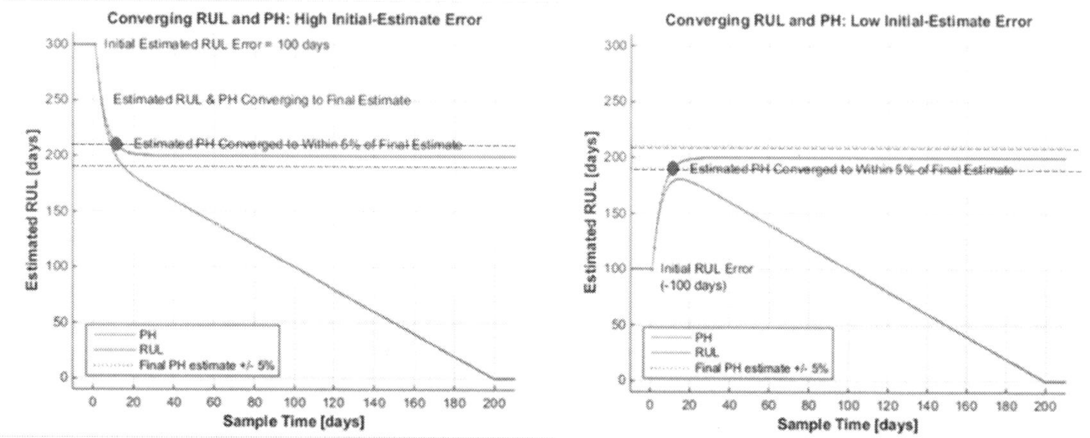

Figure 7. Ideal RUL Estimates with PH_5 (Red) Shown.

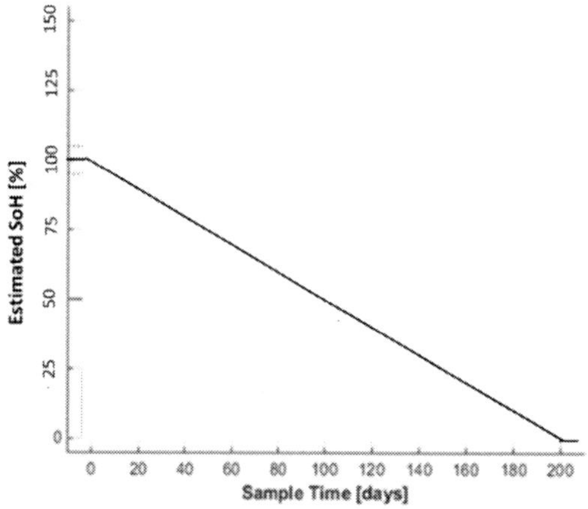

Figure 8. Ideal SoH Estimates.

3. Evaluating Prognostic Estimates

Ideal prognostic estimates are highly unlikely in a non-laboratory and/or non-simulation experiment in the presence of noise and imprecise measurements. Realistic prognostic estimates are noisy and, when first obtained, are unlikely to meet accuracy requirements.

Important Performance Metrics

Important performance metrics are the following:

- Offset errors that reduce the maximum prognostic distance: $\mathbf{BD_{OE}}$ $\mathbf{EOL_{OE}}$

- Errors other than offset errors that result in a failure to meet requirements for a minimum prognostic distance: $\mathbf{PD_{MAX}}$. An example is a requirement that the system shall meet a 72-day advance notice of failure and the degradation is detected only 20 days prior to functional failure.

- Failure to converge fast and accurately enough to meet a-accuracy requirements: χ_α, $\mathbf{PH_\alpha}$, and $\mathbf{RUL_\alpha}$.

The following accuracy requirements for our exemplary PHM system are used for illustrative purposes: (1) convergence to $\mathbf{PH_{25}}$ with $\mathbf{PD \geq 0.50\, PD_{MAX}}$ and (2) convergence to PH_{10} with $\mathbf{PD \geq 0.25\, PD_{MAX}}$.

Primary Contributors: Failure to Meet Requirements

The following are primary contributors for inaccurate prognostic information: initial RUL estimate too large (sometimes too small); offset errors that reduce maximum prognostic distance; noise with insufficient data conditioning (filtering and smoothing) and/or inadequate noise mitigation (noise margin); and insufficient linearization of data. The prediction algorithms in the exemplary PHM system converge to within 25% accuracy for the following boundary conditions: (1) at least 150 data points shall be sampled between when degradation is first detected and when functional failure occurs; (2) the required $\mathbf{PD_{MAX}}$ shall be > 100 sample periods.

Initial Estimate Error. For a specific prognostic target in a CBD-based PHM system, it is not possible to *know* exactly how long it will take for that degradation to worsen to a level where functional failure occurs.

Rather than issue a ***RUL is unknown*** in the absence of degradation, a PHM system could be designed to include a default value, such as a vendor-specified MTTF value or a value based on historical data or field experience: see PITTFF in Figure 7. An initial estimate error is probable and the prediction algorithm needs to incorporate methods to convergence.

BD Offset Error. It is not practical to use large sampling rates nor is it possible to filter out all noise and environmental drifts in any measurement of any signal. Instead, sampling is likely to result in an offset error as illustrated in Figure 9.

$$BD_{OE} = BD_{DETECTED} - BD_0 \qquad \text{Offset error} \qquad (1)$$

If the sampling period is, for example, one hour, the offset error is ~4.6 hours and might not be worrisome. But if the sampling period is one week, the offset error is ~4.6 weeks and might be considered unacceptable. One method to mitigate such an offset error is to use simple extrapolation using the BD-detected and most-previous data points. Letting *TS* represent the sampling period,

$$BD_1 = (BD_{DETECTED} - TS/2) \qquad \text{Adjusted BD time} \qquad (2)$$

$$BD_{OE\_1} = BD_1 - BD_0 \qquad \text{Period-extrapolated offset error} \qquad (3)$$

Data Conditioning and Noise. Evaluation of Figure 9 reveals that, even after simple extrapolation, there might still be a significant BD-offset error. Rather than attempting to design and development a more complex method of further reducing the magnitude of BD-offset errors, focus on methods to reduce NM: common methods include noise filtering, noise cancelation, and data smoothing. Figure 10 illustrates a reduction on offset error due to reducing NM.

EOL Offset Error. Similar to a BD offset error, sampling is likely to cause an EOL offset error.

Implied Accuracy and Precision: Effect on Sampling. Accuracy of prognostic information is often couched in the following manner: "The PHM system shall provide a failure alert at least 72 hours, ± 1 hour, prior to failure with 10% accuracy." Not explicitly stated are the following: (1) the requirement for a 1-hour precision means the data must be sampled at least twice per hour and (2) the requirement for10% accuracy increases the minimum sampling rate to 20 times per hour – a sampling period of 3 minutes or less.

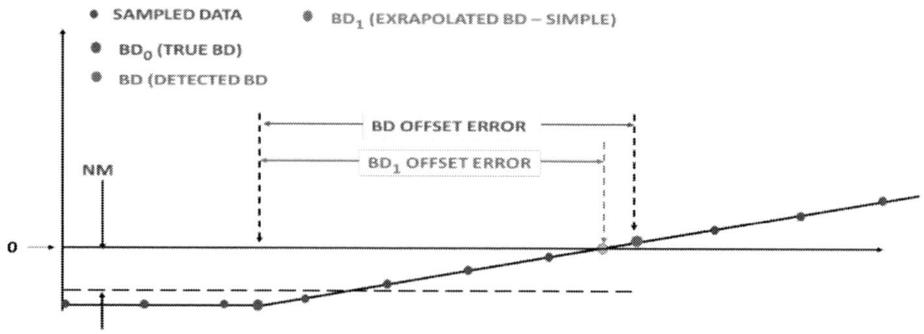

Figure 9. Offset Error in Detecting Beginning of Degradation.

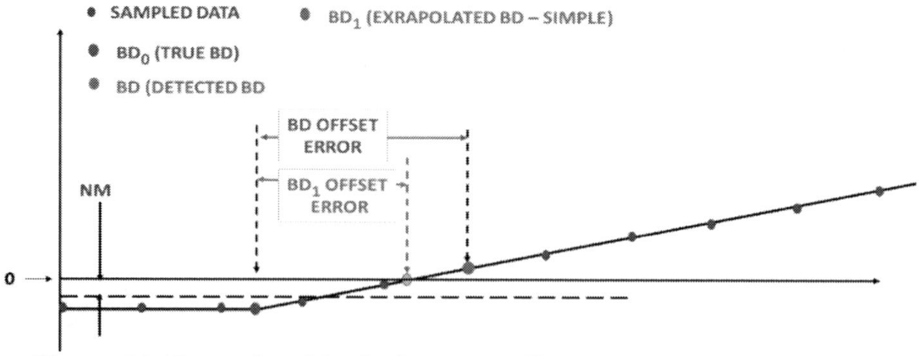

Figure 10. Example of Reducing BD-Offset Error by Reducing NM

Instead of continuously sampling 20 times or more per hour, burst-mode sampling could be employed. Referring to Figure 11, initiate data sampling four times per hour (satisfies the 1-hour precision requirement); use high-rate burst sampling for 20 times; average the burst samples; and return a scalar value (satisfies the 10% accuracy of the estimated time of failure).

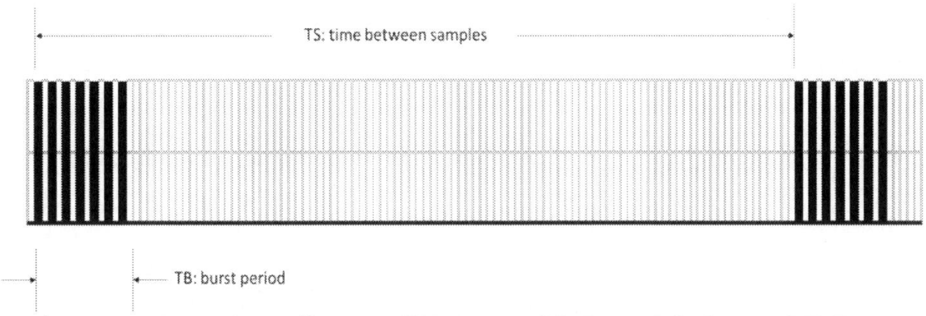

Figure 11. Data Sampling at 1/TS Rate with Burst Mode at 1/TB Rate.

Insufficient Linearization. The exemplary PHM system is designed to transform noisy CBD (Figure 12) into less noisy signature data. The CBD shown in **Figure 12** is conditioned and transformed to FFS (Figure 13, left) which was processed by the prognosing stage to produce the prognostic information shown in Figure 13, right: the red colored circles indicate the required accuracy points. See Table 2 for the evaluation results

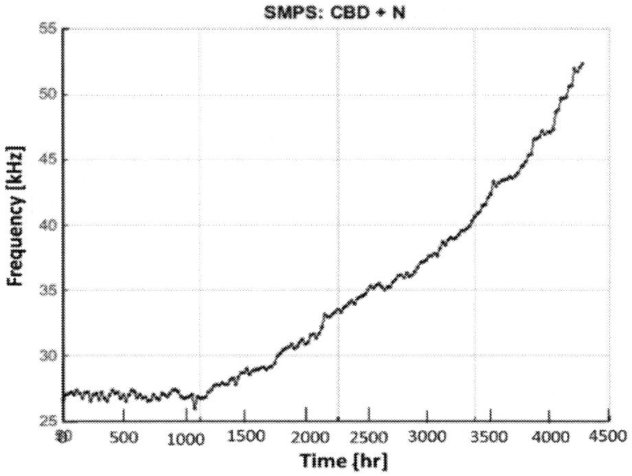

Figure 12. Experiment-Obtained CBD (Frequency).

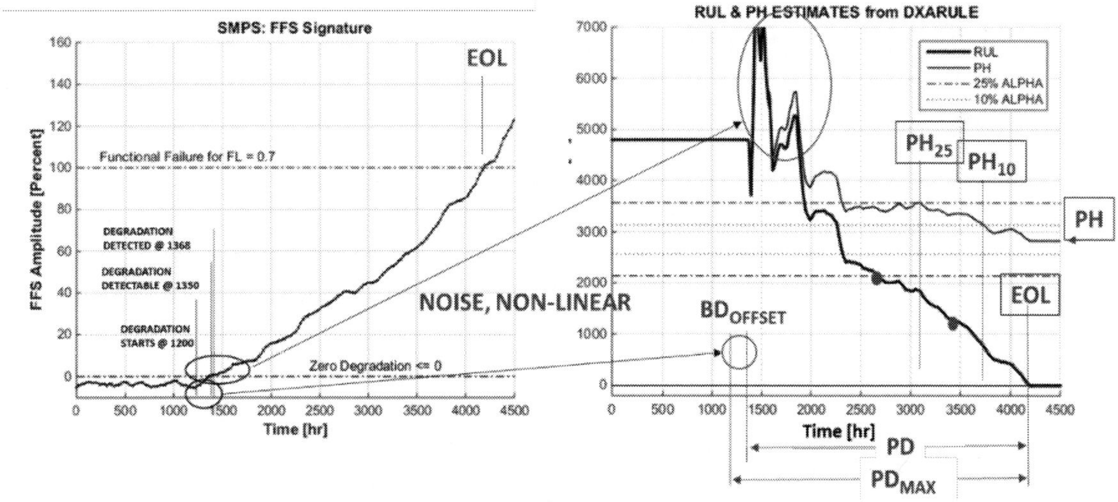

Figure 13. Insufficient Linearization (Left) – Insufficient Accuracy (Right)

A CBD to FFP to Degradation Progression Signature (DPS) to FFS series of transforms produces a more linear FFS (Figure 14, left). Further, the signature is convex rather than concave, which reduces the effect of noise at the beginning of the signature. See Table 2 for the evaluation results.

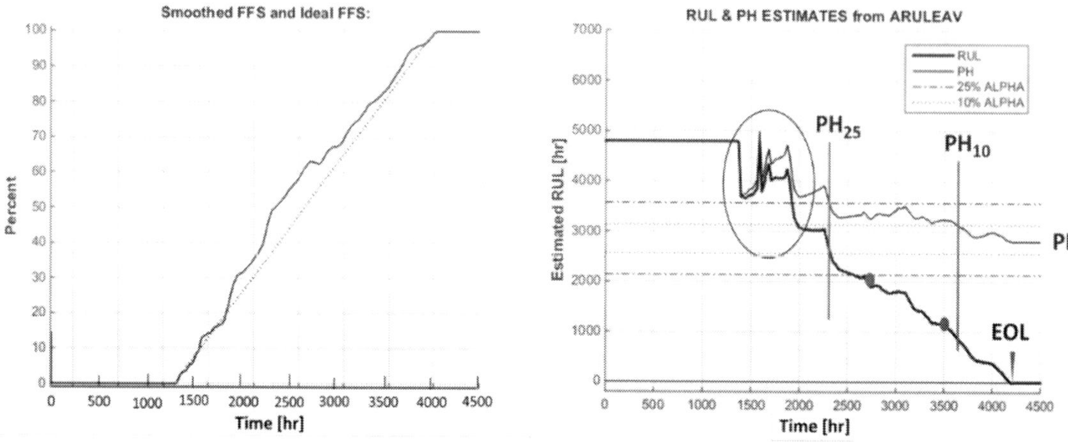

Figure 14. FFP-DPS-FFS Result (Left) and Prognostic Estimates (Right).

FFS data can be further linearized by heuristically adapting FFS data to an FFS-based model to improve prognostic accuracy (Figure 15).

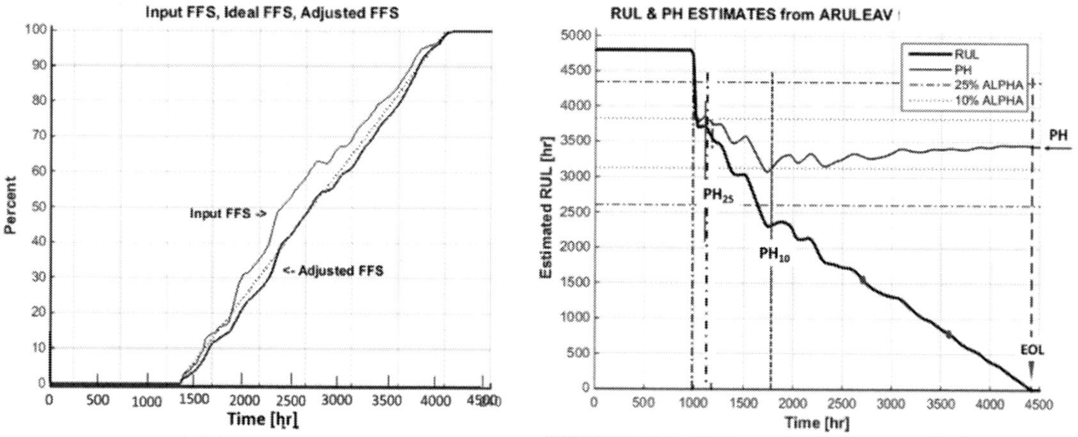

Figure 15. Heuristically Adjusted FFS (Left) and Prognostic Estimates (Right).

The heuristic method takes advantage of the FFS characteristic progression of amplitude from 0 to 100%, which is amenable to Kalman-like filtering, but instead of adapting a model to the data, the data is adapted to the model. Accuracy is significantly improved: (1) both the **PH_{25}** and **PH_{10}** points occur well before requirements, (2) the beginning of degradation was detected at ~1000 hours rather than ~1300 hours, and (3) the end of life was detected at ~4500 hours rather than ~4200 hours. See Table 2 for comparison of evaluation results.

Table 2. Accuracy Results of the Examples

#	Data Processing Used	**PH$_{25}$ at PD≥0.50 PD$_{MAX}$ $\chi_{25} \geq 0.50$**	**PH$_{10}$ at PD≥0.25 PD$_{MAX}$ $\chi_{10} \geq 0.25$**	**BD & EOL (hours)**
1	CBD to FFP FFP to FFS	$\chi_{25} \sim 0.37$ Failed to meet	$\chi_{10} \sim 0.17$ Failed to meet	~1200 & ~4200
2	CBD to FFP FFP to DPS to FFS	$\chi_{25} \sim 0.63$ **Met requirements**	$\chi_{10} \sim 0.17$ Failed to meet	~1200 & ~4200
3	CBD to FFP FFP to DPS to FFS heuristically adapted	$\chi_{25} \sim 0.94$ **Exceeded requirements**	$\chi_{10} \sim 0.77$ **Exceeded requirements**	~1000 & ~4500

4. SUMMARY

This paper presents a high-level overview of an exemplary, 3-stage PHM system as a means for describing inaccuracies in prognostic information and improvement methods. A list of performance metrics and other terms, names, and descriptions related to prognostic information are tabulated. How those terms are related to prognostic information is provided along with the models for the performance metrics.

Ideal estimates of prognostic information are discussed and is followed by methods for evaluating non-ideal estimates. Examples of common errors and methods for improving accuracy are described. Included are three examples dealing with accuracy and linearization of signatures using the same input CBD. The third example showed how, after transforming CBD to FFP to DPS data to FFS data, using heuristic adaption of FFS data improves accuracy of the prognostic information from 0.37 to 0.94 convergence for estimates within 25% accuracy, and from 0.17 to 0.77 convergence for estimates within 10% accuracy: compare Figure 13 and Figure 15.

Bibliography

[1] Goodman, Douglas L.; Hofmeister, James P.; and Szidarovszky, Ferenc; 2019 (Pending), *Prognostics and Health Management: A Practical Approach Using Conditioned-Based Data,* 1st Edition, John Wiley & Sons, Ltd, The Atrium, Southern Gate, Chichester, West Sussex, PO19 8SQ, United Kingdom, *ISBN: 9781119356653, 2019*, pp. 255-274, 300-311.

[2] Hofmeister, J., Wagoner, R. and Goodman, D. (2013) "Prognostic Health Management (PHM) of Electrical Systems using Conditioned-based Data for Anomaly and Prognostic Reasoning," Chemical Engineering Transactions, Vol. 33, pp. 992-996.

[3] Hofmeister, James P.; Goodman, Douglas, L.; and Szidarovszky, Ferenc (2018); "Transforming Condition-based Data Signatures into Functional Failure Signatures," IEEE 2018 Aerospace Conference, Big Sky, MT, 3-9 March, 2018.

[4] Saxena, A., Celaya, J., Saha, B., Saha, S. and Goebel, K. (2009) "On Applying the Prognostic Performance Metrics," Annual Conference of the Prognostics and Health Management Society (PHM09), San Diego, CA, 27 Sep.-1 Oct. 2009.

ELECTROMECHANICAL ACTUATOR CASE STUDY: MULTIVARIABLE ANALYSIS OF PHASE CURRENTS TO DETECT THREE TYPES OF FAULTS

James P. Hofmeister, Robert S. Wagoner, Douglas L. Goodman
Ridgetop Group, Inc.
3580 West Ina Road
Tucson, AZ 85741
520-742-3300
jhofmeister@ridgetopgroup.com

Abstract: This paper describes a multiple-variable analysis (MVA) methodology to detect and prognose three types of faults associated with an electromechanical actuator (EMA): (1) loading faults, such as friction, on the shaft of an EMA motor, (2) shorting faults in the stator windings of the EMA motor, and (3) on-resistance faults in one or more power-switching transistors used to convert direct voltage/current into alternating current. The presented methodology overcomes difficulties associated with typical multivariate analysis (as opposed to multiple-variable analysis) methods such as the following examples: solving simultaneous equations and performing a statistical-based analysis such as K-nearest neighbor (KNN) regression and other Euclidean-based distance methods. Examples of those difficulties are the following: (1) analysis methods that produced information suitable for classification rather than diagnosis or prognosis; (2) noisy data; (3) dependent data, rather than independent data; and (4) difficulty in processing test data to identify, extract, and use leading indicators of failure for prognostic purposes. The primary MVA solution methods included (1) noise mitigation, (2) a unique root-mean-square (RMS) of quantifying phase current values, and (3) a combination of nearest neighbor and distance methods of processing phase-current data to unequivocally identify and isolate faults and to prognose a future time at which functional failure is likely to occur.
*ARULEAV is a trademark of Ridgetop Group, Inc.

Key words: Diagnostics; electromechanical actuator; EMA; IVHM; multiple-variable analysis; MVA; prognostics; PHM

1. Introduction

A NASA-funded Small Business Innovation Research (SBIR) program on electromechanical actuators (EMAs) had the following objectives: deliver a Model-based Avionic Prognostic Reasoner (MAPR) for non-intrusive monitoring of the state of health (SoH) and remaining useful life (RUL) of electromechanical assets by using data obtained from a standard avionic data bus: see Figure 1. The following specific faults were identified: (1) loading faults, such as friction, on the shaft of an EMA motor, (2) shorting faults in the stator windings of the EMA motor, and (3) on-resistance faults in one or more power-switching transistors used to convert direct voltage/current into alternating current. [1]-[5]

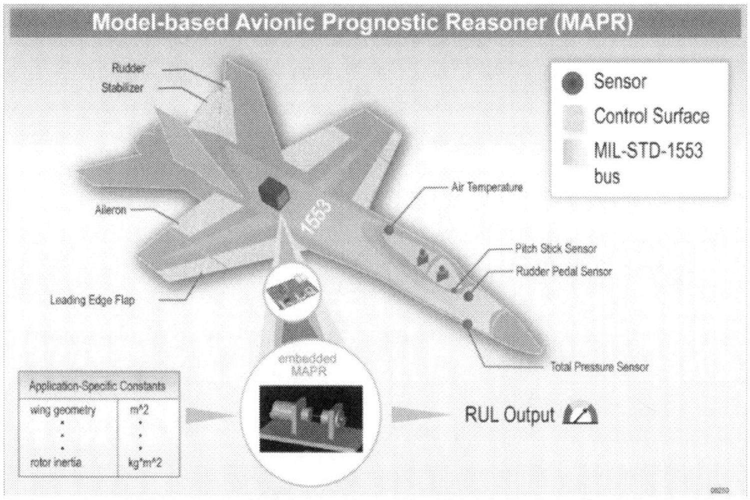

Figure 1: Concept for a Model-based Avionic Prognostic Reasoner (MAPR).

The proposed solution methodology seemed straightforward and not complicated: (1) solve a set of simultaneous equations and (2) use statistical methods, such as principal component analysis (PCA), K-nearest neighbor (KNN), and Euclidean distance methods; and (3) show how the test bed and methods would detect, isolate, and characterize EMA faults. An example of a set of models for multivariate analysis is shown in Figure 2 [5].

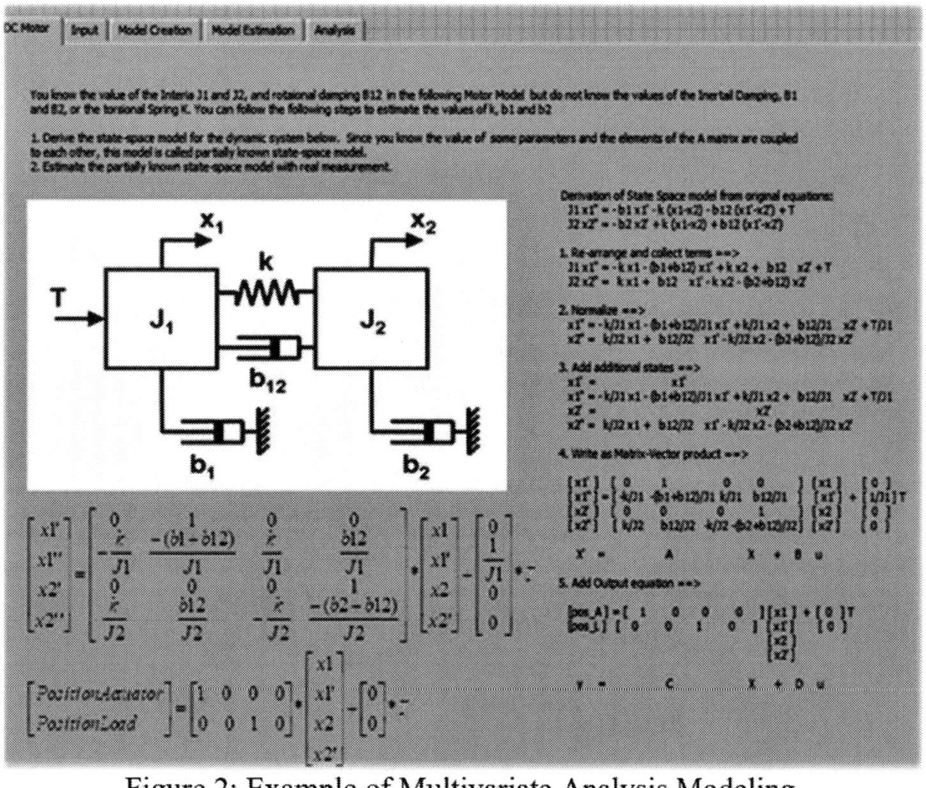

Figure 2: Example of Multivariate Analysis Modeling.

The data set was collected from a test bed (see Figure 3) for an EMA comprising a brushless DC motor (BLDC) driven by an H-bridge switching type of commutation (see Figure 4). [5]

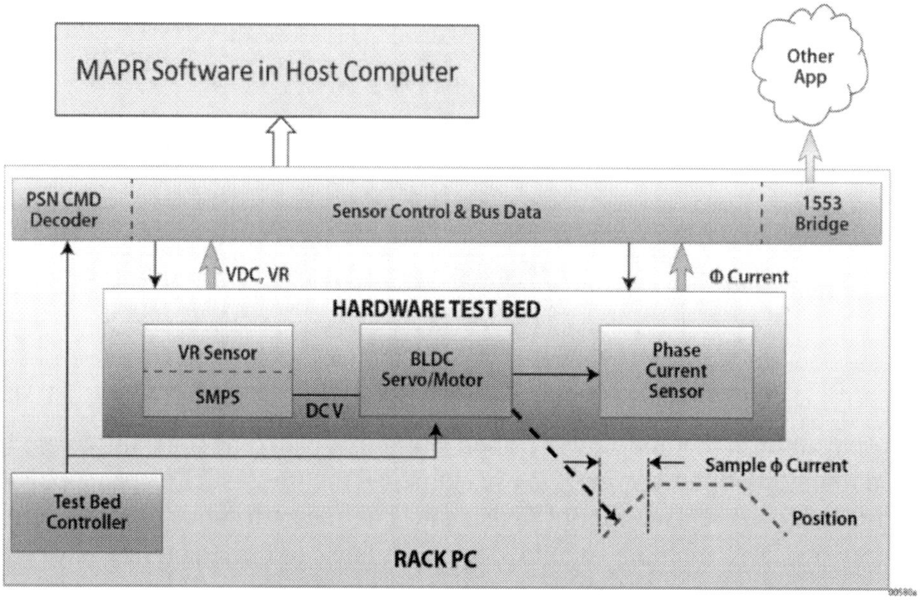

Figure 3: Test Bed Diagram.

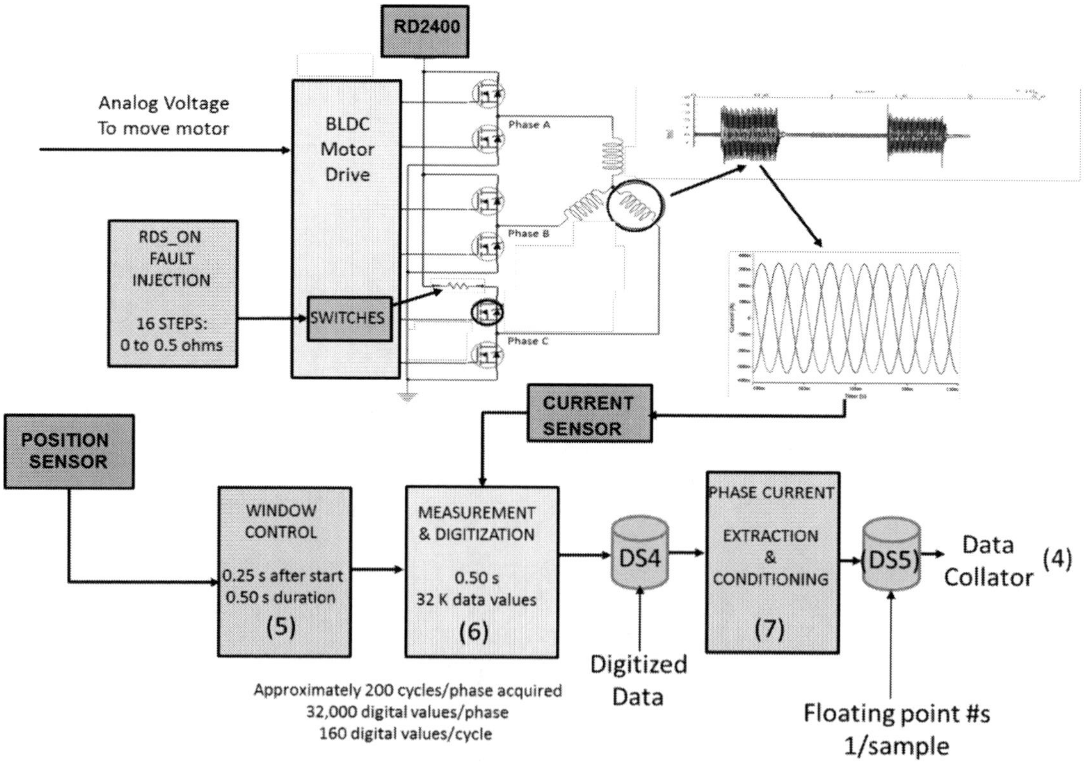

Figure 4: EMA Test Diagram.

In practice, the proposed initial methods were found not feasible because of, for example, the following: (1) analysis methods that produced information suitable for classification rather than diagnosis or prognosis; (2) noisy data (see Figure 5); (3) dependent, rather than independent, data; and (4) difficulty in processing test data to identify, extract, and use leading indicators of failure for prognostic purposes [5].

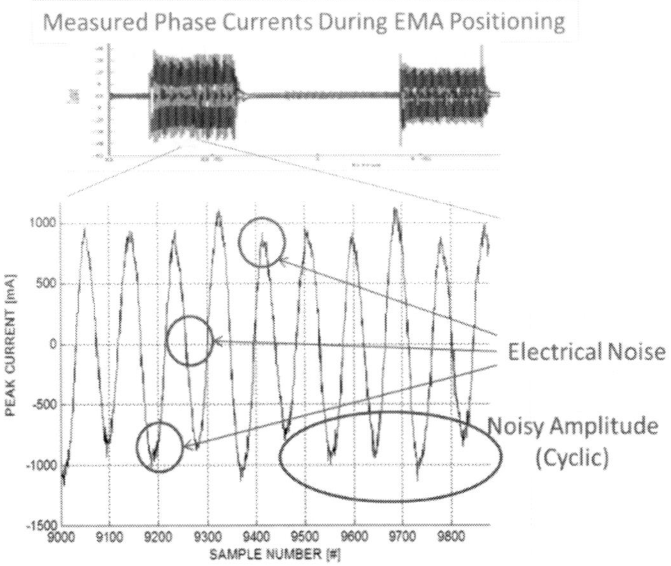

Figure 5: Example of Noisy Phase Current.

A unique root-mean-square (rms) method was created for quantifying measured values of phase currents. A methodology based on KNN and distance methods is used to process feature data extracted from condition-based data (CBD) to unequivocally identify and isolate various types of faults, and to prognose future times of functional failure [5], [6].

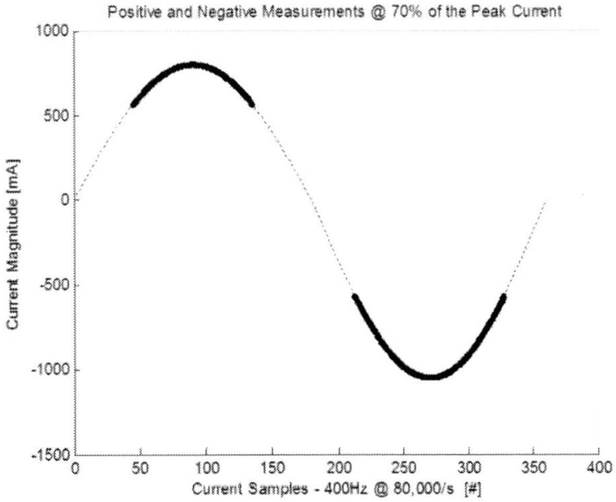

Figure 6: Peak Threshold Values for Calculating RMS.

2. MVA Solution: Approaches and Results

Fault-Classification Algorithm

PCA-based approach: An original objective of the program was to develop a fault-classification algorithm based on principal component analysis (PCA) that would model a process, program rules to produce a vector of residual errors (such as the following errors seen in Figure 7), and group those errors using machine-learning methods such as support vector machines (SVM) and a self-organizing map (SOM) to determine the best approach for fault-classification solutions in complex EMA systems [5].

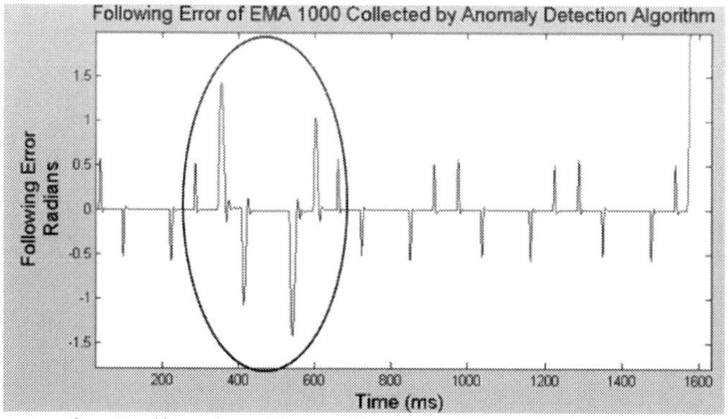

Figure 7: EMA Anomaly, Following Error Due to a Degraded Transistor in an H-bridge.

Clustering algorithms such as K-Means, Jarvis-Patrick, or Unsupervised k-Windows for fault classification were to be considered to categorize excursions. Algorithms of this type are used to group the incoming data into separate clusters based on their statistical behavior.

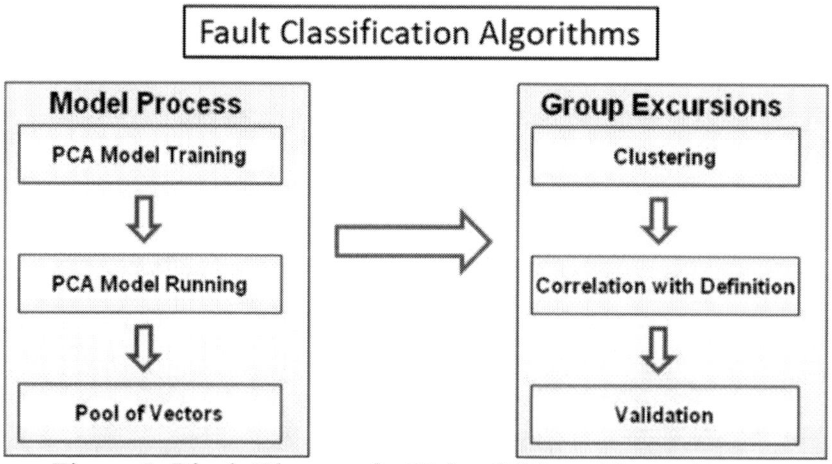

Figure 8: Block Diagram for Using PCA to Cluster Errors.

The main objective of clustering is to find similarities between events and then group similar data or events together to assist in understanding relationships or classifications that might exist among them. [7]-[9] Also, self-organizing maps (SOM), support vector machines (SVM), and Bayesian classification were researched to classify them as either feed analyzer or effluent analyzer faults, while PCA is used to model processes by providing information about the state of the process. Those methodologies will be categorized for different applications, and the strengths and weaknesses will be summarized and documented.

Result: We concluded that using a PCA-based approach to develop fault-classification algorithms for EMAs had unnecessary redundancy. The rationale for that conclusion included the following:

- PCA is a statistical procedure to convert a set of observations of possibly correlated variables into a set of uncorrelated variables (principal components). Given a EMA as a base platform, given an EMA test bed, given designed, known faults that are injected into the test bed,
- The set of principal components is known and is constrained by the specific EMA used in the test bed, the selection of the nodes at which faults are to be injected, the type of fault injected into the test bed, and the sensors (data collector) types and locations used in the test bed.

PCA and classification is useful when historical data exist and data relationships are uncertain, when it is unclear how many uncorrelated variables are in the data; and when it is not known how to group uncorrelated variables.

MVA Solutions: Modeling and Data Processing

A common MVA-solution approach for identifying and isolating a number of faults, N, is to set up and solve a set of N simultaneous equations. For example, let FD represent a feature data associated with three phase currents (A, B, & C) and the number of faults be limited to three, then basic MVA modeling yields

$$FD_A = FD_{A1} + FD_{A2} + FD_{A3} \quad (1)$$
$$FD_B = FD_{B1} + FD_{B2} + FD_{B3} \quad (2)$$
$$FD_C = FD_{C1} + FD_{C2} + FD_{C3} \quad (3)$$

Then let V represent the phase voltage, Z_W represent the impedance of the EMA-motor stator winding, Z_T represent the total impedance of the switching transistor and the connections to-from the EMA-motor stator winding, and $-Z_M$ represent the effective impedance of the EMA-motor load reflected back into the EMA-motor stator winding. EQ. (1) through (3) become

$$FD_A = V_A(1/(Z_{WA} + Z_{TA} - Z_{MA})) \quad (4)$$
$$FD_B = V_B(1/(Z_{WB} + Z_{TB} - Z_{MB})) \quad (5)$$
$$FD_C = V_C(1/(Z_{WC} + Z_{TC} - Z_{MC})) \quad (6)$$

where V_A, V_B, and V_C are of the form

$$V = V_{DC} + V_{AC}(\cos(\omega t + \theta) + V_{NOISE}) \quad (7)$$

With the restriction that the magnitude of $-Z_M$ be identical for all measured phase currents and V be identical for all measured phase currents, it was felt that using KNN and distance methods, any fault in any stator winding or switching-transistor circuit or excessive loading could be detected and isolated.

3. MVA AND NOISE

MVA Solutions: Noise is an Issue

Statistical-based MVA modeling is difficult to apply in practice, including test beds because of noise issues. CBD (sampled by sensors and processed by hardware, firmware, and software) contains both FD and noise. Here, noise is defined as anything not related to FD,

$$CBD = FD + Noise \qquad \text{model for noisy CBD} \tag{8}$$

Power-supply noise: The DC voltage from a power-supply input to an EMA is noisy: the most significant in amplitude being the damped-ringing responses due to abrupt changes in load (see Figure 9).

Actuator-movement noise: The magnitude of measured phase currents varies. One type of variation (noise) is associated with whether the actuator is lifting a load, such as wing surface, or lowering a load. Referring to Figure 8, the magnitude of the phase current is larger when a load is lifted compared to when a load is lowered.

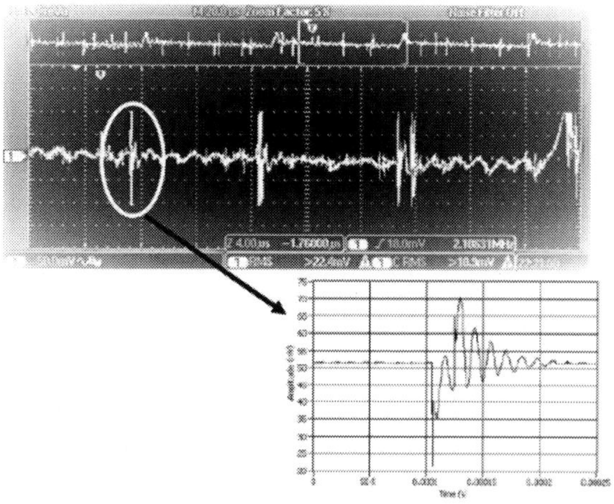

Figure 9: EMA Input Voltage – from the RD2400 (Green) Seen in Figure 4.

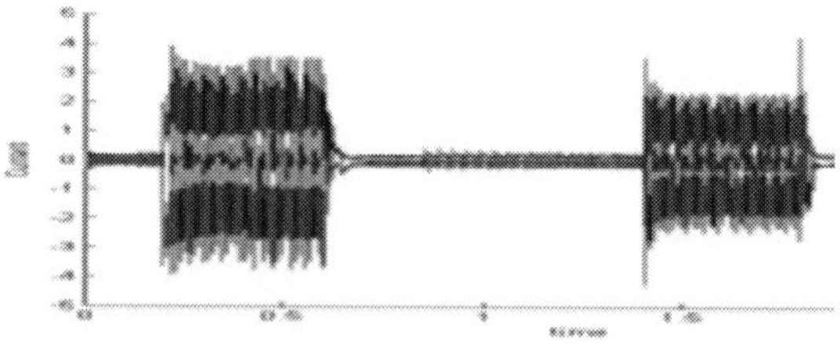

Figure 10: Phase Currents – Lifting (Left) and Lowering (Right) a Load.

Inertia and momentum noise: Move up (lift) or down (lower) commands cause an EMA to encounter load inertia and, similarly, when an actuator stops lifting or lowering a load, it encounters load momentum, which results in another type noise (Figure 11).

Normal-load and excessive-load noise: There is noise associated with the weight of a load - a normal load or an excessive load. For example, the right-hand plot in Figure 12 exhibits more harmonic distortion compared to the left-hand plot.

Electrical and cyclic-amplitude noise: Referring back to the bottom plot in Figure 5, phase current data exhibits noise marked 'Electrical Noise' and 'Noisy Amplitude' - electrical noise is primarily due to switching noise in the input DC voltage and noisy amplitude is due to level shifting because the three stator windings have a floating, common reference (refer back to Figure 4).

Noise Filtering and Mitigation Methods

Sampling and accuracy: A common method for filtering out noise is use data sampling as a low-pass filtering method: but at what sampling frequency? Since the frequency of the phase current is about 400 Hz, it might seem logical to sample at 800 Hz – to satisfy the Nyquist criteria: but that turns out to be too low of a sampling frequency.

For example, suppose you want to measure the AC component of FD with an accuracy of at least 90%: then from EQ. (8) and letting *S* represent sampling frequency, *N* represent noise, and α represent accuracy (as a ratio),

$$S \geq (2/a)N_{FREQUENCY} \tag{9}$$

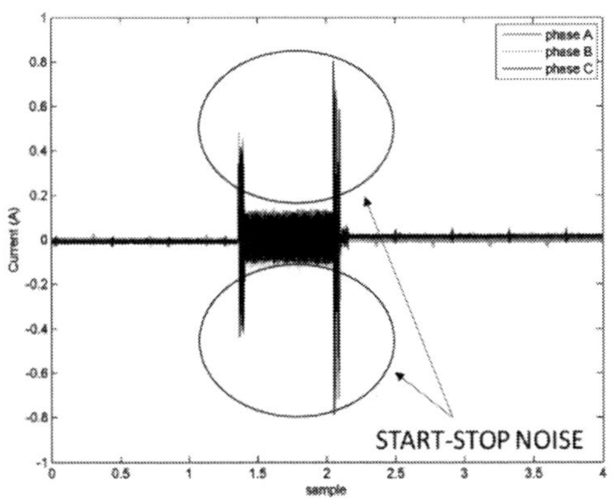

Figure 11: Noise Associated with the Start-Stop Lifting of a Load.

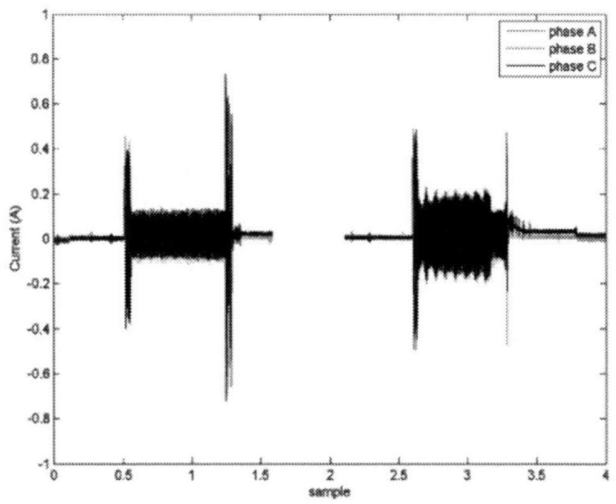

Figure 12: Normal Load (Left), Excessive Load (Right).

For a frequency of 400 Hz, EQ. (9) results in,

$$S \geq (2/0.10)400 = 8 \text{ kHz}$$

It turns out that 8 kHz is also too low because the frequency of the switching noise seen in Figure 9 is a little over 2.5 kHz, and from EQ. (9), you need a sampling frequency of more than 50 kHz to satisfy a 10% accuracy requirement.

Sampled data and loading: Since an objective of SBIR program is to detect excessive loading, such as too much weight and/or excessive friction, you need to sample data when the EMA is lifting a load rather than lowering a load. This means you need to provide a method for detecting the start of a positioning command and you need to determine whether the positioning direction is up or down. The test bed needs a positioning sensor, as shown by a green-colored block in Figure 4, and it needs firmware (or software) programming logic to setup a sampling window

Sampling Window: Evaluation of the right-side plot in Figure 12 and experimenting with EMA positioning commands leads the following window specifications (block 5 in Figure 4): (1) start sampling 0.25 s after a move-up command, (2) sample at 64 kHz for 0.50 s (32,000 data samples).

MVA Solutions: Fault Testing

Winding and Load Faults: The execution of a design of experiment when a winding type of fault (a reduced impedance in a stator winding) or a loading type of fault (increased lifting weight) is injected into the test bed yields results as predicted by the models of EQ. (4) – EQ. (6): referring to Figure 13, when a winding fault is injected into the test bed, only one of the three phase currents increase in amplitude, and when a load/friction fault is injected into the test bed, all three phase currents increase in amplitude.

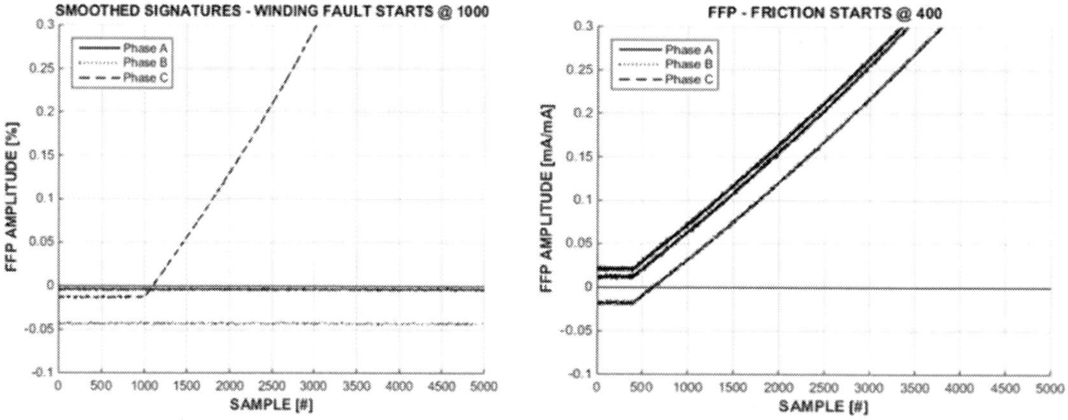

Figure 13: Phase Currents for a Winding Fault (Left) and Load (Friction) Fault (Right).

Switching Transistor Fault: A switching transistor fault is injected into the test bed by inserting a low-value resistance in series with one of the six transistors in the h-bridge. The experiment is deemed a failure because there is no significant in measured amplitude of the phase currents. This is because the on-resistance of a switching transistor, compared to the impedance of a stator winding, is too low: this is verified by experimentation and simulation. A degraded transistor in one of the six current branches in the h-bridge (Figure 4) causes a change in the current amplitude of either the positive or the negative half of one of the three phase currents. Because of a floating reference topology in the motor of the EMA (right-hand diagram in Figure 14), the reference levels of the three phase currents shift with respect to each other (left-hand plot in Figure 14).

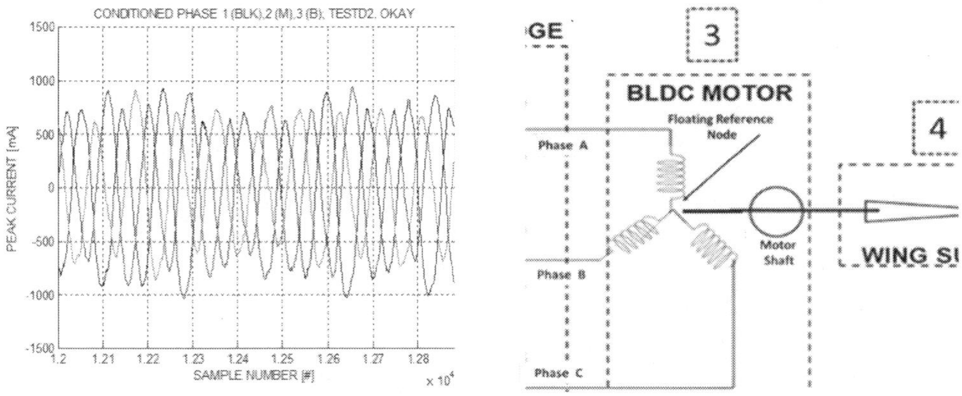

Figure 14: Data Caused by a Degraded Transistor (left) and BLDC Motor (Right).

More importantly, the impedance presented by those switching transistors is significantly smaller in comparison to the impedance of the stator windings: changes in the amplitude caused by transistor degradation are lost in the noise of all other variations in current amplitude. Even when the series impedance of a rectifying branch in the h-bridge is increased from 0.01 Ω to 1.0 Ω (a functionally-failed value), the maximum change in peak current was about 25 mA – less than 3.0% (see Figure 15).

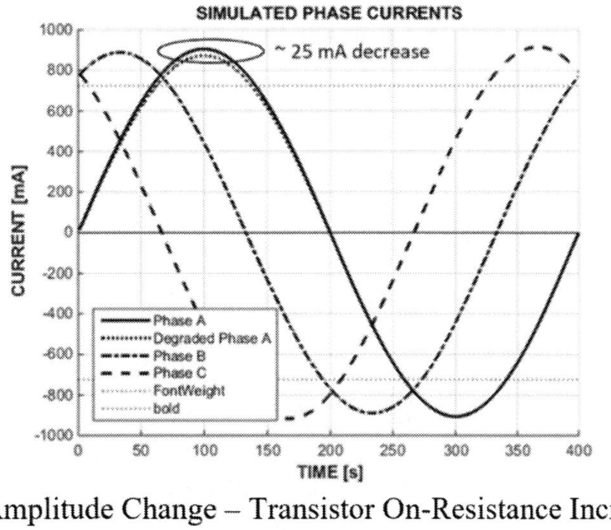

Figure 15: Amplitude Change – Transistor On-Resistance Increased by 10x.

4. MVA Solution: Special RMS

We developed a unique, innovative solution: (1) define peak-threshold levels (T_{PEAK}); (2) truncate all values of current below the positive threshold and above the negative threshold; (3) calculate the magnitudes of the positive phases (P_{RMS}) and negative phases (N_RMS) and (4) sum the magnitudes. This method emphasizes any differences between peak current values and the defined thresholds.

Algorithm: Special RMS,

$$T_{PEAK} = 0.70 \quad \text{set threshold level} \tag{10}$$

$$I_{PEAK} = 900 \quad \text{set nominal value of peak current (mA)} \tag{11}$$

$$I_{TRUNC} = T_{PEAK} * I_{PEAK} \tag{12}$$

When (□(□) > 0) && (□(□) > □□□□□□)   then letting P = count of true

$$P_{RMS} = [(1/P)\sum_1^n(I(n) - I_{TRUNC})] \quad \text{do not square} \tag{13}$$

When (□(□) < 0 && □(□) < −□□□□□□   then letting N = count of true

$$N_{RMS} = [(1/N)\sum_1^n(I_{TRUNC} - I(n))\,] \quad \text{do not square} \tag{14}$$

Take a difference summation

$$I_{RMS} = P_{RMS} + N_{RMS} \quad \text{Sum the differences} \tag{15}$$

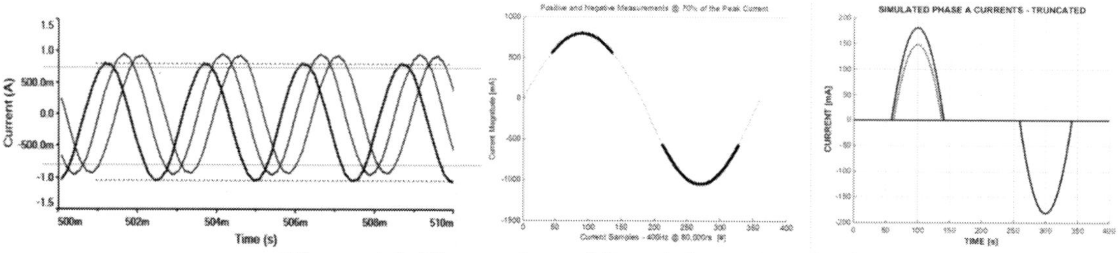

Figure 16. Illustration of Special-rms Method

Algorithm Result: Examples

The left-hand plot in Figure 17 illustrates the result of applying the special-rms algorithm to one sinusoid of a phase current in which the switching transistor in the positive branch is degraded. The algorithm was applied to all three sets of measured current data and plotted as shown in Figure 17.

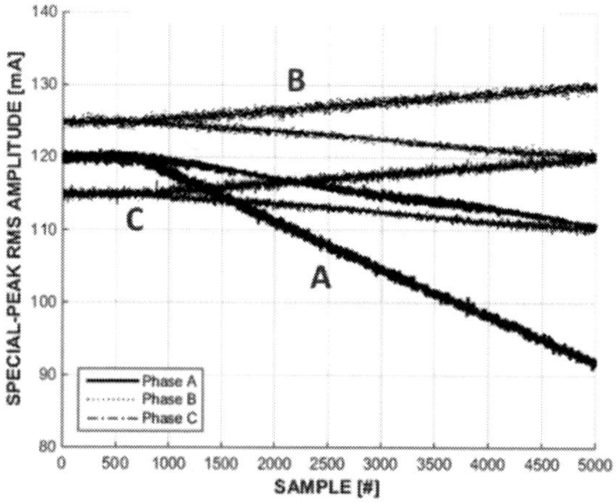

Figure 17: Plots of Test Data (3 Phases) After Special-rms Method.

The data is collected and conditioned, transformed to functional-failure signature (FFS) data, further conditioned, and input to a prediction algorithm: one data point at time, to produce (1) the conditioned data shown in Figure 18 (left) ; (2) the conditioned FFS data shown in Figure 18 (right); and prognostic information from the prediction algorithm is plotted in Figure 19.

Prognostic Information

The prediction algorithm accepts FFS input data, processes the input, and produces prognostic information – one data point at a time. The prognostic information comprises Remaining Useful Life (RUL), prognostic horizon (PH), and State of Health (SoH). The FFS data plotted in Figure 18 produced the prognostic information plotted in Figure 19.

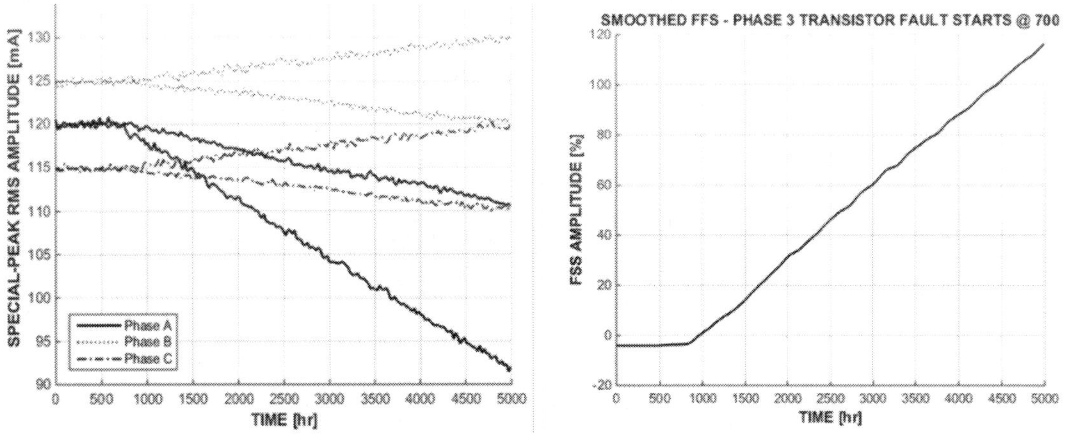

Figure 18: Conditioned Data (Left) – Transformed to Smoothed FFS (Right).

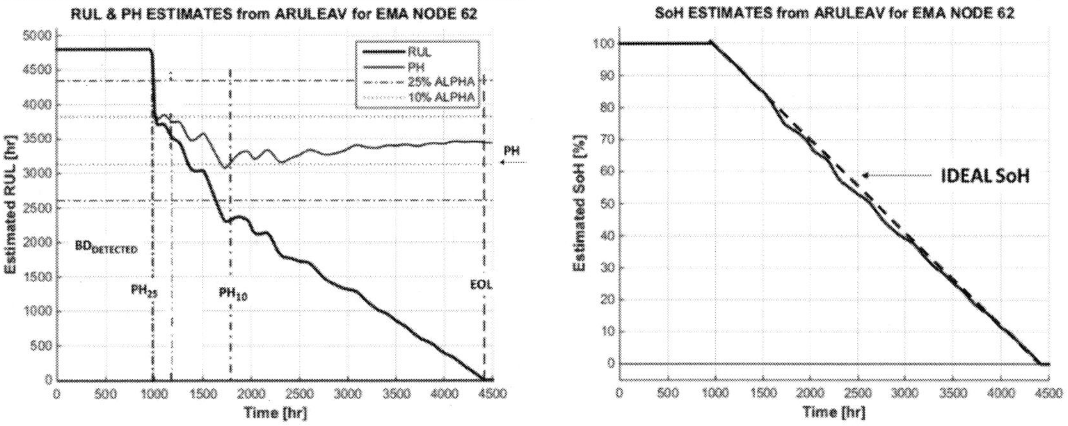

Figure 19: Prognostic Information – RUL & PH (Left) and SoH (Right).

Prognostic Accuracy

The prediction algorithms in a program called ARULEAV (Adaptive RUL Estimation – Advanced Version) employs a number of algorithms to produce very accurate prognostic estimates (refer to Figure 20):

Treat data point as a particle having momentum and inertia that tend to maintain direction and speed – use weighting and coefficients.

Data points tend to move in a data space from lower-left corner to upper-right corner – a random walk.

Data points form a characteristic curve that is represented by three data spaces concatenated together from bottom to top, left to right, at the corners – employ piece-wise solution methods.

Employ Kalman-like filtering

- Remember the time when degradation is detected (BD)
- Remember the last data point
- Use the data space models to predict where the next data point should be
- Compare predicted point to sampled next data point – the sample time (ST)
  - Compute a compromise between predicted and actual data point
  - Adjust the model lengths to correspond to the computed point
  - Use the adjusted model to estimate when functional failure (EOL) occurs
  - Compute RUL = EOL – ST
  - Compute SoH = 100% * (EOL-RUL)/EOL
  - Compute PH = ST + RUL = estimated EOL (relative to BD)

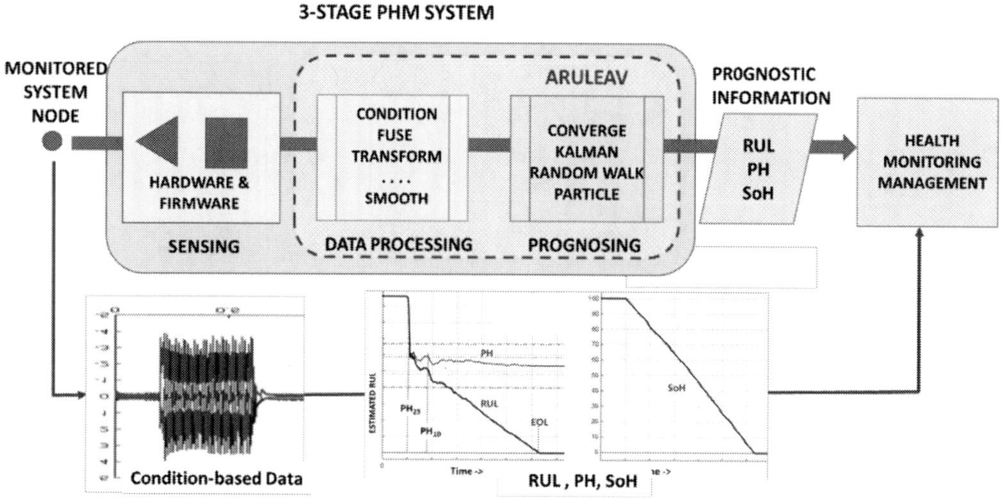

Figure 20. Diagram of a 3-stage PHM System.

ARULEAV also incorporates fast-convergence algorithms that generally produces PH estimates within an alpha (α) accuracy of 25% within 10 data points with a standard deviation of 3 points.

5. Conclusion

Statistical modeling and methods such as PCA and machine learning methods such as SVM and Euclidean distance proved not be helpful in prognostic enabling an EMA: primarily because we designed and built an EMA test bed with predetermined fault injection locations and means, and data was collected using known sensors located at specific nodes in the test bed. Therefore, the existence of or lack of correlation between the variables of the collected data was known, and the principal components of the data (voltage, ripple, phase current, and so on) was known or readily identified by examination of the data.

We did use the concepts of KNN and distance methods to identify and located faults: winding versus load/friction versus switching transistor.

The filtering and mitigation of noise was the most significant issue that had to be addressed to provide a prognostic solution for winding and load faults in an EMA. Because of the large difference in the values of the impedance of the EMA stator windings compared to the on resistance of switching transistors, a special-rms method had to be developed to detect and process phase currents.

In the end, we developed an effective solution for EMA monitoring. The prognostic information is produced by conditioning and transforming phase current data into signature data as input to prediction algorithms resulted in fast and accurate prognostic information.

References

[1] Kunst, et al., (2008) "Electronic Prognostics Solutions for Electromechanical Actuator Systems", IEEE PHM Conference, Denver, CO, October 2008

[2] Kunst, et al., (2009) "Damage Propagation Analysis Methodology for Electromechanical Actuator Prognostics", IEEE Aerospace Conference, Big Sky, MT, March 2009

[3] M. Walker, R. Kapadia, B. Sammuli, and M. Venkatesh. (2007) "A Model-based Reasoning Framework for Condition-based Maintenance and Distance Support". Automation and Controls Symposium, ASNE, Biloxi, MS. December 2007.

[4] R. Kapadia, G. Stanley, and M. Walker. (2007) "Real World Model-based Fault Management". Proc. of the 18th International Workshop on Principles of Diagnosis, Nashville, TN. May 2007.

[5] NASA Phase 2 Proposal A1 12-9941, (2012) "Physical Modeling for Anomaly Diagnostics and Prognostics," Ridgetop Group, Inc., 24 November 2012.

[6] Goodman, Douglas L.; Hofmeister, James P.; and Szidarovszky, Ferenc; 2019 (Pending), *Prognostics and Health Management: A Practical Approach Using Conditioned-Based Data,* 1st Edition, John Wiley & Sons, Ltd, The Atrium, Southern Gate, Chichester, West Sussex, PO19 8SQ, United Kingdom, *ISBN: 9781119356653, 2019*, pp. 255-274, 300-311.

[7] Ding, C., & He, X., (2004) "K-means clustering via principal component analysis," Proceedings of the 21st International Conference on Machine Learning, 2004

[8] A. Vathy-Fogarassy, A. Kiss, J. Abonyi, (2007) "Improvement of Jarvis-Patrick. Clustering Based on Fuzzy Similarity," Lecture Notes in Computer Science, 2007

[9] M. N. Vrahatis, B. Boutsinas, P. Alevizos, and G. Pavlides (2002 "The new k-windows algorithm for improving the k-means clustering algorithm," Journal of Complexity, 18:375–391, 2002

Bibliography

[1] Hofmeister, J., Wagoner, R. and Goodman, D. (2013) "Prognostic Health Management (PHM) of Electrical Systems using Conditioned-based Data for Anomaly and Prognostic Reasoning," Chemical Engineering Transactions, Vol. 33, pp. 992-996.

[2] Kumar, S. and Pecht, M. (2010) "Modeling Approaches for Prognostics and Health Management of Electronics," *International Journal of Performability Engineering*, Vol. 6, No. 5, pp. 467-476.

[3] Pecht, M. (2008) *Prognostics and Health Management of Electronics*. Hoboken, NJ: John Wiley & Sons.

[4] O'Connor, P. and Kleyner, A. (2012) *Practical Reliability Engineering*. Chichester, UK: John Wiley & Sons.

[5] Hofmeister, James P.; Goodman, Douglas, L.; and Szidarovszky, Ferenc (2018); "Transforming Condition-based Data Signatures into Functional Failure Signatures," IEEE 2018 Aerospace Conference, Big Sky, MT, 3-9 March, 2018.

[6] Saxena, A., Celaya, J., Saha, B., Saha, S. and Goebel, K. (2009) "On Applying the Prognostic Performance Metrics," Annual Conference of the Prognostics and Health Management Society (PHM09), San Diego, CA, 27 Sep.-1 Oct. 2009.

Design Approach for a PHM System: Dual Set of Electromechanical Actuator Subsystems

James P. Hofmeister, Erik Sandberg, Wyatt Pena, and Robert S. Wagoner
Ridgetop Group, Inc.
3580 West Ina Road
Tucson, AZ 86741
520-742-3300
jhofmeister@ridgetopgroup.com

Abstract: This paper describes a design for an example of a Prognostic Health Monitoring (PHM) system that is able to detect a state of degraded health and make an accurate prediction of when a resulting future failure in the system is likely to occur in a dual set of electromechanical actuators (EMA) subsystems, each comprising a switch-mode power supply (SMPS) and two identical EMAs. Examples of considerations in a design of a PHM system include the following: (1) What is the framework for the PHM system? (2) What units (targets) are to be prognostic enabled? (3) What failure modes are to be monitored? (4) What kind of data conditioning is necessary to isolate and extract condition indicators and/or leading indicators of failure from noisy condition-based data (CBD)? (5) Do special methods of data processing need to be developed and, if so, how? (6) What are the prognostic accuracy requirements? (7) What alerts and alert levels are required to support prognoses? (8) What are the requirements related to starting, stopping, resuming, and recovery of the system? (9) What architecture approach is going to be used to define the target system to the PHM system?
*RotoSense and ARULEAV are trademarks of Ridgetop Group, Inc

Keywords: condition-based data; CBD; electromechanical actuator; EMA; prognostic health monitoring; PHM; Accelerometer; gear; MEMS; rolling stock; RotoSense; signal quality; train; wheel hub

1. Introduction

An important purpose of a Prognostic Health Monitoring/Management (PHM) system is to detect a state of degraded health and accurately predict when functional failure is likely to occur [1]-[2]. This paper presents a design of a PHM system for a dual set of electromechanical actuators (EMA) subsystems, each comprising a switch-mode power supply (SMPS) and two identical EMAs. The design and development of a system to support PHM is complex and there are many approaches to do so. Even though each such system (with the possible exception of demonstrations, test beds, and experiments) is unique, there are important design considerations. Examples of considerations in a design of a PHM system include the following: (1) What is the framework for the PHM system? (2) What units (targets) are to be prognostic enabled? (3) What failure modes are to be monitored? (4) What kind of data conditioning is necessary to isolate and extract condition indicators and/or leading indicators of failure from noisy condition-based data (CBD)? (5) Do special methods of data processing need to be developed and, if so, how? (6) What are the prognostic accuracy requirements? (7) What alerts and alert levels are required to support prognoses? (8) What are the requirements related to starting, stopping, resuming, and recovery of the system? (9) What architecture approach is going to be used to define the target system to the PHM system?

There are other considerations that are not presented, including but not limited to the following: comparison and selection of sensors; the cost to design, develop, test, verify, field, and maintain a PHM solution for each prognostic target; evaluation of the cost of failure versus the cost of failure prevention; repair versus replacement cost; long-term costs associated with installing, operating, and maintaining a PHM system.

2. Design Considerations

Framework for the PHM System

The framework shown in Figure 1 is selected as the base framework.

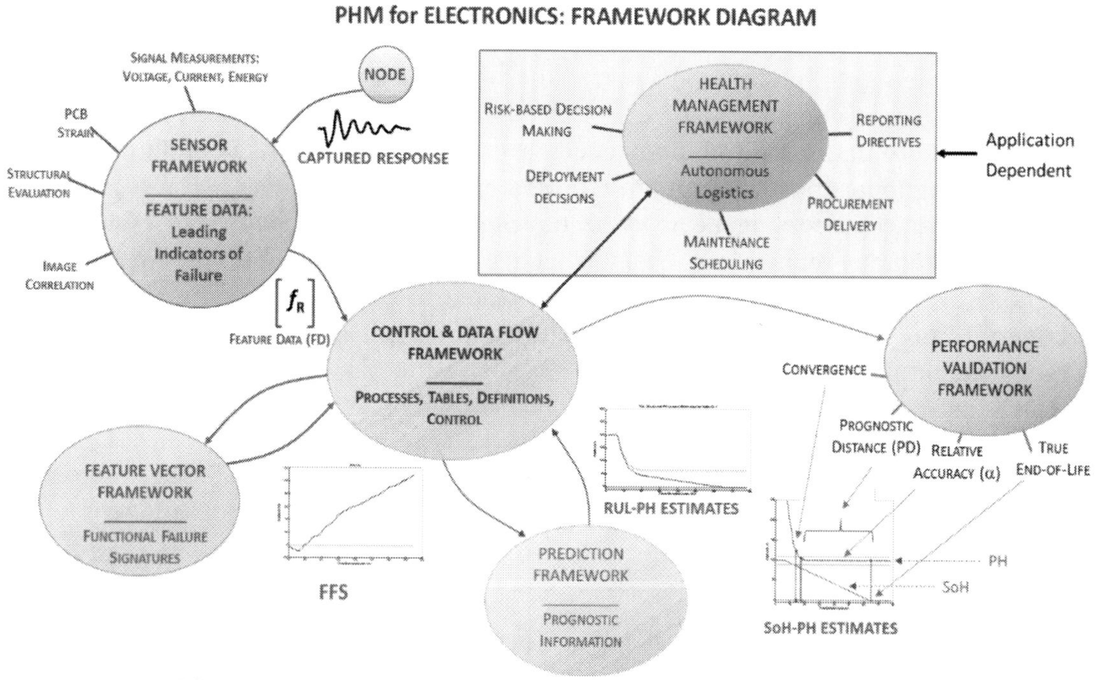

Figure 1: Framework for a PHM System (based on [3]-[5]).

Health Management Framework. Health management is very complex and includes decisions that consider risk, deployment, maintenance, reporting and directions (imperatives), procurement and delivery, load shedding, soft shutdown, and much more.

Performance Validation Framework. Because predictions are estimates, there is always a question as to how good (accurate) those estimates are. A PHM system could be designed and developed to provide means and methods to evaluate and validate the performance of the prediction algorithms. This paper does not address the design of a Performance Validation Framework.

Prognostic Targets

The selected prognostic targets are two EMA subsystems, each comprising a power supply and two EMAs as shown in Figure 2.

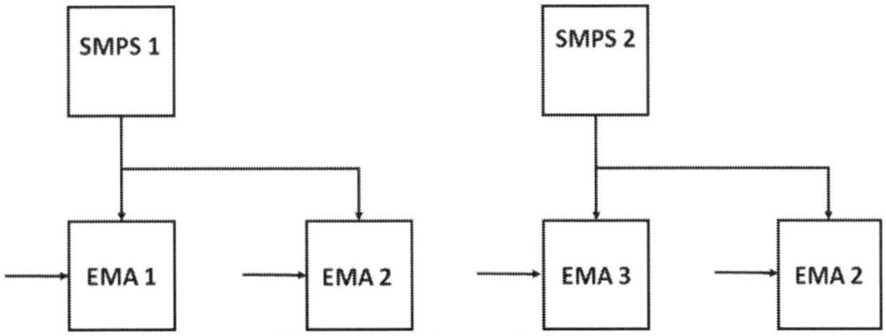

Figure 2: Diagram of Two EMA Subsystems.

Power Supply. The power supplies are switched-mode power supplies (SMPS) and the identified fault is loss of capacitance in the output filter (block 1 in Figure 3 and Figure 4). That fault is known to result in excessive noise and loss of ability to deliver power, which results in irreversible damage to both a power supply (damaged switching transistors) and a downstream EMA (block 2 in Figure 3) connected to it.

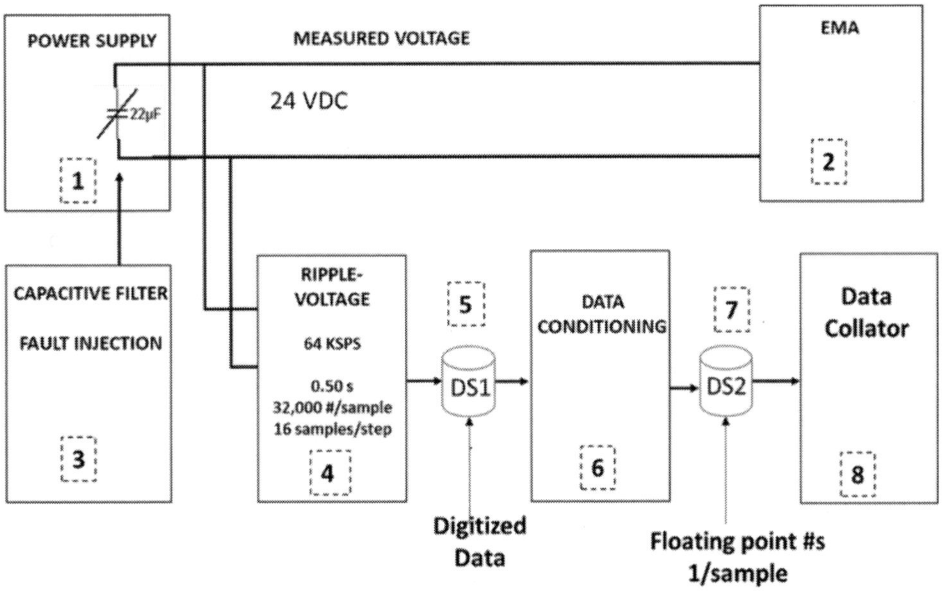

Figure 3: Block Diagram of a Prognostic-Enabled Power Supply in a Test Bed.

EMA. Each EMA (block 2 in Figure 3 and Figure 4) comprises a controller (block 11) for a brushless DC (BLDC) motor; the motor has three stator windings (block 13) and a shaft (block 14) to position a load by moving up or down; AC power is delivered by six power-switching transistors (block 12) configured as an H-bridge type of commutation.

The EMA test bed can inject a fault (increase in on-resistance) into any one of the six power-switching branches in the H-bridge. Although not shown are two other fault injection methods: one to simulate a damaged stator winding and another to increase the load on the motor shaft. Three current sensors (block 10) are used to provide CBD for detecting and prognosing faults.

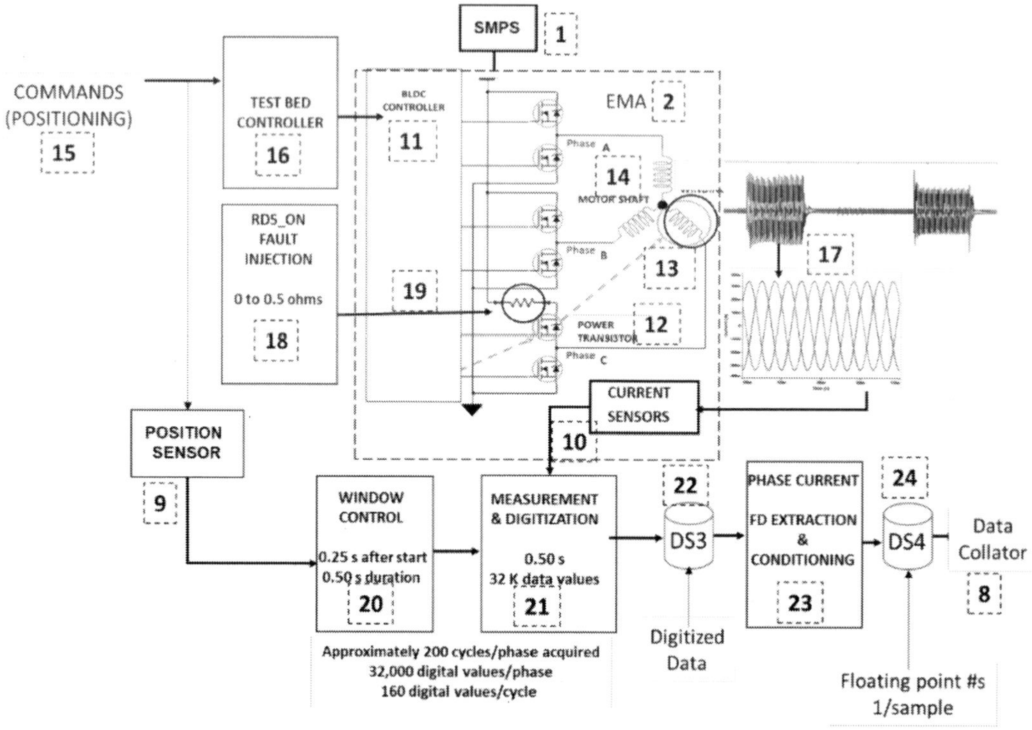

Figure 4: Block Diagram of a Prognostic-Enabled EMA in a Test Bed.

Failure Modes

The PHM system design needs to detect and prognose the failure modes (faults): (i) loss of filtering capacitance in either of the two EMA power supplies; (ii) excessive load/friction on the motor shaft (block 14 in Figure 4); (iii) damage to any of the three stator windings (block 13 in Figure 4); and (iv) damage to any of the six power-switching transistors in the H-bridge connected to the stator windings.

Data Conditioning

You need to collect and analyze sensor data, historical and test, to determine what kind of data conditioning you need – in practical applications, data conditioning, which includes data sampling, is always required.

Power supply output. The output of the power supply () comprises ripple voltage, DC voltage, and noise. Noise is anything other than a feature data of interest and includes thermal noise, switching (spikes and glitches) noise, and other features such as droop, harmonic distortion, and damped-ringing responses.

Design is to configure your ripple-voltage sensor to employ bandpass filtering when sampling: frequencies between 200 and 250 kHz.

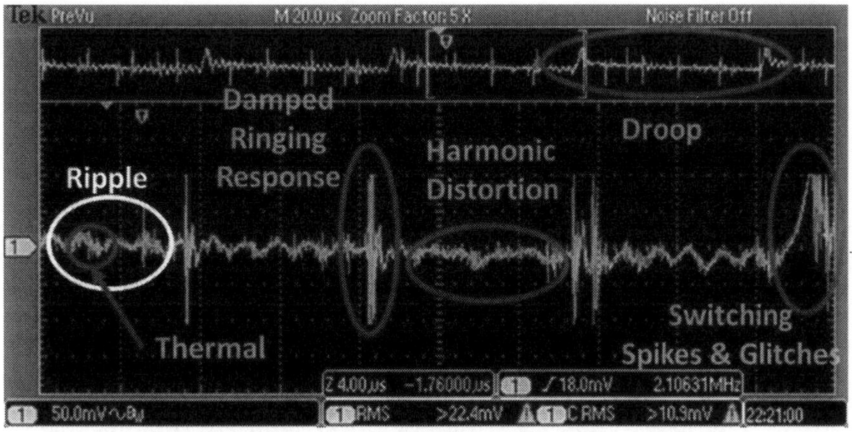

Figure 5: Noisy Voltage from the Power Supply.

Phase Currents. The phase currents of an EMA are also noisy (Figure 6), and they vary depending on positioning direction and load. There is also amplitude variation resulting from the floating reference topology of the EMA motor.

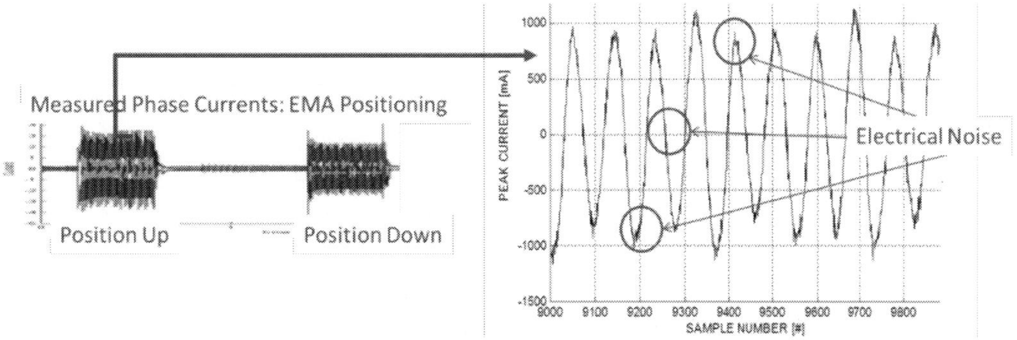

Figure 6: Phase Currents and Position Direction (Left) – Electrical Noise (Right).

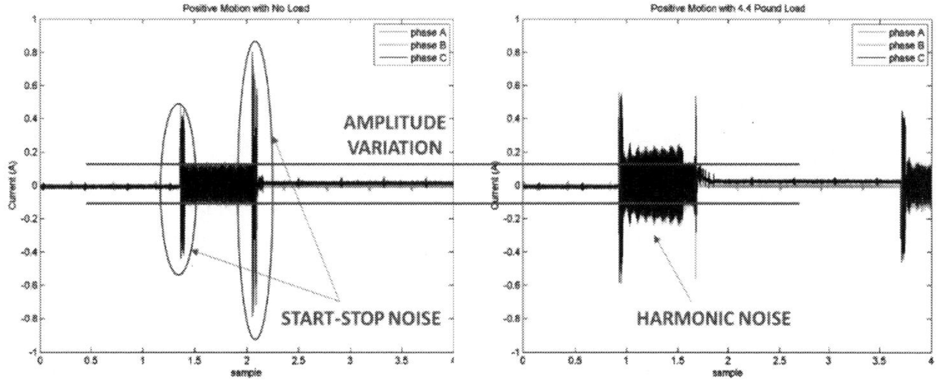

Figure 7: Noise & Amplitude – Normal Load (Left) & Heavy Load (Right).

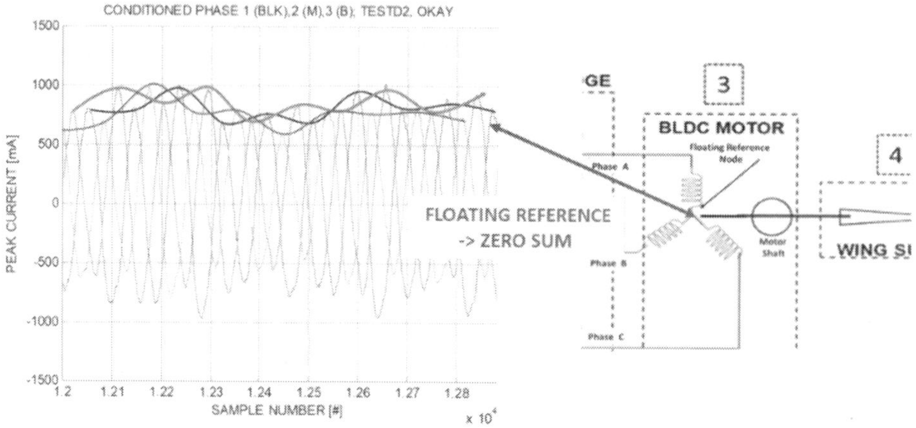

Figure 8: Floating Reference Results in Zero-Sum-Amplitude Variation.

Referring back to Figure 4, a design point is to employ windowing (Figure 9) with a positioning sensor (block 9) and a Window Control (block 20) to sample and measure currents (block 21). The window result is shown in Figure 10.

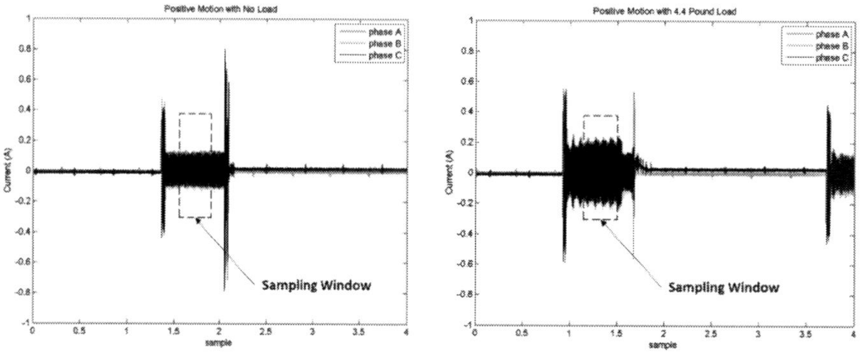

Figure 9: Sampling Window, Up Positioning for Normal (Left) and Heavy (Right) Loads.

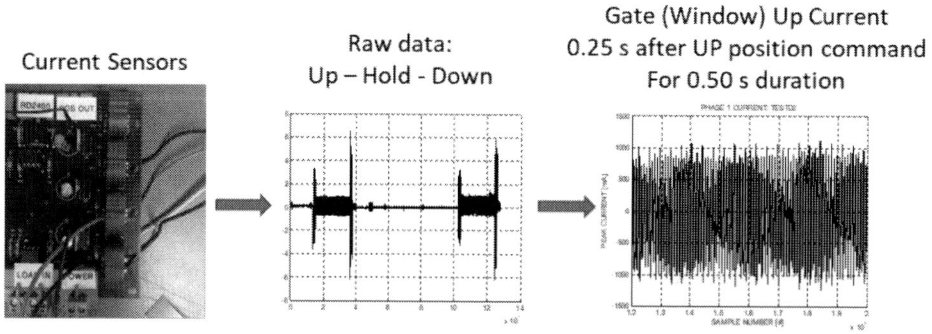

Figure 10: Window Design Employed on the EMA Test Bed.

Special Methods

Because there are 10 EMA-fault modes to be detected and isolated (six transistors, three motor windings, and a motor shaft) with only three current sensors, special methods of processing phase-current data are required to satisfy the 10 degrees of freedom.

Positive and negative halves of phase currents. One design approach is to distinguish changes in the positive halves versus changes in the negative halves of the phase currents (Figure 11, left).

For example, normal values of on-resistance of power-switching transistors range from 0.01 to 0.20 Ω, so that a value of 2.0 Ω would be defined as functional failure. Experimentation with the EMA test bed reveals that level of degradation results in decrease of only 25-mA in either the positive or the negative half of the phase current (Figure 11, left): only a 1.4% change in total current – the feature (current change) is lost in the noise (Figure 11, right). Differentiating between peak changes in the current halves improves resolution: but only from 1.4% to 2.8%.

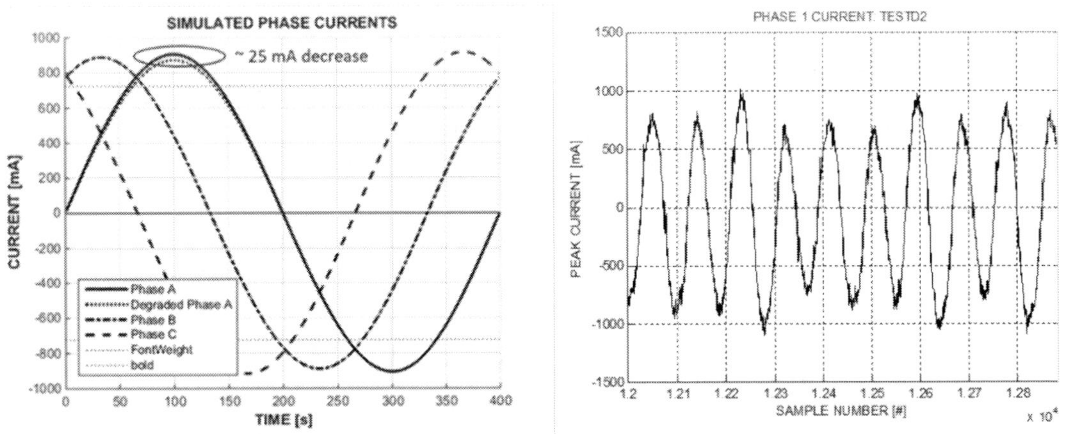

Figure 11: Positive-Negative Difference (Left) & Noisy Data (Right).

Special-rms calculations. A design solution does the following: define threshold levels, truncate all current values between the threshold levels, use the remaining values to calculate positive and negative magnitudes, and then sum the magnitudes. This method, called special-rms, emphasizes any differences between peak current values and the defined thresholds: the method overcomes the low-amplitude changes in phase currents.

$$T_{PEAK} = 0.70 \qquad \text{set threshold level} \tag{1}$$

$$I_{PEAK} = 900 \qquad \text{set nominal value of peak current (mA)} \tag{2}$$

$$I_{TRUNC} = T_{PEAK} * I_{PEAK} \tag{3}$$

When $(I(n) > 0)\ \&\&\ (I(n) > I_{TRUNC})$ then letting P = count of true,

$$P_{RMS} = [(1/P)\textstyle\sum_1^n (I(n) - I_{TRUNC})] \qquad \text{do not square} \tag{4}$$

When $(I(n) < 0\ \&\&\ I(n) < -I_{TRUNC}$ then letting N = count of true,

$$N_{RMS} = [(1/N)\textstyle\sum_1^n (I_{TRUNC} - I(n))\ ] \qquad \text{do not square} \tag{5}$$

$$I_{RMS\_DIFF} = P_{RMS} + N_{RMS} \qquad \text{Sum the magnitudes} \tag{6}$$

The special-rms method significantly improves resolution compared to the normal rms method: about a 5:1 increase for a 0.70 threshold.

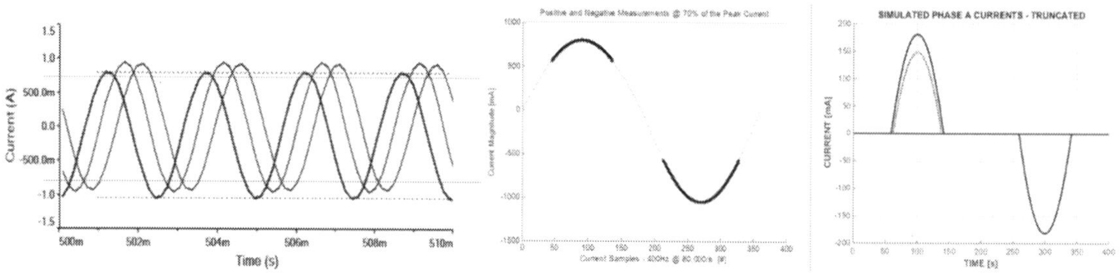

Figure 12. Graphical Illustration of Special-rms Method

Figure 13 illustrates the effectiveness of using the special-rms method for both the positive and negative halves of the three phase currents: (1) in the presence of degradation in an H-bridge transistor, there is a change in measured amplitudes of all six halves; (2) there is a significant change in the sum of the current halves for a degraded transistor, while the sums of the other two generally remains unchanged except for noise.

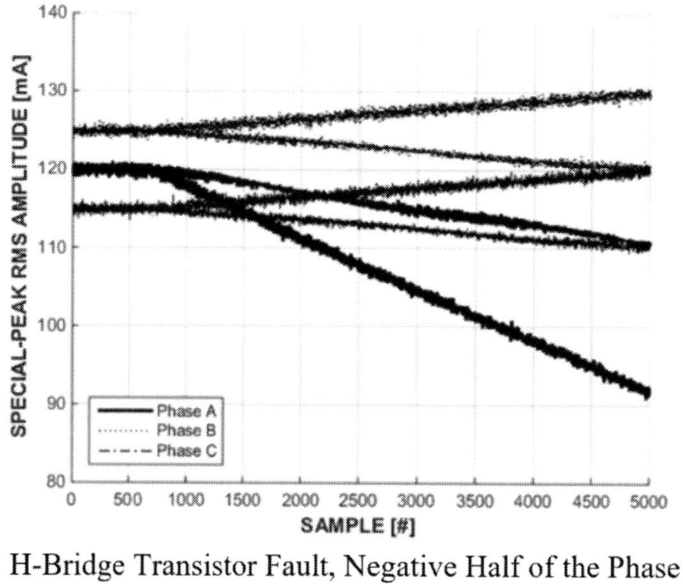

Figure 13: H-Bridge Transistor Fault, Negative Half of the Phase A Current.

Current offset. Close examination of the phase currents for a zero-degradation, no-load test (Figure 14) indicates the three phase currents are not centered about zero: this shift in the reference points of phase currents is confirmed by evaluation of plots in Figure 13. Differences in the reference levels is another form of noise of significance when comparing two sets of data, such as the magnitudes of current.

A method to mitigate current-offset noise (a source of error) is the following: (1) Sample and measure the phase currents during a calibration test (position up, no load, no degradation); (2) calculate and save the value of the mean amplitudes; (3) normalize subsequent current measurements.

Let $x = 1{:}3$ represent a phase current (A, B, C)
Let $z = 1{:}n$ represent a calibration sample to a design-defined limit

When $(I_x(n) > 0)$ then letting P = count of true,

$$I_{0ADJ\_P} = [(1/P)\textstyle\sum_1^x \sum_1^n (I(n) - I_{TRUNC})] \quad \text{do not square} \tag{7}$$

When $(I_x(n) < 0)$ then letting N = count of true,

$$I_{0ADJ\_N} = [(1/N)\textstyle\sum_1^x \sum_1^n (I_{TRUNC} - I(n))] \quad \text{do not square} \tag{8}$$

Then,

$$I_{OFF} = (P_{RMS} + N_{RMS})/2 \qquad \text{Offset adjustment values} \tag{9}$$

Subsequently,

$$I_{ADJ} = I(n) - I_{OFF} \qquad \text{Perform offset adjustment} \tag{10}$$

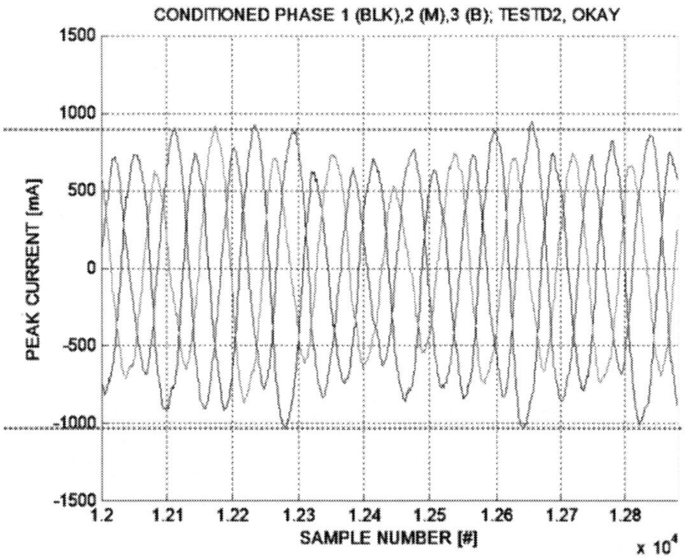

Figure 14: Phase Currents for Zero Degradation, Normal Load, Position Up.

Data Smoothing. The sensor data, pre-conditioned or otherwise, is noisy and is likely to result in noisy feature data (Figure 13). Subsequent experiments confirm that the noisy feature data ultimately results in prognostic information that does not meet accuracy requirements.

The design solution is to smooth the feature data, the transformed fault-to-failure progression (FFP) signature data, and the functional-failure signature (FFS) data that is input to a set of prediction algorithms (Figure 15). Subsequent experimentation confirms that after data smoothing, prognostic accuracy requirements are met.

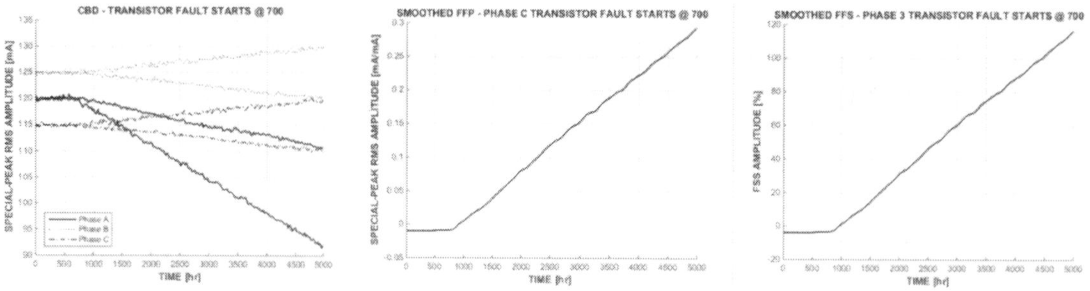

Figure 15: Smoothed Feature Data (Left), FFP Data (Middle) and FSS Data (Right).

Prognostic Accuracy

Convergence specifications, initial error. The specifications call for prognostic information to converge to within 25% when the remaining life is ≥ 50% (upper blue dots in Figure 16) and to within 10% when the remaining life is ≥ 30% (lower blue dots in Figure 16). The specifications are for cases where the initial estimated life when degradation begins is within 50% of actual.

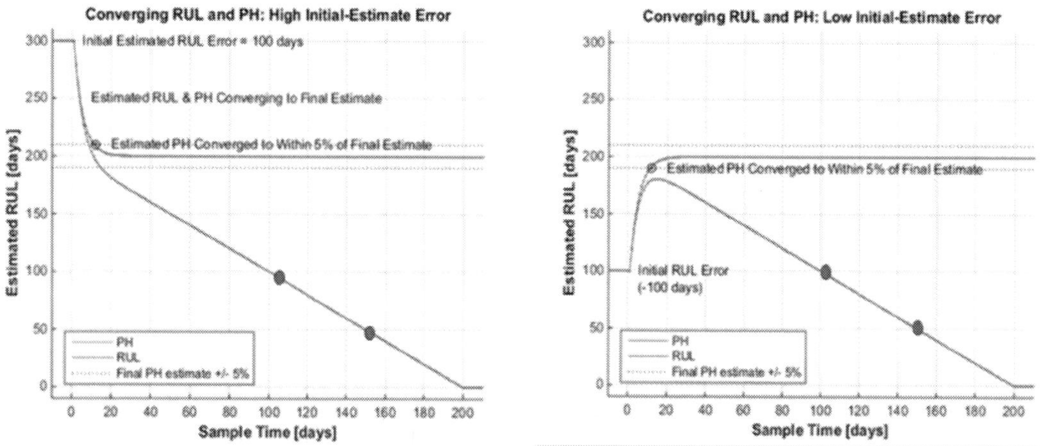

Figure 16: RUL Convergence from 50% Error – PD 50% & 10% Points Shown.

Referring to Left-hand plots in Figure 17, a 25% accuracy within 50% RUL is met, but a 10% accuracy within 25% is not met; in the right-hand plots, both the 25% and 10% accuracy specifications are met.

During design, you need to run experiments and/or simulations to indicate the design is likely to result in an implemented PHM system that meets specifications.

Alerts

The PHM system is designed to issue an alert at the SoH conditions listed in Table 1. The alerts include alert text and actions to be taken as indicated. To do so, the design needs to include the following functionality and methods: (1) prediction algorithms to produce SoH and other prognostic information; (2) support to monitor SoH values and appropriately trigger alerts; and (3) support to include relevant prognostic and failure information such as RUL, PH, and times to an alert subsystem, such as a Health Management/Services framework (see the lower left of Figure 19).

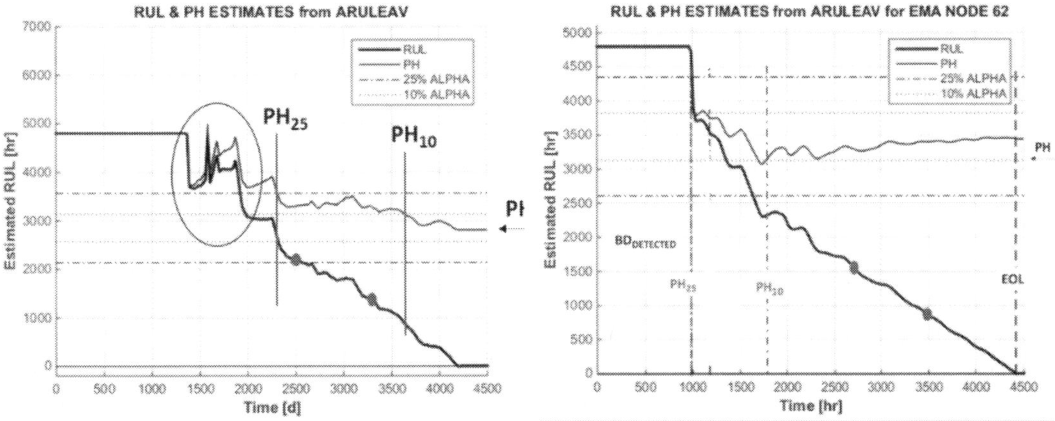

Figure 17: (Left Plots) Accuracy Within 25% is Met, Accuracy Within 10% is Not Met - (Right Plots) Accuracy Within 25% and Within 10% are Both Met.

Table 1: SoH and Alert Specifications

State of Health (SoH) Level	Level: Clear Text	Action
< 100%	Warning: xxx is degraded	Monitor
≤ 75%	Warning: xxx is ≤ 75% degraded	Next Maintenance
≤ 50%	Error: xxx is ≤ 50% degraded	Maintenance within RUL/2
≤ 25%	Error: xxx is ≤ 25% degraded	Maint/Repair within RUL/2
≤ 10%	Alarm: xxx failure within RUL	Immediate Repair
≤ 5%	Alarm: imminent failure within RUL	Immediate Repair
0%	Bells/whistles: functional failure	Immediate Repair

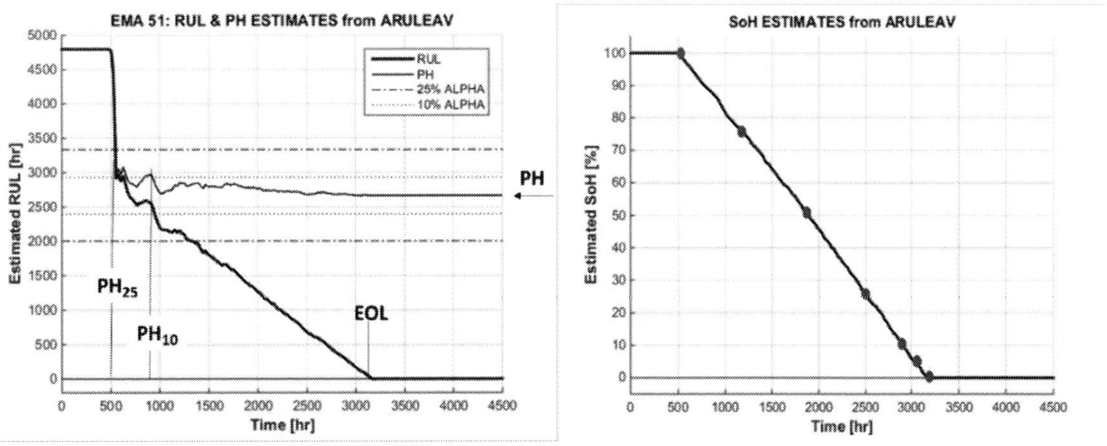

Figure 18: Example of RUL & PH Estimates (Left) and SoH Estimates with Alert Levels.

Start, Stop, Resume

The PHM system is to include the following: (1) support for stopping and restarting health monitoring; (2) support for resuming health monitoring after a planned stop or an unplanned system stop (a crash); (3) when health monitoring is resumed, the state of health of the system shall be the same as that at a checkpoint time; (4) checkpointing shall be at the prognostic-target level – specifically for each of the two power supplies and for each of the four EMAs.

A design solution is the approach shown in Figure 22.

Architecture to Define Targets to PHM System

Figure 19 is a diagram of a system architecture to support the design of the PHM system presented in this paper. The design approach uses a node-based approach. The design specifications, decisions, and architecture are subject to change as the design is refined, as the system is implemented, and as it is updated during development.

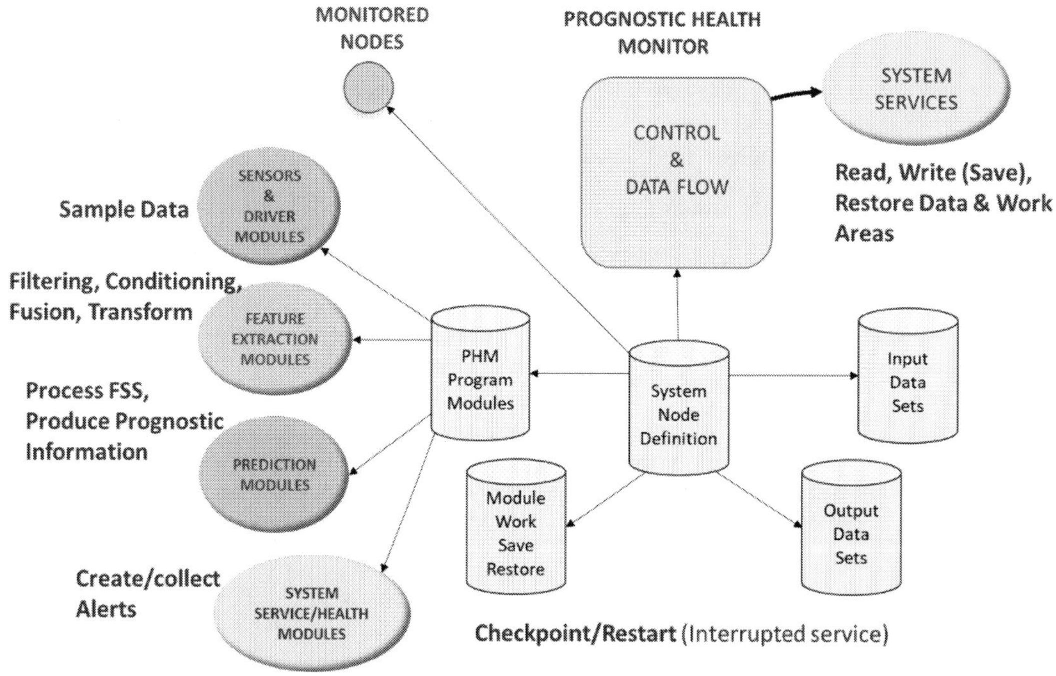

Figure 19: Design Level of Architecture.

Node-Based Approach

Figure 20 diagrams the architecture of Figure 19 to show more details of a node-based architecture as applied to the dual EMA subsystem presented in this paper.

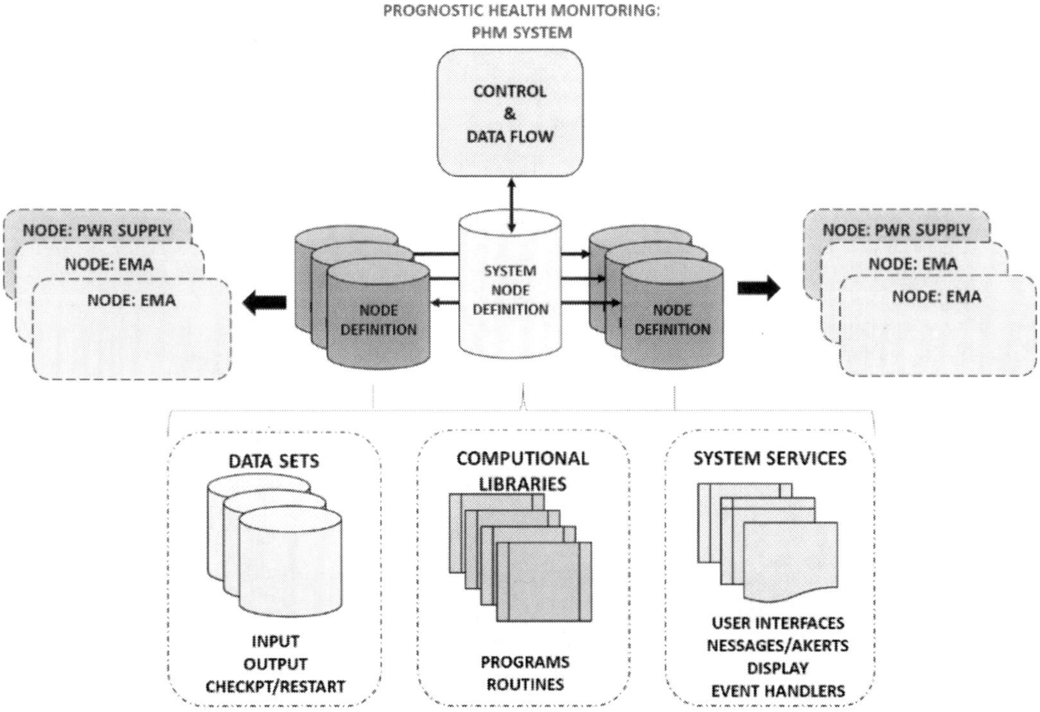

Figure 20: Node-Based Approach for Dual EMA Subsystem.

Node-Based Control and Data Flow

Figure 21 is a block diagram that illustrates the node-based control and data flow of the PHM system design.

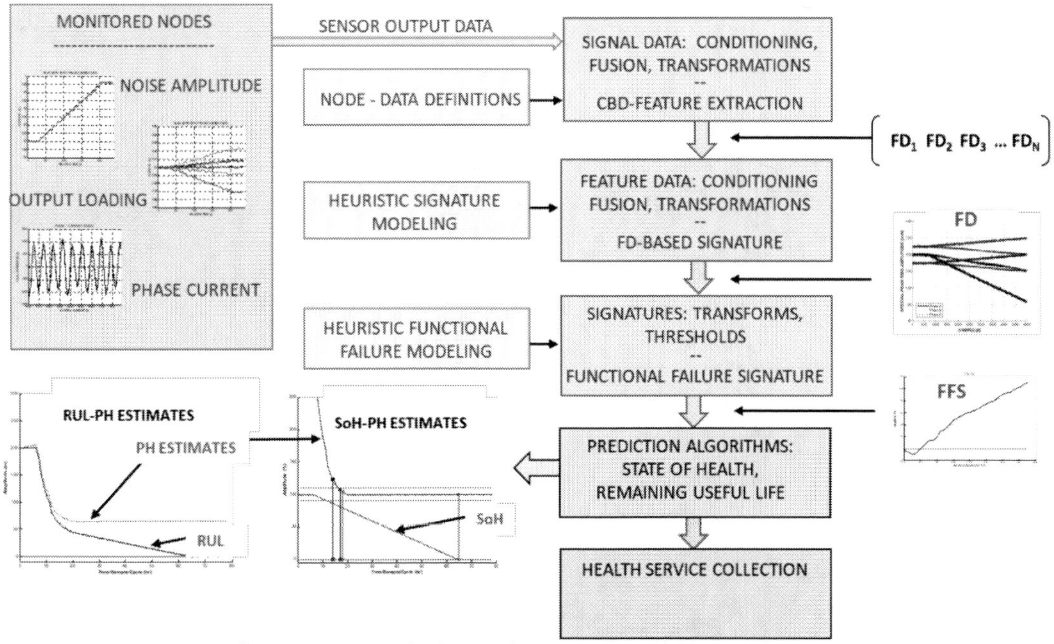

Figure 21: Node-Based Control and Data Flow.

Design Approach for Checkpoint/Restart

Figure 22 is a block diagram that illustrates a design approach to support checkpoint/restart.

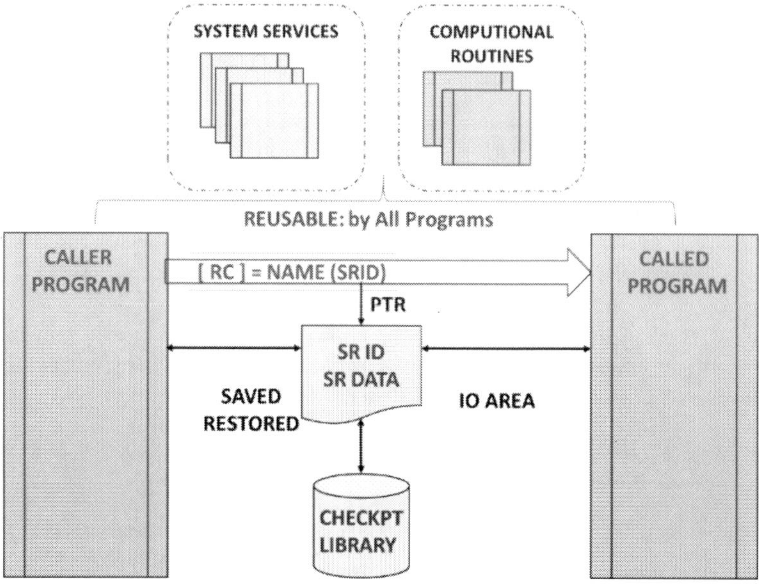

Figure 22: Design Approach to Support Checkpoint/Restart.

3. Design: Verification

A critical design process is data analyses: you need to design and run a series of experiments (DoE) using the EMA test that collects sensor data for each of the four selected faults. The data should comprise EMA operation before the onset of degradation, from the onset of degradation through increasing degradation to functional failure. Functional failure is defined as the level of degradation at which the prognostic target (power supply or EMA) no longer operates within specification.

Power Supply

Evaluation of Figure 23 and Figure 24 verifies the design for the power supply.

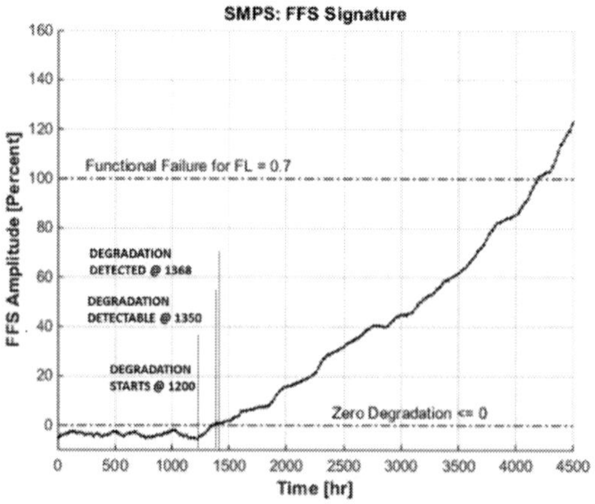

Figure 23: Power Supply FFS

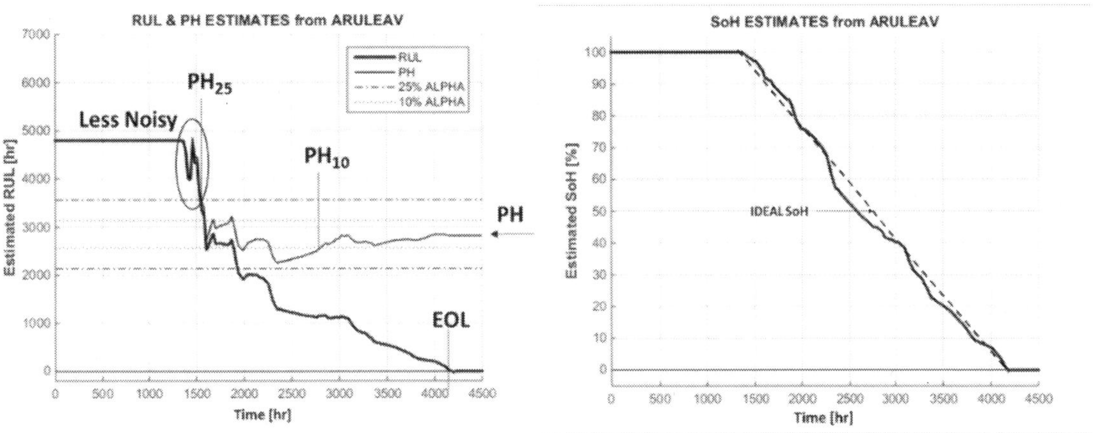

Figure 24: Power Supply RUL & PH (Left), SoH (Right)

EMA Winding, Load, Switching Transistor

Evaluation of the results plotted in Figure 25 through Figure 28 verifies the design for the EMAs.

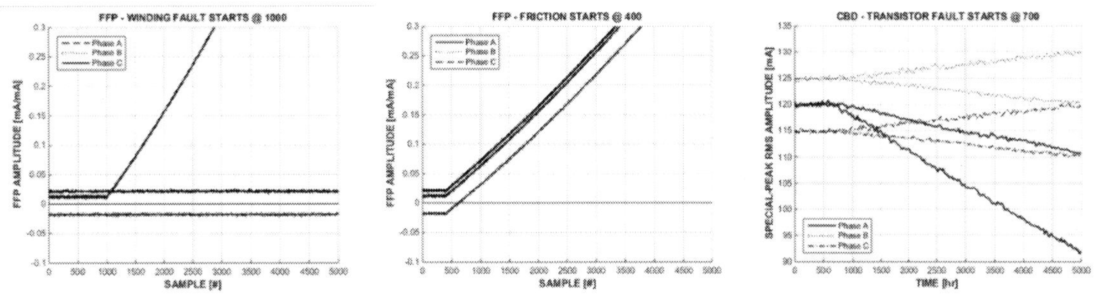

Figure 25: Fault Data for Winding, Friction/Load, and Switching Transistor

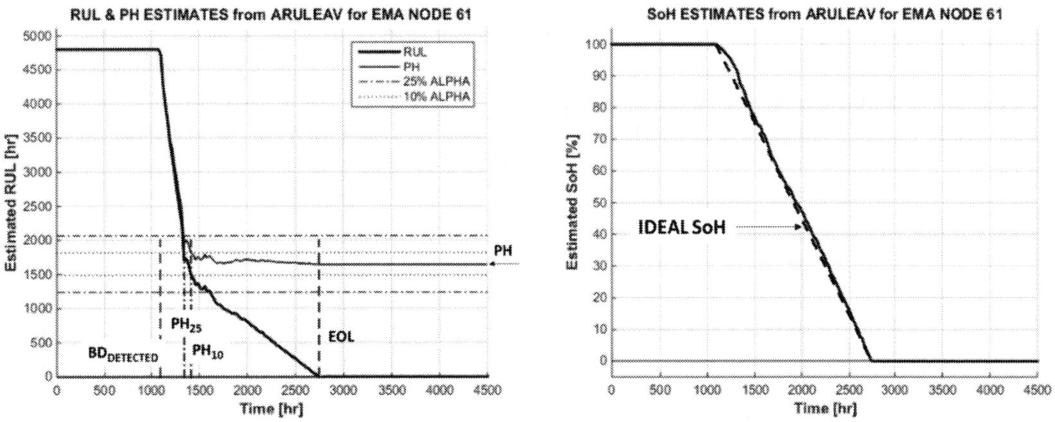

Figure 26: Winding Fault – RUL & PH (Left) and SoH (Right)

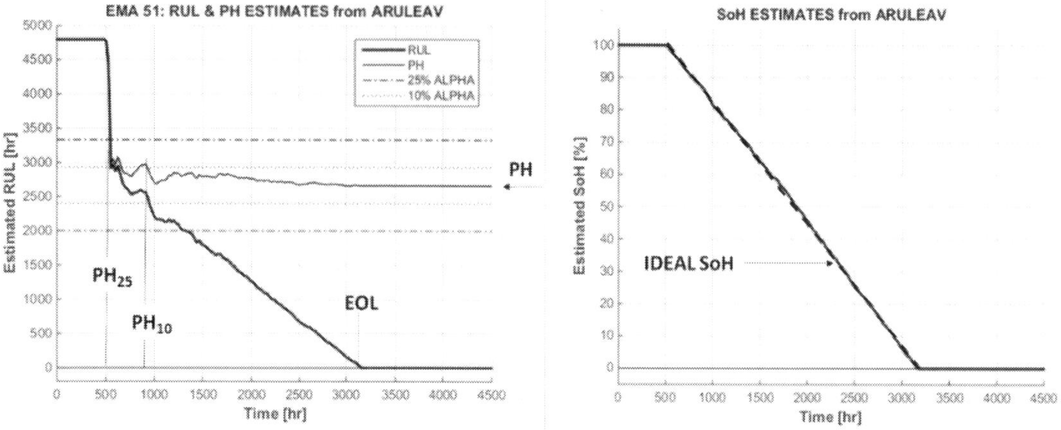

Figure 27: Loading/Friction Fault – RUL & PH (Left) and SoH (Right)

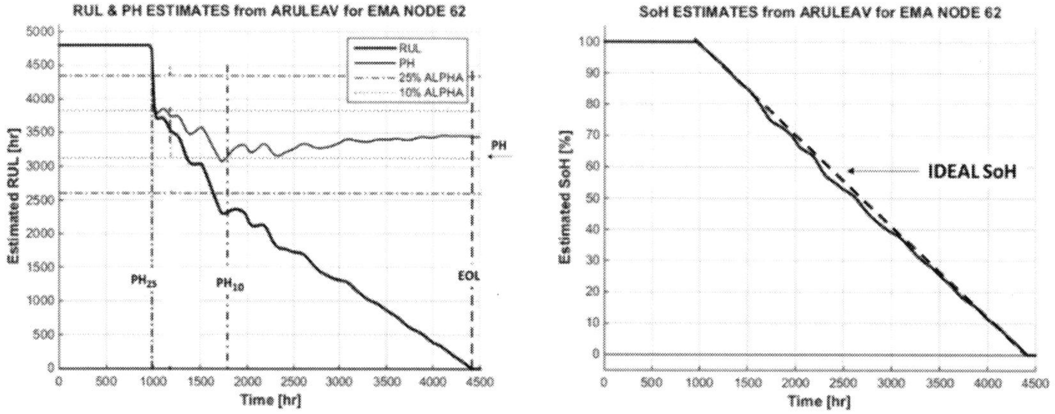

Figure 28: Transistor Fault – RUL & PH (Left) and SoH (Right)

4. Summary

This paper presented an example of a design for a robust PHM system comprising a dual set of two EMAs and a power supply. The design began with a general framework for a PHM system (Figure 1) and a block diagram description of a dual EMA system to be prognostic enabled. A critical design approach is to collect and analyze data related to the failure modes to be supported. Both the feature data related to failure and the noise in the collected data requires analyses to determine how to process the data to isolated and extract features for creating failure signatures and how to process the data to sufficiently eliminate or otherwise mitigate noise to meet specifications, especially accuracy requirements for detecting the onset of failure and predicting a time of future failure.

The following feature noise considerations were presented for the power supply: ripple voltage is the feature and significant noise such as thermal noise, harmonic distortion, and switching noise. The common noise associated with the EMAs included (1) the positioning of the EMA (up or down and (2) the start and the stop periods – the design solution was to employ a sampling window. Because of magnitude differences, a design for a special-rms method was developed, and a design to mitigate significant offset errors in current references was developed. A third design to mitigate noise was to employ data smoothing of the feature data and the transforms of that feature data into signature data.

An updated framework was presented that would support node-based sensors for monitoring, data conditioning and transformation into signatures, prediction processing of the signature data, detection of alerts and the issuing of alerts. The updated framework included an architecture to support checkpoint-restart to start, stop, resume, and recover the PHM system.

The design was verified by using an EMA testbed and injecting faults into the power supply and the EMAs. Data, from no degradation to degradation and then to functional failure was collected and processed to produce prognostic information. The prognostic information – RUL, PH, and SoH – was plotted, analyzed, and evaluated. The results indicated the design will correctly detect all four of the fault modes and will produce prognostic information that meets the design specifications.

Acknowledgments

The authors thank Naval Air, Naval Sea, U.S. Army, U.S. Air Force, and NASA research centers for their support and funding of multiple projects that led to the design and development of the prototype solutions and the results described and shown in this paper.

References

[1] Pecht, M. (2008) *Prognostics and Health Management of Electronics.* Hoboken, NJ: John Wiley & Sons.

[2] Goodman, Douglas L.; Hofmeister, James P.; and Szidarovszky, Ferenc (2019); (Pending), *Prognostics and Health Management: A Practical Approach Using Conditioned-Based Data,* 1st Edition, John Wiley & Sons, Ltd, The Atrium, Southern Gate, Chichester, West Sussex, PO19 8SQ, United Kingdom, ISBN: 9781119356653, 2019.

[3] IEEE 1856/D33 (2017), Draft Standard Framework for Prognosis and Health Management (PHM) of Electronic Systems, Mar 2017, IEEE Xplore Digital Library.

[4] CAVE3 (Accessed 2015, Nov.), Prognostics Health Management for Electronics, Auburn University, Auburn, Alabama. [Online]. Available: http://cave.auburn.edu/rsrch-thrusts/prognostic-health-management-for-electronics.html.

[5] Hofmeister, James P.; Szidarovszky, Ferenc; and Goodman, Douglas, L. (2017) "An Approach to Processing Condition-based Data for Use in Prognostic Algorithms," 2017 Machine Failure Prevention Technology, Virginia Beach, VA, 15-18 May, 2017.

[6] Hofmeister, James P.; Wagoner, Robert; and Goodman, Douglas (2019); "Electromechanical Actuator Case Study: Multiple Variable Analysis of Phase Currents to Detect Three Types of Faults," 2019 MFPT Conference, King of Prussia, PA, 16-18 May, 2019

[7] Hofmeister, James P.; Goodman, Douglas, L.; and Szidarovszky, Ferenc (2017) "PHM/IVHM: Checkpoint, Restart, and Other Considerations," 2017 Machine Failure Prevention Technology, Virginia Beach, VA, 15-18 May, 2017.

INTELLIGENT FAULT IDENTIFICATION OF PLANET BEARINGS USING DISCRIMINATIVE DICTIONARY LEARNING BASED SPARSE REPRESENTATION CLASSIFICATION FRAMEWORK

Yun Kong*, Tianyang Wang, Fulei Chu
Department of Mechanical Engineering, Tsinghua University, Beijing 100084, China
kongy15@mails.tsinghua.edu.cn

Abstract: Planet bearing fault identification is an attractive but challenging task in numerous engineering applications, such as wind turbine and helicopter transmission systems. However, traditional fault characteristic frequency identification and impulsive feature extraction based diagnosis strategies are not sufficient to resolve the problem of planet bearing fault detection, due to complex physical configurations and modulation characteristics in planetary gearboxes. In this paper, a novel discriminative dictionary learning based sparse representation classification (SRC) framework is proposed for intelligent planet bearing fault identification. Within our approach, the optimization objective for discriminative dictionary learning introduces a label consistent constraint called 'discriminative sparse code error' and incorporates it with the reconstruction error and classification error to bridge the gap between the classical dictionary learning and classifier training. Therefore, not only the reconstructive and discriminative dictionary for signal sparse representation but also an optimal universal multiclass classifier for classification tasks could be simultaneously learnt in the proposed framework. The optimization formulation could be efficiently solved using the well-known K-SVD dictionary learning algorithm. The effectiveness of the proposed framework has been validated using experimental planet bearing vibration signals. Comparative results demonstrate that our framework outperforms the state-of-the-art K-SVD based SRC method in terms of classification accuracy for intelligent planet bearing fault identification.

Key words: Planet bearing; Fault identification; Discriminative dictionary learning; Sparse representation classification; K-SVD; Orthogonal matching pursuit.

1. INTRODUCTION

Planetary gearbox possesses attractive advantages such as large transmission ratio and excellent load-bearing capacity in a relatively compact structure, which make it widely applied in many engineering applications such as wind turbine and helicopter transmissions. Failures of planetary gearbox will not only reduce the reliability of engineering system but also lead to great operation and maintenance (O&M) costs. Vibration-based fault detection technique has been proven as one of the most effective techniques for condition monitoring and O&M costs saving for planetary gearbox [1]. However, vibration signals of planetary gearbox are more complicated than the parallel gearbox due to the complex physical configuration and kinematic mechanism. As illustrated in Figure 1(a), planetary gearbox comprises four main components, including ring gear, sun gear, planet carrier and multiple planet gears. Concerning the kinematic mechanism of planetary gearbox, sun gear rotates around its fixed axis while multiple planet gears not only rotate around their own centers but also revolve around the sun gear, as shown in Figure 1(b). As a result, the vibration transmission paths between meshing points of ring-planet or sun-planet gear pairs and the fixed sensor are periodically varying due to the revolving nature of planet gears [2]. Therefore, vibration-based fault diagnosis of planetary gearbox is challenging due to the complex physical configurations and revolving planet gears inducing modulation characteristics [3].

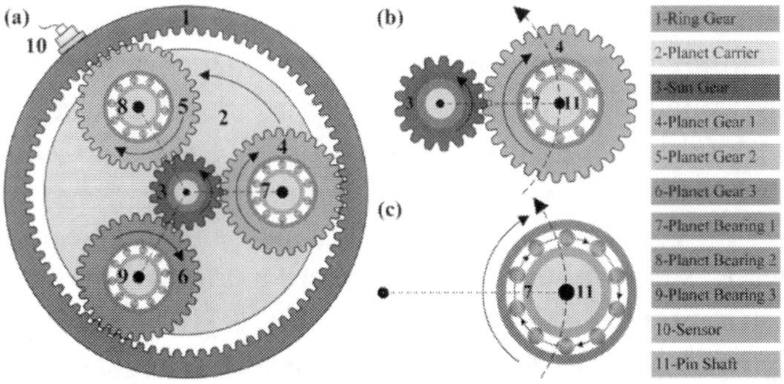

Figure 1: Physical Configurations of Planetary Gearbox and Kinematic Mechanism of Planet Bearings.

To address above challenges, many signal processing approaches have been developed for planetary gearbox fault diagnosis, such as statistical analysis [4], spectral kurtosis [5], time synchronous averaging [6], wavelet transform [7-8], demodulation analysis [9-10] and time frequency representation method [11]. Within these approaches, periodicity or time-varying characteristic frequency of fault-related features are analyzed to isolate the fault location, because the localized defect during meshing process will produce local anomalies repetitively and each gear has fault characteristic frequency proportional to the rotational frequency. To a certain extent, these contributions have successfully provided feasible means to address planetary gearbox fault detection. However, these reported literatures have been restricted to planetary gear fault detection and rarely devoted to identifying the planet bearing fault.

Planet bearing, as the most intricate component in term of kinematic mechanism, is also the most challenging component for fault identification in planetary gearbox. As illustrated in Figure1(c), the outer ring of planet bearing, which is fixed to the planet gear, not only rotates around its own center but also revolves around the sun gear. On the other hand, the inner ring stays relatively stationary to the planet carrier and only revolves around the sun gear. Therefore, vibration signature of planet bearings is complicated and significantly different from the common fixed-axis bearings due to the complex and time-varying vibration transmission paths. To understand the dynamic behavior of planetary gearbox containing localized planet bearing defects, Jain et al. first developed an analytical model to predict vibration signatures of the faulty planetary gearbox with localized planet bearing defects [12] and further predicted the fault signatures of planet bearing defects by identifying different sources of modulation sidebands [13]. Inspired by these thoughts, Feng et al. [14] developed analytical vibration signal models for defective planet bearings. Moreover, amplitude and frequency demodulation analyses were applied to identify the fault characteristic frequencies in demodulation spectrums of defective planet bearing vibration signals [15]. To locate the resonance frequency bands induced by planet bearing defects, Wang et al. proposed meshing frequency modulation index-based kurtogram [16] for filter design to detect planet bearing defects by identifying fault characteristic frequencies in envelope spectrums. These contributions have provided deep insights into understanding the vibration behavior of planetary gearbox in presence of defective planet bearings.

However, these reported literatures are only to reveal the frequency features for faulty planetary gearbox with planet bearing defects. These frequency features are too weak to identify due to the following reasons. First, the performance of the frequency feature based fault detection algorithms deteriorates easily due to planet bearing manufacturing error, misalignment and low- torque operating condition [17]. Second, vibration signatures of planet bearing are overwhelmed by gear meshing vibrations and other unwanted background noises [8, 16]. Third, the planet bearing defect-induced resonance frequency bands are very hard to determine for fault feature extraction due to heavy interferences from planetary gear meshing vibrations. Thus, these issues greatly challenge the fault characteristic frequency identification based diagnosis strategies.

Alternatively, classification-based intelligent identification approaches, which are free of detecting the weak frequency features induced by planet bearing defects, are more attractive and promising for planet bearing fault identification. Recently, dictionary learning based sparse representation classification (SRC) techniques have attracted extensive attentions in a wide range of academic communities such as image classification [18], face recognition [19] and computer vision [20]. The classical dictionary learning framework attempts to adaptively learn an overcomplete dictionary $\boldsymbol{D}\in\boldsymbol{R}^{n\times K}$ containing K signal-atoms of columns $\{\mathbf{d}_j\}^{K}_{j=1}$ so that the input signal $\boldsymbol{Y}\in\boldsymbol{R}^{n\times N}$ could be well approximated using sparse linear combinations of these atoms $\boldsymbol{Y}\approx\boldsymbol{DX}$ in terms of minimal reconstruction error $\|\boldsymbol{Y}-\boldsymbol{DX}\|_F^2$ [21]. Inspired by the merits of dictionary learning, the dictionary learning based SRC approaches have been developed for image classification [22-23] and fault identification [24]. Within the classical dictionary learning based SRC approach in [22], the main idea is to learn sub-dictionaries first for sparse representation of training signals from each category independently and then achieve the recognition tasks based on the corresponding reconstruction errors. Inspired by this idea, Zhao et al. [24] proposed the sparse representation classification for planet bearing fault identification, which achieved the recognition according to the minimal reconstruction error of testing signals with respect to these learned sub-dictionaries directly. However, these classical dictionary learning based SRC approaches only exploit the reconstruction power (best sparse representation of training signals for minimal reconstruction error) but neglect the discriminative power of the learned dictionary for recognition tasks. In contrast, a discriminative dictionary was achieved for pattern recognition by iteratively updating dictionary atoms based on the results of a linear predictive classifier [23]. In addition, the approach in [25] first learned a representative dictionary and then implemented the classifier training for recognition tasks. Nevertheless, these dictionary learning based SRC approaches treat the dictionary learning and classifier training as two separate processes, which might make the learned dictionary suboptimal for classification tasks. Therefore, learning a discriminative dictionary by identifying the discriminability of sparse codes and training the classifier model simultaneously in a supervised manner, could be promising to achieve better classification performance, which stimulated the idea of the discriminative dictionary learning based sparse representation classification framework in this paper.

In summary, to avoid the dilemmas of the frequency feature based diagnosis strategies for planet bearing fault identification and address the challenges for classical dictionary learning based SRC approaches, we propose a novel discriminative dictionary learning based sparse representation classification (DDL-SRC) framework for intelligent fault identification of planet bearings. The key features and contributions of this paper are presented as follows:

- To exploit class labels of training signals for supervised dictionary learning, the label information are associated with each dictionary atom to enforce the discriminability of sparse codes during dictionary learning process. Mathematically, we introduce a discriminative sparse code error into the optimization objective in the DDL-SRC framework.
- In addition to the reconstruction error for representative dictionary purpose, the discriminative sparse code error for discriminative dictionary purpose and the classification error for classifier training are jointly incorporated into the mixed optimization objective. Therefore, the proposed DDL-SRC framework could learn not only a discriminative dictionary but also an optimal multiclass classifier simultaneously to achieve better classification performance.
- The optimization problem in the proposed DDL-SRC framework could be efficiently solved with complexity bounded by the K-means Singular value decomposition (K-SVD) algorithm.
- In contrast to other existing pattern recognition methods, the proposed DDL-SRC framework is free of feature design and selection, which achieves the intelligent classification tasks via discriminative dictionary learning and classifier training directly from raw training signals.
- The DDL-SRC framework has been extended to achieve the intelligent fault identification of planet bearings and outperforms the state-of-the-art SRC method.

The rest of this paper is organized as follows: Section 2 presents the proposed DDL-SRC framework for intelligent fault identification, which involves discriminative dictionary learning, classification approach and intelligent fault identification for mechanical components. Section 3 validates the effectiveness of the DDL-SRC framework for intelligent fault identification of planet bearings using experimental data. Finally, conclusions are summarized in Section 4.

2. DISCRIMINATIVE DICTIONARY LEARNING BASED SPARSE REPRESENTATION CLASSIFICATION FRAMEWORK FOR INTELLIGENT FAULT IDENTIFICATION

In order to enhance the discriminative power of dictionary learning for classification tasks, learning a discriminative dictionary along with an optimal multiclass classifier simultaneously and leveraging the discriminability of sparse codes is preferred. To this end, the discriminative dictionary learning based sparse representation (DDL-SRC) framework is proposed for intelligent fault identification, which involves three procedures:

discriminative dictionary learning, classification approach and intelligent fault identification for mechanical components.

2.1 Discriminative Dictionary Learning

The discriminative dictionary learning aims to learn a reconstructive and discriminative dictionary and an optimal linear classifier simultaneously, which could bridge the gap between classical dictionary learning and classifier training in traditional SRC methods.

2.1.1 Optimization objective considering discriminative sparse codes and classifier training

In the discriminative dictionary learning, we aim to unify the dictionary learning and classifier training processes into one mixed optimization objective. In addition, the performance of linear classifier depends on the discriminability of sparse codes. As such, we prefer the supervised learning manner to leverage the label information of training signals for learning a reconstructive and discriminative dictionary. To this end, dictionary atoms should be chosen so that they represent a subset of the training signals ideally from one single class, and hence each dictionary atom will be associated with a particular label in our approach (see Figure2). Ideally, the discriminative sparse codes will be obtained over a discriminative dictionary, if we establish the explicit correspondences between dictionary atoms and the label information in a supervised fashion. The sparse representation using ideal discriminative sparse codes for optimal classification [27] is illustrated in Figure2. The ideal sparse codes $\boldsymbol{Q}$ are considered discriminative and optimal for classification tasks if $\boldsymbol{Q} = [q_1, q_2,\ldots,q_N] \in \boldsymbol{R}^{K\times N}$ where q_i is of the form of $[0,\ldots,1, 1, 1,\ldots,0]^{\mathrm{T}} \in \boldsymbol{R}^{K}$. Taking the case in Figure2 as an example, training signal matrix $\boldsymbol{Y} = [\boldsymbol{Y}_1, \boldsymbol{Y}_2, \boldsymbol{Y}_3]$ contains training signals from three classes, where $\boldsymbol{Y}_1$ contains three samples y_1, y_2, y_3, $\boldsymbol{Y}_2$ contains four samples y_4, y_5, y_6, y_7, and $\boldsymbol{Y}_3$ contains two samples y_8, y_9. Dictionary $\boldsymbol{D}$ contains three sub-dictionaries for each class and each sub-dictionary $\boldsymbol{D}_l$ has three atoms. In this case, the resulting discriminative sparse codes are illustrated in Figure 2.

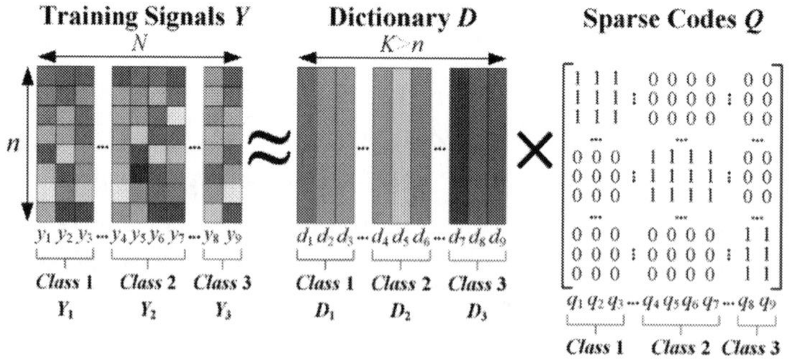

Figure 2: Sparse Representation Using Ideal Discriminative Sparse Codes for Optimal Classification.

Suppose a set of N n-dimension input signals $\boldsymbol{Y} = [y_1, y_2, \ldots, y_N] \in \boldsymbol{R}^{n\times N}$, the optimization objective for discriminative dictionary learning could be defined as follows:

$$\langle \boldsymbol{D},\boldsymbol{W},\boldsymbol{A},\boldsymbol{X}\rangle = \arg\min_{\boldsymbol{D},\boldsymbol{W},\boldsymbol{A},\boldsymbol{X}} \|\boldsymbol{Y}-\boldsymbol{D}\boldsymbol{X}\|_F^2 + \alpha\|\boldsymbol{Q}-\boldsymbol{A}\boldsymbol{X}\|_F^2 + \beta\|\boldsymbol{H}-\boldsymbol{W}\boldsymbol{X}\|_F^2 \quad s.t. \forall i, \|x_i\|_0 \le T, \tag{1}$$

where $\boldsymbol{X} = [x_1, x_2, \ldots, x_N] \in \boldsymbol{R}^{K\times N}$ are the sparse codes of the input signals $\boldsymbol{Y}$ with respect to dictionary $\boldsymbol{D} = [d_1, d_2, \ldots, d_K] \in \boldsymbol{R}^{n\times K}$. $\boldsymbol{A} \in \boldsymbol{R}^{K\times K}$ represents a linear transformation, which is designed to transform the obtained sparse codes $\boldsymbol{X}$ to be most discriminative in sparse feature space. $\boldsymbol{W} \in \boldsymbol{R}^{L\times K}$ denotes the linear classifier model parameters and $\boldsymbol{H} = [h_1, h_2, \ldots, h_N] \in \boldsymbol{R}^{L\times N}$ associates the label h_i with each training signal y_i. The first term in Equation (1) represents the reconstruction error. The second term in Equation (1) represents the discriminative sparse codes error, which enforces the sparse codes $\boldsymbol{X}$ to approximate the ideal discriminative sparse codes $\boldsymbol{Q}$. The third term in Equation (1) represents the classification error, which encourages achieving a relatively optimal classifier model parameters $\boldsymbol{W}$ for recognition tasks. Regularization parameters α and β control the relative contributions of the corresponding terms, respectively.

As a result, the above optimization objective for discriminative dictionary learning not only considers the reconstruction error, but also incorporates the discriminative sparse codes error and the classification error terms by leveraging the supervised information (label information $\boldsymbol{H}$ and discriminative sparse codes $\boldsymbol{Q}$). Therefore, this optimization objective is promising to learn a discriminative dictionary $\boldsymbol{D}$ and an optimal linear classifier jointly for classification tasks.

2.1.2 Numerical optimization algorithm

In this subsection, we will demonstrate that the proposed discriminative dictionary learning could be cast as a standard dictionary learning problem and solved using the well-known K-SVD algorithm. The proposed optimization objective in Equation (1) could be rewritten as follows:

$$\langle \boldsymbol{D},\boldsymbol{W},\boldsymbol{A},\boldsymbol{X}\rangle = \arg\min_{\boldsymbol{D},\boldsymbol{W},\boldsymbol{A},\boldsymbol{X}} \left\| \begin{pmatrix} \boldsymbol{Y} \\ \sqrt{\alpha}\boldsymbol{Q} \\ \sqrt{\beta}\boldsymbol{H} \end{pmatrix} - \begin{pmatrix} \boldsymbol{D} \\ \sqrt{\alpha}\boldsymbol{A} \\ \sqrt{\beta}\boldsymbol{W} \end{pmatrix} \boldsymbol{X} \right\|_F^2, \quad s.t. \forall i, \|x_i\|_0 \le T, \tag{2}$$

Through defining the generalized training signal matrix $\boldsymbol{Y}_{\text{new}} = (\boldsymbol{Y}^T, \sqrt{\alpha}\boldsymbol{Q}^T, \sqrt{\beta}\boldsymbol{H}^T)^T$ and dictionary matrix $\boldsymbol{D}_{\text{new}} = (\boldsymbol{D}^T, \sqrt{\alpha}\boldsymbol{A}^T, \sqrt{\beta}\boldsymbol{W}^T)^T$, the optimization objective can be reformulated as follows,

$$\langle \boldsymbol{D}_{\text{new}},\boldsymbol{X}\rangle = \arg\min_{\boldsymbol{D}_{\text{new}},\boldsymbol{X}} \|\boldsymbol{Y}_{\text{new}} - \boldsymbol{D}_{\text{new}}\boldsymbol{X}\|_F^2, \quad s.t. \forall i, \|x_i\|_0 \le T, \tag{3}$$

Consequently, this equivalent optimization is exactly the standard dictionary learning problem and can be efficiently solved by K-SVD algorithm.

Following the well-known K-SVD algorithm, we implement sparse coding and dictionary updating procedures alternately. In our approach, we utilize the OMP algorithm to accomplish the sparse coding. Within the dictionary updating procedure, we update one dictionary atom d_k^{new} and the associated nonzero sparse codes at a time for best reducing the optimization objective using SVD [18]. Let d_k^{new} be the k-th atom in dictionary $\boldsymbol{D}_{\text{new}}$ and x_R^k denotes the k-th row in sparse codes $\boldsymbol{X}$. In addition, we define $\boldsymbol{E}_k = (\boldsymbol{Y}_{\text{new}} - \sum_{j \neq k} d_j^{\text{new}} x_R^j)$, $\tilde{x}_R^k$ denotes the results of discarding the zero entries in x_R^k and $\tilde{\boldsymbol{E}}_k$ denotes the resulting $\boldsymbol{E}_k$ by discarding the corresponding columns. Hence, dictionary updating can be achieved by solving the following problem,

$$\left\langle d_k^{\text{new}}, \tilde{x}_R^k \right\rangle = \arg \min_{d_k^{\text{new}}, \tilde{x}_R^k} \left\| \tilde{\boldsymbol{E}}_k - d_k^{\text{new}} \tilde{x}_R^k \right\|_F^2, \tag{4}$$

The problem in Equation (4) could be solved by SVD for $\tilde{\boldsymbol{E}}_k$, namely, $\tilde{\boldsymbol{E}}_k = \boldsymbol{U}\Sigma\boldsymbol{V}^{\text{T}}$. Then, the updated dictionary atom d_k^{new} and the associated nonzero sparse code $\tilde{x}_R^k$ are computed as follows,

$$d_k^{\text{new}} = \boldsymbol{U}(:,1),\ \tilde{x}_R^k = \Sigma(1,1)\boldsymbol{V}(:,1). \tag{5}$$

Finally, nonzero entries in x_R^k are correspondingly replaced by $\tilde{x}_R^k$. The resulting discriminative dictionary learning algorithm is detailed in Algorithm 1, which learns $\boldsymbol{D}$, $\boldsymbol{A}$ and $\boldsymbol{W}$ simultaneously, and thus could avoid the local minima problem [23] in traditional SRC methods.

Algorithm 1 Discriminative dictionary learning

Input: $\boldsymbol{Y}, \boldsymbol{Q}, \boldsymbol{H}, K, T, \alpha, \beta$

Output: $\hat{\boldsymbol{D}}, \hat{\boldsymbol{A}}, \hat{\boldsymbol{W}}$

Initialize: compute $\boldsymbol{D}^{(0)}$, $\boldsymbol{A}^{(0)}$, $\boldsymbol{W}^{(0)}$ (See Sec. 2.1.3)

Initialize $\boldsymbol{Y}_{\text{new}} = (\boldsymbol{Y}^T, \sqrt{\alpha}\boldsymbol{Q}^T, \sqrt{\beta}\boldsymbol{H}^T)^T$, $\boldsymbol{D}_{\text{new}} = (\boldsymbol{D}^T, \sqrt{\alpha}\boldsymbol{A}^T, \sqrt{\beta}\boldsymbol{W}^T)^T$ and set $J = 1$.

while: until J reaches the preset maximum number of iterations $J_{\max}$.

- **Sparse coding:** compute sparse codes x_i while keeping $\boldsymbol{D}_{\text{new}}$ fixed.

  $x_i^* = \arg \min_{x_i} \left\| y_{\text{new}}^i - \boldsymbol{D}_{\text{new}} x_i \right\|_2^2 \quad s.t.\ \forall i, \|x_i\|_0 \leq T$

- **Dictionary update:**

  For $k = 1,\ldots,K$, repeat updating $\boldsymbol{D}_{\text{new}}$ along with $\boldsymbol{X}$ as follows,

  (1) Compute $\tilde{\boldsymbol{E}}_k$ and perform SVD operation $\tilde{\boldsymbol{E}}_k = \boldsymbol{U}\Sigma\boldsymbol{V}^{\text{T}}$;

  (2) Update d_k^{new} and $\tilde{x}_R^k$ by $d_k^{\text{new}} = \boldsymbol{U}(:,1),\ \tilde{x}_R^k = \Sigma(1,1)\boldsymbol{V}(:,1)$.

- Set $J = J+1$.

End while

Obtain the desired $\hat{\boldsymbol{D}}, \hat{\boldsymbol{A}}, \hat{\boldsymbol{W}}$ from $\boldsymbol{D}_{\text{new}}$ for classification using Equation (10).

2.1.3 Implementation details

In this subsection, several algorithm implementation details are presented, including algorithm initializations and obtaining the desired dictionary and linear classifier from $\boldsymbol{D}_{\text{new}}$.

(1) Initialization

In the proposed discriminative dictionary learning algorithm, initializations for $\boldsymbol{D}^{(0)}$, $\boldsymbol{A}^{(0)}$ and $\boldsymbol{W}^{(0)}$ are required. As for $\boldsymbol{D}^{(0)}$, we could exploit the standard dictionary learning to obtain sub-dictionaries for each class using several times of K-SVD, and then concatenate these sub-dictionaries of each K-SVD to accomplish the initialization for $\boldsymbol{D}^{(0)}$. Meanwhile, the label information $\boldsymbol{H}$ and discriminative sparse code $\boldsymbol{Q}$ could be determined according to the class information. As for $\boldsymbol{A}^{(0)}$ and $\boldsymbol{W}^{(0)}$, the multivariate ridge regression model [28] with quadratic loss and L_2-norm regularization is employed,

$$\boldsymbol{A} = \arg\min_{\boldsymbol{A}} \|\boldsymbol{Q} - \boldsymbol{AX}\|_F^2 + \lambda_1 \|\boldsymbol{A}\|_F^2 , \tag{6}$$

$$\boldsymbol{W} = \arg\min_{\boldsymbol{W}} \|\boldsymbol{H} - \boldsymbol{WX}\|_F^2 + \lambda_2 \|\boldsymbol{W}\|_F^2 . \tag{7}$$

The multivariate ridge regression models yield the solutions for $\boldsymbol{A}^{(0)}$ and $\boldsymbol{W}^{(0)}$ as follows,

$$\boldsymbol{A} = \boldsymbol{QX}^{\mathrm{T}}(\boldsymbol{XX}^{\mathrm{T}} + \lambda_1 \boldsymbol{I})^{-1}, \tag{8}$$

$$\boldsymbol{W} = \boldsymbol{HX}^{\mathrm{T}}(\boldsymbol{XX}^{\mathrm{T}} + \lambda_2 \boldsymbol{I})^{-1}. \tag{9}$$

Therefore, provided the initialized dictionary $\boldsymbol{D}^{(0)}$, we could compute sparse codes $\boldsymbol{X}^{(0)}$ of training signals $\boldsymbol{Y}$ using OMP algorithm. Then $\boldsymbol{A}^{(0)}$ and $\boldsymbol{W}^{(0)}$ can be computed by Equation (8) and (9).

(2) Obtaining the desired dictionary and linear classifier from $\boldsymbol{D}_{\text{new}}$

The solution $\boldsymbol{D}_{\text{new}}$ in Equation (3) could be efficiently achieved by K-SVD and the resulting $\boldsymbol{D}_{\text{new}}$ is L_2-norm normalized column-wise, i.e., $\forall k, \left\| \left(d_k^T, \sqrt{\alpha} a_k^T, \sqrt{\beta} w_k^T \right)^T \right\|_2 = 1$. However, the directly extracted dictionary $\boldsymbol{D} = [d_1, d_2, \ldots, d_K]$ from $\boldsymbol{D}_{\text{new}}$, will not satisfy the column normalization constraint for sparse representation. Therefore, proper transformation from $\boldsymbol{D}_{\text{new}}$ to the desired discriminative dictionary $\hat{\boldsymbol{D}}$ is required. This transformation can be achieved by normalizing the extracted dictionary $\boldsymbol{D}$ column-wise and scaling $\boldsymbol{A}$ and $\boldsymbol{W}$ correspondingly as follows,

$$\begin{aligned} \hat{\boldsymbol{D}} &= \left[\frac{d_1}{\|d_1\|_2}, \frac{d_2}{\|d_2\|_2}, \ldots, \frac{d_K}{\|d_K\|_2} \right] \\ \hat{\boldsymbol{A}} &= \left[\frac{a_1}{\|d_1\|_2}, \frac{a_2}{\|d_2\|_2}, \ldots, \frac{a_K}{\|d_K\|_2} \right] , \\ \hat{\boldsymbol{W}} &= \left[\frac{w_1}{\|d_1\|_2}, \frac{w_2}{\|d_2\|_2}, \ldots, \frac{w_K}{\|d_K\|_2} \right] \end{aligned} \tag{10}$$

Therefore, provided the learned dictionary $\boldsymbol{D}_{\text{new}}$, we could directly extract the dictionary $\boldsymbol{D}$, linear transformation $\boldsymbol{A} = [a_1, a_2, \ldots, a_K]$ and classifier model $\boldsymbol{W} = [w_1, w_2, \ldots, w_K]$ from $\boldsymbol{D}_{\text{new}}$.

Besides, the desired $(\hat{\boldsymbol{D}},\hat{\boldsymbol{A}},\hat{\boldsymbol{W}})$ could be computed using the learned $(\boldsymbol{D},\boldsymbol{A},\boldsymbol{W})$ and Equation (10).

2.2 Classification Approach

Another critical procedure for the proposed DDL-SRC framework is the classification approach for recognition tasks. The classification in our DDL-SRC framework is carried out by two steps, namely sparse coding of testing signals and label estimation using classifier $\hat{\boldsymbol{W}}$.

2.2.1 Sparse coding of testing signals

For test signal y_i, sparse coding could be achieved by solving the following problem,

$$\hat{x}_i^* = \arg\min_{x_i} \left\| y_i - \hat{\boldsymbol{D}}\hat{x}_i \right\|_2^2, \quad s.t. \forall i, \left\| \hat{x}_i \right\|_0 \le T, \tag{11}$$

The sparse coding could be efficiently solved using OMP algorithm [26].

2.2.2 Label estimation using classifier model

Estimating the label information of test signal y_i in the proposed framework is carried out according to the optimized sparse codes $\hat{x}_i^*$ and the linear classifier model $\hat{\boldsymbol{W}}$. We simply employ the linear classifier to estimate the label vector $\boldsymbol{l} = \hat{\boldsymbol{W}}x_i^*$. The predicted label j^* of y_i is determined by the index corresponding to the largest absolute value in label vector as follows,

$$j^* = \arg\max_{j} \ \text{abs}\,(l_j), \ \boldsymbol{l} = \hat{\boldsymbol{W}}x_i^* = [l_1,\ldots,l_L]^T. \tag{12}$$

2.3 Intelligent Fault Identification for Mechanical Components

In this section, we aim to extend the DDL-SRC framework for intelligent fault identification. To this end, we first propose an overlapping segmentation strategy for vibration signals and then summarize the overall procedures of the DDL-SRC framework for intelligent fault identification.

2.3.1 Overlapping segmentation strategy for vibration signals

Proper segmentation strategy is crucial to extract robust two-dimensional features from one-dimensional vibration signals. Periodic self-similarity is a robust feature in vibration signals of rotating machinery under stationary operating conditions. Hence, it is favorable to develop a wise segmentation strategy, which could take the best of self-similarity information and avoid the boundary artifacts between adjacent segments as well. To this end, we introduce a wise segmentation operator $\boldsymbol{R}$ to partition the original vibration signals into a set of local segments. The segmentation operator $\boldsymbol{R}$ is parameterized with two critical parameters, namely the segmentation window size W and overlapping rate δ. The principle of segmentation operator is illustrated in Figure 3. Let $y \in \mathrm{R}^{m}$ denotes the original vibration signal, the segmentation operator $\boldsymbol{R}: R^{m} \mapsto R^{W \times N}$ maps original vibration signals y into a signal matrix $\boldsymbol{Y} \in \mathrm{R}^{W \times N}$ as follows,

$$\boldsymbol{Y} = \underbrace{\left[\boldsymbol{R}_1\ \boldsymbol{R}_2 ... \boldsymbol{R}_{N-1}\ \boldsymbol{R}_N\right]}_{\boldsymbol{R}} y = \boldsymbol{R}(y) . \tag{13}$$

where the operator $\boldsymbol{R}_i : R^{m} \mapsto R^{W}$ represents the procedure that first takes the i-th segment from the original vibration signal y and then transposes the obtained segment, thus $\boldsymbol{R}_i(y) = y_i^{\mathrm{T}} = \boldsymbol{Y}(:,i)$.

Taking into account the multi-state vibration signals with different health conditions of mechanical components, we could employ the overlapping segmentation operator $\boldsymbol{R}$ to transform original vibration signals of each state. Then, constructing the overall training/testing signal matrix is accomplished by concatenating all signal matrix for different states. This overall overlapping segmentation strategy for overall signal matrix construction is illustrated in Figure 3.

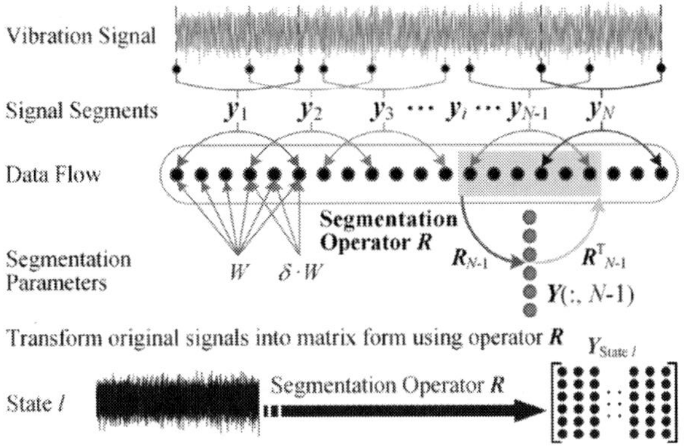

Figure 3: Overlapping Segmentation Strategy for Vibration Signals

As a result, the overlapping segmentation strategy could leverage the self-similarity feature maximally and construct the overall training/testing signal matrix, which serves as a premise for our proposed DDL-SRC framework for intelligent fault identification using vibration signals.

2.3.2 Overall procedures of DDL-SRC framework for intelligent fault identification

In this section, we summarize the overall algorithm procedures of the DDL-SRC framework for intelligent fault identification using vibration signals as follows.

Step 1 Acquire vibration signals of mechanical components under various health conditions and construct training/testing signal matrixes using the overall overlapping segmentation strategy.

Step 2 Implement the discriminative dictionary learning algorithm for learning the discriminative dictionary and an optimal multiclass linear classifier model simultaneously.
Step 3 Implement the sparse classification using the learned discriminative dictionary and predictive multiclass linear classifier model in step 2.
Step 4 Achieve intelligent fault identification of mechanical components using the results of sparse representation classification.

3. CASE STUDY: INTELLIGENT FAULT IDENTIFICATION FOR PLANET BEARINGS

In this section, experiments and vibration measurements are conducted on a planetary gearbox test bench, which are used to validate the effectiveness of the proposed DDL-SRC framework for intelligent fault identification of planet bearings. In addition, the state-of-the-art dictionary learning based sparse representation classification (DL-SRC) method in [24] is compared for demonstrating the superiority of the proposed framework.

3.1 Experiment Description

The planetary gearbox experiment setup is illustrated in Figure 4(a), which consists of a drive motor, an encoder for rotational speed measurement, a planetary gearbox, a magnetic powder brake for applying load and the vibration measurement system. To simulate the local defects in planet bearings, minor damages are introduced on surface of different planet bearing components using electrical discharge machining technique. As a result, vibration measurements could be performed on the planetary gearbox with four different health conditions (namely, healthy, outer ring defect, inner ring defect and rolling element defect), as illustrated in Figure 4(b). Detailed physical parameters of the planetary gearbox are listed in Table 1. As for the experiment measurement, the sun gear rotates at a constant frequency of 24.97 Hz and vibration signals are sampled at a frequency of 20480 Hz.

Table 1: Physical Parameters of the Experimental Planetary Gearbox.

Gear	Tooth number	Planet bearing	Size (mm)
Carrier	--	Roller diameter	9
Sun gear	13	Diameter of pitch circle	36
Ring gear	92	Number of rollers	10
Planet gear	38(3)	Contact angle (°)	0

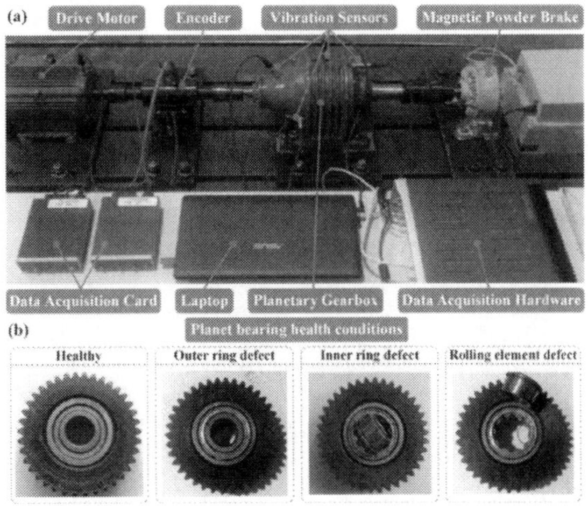

Figure 4: Planetary Gearbox Experiment Setup and Planet Bearings with Four Health Conditions.

3.2 Planet Bearing Fault Identification

Vibration signals of planetary gearbox with four different health conditions are measured in two different processes, the one of which with time duration of 60 seconds serve as the training signals and the other with time duration of 60 seconds serve as the testing signals for the proposed DDL-SRC framework. The time domain waveforms of vibration signals of planetary gearbox in presence of four different health conditions of planet bearings are illustrated in Figure 5. In order to reduce the computation burden of discriminative dictionary learning algorithm, all original vibration signals are downsampled at a sampling frequency of 1024 Hz.

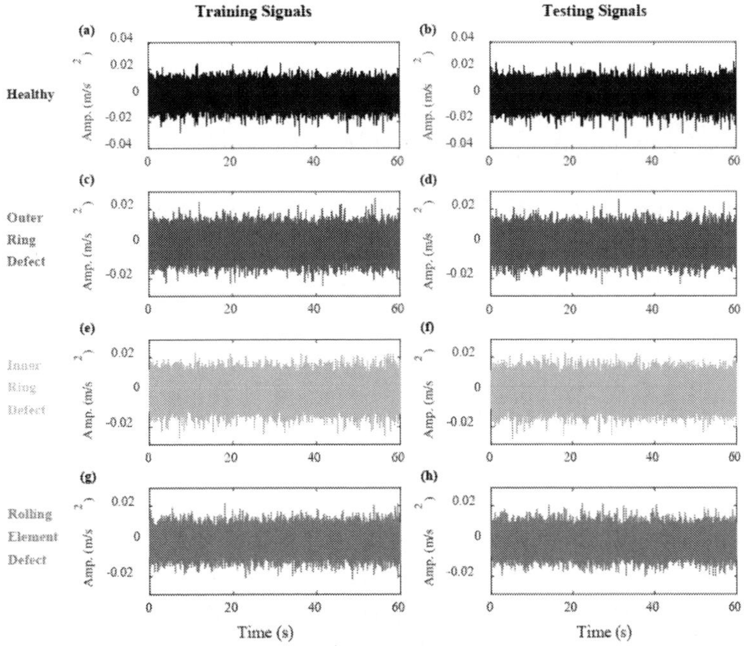

Figure 5: Vibration Signals of Planetary Gearbox with Four Health Conditions of Planet Bearings.

3.2.1 Intelligent identification of planet bearing defects

In this subsection, we evaluate the effectiveness and superiority of the proposed DDL-SRC framework over the DL-SRC method in [24] for intelligent fault identification of planet bearings.

In this experiment, dictionary size is set as 4350 such that an average of 1088 dictionary atoms corresponds to each health condition of planet bearing. The parameters for overlapping segmentation strategy are chosen that the segmentation window size W be 1800 and overlapping rate δ be 0.97. Hence, both the training and testing signal matrixes for each health condition consist of 1105 samples. Besides, the sparsity threshold T and maximum iteration number $J_{\max}$ are set as 4 and 35 for discriminative dictionary learning, respectively. The regularization parameters α=0.006 and β=0.001 are used in this case. As a result, the confusion matrix of the proposed DDL-SRC framework and classification performance including the comparison with the state-of-the-art DL-SRC method [24], are illustrated in Figure 6(a) and (b), respectively.

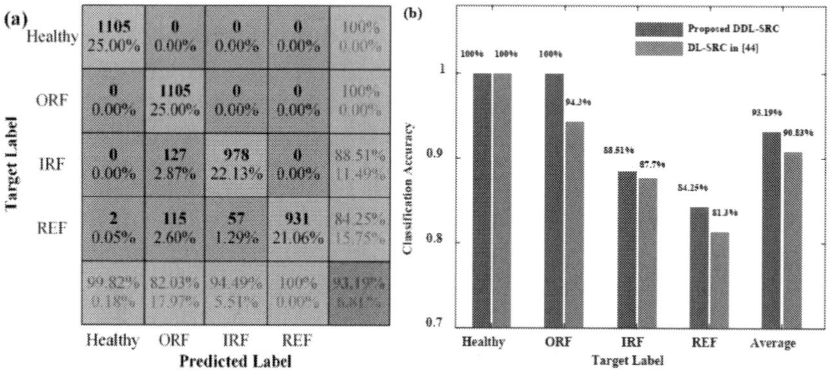

Figure 6: Confusion Matrix of The Proposed DDL-SRC Framework and Classification Performance for Intelligent Fault Identification of Planet Bearings.

As illustrated in Figure 6(a), the proposed DDL-SRC framework achieves no false prediction for healthy planet bearings and faulty planet bearings with outer ring defect. Additionally, the overall percentage of which the faulty planet bearing is wrongly predicted as the healthy one is as low as 0.045% (i.e., 2/(4*1105)). As a whole, the proposed DDL-SRC framework achieves an excellent average identification accuracy as higher as 93.19% for intelligent fault identification of planet bearings in this general case. In contrast, our proposed framework always achieve better classification accuracy than the state-of-the-art DL-SRC method [44], as illustrated in Figure6(b).

3.2.2 Parameter analysis

In this subsection, we discuss the effects of related algorithm parameters on classification performance of the DDL-SRC framework. As for the fault identification of planet bearings in this experiment, we evaluate several crucial algorithm parameters with cross-validation strategy. These parameters are divided into four different groups and each group of parameters will be discussed with other groups fixed. More specifically, we firstly analyze the effect of dictionary size K and with other parameters fixed. Then we will fix the dictionary size and check the performance of regularization parameters α and β. Next, the segmentation window size W and overlapping rate δ are discussed. Finally, the sparsity threshold T and maximum number of iterations J_{max} for discriminative dictionary learning are validated.

(1) *Dictionary Size K*. Dictionary size K determines the number of atoms in the learned dictionary. On the one hand, a higher dimension of feature subspace will contain more information so as to enhance the representational ability of dictionary. On the other hand, a higher dimension of dictionary will generally be more discriminative than a lower one for classification tasks [37]. To validate these two suggestions, dictionary size K is varied from 2400 to 4400 and the corresponding classification accuracies are illustrated in Figure 7(a). It can be concluded that increasing the dictionary size will lead to a higher classification performance and the highest classification accuracy can be achieved around the maximum considered dictionary size.

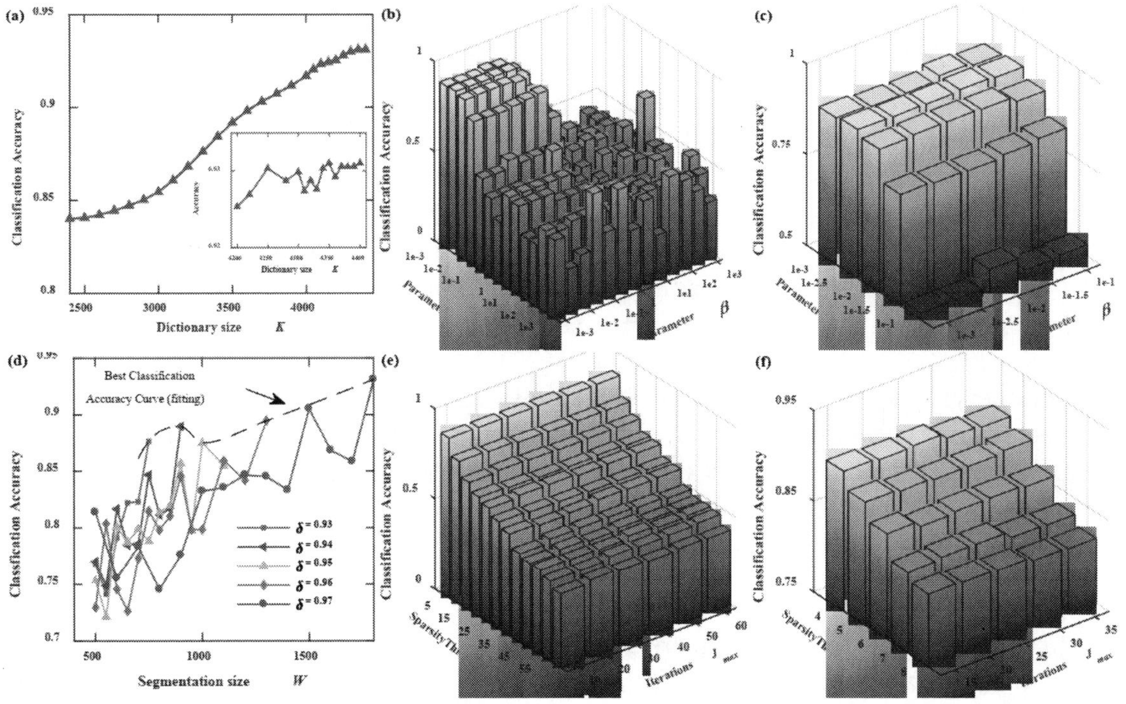

Figure 7: Parameter Analysis on Classification Performance. (a) Effect of Dictionary Size K; (B) Effect of Regularization Parameters α and β; (c) Enlarged Illustration of (b); (d) Effect of Segmentation Size W and Overlapping Rate δ; (e) Effect of Sparsity Threshold T and Maximum Number of Iterations J_{max}; (f) Enlarged Illustration of (e).

(2) *Regularization parameters α and β*. The regularization parameters determine the relative contributions of the discriminative sparse code error and classification error terms

in the optimization objective. A balanced pair of regularization parameters will lead to the desired dictionary with both reconstructive and discriminative power, and an optimal classifier model for classification tasks. To investigate their effects, both α and β are varied from 10^{-3} to 10^{3} exponentially and the corresponding results are illustrated in Figure 7(b). By checking these classification accuracies, we can get the conclusions. Firstly, regularization parameters with relatively small values could lead to higher classification performance than pairs of regularization parameters with greater values. Secondly, the regularization parameters with the order of magnitude of 10^{-3} achieve the best classification accuracy in our task, as shown in Figure 7(c).

(3) *Segmentation window size W and overlapping rate δ.* These two parameters influence the results of the overlapping segmentation strategy for vibration signals. In order to take the best of self-similarity information in vibration signals, we consider the cases with relatively large overlapping rate, namely δ=0.93:0.01:0.97, and varying segmentation sizes. The corresponding results are illustrated in Figure 7(d). By carefully checking these classification accuracies, we can get the following conclusions. Firstly, though there are few outliers like jump points, increasing the segmentation size generally could achieve a better classification performance. Secondly, the best classification accuracy using a higher overlapping rate is generally better than that using a lower overlapping rate, as illustrated by the 'best classification accuracy curve' in Figure 7(d).

(4) *Sparsity threshold T and maximum number of iterations $J_{\max}$.* Finally, the effects of sparsity threshold and maximum number of iterations on classification accuracy are studied and illustrated in Figure 7(e) and (f). These two parameters mainly affect the discriminative dictionnary learning algorithm. The following conclusions can be drawn from Figure 7(e) and (f). Firstly, smaller sparsity thresholds T could lead to better classification performance than greater sparsity thresholds. Secondly, the classification accuracy is merely promoted very little as the increase of maximum number of iterations $J_{\max}$. In our experiment, sufficiently better classification performance could be obtained with 35 iterations.

In summary, the proposed DDL-SRC framework is effective using proper algorithm parameters and outperforms the state-of-the-art DL-SRC method for fault identification of planet bearings, which indicates that it is promising to achieve intelligent machine fault diagnosis.

4. Conclusion

In this paper, a novel discriminative dictionary learning based sparse representation classification framework is proposed for intelligent fault identification of planet bearings. In contrast to traditional frequency feature based diagnostic strategies for planet bearing fault detction, the proposed framework could learn adaptive sparse features directly from raw vibration signals for reliable identification performance. The main conclusions are drawn as follows.

(1) The proposed framework bridges the gap between the reconstructive dictionary learning and classifier training in classical sparse representation classification methods, by introducing a discriminative sparse code error term and incorporating it with the

reconstruction error and classification error into one mixed optimization formulation.

(2) The proposed framework could learn a discriminative dictionary and an optimal multiclass linear classifier simultaneously for better classification performance over classical sparse representation classification methods. Additionally, the proposed discriminative dictionary learning problem could be efficiently solved by a simple extension of the K-SVD algorithm.

(3) In contrast to other conventional classifier based pattern recognition methods, the proposed framework is free of feature design and selection. It can adaptively learn adaptive features from raw vibration signals for intelligent fault identification.

(4) The proposed framework outperforms the existing state-of-the-art sparse representation classification method and achieves intelligent fault identification of planet bearings with the highest classification accuracy among all reported literatures.

References

[1] Z. Hameed, Y.S. Hong, Y.M. Cho, S.H. Ahn, C.K. Song, Condition monitoring and fault detection of wind turbines and related algorithms: A review, Renew. Sustain. Energy Rev. 13 (2009) 1–39.

[2] Z. Feng, M.J. Zuo, Vibration signal models for fault diagnosis of planetary gearboxes, J. Sound Vib. 331 (2012) 4919–4939.

[3] Y. Lei, J. Lin, M.J. Zuo, Z. He, Condition monitoring and fault diagnosis of planetary gearboxes: A review, Measurement 48 (2014) 292–305.

[4] W. Bartelmus, R. Zimroz, A new feature for monitoring the condition of gearboxes in non-stationary operating conditions, Mech. Syst. Signal Process. 23 (2009) 1528–1534.

[5] T. Barszcz, R.B. Randall, Application of spectral kurtosis for detection of a tooth crack in the planetary gear of a wind turbine, Mech. Syst. Signal Process. 23 (2009) 1352–1365.

[6] P.D. McFadden, A technique for calculating the time domain averages of the vibration of the individual planet gears and the sun gear in an epicyclic gearbox, J. Sound Vib. 144 (1991) 163–172.

[7] Y. Kong, T. Wang, Z. Li, F. Chu, Fault feature extraction of planet gear in wind turbine gearbox based on spectral kurtosis and time wavelet energy spectrum, Front. Mech. Eng. 12 (2017) 406–419.

[8] Y. Kong, T. Wang, F. Chu, Meshing frequency modulation assisted empirical wavelet transform for fault diagnosis of wind turbine planetary ring gear, Renew. Energy. 132 (2019) 1373–1388.

[9] Z. Feng, M. Liang, Y. Zhang, S. Hou, Fault diagnosis for wind turbine planetary gearboxes via demodulation analysis based on ensemble empirical mode decomposition and energy separation, Renew. Energy. 47 (2012) 112–126.

[10] Z. Feng, X. Lin, M.J. Zuo, Joint amplitude and frequency demodulation analysis based on intrinsic time-scale decomposition for planetary gearbox fault diagnosis, Mech. Syst. Signal Process. 72–73 (2016) 223–240.

[11] Z. Feng, M. Liang, Fault diagnosis of wind turbine planetary gearbox under nonstationary conditions via adaptive optimal kernel time-frequency analysis, Renew. Energy. 66 (2014) 468–477.

[12] S. Jain, H. Hunt, Vibration response of a wind turbine planetary gear set in the presence of a localized planet bearing defect, Proceedings of the ASME 2011 International

Mechanical Engineering Congress and Exposition (2011) 943–952.
[13] S. Jain, P.E. Whiteley, H. Hunt, Detection of planet bearing faults in wind turbine gearboxes, Proceedings of the International Conference on Noise and Vibration Engineering (2012) 4351–4362.
[14] Z. Feng, H. Ma, M.J. Zuo, Vibration signal models for fault diagnosis of planet bearings, J. Sound Vib. 370 (2016) 372–393.
[15] Z. Feng, H. Ma, M.J. Zuo, Amplitude and frequency demodulation analysis for fault diagnosis of planet bearings, J. Sound Vib. 382 (2016) 395–412.
[16] T. Wang, F. Chu, Z. Feng, Meshing frequency modulation (MFM) index-based kurtogram for planet bearing fault detection, J. Sound Vib. 432 (2018) 437–453.
[17] M. Inalpolat, A. Kahraman, A dynamic model to predict modulation sidebands of a planetary gear set having manufacturing errors, J. Sound Vib. 329 (2010) 371–393.
[18] Z. Jiang, Z. Lin, L.S. Davis, Label consistent K-SVD: Learning a discriminative dictionary for recognition, IEEE Trans. Pattern Anal. Mach. Intell. 35 (2013) 2651–2664.
[19] J. Wright, A.Y. Yang, A. Ganesh, S.S. Sastry, Y. Ma, Robust face recognition via sparse representation, IEEE Trans. Pattern Anal. Mach. Intell. 31 (2009) 1–18.
[20] X. Mei, H. Ling, Robust visual tracking and vehicle classification via sparse representation, IEEE Trans. Pattern Anal. Mach. Intell. 33 (2011) 2259–2272.
[21] M. Aharon, M. Elad, A. Bruckstein, K-SVD: An Algorithm for Designing Overcomplete Dictionaries for Sparse Representation, IEEE Trans. Signal Process. 54 (2006) 4311–4322.
[22] J. Mairal, F. Bach, J. Ponce, G. Sapiro, A. Zisserman, Discriminative learned dictionaries for local image analysis, IEEE Conf. Comput. Vis. Pattern Recognition, CVPR (2008) 1–8.
[23] D.S. Pham, S. Venkatesh, Joint learning and dictionary construction for pattern recognition, IEEE Conf. Comput. Vis. Pattern Recognition, CVPR. (2008) 1–8.
[24] C. Zhao, Z. Feng, X. Wei, Y. Qin, Sparse classification based on dictionary learning for planet bearing fault identification, Expert Syst. Appl. 108 (2018) 233–245.
[25] K. Huang, S. Aviyente, Sparse representation for signal classification, Advances in Neural Information Processing Systems (2007) 609–616.
[26] J.A. Tropp, A.C. Gilbert, Signal recovery from random measurements via orthogonal matching pursuit, IEEE Trans. Inf. Theory 53 (2007) 4655–4666.
[27] Y. Zhang, Z. Jiang, L.S. Davis, Learning structured low-rank representations for image classification, IEEE Conf. Comput. Vis. Pattern Recognition, CVPR. (2013) 676–683.
[28] G.H. Golub, P.C. Hansen, D.P. O'Leary, Tikhonov regularization and total least squares, SIAM J. Matrix Anal. Appl. 21 (1999) 185–194.

ROBOT MANIPULATOR DRIVE FAULT DIAGNOSTICS USING DATA-DRIVEN AND ANALYTICAL MODELLING

Michael Lipsett[1], Anthony Maltais[1], Mohammad Riazi[2], Nicolas Olmedo[1], Osmar Zaiane[2]
[1]Department of Mechanical Engineering
[2]Department of Computing Science
University of Alberta
Edmonton, AB T6G 1H9
Michael Lipsett (780)-492-9494

Abstract: Belt-driven mechatronic systems are popular for a range of applications. A modified robotic manipulator was adapted to allow different belt-drive faults to be incorporated into the mechanism, with additional sensors to characterize the compromised kinematics. Different data-driven models were used studied to detect anomalies in motor power consumption and end-effector motion; and a physics-based, lumped-parameter dynamic model was used to identify different faults. Comparative assessment metrics were sdused to compare the performance of different fault models from sets of laboratory test data.

Key words: Machinery diagnostics; fault detection; machine learning; modelling; robotics; analytics; belt drives, time-varying systems.

1. INTRODUCTION

Mechatronic systems are increasingly popular in a range of products to improve quality and production in a variety of different ways. Many mechatronic devices use belt-driven power transmission systems. Improved understanding of the failure modes and effects in belt drives would enable diagnostic and prognostic techniques for predictive maintenance. Diagnostic models can be based on an understanding of the physics of the system and the effects of progressing component failure, or on data collected from systems that incorporate the component of interest under a range of operating conditions.

Physics-based models are used to represent a system using differential equations [1]. Within these differential equations, a variety of variables can be included in the model to accurately define the state of the system. A few examples of these factors include material properties, degrees of freedom, environmental and equipment specifications. The physics-based model provides an approximation of the response of a system to these factors. If the system has many of these parameters known and are appropriately integrated into the governing equations of the system, it is possible to estimate the other parameters and thereby calibrate a model that is an accurate representation of the physical system. By using this model, a response that varies from normal operating conditions in a way that relates to a change in a component indicates a fault in the system. Such a model can be used to simulate the effect of faults that can occur in the system. Bartelmus et al. [2] showed a case study of a dynamic model for fault detection in a real-world mechanical system.

In industrial practice, fault diagnosis and maintenance are generally done based on the knowledge and expertise of the people working with the equipment [3]. From observable signs of damage or an understanding of the rate at which wear-out occurs, maintainers perform repairs based on condition or schedule. Process and condition monitoring datasets allow diagnostics to be automated, using data to determine what normal operating conditions are (with normal variability or even time-varying behaviour for non-steady state systems) and comparing new data to assess whether there is a system fault. A drawback of this method is that an observer is limited to the information that is available from sensors. Determining the appropriate type, number, and locations is critical for observing features that correlate well to developing faults [4].

Faults can affect the system subtly, making them difficult to identify in the presence of other system effects and noise. The premise of data-driven diagnostics is that it is possible to observe features of faults of interest. The purpose of this study is to demonstrate the use of classical mechanical models and data driven analytical models on a robotic manipulator drive system operating cyclically under time-varying load.

There are many ways that a belt-driven system can change over its operating lifetime. Examples of common faults include: belts that become loose or are installed too tightly, rapid changes in temperature that may affect the materials, dust or debris build up in the track, and loose nuts along the body that introduce vibrations of components that aren't secured properly. Belts themselves are not instrumented in practice; but in this study, strain gauges were installed on the motor shaft to measure the torque output from the motor, and on a custom aluminum plate introduced into the belt itself to measure the tension directly. Using these measurements, it becomes possible to observe the effort of the motor under typical operating conditions and monitor how the systems changes when a fault is created in the system. This degree of observability allows the system to be characterized, and then the diagnostic performance of reduced observability can be assessed for different combinations of features are their relationships (by not using some measured data).

2. THEORY

This section presents the formulation of the governing equations for the laboratory-scale belt-driven system of interest, and introduces some candidate machine learning techniques for anomaly detection.

2.1 LUMPED-PARAMETER DYNAMIC MODEL FORMULATION FOR MOTOR AND BELT DRIVE

A custom planar belt-drive system with one actuator was designed and developed, with the motor, structure, and idlers illustrated in Figure 1. The mechanism moves in the horizontal plane, loaded with a mass that is supported on a flat plate by an unactuated wheel.

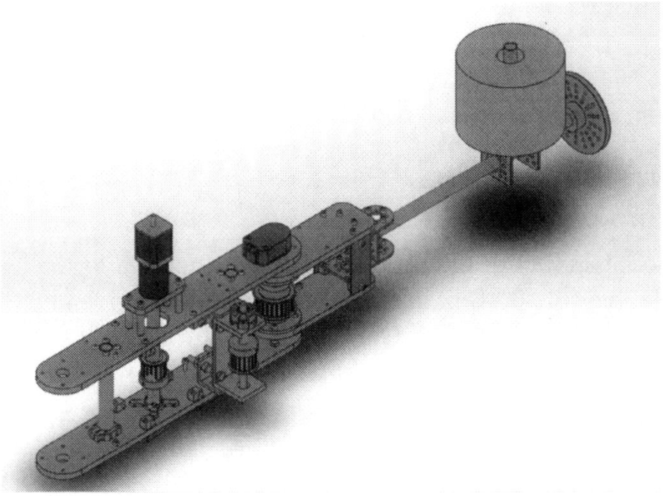

Figure 1: CAD Representation of Custom Belt-driven System (Belt Not Shown)

The robot manipulator belt drive contains three pulleys. There is a pulley attached to the motor to transmit power to the system, an adjustable idler pulley that is used to manipulate the tension in the belt and a pulley attached to a larger arm link that rolls on a steel surface. Nomenclature for variables and parameters for the governing equations are listed in Table 1.

Table 1: Nomenclature

r_m	Radius of the motor pulley
r_a	Radius of the arm pulley
r_i	Radius of the idler pulley
l_1	Center to center length on belt side 1
l_2	Center to center length on belt side 2
l_3	Center to center length on belt side 3
J_m	Polar moment of inertia of motor pulley
I_i	Mass moment of inertia of the idler pulley
I_a	Mass moment of inertia of the arm link
d_1	Effective damping coefficient of belt length 1
d_2	Effective damping coefficient of belt length 2
d_3	Effective damping coefficient of belt length 3
k_1	Effective stiffness coefficient of belt length 1
k_2	Effective stiffness coefficient of belt length 2
k_3	Effective stiffness coefficient of belt length 3
T_1	Tension in belt length 1

T_2	Tension in belt length 2
T_3	Tension in belt length 3
F_F	Coulomb friction force
τ_m	Input torque from the motor

It is assumed that there is stiffness and damping in the belt and each is a function of the belt length of each side. There are three degrees of freedom in this system, which are given by θ_m θ_i and θ_a. The governing equation for the motor pulley depends on tension forces from the belt, which are functions of their stiffening and damping properties across their respective lengths,

$$T_1 = k_1(\theta_i r_i - \theta_m r_m) + D_1(\dot{\theta}_i r_i - \dot{\theta}_m r_m) \tag{1}$$

$$T_3 = k_3(\theta_m r_m - \theta_a r_a) + D_3(\dot{\theta}_m r_m - \dot{\theta}_a r_a) \tag{2}$$

which yield the angular governing equation of motion:

$$\ddot{\theta}_m J_m = \dot{\theta}_i d_1 r_i r_m + \dot{\theta}_a d_3 r_a r_m - \dot{\theta}_m r_m^2(d_1 + d_3) + \theta_i k_1 r_i r_m + \theta_a k_3 r_a r_m - \theta_m r_m^2(k_1 + k_3) + \tau_m - F_{f_1} r_m \tag{3}$$

Similarly, the idler pulley, which is adjustable to tension the belt, has the following governing equation:

$$\ddot{\theta}_i I_i = \dot{\theta}_a d_2 r_a r_i + \dot{\theta}_m d_1 r_m r_i - \dot{\theta}_i r_i^2(d_2 + d_1) + \dot{\theta}_a d_2 r_a r_i + \theta_m k_1 r_m r_i - \theta_i r_i^2(k_2 + k_1) - F_{f_2} r_i \tag{4}$$

and the arm pulley, which is rigidly connected to the inertial load, is described by

$$\ddot{\theta}_a I_a = \dot{\theta}_m d_3 r_m r_a + \dot{\theta}_i d_2 r_i r_a - \dot{\theta}_a r_a^2(d_3 + d_2) + \dot{\theta}_m d_3 r_m r_a + \theta_i k_2 r_i r_a - \theta_a r_a^2(k_3 + k_2) - F_{f_3} r_a \tag{5}$$

The parameters for the analytical model are given in Table 2.

Table 2: Analytical Model Parameters

Parameter	Value	Units
r_m	0.0122	m
r_a	0.0122	m
r_i	0.0189	m
l_1	0.127	m
l_2	0.1524	m
l_3	0.1778	m
J_m	2.048E-9	m^4
I_i	2.445E-6	kg-m^4
I_a	0.5472	kg-m^4
d_1	0.9460	kg/s
d_2	1.135	kg/s
d_3	1.324	kg/s
k_1	3.376E+5	N/m
k_2	4.051E+5	N/m
k_3	4.726E+5	N/m

The inertia parameters for the motor pulley and idler pulley were calculated directly using the approximation equations for polar and mass inertia of hollow cylinders. The calculation of inertia for the arm link was done using a 3D model in SolidWorks as seen in Figure 2. The mass moment of inertia was calculated about the center of mass and translated to the center of the pulley using the Parallel-axis Theorem.

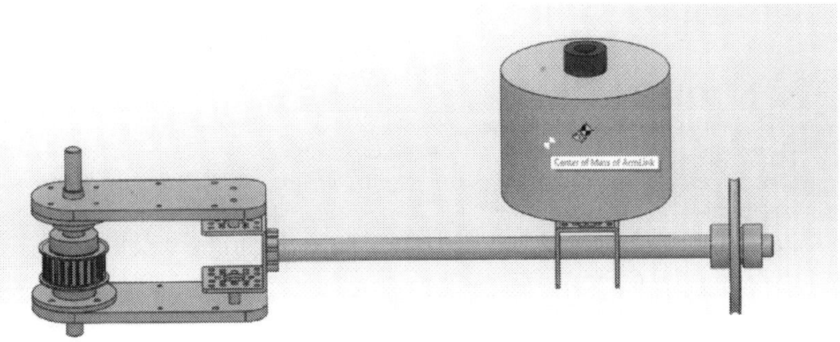

Figure 2: Arm and Driven Pulley with Support Idler

The effective inertias of the system are summarized in Table 3.

Table 3: Inertia Coefficients

Inertia Parameter	**Value**	**Units**
J_m	2.048E-9	m^4
I_i	2.445E-6	kg-m^4
I_a	0.5472	kg-m^4

The stiffness of each segment of the belt is a function of its length. A segment of belt of nominal length 0.163 m was loaded in an MTS machine to determine the stiffness per unit length at room temperature. Two trials were run, and the MTS machine generated load-displacement plots from which the average stiffness was calculated to be 43.3 kN/m. The stiffness was then scaled to determine the stiffness of the belts for each of the three lengths on the manipulator belt drive, as listed in Table 3.

Table 4: Stiffness Coefficients

	Length (m)	**Stiffness, k (kN/m)**
l_{test}	0.163	43.3
l_1	0.127	33.8
l_2	0.153	40.5
l_3	0.178	47.3

As with stiffness, the damping of the belt is a function of the length of each side. To determine the effects of damping in the belt, the 0.163 m specimen was half loaded in the MTS machine while the other side was attached to a small weight. The belt and weight were then lifted and released. The response of the belt for a typical trial is shown in Figure 3.

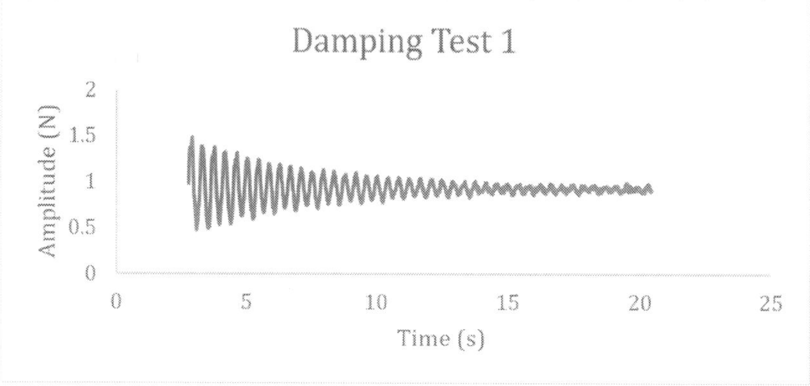

Figure 3: Decaying Exponential Response of Belt with Hanging Mass

By using the logarithmic decrement, the damping ratio ζ, can be calculated via:

$$\zeta = \frac{\delta}{2\pi} \quad (10)$$

The critical damping coefficient can be calculated via:

$$C_c = 2\left(\sqrt{k_{eff} m_{eff}}\right) \quad (11)$$

Finally, the damping coefficient, *d,* is calculated as

$$d = \zeta * C_c \quad (12)$$

As was done with the stiffness, the damping coefficient was scaled to the corresponding length of each side of the belt drive system. The results are given in in Table 5.

Table 5: Damping Coefficients

	Length (m)	Damping, *d* (kg/s)
l_{test}	0.163	1.121
l_1	0.127	0.946
l_2	0.152	1.135
l_3	0.178	1.324

The calibrated model can now be used to simulate dynamic behaviour from input of a non-conservative generalized force (motor torque) as a function of time. Differences between the simulation results and the response of an actual system are due to changes in the system dynamics (which may be faults), unmodeled dynamics, error in the model due to calibration, error in the simulation solution (solver type, step size, or implementation), or noise.

2.2 MACHINE LEARNING FOR PROGNOSTICS & HEALTH MANAGEMENT

Machine learning is a set of data analytical techniques that allow for the extraction of information from data that can help to accelerate the development of data-driven methods required in a prognostics and health management (PHM) system comprising of anomaly detection, diagnosis and prognosis methods. In general, a machine learning program can be used to "learn" from a measurable experience so that when such an experience happens again, it is recognized. In mechanical system diagnostics, a system that learns to recognize measurements of both regular operating conditions and anomalous operating conditions – and be able to tell the difference between the two.

Machine learning techniques can typically be categorized as supervised (data are labeled) and non-supervised techniques (data are un-labeled) [5]. Within these branches, there are many methods that can be used in detecting anomalies in a dataset. These include statistical methods, proximity-based methods, and deviation (or classification-based) methods.

Statistical methods assume that the data follows a specific distribution, and so the model can be created from a specified probability distribution [6]. By utilizing a probability distribution, anomalies can be detected by finding data that deviate dramatically from the distribution. Common statistical methods include Gaussian Mixture Models [7], various regression models, and Histogram-based Outlier Score models [8]. Some advantages and disadvantages are provided in Table 6.

Table 6: Statistical Methods

Advantages	Disadvantages
• If the underlying data distribution is correctly identified (assumed), these methods present a statistically proven solution for anomaly detection. • The resulting anomaly score is associated with a confidence interval. • Depending on the robustness of the distribution to anomalies within data, these methods can be used in an unsupervised fashion.	• Statistical methods assume that the data is from a specific distribution, which is normally not the case in a real world setting with high dimensional data. • Require a large sample of training data to estimate model parameters

Proximity, or cluster-based, methods for detecting data anomalies assume that similar data points can be "clustered" together. The more a data point has a certain characteristic, the closer it would appear to a parameter called a "cluster centroid". Therefore, anomalous data points can be determined by setting a threshold for its distance from a cluster centroid. These methods are typically categorized as unsupervised because they can be used to identify clusters of data points that are similar, without labels. Some commonly used cluster-based methods are k-means clustering [9] and self-organizing maps [10]. Some advantages and disadvantages of using proximity-based methods can be found in Table 7 below.

Table 7: Proximity Methods

Advantages	Disadvantages
• Can operate in an unsupervised fashion, without requiring any labeled data. • Capable of being used in incremental models, i.e., new data points. can be given and tested for anomaly. • Fast test time results due to testing against small number of identified clusters.	• Performance of these methods is very much dependent on the clustering algorithm and how successful they can identify the structure of normal data. • Some clustering methods force every data point to be part of a cluster. Hence, clustering methods that assume anomalies do not belong to clusters consider these as normal. • Some methods only work well if anomalies do not form smaller clusters amongst themselves.

Classification methods are used to classify test samples using a learned classifier by training on labeled datasets. The classifier learns by using the labeled training set which consists of labeled normal and anomalous operating conditions. It can then be tested on a second unlabeled dataset to determine the accuracy of the classifier in identifying normal and abnormal datapoints. Two commonly used classification methods are one-class support vector machines [11] and neural networks [12]. Some advantages and disadvantages of classification-based methods are given in Table 8.

Table 8

Advantages	Disadvantages
• Classification-based methods, particularly multi-class methods, can be used to distinguish between samples belonging to different normal classes. This can be useful with systems that have multiple normal operating conditions. • Test time computation is very fast due to the pre-trained model.	• Multi-class methods require accurate labeling of normal classes, which are often hard to acquire. • Some classification-based methods assign a class to samples. This may be a disadvantage as it does not provide a tangible anomaly score.

3. EXPERIMENT DESIGN

An experimental system was developed to produce a cyclical motion in a simple belt-driven system corresponding to the analytical model described above. The system was designed for reproducibility so that multiple experiments could be run and have similar results. The apparatus is shown in the photograph in Figure 4.

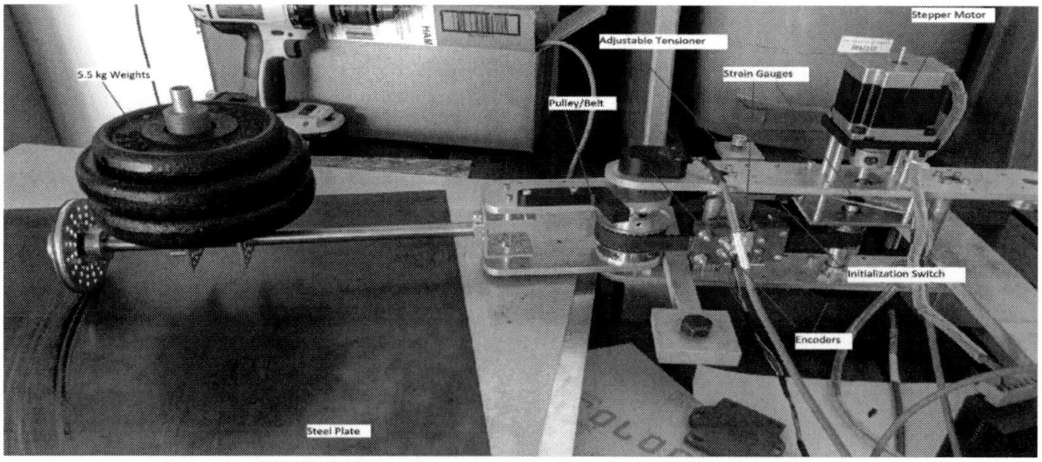

Figure 4: Side View of Arm Driven by Stepper through Belt Drive

Figure 5 shows the belt tension measurement ligament with strain gauges. A prismatic rectangular piece of aluminum is connected to the belt. A full-bridge strain gauge circuit measures the instantaneous tension in the belt, which is recorded by a microcontroller.

Figure 5: Belt Tension Measurement Sensor and Joint Sensor at Arm Base

Figure 6 shows a top view of the apparatus, with a second joint sensor that measures the arm motion (in addition to the sensor collocated on the motor shaft). Both are US Digital E2 optical encoders. A full-bridge strain gauge sensor measures torque on the motor shaft. A steel plate supports the 5.5 kg mass loading the aluminum idler at the end of the arm. Restraining clamps hold the base of the mechanism on a 3 tonne seismic platform. The drive train is fitted with an adjustable tensioning device mounted with an idler pulley, which allows the tension in the belt to be modified in different trials. The tensioner has four bolts that fasten it to the main body of the arm and the heights of the bolts can be changed to adjust the height of the idler pulley, increasing or decreasing the tension in the belt.

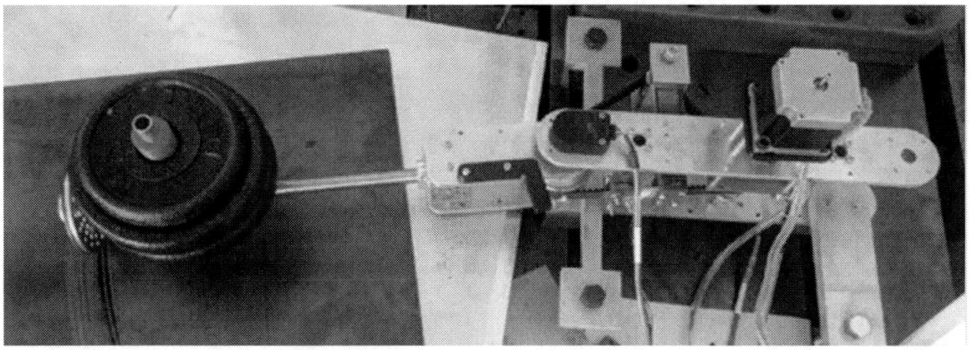

Figure 6: Top View of Apparatus

The system is controlled by a programmed Teensy 3.2 Microcontroller to actuate the motor and read the measurements from the instrumentation at a sampling rate of 100 Hz. The mechanism is actuated by a brushless stepper motor to drive the belt through a programmed range of motion (typically between 19 and 126 degrees). The motor is controlled by a Leadshine DM542 Microstep Drive. A switch on the arm is activated during the arm's first cycle. The switch tells the microcontroller that it has reached the starting point, after which the motor begins a looping script to move the arm in an oscillatory path (5 seconds counterclockwise, stop for 0.1 s, 5 s clockwise, 0.1 stop, and repeat for a preset number of cycles). The robot arm transmits power from the motor to the arm link via a timing belt and pulley system. The pulley on the shaft connected to the motor and the pulley on the adjustable tensioner are both 5GT 16-tooth pulleys; and the pulley on the joint is a 5GT 24-tooth pulley.

4. EXPERIMENTAL RESULTS

A set of cycles were run to do baseline trials of normal operating condition at a belt tension of the belt of 150 N. The torque profile over multiple cycles is shown in the plot in Figure 7.

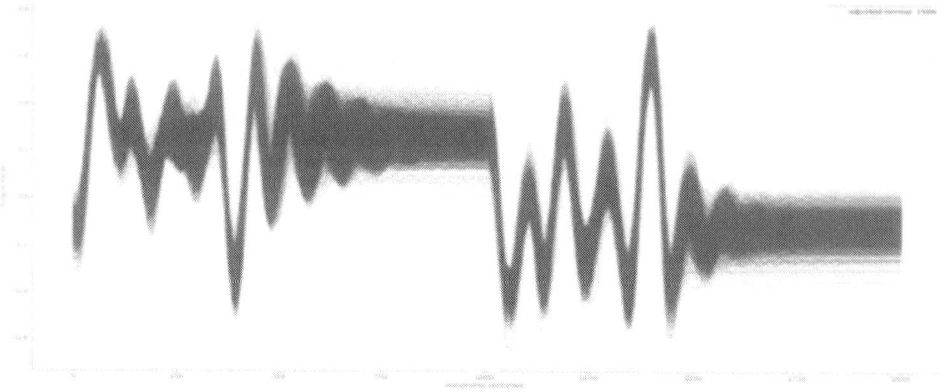

Figure 7: Normal Operation Torque Profile at 150 N Belt Tension

After a baseline torque profile was established, a variety of fault modes were individually tested on the system:

- **Loose Belt:** Using the tensioner mechanism and the instrumentation on the belt, the belt was able to be precisely loosened. The loose belt fault mode was tested at 100 N and 120 N.
- **Tight Belt:** Similarly, the belt was able to be tightened above the normal operating tightness. For this fault mode, the tension was set to 168 N.
- **Increased Friction/Sand Buildup:** For this fault mode, sand was scattered along the steel plate that the pulley was running along.
- **Extreme Temperature Change:** For this fault mode, the ambient temperature around the manipulator arm was measured by a thermocouple. The temperature was increased to approximately 40 degrees Celsius from room temperature using a heat lamp. It was observed that the belt contracted, resulting in a tension increase from 150 N to 168 N.

The experimental conditions for these fault modes are summarized in Table 9.

Table 9: Fault Modes and Number of Cycles per Trial

	TEMP	TENSION	NO. SAMPLES
Normal	~23.5	~150	7193 (22hrs)
Loose l1	~23.5	~120	182
Loose l2	~23.5	~99-100	180
Tight	~23.5	~168	194
High Temperature	~40-42	~166-169	210
Sand (friction)	~23.5	~150	183

Figure 8 on the following page shows the effects of these fault modes on the torque profile compared to the normal operating condition profile. It can be observed that while these are fault modes of the system, their impact affects the torque profile very subtly, meaning that they would be difficult to identify in a real-world system.

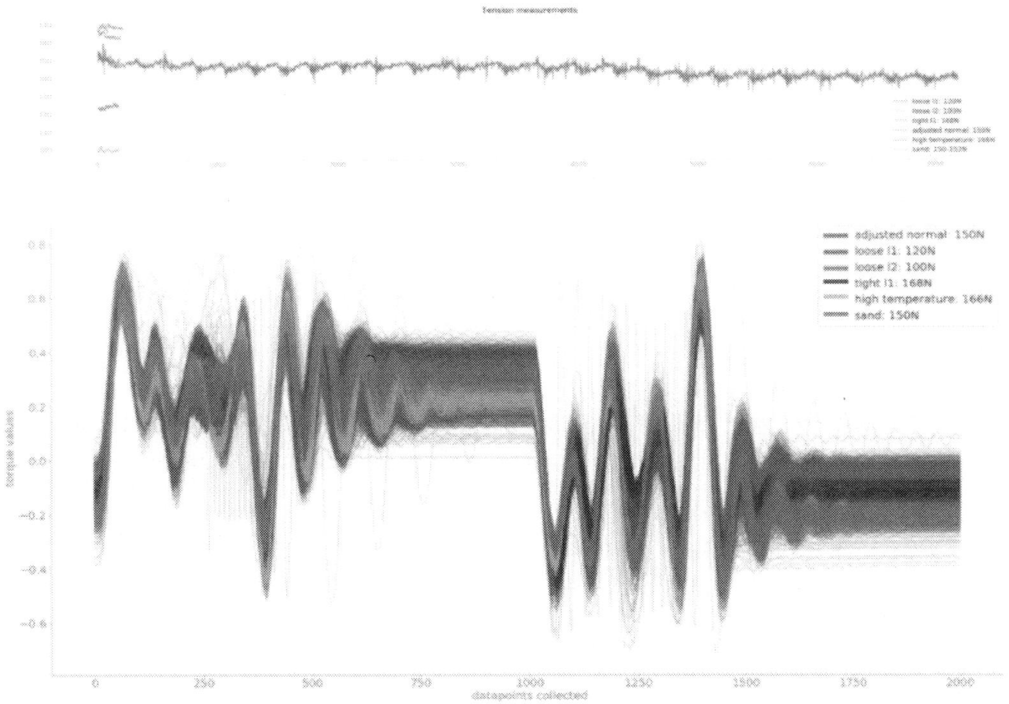

FUTURE WORK

Feature ranking and comparative assessment of a range of machine learning techniques are currently being analyzed. The anomaly detection and fault identification performance using machine learning methods will be compared with the goodness of fits for a set of physics-based models (normal, loose belt, tight belt, high friction).

ACKNOWLEDGMENTS

Funding support is gratefully acknowledged from the Natural Sciences and Engineering Research Council of Canada and the Alberta Machine Intelligence Institute.

REFERENCES

[1] Isermann, R. (2005). Model-based fault-detection and diagnosis – status and applications. *Annual Reviews in Control, 29*(1), 71-85. doi:10.1016/j.arcontrol.2004.12.002

[2] W. Bartelmus, F. Chaari, R.Zimroz and M. Maddar, Modelling of gearbox dynamics under time-varying nonstationary load for distributed fault detection and diagnosis, *"European Journal of Mechanics-A/Solids, vol. 29, no.4, pp.637—646,2010.*

[3] Potes Ruiz, Paula, Bernard Kamsu Foguem, and Bernard Grabot. "Generating Knowledge in Maintenance from Experience Feedback." *Knowledge-Based Systems*, vol. 68, 2014, pp. 4-20.

CrossRef, https://www.sciencedirect.com/science/article/pii/S0950705114000422, doi:10.1016/j.knosys.2014.02.002.

[4] Ting-Hua Yi, Hong-Nan Li, and Ming Gu. "Optimal Sensor Placement for Health Monitoring of High-Rise Structure Based on Genetic Algorithm." *Mathematical Problems in Engineering*, vol. 2011, 2011, pp. 1-12.

[5] Sathya, R., and A. Abraham. "Comparison of Supervised and Unsupervised Learning Algorithms for Pattern Classification." *International Journal of Advanced Research in Artificial Intelligence*, vol. 2, no. 2, 2013, pp. 34–38., doi:10.14569/issn.2165-4069.

[6] Clifton, David A., Peter R. Bannister, and Lionel Tarassenko. "A Framework for Novelty Detection in Jet Engine Vibration Data." *Key Engineering Materials*, vol. 347, 2007, pp. 305-310. CrossRef, doi:10.4028/www.scientific.net/KEM.347.305.

[7] G. J. McLachlan and K. E. Basford, *Mixture models: Inference and applications to clustering*. Marcel Dekker, 1988, vol. 84

[8] M. Goldstein and A. Dengel, \Histogram-based outlier score (hbos): A fast unsupervised anomaly detection algorithm," *KI-2012: Poster and Demo Track*, pp. 59-63, 2012.

[9] J. MacQueen et al., \Some methods for classification and analysis of multivariate observations," in *Proceedings of the fifth Berkeley symposium on mathematical statistics and probability*, Oakland, CA, USA, vol. 1, 1967, pp. 281-297.

[10] T. Kohonen, \The self-organizing map," *Proceedings of the IEEE*, vol. 78, no. 9, pp. 1464-1480, 1990.

[11] J. Ma and S. Perkins, \Time-series novelty detection using one-class support vector machines," in *Neural Networks, 2003. Proceedings of the International Joint Conference on*, IEEE, vol. 3, 2003, pp. 1741-1745.

[12] S. Hawkins, H. He, G. Williams, and R. Baxter, \Outlier detection using replicator neural networks," in *International Conference on Data Warehousing and Knowledge Discovery*, Springer, 2002, pp. 170-180.

AERIAL MANIPULATION FOR REMOTE ROBOTIC MACHINERY DIAGNOSTICS

MG Lipsett, Rijesh Augustine and Mark Sherstan
Department of Mechanical Engineering
University of Alberta
Edmonton, Alberta T6G 2G8
Canada
+1-780-492-9494
mlipsett@ualberta.ca

Abstract: There is a growing trend of using unmanned aerial vehicles (UAVs, or drones) for inspection of industrial facilities and infrastructure. The types of inspections currently done entail remote sensing using cameras: qualitative assessment and photogrammetry using standard cameras, as well as temperature monitoring using thermal imaging cameras. The benefits of using drones are to reduce risk to inspection personnel near operating equipment, reduced cost, and improved auditability of archived data. Machine condition monitoring can benefit from this enabling technology provided that vibration monitoring and lubricant analysis can also be done remotely. Laser vibrometers can be used for remote monitoring of equipment that does not have permanently mounted, dedicated vibration monitoring sensors; but this sensing method is capital intensive, requires a laser head and controller with a typical mass of over 5 kg, and must be pointed from an appropriate direction, which may not be accessible. Lubricant analysis requires access to a lubricant sampling port and a means for collecting and analyzing the sample. A system is described for deploying a drone with a manipulator that engages the equipment of interest for vibration data collection and employs a more versatile payload for drop-tube vacuum sampling of lubricant by inserting a tube through a fill port or dip stick port and withdrawing an oil sample from the sump cavity. The key technical challenges are collision-free navigation and hovering, followed by control of dynamic deployment of the payload that connects to the machinery. A proof-of-concept system is described with very preliminary experimental results.

Key words: Aerial manipulation; automated sampling; contact inspection; inspection; lubricant sampling; remote diagnostics; robotics; unmanned aerial vehicle

1. INTRODUCTION

The main diagnostic approaches for detecting impending machinery failure are vibration spectral features, elevated temperature trends, and anomalous components of lubricant analysis [1]. Vibration and temperature measurements are readily collected by dedicated condition monitoring systems; but for lubricant sample collection and for locations where equipment is not connected to a centralized monitoring system, manual inspection remains the norm. Worker safety is a significant issue. In 2017, approximately 50 000 Canadian workers were injured due to a fall, with 27% of the fall incidents occurring at heights [3]. Remote technologies and automation can alleviate other risks to workers, reduce inspection time, and mitigate the opportunity cost of lost production by eliminating the need to stop, safe-out process equipment, and establish safe access to inspection locations.

Access to equipment can be problematic. Wheeled and tracked robots have been used for some types of inspections, but only for extremely hazardous environments such as nuclear plant inspections and security work. Low mass thermal cameras are now easily deployed by drones, and are useful for diagnostics when excess heat generation produces persistent changes in surface temperature [4]. Prior work by our research has considered the system requirements for unmanned aerial vehicles (UAVs) to be used to collect vibration data by deploying a vibration sensor to contact a target location [5]; but the issues of target location and stable deployment of a sensor remained to be demonstrated.

Having the ability to make remote condition assessments by autonomously taking a lubrication sample has the potential benefit of quickly identifying minor lubrication-related problems, which improves the chance that the issue can be addressed before functional failure occurs [2]. Being able to set an autonomous schedule to retrieve oil samples as well as vibration and temperature would establish a baseline of performance that can lead to increased reliability of machinery. If a multirotor UAV is able to retrieve samples from heights so that a worker does not need to put themselves at risk, then the UAV is an excellent source of risk mitigation as well as cost reduction, freeing inspectors for more value-added activities.

The aerial manipulation task for lubricant sampling is much more challenging than simply pressing a sensor against a target location. Taking an autonomous lubrication sample is a repetitive task consisting of flying to a predefined setpoint, removing a cap, extracting a sample, replacing the cap, and flying back to a home location with the sample for lab analysis. Although the task is repetitive at a high level, there are many external disturbances and variations that can occur between trials. These include: environmental conditions, varying cap sizes, and the accessibility to a region of interest without collisions. Creating a robust package of hardware and software can reduce disturbances, the time to complete a task, and increase the reliability of the system, in turn reducing the downtime of the UAV.

Although there are many different ways to extract lubricant samples from machinery, a sump cavity secured by a cap was selected as the primary scenario opposed to pre-installed sampling ports. This scenario is of more general interest, as providing a torque with a UAV is a challenging control problem that is applicable to other use cases, such as opening doors to access locations, twisting process control knobs and actuating the handwheel of a process valve.

One of the most significant challenges is to compensate for reaction dynamics. When a UAV has coupled itself with the surrounding environment, the forces the UAV experiences interfere with the aerodynamics (and associated control scheme) of free flight, creating the risk of the UAV going unstable and crashing or failing its assigned task. "An analogy is to imagine a scuba diver turning an underwater valve. The diver contorts his body and aggressively kicks his legs in order to throw his body weight into turning the valve with his arms. Likewise, a UAV must 'swim' through air and may have to vector its body in order for the arms to dexterously manipulate an object with its arms" [6].

This paper will review relevant literature of aerial manipulation for autonomous diagnostics, describe potential components and systems to be used, provide an update on progress for a demonstration system, provide provisional conclusions, and an outline of future work.

2. Literature Review

Having the ability to autonomously sample and gather data is being increasingly utilized by engineers and scientists due to the efficiency and precision of the advancement of robotics [7]. Many UAVs are used for visual inspections for hard to reach or dangerous areas such as under bridges [8], various railway infrastructure [9], and high voltage power lines [10]. These use cases take advantage of both manual and autonomous flights, however they do not physical interact with the area of interest beyond imaging. Visual inspections, surveillance, monitoring and other passive tasks are now just a subset of what UAVs are capable of; and aerial robots have evolved to perform active tasks such as manipulation and physical interaction [11]. Examples include Amazon drones demonstrating the delivery of customer packages in 30 minutes or less using a hexacopter [12], two UAVs or a UAV and a ground manipulator working together to overcome payload restrictions and navigate physical barriers [13], [14], and multimodal aerial locomotion for active tool handling such as contact pipe inspections [15]. The AEROARMS Project aims to develop aerial robots with advanced manipulation capabilities for various inspections and maintenance [14]; and the reported system is highly pertinent to the challenge of interacting with machinery for diagnostics. Having various hardware and control packages such as dual arm manipulators and contact inspection platforms [16], the basic necessary components of extracting lubricant samples can be related.

One of the largest challenges with aerial manipulation is the control schemes of the UAV due to interactions with its environment [11]. UAVs interacting with the environment can be simplified into three main categories [18]: 1) Momentary coupling, interacting with objects of finite mass in the air that are not fixed to the surrounding environment; 2) Loose coupling, interacting with objects attached to the environment without securing to the environment, in tasks such as assembling, inserting, pushing, or pulling; and 3)

Strong coupling, securing or perching onto fixed objects in the environment and becoming part of it. To mitigate an unstable control system and increase the robustness due to assumptions during the modelling process, various control schemes and techniques have been proven successful such as adaptive controllers [19], adaptive sliding mode controllers [20], auto disturbance rejection (ADRC) controllers [21], gain scheduling [15] and classical proportional-integral-derivative (PID) controllers [18].

For the proposed task of autonomously extracting a lubrication sample while executing all three dynamic coupling types, there has been no detailed study found in the literature. A gas powered, ducted fan UAV design for condition monitoring of wind turbines was presented at the 32nd International Maintenance Conference in 2017 with limited details [22]. There is potential overlap within the field of aerial manipulation due to the nature of manipulators and end effectors and their multifunctional uses [16], [18]; however, the current sampling and inspection use case presented is still unique.

3. GENERAL SYSTEM DESCRIPTION

The system comprises a multirotor drone platform, a payload that interacts with fixed equipment (*i.e.,* gearbox) with an integrated sampler, and a simple manipulator to extend the payload away from the drone to engage with the equipment. All systems are electrically actuated.

Hardware

A 500 mm wheel-based MWC X-Mode Alien quadcopter frame provides a foundation for all necessary components. Four Emax XA2212 980KV brushless motors with 10 inch x 4.5 inch propellers provide a maximum of 8.6 N of thrust per motor. The motors are powered by a 3 cell LiPo battery and are controlled with Racerstar RS20A V2 20A electronic speed controllers (ESCs). One power distribution board (PDB) provides access for all electrical components and regulates voltage to the Pixhawk flight controller, Odroid C2 computer, Arduino microcontroller, and other actuation or sensing devices.

Flight Controller

To handle the control of the quadcopter a Pixhawk flight controller is utilized and comprises an inertial measurement unit (IMU), barometer, and magnetometer. The Pixhawk has a variety of interfacing ports which allows for additional components such as a global positioning system (GPS), secondary magnetometer, remote control (RC) receiver, SiK telemetry radio, and the ESC for each of the four motors. The Pixhawk runs PX4 software which is open source and allows for easy customization and deployment of new control schemes. An Odroid C2 companion computer communicates with the Pixhawk through a serial connection and handles computationally intensive tasks such as image processing. An Arduino Uno communicates with the Odroid C2 also using a serial connection for task specific applications such as analog sensing.

Coordinates and Assumptions

The axes convention to be used for describing the location and orientation of the quadcopter is North, East, Down (NED) where the right-hand rule is followed and roll φ occurs about the North axis, pitch about the East axis θ, and yaw ψ about the down axis. The convention can be observed in Figure 1.

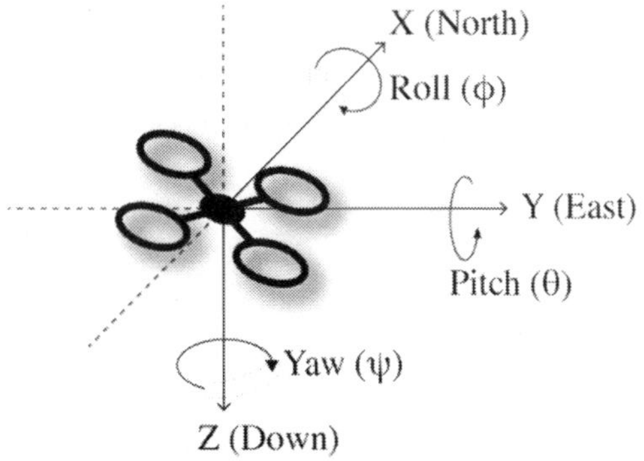

Figure 1: North East Down (NED) Reference Coordinate System

To limit the complexity of the initial modelling the following assumptions are made: 1) The quadcopter is a rigid symmetric body. 2) Thrust and drag forces produced from the propellers do not interfere with each other. 3) The ground effect is neglected.

4. Positioning the Quadcopter

The quadcopter needs to be in a precise and stable position to be able to interface with a lubricant sampling port. Depending on environmental and situational conditions, this may be difficult for a remote human operator to achieve. If the quadcopter is outside of visual line of sight or the sampling port is too far away from the human operator, then a camera system or other relative location measurement system will be required to relay information back to the operator. Factors that affect how well a human may be able position a quadcopter include the individual's depth perception, the individual's reaction time, camera latency, and control system latency. A GPS system may be an alternative to a human operator, but even then a traditional GPS unit will provide only sub-meter accuracy in the best-case scenario. A more expensive Real-time kinematic (RTK) GPS can yield sub-centimeter accuracy, but this approach requires ground station infrastructure and absolute surveying of equipment locations. The ideal scenario for positioning the UAV requires as little external infrastructure as possible. This is where a combination of GPS and visual servoing can lead to an inexpensive and reliable solution. In this scenario, the UAV moves into a collision-free position near the sampling port using GPS, and then utilizes visual servoing of a marker to tightly control the position of the UAV and any payload deployment.

The use of a marker is suggested where it can be placed anywhere in the environment, such that the UAV has an unobstructed view of it throughout the oil sampling procedure. A global shutter camera, attached to the Odroid C2 companion computer, continuously acquires images. Each image is processed by the computer in a multistep process, utilizing the ArUco library [23], [24]. The ArUco library is responsible for detecting, verifying and calculating the pose of the markers. There are many marker libraries available; but ArUco was selected because it was open source and already integrated with OpenCV. The actual steps in marker detection involve thresholding the image, finding contours in the image, analyzing the contours to see which one contain a marker ID, and then finally calculating the pose of the marker relative to the camera. Figure 2, shows an ArUco marker that is coded with a 6x6 array of bits and has a marker ID of 19.

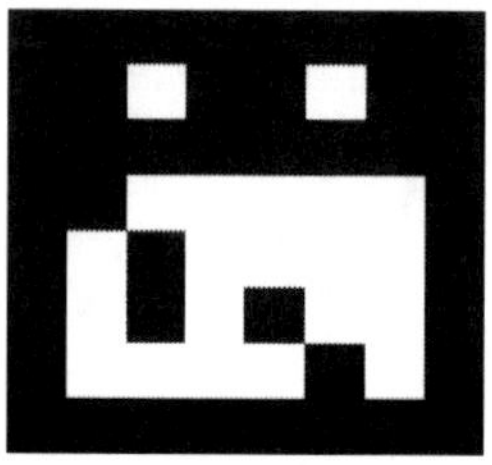

Figure 2: ArUco Marker ID 19

The accuracy of the marker pose is dependent on the number of pixels that the marker takes up. The greater the number of pixels that the marker takes up, the greater the positional accuracy. Increasing accuracy is done by either increasing the camera resolution or by increasing the size of the marker. Increasing camera resolution comes at the cost of increased processing times, whilst increasing the size of the marker makes it more difficult to install the marker and keep it in frame throughout flight manoeuvers. A trade-off is made here depending on the specific requirements of a sampling site.

After the pose of the marker is calculated, this pose is then translated from the marker coordinate frame to the quadcopter frame. The computer is preprogrammed with information of the relative location of the sampling port with respect to the marker frame of reference. This information along with the pose of the UAV is enough for the computer to calculate a trajectory. The trajectory is then sent to the PX4-based flight controller via the serial link. The quadcopter is then able to position itself to carry out the required task.

5. Lubricant Cap Removal and Replacement

The first complication of extracting a lubricant sample is the removal of the cap sealing the lubricant in its reservoir. Initial assumptions assume that the cap diameter ranges between 25 mm and 100 mm and uses a right hand thread. The largest force associated with removing the cap is overcoming the static friction posed by the threads. Once the static friction has been overcome, kinetic friction is experienced only on the threads and the torque requirement is much reduced. During the opposite process of reapplying the cap to the reservoir, the forces begin with kinetic friction and an inertial element reducing the required torque, and so cap removal is the constraining scenario requiring the most force. The greatest force application challenge is thus generating sufficient torque with adequate control. It is assumed that the UAV is able to securely attach itself to the cap with a gripper or customized spanner on a manipulator end effector.

Quadcopter Achieving Torque with Yaw

A potential approach for applying torque to the cap is to use the yaw produced from the quadcopter as the torquing force. Quadcopters achieve lift by spinning the propellers to generate thrust, with the direction of rotation of a propeller always opposite to its neighbour to produce opposing torques as displayed in Figure 3. To achieve yaw motion, two non-adjacent propellers must spin faster than the other pair of non-adjacent propeller pairs, creating a net aerodynamic torque that yields rotation of the quadcopter.

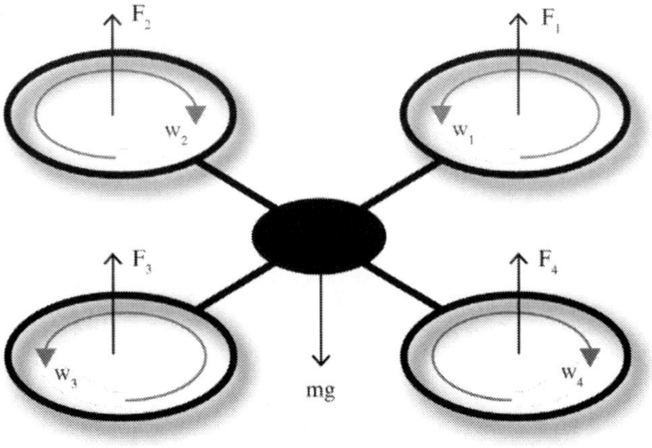

Figure 3: Free Body Diagram of Quadcopter While Hovering

In Figure 3 F_i is the thrust force produced by motor i, w_i is the angular velocity and direction of motor i, m is the mass of the quadcopter, and g is gravity. Assuming that the quadcopter is securely coupled with the cap, the following conditions must be met to generate a torque while also maintaining flight:

$$+\uparrow \sum_i^4 F_i \geq mg \quad (1)$$
$$F_1 = F_3 \quad (2)$$
$$F_2 = F_4 \quad (3)$$
$$F_1 + F_3 \geq F_2 + F_4 \quad (4)$$
$$F_1 + F_3 \leq F_2 + F_4 \quad (5)$$

To maintain level flight and controlled yaw, the UAV must produce sufficient thrust Equation (1) and the opposite pair of rotors must be producing the same amount of thrust Equation (2) and Equation (3). Referencing Figure 3, yawing clockwise tightens the cap and is achieved by the conditions in Equation (4), while unscrewing a cap is achieved by rotating counter clockwise as constrained by Equation (5). Based on the quadcopters thrust to lift ratio the torque limitation can be observed as the UAV must keep itself airborne and can only utilize half of its motors and a percentage of its thrust to achieve yaw. This was experimentally viewed in [25] where a quadcopter UAV was tasked with a strong coupling task of opening a valve. Although the valve had reduced friction and was for demonstration purposes, the UAV struggled to produce the necessary torque. It should be noted that a classical controller using a PID loop was used [23], which is promising for ease of implementation and providing a foundation to which more complex control schemes might be used with improved results.

Quadcopter Achieving Torque through Rotation of its Thrust Vector

With the limited torque available by utilizing only the yaw of the quadcopter, a rotated thrust vector is a more likely way to deliver sufficient torque to open the cap. There are two variations to produce this thrust vector.

The first alternative uses a moment arm from an extended arm. Once the quadcopter is secured to the cap with an elongated end effector, the UAV rolls about the connection point at its center of mass, transitioning the thrust vector from the upwards direction into vertical and horizontal components with an angle φ. Figure 4 displays the forces due to the roll angle φ.

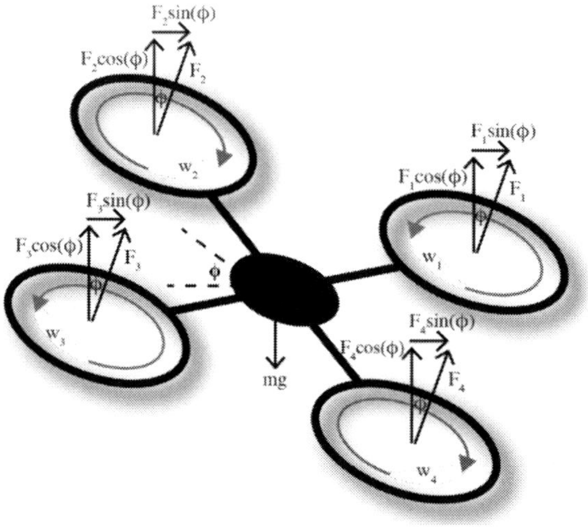

Figure 4: Free Body Diagram of Quadcopter Executing Clockwise Rolling Motion

Referring to Figure 4 and using the same variable definition as defined previously, the following conditions are required to keep a constant tangential force and the UAV airborne:

$$+\uparrow \sum_i^4 F_i \cos(\phi) \geq mg \quad (6)$$
$$F_1 = F_4 \quad (7)$$
$$F_2 = F_3 \quad (8)$$
$$F_1 + F_4 \leq F_2 + F_3 \quad (9)$$
$$F_1 + F_4 \geq F_2 + F_3 \quad (10)$$

Summing the horizontal components of the thrust vector yields a force *P* as defined in Equation (11). As the quadcopter is assumed to be symmetric, varying the distance *r* from the center of the quadcopter to the cap will increase the torquing force as observed in Equation (12) where clockwise rotation would occur with conditions from Equation (9) and counter clockwise rotation would occur with conditions from Equation (10).

$$P \overset{+}{=\!\rightarrow} \sum_i^4 F_i \sin(\phi) \quad (11)$$
$$\tau = r \times P \quad (12)$$

Maximizing φ produces the largest amount of horizontal thrust as long as the UAV produces enough vertical lift to remain airborne. Alternatively the moment arm *r* can be extended to produce a greater torque as outlined in Equation (12) and in turn the reduce the roll angle whilst providing the same amount of torque. In reality, the moment arm must be kept at a reasonable length for maneuverability of the UAV and keeping a balanced center of mass. One major aspect of this approach is the rolling motion for a quadcopter is a common and simple task to execute and control. An example relating to the proposed method can be found in [26] where a three degree-of-freedom helicopter rotating about a center point is developed for testing various control schemes. The first approach only works in a region that allows collision-free motion of the drone.

The second method for rotating the thrust vector to generate a torque for cap removal is to rotate all the rotors with respect to the frame of the UAV. Such a design turns an under-actuated quadcopter into an over-actuated machine. To model such a system six coordinate frames must be used as depicted in Figure 5. Frames one through four correspond to the respective motors and their orientation, frame B corresponds to the body frame of the UAV about its center of mass, and frame G corresponds to a ground reference frame.

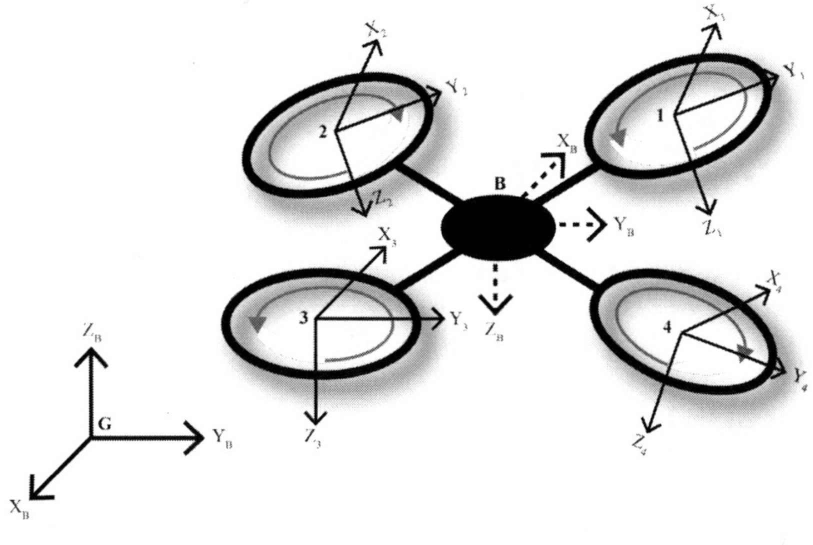

Figure 5: Rotor, Body & Ground Coordinate References for Rotating Rotor Quadcopter

The modelling of the dynamics is a non-trivial problem consisting of rotational matrices based on many different frames and coordinates. Such derivations have been achieved in both [27] and [28] where physical and numerical experiments were respectively conducted with success. Similar to the first proposed concept of thrust vector rotation, the ability to vary the rotor angle allows for the harnessing of multiple rotors thrust as long as enough lift is generated to stay airborne. Having the ability for each rotor to move independent of the others allows for an extremely maneuverable vehicle with the capability to produce high amounts of torque in a relatively small footprint. With this proposed method a UAV can achieve torque from above the cap without the need of a torquing arm. The benefits come at a cost of an increased complexity of mechanical design (and controller design); however, the methodology and ideology has been proven successful in other applications [29].

Quadcopter Achieving Torque by Coupling Itself to the Environment

To reduce the complexity of producing a torquing force with a UAV, the UAV can be coupled to the surrounding environment. Affixing the quadcopter can be achieved in a variety of different ways, such as adhering itself to the ground (*i.e.,* a steel gearbox structure) with electromagnets. Alternatively the UAV can stay airborne by securing itself to the throat of the lubricant cap with a gripper and having a secondary device provide torque to the cap from the new fixed reference. Both proposed methods and any such derivatives greatly simplify the torque limitation issue described above, as well as the problem of collision-free trajectories near equipment. The simplification comes as the UAV is stationary after approaching and landing on the structure and a secondary mechanism delivers the torque to remove the cap, as opposed to the UAV directly being part of the torquing mechanism. Such a mechanism makes many assumptions of the surrounding environment and potentially limits the autonomous capabilities of focusing on a single cap as a target; but this strategy likely increases the robustness and repeatability of the actuation task.

6. Lubricant Sampling

Once the cap has been successfully unthreaded it must be positioned so that the lubricant reservoir is accessible. Using the assumption that the cap is gripped sufficiently tightly, once freed the cap can be moved away from the access port by the end effector or by the UAV performing a slight hover to create a gap in the Z direction. With the reservoir exposed, the task becomes a pin-in-hole or loose-coupling scenario in which a sampling tube is inserted and lubricant is extracted. Based on industry standards, approximately 80 mL of lubricant is sufficient to perform the analyses necessary for machinery diagnostics [30-32]. To avoid cross contamination between samples, only one sample is taken per flight so that the sampling unit can be cleaned or swapped before the next mission (although a magazine could be employed on a larger UAV to collect multiple samples). To get a lubricant that truly reflects the state of the machinery of interest, the sample should be taken from the middle region of the sump to avoid the sludge line [33] and particle fallout [34]. Four possible designs are proposed for the drop tube sample extraction: a peristaltic pump with tubing, a vacuum pump with tubing, an evacuated bottle and tube, and a container sample lowered directly into the reservoir.

Peristaltic Pump

A peristaltic pump using positive displacement is suggested to limit contamination. Only the tubing requires cleaning or replacement between sampling missions opposed to the entire pump assembly. The cyclic motion of the peristaltic pump allows for an approximation of a fixed amount of fluid for each rotation reducing the need for fluid level sensors or other monitoring techniques. Additionally the pump can handle a range of viscosities and prevents backflow. The largest drawback from using a peristaltic pump is the pump is constantly under load increasing power consumption. Due to the UAV being a battery-operated system with a finite amount of power operating the pump for prolonged periods of time or adding a secondary power source greatly reduces the total possible mission or flight time.

Vampire Pump

A common industry solution for drop tube sampling is termed "vampire pump" [31] which is based on a piston suction pump design. A piston is actuated in a reciprocating linear motion creating a partial vacuum, which draws fluid into a sampling container. If done properly the risk of contamination is minimal, however should too much fluid be sampled the lubricant can flow into the plunger cavity creating the risk of cross contamination and requirement for cleaning. Due to the linear actuation of the pump, quantifying the fluid volumes can be easily approximated and a mechanical advantage in the form of a rack and pinion can reduce the power consumption of the system. Due to the linear nature of the pump, a large relative foot print with respective to the UAV may be required. Optimizing spatial constraints, center of mass, and total weight may pose a problem with such a design.

Evacuated Bottle

Using an evacuated bottle would eliminate the requirement for any pumping or significant actuation. Commercial products are available such as the Simplo sample bottle [35] where once a sample tube is submerged a valve on the lid is rotated (*i.e.,* by a servo) and the pressure difference draws the lubricant sample into the bottle. Although an evacuated bottle can only be used as long as its internal pressure remains below atmospheric it can handle a range of high and low viscosities. The primary difference in sampling time depends on the viscosity of the liquid with high viscosities being fast and low viscosities slow. Due to the vacuum and the sample retrieval time being a function of some unknown viscosity a fluid level indicator will have to sense when the valve should be closed again once the required quantity of fluid is acquired. With the small footprint of an evacuated bottle this is of little concern.

Container Sampling

A cup sampler method is proposed for extracting the required lubrication sample by lowering a cup or bottle into the lubricant sump. The lid separates from the sampling container to allow the lubricant to flow into the container. Once a predetermined time threshold is met the lid is replaced and the container is retrieved from the sump. The proposed method is extremely simple and robust, however comes with challenges. The outer diameter of the sampling container must not be larger than the inner diameter of the reservoir entrance. Lowering the container into the optimal sample region provides a challenge and may require rigid actuation. Upon removal of the sample, the entire container will be covered with the lubricant and must be cleaned or contained before flight to reduce spilling and contamination of the surrounding environment.

7. Vibration Monitoring and Trajectory

The hardware required for taking a vibration measurement is fairly straightforward when compared to oil sampling. An arm has been developed with an electromagnet and a vibration sensor at the end. Dampening foam is used to isolate the vibration gathering equipment from vibrations generated by the quadcopter. To ensure the center of gravity of the UAV is maintained, a counterweight is used at the other end of the arm. The vibration sensor at the end of the arm uses a ±2 g MEMS accelerometer. A Fourier transform is applied to raw accelerometer measurements to represent the vibrations in the frequency domain. Once in the frequency domain, the relative and absolute strength of vibrations across the frequency spectrum can be analyzed to detect faults.

The trajectory generation is done by scheduling setpoints. Before the UAV reaches a setpoint, a new setpoint is sent. When this scheduling is done at a high enough frequency, linear 2D motion is no longer perceived as a series of line segments, but as a smooth curve. Figure 6 shows the simulated position of the quadcopter vs the position of the setpoint. Arrows indicate the path the UAV would take if a new setpoint is not sent. This method in conjunction with the PID position controller on the autopilot, allows for a smooth trajectory that ends in a gentle touch.

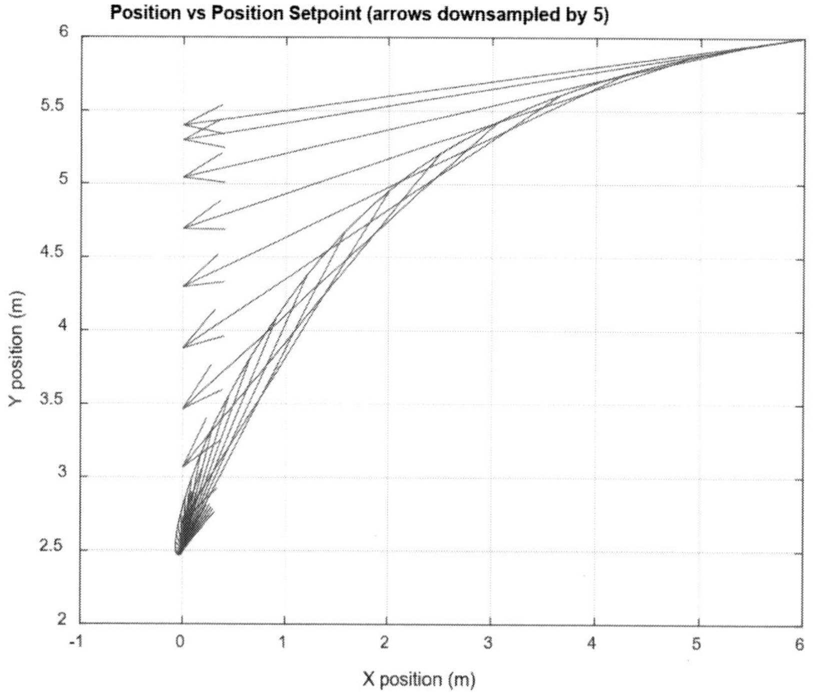

Figure 6: Position vs Position Setpoint (arrows down sampled by 5)

8. Prototype Development to Date

Quadcopter

The quadcopter previously described is not a complete kit; however, all essential components are commercially available off-the-shelf for reliability and ease of characterization. The MWC X-Mode Alien quadcopter frame does not satisfy the symmetric rigid body assumption proposed, but it does provide an extensive structural frame that can utilize large propellers for large amounts of thrust. There is ample room to organize sensors, manipulators, controllers, and batteries. The UAV with its essential elements and companion computer is displayed in Figure 7. The Pixhawk flight controller stores all flight data (such as attitude, position, and motor output) to an onboard SD card as well as wirelessly transmitting real-time data to an off-board computer using a 915 MHz SiK telemetry radio.

Figure 7: Top View of Quadcopter with Key Components and Companion Computer

Adaptive Gripper

An adaptive gripper end effector was designed based on the Robotiq 2F-140 [36] due to the wide range of tasks the gripper can perform. The gripper was 3D printed using polylactic acid (PLA) filament and assembled with standard metric hardware. An elastomeric surface was added to the inner contact surface of the gripper to increase the coefficient of friction. To actuate the device, a HS-645 MG ultra torque servo rotates two gears which are fixed to linkages. The gripper is controlled by a switch on an RC transmitter that passes a signal to the receiver onboard the UAV and then through the Pixhawk flight controller auxiliary pins. The auxiliary pins can easily be adapted into an autonomous control scheme, which allows flexibility for future development. Using camera-mounting holes located on the front of the UAV, the gripper was installed and counterbalanced with a LiPo battery to balance the UAV center of mass and minimize the effect on the controller. An image of the gripper fixed to the UAV is shown in Figure 8.

Figure 8: Adaptive Gripper Prototype Mounted on UAV

The first iteration of the gripper design was able to span from 0 mm when fully closed to 200 mm when fully open. The gripper can grasp a rigid, smooth plastic container weighing over 1 kg and successfully move it even when the payload was subjected to acceleration forces.

Flight trails were conducted to observe the effect actuating the gripper would have on an unmodified PX4 control scheme. The main concern was actuating the gripper changes the UAV center of mass and creates an inertial disturbance. Results from one flight trail are displayed in Figure 9.

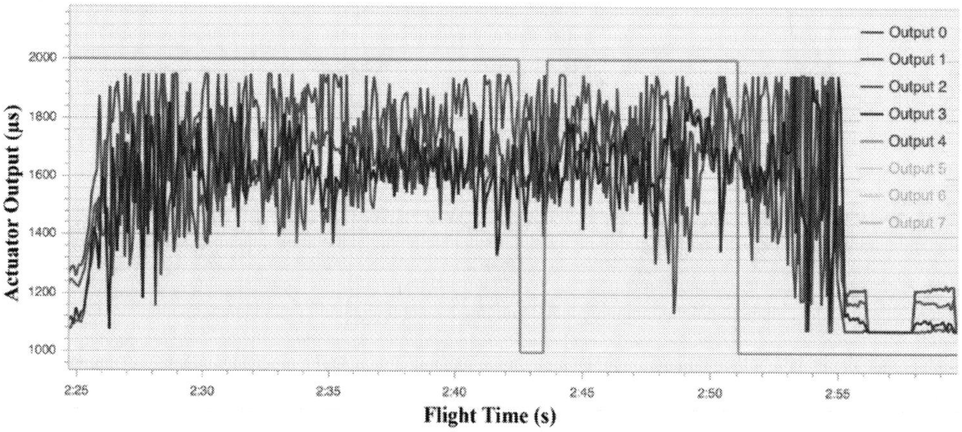

Figure 9: Main Actuator Outputs in Stabilized Flight Mode

Outputs 0 through 3 correspond to signals being sent to each of the motors in stabilized flight mode, and output 4, the grey line, is the signal that is actuating the gripper. The gripper was actuated three times: at 2:43, 2:44 and 2:51. During these three points in time, it can be observed that it had little to no effect on the flight performance, which is a promising result for the testing platform and use of the gripper for aerial manipulation.

Linear Clamp Gripper

The linear clamp gripper was constructed with the same manufacturing techniques and hardware as the adaptive gripper. The design was selected as an alternative to the adaptive gripper as this effector would allow approaching a cap from above opposed to perpendicular of the axis of rotation. This method will allow for testing the torque output produced by the yaw of the quadcopter and also provide a concept for future design considerations. A shaft is driven by a worm drive that is fixed to a continuous rotation HS-645 MG ultra torque servo attached to a gear train that outputs translational motion through a rack and pinion. Two sets of clamps are fixed to opposite racks that move in opposing directions, delivering a clamping range between 25 mm and 100 mm. To control the range of motion of the jaws, limit switches provide a physical to digital reference point. The current prototyped configuration is shown in the photographs in Figure 10.

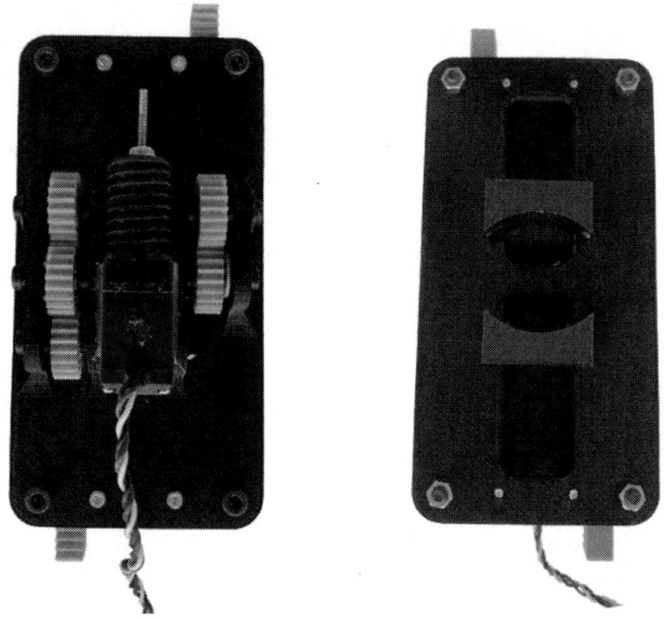

Figure 10: Linear Clamp Gripper Prototype (Top View Left, Bottom View Right)

Initial test results of the linear clamp gripper proved successful in multiple trials where the prototype successfully twisted off a moderately torqued 75 mm diameter cap with the aid of a planetary carrier rotating device. Although not quantified, the gripping force was greater than the adaptive gripper (at the cost of actuation time). As the required torque of the cap was increased, the limiting factor was the friction between the jaws and the cap. Similar to the adaptive gripper, a high friction surface was used; however, further design and experimentation will be required to characterize the response.

9. Conclusions and Future Work

In this paper the framework for an autonomous lubricant sampler is explored. The primary challenge is executing a torque to remove the lubricant cap so a sample can be extracted from the reservoir. Four solutions were proposed where torque is to be achieved using the UAV's yaw, rotating the thrust vector with a rotation joint and moment arm or with over actuated rotational rotors, and by coupling the UAV to its environment. Lubricant extraction can be achieved by using positive displacement pumps such peristaltic or vampire pumps, an evacuated bottle, or a container sampling approach. Vibration data acquisition and UAV trajectories were presented and initial prototyping and test results for the UAV and two end manipulators were then discussed.

The next testing stage is to use the actuator to couple the UAV with the environment and observe the effects. Once results and any required changes have been implemented the UAV will be used to rotate a cap with a resistant torque. Upon completion of cap removal the fluid extraction portion of the design will be more thoroughly considered.

In terms of the gripper modifications, lock nuts must be used to deal with the movement of the linkages and vibrations from the UAV. The range of actuation, geometry, and required gripping strength will need to be fully quantified and the gripper will be resized accordingly. The linear clamp gripper has yet to be tested on the UAV, but it is designed with mounting points to balance the center of mass of the UAV. Further future design considerations include reducing the size of the apparatus and decreasing the actuation time while maintaining a similar clamping force. Similar to the adaptive gripper, the forces to be experienced should be better quantified to aid the design and construction of the device and reduce size and mass. A full range of trails remains to be done for both manipulators, including detailed in air actuation, environmental coupling, and cap torquing.

The remaining future work can be summarized as quantifying a range of cap torque requirements, selecting a torquing and sampling method, developing a system and control package for cap removal and sampling, and performing experimental validation.

Acknowledgments

Dr. Duncan Elliott provided key insights into the visual navigation technique. Nicolas Olmedo provided technical and experimental support. Funding from the Natural Sciences and Engineering Research Council is gratefully acknowledged.

References

[1] M. M. Botha, "Electrical machine failures, causes and cures," 1997 Eighth International Conference on Electrical Machines and Drives (Conf. Publ. No. 444), Cambridge, UK, 1997, pp. 114-117.
[2] "Fluid Life Oil Analysis", *Fluidlife.com*, 2018. [Online]. Available: https://www.fluidlife.com/services/condition-monitoring/fluid-analysis/oil-analysis/. [Accessed: 02- Dec- 2018].
[3] The Association of Workers' Compensation Boards of Canada (AWCBC) (2017). *National Work Injury, Disease and Fatality Statistics (NWISP)*. 2015-2017. [online] The Association of Workers' Compensation Boards of Canada (AWCBC). Available at: http://awcbc.org/?page_id=89 [Accessed 11 Feb. 2019].
[4] M.G. Lipsett, R. Vaghar Anzabi (2013). "Seasonal effects on haul truck tyre life."

Int'l J Strategic Eng Asset Mgmt, Vol1, No3, pp 316 - 327
[5] M.G. Lipsett, N. Olmedo, B. Tesfay (2017). "A Wireless System for Conducting Remote Vibration Monitoring Using a Remotely Operated Deployment Robot." *Condition Monitor 365*, Inst British Non Dest Testing, August, pp 6-9.
[6] T. Ricci, "Unmanned Aerial Vehicles Soar High", *Asme.org*, 2012. [Online]. Available: https://www.asme.org/engineering-topics/articles/robotics/unmanned-aerial-vehicles-soar-high. [Accessed: 22- Jan- 2019].
[7] M. Dunbabin & L. Marques (2012). Robots for environmental monitoring: Significant advancements and applications. *IEEE Robotics & Automation Magazine*, *19*(1), 24-39.
[8] H. Yu, W. Yang, H. Zhang and W. He, "A UAV-based crack inspection system for concrete bridge monitoring," *2017 IEEE International Geoscience and Remote Sensing Symposium (IGARSS)*, Fort Worth, TX, 2017, pp. 3305-3308.
[9] K. Máthé, L. Buşoniu, L. Barabás, C. Iuga, L. Miclea and J. Braband, "Vision-based control of a quadrotor for an object inspection scenario," *2016 International Conference on Unmanned Aircraft Systems (ICUAS)*, Arlington, VA, 2016, pp. 849-857.
[10] L. F. Luque-Vega, B. Castillo-Toledo, A. Loukianov and L. E. Gonzalez-Jimenez, "Power line inspection via an unmanned aerial system based on the quadrotor helicopter," *MELECON 2014 - 2014 17th IEEE Mediterranean Electrotechnical Conference*, Beirut, 2014, pp. 393-397.
[11] F. Ruggiero, V. Lippiello & A. Ollero, "Aerial Manipulation: A Literature Review," in *IEEE Robotics and Automation Letters*, vol. 3, no. 3, pp. 1957-1964, July 2018.
[12] "Amazon.com: Prime Air", *Amazon.com*, 2019. https://www.amazon.com/Amazon-Prime-Air/b?ie=UTF8&node=8037720011 [12-Feb- 2019].
[13] H. Kim, H. Seo, C. Y. Son, H. Lee, S. Kim and H. J. Kim, "Cooperation in the Air: A Learning-Based Approach for the Efficient Motion Planning of Aerial Manipulators," in *IEEE Robotics & Automation Magazine*, vol. 25, no. 4, pp. 76-85, Dec. 2018.
[14] N. Staub *et al.*, "The Tele-MAGMaS: An Aerial-Ground Comanipulator System," in *IEEE Robotics & Automation Magazine*, vol. 25, no. 4, pp. 66-75, Dec. 2018.
[15] H. W. Wopereis *et al.*, "Multimodal Aerial Locomotion: An Approach to Active Tool Handling," in *IEEE Robotics & Automation Magazine*, vol. 25, no. 4, pp. 57-65, Dec. 2018.
[16] A. Ollero *et al.*, "The AEROARMS Project: Aerial Robots with Advanced Manipulation Capabilities for Inspection and Maintenance," in *IEEE Robotics & Automation Magazine*, vol. 25, no. 4, pp. 12-23, Dec. 2018.
[17] "Multimedia | AEROARMS", *Aeroarms-project.eu*, 2019. [Online]. Available: https://aeroarms-project.eu/category/multimedia/. [Accessed: 12- Feb- 2019].
[18] M. Orsag, C. Korpela, S. Bogdan and P. Oh, "Dexterous Aerial Robots—Mobile Manipulation Using Unmanned Aerial Systems," in *IEEE Transactions on Robotics*, vol. 33, no. 6, pp. 1453-1466, Dec. 2017.
[19] I. Palunko, P. Cruz and R. Fierro, "Agile Load Transportation : Safe and Efficient Load Manipulation with Aerial Robots," in *IEEE Robotics & Automation Magazine*, vol. 19, no. 3, pp. 69-79, Sept. 2012.
[20] S. Kim, S. Choi and H. J. Kim, "Aerial manipulation using a quadrotor with a two DOF robotic arm," *2013 IEEE/RSJ International Conference on Intelligent Robots and Systems*, Tokyo, 2013, pp. 4990-4995.

[21] Q. Wang, H. Xiong and B. Qiu, "The Attitude Control of Transmission Line Fault Inspection UAV Based on ADRC," 2017 International Conference on Industrial Informatics - Computing Technology, Intelligent Technology, Industrial Information Integration (ICIICII), Wuhan, 2017, pp. 186-189.
[22] M. Holloway, *The Use of Drone Technology for Remote Oil Sampling - 32nd International Maintenance Conference*. 2018. Available: https://reliabilityweb.com/videos/article/the-use-of-drone-technology-for-remote-oil-sampling [Accessed: 14- Feb- 2019].
[23] "Speeded up detection of squared fiducial markers", Francisco J.Romero-Ramirez, Rafael Muñoz-Salinas, Rafael Medina-Carnicer, Image and Vision Computing, vol 76, pages 38-47, year 2018
[24] "Generation of fiducial marker dictionaries using mixed integer linear programming",S. Garrido-Jurado, R. Muñoz Salinas, F.J. Madrid-Cuevas, R. Medina-Carnicer, Pattern Recognition:51, 481-491,2016
[25] M. Orsag, C. Korpela, S. Bogdan and P. Oh, *TRO2015.mp4*. 2017. Available: https://ieeexplore.ieee.org/ielx7/8860/8113882/8059875/tro-orsag-2750693-mm.zip?tp=&arnumber=8059875
[26] Mirko Brentari, Paolo Bosetti, Isabelle Queinnec, Luca Zaccarian. Benchmark model of Quanser's 3 DOF Helicopter. Rapport LAAS n° 18040. 2018. <hal-01711135>
[27] M. Kamel *et al.*, "The Voliro Omniorientational Hexacopter: An Agile and Maneuverable Tiltable-Rotor Aerial Vehicle," in *IEEE Robotics & Automation Magazine*, vol. 25, no. 4, pp. 34-44, Dec. 2018.
[28] C. Yih, "Flight control of a tilt-rotor quadcopter via sliding mode," *2016 International Automatic Control Conference (CACS)*, Taichung, 2016, pp. 65-70.
[29] "VOLIRO", *Voliro.ethz.ch*, 2018. [Online]. Available: https://www.voliro.ethz.ch. [Accessed: 13- Feb- 2019].
[30] "Guide to Submitting an Oil Sample | Same Day Testing & Results", *Apex Oil Lab*, 2017. [Online]. Available: https://apexoillab.com/submitting-an-oil-sample/. [Accessed: 17- Jul- 2018].
[31] "SAMPLING HARDWARE 38U Series Vampire Pump", *Fluidlife.com*, 2018. [Online]. Available: https://www.fluidlife.com/wp-content/uploads/2014/11/Sampling_VampirePump.pdf. [Accessed: 16- Jul- 2018].
[32] "COMPLETE TEST LISTS", *https://polarislabs.com/testing/oil-analysis/*, 2018. [Online]. Available: http://www2.eoilreports.com/CompleteTestList. [Accessed: 19- Jul-2018].
[33] "Oil Analysis Sampling Procedures | Fluid Life", *Fluidlife.com*, 2019. [Online]. Available:https://www.fluidlife.com/services/condition-monitoring/fluid-analysis/sampling-resources/sampling-procedures/. [Accessed: 24- Jul- 2018].
[34] D. Troyer and J. Fitch, "Oil Analysis Basics - Second Edition", Noria Corporation.
[35] "Oil Sampling Made Easy: Simplo sample bottle", *Relyassistllc.com*, 2018. [Online]. Available: http://www.relyassistllc.com/#simploInfo. [Accessed: 14- Feb- 2019].
[36] "2F-85 and 2F-140 Grippers - Robotiq", *Robotiq.com*, 2018. [Online]. Available: https://robotiq.com/products/2f85-140-adaptive-robot-gripper. [Accessed: 01- Feb-2019].

MOTION MAGNIFIED VIDEO APPLICATION TO MACHINERY DIAGNOSIS

William D. Marscher and Chad Pasho
Mechanical Solutions, Inc.
11 Apollo Drive
Whippany, NJ 07981
(001) 973-326-9920
wdm@mechsol.com

Abstract: Vibration detection can be performed by a meter, or a single channel spectrum analyzer. The former has quantified vibration amplitude levels since the 1950's, while the latter was able to implement the Fast Fourier Transform (FFT) since the 1960's to break vibration down into its frequency components. Multiple channel FFT analyzers enabled the process of Operating Deflection Shape (ODS) determination, which has been an important tool in visualizing the vibration of the machine and its system, including the foundation and piping networks. The input for ODS is the phase-linked signal set from a group of accelerometers, moved over often hundreds of test points. The data is superimposed onto a CAD model, and then scaled-up vibrations are animated at frequencies of interest. This process provides valuable insights, but is time-consuming and therefore expensive each time it is applied by experts, and it is error-prone. An alternative method has been developed that is based on evaluation of high resolution/ high speed videos. The method provides information equivalent to a high-sensor-count ODS, by treating each pixel as an accelerometer, using the pixel's light intensity modulation to translate information embedded in the video into vibration motion able to be observed and interpreted by human investigators. This method is known by some as Motion-Magnified Video (MMV).

Key words: Motion Magnification; Video Amplification; Operating Deflection Shape; Turbomachinery; Machinery Diagnostics; Predictive Health Management

1. Introduction

Turbomachinery, pump, and motor vibration is often useful in determining whether or not a machine is operating properly, and for diagnosis of problems if the operation appears improper, or if reliability issues have been experienced (e.g. fatigue cracking, or premature failure of bearings and seals). For several decades, a visual method called Operating Deflection Shapes (ODS) has been an important tool in getting a complete and simultaneous view of the vibration of the machine and its system (e.g. piping, foundation, and driver or driven machine). The input for ODS is the phase-linked signal spectra from a group of accelerometers, moved over often hundreds of test points while one phase-reference accelerometer is kept at a consistent location and direction. The data is superimposed onto a simplified CAD model of the machine and system, and then the exaggerated (but to-scale) vibrations are animated at frequencies where the response is sufficient to be of interest to the researcher or troubleshooter. This process takes several days for complex machinery and systems, operated at a variety of process points.

An innovative method has been developed that is based on evaluation of high resolution/ high speed video taken of the operating machinery. The method provides information equivalent to a high-sensor-count ODS, by treating each pixel as an accelerometer, using the pixel's light intensity modulation to determine local vibration displacement frequency spectrum. From this information, realistic magnification of slow-motion video footage permits microscopic vibration to be magnified and thereby observed and interpreted by the human investigator. This method may be called Moton-Magnified Video (MMV). This method is much faster than ODS, yet provides similar information at many more locations (at least in directions in the 2-D field of view), with less opportunity for error.

2. Background

ODS Methods:

About 1980, some researchers developed the concept of "seeing" vibration. Their approach was to acquire data at many locations and directions on a vibrating structure, and then allocating those motions to a stick-figure CAD model. This model was then animated on a computer video screen, or the extremes of the vibration in the model were plotted by a computer printer or plotter. This technique became known as Operating Deflection Shapes (ODS). The following outline summarizes the technique:

- It is based on the "natural excitation" frequency spectra of the rotating equipment
- The User acquires vibration data from various locations and directions on the machine (hundreds of vibration measurements)
- One sensor is always kept at the same location and direction, to provide a reference signal as other probes are "roved"
- A large database is built up of amplitude vs. frequency and phase angle
- A 3-D CAD model is constructed, and assigned motion from each individual vibration data point
- Some dedicated software is used to amplify the results filtered at specific frequencies of interest, creating animations of the CAD model of the equipment and related piping and foundation.

ODS is a powerful method for perceiving the source and relative importance of vibrations in machines and their foundations and systems. It also often provides visual "hints" concerning options to reduce any excessive vibration. However, in general ODS has its drawbacks as well. The pros and cons of ODS are summarized in the list below:

- ✓ Powerful and intuitive diagnostic tool
  - Can clearly demonstrate modes and frequencies of vibration
- ✓ Proven over decades of application
- × Time consuming
  - Data acquisition – hundreds of data points
  - Post processing – compiling database to match model points
- × Potential for bookkeeping error
  - Match all gathered data points to appropriate place and direction on the model
- × Requires proximity – not appropriate for restricted access
  - Heat, radiation, shock, accessibility/scaffolding requirements

Moton-Magnified Video (MMV) Methods:

This led researchers to wonder whether the troubleshooter could truly see vibration. Academic researchers, some of whom are listed in the References (Ref. 1-3) had this thought about 25 years ago, and have been gradually perfecting techniques. Basically, these techniques fall into two categories: 1) tracking of specific points, edges, or (as machine-vision scientists call them) "blobs", and 2) performing statistics, including signal vs. time as well as signal FFT frequency spectra, on the individual independent pixels. The former are called "Lagrangian Methods" by many researchers, and the latter are called "Eulerian Methods". A rich literature basis exists for these methods. This paper's references emphasize the Eulerian technique.

The Moton-Magnified video method may be summarized as follows:

- Uses high speed, high resolution video
  - Eulerian method is equivalent of millions of accelerometers, 1 per pixel
- Analyzes/quantifies motion
  - Frequencies
  - Displacement (2-dimensional)
- Algorithms amplify motion to human visual threshold
  - Filterable by desired frequencies, high resolution up to 3.4 kHz in the system developed by the authors
  - Amplifies up to 1000+ times the actual motion, using a statistical algorithm that can perceive and magnify motions down to less than 1% of pixel. At a 10 foot field of view, motions of 50 millionths of an inch, peak-to-peak, can be reliably detected and quantified, based on both controlled laboratory as well as practical field tests.

History:

Figure 1 shows a picture of a galloping horse, taken in 1878. The story behind the photo is that it was one of twelve photos taken by cameras staged along the track, such that a side-shot photo was taken about every quarter of a second. These photos were then imprinted onto paper, and the photos were stacked together in series into a "flip deck" of cards. By flipping the cards, once each quarter second, a nineteenth century "video" was made of the horse running. This video settled a bet between the founder of Stanford University and his friend about whether or not all of a horse's hooves are off the ground at the same time when it is at full gallop. The answer is "yes", as is clear from the video of this incident, shown during the MFPT2019 presentation of the present paper. The dynamic situation of the horse's motion was made fully evident, and looking at it today, it is also clear from the video that if there was a problem with the horse's gait, that could have been easily diagnosed as well.

Figure 1: Galloping horse video from 1878

This has led to various researchers, in the US as well as Europe, to attempt to develop such video acquisition and evaluation processes to study other dynamic phenomena.

In fluid dynamics, laser-doppler velocimetry (LDV) has been able to track streamlines, and similar processes have been applied to mechanical vibration. In a lower-tech method, strobe lights have been used on machinery for many decades to "freeze" motion. These method have been useful in interpreting dynamic behavior if relatively large motions are involved. However, displacement magnification was not achieved.

In the 1920's, advanced research organizations used relatively high speed cinematography to guide development and to evaluate the validity of new vibration analysis procedures. A good example is the work of Wilfred Campbell at GE. At the time, steam turbines driving generators were growing larger, and as they did they encountered unexplained fatigue failures. Campbell's work, as exemplified by the photo in Figure 2 (a single frame of a "movie" or video taken by his cameras), determined that bladed disks had much more complicated natural frequencies than had been predicted up to that time, including zero-vibration nodal lines in their mode shapes, now known as nodal diameters and nodal circles. When the associated natural frequencies matched a strong excitation frequency (such as the number of stator nozzles times running speed), and when simultaneously the mode shape of the natural frequency matches the lobes of a circumferential static pressure pattern (such as associated with the number of blades versus number of stator nozzles), a strong resonant response was seen to occur.

Based on this insight, GE and eventually others were able to avoid damaging resonances by either adjusting natural frequencies, or by adjusting problematic lobe patterns of the nozzle pass frequency. The Campbell Diagram, well known to turbomachinery engineers, became a graphical method of codifying this procedure.

Figure 2: Campbell's bladed disk video from 1926

This range of efforts, and others, led scientists and engineers to an appreciation for the power of video to uncover useful information in dynamic scenes, assuming that appropriate algorithms, implemented by software, were able to be used to evaluate the video footage.

In the US, several universities, including MIT [1], [2], have been actively pursuing MMV. At MIT, the particular technology which became the core of their research was an Eulerian approach [3], which essentially tracks the variation of individual pixels over time, and then exaggerates those differences. To the human eye, no matter how long and hard you stare at a machine with vibration at, say, several times the ISO limits, you would struggle to detect any motion. For a computer, however, the tiniest per-pixel fluctuations (between white and slightly-off-white, say) are easy to detect. MIT originally developed their software to measure the vital signs of neonatal babies without physical contact, but they realized that there are other, far-ranging applications. For biological applications, Eulerian MMV detects changes in skin color, as well as exaggerates movements. The steps and hardware associated with Eulerian MMV are illustrated in Figure 3.

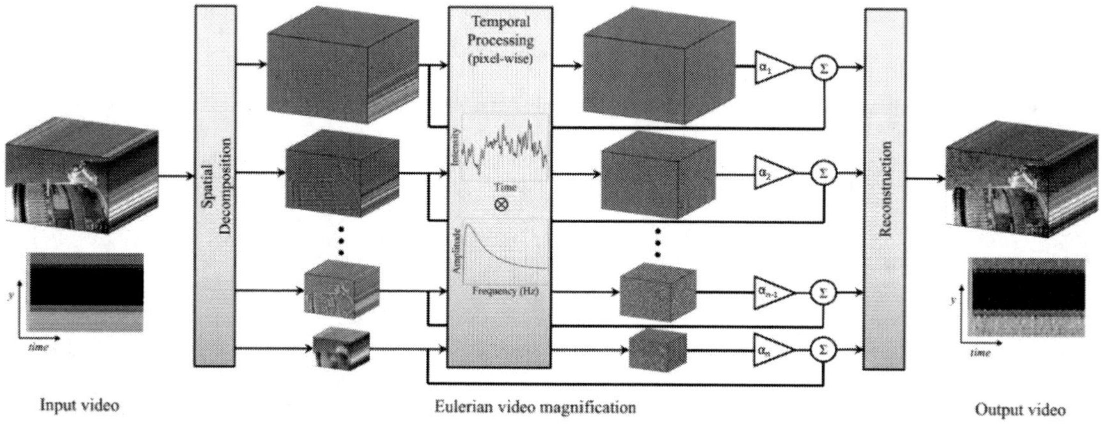

Figure 3: Illustration of the MMV process, as implemented by MIT [3]

Research by the authors:

The authors' organization initiated its detailed research into engineering uses for high speed video in 2003. The US Department of Defense contracted the authors' group to study how to make non-lethal projectiles operate reliably, ensuring that they functioned without harming humans or animals. Applications included various forms of "soft bullets". High speed video, up to 20,000 frames per second, documented their transient behavior during flight and impact.

A later application (2005) involved an inexpensive video sensor to use pattern recognition software to rapidly identify rocket-propelled grenades (RPGs) fired by a belligerent, so that defensive steps could be taken. Following this was use (2011 to present) of high speed video to determine the moment-to-moment effectiveness of passive counter-measures to fool MANPADS missiles fired by "bad guys" to bring down (for example) an airliner, providing feedback facilitating improved countermeasure design. More recently (2015 to present), the authors' group was contracted to use high speed video and pattern recognition to rapidly characterize lethal ranges of various munition types, so friendly military forces are better able to ensure sufficient buffer zones for civilians, schools, and hospitals.

During this process, the authors began applying Eulerian algorithms to measure and amplify motions of individual pixels [4]. Different procedures were tried during both government and IRAD-funded testing, and after several years a successful method was achieved. Transient motions and vibration levels as low as tenths of a mil (i.e. several microns) of machinery surfaces were demonstrated first in the laboratory, and soon after in on-site machinery troubleshooting.

The authors' methods are based on the Eulerian approach, which observes the intensity variation of individual pixels, combining their effects statistically into a full scene. Alternative methods typically keep track of motion pixel-to-pixel, the so-called Lagrangian techniques. Variations include feature tracking, and an approach called "optical flow". These Lagrangian methods are good for amplifying motions that occur over multiple pixels, e.g. motions of more than 2mm (0.080 inches) for a 3 m (10 foot) field-of-view. However, these approaches are limited in how small a displacement they can detect for a given field of view and a given camera resolution, in that interesting vibration levels can be well below such limits, particularly at higher frequencies of vibration. If higher camera resolutions are used to overcome this limitation, the computational overhead becomes excessive. Eulerian methods have been shown to work best for detection of small displacements, since motions as low as 1/1000 of a pixel have been demonstrated. Typically, as shown in Fig. 3, Eulerian methods build a time history for each pixel, and then apply signal processing techniques common to more conventional vibration detection methods, such as digital filters, time and frequency domain digitization (Laplacian and Fourier), and amplification of the subsequent amplitudes by simple multiplication. In some implementations, an FFT is useful to separate motion from individual frequencies and/ or vibration modes, to make vibration sources and/ or mode shapes more evident. Multiple frequency animations can be made from the same ten second video strip.

The benefits of MMV performed in this manner are that it becomes a powerful and intuitive diagnostic tool. It realistically demonstrates modes of vibration, and is easy for non-experts to understand. Importantly for certain applications, the technique does not require contact, so that surfaces at 1000 F, or that are radioactive, or that are 50 feet vertically above near a ceiling, can be observed without difficulty. Furthermore, items such as instrumentation wires or lubrication lines, which would change their behavior if mass-loaded with even a small accelerometer, can be evaluated in realistic manner. The results are obtained quickly, and at modest cost and effort.

Applications and Case Histories:

Several examples will be presented as videos during MFPT2019, and are available embedded in PowerPoint presentation slides from MFPT to those who were in attendance.

Gas turbine in a combined cycle plant:

The photo in Figure 4 is one frame of a 1300 frame per second video taken of a newly installed aeroderivative gas turbine in a co-gen portion of a wastewater facility. The plant was not sure how to measure vibration on such a machine, which had some sections too hot for classical measurement (e.g. accelerometers), and many instrumentation wires, thin tubes, and guide vane linkages that were difficult to instrument, and might have their natural frequencies or spectral response altered by accelerometer attachment. The MMV video quickly determined that, in this complex system, all subcomponents were vibrating in a manner that would be considered normal, and no problematic resonances were evident. Zones near bearings were quickly proved (in ten seconds of video and minutes of post-test processing and evaluation) that the turbine and generator met all ISO 10816 vibration criteria.

Figure 4: Frame of the MMV process applied to a gas turbine generator

Fatiguing thermowell sensor in a gas pipeline at a refinery:

In many applications involving pipelines and their rotating prime movers, instrumentation is attached in a manner that leads to natural frequencies excitable by higher frequency excitations (such as vane or blade passing frequency) within the variable speed operating range. Although the higher frequency excitations such as vane pass are usually relatively weak compared to running speed excitations, small bore appurtenance pipe connection natural frequencies typically have very high amplification factors (order of 100). When the weak excitation is multiplied by the high amplification factor, high enough amplitudes can occur to cause fatigue. However, fatigue can occur due to other reasons as well, such as residual stresses from rough installation, corrosion fatigue, or (low) stress corrosion cracking. Furthermore, measuring for high vibration and/ or resonance is complicated by accelerometer mass-loading that can dramatically change the vibration, or the possibility that the peak vibration is occurring someplace other than the measurement location.

The latter was true in the thermowell case provided as an example in Figure 5. There was a natural frequency resonance leading to fatigue cracking at the base of the small bore pipe attachment to the pipeline pipe subject to temperature measurement. The plant tried measuring this at the cantilever (where the large mass is) end of the device, and from this measurement had ruled out excessive vibration by relatively strong (in this case) vane pass pressure pulsations. However, strong resonance was actually present, causing vibration high enough to result in steps toward high cycle fatigue whenever the vane pass "tuned in" to a precise resonance. The MMV was able to show the problem was a second cantilever (coat-hook-shaped) mode, where the cantilever end top did not move side-to-side as much as it twisted, causing the "head" to "nod" up-and-down. One view that helped make this point was the end-on view, as shown in Figure 5, in combination with a side view. The primary reason for bore fatigue became immediately evident.

Figure 5: Frame of the MMV process applied to a gas pipeline small bore thermowell

High noise in an industrial fan:

An industrial fan was producing extremely high noise (120 dB) due to blade passing frequency (8 blades, with a frequency of about 480 Hz). Everyone including the OEM was baffled because accelerometer measurements on the fan casing, motor, and piping indicated surface velocities were too low to result in such high acoustic response. The MMV video, taken at a frame rate to permit full-screen video up to 640 Hz, clearly showed that the problem was a strong "pulsing" in-and-out of the upper sidewall, where the pulse direction was opposite on the left side versus the right side. This led to almost no motion along the vertical centerline, hence the source of the OEM's confusion. A frame from the video is shown in Figure 6.

Figure 6: Frame of the MMV process applied to an industrial fan

Vibration of a coil of copper tubing in a gas turbine application:

Strong vibration was observed in a copper coil which was part of the instrumentation system for a gas turbine. Any attachment of an accelerometer to this coil would have dramatically disturbed its vibration. MMV, as shown by the frame in Figure 7, was able to perform a dynamic study of the motion, and based on simple stress-deflection calculations performed from the video evidence, proved that the vibration was harmless.

Figure 7: Frame of the MMV process applied to an instrumentation coil of a gas turbine

Vertical turbine pump/ motor coupling imbalance:

MMV based on the Eulerian approach can also successfully document lateral vibration of rotating shafts. The frame from an MMV video shown in Figure 8 shows a vertically oriented coupling, as seen through the discharge head/ motor stand coupling access "window". The video clearly shows that the coupling is whirling, and moving more than the shafting above or below it. The displacement was found to indicate excessive coupling imbalance.

Figure 8: Frame of the MMV process applied to a vertical turbine pump/ motor coupling

MMV of a complex piping network:

The complicated piping network shown in MMV frame of Figure 9 would take days to quantify its motion, and to determine natural frequency mode shapes near important excitation frequencies. Through use of Eulerian-based MMV, both low frequency (e.g. order of several Hz) and high frequency (e.g. hundreds of Hz) character were able to be detected for all pipes, hangers, and attachments in the field of view of the camera. This data was acquired in about 10 seconds, followed by a couple hours of careful video inspection and filtering by frequency. The 1000x motion amplification made it very easy to determine which pipes were not sufficiently supported, and which pipes or hangers might be in jeopardy of eventual failure.

Figure 9: Frame of the MMV process applied to a complex piping network

Steam turbine driver with axial rotor motion at coupling:

A steam turbine driving a feed pump had been experiencing chronic fretting of the teeth of its gear coupling. As clearly shown in the MMV video, a frame from which is shown in Figure 10, the steam turbine rotor was undergoing something called "axial shuttling", with strong axial motion at an acoustic frequency. Once this motion was detected on a "full system" basis, the reason for the coupling problems became clear. Having this knowledge did not by itself solve the problem, but it presented the nature of the problem in a manner that focused further tests on acoustic issues, which in turn allowed identification and removal of the root cause.

Figure 10: Frame of the MMV process applied to steam turbine "axial shuttling"

Pump/motor offset misalignment:

Figure 11 shows a frame from an MMV video that clearly shows the out-of-phase "bounce" motion of a pump versus its motor, making the offset misalignment present evident. The Eulerian method can be "queried" to determine the frequency spectrum of this out-of-phase motion, which would show in this case a strong 2x component, along with significant axial motion, and the displacement involved. This displacement can be quantified by analyzing the degree of light modulation in high contrast pixel zones, providing the ability to quickly determine machinery that is not aligned in a chemical plants with hundreds, or in some large plants thousands, of installations.

Another issue obvious in the MMV is soft foot under the pump. Eulerian-based MMV typically does an excellent job locating soft foot, since it can measure to 0.050 mils p-p.

Figure 11: Frame of the MMV process applied to offset misalignment of a pump/ motor

3. Future Research

The following summarizes the research currently underway at the authors' organization:

1. Real-time processing and display
2. Handheld walkabout troubleshooting system
3. Specialty cameras, such as borescope attachment, underwater, and drone-carried
4. Rotating field-of-view
5. Synchronized arrays of cameras

4. Conclusions

As presented, the new video-based motion amplification procedures and equipment now on the market can be very useful for vibration-based diagnosis of machinery. It has advantages over the classic ODS method in many instances, and at least can help limit the amount of ODS required by demonstrating what is moving, and what is not. MMV also typically takes much less time and "logistics" than ODS to implement.

MMV benefits are:

- ✓ It is a powerful and intuitive diagnostic tool

- Realistically demonstrates modes and frequencies of vibration
  - Easy for non-experts to understand (management, etc.)
  - Comprehensive set of data points
- ✓ It does not requires "contact" – perfect for restricted areas
  - Heat, radiation, accessibility/scaffolding requirements
- ✓ It is fast
  - Millions of data points ready to evaluate in minutes, not days
- ✓ It is cost-effective
  - Helps focus effort for accelerometer-based ODS or strain gages.

Moton-Magnified video represents an important addition to a vibration diagnostic tool kit.

References

[1] Timoner, SJ, and Freeman, DM (2001), "Multi-image gradient-based algorithms for motion estimation", Optical Engineering 40.9, pp. 2003-2016

[2] Liu et al (2005), "Motion magnification", ACM Transactions of Graphics (TOG) – Proc. of ACM SIGGRAPH 2005, TOG Homepage, v. 24, Issue 3, July

[3] Wu, H-Y et al (2012), "Eulerian video magnification for revealing subtle changes in the world", ACM Trans. of Graphics (Proc. SIGGRAPH 2012), v. 31, August

[4] Marscher, W.D., and Onari, M.M. (2018), Video Motion Amplification vs. Operating Deflection Shapes for Machinery Diagnosis, MFPT2018, Virginia Beach, May

OPTIMAL ATTACHMENT METHODS FOR ACCELEROMETERS

William D. Marscher and Maki M. Onari
Mechanical Solutions, Inc.
11 Apollo Drive
Whippany, NJ 07981
(001) 973-326-9920
wdm@mechsol.com

Abstract: Vibration detection on stationary surfaces of machinery, such as bearing housings, is typically accomplished with accelerometers. Modern accelerometers have been engineered to have a very wide frequency range with strong sensitivity of voltage to surface acceleration. However, in order to implement this capability, the bottom surface of the accelerometer must faithfully track the actual motion of the surface that it is attached to. The ideal method would be a tight metal-to-metal connection, such as can be achieved with a well-torqued screw joint. However, this is often not practical (e.g. no surface modification allowed), or is very inconvenient (e.g. an Operating Deflection Shape requires measurements at hundreds of locations, or walk-around machinery monitoring routes on many uninstrumented machines in large plants). In such cases, diagnosticians use temporary attachment methods, such as a simple hand-pressure attachment, a magnetic base, two-sided tape of various thicknesses, a thin layer of wax, or glue-cement. The authors have studied the fidelity of each of these methods in a series of controlled experiments, and determined that all methods, up to about 1500 Hz, are adequate for characterization of machines over the typical range of significant dynamic excitation forces. For frequencies to over 10 kHz, an unexpected result was that double-sided tape provided some of the best results. All methods provided similar looking amplitude vs. frequency spectra up to about 5 kHz.

Key words: Accelerometer Attachment; Vibration Measurement; Operating Deflection Shape; Turbomachinery; Machinery Diagnostics

1. Introduction

Vibration of machinery is useful in determining whether or not a machine is experiencing distress, or perhaps has already worn certain key components such as bearings. In order to quantitatively characterize vibration, motion sensors are used. For rotor systems, these probes are typically non-contact (e.g. eddy current based) proximity probes. For stationary surfaces such as bearing housings, contact probes based on either vibration velocity ("velocimeters") or acceleration ("accelerometers") are used. Velocimeters are more limited in frequency range, and bulkier, while modern accelerometers are designed to be accurate over a wide range of about 1 Hz to over 10 kHz, with special accelerometers able to function with reasonable accuracy to as high as 50 kHz.

In all cases the accuracy and repeatability depends upon how well the base of the accelerometer is able to follow the true movement of the surface being measured. This requires a stiff attachment, one that does not allow the accelerometer surface to part from the measured surface, and that also does not allow the natural frequency of the mounted accelerometer to drift down into the intended measurement frequency range.

2. Application History

The ideal attachment in this regard has always been considered metal-to-metal contact, enforced by a screw or bolt that enforces significant compression on the interface. Testing has of course validated this as a means to achieve high fidelity of measurement, as discussed in references [1] to [7]. Depending upon the mass of the accelerometer and the stiffness of the mounting surface, this method can result in accurate and repeatable measurements to order of 50 kHz [4, 5, 6].

However, screw-on attachment of an accelerometer requires tapping of the measured surface, which may not be permitted, or may not be practical (e.g. thin surface walls). In other instances, it may be very time-consuming and inconvenient to perform such an attachment. This is the case in a typical Operating Deflection Shape (ODS) test, when hundreds of measurements on many types of surfaces are required [7]. It is also the case for walk-around routes in large plants, where technicians must measure key locations on a large number of machines in a brief period of time, as they follow their routes.

Therefore, other methods of attachment are typically used. One of the simplest and oldest is the use of hand-held pressure, by holding the accelerometer sensor base (sometimes through a needle point "stinger") firmly against the measurement location. This method is intuitively considered the least accurate attachment method, but in the authors' experience is typically more accurate than the other long-time method, a magnetic base firmly screwed onto the accelerometer base. The problem with the magnetic base is that at moderate frequency the magnetic surface can temporarily lose contact with the magnetic target surface, especially at higher acceleration G levels.

Other methods include the use of a thin layer of special sticky wax (only works for surfaces with temperatures to about 100 F), a thin layer of glue/ cement (typically cyano-acrylic "superglue"), and double-sided sticky tape on the order of 8 mils (0.2 mm) thick. Studies have been performed by various organizations, and opinion-based curves of adequate ranges for each technique are published by such standards organizations as ISO (ISO 5348[1]), Hydraulic Institute (ANSI/HI 9.6.4 [2]), and ANSI (ANSI S2.61 [3]).

3. Experimental Set-Up

As stated in the references, there is general agreement that a screw-on connection is the highest fidelity method of attaching an accelerometer to a machine for which vibration is to be measured. Finite element analysis confirms this conclusion, for flat-surface-on-flat-surface, where a torqued-up screw tightly compresses the two surfaces together sufficiently so that the separation force from acceleration F=ma, remains below the spring-like compression force across the surface, F=kx, where m is the accelerometer mass, k is the attachment interface stiffness, a is the acceleration being measured, and x is the compression displacement of the combined accelerometer base and measured surface material. Therefore, the author's testing used the screw attachment method as the standard against which all other methods were compared.

In the authors' experiments, a 12 inch long, 2 inch wide, 2 inch thick cantilever beam was attached to a stiff support over a 4 inch length at one end, firmly bolted to a massive slotted machine foundation plate. The beam retained 8 inches of cantilevered free length, as shown in Figure 1. The modal characteristics of the final assembly were determined by impact testing and an Operating Deflection Shape test, assisted by 3-D finite element analysis, and resulted in quantification of many natural frequency modes existed in the range of 100 Hz to 10 kHz. The test results were evaluated in the context of these modes (unavoidable in a real-life laboratory set-up), rather than assuming infinite stiffness of the attachment of the beam "fixed end" to ground, or the base of the exciter to ground. A screwed-on reference accelerometer was attached to the bottom of the free end of the beam, 2.5 inch from the beam end. The tested accelerometer/ attachment method in each test was attached to the beam, on the top side of the beam, precisely on the opposite side of the reference accelerometer, as also shown in Figure 1.

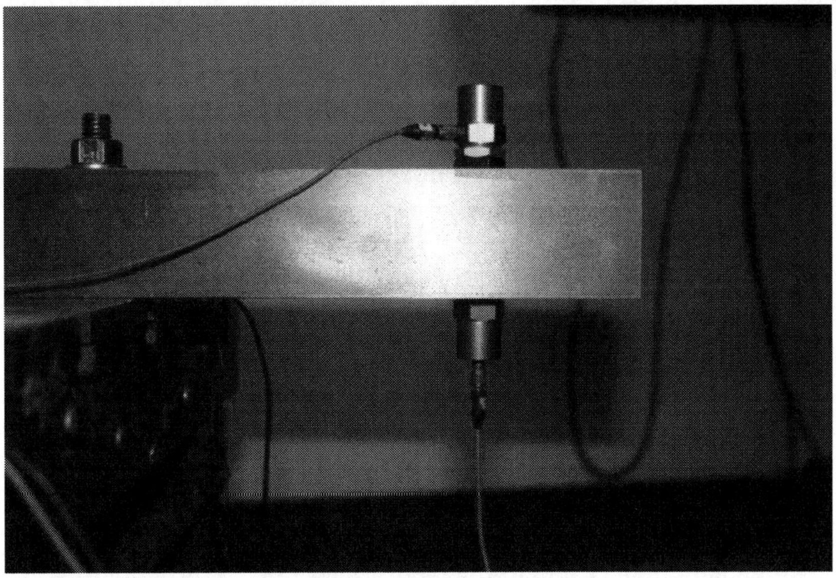

Figure 1: Acceleration attachment test set-up

The tests consisted of exciting the end of the beam with an electrodynamic shaker, with an accelerometer on its head to detect the acceleration spectrum input to the base of a stinger. For all acceleration measurements in the experimental set-up, MSI used off-the-shelf accelerometers from a major supplier. The remaining equipment consisted of a National Instruments (NI) data acquisition (DAQ) system chassis model number NI PXIe-1073 with two cards of model number NI PXIe-4497, and a Labworks ET-132-2 shaker, driven by a power amplifier PA-138, with LabView based software control custom-programed by the authors' group.

The LabView-based DAQ and FFT system was set up to read a frequency span of over 10 kHz (after anti-aliasing), a resolution of ¼ Hz, an overlap of about 90%, and an update time of 0.375 sec. The analyzer was set to Peak Averaging mode throughout a frequency sweep. The shaker was set to sweep from 100 Hz to 10 kHz frequency (later testing to 20 kHz will be reported in a future paper) at a constant rate of typically 20 Hz/s, at a nominally constant amplitude of 0.1 G at the base of the shaker stinger, where the stinger attached to the shaker head. The shaker possessed system resonances at about 4800 Hz and 8500 Hz, which resulted in increased acceleration input to the stinger of order a 1 G when the shaker was operating within several percent of those frequencies. The presence of the resonant frequencies was not observed to affect the fidelity of the results.

A 100 mV/g reference accelerometer was stud mounted to the underside of the cantilever beam. For each attachment-tested accelerometer, after acquiring data at 0.1 G excitation across the frequency range, the excitation amplitude was increased by increasing the current fed to the Power Amplifier. Three amplitudes were tested (0.1 G peak, 0.3 G peak, and 1.0 G peak) for each tested attachment and each tested accelerometer type. Specifically, the procedures were repeated with different accelerometers to ensure repeatability of the various attachment methods, or to investigate the effect of accelerometer mass on the result. Two different serial number (but same model) 100mV/g accelerometers were tested, and were then replaced with a roughly 1/10th mass 10 mV/g accelerometer. Each specific test was repeated four times to ensure consistency. The results were found to be quite consistent, and within the range of frequency and amplitude investigated, the accelerometer size had no measurable effect. Therefore, the results reported will focus only on the effect of the attachment means.

4. Experimental Results

A large volume of test data was acquired. The following figures show only representative results from the authors' tests. Initially, plots (Figure 2 to 5) show the raw frequency spectra from a given test, plotting the tested accelerometer and attachment as a blue line, and the reference accelerometer, bolted firmly in place to the beam, as the red line. These plots are log amplitude plots, which have the advantage of showing low amplitude and high amplitude responses on an equal basis, but can give the false impression of better percentage agreement than is actually present. Therefore, following the raw FFT plots, the results are re-plotted (Figures 6 to 12) from a statistical average of the repeatability testing, in this case in terms of the ratio of the tested accelerometer reading to that of the reference accelerometer, along with any phase difference between the two, over the tested range of frequency. This plotting method in turn can give a false impression of gross error, particularly at low amplitude anti-resonance "valleys". Taken together, however, in the authors' opinion the two sets of plots give a reasonable indication of the capability of each attachment method.

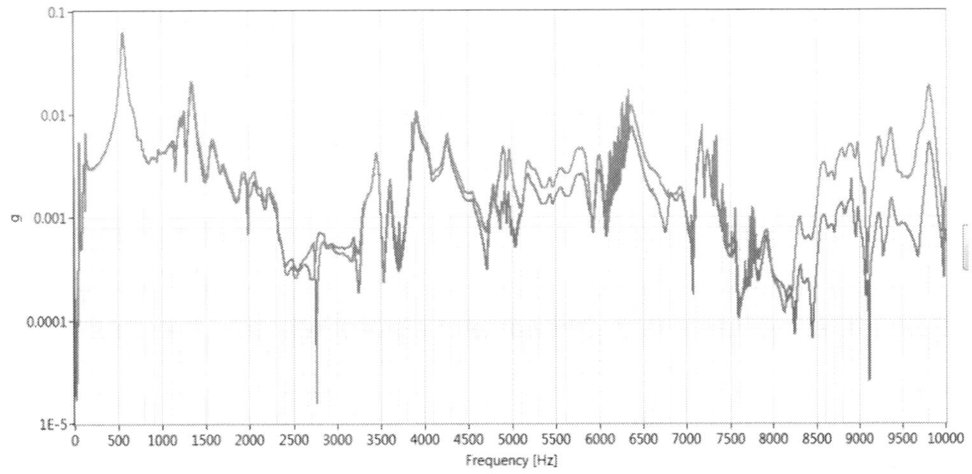

Figure 2: Acceleration attachment test raw spectra, handheld-only attachment; blue is the signal of the tested accelerometer & attachment, red is the reference accelerometer signal

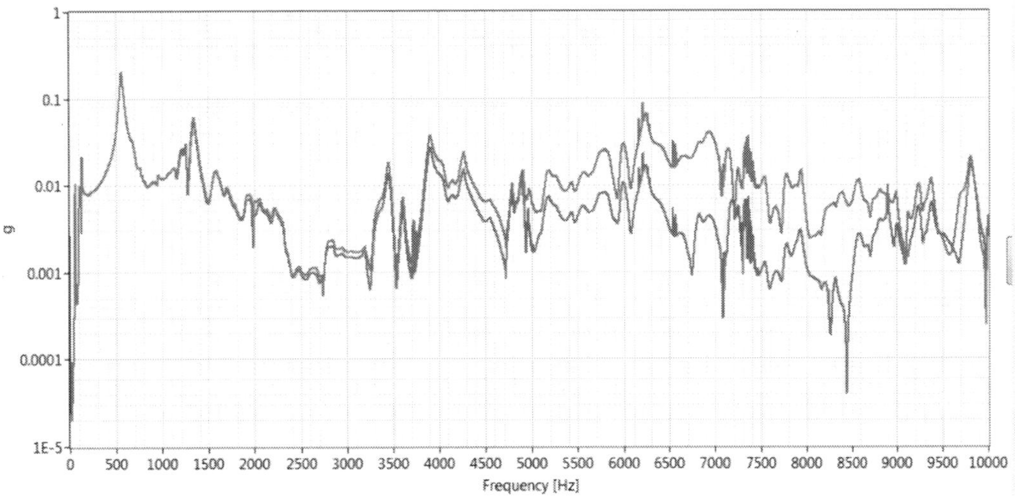

Figure 3: Acceleration attachment test raw spectra, magnet-only attachment; blue is the signal of the tested accelerometer & attachment, red is the reference accelerometer signal

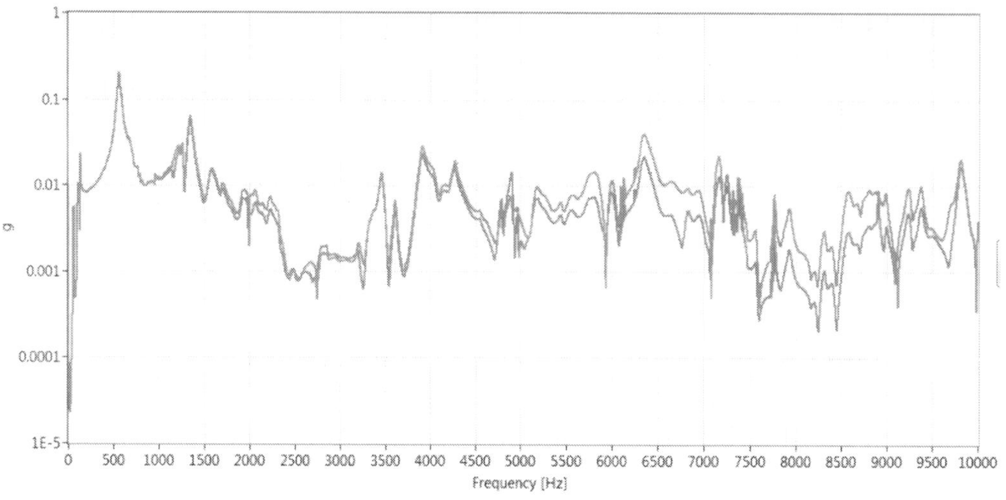

Figure 4: Acceleration attachment test raw spectra, wax-only attachment; blue is the signal of the tested accelerometer & attachment, red is the reference accelerometer signal

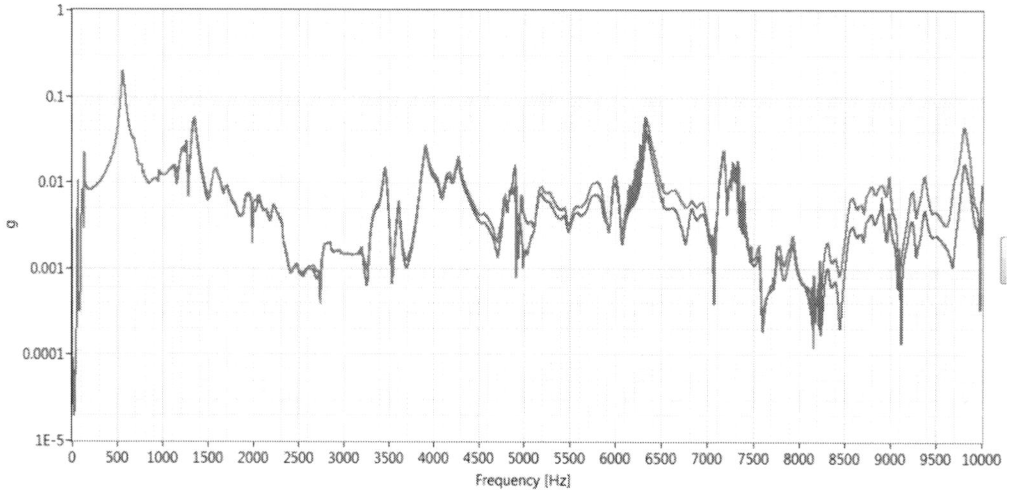

Figure 5: Acceleration attachment test raw spectra, magnet-plus-wax attachment; blue is the signal of the tested accelerometer & attachment, red is the ref. accelerometer signal

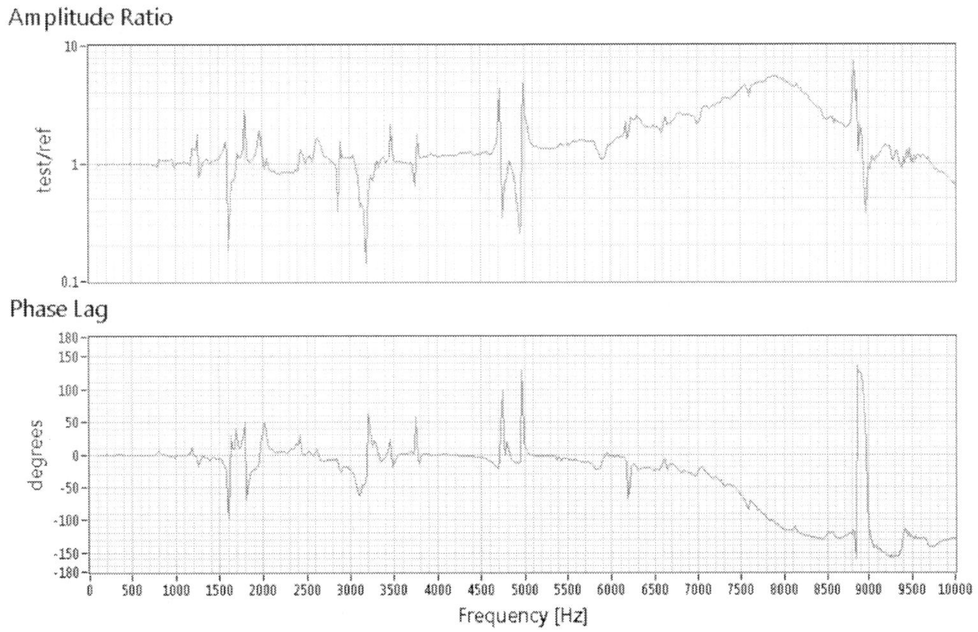

Figure 6: Acceleration attachment test comparison vs. reference, hand-held only

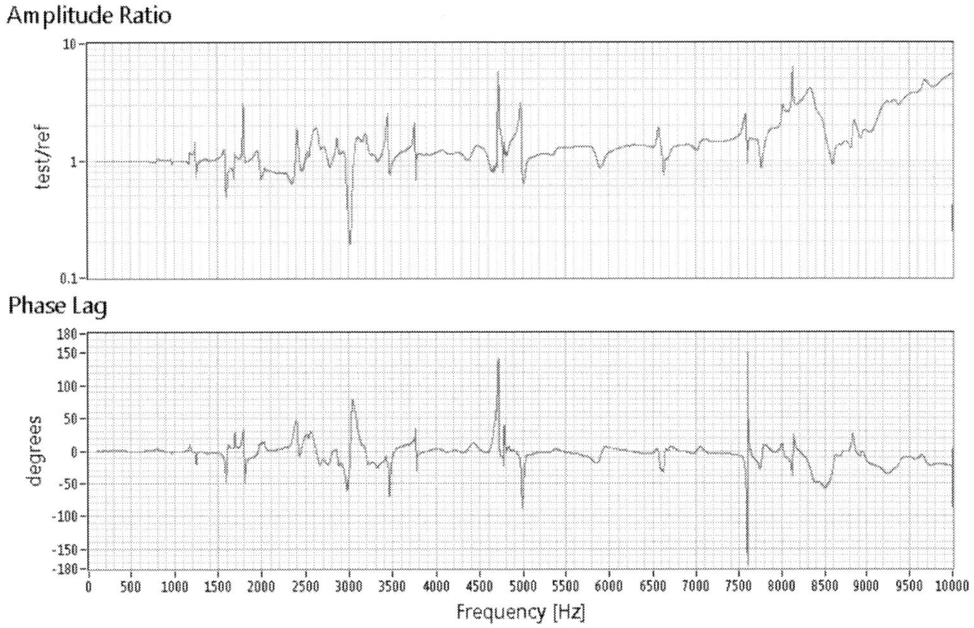

Figure 7: Acceleration attachment test comparison vs. reference, magnetic mount only

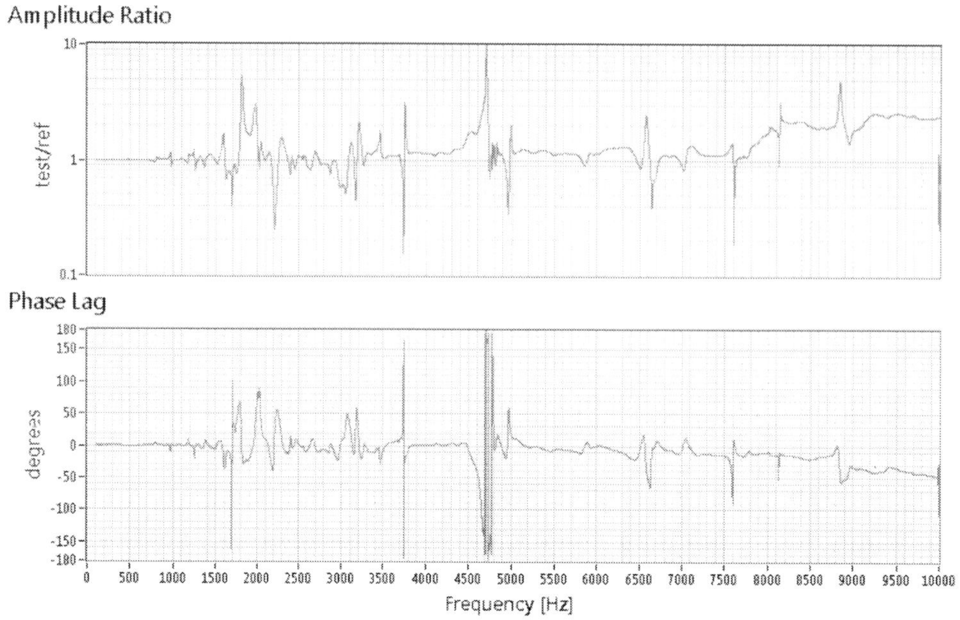

Figure 8: Acceleration attachment test comparison vs. reference, wax only

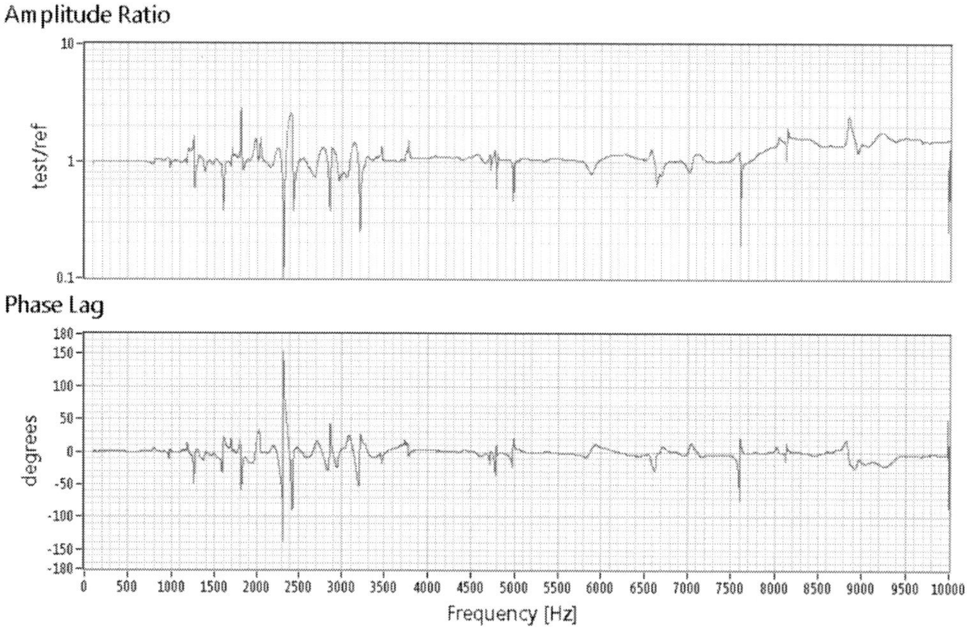

Figure 9: Acceleration attachment test comparison vs. reference, wax plus magnet mount

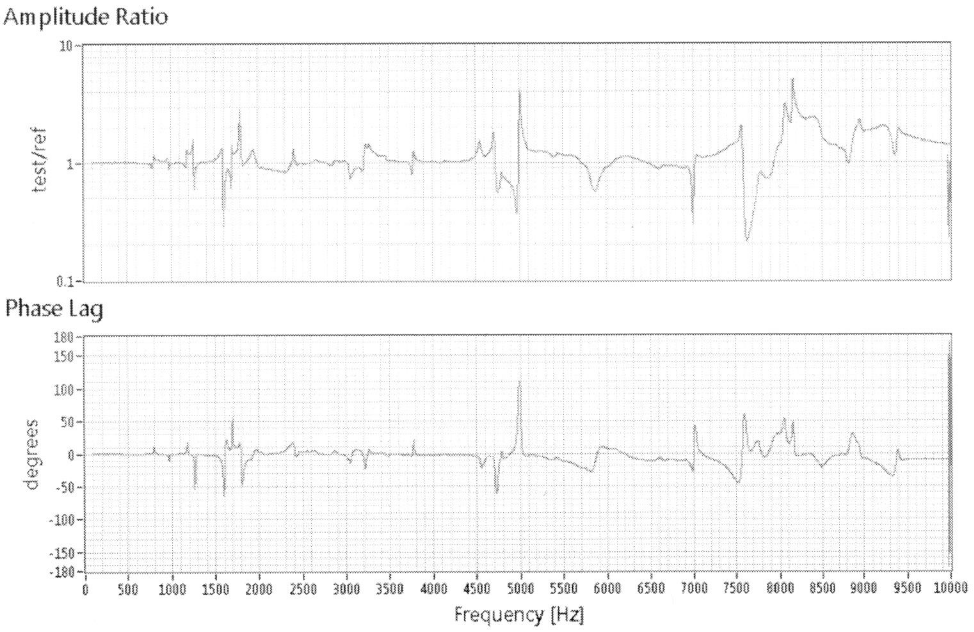

Figure 10: Acceleration attachment test comparison vs. reference, 3 mil thick tape

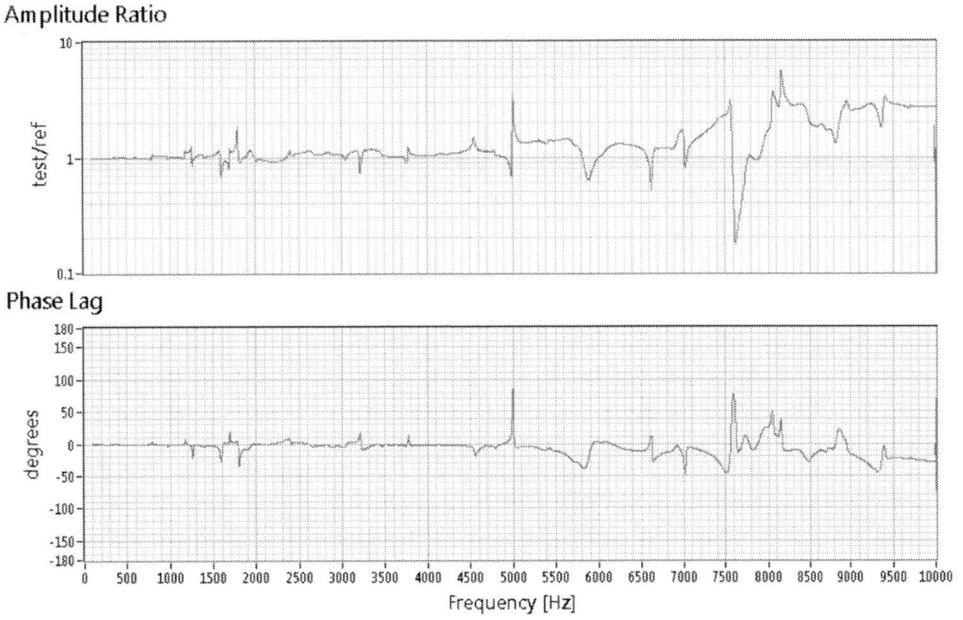

Figure 11: Acceleration attachment test comparison vs. reference, 7 mil thick tape

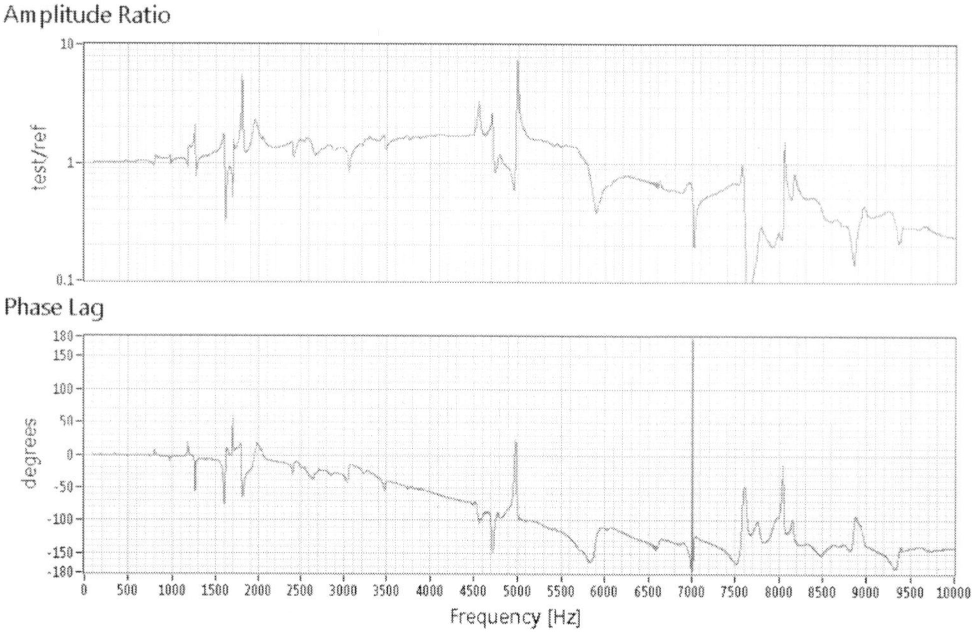

Figure 12: Acceleration attachment test comparison vs. reference, 15 mil thick tape

5. Conclusions

The test data indicated the following results:

1. In the opinion of the authors, a screw compression attachment of flat-surface-against-flat-surface is the "gold standard" of accelerometer attachment methods. This is consistent with the recommendations of international standards such as ISO and ANSI/HI.
2. All of the tested attachment techniques were found to provide accurate and repeatable results for frequencies from 1 Hz to 1.5 kHz, indistinguishable on a practical basis.
3. The best non-damaging technique of accelerometer attachment was found to be moderate thickness (7 mil, or 0.175 mm) double-sided fiber-reinforced paper tape. 50% thinner tape did not produce as high fidelity results, nor did 2x thicker tape. The 7 mil thick tape attachment method provided reasonable fidelity from 1 Hz to 7 kHz, apparently being thick enough to better fill in surface undulations than the 3 mil thick tape, but not be so thick as to lose significant stiffness though the thickness of the tape as in the case of the 15 mil thick tape. Some fidelity was lost at specific natural frequency peaks and anti-resonance "zeros", such that the reference accelerometer might report higher or lower vibration amplitude within +/-1% of such zones. Except for the highest amplification factor natural

frequencies (e.g. Q>30), the 7 mil tape provided good FFT curve shape fidelity, and practically acceptable quantitative amplitude results to 19 kHz.

4. The use of wax by itself provided results that were inferior to wax plus a magnet.
5. The lowest fidelity method was shared by magnet-alone, hand-held alone, and wax-alone. The magnet-alone was dependent on the flatness of the attachment surface. Handheld-alone was generally superior, but dependent on the firmness and steadiness of the hand holding the accelerometer. Generally, any unsteadiness in the hand holding affected frequencies at 30 Hz and below. The wax attachment was dependent to some extent on the thickness of the wax coating between the accelerometer base and the target surface.

6. Acknowledgments

The authors would like to thank Senior Engineer Joseph Gruener, and co-op students Daniel Pruess and Alison Jago for their excellent work in setting up the experimental hardware, and acquiring the data associated with this IRAD project.

References

[1] ISO 5348 [2018], "Mechanical Vibration and Shock- Mechanical Mounting of Accelerometers", International Standards Organization, Geneva Switzerland

[2] ANSI/ HI 9.6.4 [2016], "Rotordynamic Pumps for Vibration Measurement and Allowable Values", Hydraulic Institute, Parsippany NJ

[3] ANSI S2.61-1989 (ASA 78-1989) [1989, Reaffirmed 2010], "Guide to the Mechanical Mounting of Accelerometers", American National Standards Institute and ASA, New York, NY

[4] Bowers, S.V., Piety, K.R., and Piety, R.W. [1990], "Real-World Mounting of Accelerometers for Machinery Mounting", 4th Incipient Failure Detection Conference, EPRI, Philadelphia PA, October 1990

[5] Peters, J. [2007], "Restrained by Resonance", Uptime Magazine, May 2007

[6] Corelli, D.A. [2015], "How Sensor Mounting Affects Measurements", Vibrations Magazine, Vibration Institute, Oak Brook IL, May 2015

[7] Marscher, W.D., and Onari, M.M., [2019], "Vibration Problems and Solutions for Turbomachinery and Centrifugal Pumps", Short Course at MFPT2019, King of Prussia PA, May 2019

FAULT DIAGNOSIS OF EXCESSIVE PIPE VIBRATION DUE TO BEATING PHENOMENON

Meng Hee Lim, Kar Hoou Hui, and Mohd Salman Leong

Institute of Noise and Vibration
Universiti Teknologi Malaysia Kuala Lumpur
Jalan Sultan Yahya Petra
54100 Kuala Lumpur, Malaysia
Telephone: (60)3-2615 4413
mhlim.kl@utm.my

Abstract: This paper presents a case study in diagnosing an excessive pipe vibration due to beating phenomenon. The outdoor process pipes in a sewage plant were found to vibrate viciously and resulted in emitting loud hamming noise that affecting the surrounding community. The process pipes were connected to two identical blower units, each driven by a motor via belt and pulley system. Besides the loud noise, the excessive pipes vibration had also posed a concern to the plant personnel that a possibility of premature machine failures may occur if the problem persists any longer. A comprehensive vibration investigation was conducted to map-out the vibrations of the entire machine train that includes pipes, blowers, motors, skid, plinth and floor slab of the blower house. Vibration investigation found that pipes vibration was most severe when the two units of blowers were operated simultaneously. It was found that the root cause of the excessive pipe vibration was caused by beating phenomenon of which two adjacent machines operated under slightly different speeds. In this case, the two blowers were operated at 41.88Hz and 41.72Hz respectively. Beating is a phenomenon of constructive and destructive interference of two identical waveforms with slightly different frequency. As such, the remedy measure undertaken was thus to fine tune the operating speed of the two blowers. It was found that pipes vibration had subsided considerably when the two blowers' speeds were adjusted to be 7.5 Hz apart. As a result, the loud humming noise emitted from the pipes was noted to be completely mitigated with the remedial action taken.

Keywords: Pipe; beating; vibration

1. Introduction

This paper presents a case study on noise and vibration investigation that had conducted at a sewage treatment plant in Malaysia. The outdoor process pipes in the sewage plant were found vibrated viciously and thus resulted in emitting loud hamming noise affecting the residential houses nearby. Besides the loud noise, the excessive pipes vibration had also posed a concern to the plant personnel that a possibility of premature machine failures may occur if the problem persists.

Our initial investigation found that the noise originated from the sewage treatment plant has spilled to the environment primarily via the severely vibrated pipe that connected to the two blower units inside the blower house. The process pipes that connected to the two identical blower units were driven by motors via a pulley system. Besides this, the air-borne noise from the operation of the machines inside the blower house was also noted to have escaped and spilled to the environment via the exhaust and fresh air openings of the plantroom. However, the air-borne noise originated from the internal of the plantroom was found to be less severe or prominent as compared to the humming noise from the outdoor process pipe.

The objective of the investigation is to identify the root causes of the severe pipe vibration and thus provided recommendations on the remedial actions needed to resolve the problem.

2. Measurement Strategy and Procedure

Noise and vibration measurements were conducted to confirm the characteristics of the prevailing pipes vibration before any remedial works were to be performed at site. The noise and vibration measurement were carried out at the interior and the exterior of the blower house. The locations of measurement at the blower house and at the blower units itself were shown in Figure 1 and Figure 2 respectively. In short, the vibration measurement undertaken measured vibration levels of the blower house walls and floor, pipes that connected to the blowers, blower units, motors, skid, and plinth.

The measured vibration spectrum of the machines were used to identify the dominant vibration frequency peaks associated to the blowers and piping. The vibration investigation focused on identifying the root cause of excessive pipes vibration when two units of blowers (Blower 2 and 3) were ran at the same time.

Controlled tests by switching on and off the different units of blowers, and by changing the operating speed of the blowers were also carried out on site. This is conducted in order to understand the cause and effect of the changes in blower operation to the resulted vibration (and noise) levels of the machines and its connected pipes.

On top of this, ambient noise levels at the property boundary of the plant were also measured during the controlled tests. Noise measurements using a Class 1 sound level meter was deployed throughout the entire investigation work.

Point	Description
1	Wall
2	Wall
3	Wall
4	Wall
5	Floor
6	Floor
7	Floor
8	Floor
9	Pipe (Large)
10	Pipe (Small)

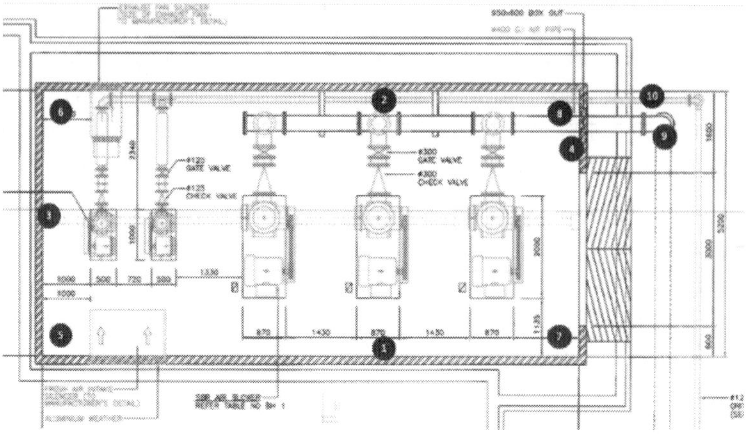

Figure 1: Vibration measurement locations at the blower house

Point	Description
1	Motor Footing
2	Blower Footing
3	Before Air Bellow (Blower Side)
4	After Air Bellow
5	Pipe (U-clamp Support)
6	Pipe (Flange)
7	Plinth
8	Raised Floor
9	Floor

Figure 2: Vibration measurement locations at the blower units

3. Results and Discussion

Table 1 tabulates the noise levels measured before the remedy works were performed at site. The noise level at the SBR Tank area was noted to have increased by 16 dB when two blowers were simultaneously ran as compared to the case with only one blower ran at a time. The increase of 16 dB can be interpreted as three times louder as perceived by the receivers.

Table 1: Noise levels at interior and exterior of the blower plantroom

Location	Single Blower in Operation	Two Blowers in Operation
Blower House (Exterior)	69 dBA	82 dBA
SBR Tank	67 dBA	83 dBA
Plant Boundary	63 dBA	71 dBA

Figure 3 illustrates the vibration spectrum measured when Blower 2 and 3 were set to run discreetly or one at a time. It was found that when the two machines were set to run discreetly, the vibration peaks as seen in Figure 3 were mainly governed by the harmonics of its own operating frequency that is at 41.88Hz and 41.72Hz respectively.

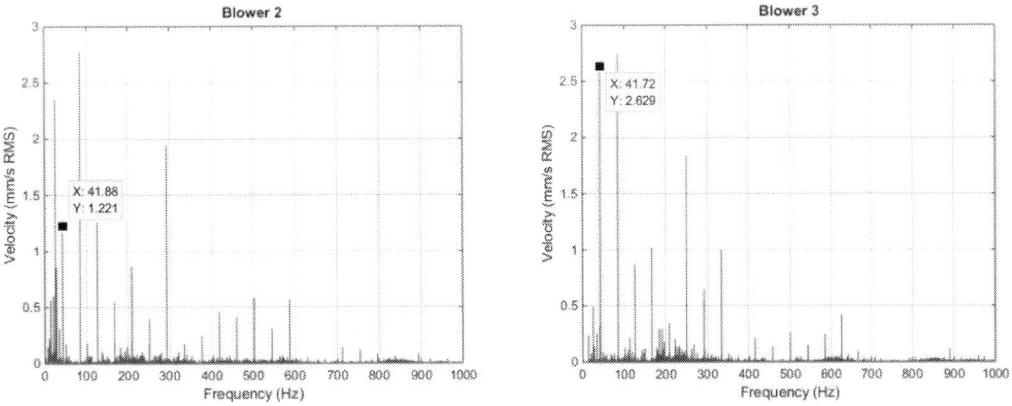

(a) Vibration spectrum of Blower 2 b) Vibration spectrum of Blower 3

Figure 3: Vibration spectrum measured at (a) Blower 2 and (b) Blower 3

Figure 4 illustrates the vibration spectrum of outdoor pipe (Point 9) with single blower and two blowers ran simultaneously. It was found that the zoom-in vibration peak at 84Hz (the second harmonics of the blower's operating speed) was actually made out of two discreet vibration peaks of 83.26Hz and 83.59Hz when two blowers were ran simultaneously.

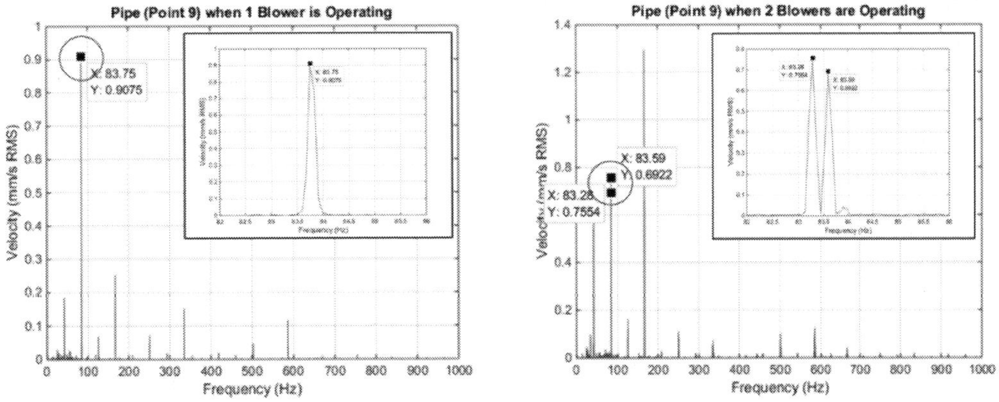

(a) Single blower ran b) Two blowers ran

Figure 4: Vibration spectrum of pipe with single and two blowers ran simultaneously

Beating phenomenon can be observed when two machines are operating at approximately but not exactly the same speed. In this case, the two blowers (Blower 2 and 3) were ran at 41.88Hz and 41.72Hz respectively. Beating is known to cause by the constructive and destructive interference of two sinewaves due to slight phase difference. Figure 5 illustrates an example of a beating wave. Beating phenomenal is widely reported in many engineering structures and machinery such as in large floating structure [1] and also in a smaller machinery such as in motor [2]. The characteristics and phenomenon of beating was also well documented in many vibration handbooks [3] and machinery diagnostic books [4].

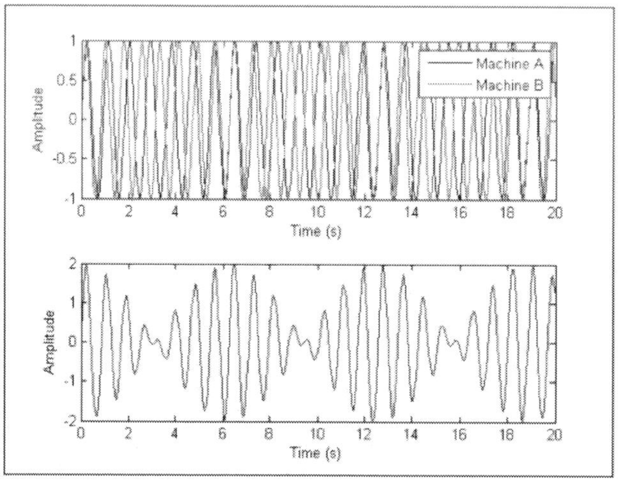

Figure 5: Example of a beating wave.

A control test was conducted by deliberately switched on Blower 2 and 3 by stages and at the same time the vibration response of the outdoor pipe (Point 9) was measured. Figure 6 depicts the vibration spectrums measured during the transition period of a single blower ran to two blower ran simultaneously. Beating wave can be clearly observed after the second blower was switched on and stabilised. The waveform of the two-blowers-ran was noticed to exhibit the characteristics of beating waveform as showed in Figure 5. Figure 6 illustrates the vibration spectrum of the beating waveform. A zoomed-in displays of the individual peak in the vibration spectrum revealed that there were actually formed by two or more individual peak which corresponding to the harmonics of the operating frequencies that triggered the beating process in the pipes.

Figure 5: Vibration spectrums of the transition of single blower to two blowers' operation

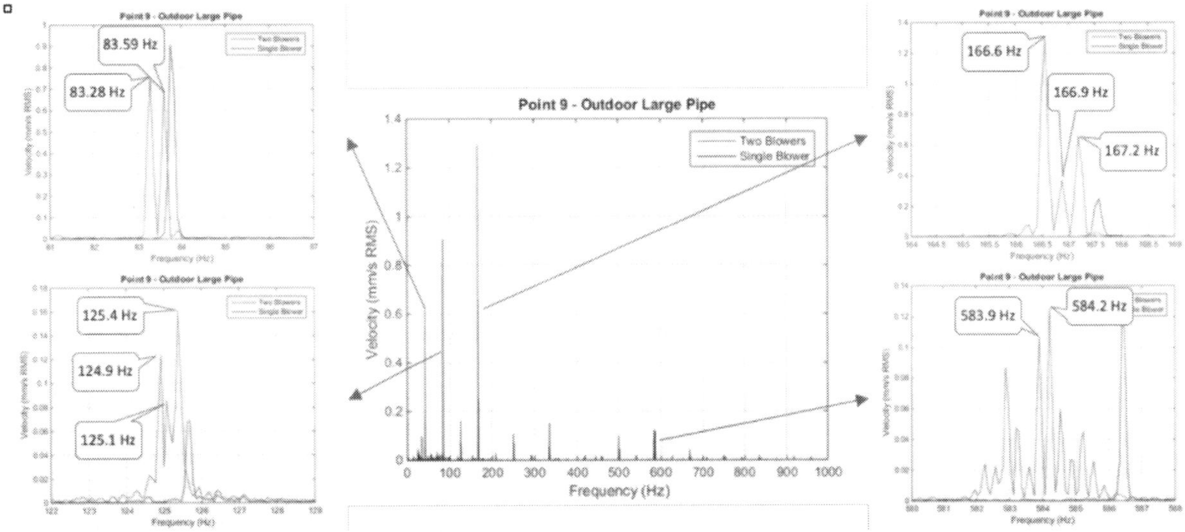

Figure 6: vibration spectrum of the beating waveform with zoomed in displays

4. Remedial Actions Proposed and Implemented

Based on these evidences, beating phenomenon was confirmed to be the root cause of the excessive noise and vibration occurred to the process pipes outdoor.

Beating can be mitigated by avoiding the presence and interaction of two or more closely spaced frequency peaks that transmitted to the same structure. As such, the remedial measure proposed in this study was to fine-tune the operating speeds of the two blowers to be either perfectly matched or reasonably far apart to eliminate the sources of the vibration waveforms that enable beating.

Subsequently, a second controlled test was carried out in the attempt to fine-tune the operating speeds of the two blowers. In short, the operating speed of Blower 2 were gradually increased from 40Hz to 50Hz while the operating speed of Blower 3 was maintained at 50Hz (and in one case at 45Hz). The corresponding vibration response of the outdoor pipe was measured and recorded for comparisons.

Table 2 provides the resulted pipe vibration levels in correspondence to the changes in blower operating speeds. It was noted that piping vibration reduced significantly under the most optimised operating speeds. It was found that the most optimised operating speed of the two blowers was when the difference of the operating speed were 7.5Hz apart or the two blowers were set to run at 42.5Hz and 50Hz respectively. The vibration levels of pipes reduced from 23.31 mm/s under the worst case scenario to only 10.91 mm/s under the most optimised operating condition.

Table 2: Pipe vibration levels in correspondence to the changes in operating speeds

Blower 2 and 3 Operating Speeds	pk-pk Vibration (mm/s)
50.0Hz & 50.0 Hz (Beating)	23.31
42.5Hz & 45.0Hz	14.10
42.5Hz & 50.0Hz	10.91
43.5Hz & 50.0Hz	12.15
44.0Hz & 50.0Hz	51.70
45.0Hz & 50.0Hz	12.64

5. Noise and Vibration Levels After the Remedial Actions Taken

Noise levels of before and after the remedial action at the exterior of the blower house were given in Table 3. It was found that the noise levels measured before remedial action was about 82 dBA. After the remedial action (by operating the Blower 2 and 3 at 42.5Hz and 50Hz), the resulted noise levels were about 73 dBA. This represents a significant 9 dB reduction, in which can be subjectively perceived as twice as quiet based on the subjective perception of loudness for human being.

Table 3: Comparison of noise levels before and after remedial actions

	Before Remedy Action	After Remedy Action
Noise Level at Blower House (Exterior)	82 dBA	73 dBA

Beside this, the vibration level of the pipes with two simultaneous blowers run was also noted to reduce significantly and matched the vibration level of a single discreetly run blower after the remedial action taken.

6. Conclusion

This paper provides a case study on beating phenomenon that occurred in process pipes which connected to two identical machines that ran under a slightly different operating speeds. It was shown that beating phenomenon can be identified with a comprehensive vibration investigation study. As beating is a phenomenon of constructive and destructive interference of two identical waveforms with slightly different frequencies, the remedy action undertaken was thus to fine-tune the operating speeds of the machines to eliminate the sources of the vibration waveforms that enable beating. This paper showed that beating in pipes can be effectively mitigated by fine-tuning the operating speeds of the machines.

Acknowledgments

The authors would like to extend their deepest gratitude to the Institute of Noise and Vibration UTM for funding this study under the Higher Institution Centre of Excellence (HICoE) Grant Scheme (R.K130000.7809.4J226, R.K130000.7843.4J227 and R.K130000.7843.4J228).

References

[1] H. Endo, H. Suzuki, "Beating vibration phenomenon of a very large floating structure", Journal of Marine Science and Technology, September 2018, Volume 23, Issue 3, pp 662–677.

[2] A. Biernat, P. Góralski, "Analysis of state of operation of asynchronous motor with stator slot frequency beat vibration". IEEE, International Symposium on Electrical Machines (SME), Naleczow, Poland, DOI: 10.1109/ISEM.2017.7993560, 2017.

[3] A. G. Piersol, C. M. Harris, "Harris' Shock and Vibration Handbook", McGraw-Hill Education; 6 edition, November 9, 2009.

[4] R. T. Buscarello, "Practical Solutions to Machinery and Maintenance Vibration Problems", Update International, Forth Revision, May 2002.

TOWARDS REPLACEMENT OF FAILED PARTS ON THE BATTLEFIELD VIA METAL CASTING IN 3D-PRINTED DESERT SAND MOLDS

Marc Pepi[a]
Jerry Thiel[b]
Nathaniel Bryant[b]
Brandon McWilliams[a]
Andelle Kudzal[a]
Jennifer Sietins[a]

[a]-Army Research Laboratory, Aberdeen Proving Ground, MD
[b]-University of Northern Iowa, Cedar Falls, IA

Abstract

The research was performed to show the feasibility and scale-up potential of production of replacement parts for long lead time DoD critical components utilizing a sand 3D-printer with indigenous desert sand to produce casting molds from a digital drawing of the actual part. Replacement and/or spare parts could then theoretically be manufactured by pouring molten metal into these molds. Success was achieved in producing parts in a laboratory setting by pouring molten aluminum into 3D-printed desert sand molds of smaller components, but scale-up proved difficult, as the fine desert sand did not fully adhere to itself, and the mold strength was less than optimal. Lessons learned could potentially be utilized to further this effort in the use of different sand, for example beach sand, which may be easier to process.

Introduction

Additive manufacturing has changed the face of conventional metal forming technologies. Conventional metal casting involves the manufacture of solid forms or patterns that sand is formed around to develop the shape of the cavity that is filled with liquid metal. The patterns or tooling for castings will often cost as much as 100 times that of the needed casting and require weeks to months to produce. The tools must then be stored, maintained and repaired between uses. Because of the amount of space required and the length of time required for producing patterns, the technology is not conducive to a mobile manufacturing platform. Additive manufacturing and digital part creation has allowed the production of castings within days rather than weeks or months. It has reduced the need for the lengthy production of costly tooling with requirements for storage space. With the proper tools and training, additive manufacturing for metal casting could allow replacement of critical cast components and provide for weapon system sustainability on or near the battlefield. This approach allows for the geometric freedom provided by 3D-printing as well as the flexibility of traditional casting [1,2]. The vision of this work was to replace the traditional sand mold equipment in the Tactical Fabrication (TACFAB) conex with a sand 3D-printer, with the hopes of using desert sand and a binder to create casting molds at the point-of-need.

Background

It would be advantageous for our warfighters to have access to technology that uses locally available indigenous materials. Several favorable outcomes of this scenario would include reduction of energy and material costs related to transportation, especially for well-established industries and support of local businesses and resource bases [3]. Shrinking the logistics tail is an important benefit of utilizing indigenous materials in-theater [4]. The 2012 Army Sustainability Report, released 31 October 2012, outlines the Army's desire to reduce the number of convoys required to resupply troops on the battlefield [5]. Reducing vulnerable convoys not only saves materiel and lives, but troops assigned to guard these convoys can actually be utilized for their intended purpose – engaging the enemy. The charter to reduce the tail in the combat zone is deemed critical to the success of the overall Army transformation, with relevance to Army future missions [6]. The Army R&D, as well as the sustainment community, should consider the reduction in the logistics footprint a principal goal. As stated in Logistics Transformation – Reducing the Logistics Footprint by Ransom et al., "Technology will be one of the primary enablers to reduce the logistic footprint, and the reduction of the logistic footprint is clearly a key element of the future battlefield" [7]. In addition, the armed forces are increasingly playing humanitarian roles in the context of assisting citizens who have lost their assets in a natural disaster and/or live in parts the world where there is no infrastructure for creating buildings, roads, bridges, or manufacture materials that can clean water, create energy, or repair machines. The ability to build and make materials with indigenous materials dually serves both the armed forces and society's needs.

State of the Technology

The intent of this program was to further 3D sand printing technology with the incorporation of a printer that utilizes CAD drawings of key components and the use of desert sand (instead of the OEM sand that is intended for use with this printer). This capability would allow the manufacture of replacement parts in-theater, which can be used as "Band-Aids" until the actual replacement parts are delivered to the forward operating bases. One-to-one replacements (where actual replacement parts would not be needed) would reduce the military logistics tail and enable point-of-need production. Work has been done in the area of using sand 3D-printers for metal casting, but none of this research has used actual desert sand. It is reported that sand is the most widely used 3D printed material over metals and plastics [8]. This prior work includes research performed by UNI through America Makes entitled "Additive Manufacturing for Metal Casting" which concluded that 3D-printing casting molds with sand made casting faster and more affordable for the automotive industry within the small and medium sized companies that make up the domestic metal casting supply chain [9]. In similar work, it was determined that binder jet sand 3D printing could be used to manufacture metal components with similar properties to those of traditional sand casting processes [10]. Comparable work has been performed at Indiana University – Purdue University Indianapolis (IUPUI) [11]. Their study concluded that a significant amount of sand could be saved by using 3D printed sand molds versus conventional sand casting. This study also concluded that the surface finish of 3D printed sand molds was higher than conventionally cast parts, and that closer dimensional tolerances could be achieved with 3D printed sand molds versus conventional sand casting. Almaghariz, et. al. claimed that 3D sand printing leads to significantly reduced lead times [12]. A lead time reduction from 10 weeks down to 12 days was provided by Connor [13] and a reduction from four months to four weeks in a case study found in Donaldson et al. [14]). Additionally, reduced labor costs (a savings of $14,000 was found by Connor [13]) as well as improved flexibility, reduced tooling and enhanced mold designs were found to be benefits from using 3D printed sand molds. When it comes to intricate designs, Singh Gill et. al. [15] concluded that 3D printed sand molds were good for parts that were thin-walled.

Approach – Proof-of-Concept with Mock Component

To determine the feasibility of using desert sand for 3D-printing casting molds, a mock component was chosen. A drawing of this mock component was converted to an .stl file by the University of Northern Iowa (UNI) and subjected to casting software which determined the optimum positioning of gates and risers needed to cast the part.

Sand Analysis / Preparation for 3D-Printing

For this program, UNI received sand from the National Training Center, Ft. Irwin, Mojave Desert, California. This sand was nicknamed "moon dust" by the Ft. Irwin staff, but the four 55-gallon drums did contain various distributions of rocks. This sand was screened by UNI to a usable US mesh size of 50 – 140 (~297 to 105μm average diameter size).

The desert sand was subjected to preliminary chemical analysis at UNI, as well as energy dispersive spectroscopy (EDS) at ARL to determine composition. Table 1 contains the results of analysis at UNI. EDS was able to detect carbon (most likely contaminant), oxygen, sodium, magnesium, aluminum and silicon. The other elements detected via chemical analysis were in too low of a quantity to be detected via EDS.

Table 1

Desert sand chemical analysis

Mass%

SiO_2	Al_2O_3	Fe_2O_3	Cr_2O_3	TiO_2	CaO	MgO	Na_2O	K_2O
91.13	4.564	0.707	0.001	0.196	1.044	0.031	1.496	0.833

UNI also calculated the pH and acid demand value (ADV) of the as-received sand, versus the same sand that had been screened to 70, 100 and 140 mesh size, respectively. As shown in Tables 2 and 3, the pH and ADV was lowered after sieving, making it more suitable for 3D-printing. Chemically, the desert sand underwent baseline testing at UNI according to AFS standard procedure. The pH test determines the basicity or acidity of the base material. This test is often associated with the ADV, or acid demand value. The ADV is used to determine how much acid is needed to cure the resin in chemically bonded sand systems. Lastly, the moisture test is conducted to determine how much water is withheld in the sand prior to experimentation. The pH was determined to be 9.57, which means the material is slightly basic. This is further supported by a high ADV of 24.5. The sand was determined to have a moisture content of 0.41%.

Table 2
pH, Acid Demand Value (ADV) and Moisture Content of
As-Received Mojave Desert Sand

pH	Acid Demand Value	Moisture
Average = 9.570	Average = 24.500	0.41

Table 3
pH and Acid Demand Value (ADV)
Screened Mojave Desert Sand

pH		Acid Demand Value	
70 mesh size	8.240	70 mesh size	10.150
100 mesh size	8.410	100 mesh size	10.200
140 mesh size	8.440	140 mesh size	9.950
Average = 8.363		Average = 10.100	

Scanning electron microscopy of the Mojave Desert sand showed a layer of dust/clay on the particles. Image J® software calculated the average diameter at 61 +/- 35μm from analysis of 50 random particles.

The commercial sand (silica) that is typically used with the ExOne® S-Max 3D-printer was compared to the indigenous sand from the Mojave Desert. Of note were differences in morphology, particle size and composition. Of these differences, particle size would not matter due to sieving that would result in particles of useful diameter for 3D-printing. Morphology may pose a problem, as particles may range from sharp and acicular to spherical in geometry, affecting packing density. Composition could also have an effect on the as-cast part, if metals such as steels were cast. For the light alloy being cast for this program (A356 aluminum), this composition would unlikely affect the physical and mechanical properties of the as-cast part.

<u>Printing of Casting Molds from Desert Sand</u>
UNI utilized the ExOne® S-Max 3D-printer to create the casting molds, which employs 8 kg of sand in two feeders. A sulfonic acid binder was mixed with the sand for 60 seconds in preparation for printing. Within the equipment, alcohol is coated onto the deposited sand/binder mixture to harden (12-minute chemical-bonded cure). The machine is capable of depositing a 0.28mm thick layer of sand in 40 seconds.

Casting molds for the mock part were made of both 3D-printed commercial sand and indigenous desert sand. The parts were subjected to micro-computed tomography (CT) inspection and stylus surface roughness measurements.

<u>Computed Tomography of Mock Components</u>
The two mock components were subjected to X-ray CT for the purpose of comparing the external surfaces and examining internal defects. A Northstar X5000 unit (with a scan power of 200kV, 1700μA and a resulting voxel size of 146.8μm) with efXCT and Volume Graphics software was utilized for this inspection. The exterior of each looked visually similar, with the part made from

the indigenous sand mold looking perhaps slightly rougher. It appeared that more internal defects were found within the actual part which was printed with indigenous sand, compared to the part made with the commercial sand, where the defects appeared to be more concentrated in the risers.

Surface Roughness of Mock Components

The surface roughness of a section from each part was measured using a Keyence laser profilometer. A sample was removed from each specimen in the same area. The average of 4 scans (2mm x 2mm) confirmed that the sample cast in commercial sand was smoother than those from the part cast in desert sand.

Metallography

Small samples of each of the mock components (made with commercial sand and with desert sand molds) were subjected to metallographic inspection. The samples were ground and polished and etched with a modified Murakami reagent. Scanning electron microscopy of the resultant structures showed that the type of sand mold used in this study had no obvious effect on microstructure.

Scale-Up – Sand Molds for an Actual USMC Amphibious Assault Vehicle (AAV) Component

Based on the success of producing a part from a mold 3D-printed with indigenous sand, scale-up to an actual weapon system component was subsequently attempted. The part chosen was a long lead-time component form a USMC AAV. The part is required to be made from casting aluminum 355 or 356 tempered to the T6 condition per QQ-A-601 (superseded by ASTM B26). It was decided that, concurrently, this component will also be additively manufactured using a 3D Systems ProX 300, using alloy Al12Si for comparison.

Desert Sand Challenges

It is understood that desert sand would most likely only be useful for low temperature alloys such as aluminum, since traditional casting uses molds comprised of granular refractory materials such as pure silica, olivine, chromite or zircon sands [16]. Reference 16 also lists other properties of traditional sand molds that could potentially be compromised by using indigenous desert sand:

- Strength enough to sustain the weight of the molten metal
- Allows for gases formed in the mold to escape
- Resistance to the erosive action of the molten metal
- Allows for unhindered metal constricting upon cooling
- Able to cleanly strip away from the cooled casting

Although a mock component was made from a mold comprised of 3D-printed indigenous sand, it turned out the sand received from Ft. Irwin, CA contained too many large pebbles and rocks, as well as a large amount of "fines". This meant that collecting enough usable sand to make the molds for the AAV component was very difficult. The ExOne S-Max 3D-printer needs a fairly normal distribution of sand particle size from 70 to 200 mesh to operate efficiently. Methodology for sieving and screening while obtaining a usable yield provided a challenge, and it was unknown initially whether there would be enough usable sand from the amount received to make the intended casting molds. It was decided to screen the base material using a US-8 mesh riddle. This eliminated large, unusable particles from the sand. The pre-screened sand distribution was then determined by sieve analysis according to AFS-1105-00-S.

The pre-screened sand distribution was then manipulated using an automatic double-deck screener, which has three chambers that retain particular grain sizes and then exports that material into individual containers. Large particles are retained in the top chamber while fines fall through to the bottom. The sand retained in the middle chamber was collected for sieve analysis. The screener effectively eliminated coarse particles up to the US-100 mesh screen; however, it was determined by this data that there was an accumulation of fine particles in the system. It was then observed that the bottom screen was blinded by the dust present in the sand. The dust obstructed the flow of fine particles which allowed the excess fines to be retained in the middle chamber.

It was determined that the fine particles needed to be separated prior to being sent through the automatic screener, since they would be detrimental to the printing process. To accomplish this, a fluidized chamber was constructed. A fluidized chamber used air to convert granular materials from a static solid-like state to a dynamic fluid-like state. The bottom of the chamber contained air that is blown in to create pressure. The top channel contained the sand. They were separated by a porous fabric that allows air to travel through freely while still retaining the material in the top channel. An industrial blower used for a gas-fired furnace was needed to provide enough pressure for fluidization. With the equipment ready for operation, pre-screened sand was loaded into the fluidized chamber and energized. After five minutes of fluidization a sample was collected for sieve analysis.

Finalization of Sand Processing

Due to the low sand yields (approximately 35%), an additional 55-gallon drum of the Mojave Desert "Moon Dust" was forwarded to UNI. The equipment was further modified in an effort to reduce not only the larger aggregates, but also the large amount of "fines". The modified equipment was tilted to various angles to determine the optimal screening.

After receiving this fifth drum of sand, 9 barrels were fully processed to reach the finalized screen distribution described in Figure 1. There was a total yield of 884lb (401kg) of printable material as a result of the sand processing methodology from the approximately 2,500lb (1,134kg) of sand received. Since the sand distribution and processing were finalized, several baseline tests were conducted to understand the printability of the aggregate.

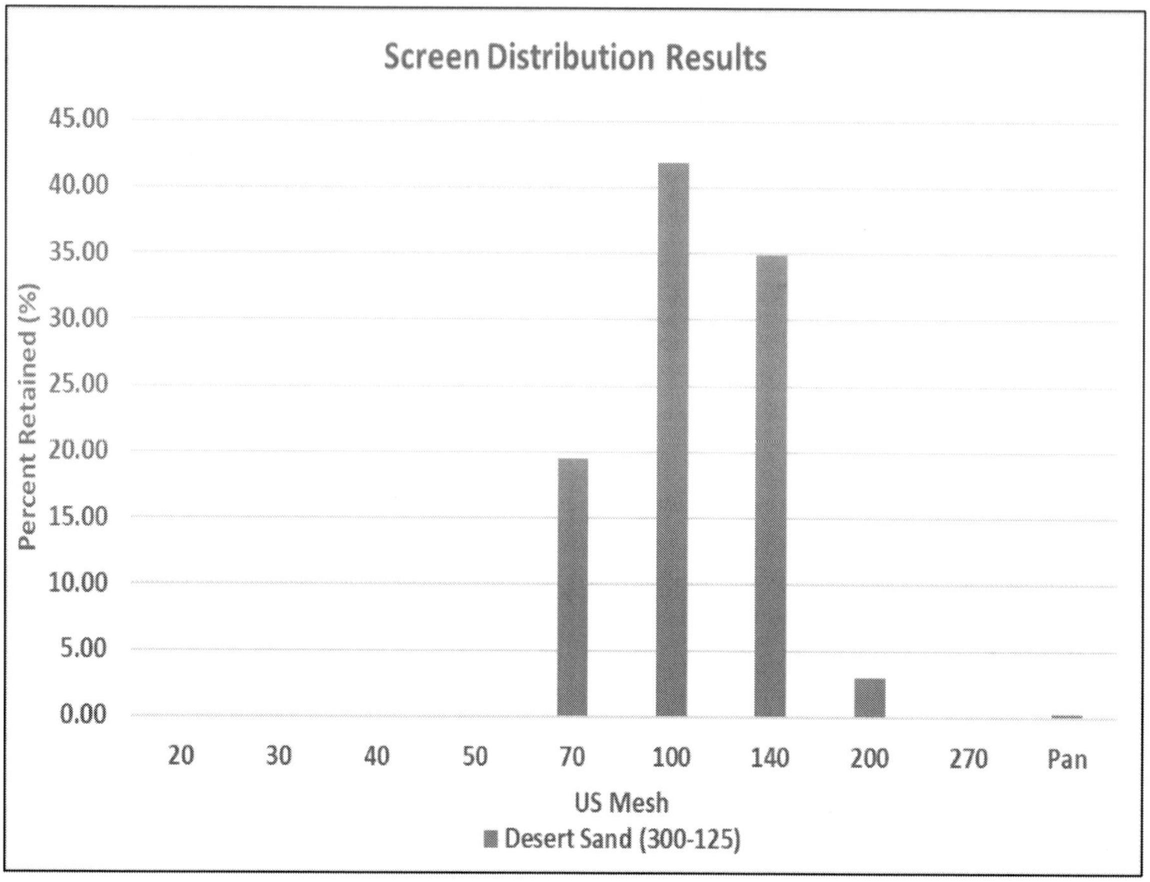

Figure 1 Accepted screen distribution of desert sand for printing.

<u>Flowability Trials</u>

An important consideration for sand printing is the flow characteristics of the preactivated sand. During the printing process, virgin material is siphoned into a hopper. A predetermined quantity of sand is fed into the mixing chamber where a liquid activator, toluenesulphonic acid, is introduced and the mixture is homogenized. The acid addition is calculated based on sand weight, and the printer is capable of printing in a range from 0.1 to 1% activator. When the mixer finishes a cycle, the coated sand is deposited into the recoater (the component that creates the sand layer). Several factors, including print environment humidity, activator %, and sand surface area directly influence the flow characteristics of the media. To understand the flowability of the desert sand, a granulate/powder flowability tester (powder rheometer) was used to measure how different activator levels change the sand flowability. The funnel has an interchangeable disc that is placed at the tapered end with a specific size hole placed concentrically. The hole diameters ranged from 4 to 36mm (0.16 to 1.42in).

Several common materials used for sand printing were tested alongside the desert sand for direct comparison. The sand was coated with toluenesulphonic acid at four different weight percentages (0.25%, 0.35%, 0.45% and 0.50%) and loaded into the funnel. The hatch was dropped, and if sand didn't flow out of the funnel without irritation, progressively larger hole diameters were tested until the sand did freely flow. The recorded value was the minimum hole diameter for the coated sand to flow freely. Flow was measured by starting with a small hole diameter (3mm) and loading the funnel with sand (with a specified weight percentage of activator). The hatch was dropped, and it was noted if the sand freely flowed through the hole without any irritation. If the sand did not flow, progressively larger hole diameters were tried until it did flow. For example, in Figure 2, for W730, a commercially available silica sand, the sand would flow out once the hole diameter was 20mm if it contained 0.10% TW40, but not before. After increasing the activator content to 0.25%, the sand would not freely flow out until the hole diameter was 30mm.

Figure 2 shows that the slope of the desert sand trend line was significantly lower than the other materials. This phenomena could be due to contaminants absorbing the moisture from the liquid acid. There is a much more noticeable rate of change increase in the other materials at significantly lower activator percentages. It may be worth mentioning that the other materials used are high purity silica sand and engineered ceramic materials with very low contaminants. The surface area of each sand was calculated based on the screen distribution and grain shape of the material and compared to the slope. The trend shows that higher surface area decreases the rate of sand adhesion with increasing activator content.

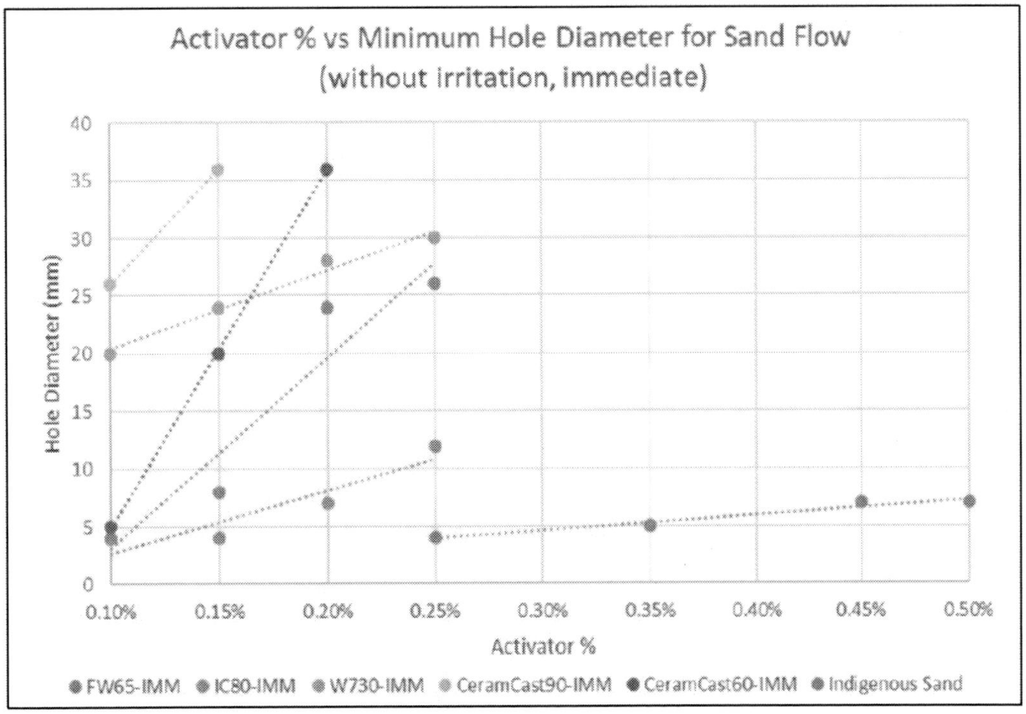

Figure 2 Flow characteristics for various sands

Another important consideration when printing sand is the acid demand value of the base sand. In acid catalyzed binder systems, such as furan, there is a quantity of acid that is needed to neutralize any basic components present in the aggregate. This gives the operator an idea of how the sand bonds. ADV trials were conducted on the desert sand to understand the bonding characteristics of the material. There were interesting results found from these experiments. When certain acids were added to the desert sand, the acid demand value increased, which is counter-intuitive. It was concluded that some sort of reaction between contaminants in the sand and the acid released more basic components into the sand; therefore, the acid demand value increased. This meant that certain acids could not be used to neutralize the aggregate. After several trials with various recipes there were no successful bonds formed with the desert sand. It also turned out that it was much harder to retain the desert sand within the recoater of the sand 3D-printer than with commercial sands

Initial Bonding Trials

As mentioned, the indigenous sand was found to have a much higher acid demand value (ADV) than the commercial sand typically used for printing (17.67 vs. 1.4 ADV, respectively). The ADV of sand is indicative of the alkalinity of the sand, and represents the chemical balances of the sand system [17]. ADV values measure the amount of milliliters of sodium hydroxide necessary to neutralize 100 grams of sand. An ADV of 17.67 indicates that 17.67 ml of NaOH are needed to neutralize 100g. The ADV has an important effect on the catalyst requirements of cold-setting acid-catalyzed binders [18]. This reference also indicates that sands with ADVs greater than 15 are not suitable for acid catalyzed binder systems, and ADVs less than 6 are preferred. Table 4 lists the many bonding trials attempted with the indigenous sand in efforts to lower the ADV. It was found that sulfuric additions were able to lower the ADV of the indigenous sand, but for the most part, none of these recipes were considered "successful". This study found that the ADV of the indigenous sand could be lowered to zero with an addition of 0.4% sulfuric acid. Bond testing was performed on the indigenous sand mixed with 0.4% sulfuric acid, 5% dry acid and 2.5% FNB resin. As shown in Table 5, the results were much lower than the nominal 80+ psi needed for casting purposes.

To increase strength, blending of the indigenous sand with commercial sand was attempted. As Table 6 shows, even with a 50-50 mix of indigenous and commercial sand, the tensile strength was still less than half of that required for printing molds.

Table 4
Initial Bonding Trials

Trial #	Resin %	Acid Catalyst %
1	1.50	0.2% toluenesulphonic acid
2	1.50	0.5% toluenesulphonic acid
3	1.50	1% toluenesulphonic acid
4	2.00	1% toluenesulphonic acid
5	1.50	1% fluoroboric acid
6	1.50	0.5% toluenesulphonic acid + 1% fluoroboric acid
7	1.50	0.325% toluenesulphonic acid + 0.175% xylenesulphonic acid 1% fluoroboric acid
8	2.00	0.65% toluenesulphonic acid + 0.175% xylenesulphonic acid
9	1.50	0.2% sulfuric acid
10	1.50	0.5% sulfuric acid
11	1.50	0.5% sulfuric acid + 1% toluenesulphonic acid
12	1.50	2% dry acid
13	1.50	4% dry acid
14	1.50	5% dry acid
15	2.00	0.2% water + 5% dry acid
16	2.00	0.2% water + 5% dry acid + 0.5% sulfuric acid

Table 5
Indigenous Sand Bonding Strength Testing

Sample #	Tensile Strength (psi)
1	11.3
2	11.2
3	5.7
4	10.2
5	10.2
6	10.6

Table 6
Blended Sand Bonding Strength Testing

Sand	TW40%	Dry Acid%	E3D%	Tensile Strength (psi) - Average
100% indigenous	0.5	5	2.5	9.9
75% indigenous / 25% FW80	0.5	2	1.5	4.5
75% indigenous / 25% FW80	0.5	3	2.0	10.8
75% indigenous / 25% FW80	0.5	5	2.5	16.4
50% indigenous / 50% FW80	0.5	5	2.5	37.3

Improved Binder

Due to the presence of increased fines within the screened Mojave desert sand, an improved binder was sought to deal with the higher surface area exhibited by these fines. Furan resin weight percentages that were used were already above the recommended range of the supplier to account for these increased fines. Batch mixing trials were conducted on the indigenous sand with a new, cement-based experimental binder provided by Rutgers University. Upon receipt of the material, four recipes (Table 7) were tried using a modified mixing methodology. The results were promising and tensile strengths as high as 65 psi were achieved with this new binder (Table 8). However, at such high binder concentrations, the curing occurred quickly, reducing the time that one can work with the sand mixture. Test cores (called "Gertzman" cores) were also created with these recipes and the resultant casting images are supplied within. These test cores are used to assess and evaluate metal penetration, and potential casting-related defects on a test scale, without making full-scale castings. A surface plate casting was also created to measure the surface roughness of metal cast against the indigenous material. It was found that it behaved most similarly to a commercial, 44 AFS-GFN silica sand.

Testing Methodology

Small Batch Mixing - A representative, 3500g batch of pre-screened indigenous sand was placed in a small batch mixing container. The first part of the experimental binder system, the "N-400" powder, was then introduced at the specified weight percentage defined by the recipes in Table 7. The result was mixed for 3 minutes, but the sand/powder mixture was flipped every 1.5 minutes. Next, either the specified weight of "X2" or "X11" binder was added and mixed for 1 minute.

This sand was used to create standard tensile specimens, Gertzman test cores, and a surface plate test core.

Melting Operations

The Gertzman test molds were cast using class 30 gray iron, with a pouring temperature of 1440°C (2625°F). After 24 hours, shakeout occurred, and the resultant castings were sectioned and the evaluation areas were imaged using a high-resolution camera. Comparisons were drawn between both sand type and the presence of a graphite-based refractory mold wash. A surface plate mold made from desert sand utilizing recipe 4 was also cast using A356 aluminum to measure the metal surface roughness. Using a Taylor-Hobson stylus profilometer, eight measurements were gathered on the resultant casting and the results were compared to data from a previous investigation [19].

Tensile Testing

Table 7 also lists the tensile strength results from the small batch mixing recipes. The first attempt, recipe 1, did not achieve any strength; however, upon increasing the overall addition rates of both the N-400 powder and X2 liquid material strengths of approximately 12.5 psi were achieved after 24 hours (the typical strength development period for foundry binders). Recipes 3 and 4 utilized the X11 liquid material rather than the X2, and the addition rates were increased. It can be observed that 24 hour strengths of approximately 65 psi were obtained using the last recipe, but it was noted that the work and strip time of the material was quite fast (as low as 3 minutes), and any addition rates higher than those tried in this recipe would cure too quickly. X11 exhibited a much faster curing rate than X2, so higher addition rates may be tried utilizing X2 to potentially achieve higher maximum strengths. Rutgers mentioned that the X11 would have higher strength than the X2, based on the chemistry.

Table 7

24-hour tensile strength results from each small batch mixing trial conducted

Recipe ID	Composition (wt% BOS)	24Hr Tensile Strength (psi)
Recipe #1	4% N-400, 5% X2, Desert Sand	0.0
Recipe #2	6% N-400, 7.5% X2, Desert Sand	12.5
Recipe #3	6.2% N-400, 7.5% X11, Desert Sand	28.7
Recipe #4	9% N-400, 10.9% X11, Desert Sand	64.9

Gertzman Casting Results

The Gertzman casting result for a commercial, 57 GFN silica sand utilizing 4% N-400 powder and 5% X2 binder without the application of the graphite-based refractory coating showed metal penetration defects in the evaluation area. This is typical for baseline silica sand and were effectively eliminated when the graphite-based refractory coating was applied.

The first Gertzman casting result for the indigenous sand was created using 6% N-400 and 7.5% X2 binder. This core was cast without any refractory mold coating, and as a result, there was severe metal penetration in the evaluation area. Large particles could be seen adhering to the casting surface, which were irremovable by sand blasting at 80psi. When the graphite-based refractory mold wash was used with the indigenous sand, the overall penetration was reduced but there were still a significant number of defects present, including embedded large particles on the evaluation surface.

Surface Roughness Results

Table 8 presents the results from the surface roughness measurements. The desert sand yielded an average surface roughness of 282.5 RMS, or root mean squared. When comparing this number to data gathered for silica sand during a prior investigation [19], it was most similar to a commercial, 44 GFN material.

Table 8

Surface roughness results measured from indigenous sand casting.

Measurement #	Result (RMS - μin)
1	300 (273 Ra – converted)
2	280 (255 Ra – converted)
3	240 (218 Ra – converted)
4	240 (218 Ra – converted)
5	310 (282 Ra – converted)
6	320 (291 Ra – converted)
7	340 (309 Ra – converted)
8	230 (209 Ra – converted)
Average	*282.5* (257 Ra – converted)

Process Simulation

Magmasoft® software was utilized to simulate the pouring and solidification of the AAV component made from A356 aluminum alloy. This software is commonly used to reduce/eliminate the porosity, shrinkage and other deleterious casting defects, and suggests optimal location of gates, risers and chills. Initial software results of casting the AAV component was optimized with strategically placed gates and risers, which pushed potential defects out of the actual component.

Mold from Indigenous Sand with Improved Binder

An attempt was made to pack indigenous sand (with the improved binder) into molds 3D-printed with the IC80 commercial sand (silica). Unfortunately, even with the new binder, the strength was not sufficient, and the indigenous sand mold broke upon removal.

3D-Printing Casting Mold from Commercial Sand

As a proof-of-concept, cope and drag patterns were 3D-printed with commercial sand based on the optimized results of the Magmasoft® software. IC80 commercial sand was utilized for these pieces. This was 3D-printed from the IC80 commercial sand that was coated with polyurethane. The resultant part is shown in Figure 3.

Recommended Post-Processing

As mentioned previously, the USMC AAV component is required to be fabricated from 355 or 356 aluminum casting alloy and tempered to the T6 temper condition, meaning an additional heat treatment would be needed for any AAV parts made from molds 3D-printed from indigenous sand. The T6 temper consists of solution heating following by an artificial age. The artificial age (referred to as precipitation heat treatment) is a lengthy, low-temperature process where temperature control/uniformity is critical [20]. The required heat treatments for these alloys are listed in Table 9, as found in the governing specification (ASTM B917 [21]).

Figure 3 Cast A356 AAV part from 3D-printed molds using commercial sand.

Table 9

Required Heat Treatments for Aluminum Casting Alloys 355 and 356 [21]

Alloy	Solution Heat Treatment [1,2]		Precipitation Heat Treatment [3] (Temper = T6)	
	Metal Temperature +/- 12°C (10°F)	Time at Temperature (h)	Metal Temperature +/-12°C (10°F)	Time at Temperature (h)
355	527 (980)	6 to 12	154 (310)	3 to 5
356	538 (1000)	6 to 12[4]	154 (310)	3 to 5

1 – Quench in water at 66 to 100°C (150 to 212°F).
2 – Time at solution temperature may be increased for section thickness over 1 in.
3 – No quenching required. Cool in still air outside the furnace.
4 - Solution time may be reduced when the silicon eutectic has been well modified such as when modified with Sr or Na.

AM Builds of Component Using Aluminum Alloy

The intent of this project was to compare properties of an AAV part cast in a mold made from indigenous sand to those of a part built with metal additive manufacturing. The following outlines the work performed in building this component with aluminum powder via additive manufacturing.

Aluminum Powder - ARL also AM built three AAV parts with the 3D Systems ProX 300 3D-printer, using manufacturer recommended Phenix PS2585-18 casting aluminum alloy AlSi12 powder, the chemical analysis of which is listed in Table 10. The powder was found to be free flowing with a tap density of 43.3%. Tap density is the packing density of the powder without any additional applied pressure/compaction and is the mass divided by the final volume of the powder which is normalized by the theoretical density of 2.6 g/cc.

Table 10

Chemical analysis of AlSi12 powder used for AM build [22]

Element	Result (wt %)
Carbon	0.003
Sulfur	<0.001
Oxygen	0.178
Nitrogen	<0.001
Iron	0.22
Silicon	12.0
Manganese	0.020
Chromium	0.016
Gallium	0.013
Nickel	0.021
Lead	0.028
Tin	0.013
Titanium	0.036
Zinc	0.063
Aluminum	remainder

Notes: No other elements detected in excess of 0.01% wt %

Methods: carbon & sulfur: combustion infrared detection per ASTM E1019-11
hydrogen: inert gas fusion per ASTM E1447-09
oxygen & nitrogen: inert gas fusion per ASTM E1019-11
all others: direct current plasma emission spectroscopy per ASTM E1097-12

Scanning electron microscopy of the AlSi12 powder was performed on a Hitachi S4700 Field Emission SEM. Figure 4 shows the powder at 500x magnification. For the most part, the powder was spherical, with a mix of oblong particles.

Figure 4 SEM of the 3D Systems AlSi12 powder at 500x magnification.

Additive Manufacturing Build - The printer utilized was the 3D Systems ProX300®. The height of this part was close to the limit that this machine can build, and was the tallest build by ARL on this equipment, up to this point. After pre-processing, the parts were completed after 100-hours (machine time) of build time. A total of 5,000 40μm layers were utilized to complete these parts. Although exterior supports were used, there were no internal supports associated with this build.

The processing parameters used to build these parts are listed in Table 11. The parameters are listed not only for the actual parts, but for the structure that supported the part during the build. The part is shown in Figure 5 after removal from the build plate, and support material.

Table 11
Processing parameters used for AM builds

Parameter	Build	Support
Laser power (%)	45	80
Laser power (W)	225	400
Laser speed (mm/s)	1200	2500
Layer thickness (μm)	40	40
Hatch spacing (μm)	70	70
Interlayer rotation angle (deg)	45	45
Hexagon radius (μm)	5000	25000
Hexagon overlap (μm)	100	100

Part Characterization

Dimensional Verification - The AM parts were measured in a metrology laboratory located at the Army Research Laboratory to determine conformance to the governing 2-dimensional drawing. There were some glaring discrepancies noted on the first three AM-built parts. The wall thickness of the part went from 0.792cm (0.312-inch) to 0.254cm (0.1-inch) when converting to a .stl file. Additionally, the inner diameter was out of tolerance and there was some overlap on one of the lips. Sagging of one of the corners was also noted, most likely due to the weight of the build. Finally, one of the bolt through-holes was out-of-round and the centerline distance between the bolt holes was 6.6cm (2.6-inch), instead of the required 6.7 cm (2.65-inch). The .stl file was amended and an AM build (two per plate) was accomplished. One of these parts was also subjected to dimensional verification which showed conformance to all dimensions.

Figure 5 AM-built part using amended .stl file ensuring proper dimensions.

Computed Tomography of AM Built AAV Component - The AM parts were subjected to X-ray computed tomography (CT) in order to inspect for internal defects. A Northstar X5000 unit (with a scanning power of 180kV and 140μA and a voxel size of 129.5 μm) with efXCT, CTVox, and CTAn software was utilized for this inspection. Evidence of witness lines can be seen on the outer diameter which was caused by a pause in the build due to a power outage. The lines were horizontal to the build direction. CT inspection for porosity both within the support material and the as-built material determined that the support material exhibited a porosity of 0.48% while the as-built part had 0.59% porosity.

CT inspection was also used to determine defect sphericity for both the support and as-built material. Sphericity is a measure of porosity shape, as this defect is formed by trapped gas bubbles, and is generally round in morphology. It was found that there was not a strong change in the sphericity with the change in processing parameters. In addition, it was noted that support and build defect volumes were comparable for defects below 10,000 μm^3, but slightly larger defects were found in the as-built parts.

Density Measurements - The density of the AM built parts (using tensile specimens) was determined with the Archimedes method:

$$\rho = \frac{(m_{dry} * \rho_{water})}{(m_{dry} - m_{wet})}$$

Table 12 contains the results of this analysis. No appreciable density difference was noted between the as-built specimens which were also similar to the supports. These results compared favorably with the 2.66 g/cc theoretical bulk density reported in literature [23].

Table 12

Density Measurements for As-Built AlSi12 Powder

				$AlSi_{12}$ B2 As Built										
				Dry mass (g)				Wet Mass (g)						
Specimen	Sample Label	T_{water} (°C)	ρ_{water} (g/mL)	1	2	3	AVG	1	2	3	AVG	ρ (g/mL)		
1	b21	22.4	0.998	10.850	10.850	10.850	10.850	6.725	6.726	6.727	6.726	2.625	ρ_{AVG}	2.63
2	b22	22.4	0.998	10.884	10.884	10.884	10.884	6.748	6.749	6.749	6.749	2.626		
3	b23	22.4	0.998	10.957	10.957	10.957	10.957	6.819	6.821	6.820	6.820	2.643		
				$AlSi_{12}$ B3 As Built										
				Dry mass (g)				Wet Mass (g)						
Specimen	mple Lab	water (°C	ater (g/m	1	2	3	AVG	1	2	3	AVG	ρ (g/mL)		
4	b31	22.4	0.998	10.799	10.799	10.799	10.799	6.673	6.674	6.674	6.674	2.612	ρ_{AVG}	2.61
5	b32	22.4	0.998	10.833	10.833	10.833	10.833	6.697	6.699	6.699	6.698	2.614		
6	b33	22.4	0.998	10.801	10.801	10.801	10.801	6.675	6.676	6.677	6.676	2.612		
				$AlSi_{12}$ B3 Support										
				Dry mass (g)				Wet Mass (g)						
Specimen	mple Lab	water (°C	ater (g/m	1	2	3	AVG	1	2	3	AVG	ρ (g/mL)		
7	b3s1	22.4	0.998	11.441	11.441	11.441	11.441	7.148	7.149	7.149	7.149	2.659	ρ_{AVG}	2.66
8	b3s2	22.4	0.998	11.490	11.490	11.490	11.490	7.178	7.180	7.179	7.179	2.659		
9	b3s3	22.4	0.998	11.442	11.442	11.442	11.442	7.148	7.150	7.150	7.149	2.659		

Surface Roughness - The surface roughness of a section from the interior and exterior of an AM-built part was measured using a Keyence laser profilometer. The average of 4 scans (2mm x 2mm) showed that the interior surface of an AM-built part was smoother than the exterior surface of an AM-built part.

Hardness Testing - Vickers microhardness (major load of 500 grams) traverses were performed on a representative as-built piece of material. The average hardness of the part was 117.9 +/- 16.6 HV (average of 171 readings).

Heat Treatment of AM Build – Although post processing heat treatment of this powder was not recommended by the manufacturer, a heat treatment was performed on a few specimens, with parameters found in open literature for this alloy (500°C for 2 hours). This treatment differed greatly from the T6 treatment recommended for a cast 356 aluminum component (shown in Table 10). Research after the fact showed that literature exists that provides recommended heat treatment schedules for strengthening aluminum powder AM builds. For instance, one reference for heat treatment of AlSi10Mg alloy (similar to the one used in this study) lists a schedule of solid solution at 535°C (995°F) and artificial aging at 158°C (316°F) [24], which are almost the same as listed in Table 12 for cast A355 and A356 aluminum alloy. Future work could focus on trying this alternative heat treatment on AlSi12 AM builds to determine if mechanical properties could be optimized.

Tensile Testing - Tensile tests were conducted on specimens built with the actual parts using both the build and support parameters (shown previously in Table 12). Testing was accomplished in accordance with ASTM E8M-09 and with dynamic image correlation (DIC) VIC software to measure elongation. The specimens were sub-sized flat dogbones, and were tested at 1.25 mm/min in tension. The results showed that it appeared that the specimen representing the support structure parameters had the best combination of strength and elongation, slightly better than the specimens that had used the build parameters. Heat treatment degraded the mechanical properties, suggesting that an alternate heat treatment needs to be studied for improved post-treatment properties.

Metallography - Samples utilizing the as-built parameters, the support structure parameters, and heat-treated were subjected to metallographic inspection. The samples were ground and polished, and etched with a modified Murakami reagent. Figures 6, 7, and 8 are scanning electron micrographs of the support structure parameters representative material at low (figure 6) and high (figure 7) magnifications. In figure 6, note the "weld bead-type" patterns representing the laser melted regions that had solidified, common with DMLS processes. Also note the porosity (entrapped gas bubbles) within the sample. Figure 7 shows the dendrites (feathery structure) indicative of the aluminum cooling and solidifying from a molten state. Figure 8 shows a cavity-like defect that appears to be located along a witness line within the structure of the sample.

After heat treatment, the dendritic structure was no longer evidenced, nor was the weld bead-type structure (see low magnification example in Figure 9). As this figure shows, heat treatment had formed many precipitates, and what appears to be increased porosity. This porosity could be artifacts caused by pullouts of the precipitates during metallographic preparation and were not studied further.

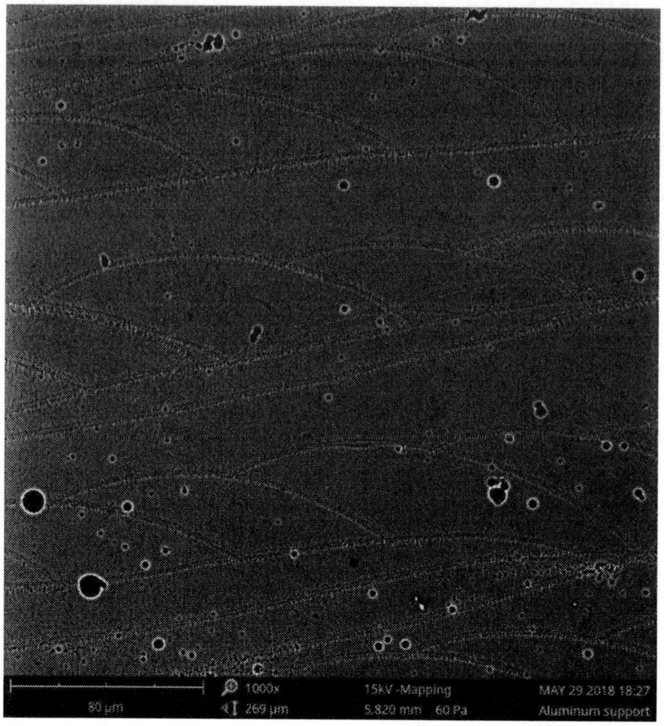

Figure 6 Scanning electron micrograph showing the microstructure of the DMLS specimen using support-structure parameters (representative). Etchant: modified Murakami's etch.

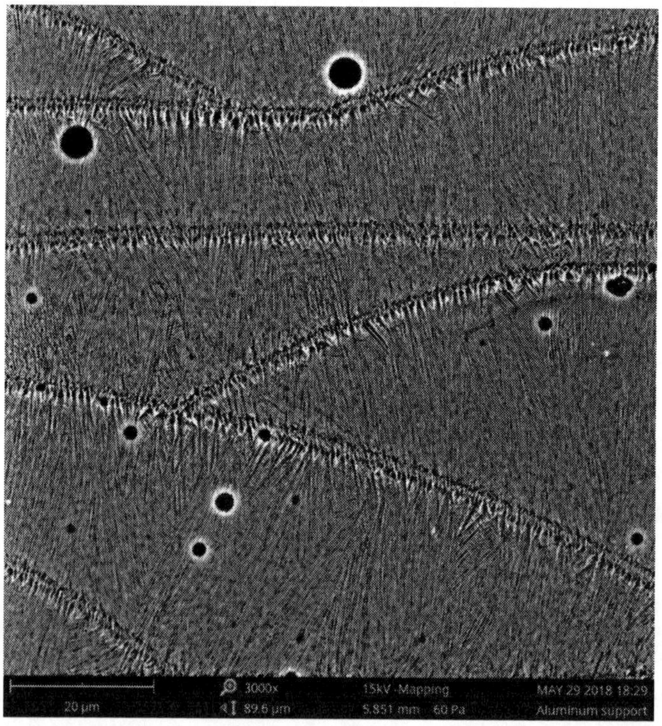

Figure 7 Higher magnification of the sample shown in Figure 6 – note dendrites (feathery structure) representing solidification of molten aluminum. Etchant: modified Murakami's etch.

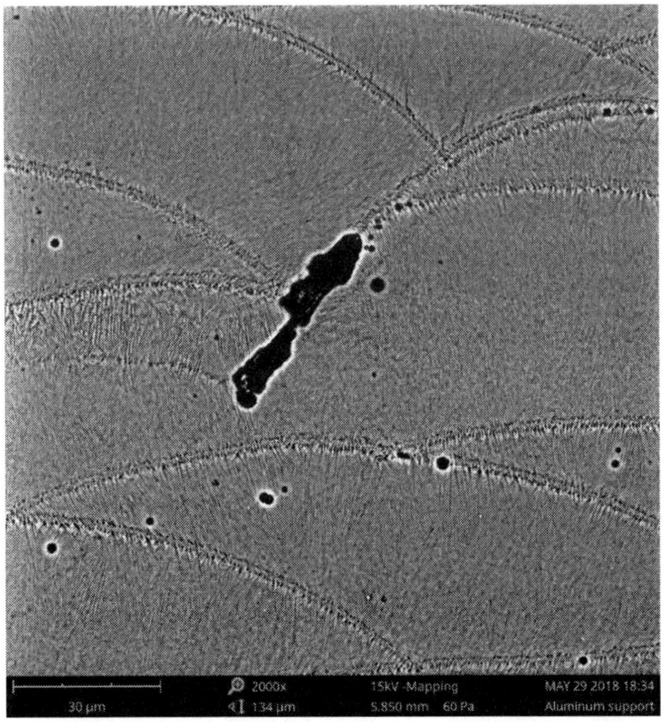

Figure 8 A cavity-like defect located on what appears to be a witness line within the structure of the sample. Etchant: modified Murakami's etch.

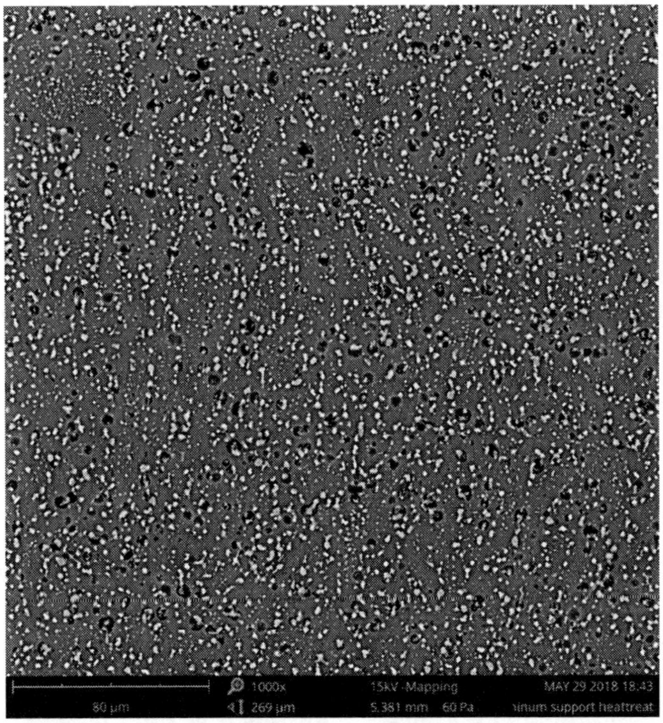

Figure 9 Structure of the sample representing the AM support structure parameters after heat treatment. Note the formation of precipitates, and the lack of any evidence of the dendrites and weld bead-type features noted prior to heat treatment. Etchant: modified Murakami's etch.

Business Case / Cost Comparison

Based on the amount of work needed to process indigenous sand for use as a viable 3D-printed mold material at the point-of-need (and the choice herein of desert versus beach sand), a true cost comparison could not be made. Listed below, however, are the costs needed to build these parts using metal additive manufacturing in a lab setting:

- 4 hours pre-processing time (human time)
- 100 hours print time (machine time)
- 4-6 hours post processing time (mostly machine time)
- Power requirement: 400 V / 15 KVA / 3 phase
- Powder cost: ~$800 per part (a lot of this powder is excess and can be reused though so actual cost would decrease with increased quantity)
- Al build platform cost: $280 (these are reusable)

In addition, thermal post-processing may be necessary, which would increase the total cost of AM production.

Lessons Learned

- 3D-Printing with Desert Sand:
    - Effect of clay contamination within the desert sand…will this affect the mold build?
    - Use of ultrafine desert sand clearly the most difficult path; beach sand with less fines may have been easier to work with.
    - Desert sand will require a modification to the 3D-printing process.
    - Developed a new screening process to remove large aggregates and most fines.
    - Alternative binder material was found to show improvement in bond strength, but further enhancement is needed.
    - Dry acid catalyst was found to work well with non-optimal, sub-par materials, and provided ability to print with fine materials.
    - Fast tooling was shown to be possible with sand printed molds.
- AM Builds
    - Ensure digital file is accurate prior to printing.
    - Choose an appropriate heat treatment that optimizes final properties.
    - As shown through CT, uninterrupted builds are key to better properties.

Future Work

Indigenous sand screening - The screening of the sand to the appropriate/acceptable particle size distribution turned out to be a monumental task. Dr. Peter Lucon [25], Assistant Professor of Mechanical Engineering, Department of Mechanical Engineering, Montana Tech University (and holder of many patents in the area of resonant acoustic mixing when employed by Resodyn Corp.) believes the RAM technology may offer a solution. Dr. Lucon mentioned that the benchtop RAM would be able to sieve 100 grams of sand in less than a minute. Research would be needed to upgrade the RAM system into a continuous-feed system to provide sieving capability to the size distribution needed and to determine if it can not only screen out the large aggregate but also the fines.

Improved indigenous sand binder - The current commercial resin is not strong enough to work with the fine particles of desert sand that cannot be removed. The binder used was well above the range specified by the supplier (and outside the range that is currently able to be deposited by a 3D printer). Trying to maintain a recipe that was usable in a 3D printer was the main constraint, and anything higher than what was tried wouldn't be feasible. As mentioned, foundry binder strength development is generally complete after 24 hours, and will not increase after that point. The inorganic resin used herein showed promise in increasing the strength of the mold 3D-printed from indigenous sand. This is most likely due to it utilizing the fine particles for bonding. Further work in this area would be beneficial, for successful, this would represent a major step in using indigenous materials for additive manufacturing.

Improved AM build thermal post-treatment - As listed herein, the heat treatment utilized on the AM-built samples actually lowered the strength of the material. Work could be performed in this area in an attempt to find an optimal thermal post treatment (if any) for maximized mechanical properties of AM builds using the AlSi12 alloy powder.

Acknowledgements

The authors wish to thank Dr. Rik Riman (Rutgers University) for his guidance on (and sample of) low-temperature solidified ceramic binders, and Tony Delgado and Eric Kirchner (DLA) for their funding of this project, and their continued inputs along the way.

References

[1] "Digital Metal: AM for Casting Metal Parts in Architecture", M. Meibodi, TCT Magazine, Volume 4, Issue 4.

[2] "Aerospace Project Pairs 3D Printing with Investment Casting", S. Hendrixson, Additive Manufacturing, November 2017, p. 64.

[3] "http://earth.sustainablesources.com, referenced 12/8/11.

[4] "Force Projection Research and Development: The Key Enabler for Army Transformation", Col Genaro J. Dellarocco, 10 April 2001.

[5] "US Army Outlines Push to Save Resources, Lives", Defense News, 11/19/13.

[6] "Shrinking the Logistics Tail in the Combat Zone", Col Larry W. Jameson, 9 April 2002.

[7] "Logistics Transformation – Reducing the Logistics Footprint", LTC Darrell S. Ransom, 5 April 2002.

[8] "Sand Printing Expands Metal Possibilities, Too", Additive Manufacturing, November 2016, p. 37.

[9] "4063 Additive Manufacturing for Metal casting (AM4MC)", presentation by R. Lonardo and N. Pramanik for America Makes, referenced on 7 March 2017.

[10] "The Effects of 3D Printed Molds on Metal Castings", D. Snelling, et. al., Proceedings of the Solid Freeform Fabrication Symposium (2013), pp. 827-845.

[11] "Additive Manufacturing of Molds for Sand Casting", N. Hawaldar and J. Zhang, Poster at the Powdermet 2017 conference, Las Vegas, NV, 14 June 2017.

[12] "Quantifying the Role of Part Design Complexity in Using 3D Sand Printing for Molds and Cores", E. Almaghariz et. al., International Journal of Metalcasting, Volume 10, Issue 3, 2016.

[13] "Paradigm Shift", B. Connor, Defense AT&L Journal, Special Issue – Additive Manufacturing (3D Printing), November-December 2016.
[14] "AM Answers its Casting Call", B. Donaldson, Additive Manufacturing, November 2017.
[15] "Comparative Study of 3D Printing Technologies for Rapid Casting of Aluminum Alloy", S. Singh Gill et. al., Journal of Materials and Manufacturing Processes, 24, pp. 1405-1411, 2009.
[16] "Guide to Casting and Molding Processes", 2015 Metal Casting Design and Purchasing Journal.
[17] "Reclamation and reusability studies on CO2 sodium silicate bonded sands for ferrous foundries", SM Nazirudeen, 13_Chapter 4, 1999.
[18] Foseco Ferrous Foundryman's Handbook, edited by John Brown, Butterworth-Heinemann, 2000.
[19] "Critical Characteristics Affecting the Surface Finish of Castings", N. Bryant and J. Thiel, AFS Transactions, Paper 17-094 v125, 2017.
[20] "Aluminum Heat Treatment Revisited: Common Concerns and How to Resolve Them", D. Herring, Industrial Heating, February, 2018, p. 18.
[21] ASTM B917, "Standard Practice for Heat Treatment of Aluminum-Alloy Castings from All Processes", 2012, ASTM, Conshohocken, PA.
[22] Luvak Inc. Analytical Report No. 0-82641, 10/5/17, provided by Andrew Burke, Quality Manager.
[23] "Influence of Powder Characteristics on Processability of AlSi12 Alloy Fabricated by Selective Laser Melting", R. Baitimerov, et. al., MDPI, Materials, 5/7/18.
[24] "Using Two-Stage T6 Heat Treatment to Tailor the Mechanical Properties of 3D Printed Aluminum AlSi10Mg Alloy, S. Saunders, 3Dprint.com, 8/20/18.
[25] Personal correspondence with Dr. Peter Lucon, Assistant Professor of Mechanical Engineering, Department of Mechanical Engineering, Montana Tech University on 2/7/18 and follow up electronic correspondence on 2/20/18.

Advances in Comprehensive Fluid Analysis

J. Reintjes[1], J. E. Tucker[2], P. F. Henning[3], Paul Howard[4]

[1] KeyW Corp, Hanover, MD, USA
[2] Naval Research Laboratory, Washington, DC, USA
[3] AMETEK Spectro Scientific, Chelmsford, MA, USA
[4] Paul L. Howard Enterprises Newmarket, NH

Abstract

We describe advanced forward deployed comprehensive fluid analysis for wear debris and fluid condition of lubricating oil, hydraulic oil and fuel. Examples from mechanical systems are given for detection of wear-related and fluid condition-related faults.

Introduction

We describe advances in comprehensive forward deployed fluid analysis based on fluid condition assessment with a FluidScan® analyzer and a viscometer and contaminant and wear debris analysis with the LaserNet Fines® (LNF) optical particle analyzer. The combined technologies provide comprehensive analysis of fluid condition and machinery health, and are implementable either as a forward-deployed portable instrument or a continuous, autonomous online monitor. The combined technologies provide comprehensive immediate on-the spot actionable information at the operational level about the current condition of platforms and machinery, readiness for deployment or usage, and advanced warning about impending failures or need for specific maintenance. A forward deployed portable instrument, which can analyze lubricating oils, hydraulic fluids and fuels is of value for monitoring fleets of vehicles or platforms with a single instrument using bottle sampling. An online monitor [1] is of value in providing continual assessment of the condition of critical, high-value equipment where serious faults can develop between bottle sample intervals, and can help support reduced manpower or workload environments.

The first generation Portable Fluid Analyzer (PFA) [2] is shown in Fig. 1.

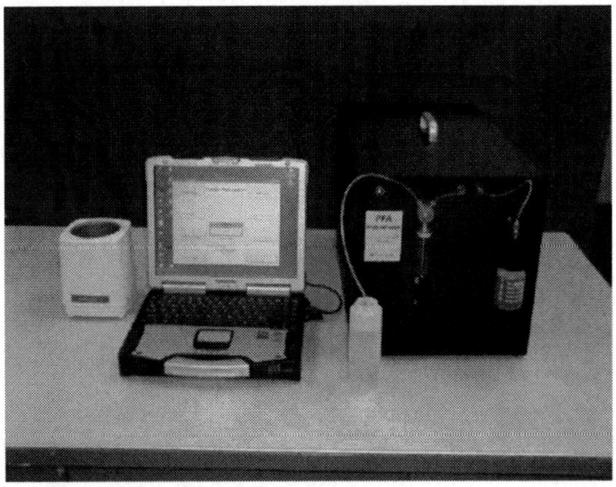

Fig. 1. First generation PFA with laptop computer

Analysis of a fluid sample is accomplished by drawing fluid from a sample bottle into the PFA, which contains LNF, FluidScan and viscometer analyzers arranged in series along the fluid path. Raw data from each device is transmitted to a laptop computer. Software on the laptop computer controls the PFA instrument, analyzes the results of measurements, and displays results to the user. The instrument also automatically flushes the internal lines between samples with solvent. Spent fluid is collected in a separate bottle for disposal

Case Studies
Here we will illustrate the capabilities of the PFA instrument in detecting mechanical and fluid faults with two case studies as examples. Our examples will show the benefits of immediate on-site fluid analysis, as well as combined analysis of fluid condition and wear debris, allowing immediate assessment of equipment readiness and improved ability for remedial action.

The PFA wear debris measurements from transmissions in a fleet of six vehicles over a period of nine months are shown in Fig. 2. For this study, samples were taken on an "as available" basis as opposed to prescribed intervals driven by calendar time or operating hours.

A baseline of "normal" operation is established by assuming that the vehicles have all been subjected to the same operational conditions historically, and are all operating normally at the start of the monitoring period. Over the period of study, the particulate concentration in one of the PFA particle categories (fatigue particles between 25 and 50 microns in size) varies randomly over a small range at a relatively low level for five of the six vehicles. This range can be taken to be indicative of the concentration of fatigue particles for normal operation.

Near the start of the study the concentration of particles in vehicle 4 (V4 TR) starts to increase and continues to grow, slowly at first and then more quickly. At the maximum level detected in the study (Point "A") a catastrophic failure occurred for vehicle 4 with the particle filter plugged and the transmission inoperable. At this point, not only was the fatigue particle concentration significantly elevated in the 25-50 micron size range (Fig 2b), but the size distribution shows an peak in that size interval, as shown in the histogram in Fig 2c. Following the failure event, the transmission was replaced. Analysis of samples taken immediately after replacement (V4 TR2) continued to show high levels of fatigue wear debris because of incomplete flushing of the transmission sump. Subsequent measurements made after a complete flush of the transmission (Point "B") show that the wear particle concentration returned to the "normal" range (Fig 2d) and, more importantly, the size distribution changed to one that falls rapidly with increasing particle size (Fig 2e).

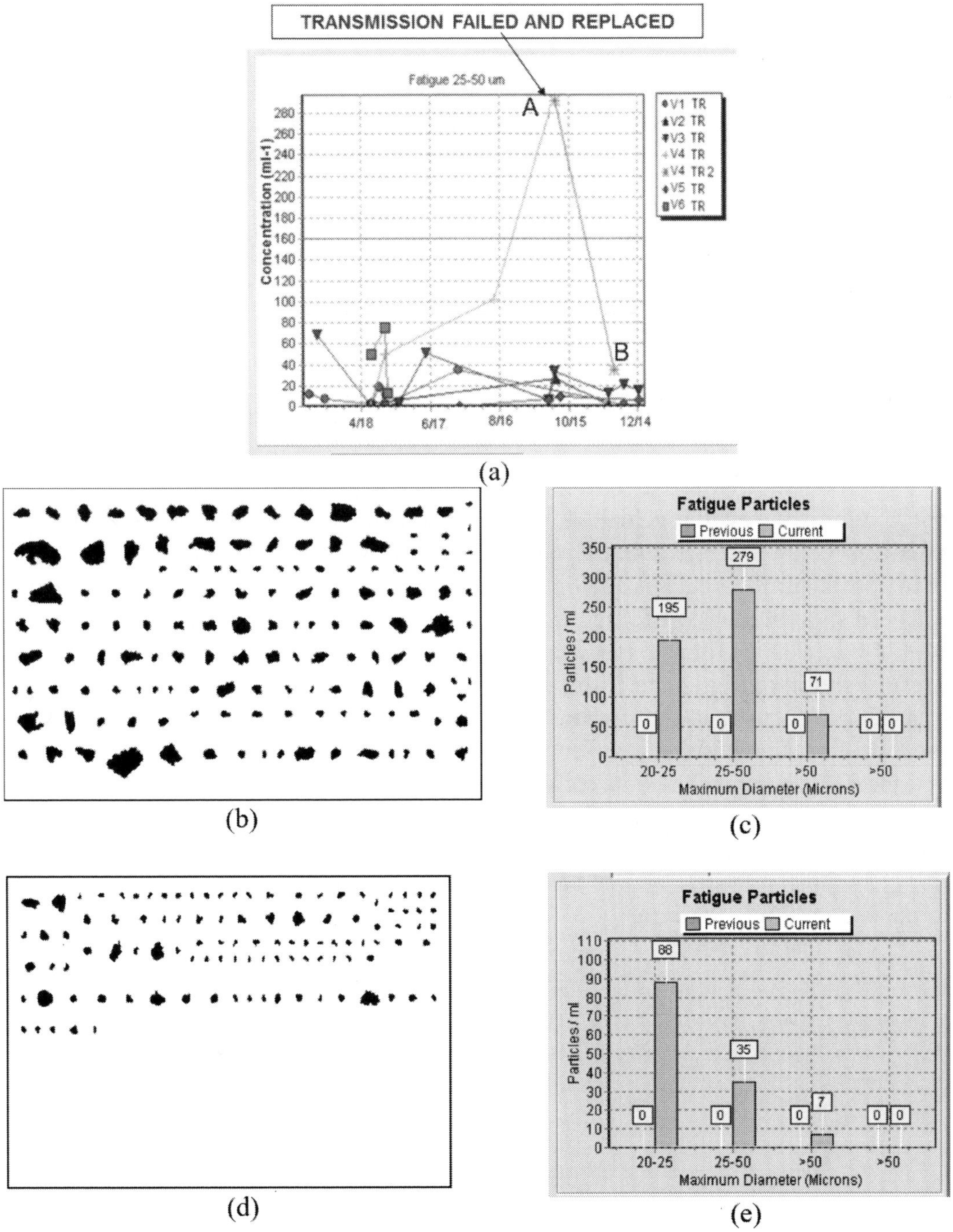

Fig. 2. Wear particle analysis by PFA on the transmissions (TR) a fleet of six vehicles. At point "A", the transmission of vehicle 4 failed and was replaced. After replacement, normal operation resumed at point "B". (a) fatigue particle concentrations in the 25-50 µm size range. (b) Particle image map of vehicle 4 at point A (1 of 6 pages of fatigue particles). (c) Fatigue particle size distribution for vehicle 4 transmission at point A. (d) Complete fatigue particle image map of vehicle 4 at point B. (e) Fatigue particle distribution for vehicle 4 at point B.

These results indicate the benefits and power of debris analysis that combine morphological analysis with quantitative measures. Morphological analysis allows us to separately identify those particle associated with specific types of faults. Typically, these are a small fraction of the total population of particles contained in the lube oil. By singling them out for individual analysis, it is possible to detect faults at their earliest stages. Simultaneous quantitative analysis allows us to identify the transition from normal or early wear to the onset of fault-related excess wear more reliably. This transition can be judged not only by an increase in particle concentration, which can happen for a number of reasons, but by changes in the size distributions within a class, which is generally associated with the progression of fault conditions.

Additional information can be obtained by combining the wear debris results with fluid condition analysis. The same particle data in Fig 2a is shown in Fig. 3a, along with water concentration measurement over the same period in Fig 3b.

Here it can be seen that for several months prior to the catastrophic transmission failure (Point A) the water level in vehicle 4 was elevated at a value of the order of 0.08%. This level is well below the maximum allowed level of 0.2% for this vehicle. However, it is known from controlled studies [3] that water levels in the range of 0.08% for extended periods can shorten bearing life significantly, as shown in Fig. 4.

The results shown in Figs 2 and 3 illustrate the value of on-site comprehensive monitoring. The particulate data shows that five of the six vehicles in the fleet remained fit for service, but vehicle 4 became unfit near the end of the sixth month in the study. Such information is vital for field commanders, both military and civilian, to obtain optimal effective service from their fleets. An addition, the fluid condition results strongly indicates that the failure near the end of month 8 of the study could be traced to the elevated water levels occurring earlier. Suitable remedial action based on these results could well have averted the subsequent catastrophic failure.

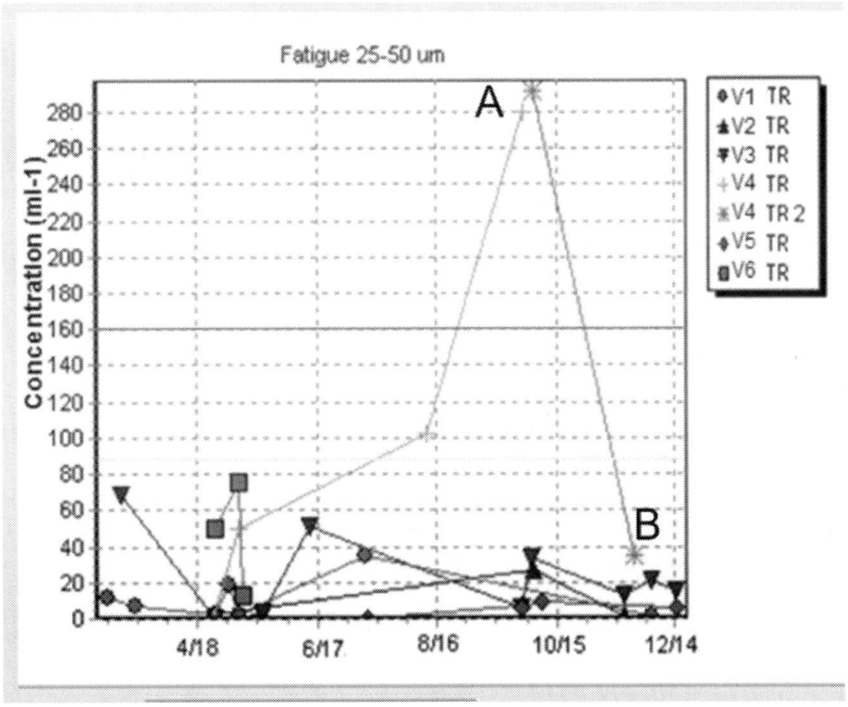

(a)

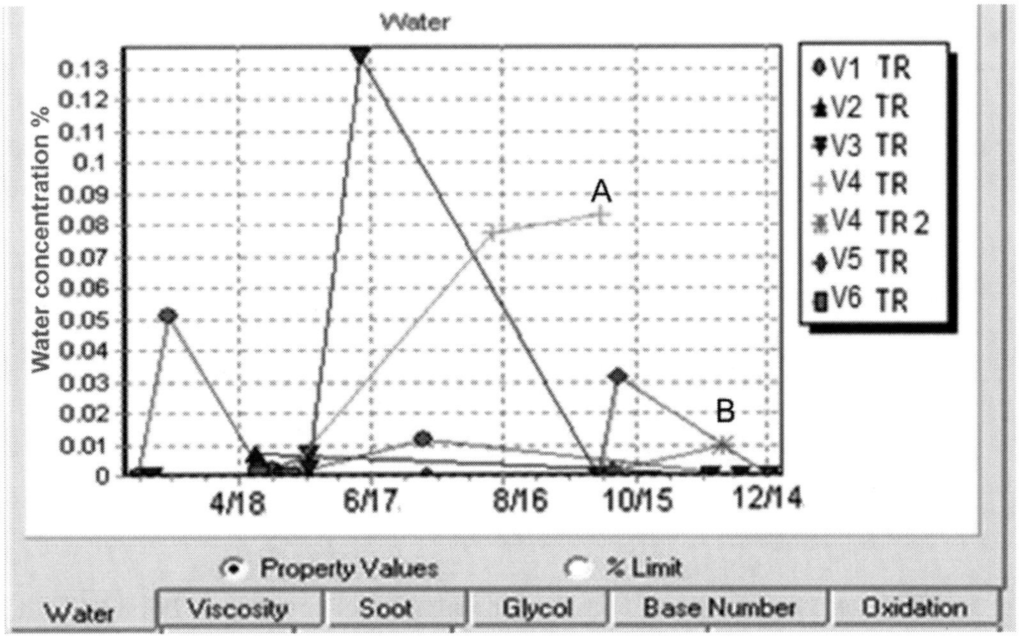

(b)

Fig. 3. PFA (a) fatigue particle and (b) water concentration measurements for a fleet of six vehicles. Point A corresponds to the point where vehicle 4 transmission failed. Point B corresponds to the point where the transmission was replaced and the system flushed.

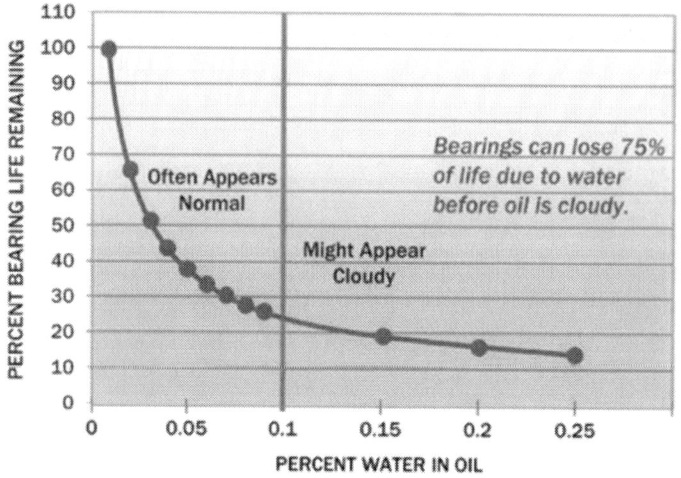

Fig. 4. Dependence of bearing life on concentration of water in lubricating oil as described in Ref 3. Reprinted with permission.

An additional example of the value of immediate on-site information is illustrated in Fig. 5.

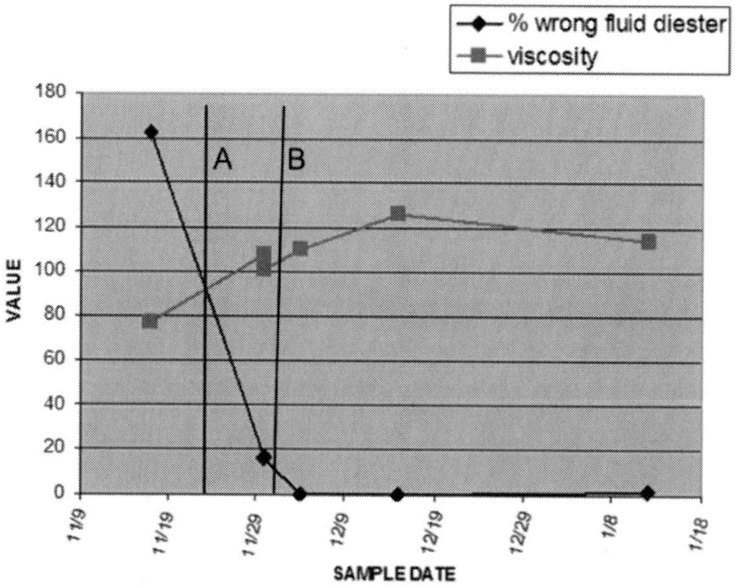

Fig. 5. Measures of viscosity and wrong fluid in a diesel engine. At Point "A", the engine was replaced based on PFA analysis. At that time, the system oil was replaced. At Point "B", the engine oil was replaced a second time.

Here we show results of measures of viscosity and wrong fluid (synthetic diester in mineral lube oil) in a diesel engine. During routine analysis, one of the samples showed a significant concentration of synthetic diester (wrong fluid) in the mineral lubricating oil, along with a reduction in viscosity from a normal value of 120 cSt down to 78 cSt. The elevated diester level was subsequently traced to a broken seal in the engine between the high-pressure hydraulic system and the lower pressure lube oil system. Corroborating this finding is a drop in viscosity, although the viscosity remained at an acceptable value and would not have, in itself, indicated a problem. In this situation, the broken seal was replaced before the engine failed.

Fuel Analysis

We have also developed algorithms for fuel analysis using mid infrared spectroscopy with an FTIR spectrometer. The purpose of the algorithms is to determine if an unknown fluid is a fuel, if so, its type and multiple physical properties relevant to the fuel type. The approach is based on Soft Independent Modeling by Class Analogy (SIMCA). We have developed the ability to distinguish the broad classes of gasoline, diesel and aviation fuels, and also several subtypes within each class. The truth table showing the SIMCA classification of over 1400 samples of various fuel types compared to ground truth in Fig. 6.

CLASSIFIER

GROUND TRUTH	Unknown	JP-8	JP-5	Diesel	Gasoline	fraction correct
Unknown	1	0	0	0	0	1
JP-8	1	161	0	1	0	0.99
JP-5	2	0	18	0	0	0.90
Diesel	18	0	0	779	0	0.98
Gasoline	17	0	0	0	631	0.97

CONSOLIDATED FUEL CLASSES

Fig. 6. Summary truth table showing the number of fuel samples classified in each category by SIMCA compared to ground truth.

Once a fuel type has been identified, chemometric analysis of the spectra is used to identify multiple appropriate physical properties as indicated in Fig. 7.

Property	gasoline	biodiesel	JP-8	diesel
Density	X	X	X	X
Distillation – Initial Boiling Point	X	X	X	X
Distillation - 10%		X	X	X
Distillation - 50%		X	X	X
Distillation -90%		X	X	X
Distillation – Final Boiling Point	X	X	X	X
FAME (%)		X		X
Flash Point	X	X	X	X
Cetane Index		X		X
Cloud Point		X		X
Octane Number	X			
Freezing Point			X	
FSII			X	
Water	X	X	X	X
Sediment	X	X	X	X

Fig. 7. Physical properties calculated for different fuel types

The physical properties determined as shown in Fig 7, combined with specified limits and fuel type identity, allows the use of these results to determine in the field if an unknown fluid is a fuel and, if it is a fuel, what type it is and whether it is usable. The fuel analysis capability will be incorporated in a future version of the PFA.

Summary

We have described an approach to the development of forward deployed comprehensive fluid analysis for analysis of lube oil, hydraulics and fuel. The instrument can be used by non-trained personnel in the field to obtain immediate actionable information about the usability of the equipment and the presence of specific faults, allowing remedial action or planned maintenance.

Acknowledgement: The authors acknowledge the support of the Office of Naval Research

References

[1] Sølve Fjerdingstad, J. Reintjes and J. E. Tucker, "On Line Condition Monitoring for Subsea Applications", MFPT Proceedings 2016, Dayton, OH.

[2]. J. E. Tucker, J. Reintjes, T. Sebok and P. F. Henning, "Comprehensive, Field-Deployable Portable Fluid Analyzer", Presented at 61st Meeting of the Society Machinery Failure Prevention Technology April 2007.

[3]. Loren Green, "Why Clear and Bright Oil Samples Are Not Good Enough", in Machinery Lubrication, June, 2014 (Noria Corporation pub).

RAPID IMPACT™ TESTING OF ANY SIZE STRUCTURE

Brian Schwarz, Patrick McHargue, Mark Richardson
Vibrant Technology, Inc.
Centennial, CO 90112

Abstract: One of the limitations of conventional modal testing using a roving impact hammer is that the ***reference sensor*** (usually an accelerometer) ***must remain fixed*** throughout the test. Since the accelerometer must typically be connected by a wire to the data acquisition system. a ***very long wire may be required*** when testing a large structure. Furthermore, ***better quality signals*** are possible if each ***impact force is applied closer to the response accelerometer***. Because it does not require ***a fixed reference sensor*** throughout the test, a **Rapid Impact™** test is faster and easier to perform on ***any size structure***.

Key words: Experimental Modal Analysis (EMA); Frequency Response Function (FRF); Impulse Response Function (IRF); Curve Fitting; Modal Residues; UMM Mode Shape; Multi-Input Multi-Output (MIMO) Modeling & Simulation; Modal Participation.

1. Introduction

In a conventional **Roving Impact** test, the ***accelerometer must remain fixed*** while the structure is impacted at different DOFs (points & directions). In a conventional **Roving Response Impact** test, the structure ***must be impacted at the same DOF*** while one or more accelerometers are moved to different points.

In **Rapid Impact™** testing, both the impact DOF and the accelerometer location can be changed between acquisitions. Based on their DOFs, a ***chain of acquisitions*** is made during a Rapid Impact™ test. A ***chain of FRFs*** is then calculated from the acquired data.

An FRF chain is not a conventional set of single-reference FRFs, yet it can be curve fit using single-reference curve fitting methods. However, the modal residues resulting from curve fitting the FRFs do not form a mode shape of residues. Instead, the residues must be further processed to obtain mode shapes. The residue post-processing is based on the ***relationship between modal residues and mode shapes***. This relationship is illustrated in Figure 13.

2. Roving Impact Test

A conventional **Roving Impact Test** is depicted in Figure 1. In this test, the accelerometer must remain fixed throughout the test, and the structure is impacted at a different DOF with each acquisition of data. Each set of impact & response data must be ***simultaneously acquired***. Each measurement set of ***simultaneously acquired data*** is used to calculate a **Frequency Response Function (FRF)**.

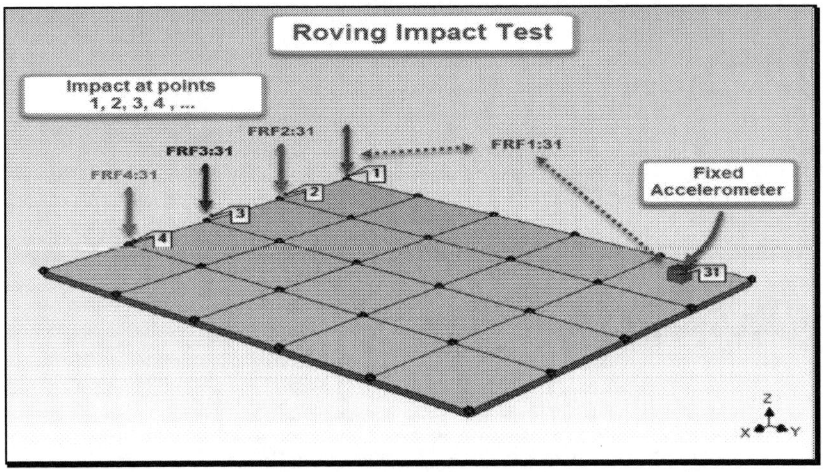

Figure 1. Roving Impact Test

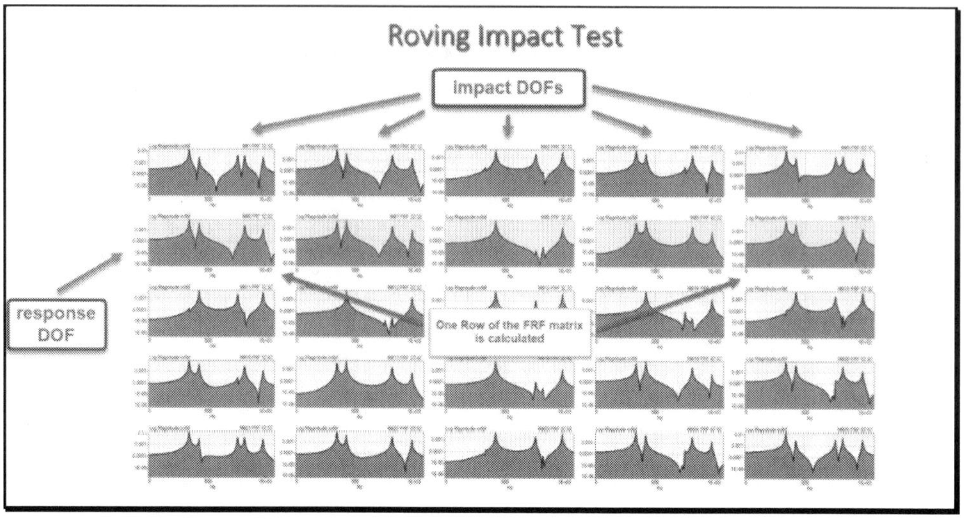

Figure 2. FRFs from a Roving Impact Test

The series of FRFs calculated from each measurement set of acquired data fill in a ***row of the matrix of possible FRFs***, as shown in Figure 2.

3. Roving Response Impact Test

The dynamics of the grating were originally captured with a ***conventional multi-reference roving accelerometer test***. Several photos of this test are shown in Figure 4. During this test, the grating was impacted in ***three directions at one corner*** (**DOFs 1X, 1Y, -1Z**), and ***tri-axial accelerometers*** were attached to points on the grating where mode shape data was desired.

During this test, calibrated FRFs were calculated by entering the sensor sensitivities into the acquisition system so that each signal was converted from a voltage to its correct engineering units. Log magnitudes of several FRFs derived from this ***multiple-reference impact test*** are shown in Figure 3.

The properties of the multi-reference FRFs are listed in the **M#s spreadsheet** to the right of the log magnitude display. The FRFs have engineering units of (**g/lbf**). The log magnitudes show the resonance peaks of ***five modes of vibration***. The mode shapes of these five modes were used to model the dynamics of the grating during a ***simulated*** Rapid Impact™ test of the grating. The results of this ***simulated*** Rapid Impact™ test are presented in the remainder of this paper.

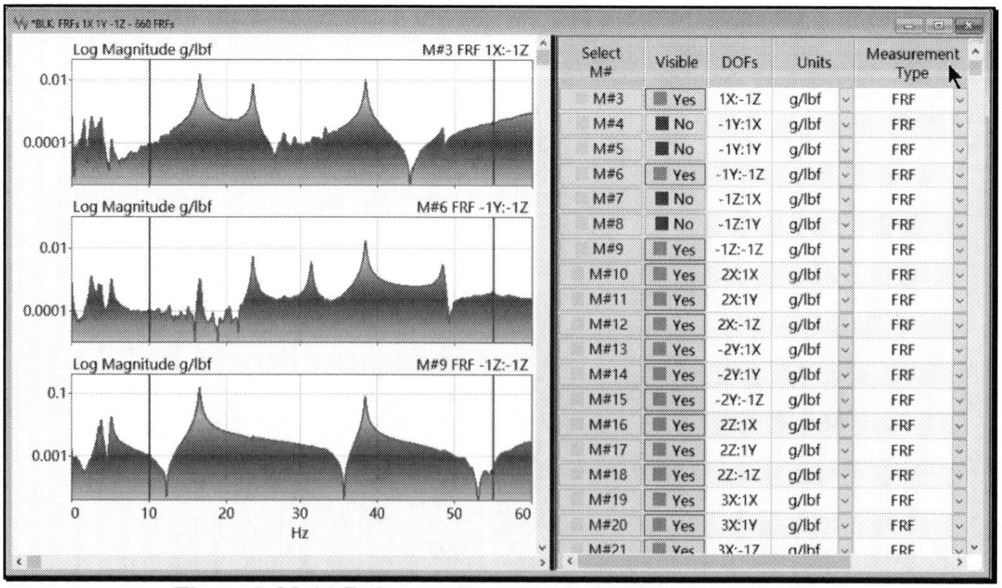

Figure 3. Multi-Reference Experimental FRFs of the Grating

Figure 4. Multi-Reference Roving Accelerometer Test of the Grating

4. Mode Shapes of The Grating

The multi-reference FRFs acquired from the grating were curve fit using ***multi-reference curve fitting***. Because the FRFs were calibrated, the dynamic properties (mass, stiffness & damping) of the structure were preserved in the FRFs. The multi-reference modal residues obtained by curve fitting the FRFs also preserves the dynamic properties. Finally, the modal residues were used to create a **Modal Model (a set a of UMM mode shapes)** for the grating.

The ***truncated dynamic model*** of the grating consists of five **UMM mode shapes**. **UMM** stands for **Unit Modal Mass**, a special scaling of the mode shapes that also ***preserves the dynamic properties*** of the structure. This **Modal Model** was used to represent the dynamics of the grating during a ***simulated*** Rapid Impact™ test of the grating.

5. Modal Participation

One of the mode shapes of the grating is displayed in Figure 5. Also, the modal participations of each mode shape in each of the three references (1X, 1Y, -1Z) are listed in Figure 5 for the five mode shapes.

Modal participation indicates the direction of ***the dominant motion of a mode shape***. Modal participations are calculated during multi-reference curve fitting of a set of multi-reference FRFs. Modal participations have values ***between 0 & +-1***. **Participation = +-1** indicates ***dominate motion*** of the mode shape in the indicated direction. **Participation = 0** means the mode shape has ***no motion*** in the indicated direction.

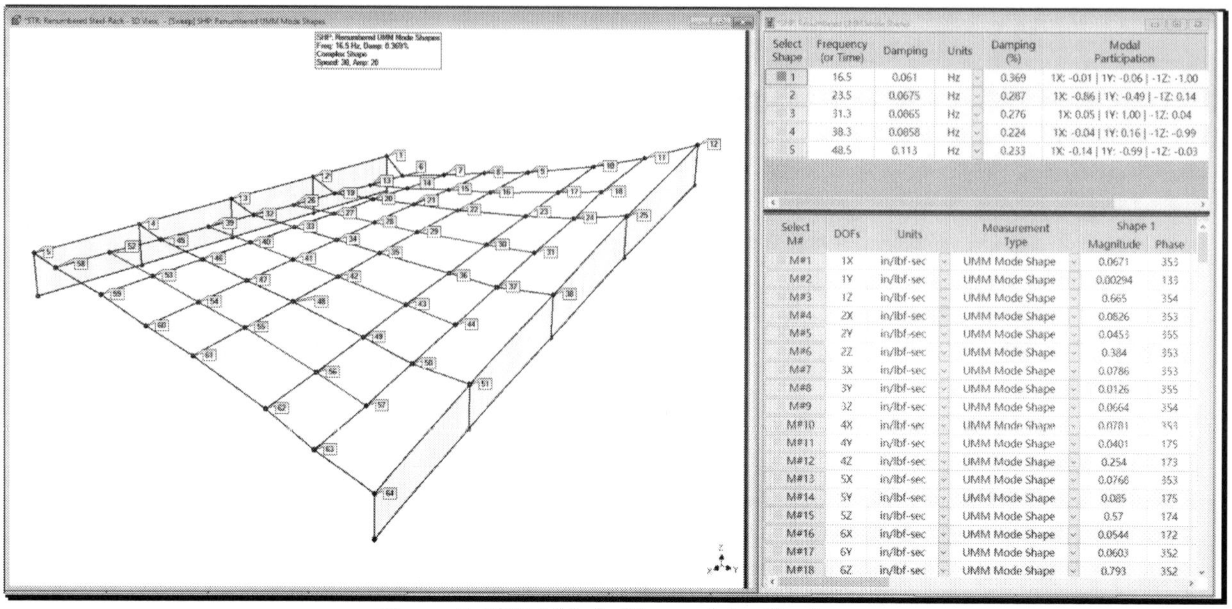

Figure 5. UMM Mode Shape of the Grating

The modal participations in Figure 5 indicate that ***modes 1 & 4 participate mostly in the Z-direction. Mode 2*** participates ***mostly in the X-direction*** & ***Y-direction***, while ***modes 3 & 5 participate mostly in the Y-direction***.

Without knowing the dominant motion of each mode shape prior to a test, impact points & directions (DOFs) must be chosen to excite ***as many modes as possible***. In the original multi-reference impact test, the grating was excited in three directions at point #1, at DOFs (1X, 1Y, -1Z).

6. Simulated Rapid Impact Test™

The five UMM mode shapes of the grating were used to simulate a Rapid Impact test of the grating. Data acquisition from the grating ***was simulated*** as if it were acquired using an ***impact hammer,*** a ***tri-axial accelerometer*** and a ***4-channel acquisition system***.

Three random impact forces were applied in succession to the Y-direction & Z-direction at five points (numbered 1 to 5 and shown in the Figure below along one edge of the grating. The 3D response of the grating to the impacts was calculated for accelerometer positions in ***rows of points closest to each impact point***. The points in each row are numbered in succession, as shown in the Figure.

- When impacted at point 1, accelerations were calculated for points 6 to 18
- When impacted at point 2, accelerations were calculated for points 18 to 31
- When impacted at point 3, accelerations were calculated for points 31 to 44
- When impacted at point 4, accelerations were calculated for points 44 to 57
- When impacted at point 5, accelerations were calculated for points 57 to 64

Each response was calculated using MIMO Modeling & Simulation. This calculation is depicted in Figure 8. The simulation uses the **Modal Model of UMM mode shapes** to synthesize an FRF between each impact & response DOF. The first several impact impact measurements of the ***simulated*** Rapid Impact™ test are shown in Figure 6.

Four impact-response pairs are shown in Figure 9. Each impact-response pair shows ***three impacts*** and their resulting ***three impulse responses*** in the X, Y & Z-directions. As expected, a small impact caused a small response.

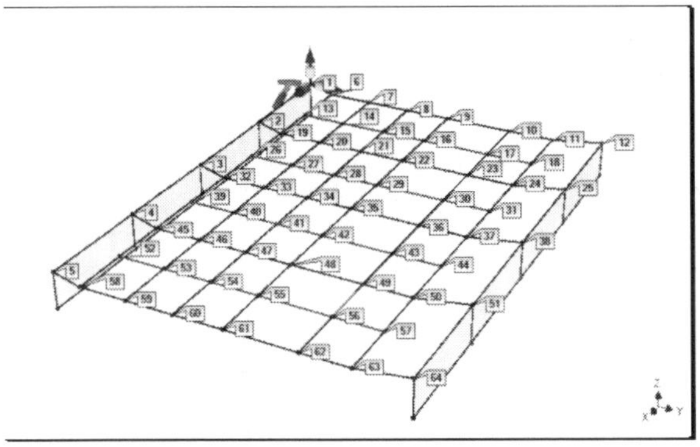

Figure 6a. First Acquisition

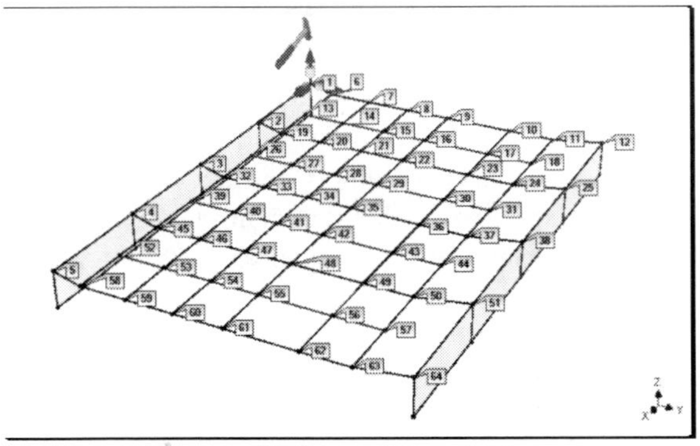

Figure 6b. Second Acquisition

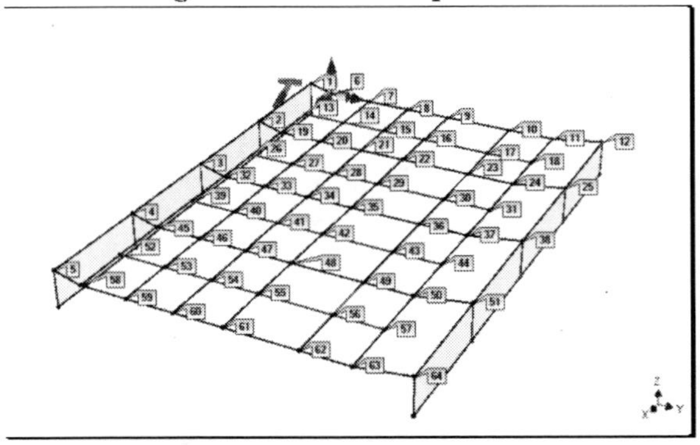

Figure 6c. Third Acquisition

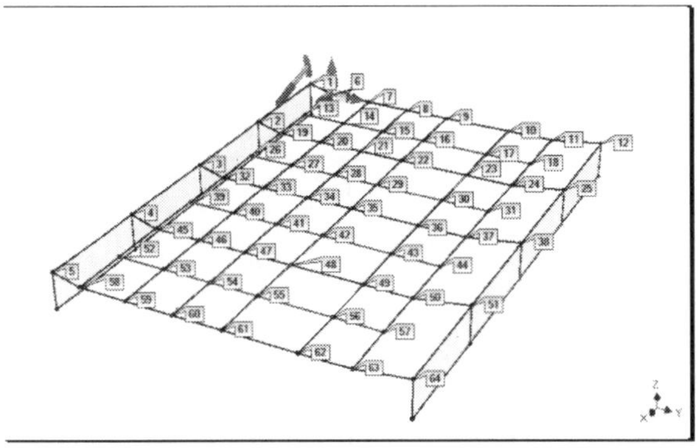

Figure 6d. Fourth Acquisition

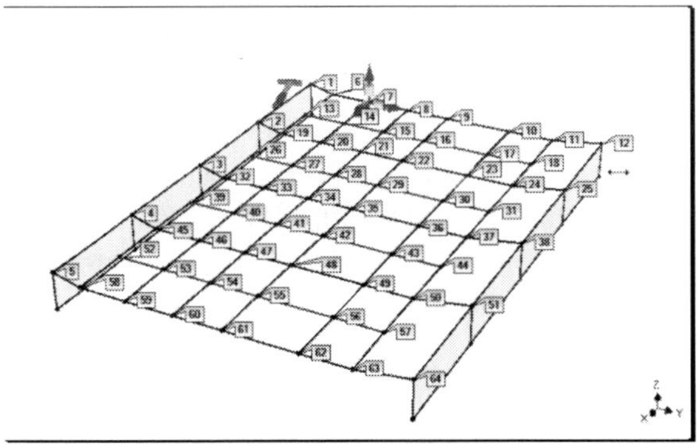

Figure 6e. Fifth Acquisition

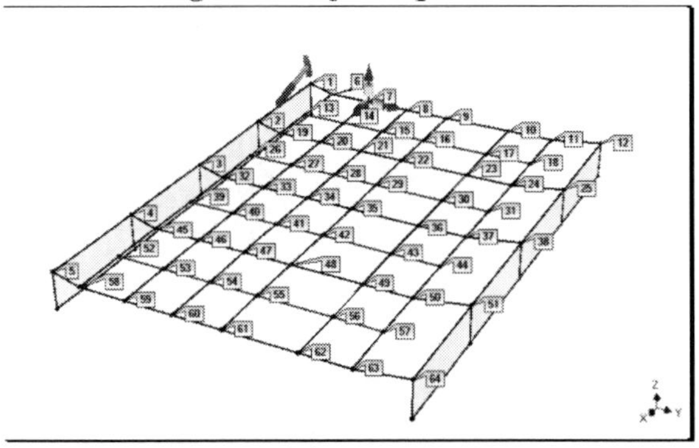

Figure 6f. Sixth Acquisition

During a Rapid Impact test, acquisitions can be made in ***any desired manner*** provided that a **chain of FRFs** can be calculated from the acquired data. An **FRF chain** is formed when each FRF has a ***Roving or Reference DOF that matches a DOF in another FRF***.

For this simulated Rapid Impact test, an FRF chain was calculated in three steps,

1. A sequence of three random impact forces was created for applying impacts in the Y-direction & Z-direction at each of five impact points along one edge of the grating. The random impacts simulate a real-world impact test. Several of these impact sequences are shown in Figure 7. Each impact force sequence contains **6144 samples** of time waveform data, enough to calculate three Fourier spectra with **1024 samples each**.

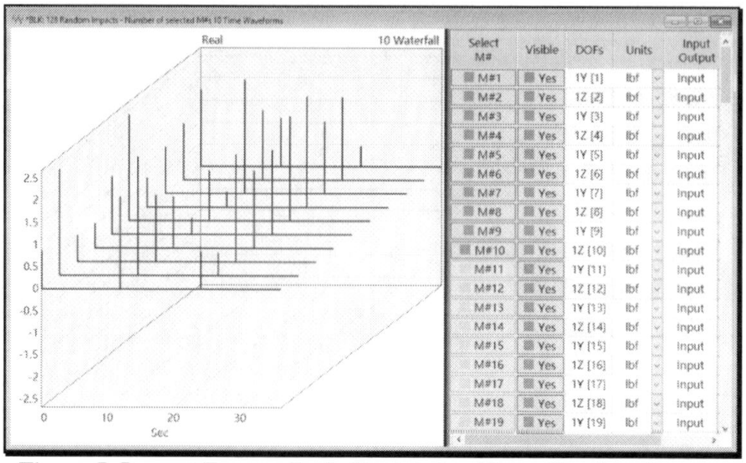

Figure 7. Impact Forces, Each Containing Three Random Impacts

2. Acceleration responses to each impact force were calculated in the X-direction, Y-direction & Z-direction. This calculation used the **Modal Model of UMM mode shapes** to represent the dynamics of the grating. FRFs were synthesized for each impact-response pair (input-output pair) using the UMM mode shapes. The MIMO calculation is depicted in Figure 8. Several impact-response pairs are shown in Figure 9. Each pair shows ***three impact forces*** and three corresponding ***impulse responses***. The responses are in the X, Y & Z-directions.
3. Data was ***"acquired"*** from the impact-response pairs shown in Figure 9. Each impact-response pair is given a ***unique*** **[Measurement Set number]** to define it as a ***simultaneous acquisition of 4 channels of data***.

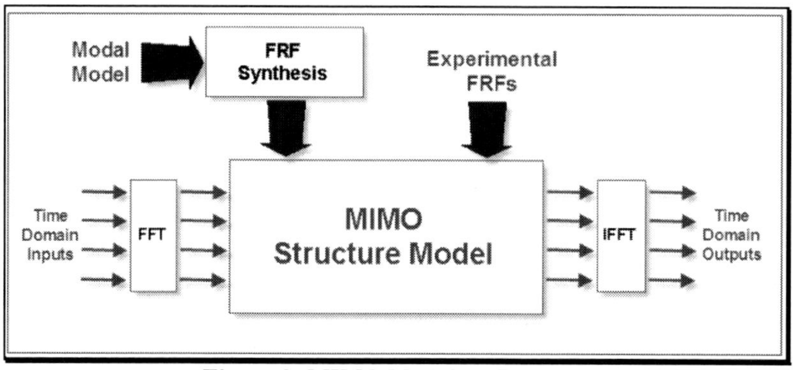

Figure 8. MIMO Modeling Diagram

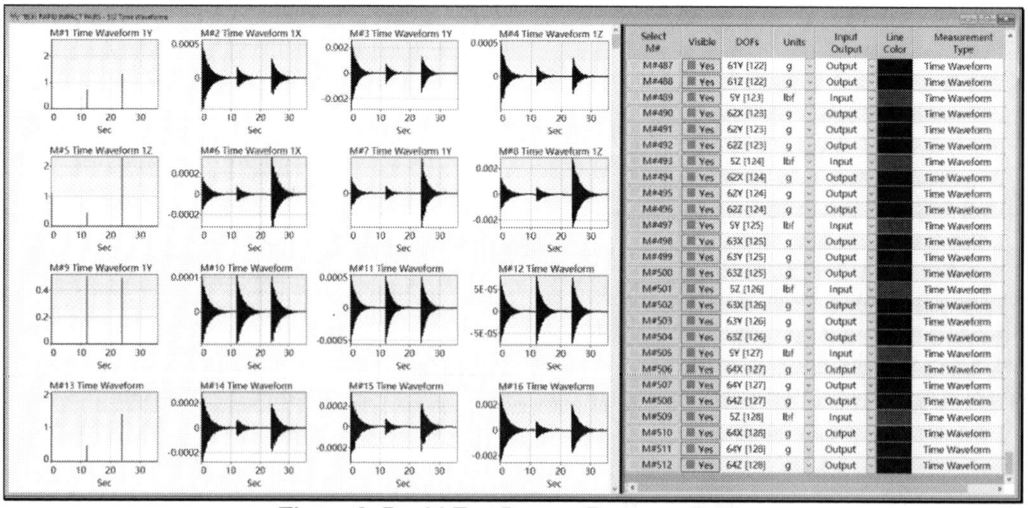

Figure 9. Rapid Test Impact-Response Pairs

7. Rapid Impact FRFs

To calculate FRFs, each **[Measurement Set]** of time waveforms was ***"acquiring"*** from the time waveforms shown in Figure 9. Three spectral estimates were averaged together using Stable (Linear) averaging, and **Coherence** was calculated together with each FRF. Each impact point & direction on the grating model is depicted with a hammer. Each accelerometer location is depicted with three arrows indicating the three responses of the tri-axial accelerometer. The acquisition process is depicted in Figure 10.

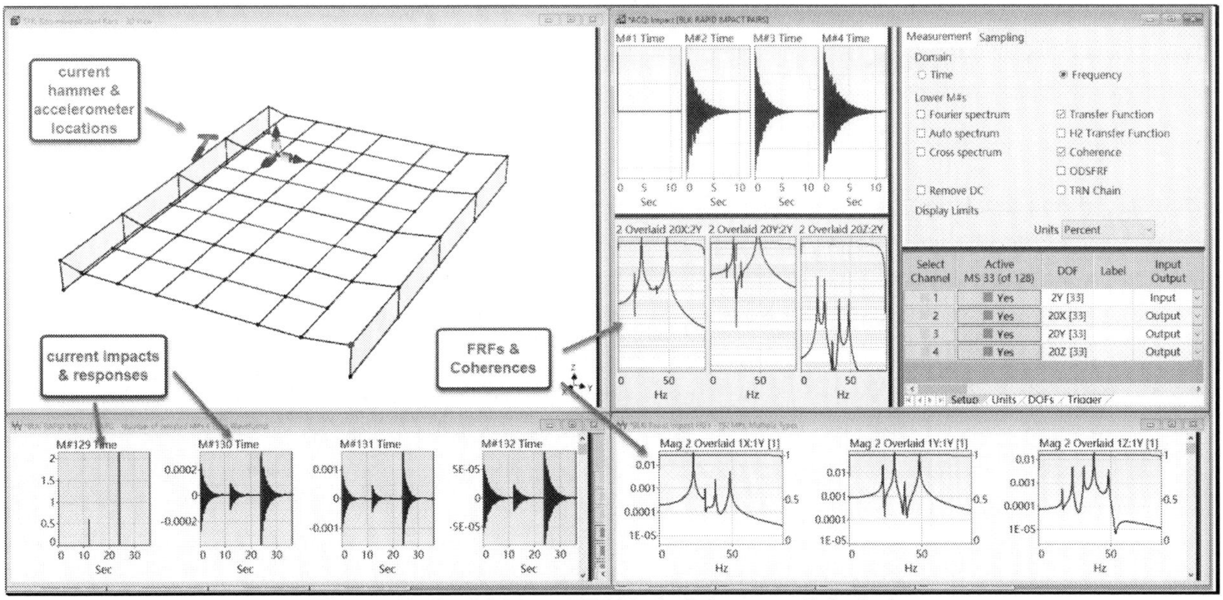

Figure 10. Rapid Impact FRFs & Coherence Calculation

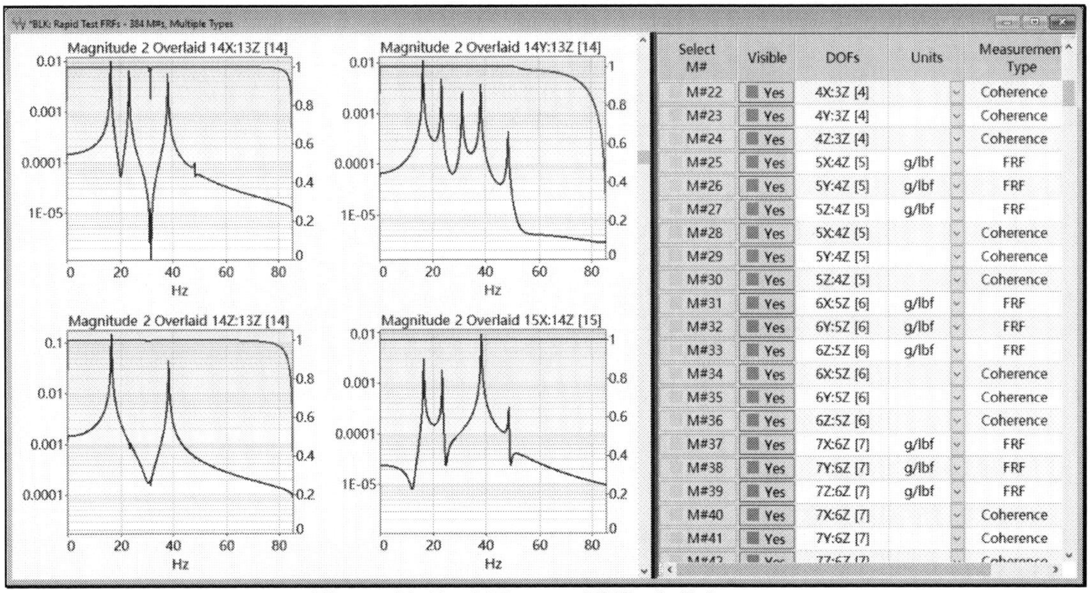

Figure 11. Rapid Impact FRFs & Coherence

408 FRF-Coherence pairs were calculated from the random impact-response pairs. Several FRFs, each overlaid with its corresponding Coherence, are shown in Figure 11. The FRFs represented measurements that would be calculated between impacts in the Y-direction & Z-direction (at 5 points along one edge of the grating), and 64 tri-axial accelerometer responses on the grating. The points are labeled in Figure 6.

Coherence ***values will drop below*** **"1"** at ***anti-resonances*** (low FRF values), ***near DC*** (zero frequency), and near the highest frequency in the FRFs. Figure 11 shows that very few Coherence values ***dropped below "1"*** indicating that the FRFs accurately measured the ***linear relationship*** between each impact force and one of its acceleration responses.

No special time domain windowing was required because the impact & response signals are both ***completely contained*** within each sampling window of time domain data (**2048 samples**). Hence, each time waveform is ***periodic in its sampling window***.

8. FRF Curve Fitting

The Rapid Impact FRFs were curve fit using the single-reference **Quick Fit** command. An example of a Quick Fit is shown in Figure 12.

Multi-reference curve fitting can only be used on FRFs that are ***multiple sets*** of ***single-reference FRFs***. To use multi-reference curve fitting, the FRFs for each reference DOF must contain the ***same Roving DOFs*** as all other FRFs with a different reference DOF. Stated differently, in order to use multi-reference curve fitting, the FRFs must fill ***multiple rows or columns*** of the FRF matrix of possible FRF measurements. Figure 2 shows an FRF matrix.

Following a curve fit, a **red Fit Function** is overlaid on each FRF. The **Fit Function** in Figure 12 closely matches the FRF, also indicated by **FRAC=1** on the ***upper right*** of the graph. **FRAC** is the frequency response version of **MAC** [4] that measures the ***co-linearity*** between an FRF and its **Fit Function**.

Each modal residue has the same DOFs as its corresponding FRF. Modal residues are displayed on the right side of Figure 12. **408 FRFs** were curve fit, so each mode has **408 modal residues.** Each modal residue has DOFs that match the DOFs of the FRF from which is was derived through curve fitting.

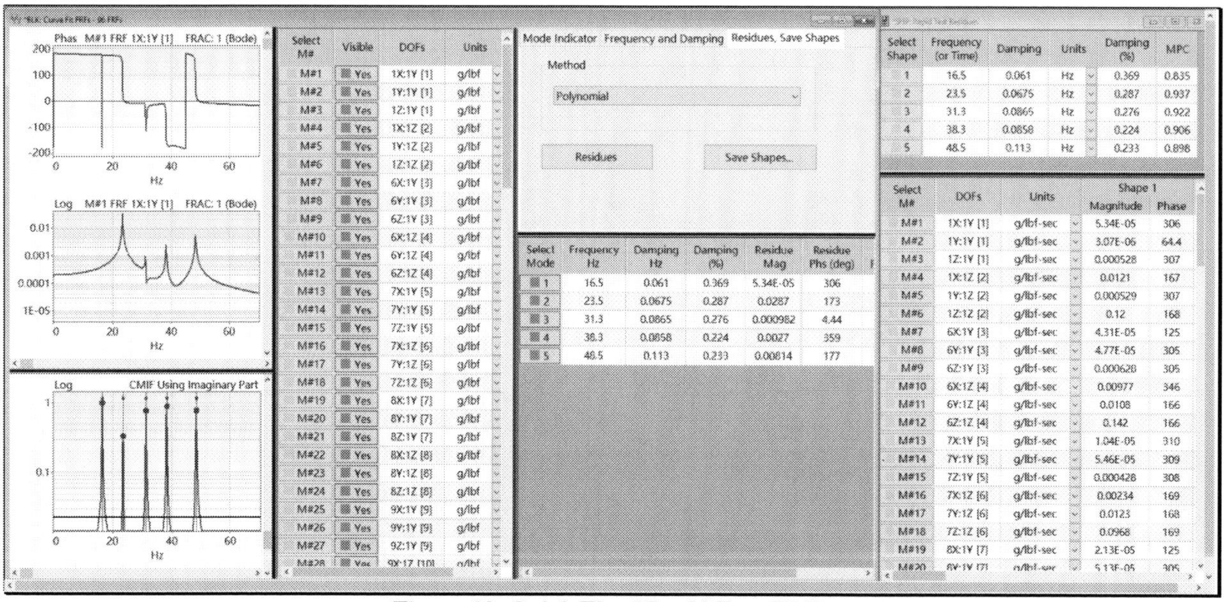

Figure 12. Quick Fit of Rapid Test FRFs

9. Modal Residues & Mode Shapes

Rapid Impact testing takes advantage of the mathematical relationship between modal residues and mode shapes. Figure 13 illustrates the relationship between modal residues and a mode shape.

Rapid Impact FRFs do not correspond to a ***row or column*** of FRFs in an FRF matrix, and therefore the modal residues obtained by curve fitting are ***not Residue Mode Shapes***. These modal residues must be further processed to obtain mode shapes, using the relationship "***Each modal residue is the product of two mode shape components***".

Figure 13 shows how modal residues obtained by curve fitting Rapid Impact FRFs are converted to a mode shape. In order to begin the conversion, a starting mode shape component is required.

The starting mode shape component is obtained either from a **driving point residue** or from **triangle point residues**. Using the starting mode shape component, all other mode shape components are computed using the relationship between residues and mode shape components.

Residues versus Mode Shape Components

$$\text{Residues} = \begin{Bmatrix} RES(1{:}2) \\ RES(1{:}3) \\ RES(2{:}3) \end{Bmatrix} \qquad \text{Mode Shape} = \begin{Bmatrix} MSC(1) \\ MSC(2) \\ MSC(3) \end{Bmatrix}$$

RES(1:1) = MSC(1) x MSC(1)

RES(1:2) = MSC(1) x MSC(2)

RES(1:3) = MSC(1) x MSC(3)

RES(2:3) = MSC(2) x MSC(3)

Single-Reference Modal Test

$$\text{Residue Mode Shape} = \begin{Bmatrix} RES(1{:}1) \\ RES(2{:}1) \\ RES(3{:}1) \end{Bmatrix} = \begin{Bmatrix} MSC(1) \times MSC(1) \\ MSC(2) \times MSC(1) \\ MSC(3) \times MSC(1) \end{Bmatrix}$$

Converting Rapid Test Residues to a Mode Shape

Starting with a Driving Point Residue

$$MSC(1) = \sqrt{RES(1{:}1)} = \sqrt{MSC(1) \times MSC(1)}$$

MSC(2) = RES(1:2) / MSC(1)

MSC(3) = RES(1:3) / MSC(1)

MSC(3) = RES(2:3) / MSC(2)

MSC(4) = RES(3:4) / MSC(3) = RES(4:3) / MSC(3)

Starting with Triangle Point Residues

$$MSC(1) = \sqrt{\frac{RES(1{:}2) \times RES(1{:}3)}{RES(2{:}3)}}$$

$$= \sqrt{\frac{MSC(1) \times MSC(2) \times MSC(1) \times MSC(3)}{MSC(2) \times MSC(3)}}$$

Figure 13. Modal Residues & Mode Shapes

10. Animated Mode Shape Comparison

As a ***"round trip"*** comparison, the Rapid Test mode shapes were compared in animation with the original UMM mode shapes. A comparison of a Rapid Test mode shape on the left and a UMM mode shape on the right is shown in Figure 14.

MAC & SDI Bars: Notice that **Modal Assurance Criterion (MAC)** and **Shape Difference Indicator (SDI)** bars are also displayed with each shape pair. These two correlation coefficients have the following properties,

- MAC is a measure of the ***co-linearity*** between two shapes [4]]
- SDI is a measure of the ***difference*** between two shapes [3]
- Both MAC & SDI have values between 0 &1
- A MAC or SDI value ***greater than 0.9*** indicates a ***strong correlation*** between two shapes
- A MAC or SDI value ***less than 0.9*** indicates a ***weak correlation***

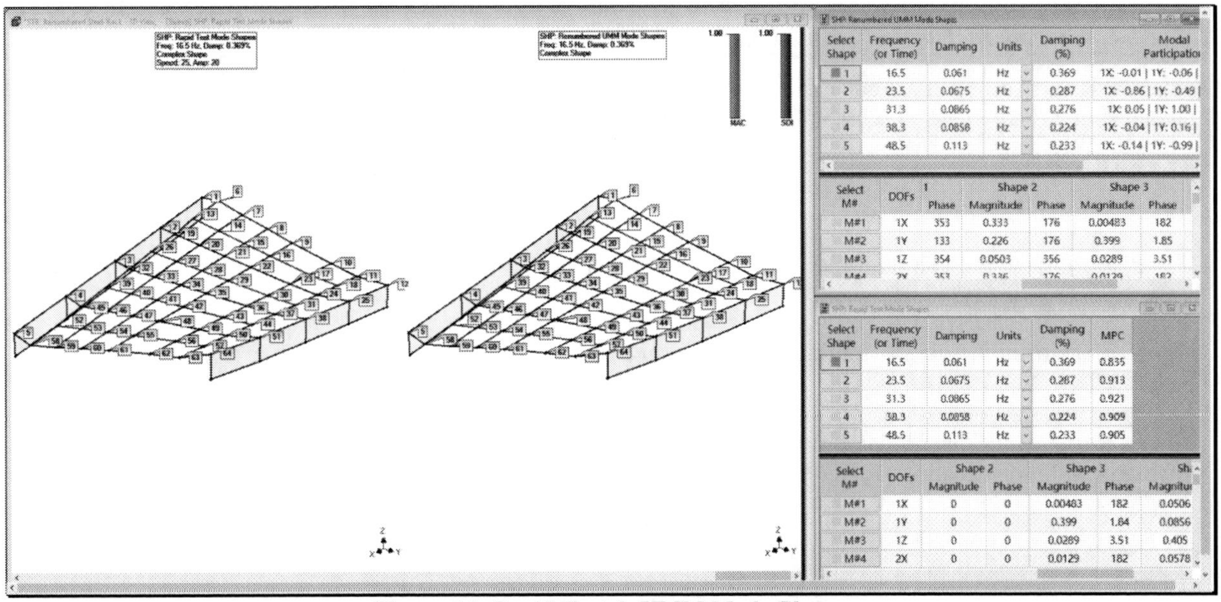

Figure 14. Rapid Test & UMM Mode Shapes

From the animated shape comparison (including the MAC & SDI bars), it was evident that the modal parameters (frequency, damping & mode shape) of all five Rapid Test modes ***matched the parameters of the original UMM modes very closely***.

11. Summary of The Rapid Impact™ Test

A Rapid Impact™ test ***was simulated*** on a grating from a water treatment plant. It was assumed that an **impact hammer**, **tri-axial accelerometer**, and **4-channel simultaneous data acquisition** were used for the test. The following steps were carried out in this simulation,

Step [1] - 136 random impact forces were created to simulate real-world impacting of the grating in the Y-direction & Z-direction at 5 points

Step [2] - Impact-response pairs were calculated using **MIMO Modeling & Simulation** and a **Modal Model of five UMM mode shapes** to represent the dynamics of the grating between pairs of impact & response DOFs

Step [3] – **Frequency Response Functions (FRFs)** and their corresponding **Coherences** were calculated from each impact-response pair. 408 FRF & Coherence pairs were calculated

Step [4] - The Rapid Impact™ FRFs were curve fit to obtain the modal frequency, damping, and **modal residues** for five modes of the grating

Step [5] - The Rapid Impact **modal residues** were converting into **mode shapes** of the grating

Step [6] - The Rapid Test **mode shapes** were compared in animation with the original **UMM mode shapes**

Rapid Impact™ testing offers several advantages over conventional single-reference or multi-reference impact testing,

1. In a conventional impact test, the ***reference sensor*** (either the impact DOF or the accelerometer location) ***must remain fixed throughout the test***. Since the reference sensor must be connected by a wire to the data acquisition system. a ***very long wire may be required*** when testing a large structure.
2. ***Better quality signals*** *are possible if each impact force* ***is applied closer to the response accelerometer***
3. In a Rapid Impact™ test, either the ***impact hammer or the accelerometer can be moved*** to a different DOF between acquisitions of data. One sensor can remain fixed while the other one is moved between acquisitions, thus forming a ***chain of acquisitions*** based on their DOFs
4. A Rapid Impact™ test is faster and more convenient to use on ***any size structure***

During a Rapid Impact™ test, a chain of acquisitions is formed based on their DOFs. A chain of FRFs can then be calculated from the chain of acquired data.

Each FRF has two DOFs associated with it. An ***FRF chain*** is formed when the ***Roving or Reference DOF of each FRF has the same DOF as another FRF in the chain***.

After the Rapid Test FRFs are curve fit, the modal residues can be converted to mode shapes using the special relationship, ***"Each modal residue is the product of two mode shape components"***.

REFERENCES

[1] - B. Schwarz, M. Richardson, "Linear Superposition and Modal Participation" IMAC XXXII, February 3-6, 2014

[2] - P. McHargue, B. Schwarz, M Richardson, "ODS & Modal Testing Using a Transmissibility Chain", IMAC XXXVI, February 2017

[3] - S. Richardson, J. Tyler, P. McHargue, B. Schwarz, M Richardson, "A New Measure of Shape Difference", IMAC XXXII, February 3-6, 2014

[4] - R. J. Allemang "The Modal Assurance Criterion (MAC): Twenty Years of Use and Abuse", Proceedings of the International Modal Analysis Conference, 2002

LOW POWER WIRELESS TRIAXIAL VIBRATION SENSOR DESIGN - PROTOTYPE REVIEW

Shannon Jelken
Emerson Automation Solutions, Inc.
301 S. 1st Ave
Marshalltown, IA 50158
641-754-2622
shannon.jelken@emerson.com

Joel Bland
8653 Riggs St.
Overland Park, KS 66212
620-757-3639
BlandJoel@gmail.com

Haibo Wang
Department of Electrical and Computer Engineering
Southern Illinois University
1230 Lincoln Drive,
Carbondale, IL 62901
618-453-1522
zhwang@siu.edu

Ed Spence
The Machine Instrumentation Group, LLC
535 Washington St
Dedham, MA 02026
781-439-1277
ed.spence@machineinstrumentation.com

Abstract: Attend a maintenance or reliability conference today and you can't fail to miss the many vendors offering wireless sensors for condition monitoring. Wireless sensor networks offer a way to expand the practice of CBM by making more condition data from more machines available for analysis. The recent advancements of MEMS accelerometers, energy harvesting techniques, and low power wireless communication facilitate the development of specialized vibration sensors for machine maintenance applications. This presentation describes a low power MEMS accelerometer based vibration sensor with energy harvesting and wireless communication capabilities.

Key words: Accelerometer; MEMS; Sensor; Vibration; Wireless; Energy harvesting; TEG; Low power

1. Introduction

The use of vibration based condition monitoring for rotating equipment is a relatively well development application, with a mature eco-system of sensor and instrumentation vendors, service providers and signal processing techniques for monitoring and diagnostics. Less well served is the emerging trend by machine makers and OEMs to instrument their own equipment, developing the components necessary for their particular machine along with new business models to service their customers. In many cases, the machines require a more specialized approach to all aspects of the system – sensor, interface, network, data processing, storage and display, with implications for the service or business model. We describe the results of an OEM program to develop a diagnostic and prognostic regime for machine faults that were previously diagnosed only by field specialists. The goal of the program is to automate the identification and diagnosis of machine faults, both old and new, with a suite of sensors envisioned to monitor aspects of the machine behavior currently not monitored.

The vibration sensor is defined to be a low-power device, with a battery that will need replacing on the order of 8 -10 years. This life span may require the sensor itself to supply at least some of its own power needs with harvested energy, potentially from more than one source. The maximum sample and transmission rate is expected to be 4 times per hour, but may often be as infrequent as once per day (depending on the measurand). The measurements will be wirelessly transmitted via Bluetooth Low-Energy (BLE) to a local 'hub' for processing with other measurands, with modest transmission distances measured in single digit meters. With the assumption that radio transmission dominates the power budget, local signal processing will be considered to reduce the amount of transmitted data, if it saves power. Research for the application reveals that vibration measurements are to be taken in all three axes of orientation in a frequency band from 10Hz to 3.5kHz.

2. Signal Processing Decisions and BOM

After early consultations with a sensor vendor the ADXL356 by Analog Devices is evaluated, validating the sensor performance for this application. The bandwidth of the sensor output is high enough to cover the vibration frequency range of interest. The output signals are in analog format which requires an ADC with MUX to digitize the analog outputs. Due to the relatively high accelerometer output resistance, additional ADC driver circuits are implemented. Since the ADXL356 outputs are ratiometric values with respect to its analog power supply voltage, the analog power supply of the ADXL356 is used as the ADC reference voltage. Further, experiments indicate that a voltage buffer is required to drive the ADC reference input in order to avoid SNR degradation when digitizing signals with relatively high frequencies.

Another early decision made was regarding the signal processing. The design team opted to transmit time series data instead of frequency spectrum. This allows the data to be processed post transmit in a way that is not power constrained (for example in the data hub or gateway). In addition, transmitting time series data prevents information loss, whereas frequency spectral data may reduce resolution of the content.

Transmitting frequency spectral data after an FFT also doubles the sample size in order to preserve phase information, hence the FFT may not necessarily have a smaller data packet.

The first prototype was constructed with 4MB of RAM, with a controller external to the radio module for signal processing flexibility. The data acquisition design centers around the ADS8331 16b 500ksps 4/8-channel input ADC with the PIC24FV32KA302-I_SN low power 16bit microcontroller (MCU).

Due to the relatively modest transmit distances and data transmission, a TI CC2640r2 radio chip was chosen for the first prototype due to availability of modules with BLE4.2 software stacks, with relatively high data transmission rates, the SimpleLink™ LaunchPad™ development environment, and a migration path to radios with more memory and BLE5.0 support. The CC2640r2 is equipped with a 32-bit ARM® Cortex®-M3 processor that runs at 48 MHz.

Figure 1. CC2640r2 radio modules from TI were chosen for the prototypes, which support the higher bit rates of BLE4.2. Radio modules that support BLE5.0 may be used for subsequent prototypes.

3. Prototype Design

Current prototype is configured to wake up, take 512 data points at 16 Ks/s sampling rate for each x, y, z axis of orientation and immediately transmit to the BLE radio developer board, repeating the measurement and transmission cycle once per minute. For data display and evaluation, the team is using the MATLAB GUI. The prototype card size is 2.5 x 4.0 inches. Size reduction approaches to the 2nd prototype may include using the 12-bit ADC integrated in the controller. We will also probably migrate to a two card stack solution, as there is more available space in the application for height than there is footprint. Architecturally, it also makes more sense to separate the sensor plus DAQ components from the radio, making the design more modular for future modifications and component replacements.

Figure 2. The first prototype, with the radio card shown upper left, the supercap at the edge on the right, and the MEMS accelerometer in the lower right.

Many of the problems encountered early in prototype development are probably common to new designs with a multiplexed sampling input (necessary for sampling the 3 outputs of the triaxial accelerometer). Prototypes were tested on a shaker table with a reference accelerometer from PCB, sampled at 30kHz.

4. Limitations on Sample Rate

Early designs ran into issues achieving the target sample rate, leading to erroneous samples. Some of the more mundane problems to track down included firmware bugs (thought to be internal CLK frequency inaccuracy).

Analysis of the timing signals reveals that, although designed to sample at 16Ks/s, the system was actually sampling at 13.4 Ks/s instead. The first suspect was an inaccurate RC oscillator built in to the microcontroller used to set the timing, resulting in the unexpected data capture rate, but the error was tracked down to a firmware bug.

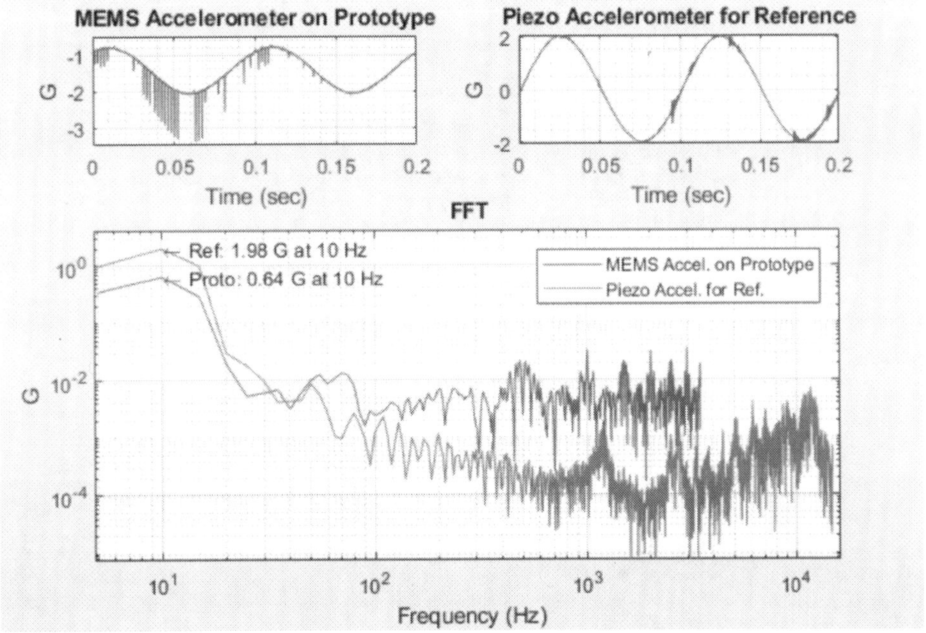

Figure 3. Time series plots and frequency spectrum of the MEMS based prototype compared to a reference piezo accelerometer, both mounted on a shaker table. The sample errors are evident in the time series plot, with the resulting noise in the frequency spectrum.

The prototype samples one axis and then switches to the next axis until 1536 samples were collected for all 3 axes. In the time series the sampling errors manifest as downward spikes, and in the FFT the noise floor is a couple orders of magnitude greater than the reference. The sensor vendor (ADI) recommended including a low-power nearly zero footprint op amp in the design that all but eliminated any potential switching noise by the MUX. The simultaneous sample rate is now only limited by the microcontroller OS.

It was also found that while most of the switching transient amplitude has been eliminated, a small residual capacitance affects the simultaneously-sampled measurement. This can be observed as an increase in roughly 0.04g compared to the single-channel measurement. This is well within acceptable measurement error and was not considered a problem.

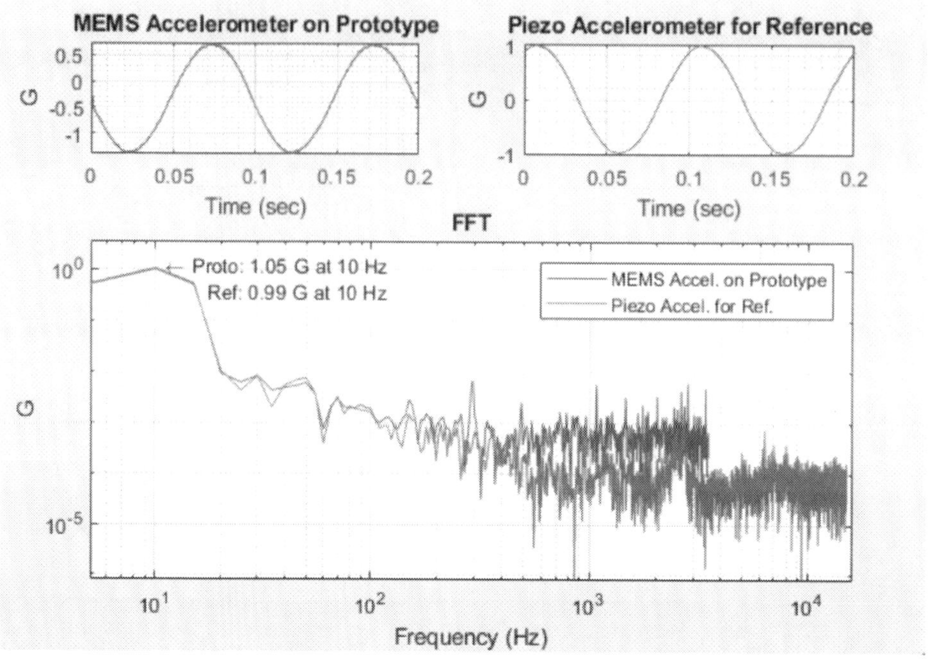

Figure 4. The improved performance is evident once the errors induced in the sampled data are addressed.

5. Energy harvesting

Both Thermal Energy Generator (TEG) and RF Energy harvesting technologies are under investigation. The power management circuit is designed to accept additional current from any source to charge a 1F, low leakage super capacitor. For this prototype, a TEG generator is added to the design for experimenting with the amount of power that might be generated. Several experiments are performed to test the TEG charging capabilities.

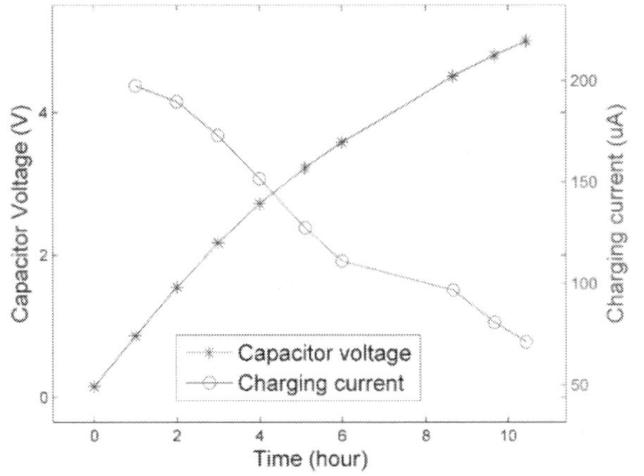

Figure 5. In this case, a temperature differential of 13C is measured across the electrodes of the TEG, presenting 67mV at the input of the charging circuit. The circuit took 10 hours to charge the cap to +5Vdc.

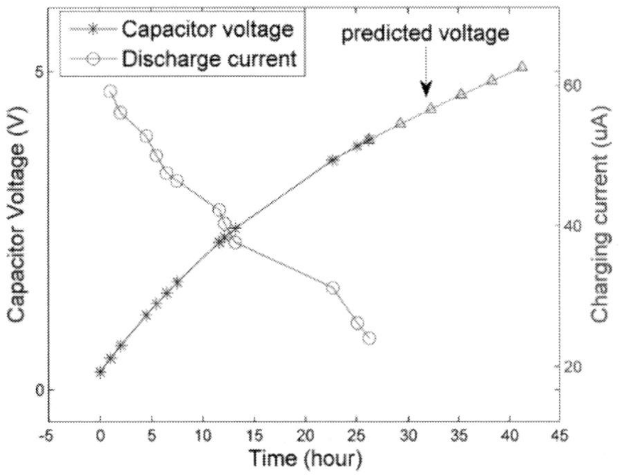

Figure 6. In a second experiment, a temperature differential of 9.6C is measured across the electrodes of the TEG, presenting 40mV at the input of the charging circuit. The circuit took 48 hours to charge the cap to +5Vdc.

Once the capacitor was charged, measurements were taken with no load to estimate the impact of circuit leakage.

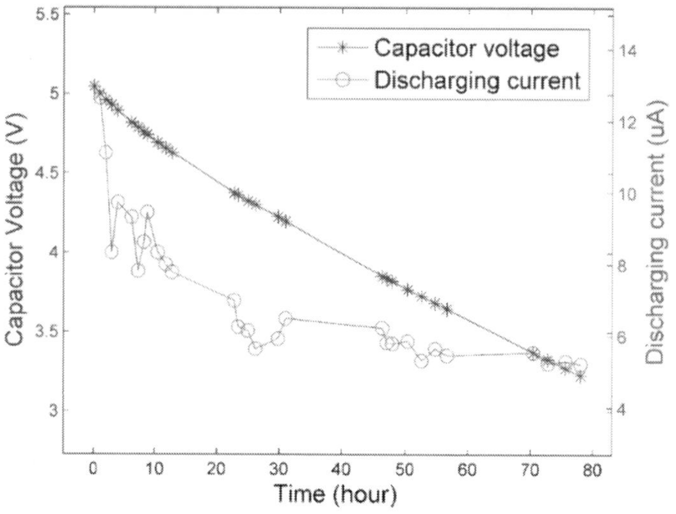

Figure 7. The leakage current for the circuity is about 6 uA. After losing the energy source, the circuit may remain on standby for approximately three days.

During data sampling, the cap discharges 3 mV for wakeup, 512 measurements on all 3 axes, and a write cycle to flash memory (or 0.06% of charge).

Once the initial BLE communications were up and running, it was found to discharge 250 mV for new data and transmit (we believe that this can be reduced in the future). Each data sample and transmission cycle discharges roughly 0.3V from the supercap, demonstrating our expectation that the radio transmission would dominate the power budget. The device continues to function until the power supply drops below 3.3V. In practice, 6-7 transmissions are likely without recharging the TEG energy source.

Improvements in the charging performance of a TEG may be to position the generator such that airflow from a local source will help to reduce the temperature of the sink side, increasing the average temperature differential, hence the power harvested over a given period of time.

6. Next Steps and Future Work

The 2nd revision will reduce the size further, and the design team may consider a 2-stack PCB to reduce the footprint at the expense of height. Low hanging fruit might include replacing the external 16-bit ADC with the internal 12-bit ADC of the microcontroller. Some testing with the integrated converter will be required to ensure adequate resolution for the measurements.

Also, a sound partitioning strategy would be to separate the sensor plus DAQ from the Radio, to make it easier for future component replacements based on results achieved. We will also probably continue to keep a separate MCU (plus memory) for flexibility regarding sensor signal processing. This provides flexibility to provision with capable components, and avoids problems and risks of trying to use the radio chip resources, such as filling the memory, counting on vendor support and trying to understand protocol code.

Other sensor features to add include:

Configuring the sensor MCU for selectable data collection intervals, such as taking a data sample every 1, 4, 6, 12, 24 hours.

Investigate just how much flash memory storage is enough. The target sample size is defined as 0.1 seconds with at least 9ksps sample rate (2.56x the max frequency of interest is considered a minimum oversampling rate), with the ability to buffer multiple sets of data (to manage data transmission intervals). The current design uses 4M bit of Flash memory. A decision needs to be made regarding how many datasets to buffer, as too much data requires excessive power to transmit in addition to replicating calculations.

Adding elements of local control to the sensor, such as user initiated device wake up and data capture, which would be useful for research and debugging. This might be implemented with an extremely low-power, low-accuracy MEMS accelerometer to use purely for motion-based event detection wake-up (ie ADXL362). Another option may be to consider a very low power Wake Up Radio (WuR). Other event criteria to be defined.

Regarding the radio and protocol, we will consider transitioning to BLE5.0, for higher speed transmission and smaller size data packets, if necessary. Ideally, this progression could be implemented with minimal impact on recoding the stack. We have also starting working with an IIoT software vendor to develop the sensor firmware further for interface to a COTS gateway, in order to facilitate deployment in the field for testing data transmission and to collect data for further analysis.

7. CONCLUSION

The high value outcome of this project is in the diagnosis and prognosis approaches developed from analysis of the collected data from the field. The wireless sensor network enables us to deploy in friendly plant without risk of disrupting existing networks or process controls. Working in partnership with a data engineering company (Predictronics Inc), the OEM Subject Matter Experts (SME's) have been developing custom condition indicators for several months and the work is on-going. The early results look very promising – there is potential for automated monitoring of emerging faults that are now identified visually by an experienced specialist, with new ways to detect new faults, and the potential to build a prognostic outlook for earlier identification of problems.

Bibliography

[1] M. Alhawari, B. Mohammad, H. Saleh and M. Ismail, "A survey of thermal energy harvesting techniques and interface circuitry," 2013 IEEE 20th International Conference on Electronics, Circuits, and Systems (ICECS), Abu Dhabi, 2013, pp. 381-384.

[2] P. Nintanavongsa, "A Survey on RF Energy Harvesting: Circuits and Protocols," Energy Procedia, Vol. 56, 2014, pp. 414-422.
[3] C. Wang, C. Hung, W. Chen, "Design of heat sink for improving the performance of thermoelectric generator using two-stage optimization, Energy," Energy, Vol. 39, No. 1, 2012, pp.236-245.
[4] D. Li, "Design Challenges and Improvement Techniques for SAR ADC Circuit," TI Application Report, SLAA571, [Online], accessed 2018, http://www.ti.com.cn/cn/lit/an/slaa571a/slaa571a.pdf.
[5] D. Giovanelliy, B. Milosevicy, D. Brunelli, E. Farella, "Enhancing Bluetooth Low Energy with wake-up radios for IoT applications", 2017 13th International Wireless Communications and Mobile Computing Conference (IWCMC), June 2017, accessed 2017 from IEEE

DEVELOPMENT OF AN ACCELERATED DEGRADATION TESTBED FOR INTEGRATED VIBRATION MONITORING AND ONLINE OIL ANALYSIS OF VEHICLE DIFFERENTIAL BEARINGS

Cody M. Walker, Alec B. Salakovich, David K. Irick, Jamie B. Coble, and Cyrus Smith
Nuclear Engineering
Mechanical, Aerospace, and Biomedical Engineering
PYA Analytics
University of Tennessee, Knoxville
Nuclear Engineering Department
421 Nuclear Engineering Building
Knoxville, TN 37996
Telephone: (931) 797-5403
Email: cwalke42@utk.edu

Abstract: This project, sponsored by the Office of Naval Research, integrates oil and vibration analysis to develop a more complete diagnostic and prognostic model for monitoring bearing degradation and performance in the field. The results of this project will be used to optimize repair schedules and minimize the risk of catastrophic failure of machine components for military personnel during a mission through enhanced condition-based maintenance. Each test bearing is first introduced with an outer race fault. The fault's degradation is accelerated by running the bearing under overloaded conditions in the boundary lubrication regime within a large test rig designed for these conditions. Accumulated damage to the bearing is characterized in another clean, low-noise test rig to collect vibration signatures. Finally, the bearing is implemented into the differential for in-situ testing to emulate data in the field. Dynamic characterization of the designed test machines was performed by long runs spanning multiple hours under rated conditions to determine any wear-in effects and ramp-up tests to distinguish order-based bearing frequencies and structural resonances. Modal analyses were performed on the static system to provide additional evidence of structural resonances within the machines. This paper will discuss the design challenges and solutions for creating a test bed to monitor bearings in accelerated-degradation conditions and in realistic operating conditions.

Key words: Bearings; boundary lubrication; diagnostics; frequency response function; health monitoring; prognostics;

Introduction:

Bearing are an integral component to most power transport devices including motors, drive fans, compressors, pumps and conveyors. Bearing fault diagnostics typically rely on vibration analysis to identify defects located on the bearing's inner race, outer race, rolling element, or cage. Bearing vibration analysis is used to determine the type of fault, if any, that is present, estimate the size of the fault or defect, and calculate the remaining useful life (RUL) of that component.

Prognostics aim to minimize in-use component failures by calculating the remaining useful life of essential components. Prognostic modeling requires run-to-failure data, but many industries replace their systems at the first indication of a fault. For example, a reactor coolant pump should not be run-to-failure while in the system. This is why run-to-failure data is a premium. There are few sources of such data for prognostics data sets. The National Aeronautics and Space Administration has set up a prognostics data repository to allow the exchange of publicly available data sets to work with (Administration 2019). Some industries, such as paper mills, have systems where faults develop slowly over months and even years (James C. Robinson 2018). Exactly recreating such faults in a laboratory setting would not be time efficient. Accelerated degradation testing provides realistic run-to-failure data for each stages of degradation seen in these systems, while reducing the time required to days or weeks (Nectoux et al. 2012). This is accomplished by overstressing the component of interest. By running under boundary lubrication rather than full fluid-film lubrication, the sliding friction between the rollers and the races increases. Boundary lubrication occurs when the lubrication layer is so thin that its viscosity no longer plays a significant role in the friction coefficient (Snow 2001). The lubricant and the nature of the surface determine the friction coefficient. When boundary lubrication exists, frictional wear increases and the lubrication can become contaminated with wear particles as the high spots of surfaces come into contact (Snow 2001). This type of friction will accelerate the degradation of the bearing and reduce its overall life. By understanding how the bearings degrade, maintenance can be scheduled based on the condition of the bearing rather replacing them at arbitrary intervals.

Condition-based maintenance is implemented in the military-based programs such as the System Health Management (SHM) program and its subset the Health and Usage Monitoring Systems (HUMS) toolkit (Team 2013). These initiatives have attempted to utilize existing or add-on steady state sensors for both diagnostics and prognostics. The SHM programs improve safety, increase reliability, and enhance situational awareness for crews. It is imperative that Humvees and helicopters do not burn out mid-mission as the consequences could be far worse than just loss of revenue. This project is working towards monitoring bearings within the differential of wheeled land vehicles. Run-to-failure data can be produced by seeding the bearings with faults then degrading them under accelerated conditions. The ultimate goal of this research is to provide early warning of incipient faults and accurate prognosis of failures throughout the component life.

The Bearing Test Rig was designed to assess the vibrational characteristics of a damaged bearing in a low-noise environment before taking measurements in the differential where it is used. The differential is likely to have a far noisier environment due to the extensive inner workings. Characterizing the damage with a low-noise rig will provide measurements that will be used to help distinguish the bearing fault frequencies from other vibrations within the differential. While the test rig is a low-noise environment, it is still not without its own characteristics that will show up in the measurements. Structural resonances of the test rig were investigated via impact hammer testing to develop a frequency response function (FRF). Such testing is important as the cyclical inputs provided by the shaft and bearings may excite natural frequencies of the structure and such resonances can cause unrealistically high magnitudes at bearing fault frequencies. The Bearing Test Rig was placed onto isolation pads to decouple the rig from the table that it rests on. Without the isolation pads, the table will resonate in response to the Bearing Test Rig's vibrations causing a noisier environment. These kinds of effects can be seen with a FRF and impact hammer testing.

The natural frequency of an object describes the frequency at which an object will dissipate the energy through vibration when an excitation is introduced. A common example of this effect is when a bell or tuning fork is struck. The magnitude of the strike affects the volume of the subsequent ring-down, but the tone is a function of the structure of the object and therefore is always consistent. Structural resonances such as these are important when performing vibration analysis work. If they exist near a frequency of interest, they can resonate and amplify the vibration response, causing either damage or affecting the credibility of sensor readings.

Mathematical modeling of this phenomenon is most simply expressed by a single degree of freedom (SDOF) system. In the SDOF system, the vibration characteristics are represented using four idealized components: mass, spring, damper, and excitation. Figure 1 shows the configuration for the theoretical system.

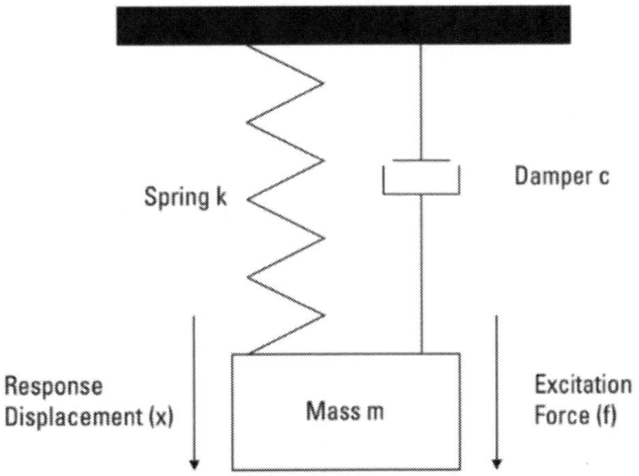

Figure 1: Configuration of Single Degree of Freedom System (Technologies 1997)

In this ideal system, the mass (m), spring constant (k), and damping constant (c) are all fixed values and the output (x) is only changed as a function of the input variable (f). The dynamics of this system are modeled by the equation of motion, shown as Equation 1.

$$\frac{d^2x(t)}{dt^2} + 2\xi\omega_n\frac{dx(t)}{dt} + {\omega_n}^2x(t) = \frac{f(t)}{m} \quad (1)$$

$$\text{where } 2\xi\omega_n = \frac{c}{m}, \omega_n^2 = \frac{k}{m}$$

Equation 1 is then manipulated with a Laplace transform and the ratio of the input and output, known as the transfer function, is shown in Equation 2.

$$\frac{x(s)}{f(s)} = \frac{1}{m(s^2+2\xi\omega_n s+\omega_n^2)} \quad (2)$$

$$\text{with roots: } s_{1,2} = -\xi\omega_n \pm \omega_n\sqrt{\xi^2 - 1}$$

The frequency response of the system can be plotted in polar coordinates to provide a visual representation of the system. Equations 3 and 4 show the formulas for magnitude and phase, respectively, of the response (Technologies 1997).

$$H(\omega) = \frac{1/m}{\sqrt{(\omega_n^2-\omega^2)^2+(2\xi\omega\omega_n)^2}} \quad (3)$$

$$\theta(\omega) = tan^{-1}(\frac{2\xi\omega\omega_n}{\omega_n^2-\omega^2}) \quad (4)$$

However, very few systems behave as ideally as a mass on a spring. Many structures have multiple degrees of freedom (MDOF) and exhibit many modes of excitation. Such systems are modeled as a weighted sum of multiple SDOF systems. Equation 5 shows the curve fitting function for a MDOF system.

$$H(\omega) = \sum_{r=1}^{n} \frac{\phi_r/m}{\sqrt{(\omega_n^2-\omega^2)^2+(2\xi\omega\omega_n)^2}} \quad (5)$$

Where ϕ_r is the modal participation factor and "is a function of excitation and mode shape coefficients at the input and output degrees of freedom (Technologies 1997)."

In order to accelerate the degradation in the bearing after characterization has been completed, operation will take place in oil-starved conditions, or more specifically, boundary lubrication. The boundary lubrication regime is where most of the load in the bearing is being supported on the asperities, or the peaks associated with the surface roughness of the interface, rather than being supported by a layer of lubrication (Hironaka 1984). Figure 2 below shows the Stribeck curve, which shows the various lubrication regimes and how the load is supported relative to the regime. In the equations shown in Figure 2, f represents the coefficient of friction, η is lubricant viscosity, V is the velocity of sliding surfaces, h is the oil film thickness and F_N is the load.

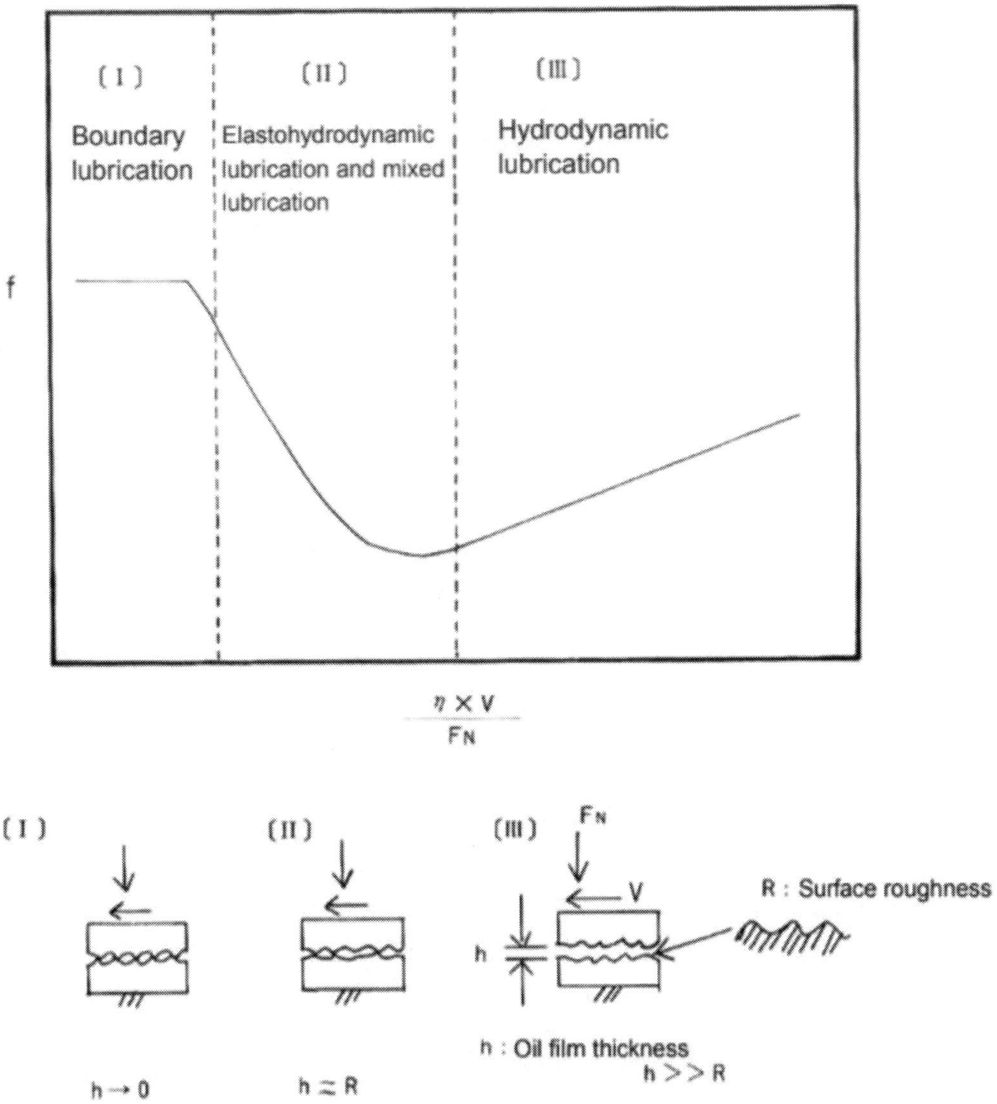

Figure 2: Stribeck Curve and Lubrication Regimes(Hironaka 1984)

Due to the higher coefficient of friction and higher temperatures acting on thinner layers of oil, oxidation reactions occur more readily within the oil. Initially, 75W-90 gear oil was intended for the tests as the chemical characteristics revealed in the oil analysis would represent the type of oil used within the differential. However, the kinematic viscosity of gear oil is 14.6 cSt at 100°C and the high viscosity makes it difficult to enter the boundary lubrication regime ('Mobil 1TM Synthetic Gear Lube LS | MobilTM Motor Oils.'). Aeroshell Turbine Oil 555 (also referred to as "triple-nickel") was used in tests due to its far lower viscosity. Figure 3 shows the temperature-viscosity curve for various Aeroshell oils with 555 emphasized in red.

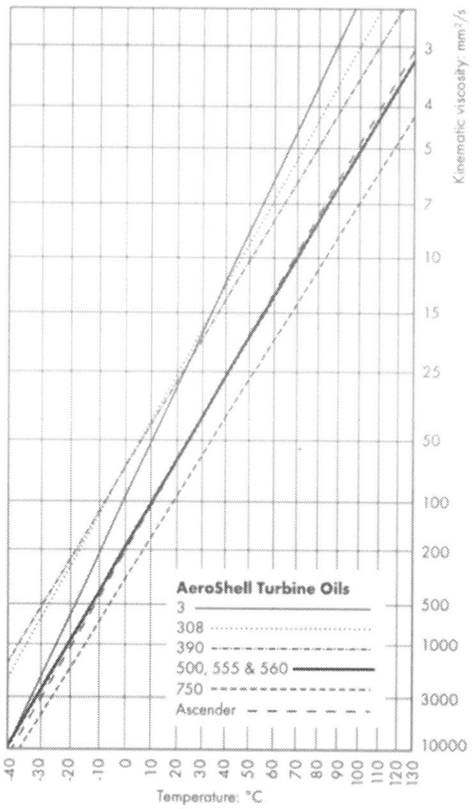

Figure 3: Typical Temperature-Viscosity Curves of Aeroshell Turbine Oils (1 cSt = 1 mm²/s)(Aeroshell Turbine Engine Oils 2012).

At 100°C, the viscosity of Aeroshell 555 is roughly one-third that of the 75W-90 gear oil. This lower viscosity results in a thinner oil layer between contact surfaces, thus making boundary lubrication more easily achieved. Requiring supervised operation, using the Aeroshell 555 reduces the time to reach boundary lubrication and more time of a single run can be used degrading the bearing.

Bearing skidding is an unintentional way to damage the bearing, which was seen at times during testing. Skidding occurs when the resistance of the roller or the cage overcomes the forward force from the inner race (Solutions 2019). This prevents the roller from adequately rolling, thus sliding. This sliding effect can form streaks and smears along the rollers and races leading to excessive pitting formation (Solutions 2019). Bearing skidding can be diagnosed through vibration analysis. It characteristically appears as an elevated noise floor in higher frequencies with protruding peaks spaced at frequencies relating to the bearing's outer or inner race fault frequency (Institute 2019). Skidding was seen in the Bearing Test Rig due a locking nut becoming loose during operation. This locking nut pressed the inner race into the rollers and outer race. Without proper compression, a small clearance formed, which allowed the rollers to skid rather than roll.

Methodology:

Three tests were used in the characterization of the Bearing Test Rig. First, impact hammer testing on the static system is used to identify structural resonances that may interfere with the collection of meaningful data through analysis of the system's response to an impulse excitation, otherwise known as the Frequency Response Function (FRF). The second test is a long run that provides a wear-in period for new bearings and gives insight into the steady state conditions achieved during extended operation. Finally, a ramp-up test that acquires a data sample at each operating frequency of interest to show how bearing and resonance frequencies move throughout the operating range. Table 1 shows the bearing defect frequencies for the bearings used in these experiments.

Table 1: Bearing Fault Frequencies (Generated from the SKF Engineering Calculator)

			Bearing Fault Frequencies (Order)			
Bearing Role	Product Number	Bearing Type	BPFI	BPFO	Cage	BSF
Test	JLM506849A	Tapered Roller	12.239	9.761	0.444	4.263
Support	7315WN	Angular Contact	7.056	4.944	0.412	2.11

Typically, data were sampled at 10,240 Hz and in the case of dynamic tests; data were sampled over 15 second intervals and all bearings were under well-lubricated conditions. The presented spectra are generated by averaging the spectra for each second within the sampling interval to minimize signal noise. The results from these tests are used to identify desirable operating conditions for data collection on bearings with known faults.

Figure 4 shows the design of the Bearing Test Rig. The Bearing Test Rig has a 3 hp motor that is controlled with a variable frequency drive (VFD). The motor is attached to a pulley, which spins the shaft. The two test bearings are located towards the ends of the shafts, while two greased-filled support bearing are closer to the pulley on the middle. The test bearings are lubricated with a once-through cycle. The oil that flows through the bearings is collected via the downward-facing spouts and will undergo further testing later. The test bearings are radially loaded with two hydraulic pumps capable of applying 20,000 lb of force. This force is applied in the same direction as the red arrows in Figure 4.

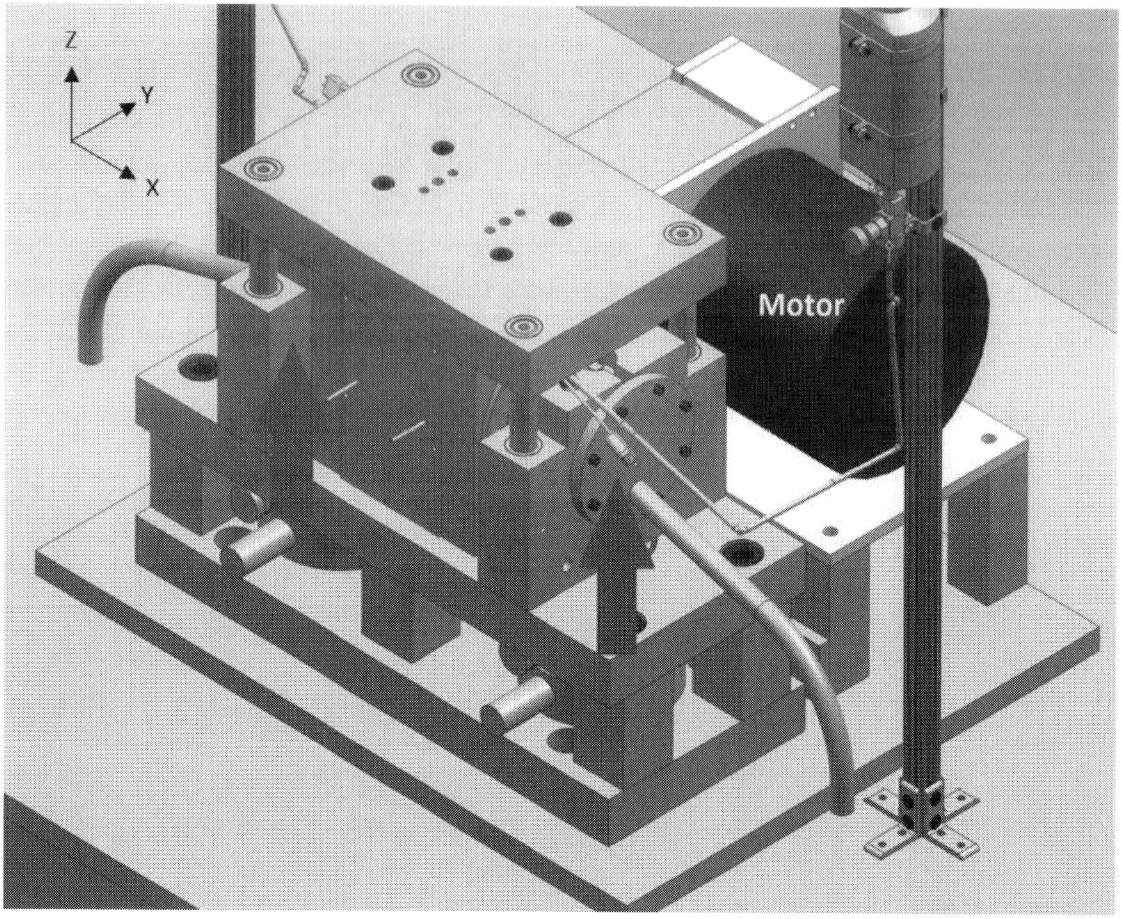

Figure 4: CAD model of the Bearing Test Rig. The red arrows indicate the location of the two test bearings.

2.1: Frequency Response Tests

To determine the frequency response function of the test rig, it was struck with an impact hammer to excite the structural resonances and the input-output relationship was recorded. When carrying out the test, the rig was struck in the same location three times with ample time between blows to allow the rig to ring down. During these periods, data were continuously collected until the final ring-down had decayed. The data were then post processed in MATLAB by choosing a window length large enough to capture the entire ringdown from each blow. The successive blows were then concatenated as shown in Figure 5 and evaluated using the "modalfrf.m" function in MATLAB, which uses Welch's averaged, modified periodogram method to estimate the frequency response function as the cross-power spectral densities of the input and output.

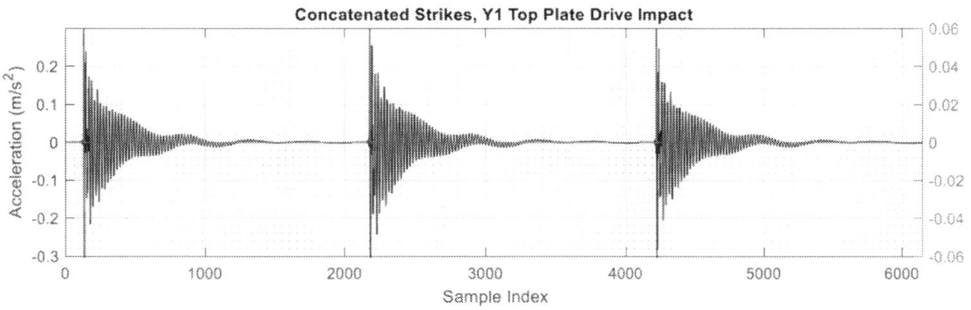

Figure 5: Preprocessed Data Sample for use in "modalfrf.m" (2048 samples per window)

There are three different components to consider when evaluating an FRF plot: magnitude, phase, and coherence. Magnitude and phase, when used together, are a great indicator of the frequencies at which resonances exist in a system. Magnitude is essentially the intensity of the response and, when shown in the frequency domain, allow for easy identification of a resonance mode. Phase is a value that exists between -180° and 180° and is also useful for identification of a resonances as the response will lag the input by 90°. Coherence is a value between 0 and 1 that is essentially used to show the goodness of the test across the frequency domain. Values near 1 indicate good repeatability and values close to 0 indicate errors in data collection. Figure 6 shows how the resonance for an idealized SDOF system would look in with respect to magnitude and phase.

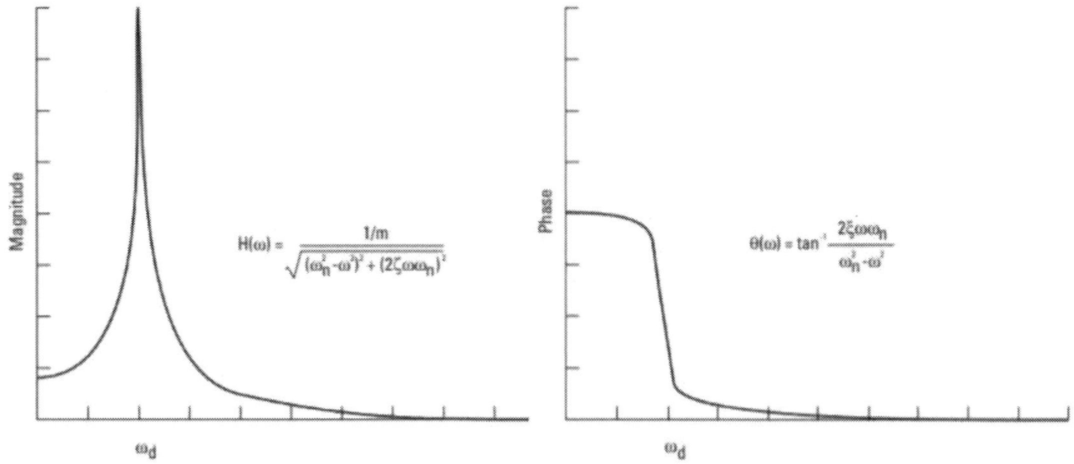

Figure 6: Magnitude and Phase for Resonance of a SDOF System

2.2: Long Run Tests

Long tests were designed to wear-in the test bearing as well as characterize the steady state nature of the Bearing Test Rig under extended operation. The tests also were used to determine the differences between unloaded and loaded scenarios at constant speeds. Tests are carried out such that the motor speed is held constant for four hours. During this time, the Bearing Test Rig is unloaded for the first two hours, then 1500 lb of radial load is applied for the second two hours. The motor speed is then increased for the next four-hour block to increase the breadth of data.

2.3: Ramp-up Tests

The motivation behind a ramp-up test is to collect vibration data at various motor speeds to identify how vibration signatures change as a function of shaft speed. The test starts by running the Bearing Testing Rig at a shaft speed of 15 Hz with 1500 lb radial load on each test bearing for 30 minutes to heat up the grease in the support bearings. This ensures that the support bearings are warm and well lubricated, while minimizing the risk of skidding in the bearings. Once 30 minutes have passed and the bearing temperatures have reached about 45 °C, the speed is reduced to 5 Hz and data collection begins. The shaft speed is increased by 1 Hz after each data collection period until the final speed of 20 Hz is reached.

2.4: Accelerated Degradation Testing

Before further testing, the test bearing's outer race was seeded with a fault using the brinelling drop-weight fixture shown in Figure 7. This fixture held the bearing's outer race in place, while an impact load was delivered to form a brinell on the test bearing's outer race.

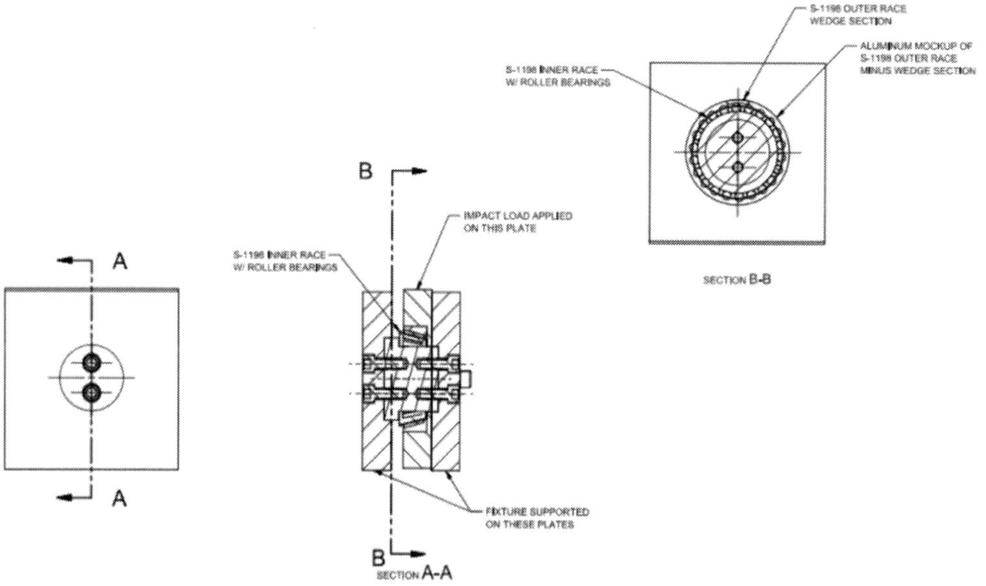

Figure 7: Brinelling Drop-weight Fixture

A progressive loading scheme and initial lubricant dose of 6ml were determined for entering the boundary lubrication regime. Lubricant was added using a syringe then the machine was brought up to an operating speed of 20 Hz. Incremental loads of 2,500 lb, 4,000 lb, and 6,000 lb were applied to the bearing. Each time, the bearing was allowed to reach a constant temperature before applying an additional load. These temperatures were 40°C, 60°C, and 75°C, respectively. The bearing is rated for 6,340 lb. Once the bearing was at 75°C, the bearings were overloaded to 8,000 lb. The temperature rise will level off until boundary lubrication is exited, which can be observed by a far more rapid increase in temperature.

Results:

This section will feature visual representations of collected data as well as highlight the features of importance in each test.

3.1: Frequency Response Results

Initially, the Bearing Test Rig was on isolation pads, while the motor was fixed to the table. This configuration showed a clean FRF with respect to the coherence, however, strong resonances were observed between the rig and the motor. Figure 8 shows the FRF in the Y-direction, parallel to the earth and normal to the shaft of the rig. The data were recorded at a rate of 10,240 Hz using a wax-mounted, single-axis accelerometer.

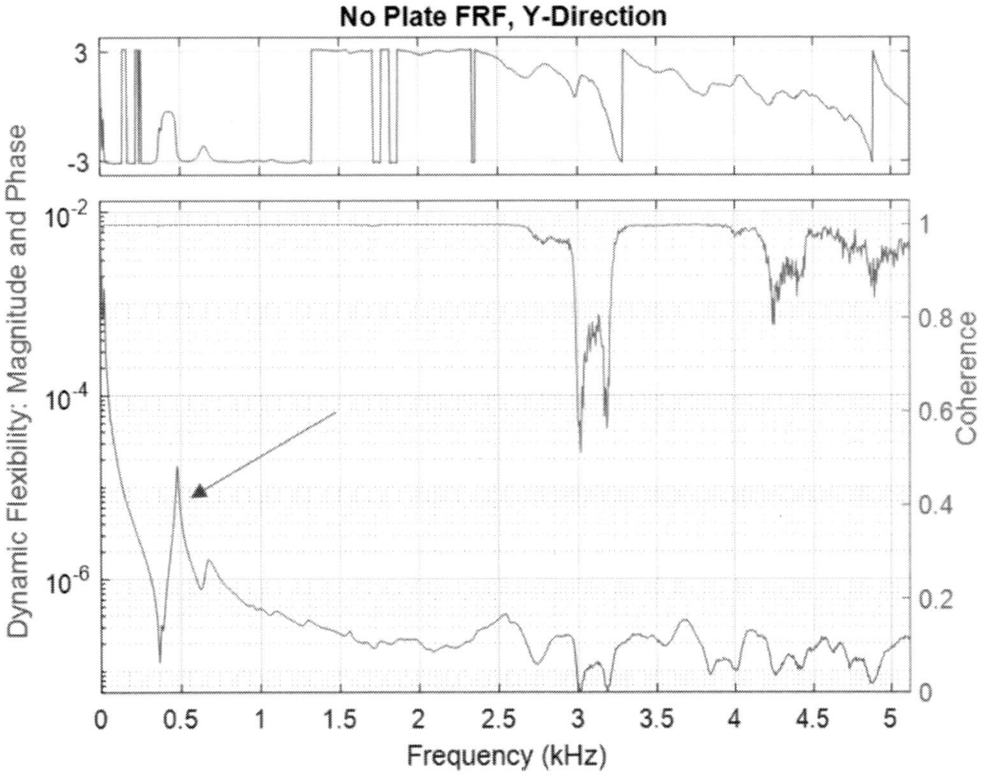

Figure 8.Frequency Response Function, No Plate, Y-Direction

Figure 8 shows good coherence up to 2700 Hz. A single, strong resonance can be seen at approximately 500 Hz. To reduce the effects of this resonance, the motor and Bearing Test Rig were mounted on a common plate to provide additional stiffness to the system. Figure

9 shows the results for the FRF when the motor and Bearing Test Rig were attached to a common plate.

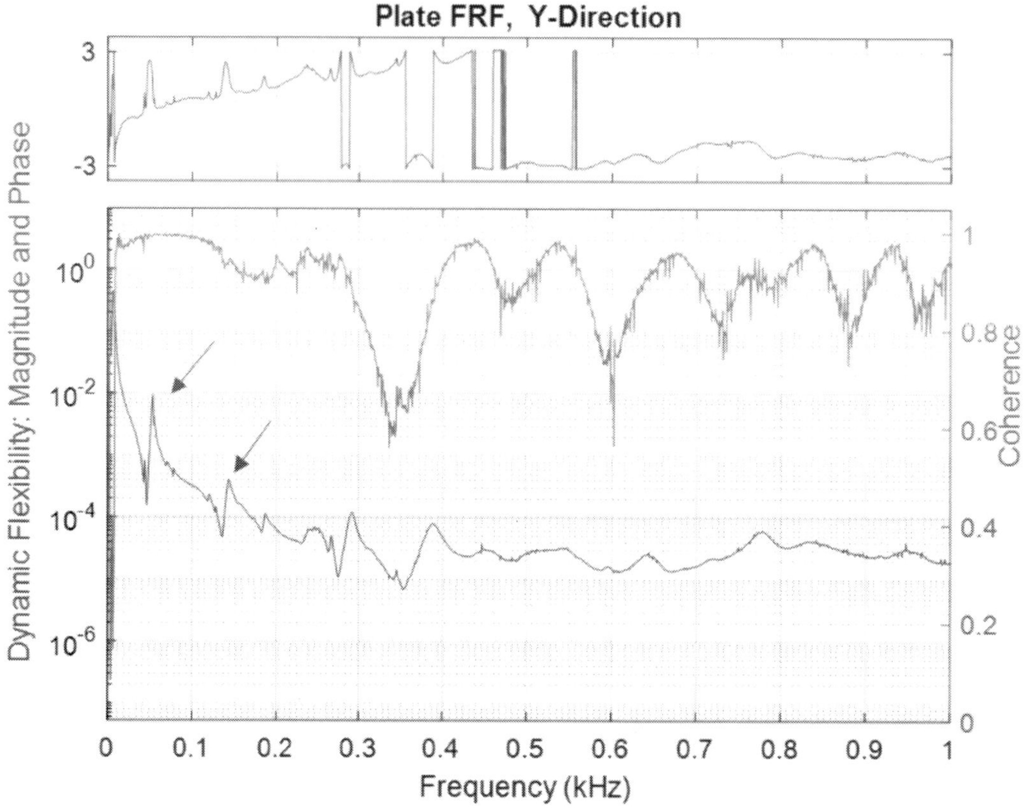

Figure 9. Frequency Response Function, Plate, Y-Direction

With the common plate, notable resonances can be observed at 50 and 150 Hz. While the resonances have a higher magnitude, the additional stiffness of the plate suppresses these resonances from occurring when operation speeds keep the bearing fault frequencies away from 50 Hz. While the coherence was near 1 at the resonances, higher frequencies show a drop in the coherence. This is also due to the additional stiffness introduced by the common plate as it provides a more complex geometry and additional pathways for vibrations to travel. Operation of the bearing test rig was found to have quieter operation in the case where the motor and the Bearing Test Rig were fixed to a common plate. In the case where the Bearing Test Rig was free from the motor, the 500 Hz vibration would occur as part of normal operation as it was excited by the elasticity of the drive belt, raising the overall RMS of the signal regardless of operating speed.

3.2: Long Run Results

The long run consisted of two different shafts speeds in combination with two different loads over several hours. This test was used to wear-in and characterize the bearings under different loading schemes. Figure 10 shows the vibration spectra for two shaft speeds: 6 Hz for samples 1 to 18 and 10 Hz for samples 19 to 37. A 1500 lb load was placed on the bearings for samples 9-18 and samples 28-37, respectively. Each combination was ran for two hours.

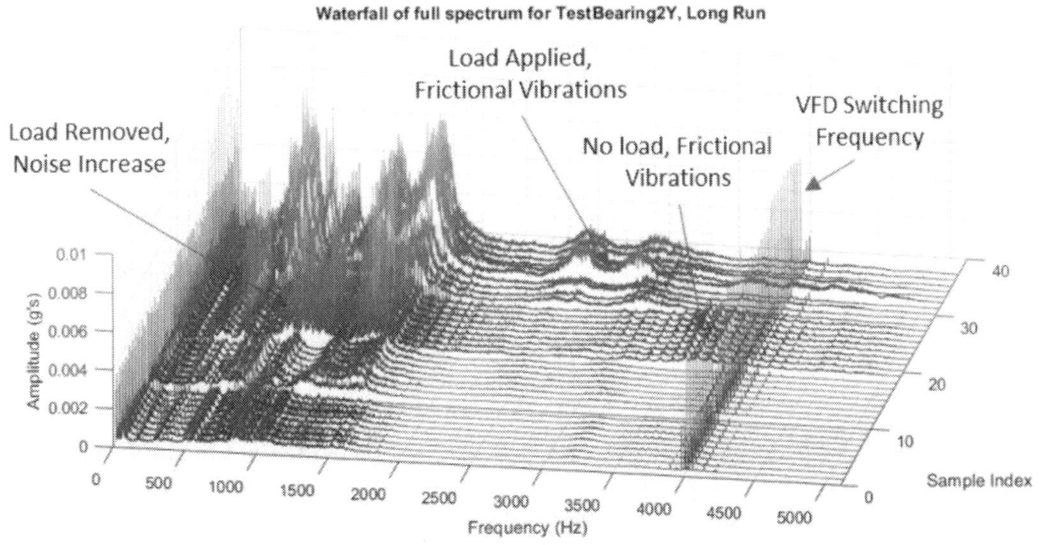

Figure 10: Evidence of Skidding, Long Run

A strong peak at 4000 Hz is observed throughout the spectra. This is due to the switching frequency in the VFD. This peak was later minimized through a sine wave filter, located between the VFD and motor, and grounding of the rig. A large increase in noise is seen in the spectra at index 19, when load was taken off of the bearings and the motor speed was increased to 10 Hz. Most of the increase in noise occurs in the 500-1500 Hz range, while frictional vibrations began to be observed in the 3000-5000 Hz range. When the load was reapplied at index 28, an increase in noise can be seen in the 2500-3500 Hz range. There is a rise in the noise floor with peaks spaced at the bearing outer race fault frequency, which points to signs of bearing skidding (Institute 2019). This skidding might be caused if the races were not sufficiently pressed there. The forward force from the inner race would not be enough to overcome the resistance from the cage causing the rollers to skid. Disassembling the Bearing Test Rig confirmed the existence of mechanical looseness in an adjacent support bearing through the discovery of a loose bearing lock nut. This bearing lock nut was used to press the inner race into the rollers and outer race.

3.3: Ramp-Up Results

This test is used to help differentiate the Bearing Test Rig's structural resonances from dynamic bearing frequencies. Figure 11 shows the vibration spectra from one such test as a waterfall plot.

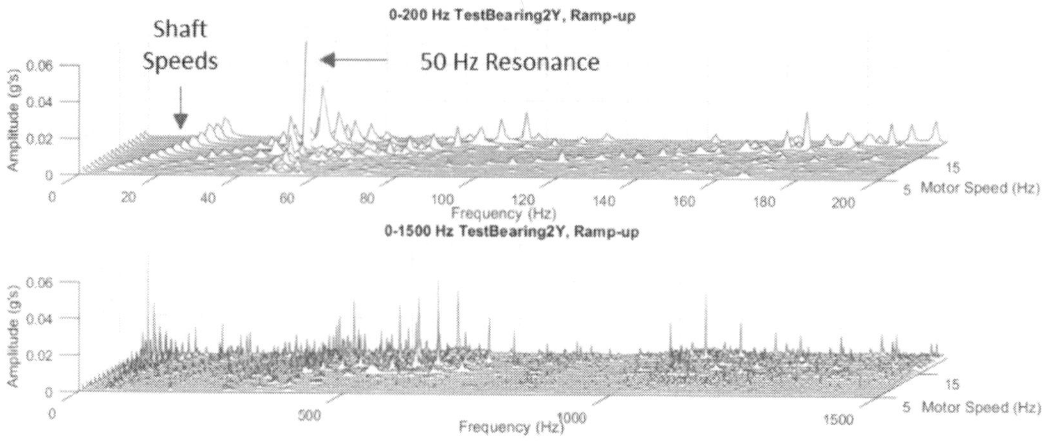

Figure 11: Ramp-up Test Results, Y-direction

In the top plot, the frequencies can be seen increasing as a function of shaft speed, which can be observed by following the first peak in each spectrum. Another notable feature of the top plot is the strong peak seen around 50 Hz. This is a result of a bearing frequency resonating with a structural resonance of the Bearing Test Rig. In the lower plot, additional noise in the spectrum can be seen at the higher frequencies. This was determined to be the onset of skidding due to mechanical looseness as the "self-locking" bearing lock nuts began to loosen. Further evidence of can be seen in the results for the long-run test that took place a few days later. A peak at 4000 Hz that corresponds to the switching frequency of the motor variable frequency drive (VFD) was the only significant peak above 1500 Hz. This peak was seen during the long run in Figure 10.

3.4: Accelerated Degradation Testing Results

As the approach to boundary lubrication is being made, temperature will level off for quite some time around 90-100°C. During this time, the oil within the bearing provides enough lubrication to keep the contact surfaces in elastohydrodynamic lubrication, while the lubricant slowly drips out. Eventually boundary lubrication will be achieved, but the indicator to observe is when the regime is exited, and dry contact begins to take place. This is observed when the temperature begins to climb again as shown by the white line for Test Bearing 1 temperature in Figure 12.

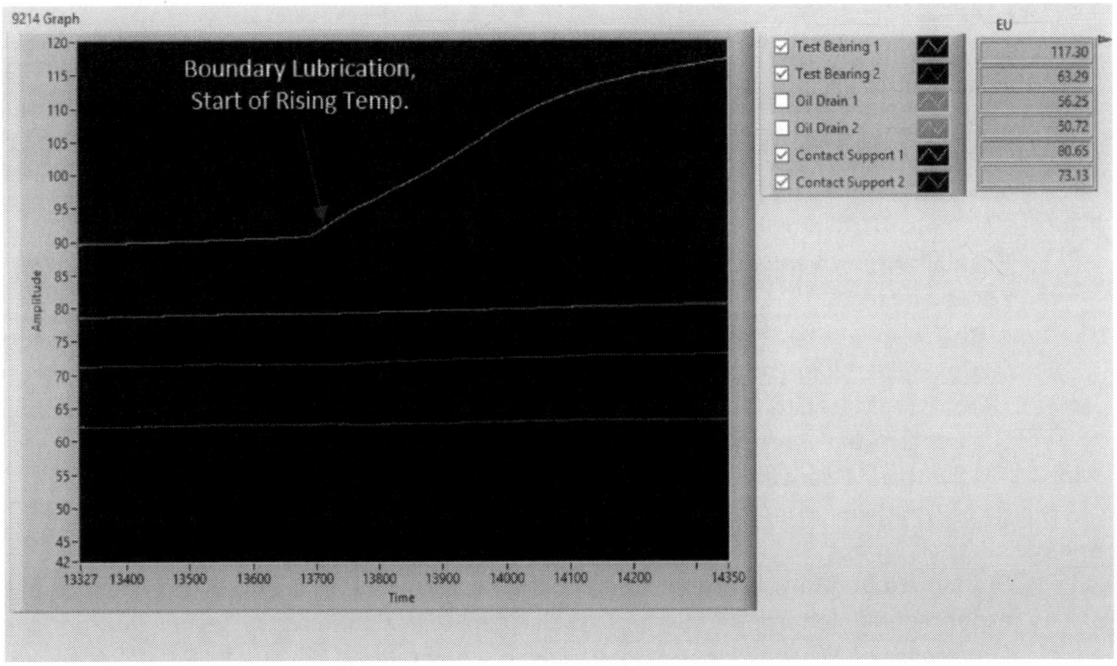

Figure 12.Bearing Temperature, Boundary Lubrication Indication

Further testing is to be carried out to determine the volume and frequency of oil additions to maintain the boundary lubrication conditions for extended runs where bearing degradation will occur. Oil samples will be extracted to determine fluid properties, contamination, and wear debris.

Ongoing work:

Bearings will be monitored inside a wheeled land vehicle's differential to provide realistic data and noise that would only be apparent inside an actual system. A differential is a key component of the drive train in a vehicle. This system allows the rotational speed of one wheel to vary from the other. This is useful during a turn, where the outer tire needs to rotate faster than the inner tire. The carrier bearings inside the differential will be monitored through vibration analysis, while samples will be regularly taken from the splash lubrication system inside the differential.

The test plan for the differential setup is a multistep process. First, the bearing of interest in brinelled using the brinelling drop-weight fixture. The bearing is then monitored and accelerated to a known level of degradation under boundary lubrication. The bearing is inserted into the differential where it is monitored in-situ. The bearing is removed from the differential and again accelerated under boundary lubrication conditions. This process of degradation acceleration then in-situ monitoring is repeated until bearing failure. Oil samples are taken throughout this process during each step. Once a population of bearings have failed, work will begin on creating accurate and robust prognostic models from actual run-to-failure data.

References:

Administration, National Aeronautics and Space. 2019. 'Prognostic Center Data Repository', Accessed Jan 24. https://ti.arc.nasa.gov/tech/dash/groups/pcoe/prognostic-data-repository/.

Aeroshell Turbine Engine Oils. 2012.

Hironaka, S. 1984. 'Boundary Lubrication and Lubricants', *Three Bond Technical News,* July.

Institute, Mobius. 2019. 'Rolling element bearings: Rolling elements skidding', Mobius Institute - Reliability Improvement & Vibration Analysis Training And Certification Per ISO 18436, Accessed Jan 25. https://www.mobiusinstitute.com/site2/item.asp?LinkID=8016&iVibe=1&sTitle=Rolling%20element%20bearings.

James C. Robinson, Meredith Christman. 2018. 'Vibration Monitoring of Paper Mill Machinery', *Sensors for Machinery Health Monitoring*.

'Mobil 1TM Synthetic Gear Lube LS | MobilTM Motor Oils.'. Accessed 11-Jan-2019. https://mobiloil.com/en/gear-lubricants/mobil-1-synthetic-gear-lube-ls.

Nectoux, Patrick, Rafael Gouriveau, Kamal Medjaher, Emmanuel Ramasso, Brigitte Chebel-Morello, Noureddine Zerhouni, and Christophe Varnier. 2012. "PRONOSTIA: An experimental platform for bearings accelerated degradation tests." In *IEEE International Conference on Prognostics and Health Management, PHM'12.*, 1-8. IEEE Catalog Number: CPF12PHM-CDR.

Snow, Dennis A. 2001. *Plant engineer's reference book* (Elsevier).

Solutions, Integrated Reliability. 2019. 'Bearing Skidding', Accessed 2-13-2019. https://www.lubrication.com.au/case-studies/bearing-skidding/.

Team, U.S. Joint Helicopter Safety Implementation. 2013. 'Health and Usage Monitoring Systems Toolkit', International Helicopter Safety Team. http://www.ihst.org/portals/54/toolkit_hums.pdf.

Technologies, Agilent. 1997. "The Fundamentals of Modal Testing." In, 14-29. Hewlett-Packard Co.

PIEZOELECTRIC VS MEMS: FUTURE OF THE VIBRATION SENSORS.

George Zusman, Ph.D., D.Sc.
Vibration Measurement Solutions, Inc.
zusger@gmail.com

The paper presents a detail comparison between the traditional piezoelectric and MEMs based vibration sensors. The future prognoses of what technology and where will be used in the future is discussed.

1. Main MEMs parameters important for vibration measurements and diagnostics

There is exit a group of parameters which could be used to define will particular accelerometer is perspective to use for vibration measurements and diagnostics. That group consists of the follow parameters: 1) cost (parts and assembly), 2) required supply power, 3) working frequency range, 4) working temperature range and 5) output noise level. The comparison of the piezoelectric and MEMs based accelerometers by the parameters 1) to 4) is obvious. For instance it is known that for the vibration machinery protection the frequency range of 10-1000 Hz is commonly used, but for diagnostics goals the up frequency is required to be 10-20 kHz. In the same time the parameter 5) – output noise level – is not always clear known for particular application.
Diagram Noise vs Max Working Frequency presented in Figure 1.

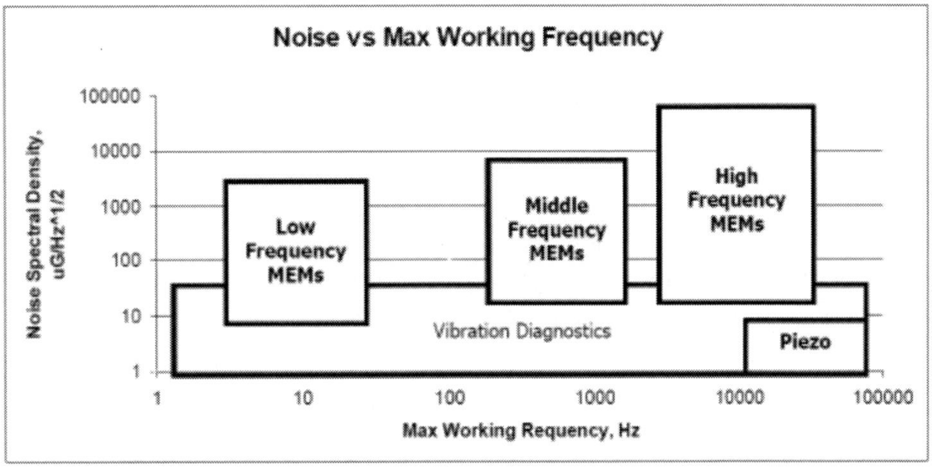

Figure 1. Diagram Noise vs Max Working Frequency

There are some rectangles crosses area usually used for vibration measurement and diagnostics. It is possible to see that some MEMs with noise going down and frequency range being wider could be used in the vibration field. The typical noise and frequency response plots from modern MEMs accelerometer is shown at Figure 2.

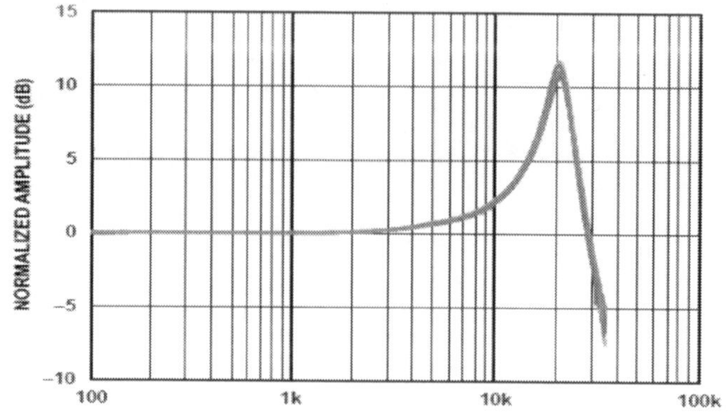

a)

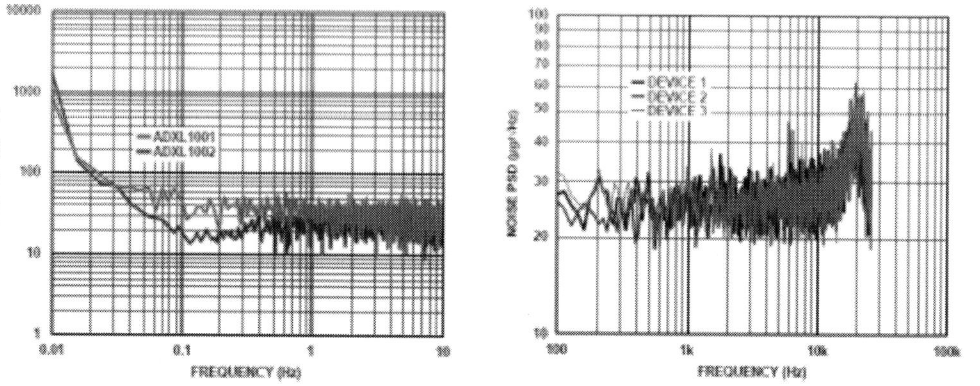

b)

Figure 2. Example of modern MEM accelerometer frequency response a) and noise density b) vs frequency

Bellow will be discussed the noise level of the MEMs based Accelerometers and Velocity transmitters.

2. Noise of MEMs accelerometers.

The principle of MEMs accelerometer is illustrated by Figure 3. The external vibration is activates a movement of inertial mass. The integrated electronics provides the output in analog or digital form proportional to that movement by measuring the capacitance between movements and fixes sensor parts. That construction might be considered as mechanical resonator. Then the equation described it movement could be written as follow:

$$M\frac{\partial^2 x}{\partial t^2}+\gamma\frac{\partial x}{\partial t}Kx=F$$

where M – value of inertial mass, x – displacement of inertial mass, K – coefficient of spring elasticity, γ - spring damper coefficient, F – external forces applied to the inertial mass and t – time.

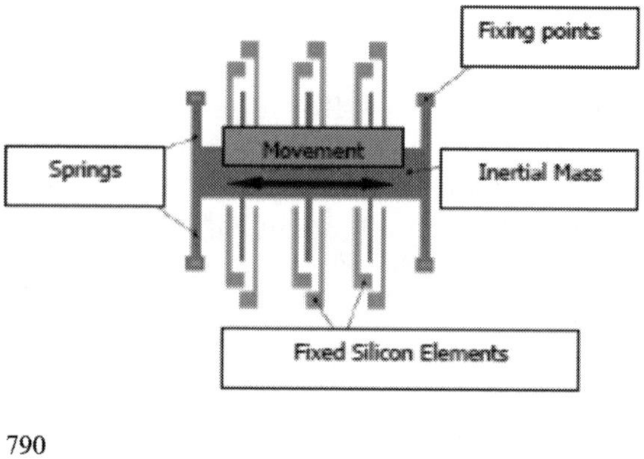

Figure 3. Diagram of the of MEMs accelerometer working principle

This equation could be transformer to the follow $pMv + \gamma v + K / p = F$, where $v = \frac{\partial x}{\partial t} = px$ and p – Laplace operator.
Thus, the square of velocity of noise generation could be equal to:

$$v_{RMS}^2 = \frac{F_{RMS}^2}{\gamma^2 + (\omega M - K / \omega)^2}$$ where ω - circular frequency of inertial Q- mass movement. Using resonance frequency $\omega_0 = \sqrt{K / M}$ and Q-factor $Q = \omega_0 M / \gamma$ the above equition could presented as

$$v_{RMS}^2 = \frac{1}{\gamma^2} \frac{F_{RMS}^2}{1 + Q^2(\omega / \omega_0 - \omega_0 / \omega)^2} .$$

The the kinetic energy accumulated in resonator could presented as

$$E = \frac{1}{2} M v_{RMS}^2 = \frac{1}{4\pi\gamma} \int_0^{\infty} \frac{F_{RMS}^2 Q d(f / f_0)}{1 + Q^2(f / f_0 - f_0 / f)^2}$$

where f=$\omega / 2\pi$ и f$_0$=$\omega_0 / 2\pi$. From above equation it is possible to get that $E = F_{RMS}^2 / 8\gamma$ and $F_{RMS}^2 = 4kT\gamma$, where k – is Boltzmann constant and T – is absolute temperature. Then the minimum noise level of the accelerometer is $\sqrt{4kT\omega_0 / MQ}$. At the reasonable values of the Q it should be possible to get minimum spectral noise density for the MEM accelerometer about 1 $\mu G / \sqrt{Hz}$ as of today the best results is about 7-10 times higher. Last years progress of reducing the MEMs accelerometers noise allow us proposing of getting the MEMs accelerometers suitable for any vibration measurements in the temperature range up to 125 degree C.

3. Noise of MEM based velocity transmitters

Vibration velocity in a range of 10 to 1000 Hz is commonly used and standard parameter for machinery protection. There are norms for different types of machinery used in the industries for emergency machinery shutdown when vibration velocity up over the recommended levels. It is typical request that the minimum velocity have to be measured lay bellow 0.01 IPS PK (about 0.1 mm/s RMS). Then it will be nassery to define what the noise level of MEM based accelerometer allows to provide such measurements. Let us do that analyzes with assumption that spectral noise density is constant in a range of 10 to 1000 Hz, which practically truth for almost all MEMs accelerometers (see Figure 2 a). So, accelerometer acceleration noise density A(ω)=ND [$\mu G / \sqrt{Hz}$]. The integrator uses to get the velocity from the acceleration with the transfer function:,
$K(\omega) = K_{INT} / \sqrt{\omega_{INT}^2 + \omega^2}$, where K$_{INT}$ - is integrator gain and ω_{INT} - is integrator low cutoff frequency. Then the velocity spectral density will be:

$V(\omega) = \text{ND } K_{INT} / \sqrt{\omega_{INT}^2 + \omega^2}$ and noise velocity RMS in the output of velocity transmitter will be:

$$V_{RMS} = NDK_{INT} \sqrt{\int_{\omega_L}^{\omega_H} d\omega / (\omega_{INT}^2 + \omega^2)} \approx NDK_{INT} \sqrt{\frac{1}{\omega_{INT}} arctg \frac{\omega_H}{\omega_{INT}}}$$

where ω_L - is low frequency of the vibration measurement range and ω_H - is high frequency of the vibration measurement range.

Based on the above equation plots of the velocity noise vs integrator cutoff frequency at different levels of acceleration noise density are presented at Figure 4.

Figure 4 shows that noise density bellow 20 $\mu G / \sqrt{Hz}$ is suitable for most application required vibration velocity measurements.

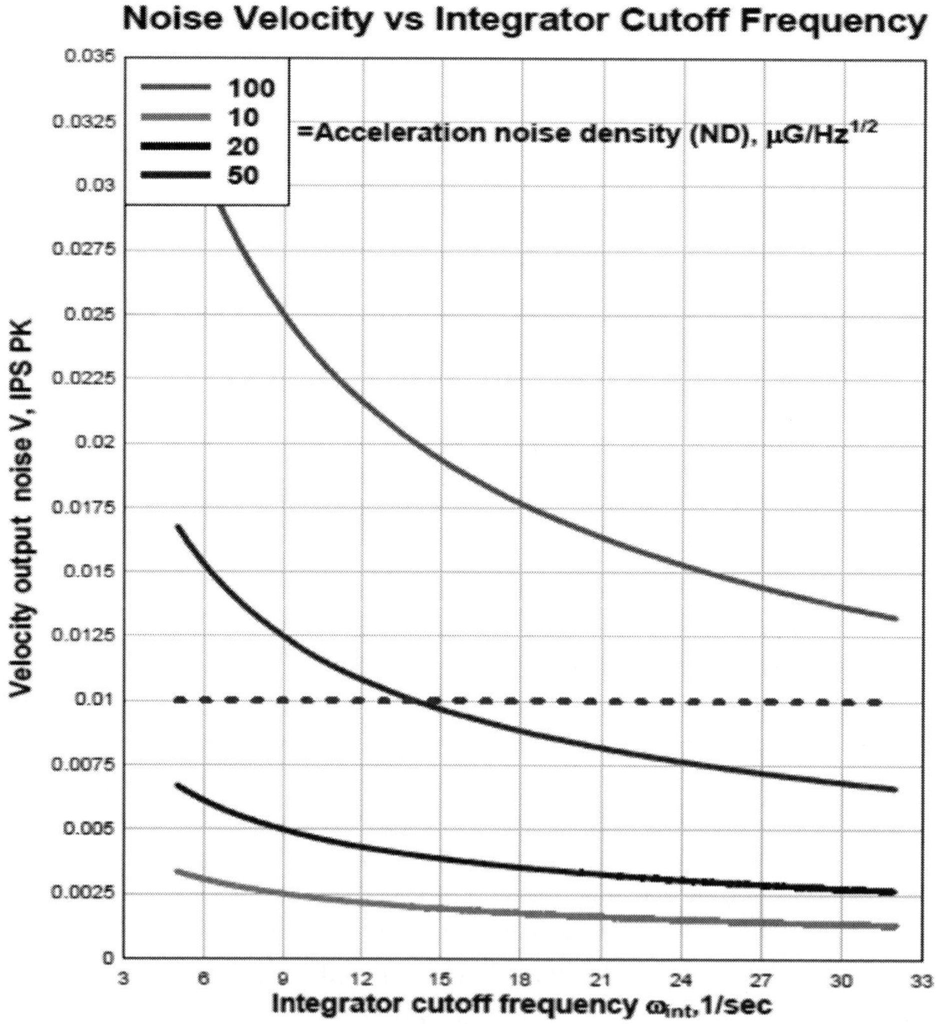

Figure 4. Velocity noise vs integrator cutoff frequency

4. Conclusion

MEMs accelerometers and velocity transmitters have a good chance to take a quite large percent of the niche for vibration machinery measurements and protection in the temperature range up till 125 degree C. Piezoelectric based accelerometers and transmitters will remain for the high temperature applications and the measurements with vary low noise or/and power requirements.

UNIVERSAL VIBRATION SENSOR MOUNT

George Zusman, Ph.D., D.Sc.
Vibration Measurement Solutions, Inc.
zusger@gmail.com

The present paper describes universal vibration sensor mount devices for attaching sensors to the machinery. Some types of machinery vibration sensors need to be mounted in a particular orientation relative to the machine being monitored. One example is two-axis and three-axis MEMS accelerometers used to detect machine-critical acceleration vector components. Another example is sensors with side connectors or cables whose orientation is restricted by space limitations. In many instances, vibration machinery sensors constructed as could be mounted using a central bolt that going throw the sensor body threads into a bore on the machine body. Such sensors are usually more expensive that similar sensors with solid body. In some cases, proper sensor orientation is achieved by simply gluing the sensor to the machine body. This makes it difficult to remove the sensor if it needs to be replaced. There are also proprietary sensor mounts designed for specific sensors, but these lack versatility. We designed the universal sensor mount that can be used to mount a wide variety of machinery sensors at precise orientations. The idea here is that sensor module mounting member is locked in a selected rotational position using the adjustable locking member as shown on the Figure 1. The sensor module is attached to the universal sensor mount by threading the sensor module onto the sensor module mounting member until tight. A determination is made whether the sensor module is substantially aligned with one or more reference axes of the machine. If not, the sensor module is detached from the universal sensor mount. The sensor module mounting member is unlocked and rotated to another selected rotational position that will substantially align the sensor module with the one or more machine reference axes as shown on Figure 2. The sensor module mounting member is then relocked in another selected rotational position and sensor module is reattached to the universal sensor mount.

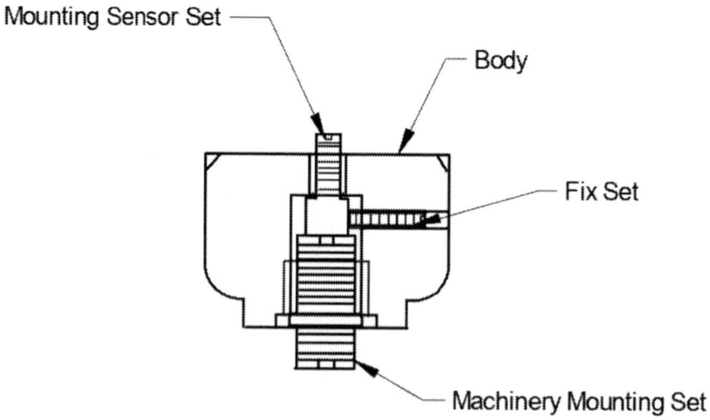

Figure 1. Structure diagram of universal sensor mount

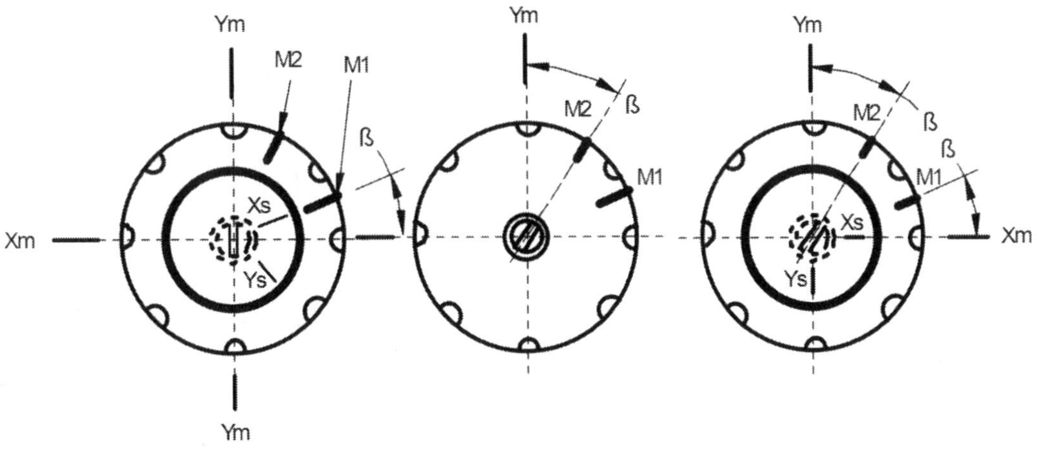

Figure 2. Illustration of the sensor mounting technology

When the mount fixed in a machinery body permanently the mounting sensor set flexes slit should oriented in the one of machinery axis (Ym in the Figure 2). Then the side set screw (fix set) fixes the sensor mounting set and sensor should be screwed on it top completely. The mark then applied in the mount side where the particular sensor axis stops (M1 in the Figure 2). After that sensor should be took out, mounting set flex and it slit turn on the watch hand direction in the particular angle (β in the Figure 2) between mark and machinery axis in parallel with which the particular sensor axis should be. Extra mark might be used to support that operation (M2 in the Figure 2) Then mounting set should fixed by side set screw and sensor should be mounted back. Now sensors axis will be directed in parallel with machinery axis. In a case sensor needed to be replaced the mount remains in the machinery and just the operation of adjusting the mounting set slit orientation should be repeated.

AUTHOR INDEX